MTI
and Pulsed Doppler Radar

For a complete listing of the *Artech House Radar Library*,
turn to the back of this book . . .

MTI and Pulsed Doppler Radar

D. Curtis Schleher

AIL Systems, Inc.

Artech House
Boston • London

Library of Congress Cataloging-in-Publication Data

Schleher, D. Curtis, 1932-
MTI and pulsed doppler radar / D. Curtis Schleher
p. cm.
Includes bibliographical references and index.
ISBN 0-89006-320-6
1. Moving target indicator radar. 2. Doppler radar. 3. Pulse compression radar. I. Title.
TK6592.M67S35 1991 90-26020
621.3848'5--dc20 CIP

685 Canton Street
Norwood, MA 02062

International Standard Book Number: 0-89006-320-6
Library of Congress Catalog Card Number: 90-26020

10 9 8 7 6 5 4 3 2 1

Contents

MTI and Pulsed Doppler Radar
by D. Curtis Schleher

FIGURE ATTRIBUTIONS

The author and the publisher wish to thank the authors and publishers of the following works for permission to reprint selected figures and tables in this book.

CHAPTER 1

Figure 1.1 Skolnik, M.I., ed., *Radar Handbook,* 1st Ed., 1970, McGraw-Hill Book Company, p. 17-3.

Figure 1.2 Schleher, D.C., *Introduction to Electronic Warfare,* 1986, Artech House, Inc., p. 30.

Figure 1.3 Skolnik, M.I., ed., *Introduction to Radar Systems,* 1st Ed., 1962, McGraw-Hill Book Company, p. 145.

Figure 1.4 Raytheon, *Electronic Progress,* 1987, Vol. XXVIII, No. 1, p. 37.

Figure 1.5 Schleher, D.C., MTI Performance in Rayleigh and Log-Normal Clutter, April 1980, *IEEE International Radar Conference,* p. 300.

Figure 1.6 Skolnik, M.I., ed., *Introduction to Radar Systems,* 1st Ed., 1962, McGraw-Hill Book Company, p. 133.

Figure 1.9 Report No. FAA-RD-76-190, MIT Lincoln Laboratories, 8 March, 1977, p. 40.

Figure 1.10 Povejsil, D.J., Raven, R.S. and Waterman, P., *Airborne Radar,* 1961, D. Van Nostrand Company, Inc., p. 321.

Figure 1.11 Povejsil, D.J., Raven, R.S. and Waterman, P., *Airborne Radar,* 1961, D. Van Nostrand Company, Inc., p. 322.

Figure 1.13 Skolnik, M.I., ed., *Radar Handbook,* 1st Ed., 1970, McGraw-Hill Book Company, p. 19-8.

Figure 1.14 Povejsil, D.J., Raven, R.S. and Waterman, P., *Airborne Radar,* 1961, D. Van Nostrand Company, Inc., p. 323.

Figure 1.15 Sugie, M., Tomita, Y. and Hagisawa, T., Advanced Air Route Surveillance Radar With Various Unique Clutter Elimination Capabilities, *NEC Research and Development,* No. 39, October 1975, pp. 71–83

Figure 1.16 Sugie, M., Tomita, Y. and Hagisawa, T., Advanced Air Route Surveillance Radar With Various Unique Clutter Elimination Capabilities, *NEC Research and Development,* No. 39, October 1975, pp. 71–83

Figure 1.17 Raytheon, *Electronic Progress,* 1987, Vol. XXVIII, No. 1, p. 29.

Figure 1.18 Cosgriff, R.L., Peake, W.H. and Taylor, R.C., *Terrain Scattering Properties for Sensor System Design (Terrain Handbook II),* 1960, Ohio State University, p. 171.

Figure 1.19 Long, M.W., *Radar Reflectivity of Land and Sea,* Artech House, Inc., p. 159.

Figure 1.20 Fishbein, W., Graveline, S. and Rittenbach, O., Clutter Attenuation Analysis, *Technical Report ECOM-2808,* March 1967, U.S. Army Electronics Command, NJ.

Figure 1.21 Billingley, *Measured Spectral Extent of L- and X-band Radar Reflections from Windblown Trees,* MIT Lincoln Laboratories Report CMT-57, 1987, p. 14.

Figure 1.22 Marple, L., *Digital Spectral Analysis with Applications,* Prentice-Hall, Englewood Cliffs, NJ, p. 176.

Figure 1.23 Skolnik, M.I., ed., *Radar Handbook,* 1st ed., 1970, McGraw-Hill Book Company, pp. 17-12, 17-13.

Figure 1.25 Harvard Graphics Plot.

Figure 1.27 *Report No. FAA-RD-76-191,* MIT Lincoln Laboratories, p. 66.

Figure 1.28 *Report No. FAA-RD-76-191,* MIT Lincoln Laboratories, p. 7.

Figure 1.29 *Report No. FAA-RD-76-190,* MIT Lincoln Laboratories, p. 5.

Figure 1.30 *Report No. FAA-RD-76-190,* MIT Lincoln Laboratories, p. 35.

Figure 1.31 Taylor, J.W., Jr., Sacrifices in Radar Clutter Suppression Due to Compromises in Implementation of Digital Doppler Filters, *IEE International Radar Conference,* October 1982, The Institution of Electrical Engineers, p. 42.

Figure 1.32 O'Donnell, R.M. and Muehe, C.E., Automated Tracking for Aircraft Surveillance Radar Systems, *IEEE Transactions on Aerospace and Electronic Systems,* July 1979, Vol. AES-15, No. 4, p. 512.

Figure 1.33 O'Donnell, R.M. and Muehe, C.E., Automated Tracking for Aircraft Surveillance Radar Systems, *IEEE Transactions on Aerospace and Electronic Systems,* July 1979, Vol. AES-15, No. 4, p. 512.

Figure 1.34 Barton, D.K., *Modern Radar System Analysis,* 1988, Artech House, Inc., p. 259.

Figure 1.36 Nitzberg, R., Clutter May CFAR Analysis, *IEEE Transactions on Aerospace and Electronic Systems,* July 1986, Vol. AES-22, No. 4, p. 421.

Figure 1.37 Khoury, E.N. and Hoyle, J.S., Clutter Maps: Design and Performance, *Proceedings of the 1984 IEEE National Radar Conference,* p. 56.

Figure 1.38 *Report No. FAA-RD-76-190,* MIT Lincoln Laboratories, p. 126

CHAPTER 2

Figure 2.1 Skolnik, M.I., ed., *Radar Handbook,* 1st Ed., 1970, McGraw-Hill Book Company, p. 3-24.

Figure 2.2 Roulston, J.F., Design Decisions Guide Airborne Radar, *Microwaves & RF,* Vol. 24, No. 3, March 1985, Hayden Publishing Co., Inc., p. 132.

Figure 2.3 Nathanson, F.E., *Radar Design Principles,* 1969, McGraw-Hill Book Company, p. 395.

Figure 2.4 Schleher, D.C., *MTI Performance Analysis,* Internal Report for Airborne Instruments Laboratory, 1966, Cutler-Hammer, Inc., p. 1–64

Figure 2.5 Schleher, D.C., *MTI Performance Analysis,* Internal Report for Airborne Instruments Laboratory, 1966, Cutler-Hammer, Inc., p. 1–64

Figure 2.6 Stimson, G.W., *Introduction to Airborne Radar,* 1983, Hughes Aircraft Company, El Segundo, California, p. 431.

Figure 2.7 Stimson, G.W., *Introduction to Airborne Radar,* 1983, Hughes Aircraft Company, El Segundo, California, p. 422.

Figure 2.8 Stimson, G.W., *Introduction to Airborne Radar,* 1983, Hughes Aircraft Company, El Segundo, California, p. 428.

Figure 2.9 Schleher, D.C., *Introduction to Electronic Warfare,* 1986, Artech House, Inc., p. 283.

Figure 2.10 Stimson, G.W., *Introduction to Airborne Radar,* 1983, Hughes Aircraft Company, El Segundo, California, p. 468.

Figure 2.11 Stimson, G.W., *Introduction to Airborne Radar,* 1983, Hughes Aircraft Company, El Segundo, California, p. 442.

Figure 2.12 Aronoff, E. and Greenblatt, N.M., Medium PRF Radar Design and Performance, *20th Tri-Service Radar Symposium,* 1974, p. 272.

Figure 2.13 Stimson, G.W., *Introduction to Airborne Radar,* 1983, Hughes Aircraft Company, El Segundo, California, p. 443.

Figure 2.14 Stimson, G.W., *Introduction to Airborne Radar,* 1983, Hughes Aircraft Company, El Segundo, California, p. 509.

Figure 2.15 Kennedy, P.D., FFT Signal Processing for Non-Coherent Airborne Radars, *Proceedings of the 1984 IEEE National Radar Conference,* p. 81.

Figure 2.17 Schleher, D.C., *MTI Performance Analysis,* Internal Report for Airborne Instruments Laboratory, 1966, Cutler-Hammer, Inc., pp. 1–64.

Figure 2.18 Stimson, G.W., *Introduction to Airborne Radar,* 1983, Hughes Aircraft Company, El Segundo, California, p. 423.

Figure 2.19 Skolnik, M.I., ed., *Radar Handbook*, 1st Ed., 1970, McGraw-Hill Book Company, p. 17-6.

Figure 2.20 Skolnik, M.I., *Introduction to Radar Systems*, 2nd Ed., 1980, McGraw-Hill Book Company, p. 105.

Figure 2.21 Skolnik, M.I., *Introduction to Radar Systems*, 2nd Ed., 1980, McGraw-Hill Book Company, p. 106.

Figure 2.22 Skolnik, M.I., *Introduction to Radar Systems*, 2nd Ed., 1980, McGraw-Hill Book Company, p. 107.

Figure 2.24 Skolnik, M.I., *Introduction to Radar Systems*, 2nd Ed., 1980, McGraw-Hill Book Company, p. 110.

Figure 2.25 Andrews, G.A., Jr., Optimal Radar Doppler Processors, *NRL Report 7727*, May 1974, Naval Research Laboratory, Airborne Radar Branch, Radar Division, Washington, D.C., p. 8.

Figure 2.26 Skolnik, M.I., *Introduction to Radar Systems*, 2nd Ed., 1980, McGraw-Hill Book Company, p. 112.

Figure 2.27 Urkowitz, H., Analysis and Synthesis of Delay Line Periodic Filters, *IRE Transactions on Circuit Theory*, Vol. CT-4, No.2, June 1957, pp. 41–53.

Figure 3.39 Goldman, S., Oscillator Phase Noise Proves Important to Pulse-Doppler Radar Systems, *Microwave Systems News*, Vol. 14, No.2, February 1984, EW Communications, Inc., p. 92.

CHAPTER 4

Figure 4.1 Barton, and Shrader, W.W., Interclutter Visibility in MTI Systems, *IEEE EASCON Record*, 1969.

Figure 4.2 Long, M.W., *Radar Reflectivity of Land and Sea*, 1983, Artech House, Inc., p. 32.

Figure 4.3 Long, M.W., *Radar Reflectivity of Land and Sea*, 1983, Artech House, Inc., p. 213.

Figure 4.4 Long, M.W., *Radar Reflectivity of Land and Sea*, 1983, Artech House, Inc., p. 278.

Figure 4.5 Barton, D.K., *Modern Radar System Analysis*, 1988, Artech House, Inc., p. 130.

Figure 4.6 Barton, D.K., *Modern Radar System Analysis*, 1988, Artech House, Inc., p. 130.

Figure 4.7 Barton, D.K., *Modern Radar System Analysis*, 1988, Artech House, Inc., p. 131.

Figure 4.8 Schleher, D.C., ed., *MTI Radar*, 1978, Artech House, Inc., p. 53.

Figure 4.9 After *IITRI Report*, Greenstein, Overland Technology Program Report, September 1969, p. 9.

Figure 4.10 After *IITRI Report*, Greenstein, Overland Technology Program Report, September 1969, p. 10.

Figure 4.11 Schleher, D.C., ed., *MTI Radar*, 1978, Artech House, p. 54.

Figure 4.12 After *IITRI Report*, Greenstein, Overland Technology Program Report, September 1969, p. 11.

Figure 4.13 Barton, D.K., ed., *Radar Clutter*, 1977, Artech House, Inc., p. 160.

Figure 4.14 Currie, N.C, Millimeter Wave Land Clutter Model, *International Conference, Radar-82*, 18–20 October 1982, The Institution of Electrical Engineers, p. 388.

Figure 4.15 Currie, N.C, Millimeter Wave Land Clutter Model, *International Conference, Radar-82*, 18–20 October 1982, The Institution of Electrical Engineers, p. 388.

Figure 4.15 Currie, N.C, Millimeter Wave Land Clutter Model, *International Conference, Radar-82*, 18–20 October 1982, The Institution of Electrical Engineers, p. 388.

Figure 4.16 Currie, N.C, Millimeter Wave Land Clutter Model, *International Conference, Radar-82*, 18–20 October 1982, The Institution of Electrical Engineers, p. 388.

Figure 4.17 Schleher, D.C, ed., *Automatic Detection and Radar Data Processing*, Artech House, Inc., p. 445.

Figure 4.19 Nathanson, F.E., *Radar Design Principles*, 1969, McGraw-Hill Book Company, p. 261.

Figure 4.20 Nathanson, F.E., *Radar Design Principles*, 1969, McGraw-Hill Book Company, p. 274.

Figure 4.21 Long, M.W., *Radar Reflectivity of Land and Sea*, 1983, Artech House, Inc., p. 276.

Figure 4.22 Blake, L.V., *Radar Range Performance Analysis*, 1986, Artech House, Inc., p. 316.

Figure 4.23 Blake, L.V., *Radar Range Performance Analysis*, 1986, Artech House, Inc., p. 317.

Figure 4.24 Nathanson, F.E., *Radar Design Principles*, 1969, McGraw-Hill Book Company, p. 255.

Figure 4.25 Long, M.W., *Radar Reflectivity of Land and Sea*, 1983, Artech House, Inc., p. 185.

Figure 4.26 Trunk, G.V., Non-Rayleigh Sea Clutter: Properties and Detection of Targets, *NRL Report 7986*, June 1976, p. 8.

Figure 4.27 Schleher, D.C., ed., *MTI Radar*, 1978, Artech House, Inc., p. 43.

Figure 4.28 Schleher, D.C., ed., *MTI Radar*, 1978, Artech House, Inc., p. 49.

Figure 4.29 Jakeman, E., and Pusey, P.N., Statistics of Non-Rayleigh Microwave Sea Echo, IEE Radar-77, 25–28 October 1977, *Conference Publication Number 155*, 1977, Institution of Electrical Engineers, p. 108.

Figure 4.30 Jakeman, E., and Pusey, P.N., Statistics of Non-Rayleigh Microwave Sea Echo, IEE Radar-77, 25–28 October 1977, *Conference Publication Number 155*, 1977, Institution of Electrical Engineers, p. 108.

Figure 4.31 Nathanson, F.E., *Radar Design Principles*, 1969, McGraw-Hill Book Company, p. 243.

Figure 4.32 Nathanson, F.E., *Radar Design Principles*, 1969, McGraw-Hill Book Company, p. 244.

Figure 4.33 Nathanson, F.E., *Radar Design Principles*, 1969, McGraw-Hill Book Company, p. 248.

Figure 4.34 Long, M.W., *Radar Reflectivity of Land and Sea*, 1983, Artech House, Inc., p. 145.

Figure 4.35 Barton, D.K., *Radar System Analysis*, 1976, Artech House, Inc., p. 97.

Figure 4.36 Ward, H.R., A Model Environment for Search Radar Evaluation, *IEEE EASCON Convention Record*, 1971.

Figure 4.37 Barton, D.K., *Modern Radar System Analysis*, 1988, Artech House, Inc., p. 134.

Figure 4.38 Schleher, D.C., *Introduction to Electronic Warfare*, 1986, Artech House, Inc., p. 287.

Figure 4.39 Nathanson, F.E., *Radar Design Principles*, 1969, McGraw-Hill Book Company, p. 207.

Figure 4.40 Nathanson, F.E., *Radar Design Principles*, 1969, McGraw-Hill Book Company, p. 211.

Figure 4.41 Eaves, J. and Reedy, E., *Principles of Modern Radar*, Van Nostrand Reinhold, p. 300.

Figure 4.43 Schleher, D.C., *Introduction to Electronic Warfare*, 1986, Artech House, Inc., p. 186.

Figure 4.44 Schleher, D.C., *Introduction to Electronic Warfare*, 1986, Artech House, Inc., p. 187.

Figure 4.45 Barton, D.K., *Modern Radar System Analysis*, 1988, Artech House, Inc., p. 137.

Figure 4.46 Kerr, D., *Propagation of Short Radio Waves*, 1951, McGraw-Hill Book Company, Inc., p. 567.

Figure 4.47 Nathanson, F.E., *Radar Design Principles*, 1969, McGraw-Hill Book Company, p. 208.

Figure 4.48 After *IITRI Report*, Greenstein, A Comprehensive Ground Clutter Model for Airborne Radars, September 1969, p. 544.

Figure 4.49 Schleher, D.C., ed., *MTI Radar*, 1978, Artech House, Inc., p. 43.

Figure 4.50 Schleher, D.C., ed., *MTI Radar*, 1978, Artech House, Inc., p. 44.

Figure 4.51 Schleher, D.C., ed., *MTI Radar*, 1978, Artech House, Inc., p. 46.

Figure 4.52 Schleher, D.C., ed., *MTI Radar*, 1978, Artech House, Inc., p. 55.

Figure 4.53 Schleher, D.C., ed., *MTI Radar*, 1978, Artech House, Inc., p. 55.

Figure 4.54 Schleher, D.C., ed., *MTI Radar*, 1978, Artech House, Inc., p. 57.

Figure 4.55 Schleher, D.C., ed., *MTI Radar*, 1978, Artech House, Inc., p. 58.

Figure 4.56 Ward, H.R., A Model Environment for Search Radar Evaluation, *IEEE EASCON Convention Record*, 1971.

Figure 4.57 Ward, H.R., A Model Environment for Search Radar Evaluation, *IEEE EASCON Convention Record*, 1971.

CHAPTER 5

Figure 5.11 Schleher, D.C., MTI Detection Performance in Rayleigh and Log-Normal Clutter, *IEEE 1980 International Radar Conference*, p. 302.

Figure 5.12 Schleher, D.C. and Schulkind, D., Optimization of Non-Recursive Digital MTI, IEE International Radar Conference, October 1977, *Conference Publication No. 155*, Institution of Electrical Engineers, p. 176.

Figure 5.13 Andrews, G.A., Jr., Comparison of Doppler Radar Filtering Techniques, *NRL Report 7811*, October 1974, Naval Research Laboratory, Airborne Radar Branch, Radar Division, Washington, D.C., p. 6.

Figure 5.14 Andrews, G.A., Jr., Performance of Cascaded MTI and Coherent Integration Filters in a Clutter Environment, *NRL Report 7533*, March 1973, Naval Research Laboratory, Airborne Radar Branch, Radar Division, Washington, D.C., p. 5.

Figure 5.15 Andrews, G.A., Jr., Performance of Cascaded MTI and Coherent Integration Filters in a Clutter Environment, *NRL Report 7533*, March 1973, Naval Research Laboratory, Airborne Radar Branch, Radar Division, Washington, D.C., p. 9.

Figure 5.28 Schleher, D.C., Performance Comparison of MTI and Coherent Doppler Processors, *International Radar Conference*, October 1982, The Institution of Electrical Engineers, p. 157.

Figure 5.29 Schleher, D.C., Performance Comparison of MTI and Coherent Doppler Processors, *International Radar Conference*, October 1982, The Institution of Electrical Engineers, p. 158.

Figure 5.30 Schleher, D.C., Performance Comparison of MTI and Coherent Doppler Processors, *International Radar Conference*, October 1982, The Institution of Electrical Engineers, p. 158.

Figure 5.31 Schleher, D.C., Performance Comparison of MTI and Coherent Doppler Processors, *International Radar Conference*, October 1982, The Institution of Electrical Engineers, p. 158.

CHAPTER 6

Figure 6.1	Barton, D.K., *Modern Radar System Analysis*, 1988, Artech House, Inc., p. 238.
Figure 6.3	Skolnik, M.I., *Introduction to Radar Systems*, 2nd Ed., 1980, McGraw-Hill Book Company, p. 110.
Figure 6.4	Barton, D.K., *Modern Radar System Analysis*, 1988, Artech House, Inc., p. 238.
Figure 6.5	Blackband, W.T., *Radar Techniques for Detection Tracking and Navigation*, 1966, Gordon and Breach Science Publishers, Inc., p. 155.
Figure 6.6	Blackband, W.T., *Radar Techniques for Detection Tracking and Navigation*, 1966, Gordon and Breach Science Publishers, Inc., p. 156.
Figure 6.7	Nathanson, F.E., *Radar Design Principles*, 1969, McGraw-Hill Book Company, p. 327.
Figure 6.8	Schleher, D.C., *MTI Performance Analysis*, Internal Report for Airborne Instruments Laboratory, 1966, Cutler-Hammer, Inc., p. 202.
Figure 6.9	Skolnik, M.I., *Radar Handbook*, 1st Ed., 1970, McGraw-Hill Book Company, p. 17-54.
Figure 6.10	Rhys, T. and Andrews, G., AEW Radar Antennas, 1974, AGARD Conference Preprint No. 139, *Antennas for Avionics*, pp. 12-1–12-16.
Figure 6.11	Rhys, T. and Andrews, G., AEW Radar Antennas, 1974, AGARD Conference Preprint No. 139, *Antennas for Avionics*, pp. 12-1–12-16.
Figure 6.12	Rhys, T. and Andrews, G., AEW Radar Antennas, 1974, AGARD Conference Preprint No. 139, *Antennas for Avionics*, pp. 12-1–12-16.
Figure 6.13	Rhys, T. and Andrews, G., AEW Radar Antennas, 1974, AGARD Conference Preprint No. 139, *Antennas for Avionics*, pp. 12-1–12-16.
Figure 6.16	Andrews, G.A., Jr., Performance of Cascaded MTI and Coherent Integration Filters in a Clutter Environment, *NRL Report 7533*, March 1973, Naval Research Laboratory, Airborne Radar Branch, Radar Division, Washington, D.C., p. 2.
Figure 6.17	Urkowitz, H., Analysis and Synthesis of Delay Line Periodic Filters, *IRE Transactions on Circuit Theory*, Vol. CT-4, No. 2, June 1957, pp. 41–53.
Figure 6.18	Andrews, G.A., Jr., Optimal Doppler Radar Processors, *NRL Report 7727*, May 1974, Naval Research Laboratory, Airborne Radar Branch, Radar Division, Washington, D.C., p. 8.
Figure 6.19	Urkowitz, H., Analysis and Synthesis of Delay Line Periodic Filters, *IRE Transactions on Circuit Theory*, Vol. CT-4, No. 2, June 1957, p. 41–53.
Figure 6.20	Andrews, G.A., Jr., Optimal Doppler Radar Processors, *NRL Report 7727*, May 1974, Naval Research Laboratory, Airborne Radar Branch, Radar Division, Washington, D.C., p. 14.
Figure 6.21	Urkowitz, H., Analysis and Synthesis of Delay Line Periodic Filters, *IRE Transactions on Circuit Theory*, Vol. CT-4, No. 2, June 1957, pp. 41-53.
Figure 6.22	Linden, D.A. and Steinberg, B.D., Synthesis of Delay Line Networks, *IRE Transactions on Aeronautical and Navigational Electronics*, Vol. ANE-4, No. 1, March 1957, pp. 34–39.
Figure 6.23	Linden, D.A. and Steinberg, B.D., Synthesis of Delay Line Networks, *IRE Transactions on Aeronautical and Navigational Electronics*, Vol. ANE-4, No. 1, March 1957, pp. 34–39.
Figure 6.24	Skolnik, M.I., ed., *Radar Handbook*, 1st Ed., 1970, McGraw-Hill Book Company, p. 17-26.
Figure 6.25	Schleher, D.C., *MTI Performance Analysis*, Internal Report for Airborne Instruments Laboratory, 1966, Cutler-Hammer, Inc., pp. 1–64.
Figure 6.26	Skolnik, M.I., *Radar Handbook*, 1970, McGraw-Hill Book Company, Inc., p. 17-25.
Figure 6.27	White, W.D. and Ruvin, A.E., Recent Advances in the Synthesis of Comb Filters, *IRE National Convention Record*, 1957, pp. 186–199.
Figure 6.28	White, W.D. and Ruvin, A.E., Recent Advances in the Synthesis of Comb Filters, *IRE National Convention Record*, 1957, pp. 186–199.
Figure 6.29	Schleher, D.C., *MTI Performance Analysis*, Internal Report for Airborne Instruments Laboratory, 1966, Cutler-Hammer, Inc., pp. 1–64.
Figure 6.30	Schleher, D.C., *MTI Performance Analysis*, Internal Report for Airborne Instruments Laboratory, 1966, Cutler-Hammer, Inc., pp. 1–64.
Figure 6.32	Schleher, D.C., *MTI Performance Analysis*, Internal Report for Airborne Instruments Laboratory, 1966, Cutler-Hammer, Inc., pp. 1–64.

Figure 6.33 White, W.D. and Ruvin, A.E., Recent Advances in the Synthesis of Comb Filters, *IRE National Convention Record*, 1957, pp. 186–199.
Figure 6.34 White, W.D. and Ruvin, A.E., Recent Advances in the Synthesis of Comb Filters, *IRE National Convention Record*, 1957, pp. 186–199.
Figure 6.37 Andrews, G.A., Jr., Performance of Cascaded MTI and Coherent Integration Filters in a Clutter Environment, *NRL Report 7533*, March 1973, Naval Research Laboratory, Airborne Radar Branch, Radar Division, Washington, D.C., p. 9.
Figure 6.38 Andrews, G.A., Jr., Performance of Cascaded MTI and Coherent Integration Filters in a Clutter Environment, *NRL Report 7533*, March 1973, Naval Research Laboratory, Airborne Radar Branch, Radar Division, Washington, D.C., p. 13.
Figure 6.39 Skolnik, M.I., *Introduction to Radar Systems*, 2nd Ed., 1980, McGraw-Hill Book Company, p. 115.
Figure 6.43 Skolnik, M.I., ed., *Radar Handbook*, 1st Ed., 1970, McGraw-Hill Book Company, p. 17-40.
Figure 6.44 Skolnik, M.I., ed., *Radar Handbook*, 1st Ed., 1970, McGraw-Hill Book Company, p. 17-41.

CHAPTER 7

Figure 7.2 Stimson, G.W., *Introduction to Airborne Radar,* 1983, Hughes Aircraft Company, El Segundo, California, p. 456.
Figure 7.3 Goetz, L.P. and Albright, J.D., Airborne Pulse-Doppler Radar, *IRE Transactions on Military Electronics*, Vol. MIL-5, No. 2, April 1961, IRE.
Figure 7.4 Stimson, G.W., *Introduction to Airborne Radar,* 1983, Hughes Aircraft Company, El Segundo, California, p. 571.
Figure 7.5 Stimson, G.W., *Introduction to Airborne Radar,* 1983, Hughes Aircraft Company, El Segundo, California, p. 447.
Figure 7.6 Povejsil, D.J., *et al.*, Airborne Radar, 1961, D. Van Nostrand Company, Inc., p. 325.
Figure 7.7 Stimson, G.W., *Introduction to Airborne Radar,* 1983, Hughes Aircraft Company, El Segundo, California, p. 504.
Figure 7.8 Stimson, G.W., *Introduction to Airborne Radar,* 1983, Hughes Aircraft Company, El Segundo, California, p. 570.
Figure 7.9 Schleher, D.C, *Introduction to Electronic Warfare*, 1986, Artech House, Inc., p. 281.
Figure 7.10 Nathanson, F.E., *Radar Design Principles*, 1969, McGraw-Hill Book Company, p. 414.
Figure 7.11 Nathanson, F.E., *Radar Design Principles*, 1969, McGraw-Hill Book Company, p. 419.
Figure 7.12 Long, M.W., Medium PRF for the AN/APG-66 Radar, *Proceedings of the IEEE*, Vol. 73, No. 2, February 1985, p. 303.
Figure 7.13 Long, M.W., Medium PRF for the AN/APG-66 Radar, *Proceedings of the IEEE*, Vol. 73, No. 2, February 1985, p. 303.
Figure 7.14 Stimson, G.W., *Introduction to Airborne Radar,* 1983, Hughes Aircraft Company, El Segundo, California, p. 435.
Figure 7.15 Long, M.W., Medium PRF for the AN/APG-66 Radar, *Proceedings of the IEEE*, Vol. 73, No. 2, February 1985, p. 303.
Figure 7.16 Willems, Airborne Radar and the Three PRFs, *Microwave Journal*, 1983, Horizon House Publications, Inc., p. 133.
Figure 7.17 Willems, Airborne Radar and the Three PRFs, *Microwave Journal*, 1983, Horizon House Publications, Inc., p. 133.
Figure 7.18 Stimson, G.W., *Introduction to Airborne Radar,* 1983, Hughes Aircraft Company, El Segundo, California, p. 449.
Figure 7.19 Stimson, G.W., *Introduction to Airborne Radar,* 1983, Hughes Aircraft Company, El Segundo, California, p. 448.
Figure 7.20 Stimson, G.W., *Introduction to Airborne Radar,* 1983, Hughes Aircraft Company, El Segundo, California, p. 446.
Figure 7.21 Stimson, G.W., *Introduction to Airborne Radar,* 1983, Hughes Aircraft Company, El Segundo, California, p. 448.

Figure 7.22 Long, M.W., Medium PRF for the AN/APG-66 Radar, *Proceedings of the IEEE*, Vol. 73, No. 2, February 1985, p. 307.

Figure 7.23 Stimson, G.W., *Introduction to Airborne Radar,* 1983, Hughes Aircraft Company, El Segundo, California, p. 465.

Figure 7.24 Hovanessian, S.A., *Radar System Design and Analysis*, 1984, Artech House, Inc., p. 197.

Figure 7.25 Long, M.W., Medium PRF for the AN/APG-66 Radar, *Proceedings of the IEEE*, Vol. 73, No. 2, February 1985, p. 309.

Figure 7.26 Stimson, G.W., *Introduction to Airborne Radar,* 1983, Hughes Aircraft Company, El Segundo, California, p. 452.

Figure 7.27 Hovanessian, S.A., *Radar System Design and Analysis*, 1984, Artech House, Inc., p. 116.

Figure 7.28 Skillman, W.A. and Mooney, D.H., Multiple High-PRF Ranging, *IRE 5th National Convention on Military Electronics*, 1961, p. 44.

Figure 7.29 Skillman, W.A. and Mooney, D.H., Multiple High-PRF Ranging, *IRE 5th National Convention on Military Electronics*, 1961, p. 44.

Figure 7.29 Skillman, W.A. and Mooney, D.H., Multiple High-PRF Ranging, *IRE 5th National Convention on Military Electronics*, 1961, p. 45.

Figure 7.31 Hetrich, G.R., Frequency Modulation Techniques as Applied to Pulse Doppler Radar, *IRE Convention Record*, 1962, p. 86.

Figure 7.32 Hetrich, G.R., Frequency Modulation Techniques as Applied to Pulse Doppler Radar, *IRE Convention Record*, 1962, p. 86.

Figure 7.33 Stimson, G.W:, *Introduction to Airborne Radar,* 1983, Hughes Aircraft Company, El Segundo, California, p. 233.

Figure 7.24 Stimson, G.W., *Introduction to Airborne Radar,* 1983, Hughes Aircraft Company, El Segundo, California, p. 234.

Figure 7.35 Stimson, G.W., *Introduction to Airborne Radar,* 1983, Hughes Aircraft Company, El Segundo, California, p. 235.

Figure 7.36 Stimson, G.W., *Introduction to Airborne Radar,* 1983, Hughes Aircraft Company, El Segundo, California, p. 236.

Figure 7.37 Stimson, G.W., *Introduction to Airborne Radar,* 1983, Hughes Aircraft Company, El Segundo, California, p. 236.

Figure 7.38 Stimson, G.W., *Introduction to Airborne Radar,* 1983, Hughes Aircraft Company, El Segundo, California, p. 462.

Figure 7.39 Stimson, G.W., *Introduction to Airborne Radar,* 1983, Hughes Aircraft Company, El Segundo, California, p. 463.

Figure 7.41 After Holbourn, "Performance Analysis of Airborne Pulse Doppler Radar," IEEE *International Radar Conference*, 1985, IEEE, p. 13.

CHAPTER 8

Figure 8.1 Ivanov, A., Semi-Active Radar Guidance, *Microwave Journal*, September 1983, Horizon House Publications, Inc., p. 106.

Figure 8.3 Shohat, M., Doppler Beam Sharpened Attack Radar Uses Magnetron, *Military Electronics/Countermeasures*, October 1979, p. 60.

Figure 8.4 Shohat, M., Doppler Beam Sharpened Attack Radar Uses Magnetron, *Military Electronics/Countermeasures*, October 1979, p. 60.

Figure 8.9 Hanle, E., Survey of Bistatic and Multistatic Radar, *IEE Proceedings-F, Communications, Radar and Signal Processing*, Vol. 133, Part F, No.7, December 1986, Institution of Electrical Engineers, p. 589.

Figure 8.10 Hanle, E., Survey of Bistatic and Multistatic Radar, *IEE Proceedings-F, Communications, Radar and Signal Processing*, Vol. 133, Part F, No.7, December 1986, Institution of Electrical Engineers, p. 591.

Figure 8.11 Hanle, E., Survey of Bistatic and Multistatic Radar, *IEE Proceedings-F, Communications, Radar and Signal Processing*, Vol. 133, Part F, No.7, December 1986, Institution of Electrical Engineers, p. 591.

Figure 8.14 After Griffiths, H., and Carter, S., Provision of MTI in an Independent Bistatic Radar Receiver, *The Radio and Electronic Engineer*, Vol. 54, No. 7-8, August 1984, p. 339.

Figure 8.15 Ziemer, R.E., *et al.*, "Degradation Analysis of Pulse Doppler Radars Due to Signal Processing," *NAECON 1977 Record*, 1977, IEEE, p. 945.
Figure 8.16 Long. J.J. and Ivanov, A. Radar Guidance of Missiles, *Electronic Progress*, Vol. 16, No. 3, 1974, Raytheon Company, p. 24.
Figure 8.18 Stimson, G.W., *Introduction to Airborne Radar,* 1983, Hughes Aircraft Company, El Segundo, California, pp. 253 and 466.
Figure 8.20 Long. J.J. and Ivanov, A., Radar Guidance of Missiles, *Electronic Progress*, Vol. 16, No. 3, 1974, Raytheon Company, p. 22.
Figure 8.25 Skolnik, M.I., *Introduction to Radar Systems*, 2nd Ed., 1980, McGraw-Hill Book Company, Inc., p. 518.
Figure 8.24 Ivanov, A., Semi-Active Radar Guidance, *Microwave Journal*, September 1983, Horizoı House Publications, Inc., p. 118.
Figure 8.26 Skolnik, M.I., *Introduction to Radar Systems*, 2nd Ed., 1980, McGraw-Hill Book Company, p. 518.
Figure 8.27 Wehner, D.R., *High Resolution Radar*, 1987, Artech House, Inc., p. 275.
Figure 8.28 Wehner, D.R., *High Resolution Radar*, 1987, Artech House, Inc., p. 275.
Figure 8.29 Stimson, G.W., *Introduction to Airborne Radar,* 1983, Hughes Aircraft Company, El Segundo, California, p. 575.
Figure 8.30 Wehner, D.R., *High Resolution Radar*, 1987, Artech House, Inc., p. 276.
Figure 8.31 Ausherman, D.A., *et al.*, Developments in Radar Imaging, *IEEE Transactions on Aerospace and Electronic Systems*, July 1984, Vol. AES-20, No. 4, July 1984, p. 364.
Figure 8.32 Wehner, D.R., *High Resolution Radar*, 1987, Artech House, Inc., p. 286.
Figure 8.33 Wehner, D.R., *High Resolution Radar*, 1987, Artech House, Inc., p. 288.
Figure 8.34 Wehner, D.R., *High Resolution Radar*, 1987, Artech House, Inc., p. 289.
Figure 8.35 Wehner, D.R., *High Resolution Radar*, 1987, Artech House, Inc., p. 290.
Figure 8.36 After *Aviation Week*, September 7, 1987, p. 88.
Figure 8.37 Ausherman D.A., *et al.*, Developments in Radar Imaging, *IEEE Transactions on Aerospace and Electronics Systems*, July 1984, Vol. AES-20, No. 4, July 1984, p. 383.
Figure 8.38 Ausherman, D.A., *et al.*, Developments in Radar Imaging, *IEEE Transactions on Aerospace and Electronics Systems*, July 1984, Vol. AES-20, No. 4, July 1984, p. 386.
Figure 8.39 Ausherman, D.A., *et al.*, Developments in Radar Imaging, *IEEE Transactions on Aerospace and Electronics Systems*, July 1984, Vol. AES-20, No. 4, July 1984, p. 387.
Figure 8.40 Ausherman, D.A., *et al.*, Developments in Radar Imaging, *IEEE Transactions on Aerospace and Electronics Systems*, July 1984, Vol. AES-20, No. 4, July 1984, p. 387.
Figure 8.41 Brookner, E., ed., *Aspects of Modern Radar*, 1988, Artech House, Inc., p. 434.
Figure 8.43 Brookner, E., ed., *Aspects of Modern Radar*, 1988, Artech House, Inc., p. 436.
Figure 8.44a Brown, R., *et al.*, *Clutter Considerations in Modern Radar Systems*, RADC Report TR-88-252, October 1988, p. 11.
Figure 8.44b Brown, R., *et al.*, *Clutter Considerations in Modern Radar Systems*, RADC Report TR-88-252, October 1988 p. 16.
Figure 8.44c Brown, R., *et al.*, *Clutter Considerations in Modern Radar Systems*, RADC Report TR-88-252, October 1988 p. 17.
Figure 8.45 Sugie, M., *et al.*, Advanced Air Route Surveillance Radar with Various Unique Clutter Elimination Capabilities, *NEC Research and Development*, No. 39, 1975, NEC, p. 76.
Figure 8.46 Sugie, M., *et al.*, Advanced Air Route Surveillance Radar with Various Unique Clutter Elimination Capabilities, *NEC Research and Development*, No. 39, 1975, NEC, p. 76.
Figure 8.47 Wehner, D.R., *High Resolution Radar*, 1987, Artech House, Inc. p. 139.
Figure 8.48 Skolnik, M.I., ed., *Radar Handbook*, 1st Ed., 1970, McGraw-Hill Book Company, p. 17-51.
Figure 8.49 Klauder, J.R., *et al.*, The Theory and Design of Chirp Radars, *The Bell System Technic Journal*, July 1960, AT&T, p. 781.
Figure 8.50 Skolnik, M.I., ed., *Radar Handbook*, 1st Ed., 1970, McGraw-Hill Book Company, p. 17-53.
Figure 8.51 Barton, D.K., *Modern Radar System Analysis*, 1988, Artech House, Inc., p. 230.
Figure 8.52 Brookner, E., *Aspects of Modern Radar*, 1988, Artech House, Inc., p. 488.

Figure 8.53 Brookner, E., *Aspects of Modern Radar*, 1988, Artech House, Inc., p. 496.
Figure 8.54 Brookner, E., *Aspects of Modern Radar*, 1988, Artech House, Inc., p. 497.
Figure 8.55 Brookner, E., *Aspects of Modern Radar*, 1988, Artech House, Inc., p. 502.
Figure 8.56 Brookner, E., *Aspects of Modern Radar*, 1988, Artech House, Inc., p. 503.
Figure 8.57 Brookner, E., *Aspects of Modern Radar*, 1988, Artech House, Inc., p. 506.
Figure 8.58 Zrnic, D.S., Spectrum Width Estimates for Weather Echoes, *IEEE Transactions on Aerospace and Electronic Systems*, September 1979, Vol. AES-15, No. 5, p. 617.
Figure 8.59 Zrnic, D.S., Spectral Moment Estimates from Correlated Pulse Pairs, *IEEE Transactions on Aerospace and Electronic Systems*, Vol. AES-13, No. 4, July 1977, IEEE, p. 345.
Figure 8.60 Strauch, L., Improved Doppler Velocity Estimates by the Poly-Pulse-Pair Method, *Meteorological Conference*, 1978, p. 376.
Figure 8.61a Strauch, L., Improved Doppler Velocity Estimates by the Poly-Pulse-Pair Method, *Meteorological Conference*, 1978, p. 376.
Figure 8.61b Strauch, L., Improved Doppler Velocity Estimates by the Poly-Pulse-Pair Method, *Meteorological Conference*, 1978, p. 377.
Figure 8.62 Brookner, E., *Aspects of Modern Radar*, 1988, Artech House, Inc., p. 500.
Figure 8.63 After Doviak, R. and Zrnic, D., The Weather Equation — Receivers Bandwidth Considerations, *Meteorological Conference*, 1978, p. 487.
Figure 8.64 After Doviak, R, and Zrnic, D., The Weather Equation — Receivers Bandwidth Considerations, *Meteorological Conference*, 1978, p. 487.
Figure 8.6 5 After Doviak, R. and Zrnic, D., The Weather Equation — Receivers Bandwidth Considerations, *Meteorological Conference*, 1978, p. 488.
Figure 8.66 After Groginsky, H., and Glover, K., Weather Radar Canceller Design, *Meteorological Conference*, 1987, p. 192.

APPENDIX B

Figure B.5 Helstrom, C.W., *Statistical Theory of Signal Detection*, 1968, Pergamon Press, p. 106.
Figure B.13 Helstrom, C.W., *Statistical Theory of Signal Detection*, 1968, Pergamon Press, p. 170.

APPENDIX C

Table C.2 Urkowitz, H., Analysis and Synthesis of Delay Line Periodic Filters, *IRE Transactions on Circuit Theory*, Vol. CT-4, No. 2, June 1957, p. 52-54.
Figure C.1 Truxal, J., *Automotive Feedback Control System Synthesis*, p. 504.
Figure C.2 Skolnik, M.I., ed., *Introduction to Radar Systems*, 2nd Ed., 1980, McGraw-Hill Book Company, p. 118.
Figure C.3 Urkowitz, H., Analysis and Synthesis of Delay Line Periodic Filters, *IRE Transactions on Circuit Theory*, Vol. CT-4, No. 2, June 1957, p. 43.
Figure C.4 Urkowitz, H., Analysis and Synthesis of Delay Line Periodic Filters, *IRE Transactions on Circuit Theory*, Vol. CT-4, No. 2, June 1957, p. 43.
Figure C.5 Urkowitz, H., Analysis and Synthesis of Delay Line Periodic Filters, *IRE Transactions on Circuit Theory*, Vol. CT-4, No. 2, June 1957, p. 48.
Figure C.6 White, W.D. and Ruvin, A.E., Recent Advances in the Synthesis of Comb Filters, *IRE National Convention Record*, 1957, p. 194.
Figure C.7 White, W.D. and Ruvin, A.E., Recent Advances in the Synthesis of Comb Filters, *IRE National Convention Record*, 1957, p. 196.
Figure C.8 Urkowitz, H., Analysis and Synthesis of Delay Line Periodic Filters, *IRE Transactions on Circuit Theory*, Vol. CT-4, No. 2, June 1957, p. 48.
Figure C.9 Berkowitz, R.S., *Modern Radar — Analysis, Evaluation, and System Design*, 1965, John Wiley & Sons, Inc., p. 515.
Figure C.10 Skolnik, M.I., ed., *Radar Handbook*, 1st Ed., 1970, McGraw-Hill Book Company, p. 35-5.

Preface

A *doppler radar* is defined by the IEEE Standard Radar Definitions as one which utilizes the *Doppler effect* to determine the radial component of relative radar-target velocity or to select targets having particular radial velocities. When a doppler radar uses pulsed transmissions, it is called a *pulsed doppler* (PD) radar. Further, *moving target indication* (MTI) is defined as a technique that enhances the detection and display of moving radar targets, where doppler processing is one method of implementation. These formal definitions point to the central purpose of MTI and PD radars, which is to use the Doppler effect to separate small, moving targets from competing extraneous returns, called clutter, which typically result from reflections obtained from the ground, sea, rain and other hydrometers, chaff, birds, insects, and aurora.

The ability of MTI and PD radars to detect moving targets provides a critical function for the operation of most modern radars. A typical civil application is in air surveillance, where small nontransponder-equipped aircraft must be detected over land in heavy ground and weather clutter for air traffic control purposes. Military applications range from the detection of low flying aircraft or missiles from a surface-based radar to airborne applications such as AWACS, airborne early warning (AEW), and airborne intercept (AI) radars, which involve looking down into a heavy surface clutter environment.

Although similar in nature, there is a distinct theoretical difference between MTI and PD radars. The MTI radar ideally functions to eliminate clutter through the use of a comb filter, the stopband of which is placed in regions of heavy clutter concentrations. Moving targets are passed through those velocity regions not occupied by the clutter. PD radars function to resolve and enhance targets within a particular velocity band while rejecting clutter and other echoes outside the velocity band of interest. PD radars generally use a contiguous bank of doppler filters matched to the target response through the velocity range of interest. The bank of doppler filters serves to integrate the target's response coherently with respect to noise, and hence the PD radar provides the potential for enhanced detection performance as compared to the MTI radar.

MTI and Pulsed Doppler Radar provides an in-depth treatment of the theoretical and practical considerations involved with the system design and performance analysis of such radars. The book emphasizes a basic understanding of radar clutter properties and how an MTI or PD radar functions to extract moving radar targets from a clutter background. A detailed theoretical basis is established for the optimum detection of moving radar targets in clutter which is related to MTI and PD implementations. Detailed derivations are given for the various theoretical relationships used in these systems, followed by practical examples to illustrate their use.

This book serves as a sequel to an anthology edited by the author, *MTI Radar,* published in 1979, which at that time brought together much of the significant material on this subject. The earlier volume was composed mostly of selected reprints, liberally interspersed with comments relative to the significance of each article and also contained a section on clutter models and MTI formulas. The present book expands on much of the material in *MTI Radar,* gives detailed derivations of the theoretical results and examples, while adding extensive material on PD radars. In addition, a chapter is provided to cover specialized applications of pulsed doppler radars, including such diverse topics as inverse synthetic aperture radars (ISAR), doppler weather radars, and the use of MTI in bistatic radars.

Chapter 1 covers the fundamental principles of doppler processing, and gives an overview of significant results relating to MTI and PD radars. The overall radar detection problem in clutter and a receiver noise background is discussed from the viewpoint of a radar resolution cell. Optimal MTI and PD radar solutions are described. The properties of various clutters are described through the use of the environmental diagram. This leads to a discussion of both Gaussian and non-Gaussian clutter models, which are useful for doppler radar analysis. The need for adaptive clutter filtering that uses modern spectrum analysis techniques is described. The chapter ends with an in-depth description of the moving target detector (MTD), which is one of the more powerful implementations of modern doppler processing in air surveillance radars. The MTD is an example of a low-PRF (pulse repetition frequency) PD radar processor, which provides almost optimal detection performance against moving targets while incorporating various false alarm control mechanisms such as doppler filter CFAR (constant false alarm rate) and digital clutter maps.

Chapter 2 describes waveform design and doppler filtering principles as they apply to MTI and PD radars. In general, the waveform for a PD radar can be classified as either a low-PRF (unambiguous range), high-PRF (unambiguous doppler), or medium-PRF (ambiguous in both range and doppler) design. Each type of waveform interacts with the clutter in a different way and requires a different form of signal processing. The choice of the waveform is strongly influenced by the application and performance requirements. Doppler filter requirements are generally dependent on the type of waveform utilized. The transversal type of doppler filter is optimum, but also hardware-intensive. Comb-type filters (i.e., delay line cancellers) are used in MTI applications or in front of doppler filter banks as a clutter notch filter. Contiguous banks of doppler filters are often implemented digitally using the FFT algorithm. Modern radars use digital signal processors, which are discussed in detail at the end of that chapter.

Chapter 3 discusses the detection performance that can be achieved by using PD and MTI radars. Performance is related to the usual probability of detection (P_d) and probability of false alarm (P_{fa}) criteria generally used in radar analysis. These are related to the MTI improvement factor, which establishes a standard measure of performance for doppler radars. The improvement factor is related to two distinct measures. The first is the action of an ideal doppler processor on the clutter spectrum. The second is due to system instabilities such as transmitter and stable local oscillator (STALO) noise. In general, system instabilities form a performance floor, which is achieved if appropriate radar and doppler filter parameters are designed into the system.

Chapter 4 discusses the subject of radar clutter. We describe the average backscattering coefficient and spectra for land, sea, weather, and chaff clutter, as determined from various experimental programs. Parametric approximations are given for the data. Various clutter models such as the Gaussian, Rician, log-normal, Weibull, and K-distribution are described in detail. Several environmental models that can be used for computer simulations of clutter are also described.

Chapter 5 discusses the theoretical bases of MTI and pulsed doppler radars. This treatment starts from the general problem of detecting *n* doppler-shifted coherent signals with random initial phase in a background of Gaussian clutter and receiver noise. The resulting linear transversal filter with complex weights followed by an envelope detector is shown to be the optimal processor. The MTI solution is shown to apply when all target velocities are equally likely, while the PD solution applies when the target doppler signature is known *a priori*. The performance of the optimal processor is then compared with that of a quasioptimal processor, which consists of an MTI followed by a coherent integrator (e.g., fast Fourier transform, FFT), and further with an MTI followed by an incoherent integrator.

Chapter 6 treats various aspects of MTI system design. We describe different types of coherent and noncoherent MTIs. Performance drivers for surface-based and airborne types of systems are quantified. We include a section on the MTI performance limitations of a nonlinear receiver, using receiver distortion theory. Recursive and nonrecursive MTI filter design theory is given with design examples. The generally ambiguous velocity response of MTI systems results in blind speeds about the radar's PRF, which can be eliminated by staggering the PRF. Staggering the radar's PRF, in turn, results in a modified MTI velocity response, which is a function of the stagger pattern and the order of the canceller employed.

Chapter 7 discusses various aspects of pulsed doppler system design. General PD configurations and waveforms are described with emphasis on performance limitations. The problem of resolving range ambiguities in high-PRF designs through the transmission of multiple PRFs or FM ranging waveforms is discussed in detail. The chapter concludes with a discussion of main-lobe clutter performance requirements. This becomes the critical design parameter in many high-PRF PD radars due to the folding of clutter from multiple ranges into a single range cell, which hence competes with the target response.

Chapter 8 treats a number of specialized uses of the Doppler effect in radar systems. The discussions are self-contained and provide sufficient detail to allow a clear understanding of the principles involved. The topics covered are doppler beam sharpening, bistatic radar MTI, semiactive doppler missile guidance, ISAR, low observable target MTI, dual-beam antennas for surveillance radars with MTI, and pulse compression with MTI and doppler weather radar. The last topic provides an interesting situation, where the PD radar uses the Doppler effect both to measure pertinent characteristics of the target response (i.e., the velocity and turbulence of the weather) and also to reject ground clutter that will contaminate this measurement.

The material in the first three chapters and part of the fourth chapter has been used to teach a three-day short course on MTI and PD radars at the George Washington University. The course starts with an exposition of the properties of clutter and then describes the processing effects of doppler filters on the clutter. Covered next are performance calculations on various types of MTI and PD radars, using the MTD processor as a primary example. The course concludes with a discussion of optimal doppler processing theory and comparisons with quasioptimal systems by using FFT-based processors. Also discussed is doppler weather radar pulse-pair processing due to its contrast with conventional doppler processing.

As is appropriate, the author acknowledges a number of people who provided indispensable help in preparing the manuscript for publication. My wife, Carol, typed the entire manuscript and its many revisions, including the detailed equations that appear throughout the book. David K. Barton constructively reviewed the entire manuscript and suggested many useful revisions, which were incorporated into the text. Also, over the years, the author has received useful comments from many people in the field about *MTI Radar,* which were incorporated into the present text.

D. Curtis Schleher
July 1990

Chapter 1
Fundamental Principles and Overview

Moving target indication (MTI) and *pulsed doppler* (PD) radars provide the capability to detect target reflections having differential radial motion with respect to an interfering background (called *clutter*) that might typically consist of reflections from terrain, sea, weather, or chaff. MTI and PD radars can be operated from either a fixed based or a moving platform such as on a ship, aircraft, or satellite [1–4].

A typical application of these radars is surveillance to detect low-flying aircraft moving over terrain through possible weather disturbances. The function of the MTI or PD radar in this situation is to reject the returns from terrain and weather while retaining the return from the aircraft, thereby allowing target detection.

When the radar is surface-based, the terrain reflections, called *ground clutter,* usually are stronger than expected target returns. The ground clutter extends out to a range where terrain features that cause the clutter are masked due to the earth's curvature. This can be for ranges on the order of 10–20 nmi over relatively flat terrain to over 50 nmi, where cities and mountainous terrain are within the radar's field of view. When the radar is airborne, the ground clutter generally extends to the full operating range of the radar.

The intensity of weather clutter is in proportion to the fourth power of frequency from low frequencies up to the millimeter-wave region. This type of clutter becomes significant at L-band frequencies and above, and it differs from ground and sea clutter in that a volumetric distribution of scatterers is involved, rather than a surface distribution. Weather clutter generally extends to the maximum operating range of the radar.

The physical mechanism that allows detection of the aircraft in the surveillance radar application is a unique transformation that shifts the frequency of the reflected return from each scatterer within the radar's *field of view* (FOV) in direct proportion to its relative radial motion with respect to the radar. The transformation is due to the compression or extension of the radar wave caused by the relative radial motion of the scatterer with respect to the radar, and is generally known as the *doppler effect*. The magnitude of the doppler shift frequency can be computed by noting that the phase of the return from a scatterer, when received at the radar, is given by $\theta = 4\pi r/\lambda$ radians, where r is the range to the scatterer and λ is the radar's apparent wavelength. The instantaneous frequency from a scatterer moving radially with respect to the radar is then given by $f_d = d\theta/2\pi\, dt = 2v_t/\lambda$, where v_t is the scatterer's radial velocity. Doppler signatures from aircraft will generally be different from those of ground, sea, or weather clutter, thus allowing detection through use of appropriate doppler filters.

Figure 1.1 illustrates effective operation of an MTI radar. The normal video display shows ground clutter out to a range of 40 nmi. The MTI display shows elimination of the ground clutter while retaining aircraft target reports. The moving targets are recorded for three antenna revolutions, resulting in the triple blips shown [5].

The function of most radars is to supply position location data on targets of interest. In MTI and PD radars, this is a two-stage process whereby the radar first uses doppler filtering to separate the target signal from its surroundings, which consist

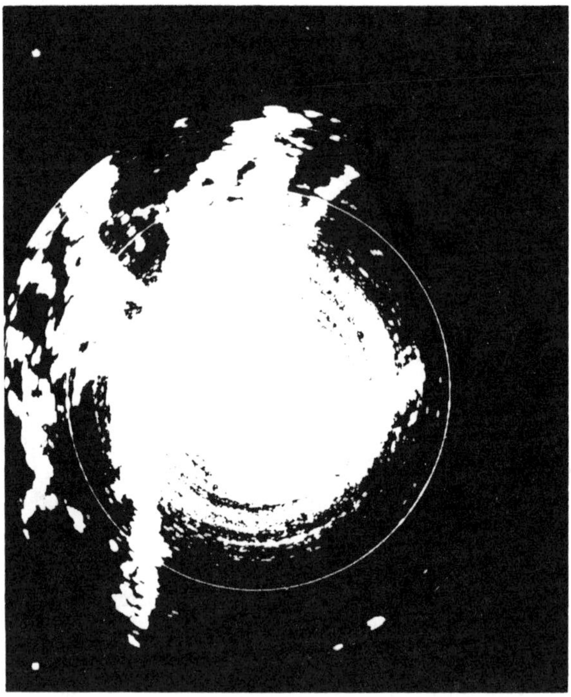

(a)

Figure 1.1 Effective MTI radar performance.

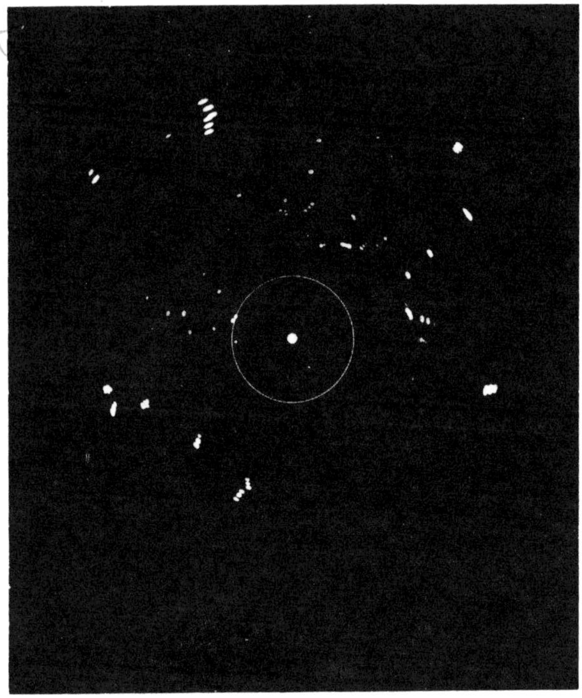

(b)

Figure 1.1 continued.

of reflections from terrain, weather, chaff, and other extraneous targets, and then performs the detection and location measurement on the extracted target. The focus in MTI and PD radars is generally on the doppler filtering process. However, the overall performance of the radar involves a consideration of the overall detection and location process.

1.1 DOPPLER PROCESSING FUNDAMENTALS

1.1.1 Resolution Considerations [6,7]

The performance of an MTI or PD radar is intimately related to its resolution properties. Resolution is defined as the ability of the radar to distinguish between two closely spaced targets in at least one of the state variables used to describe the target vector. State variables associated with radar targets generally form a four-dimensional space consisting of azimuth angle, elevation angle, range, and radial velocity (doppler).

The angular states are generally decoupled from the range-velocity states, thereby allowing the resolution properties of the radar to be described by two vectors: one consisting of the angle dimensions (azimuth and elevation angles) and the other the temporal-spectral (range-doppler) dimensions.

Radar waveform design is concerned with resolution in the range-doppler dimensions. The radar ambiguity function, which displays the complex envelope of the response of a matched filter receiver to the radar transmitter waveform reflected from a point scatterer, is useful for visualizing the resolution properties of the radar waveform. In a three-dimensional plane, this diagram displays the response to time-delay (range) in one dimension, doppler-shift (frequency shift) in an orthogonal direction, and the magnitude of the receiver response in the third, orthogonal, dimension. The desirable thumbtack type of response produced by a noiselike waveform, as shown in Figure 1.2, depicts the features of the ambiguity function response.

Resolution in the range-doppler domain can be visualized by superimposing two ambiguity diagrams displaced by the range-doppler spacing of the scatterers. With the idealized diagram of Figure 1.2, which is characterized by an isolated central spike, we can see that the range resolution is enhanced by transmitting a greater bandwidth (B) because it compresses the extent in the range plane. Doppler resolution is increased by using a longer waveform (T) because it compresses the extent in the doppler plane. However, the volume under the ambiguity diagram is fixed so that

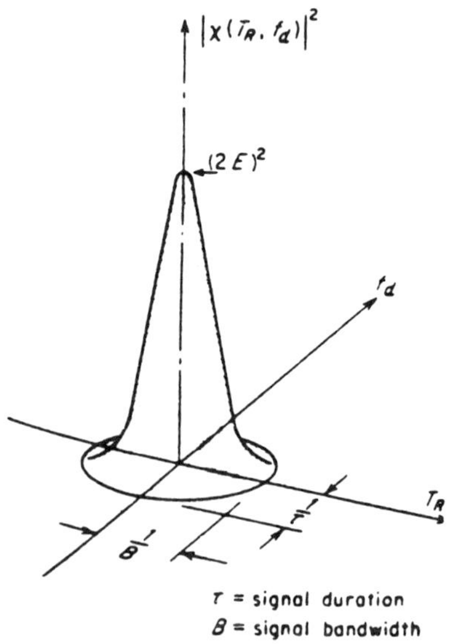

Figure 1.2 Thumbtack radar ambiguity diagram.

the range and doppler resolution must be balanced against each other (i.e., the radar waveform uncertainty principle).

MTI and PD radars represent two alternative waveform strategies for detecting targets based upon doppler filtering principles. The distinction between MTI and pulsed doppler radar is not the use of low, medium, or high *pulse repetition frequency* (PRF), but the use of a passband-stopband filter for MTI as compared to a bank of coherently integrating filters for pulsed doppler radars. Hence, there are medium-PRF MTI systems and low-PRF pulsed doppler systems (e.g., the moving target detector). These distinctions are recognized in Chapter 2.

1.1.2 MTI Doppler Filters in a Surface Clutter Environment

MTI radars use a low-PRF waveform design that is unambiguous in range, but highly ambiguous in doppler response. For surface clutter (sea or ground clutter), this means that only the clutter returned from scatterers within the azimuth beamwidth and effective pulsewidth dimensions compete with the target. The total energy returned from the clutter is minimized by matching the azimuth beamwidth and effective pulsewidth as closely as possible to the corresponding target dimensions. This is usually impossible in the azimuth dimension (a 45-ft aperture at L-band provides a 1° beamwidth), but can be approached in the range dimension using pulse compression to balance resolution and energy. The resulting resolution cell for a low-PRF radar is usually much larger than the target dimension, which allows an appreciable amount of ground clutter energy into the receiver.

Figure 1.3 depicts the resolution cell associated with a surface-based low-PRF radar superimposed on a field of randomly oriented, elemental, stationary point scatterers. When illuminated by the radar transmitter, reflections from those scatterers contained within the resolution cell will form the ground clutter return. Under conditions where the scatterers are numerous and no single scatterer predominates, we can show

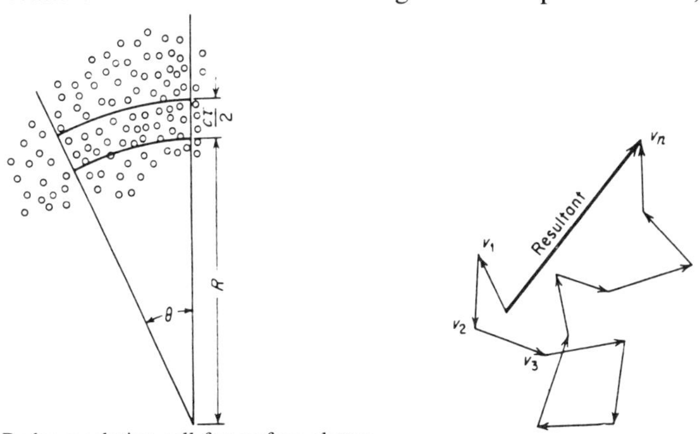

Figure 1.3 Radar resolution cell for surface clutter.

that the ground clutter signal is a Gaussian distributed random process. The power spectral density of the process is determined by the internal radial motion of the clutter scatterers (due to wind-driven vegetation) and antenna scanning motion, which induces spectral spreading into the scatterers' return [1].

A simple target can be represented as a point reflector moving with respect to the elemental clutter scatterers. The target's radial motion causes its return to be shifted in frequency with respect to the clutter scatterers, the return signal spectrum of which is centered about the transmitter's carrier frequency. The target's spectral width, and hence its power spectral density, is determined by antenna scanning motion, which induces spectral spreading onto the target.

The target and ground clutter frequency spectra are depicted in Figure 1.4, where both the underlying (equivalent *continuous-wave,* CW) and sampled-data versions are shown. Based on the somewhat idealized target and clutter model just described, to synthesize an optimum MTI detection processor is possible by using statistical detection theory (see Appendix B).

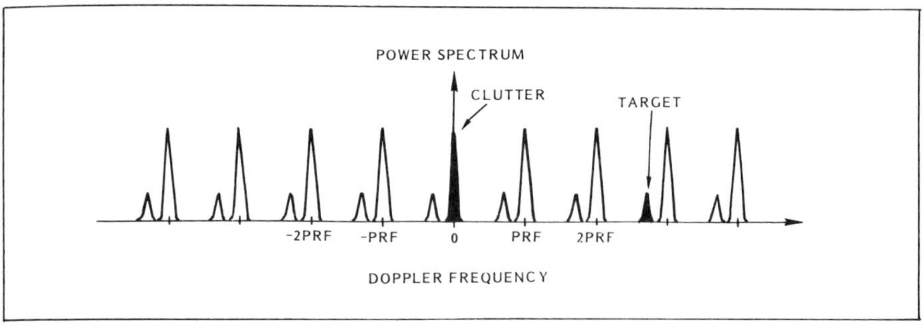

Figure 1.4 Target and clutter frequency spectra.

1.1.2.1 Optimum MTI Filter Theory

The fundamental doppler filtering problem to be solved is the detection of *n* coherently doppler-shifted target pulses imbedded in a background of Gaussian distributed interference consisting of clutter and receiver noise. The initial RF phase of the coherent pulse train returned from the target is uncertain, and hence will be randomly distributed with respect to the phase of the coherently transmitted waveform.

The detailed solution to this problem is discussed in Chapter 5. However, at this point, we will discuss some general features of the solution. First, the general architecture of the optimum doppler processor consists of a linear transversal filter using complex weights (amplitude and phase) followed by an envelope detector (see Section 5.1). The linear filter structure results from the assumed Gaussian statistics of the interference and the envelope detector results from the initial random phase of the coherent pulse train. The complex weights of the transversal filter are determined by the interference (clutter and receiver noise) covariance matrix and the assumed distribution of the signal's complex doppler distribution.

The MTI solution to the optimum doppler processor is obtained when there is no *a priori* knowledge of the doppler shift of the target return. In other words, all possible target doppler shifts are equally likely. As shown in Figure 1.5, the resulting MTI processor consists of two identical quadrature channel transversal filters using real weights, where one channel processes the in-phase (I) component and the other processes the quadrature (Q) channel. The signals from each channel are then envelope-detected, squared, and summed to form the MTI processor output. Note that the MTI processor provides no detection improvement against receiver noise.

The resulting MTI doppler filter response has a comblike shape that attenuates regions (stopband) of heavy clutter concentration and provides minimal attenuation (passband) in regions with little or no clutter energy. A typical single-channel MTI filter response is shown superimposed upon the target and ground clutter spectra in Figure 1.6.

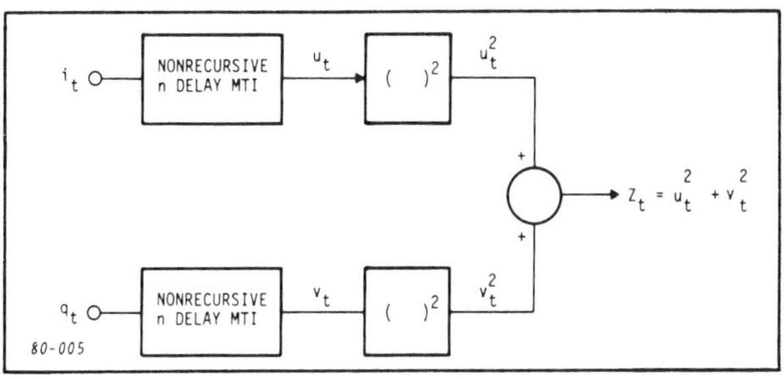

Figure 1.5 Block diagram of quadrature channel MTI.

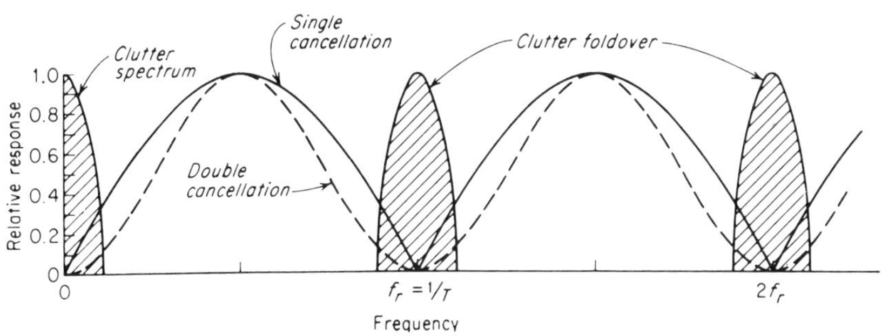

Figure 1.6 MTI filter response and clutter spectra.

A feature of the low-PRF MTI design is that range measurements are unambiguous, whereas doppler measurements are generally ambiguous, creating so-called blind speeds, which occur in small regions about the radial target velocities corresponding to multiples of the radar's PRF. To allow detection of targets in the blind regions, staggering or jittering the MTI radar's PRF is generally necessary, which has the effect of "filling in" the MTI's velocity response to cover the region of interest.

1.1.3 MTI Doppler Filters in a Surface and Volume Clutter Environment

The resolution states that apply to weather or chaff returns (sky or volumetric clutter) are the radar's azimuth and elevation angles and effective pulsewidth. Figure 1.7 depicts how these states are formed into a three-dimensional resolution cell, where orthogonal dimensions consist of the arc subtended by the radar's azimuth and elevation beamwidths and the range equivalent to the radar's effective pulsewidth.

The reflections from all weather disturbances (rain, snow, fog, and clouds) and chaff particles contained within the three-dimensional volumetric resolution cell combine to form the weather clutter or chaff return. Thus, the magnitude of sky clutter is weighted by the effect of both the azimuth and elevation antenna patterns. In addition to this difference, sky clutter has several other distinctive features that distinguish it from ground clutter. Notable among these are the variable mean velocity (doppler frequency), which is proportional to wind speed, and the considerable spectral spreading due to the effects of wind shear and turbulence.

The design of an optimum MTI doppler filter that provides detection of a moving target in the presence of sky clutter follows the same general principles as those described for ground clutter. However, the variability of the sky clutter (weather or chaff) spectra requires some form of adaptive control of the MTI's frequency response. This can be accomplished by monitoring the clutter returns in real time and adjusting the MTI's complex filter weights to minimize its clutter output. The maximum entropy

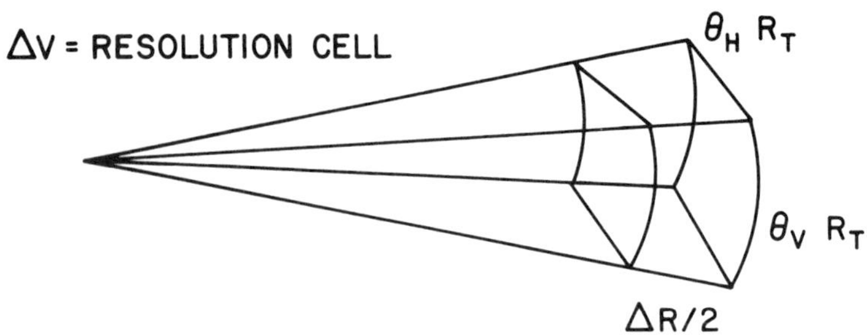

Figure 1.7 Weather clutter or chaff volumetric resolution cell.

algorithm is sometimes used to estimate the characteristics of the clutter spectrum for this purpose [8].

A difficult MTI design problem results when both surface and sky clutter are present simultaneously. Figure 1.8 depicts an approach that uses dual-offset MTI filters to process simultaneously both types of clutter. Adaptive control of the MTI's filter weights is necessary to provide reasonable performance in this situation.

TYPE OF MTI SYSTEM	NO. PULSES NEEDED IN FILTER	CLUTTER REJECTION		CONCLUSION
		GROUND	SKY	
ZERO REF CANCELLER 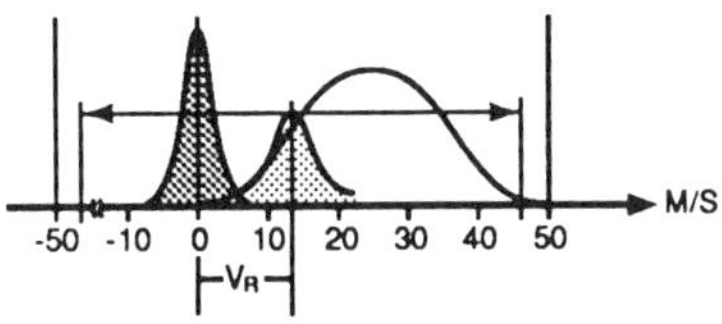	3	VERY GOOD	POOR	APPLY WHEN ONLY GROUND CLUTTER EXISTS
OFFSET CANCELLER 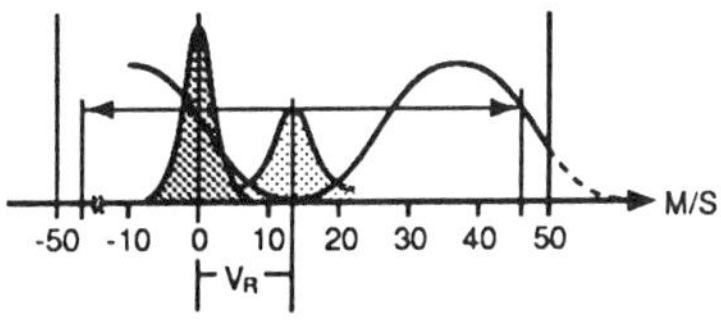	3	POOR	FAIR	APPLY WHEN ONLY MILD SKY CLUTTER EXISTS
DUAL OFFSET CANCELLERS 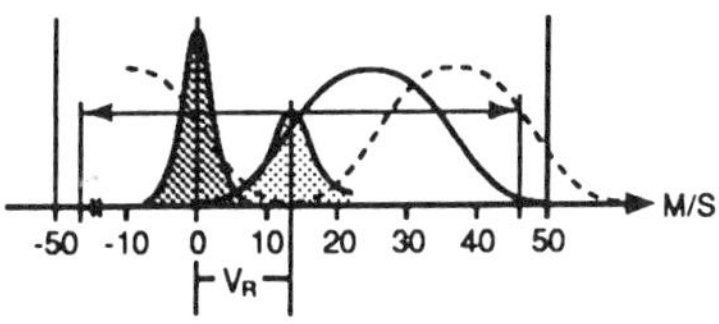	5	VERY GOOD	FAIR	APPLY FOR GROUND AND MILD SKY CLUTTER

Figure 1.8 MTI processing of weather clutter or chaff in the presence of ground clutter.

The mean doppler frequency associated with sky clutter often exceeds the radar's PRF when an MTI design is used. This causes an *aliasing* (spectral foldover) effect, whereby the doppler frequency spectrum of the clutter is folded into the region between zero and the first PRF line. The same effect occurs for targets, which often are also aliased. This results in a situation where the aliased target spectrum may overlap the sky clutter spectrum in the MTI, although the target has a different true doppler than the clutter. The general solution to this problem is to change the radar's PRF, thereby causing the aliased spectra to change their relative positions within the first PRF spectral region.

This general approach is employed in the *moving target detector* (MTD) [1]. An early version of the MTD used an MTI to suppress ground clutter, followed by a *fast Fourier transform* (FFT) doppler filter bank, the function of which was both to suppress sky clutter and to provide coherent integration. The FFT-generated filter bank, as depicted in Figure 1.9, divided the MTI passband into eight contiguous filters. The filters containing weather clutter were suppressed by sensing the heavy clutter content in these filters using a cell-averaging *constant false-alarm rate* (CFAR) to control their detection thresholds. A second PRF was then employed to shift the aliased targets into different doppler filters, thereby allowing them to be detected. The MTI-FFT doppler filter bank combination is replaced in later designs by a more efficient eight-pulse transversal filter design to provide the same function.

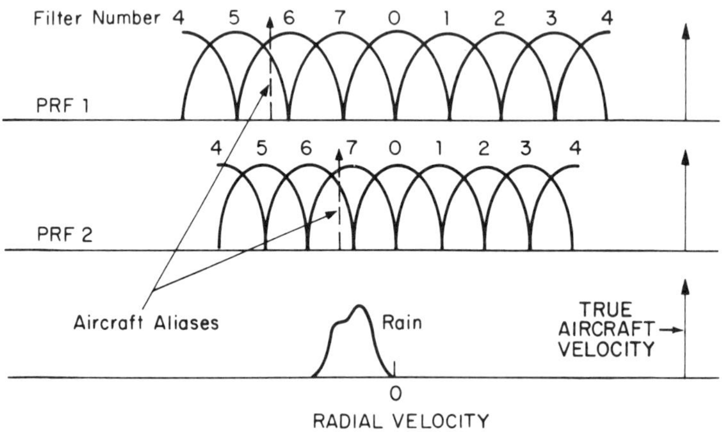

Figure 1.9 MTD processing of weather clutter.

1.1.4 Pulsed Doppler Fundamentals [2]

The pulsed doppler radar operates on the same general principle as an MTI radar, but the objectives and implementation are somewhat different. One basic difference is that the high-PRF pulsed doppler radar operates at a PRF that provides unambiguous doppler frequency returns from the target. This involves using a PRF that is at least twice the doppler frequency, corresponding to the maximum closing speed between the radar and target. Pulsed doppler radars are usually highly ambiguous in the range dimension.

The unambiguous doppler allows implementation of a signal processing filter, which provides a high degree of clutter rejection. The ambiguous range causes clutter returns from many radar resolution cells to fold into the target cell, thereby increasing the clutter power that competes with the target. Also, the resolution of the many possible target ranges is not a trivial problem and usually involves transmitting multiple PRFs with associated computer processing to determine the correct target range [3,9].

Pulsed doppler radars are used in airborne applications (e.g., AWACS) where low-PRF MTI radars are impractical, except at very low transmitter frequencies, due to spectral-spreading effects from platform motion. High-PRF pulsed doppler radars are also used in ground-based applications against low-flying, penetrating targets, where modest ranges are required. High-PRF pulsed doppler radars are particularly effective against chaff and weather clutter.

1.1.4.1 Pulsed Doppler Waveform and Clutter Considerations [2,4]

A block diagram of a simple pulsed-doppler radar and its transmitted waveform are depicted in Figure 1.10. The waveform consists of a coherent RF pulse train (the starting RF phase of successive pulses is separated by an integral number of

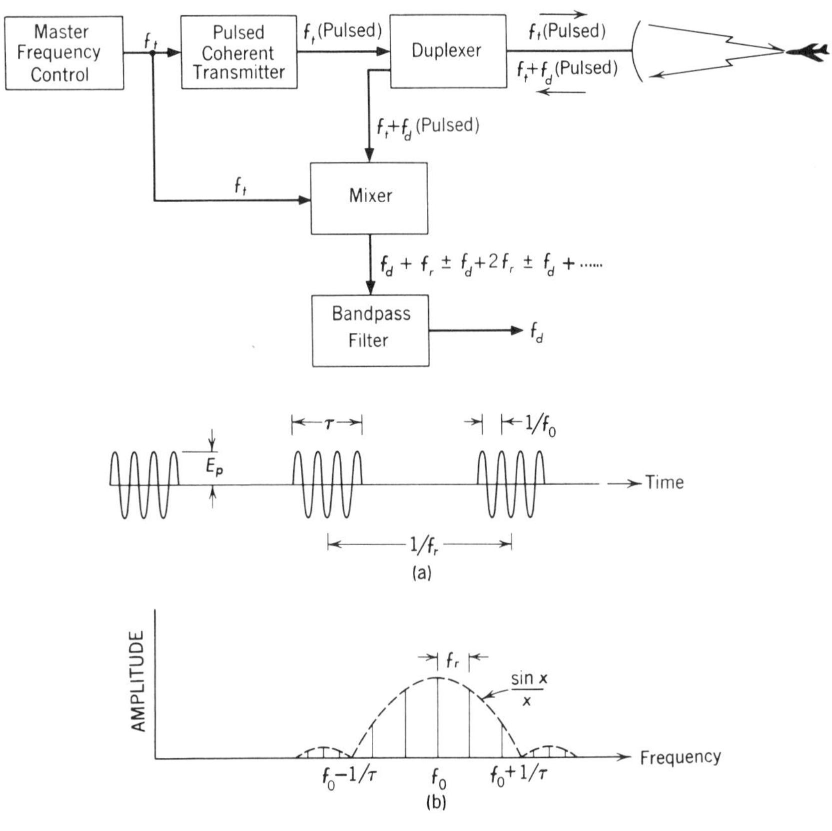

Figure 1.10 Elemental pulsed doppler system.

wavelengths) resulting in a line spectra where adjacent spectral lines are separated by the PRF (f_r). Target and clutter spectra for a single-spectral-line transmission from a moving platform are shown in Figure 1.11. Because of the sidelobes, the frequencies of the fixed clutter returns can vary $\pm 2v_r/\lambda$ from the transmitter frequency. The clutter possesses a high peak, resulting from clutter return in the main beam. For closing targets ($v_t > v_r$), the returns will be shifted in frequency by $2v_t/\lambda$, and therefore appear in the clutter-free portion of the frequency spectrum.

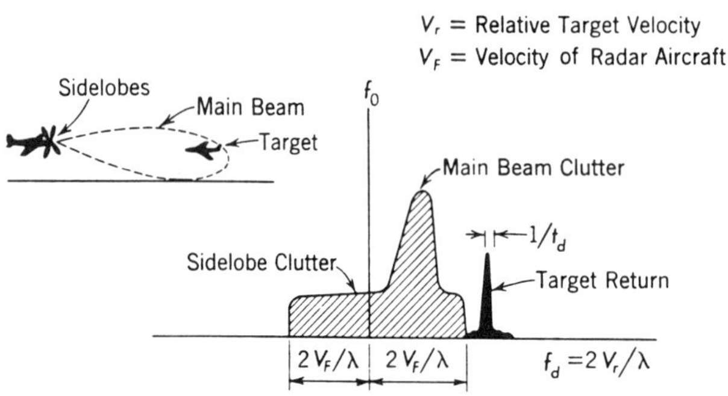

Figure 1.11 Target and clutter spectra for a single-spectral-line transmission.

When the return is mixed with the coherent "sidestep" oscillator, as shown in the block diagram, the output video spectrum is folded over about one-half the PRF. When processed through a bank of contiguous doppler filters, the resulting target doppler output, which appears in the appropriate doppler filter, is simply the doppler return associated with the central line of the transmitter spectrum. In practice, some form of the single-sideband detection process would be employed instead of the simple mixing system shown in Figure 1.10. With this system, the minimum PRF is given by $f_r \geq 2(v_r + v_t \max)/\lambda$ where v_t max is the maximum relative target closing velocity.

The bandwidth (B_D) of the doppler filter generally determines the doppler resolution of the PD radar. In a scanning radar, the bandwidth is often selected so that all of the target returns received during the time that the radar dwells on the target are collected within a single doppler filter. For this situation, the approximate relationship $B_D = 1/T_d$ is generally applicable, where T_d is the dwell time of the main beam on the target. This results in $n = f_r/B_D$ target returns being coherently integrated in the filter. The PD radar thus approximates a matched filter that provides optimum detection in a noise environment.

A variation of the pulsed doppler waveform is the uniform pulse burst (see Figure 1.12), which consists of a finite train of n pulses transmitted at an interpulse period T, where generally $nT \leq T_d$. The period nT is the *coherent processing interval*

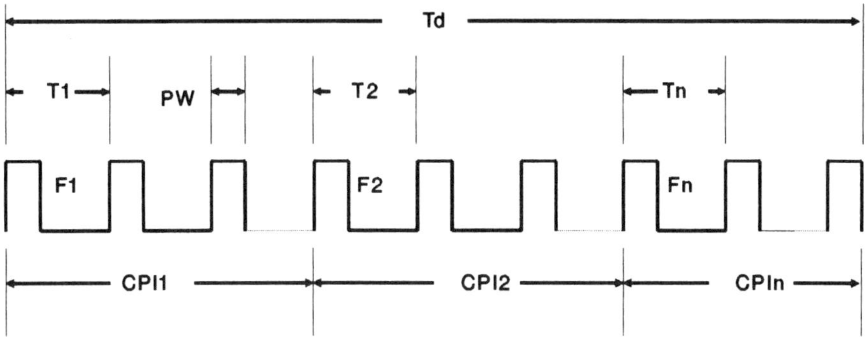

Figure 1.12 The uniform pulse burst waveform.

(CPI). The dwell time (T_d) is filled with several CPIs. By using mixed predetection (coherent) and postdetection integration (matched to T_d), to provide good overall detection performance for the PD radar is possible while changing the transmitted frequency or PRF on a block-to-block (i.e., CPI) basis. The block-to-block frequency changes are useful in applying frequency agility to the PD radar and PRF changes are useful in resolving range ambiguities.

The calculation of the ground clutter spectrum returned to an airborne PD radar is a particularly complex procedure due to the multiple range ambiguities and the spread of the clutter spectrum due to platform motion. Insight into the ground clutter spectrum can be gained from Figure 1.13, which shows the lines of constant doppler returns, called *isodops,* obtained from a horizontal moving platform on which concentric rings are superimposed, representing the ground scatterer returns collected within a particular range gate. The isodops represent the intersection between the ground and the surface of a cone about the radar's velocity vector. Because the angle between this vector and every point on the cone is the same, the return from every point along this hyperbolic line has the same doppler frequency.

The ground clutter spectrum associated with a particular doppler frequency increment is obtained by summing those returns from the incremental shaded areas (shown in Figure 1.13) between two isodops separated by the incremental frequency difference. The ground clutter *power spectral density* (psd) for a particular frequency increment can then be obtained through the following procedure: (1) computing the incremental surface area represented by each shaded area, (2) multiplying each incremental surface area by the backscattering coefficient (σ° with dimensions m^2/m^2) for the appropriate terrain and grazing angle to obtain the equivalent radar cross-sectional area, (3) substituting this cross-sectional area into the radar equation with appropriate weighting for the two-way antenna pattern and variable slant range to obtain the incremental power returned from each elemental area, (4) summing the

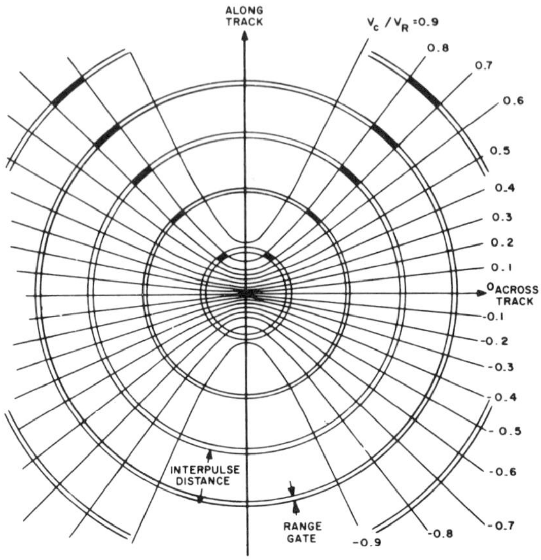

Figure 1.13 Ground clutter isodops from an airborne pulsed doppler radar.

returns from all range-ambiguous incremental areas to obtain the overall incremental psd, and (5) repeating the process for all incremental frequencies over f_d (max) = $\pm\ 2v_r/\lambda$ to obtain the complete spectrum.

The resulting shape of the psd for ground clutter associated with an airborne PD radar is generally as shown in Figure 1.14. Note that for a stationary PD radar the ground clutter spectrum becomes concentrated about each spectral line, which is similar to that of an MTI radar (recall Figure 1.4), except that the power must be summed over the various range ambiguities. The width of the spectrum for the stationary case is due to internal motion and scanning-induced spectral spreading.

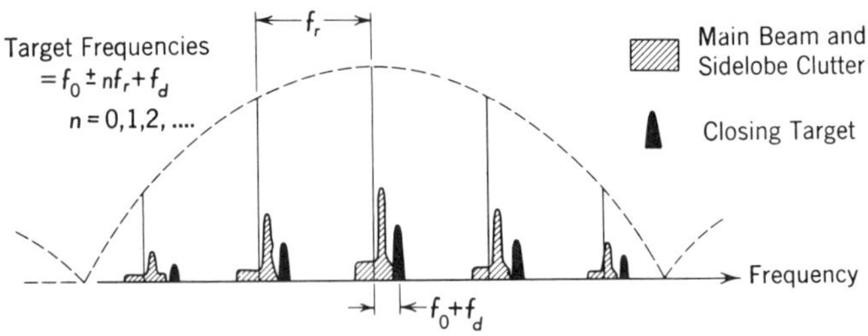

Figure 1.14 Power spectral density for an airborne pulsed doppler radar.

1.1.4.2 Optimum Pulsed Doppler Filter Theory [10]

The general structure of an optimum PD filter is that of a linear complex-weight transversal filter followed by an envelope detector, which is identical to that of the optimum MTI filter previously discussed. However, the weights of the linear transversal filter are optimized under the assumption that the doppler frequency of the target is precisely known at a particular point in the doppler domain. The range of possible target dopplers is then filled with a suitable number of point-optimized complex-weighted transversal doppler filters. The doppler filters are usually separated by a frequency equal to PRF/n, where n is the number of pulses in a CPI.

A quasioptimal pulsed doppler processor used in a number of applications is the cascaded MTI-coherent integrator, where the coherent integrator is implemented by using the FFT algorithm. This processor generally provides less detection performance than the optimal transversal filter, but utilizes considerably less hardware.

1.2 THE ENVIRONMENTAL DIAGRAM [1,11]

The principal means of discerning target echoes from clutter returns in a PD or MTI radar is through doppler processing. Consequently, the way in which the clutter is distributed over the band of possible frequencies, the clutter's doppler spectrum, and the relationship of the spectrum to the doppler frequencies of anticipated targets critically influence the radar's design. One way to examine the clutter and target's frequency distribution is by using the environmental diagram, which provides a pictorial representation of the clutter model.

An environmental diagram for a low-PRF L-band (1.3 GHz) *air traffic control* (ATC) radar is depicted in Figure 1.15 [1]. The limits on the radial velocity of various types of clutter are given on the ordinate, and the range extent is indicated along the abscissa. In the diagram, land clutter is shown to extend from the origin to 20 nmi, which represents the radar horizon for this particular location. The vertical extent of the bar is the standard deviation of the velocity spectrum for ground clutter returns from undeveloped wooded land with a 30-knot wind. Ground clutter is depicted by a solid bar to indicate that it has the highest amplitude return of the clutters shown.

Sea clutter is modeled for a sea-state 4 condition, where a mean velocity of 3.4 m/s and a standard deviation of 1 m/s are representative. The mean velocity considers that the radar is looking directly into the wind or waves. The cross-hatching is more dense near the radar, indicating that the reflectivity is higher near the radar than at lower grazing angles near the horizon. As with ground clutter, sea clutter extends to the radar horizon.

Rain clutter is modeled with an initial mean velocity of 28 m/s, which is appropriate for a moderate altitude with the wind driving the rain directly toward the radar. Like sea clutter, the mean velocity of the precipitation clutter will drop

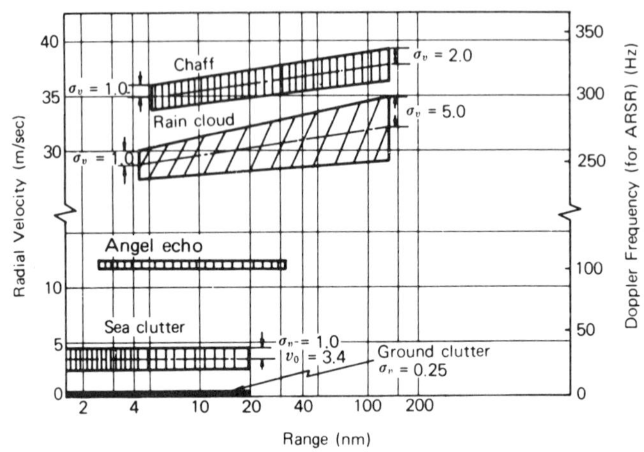

Antenna height : 20 m
Ground clutter : open land with 30 knot wind
Sea clutter : rough sea state

Figure 1.15 Environmental diagram for an air traffic control radar.

to zero under crosswind conditions. A linear gradient of wind velocity with altitude (called *wind shear*) is postulated. This causes the mean velocity of the rain clutter to increase (a linear model is assumed) with range due to the increased height intercepted by the center of the radar beam. The wind shear effect also causes the standard deviation of the rain clutter to increase with range due to the increasing vertical extent subtended by the radar beam at longer ranges. The rain clutter is shown extending to the full operating range of the radar. The standard deviation of the rain clutter spectrum varies from 1 m/s to 5 m/s at the radar's maximum range.

A high-altitude chaff corridor is modeled with characteristics analogous to those of rain clutter. The amplitude is assumed to be greater than that of the rain clutter, but with reduced vertical extent. The standard deviation of the chaff spectrum varies from 1 m/s to 2 m/s at the radar's maximum range.

Angel clutter is generally due to reflections returned from flocks of migrating birds. The return is primarily from the bird's body, which for large birds is resonant at L-band [1]. The spectral spread is thus due to the differential radial velocity among the various birds forming the flock, which is assumed to be relatively narrow. The mean velocity is modeled at 15 m/s, which assumes that the birds are generally moving in a radial direction with respect to the radar. The angel return is shown to a radar horizon of about 30 nmi, which corresponds to a target height of 300 ft for a 20 m antenna height.

The environmental diagram shows the various radial velocity parameters expressed in meters per second, which can be converted to doppler frequency (as in Figure 1.15) through multiplying by the factor $2/\lambda$, where λ is the radar's wavelength in meters. The cross-hatching indicates the normalized reflectivity per unit frequency, rather than received power spectral density. Thus, the cross-hatching for the rain and chaff shown in Figure 1.15 has a lower density in the shear region, indicating that the power spectral density of these echoes is lower for a given reflectivity than for signals with a narrow spectral width, such as ground clutter. This may be significant for pulsed doppler radars having a doppler bandwidth that is less than the standard deviation of the clutter spectral spread.

Figure 1.16 depicts typical received power as a function of range for the various clutters shown in the environmental diagram [1]. Also shown are the signal strength of typical aircraft targets at altitudes of 10,000 ft and 30,000 ft as received by the radar's dual-beam antenna system. Note that the returns from the various clutters generally would prevent detection of the targets without the use of some form of doppler processing.

The ground clutter return is the strongest shown in the diagram. The return exceeds the receiver noise level by more than 100 dB and the target returns by 20 to 50 dB. Precipitation returns are comparable to target returns, but will increase as the fourth power of frequency if the radar's L-band transmitting frequency is increased. Sea clutter returns generally exceed target returns, but require less suppression than

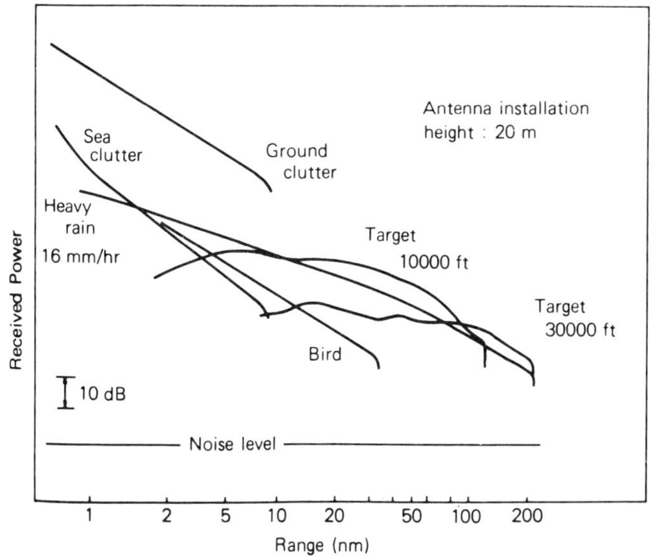

Figure 1.16 Typical received power from clutter and targets.

ground clutter to allow target detection. At short ranges, angel clutter returns are comparable to target returns and generally exceed receiver noise levels.

An alternative representation of the environmental diagram is shown in Figure 1.17 [12]. In this three-dimensional presentation, the power level of the return is shown as the ordinate and the range-velocity matrix is shown in the orthogonal horizontal plane. Thus, this form of the environmental diagram presents the information of Figures 1.15 and 1.16 in a single diagram.

The coordinate system used by both the alternative environmental diagram and the radar ambiguity diagram (see Figure 1.2) are the same. By drawing both diagrams using the same scale, we are able to gain insight into the clutter response of different waveform designs. Preferably, with this method, the ambiguity diagram is drawn as a transparent overlay so that its origin can be superimposed on the positions of the various clutter scatterers in the environmental diagram [11]. This presentation could then reveal an intersection of a high sidelobe of the radar waveform with, for example, a low-intensity rain clutter return, or, alternatively, the intersection with a high-intensity ground clutter return, which would reveal a problem.

The environmental diagram reveals the wide variability of clutter in the range, velocity, and amplitude dimensions. One characteristic is the wide dynamic range of clutter, which, coupled with the desirability of linear processing, presents a formidable problem to achieving the optimal doppler filter processing described in the previous sections. This has led to a number of clutter processing procedures to

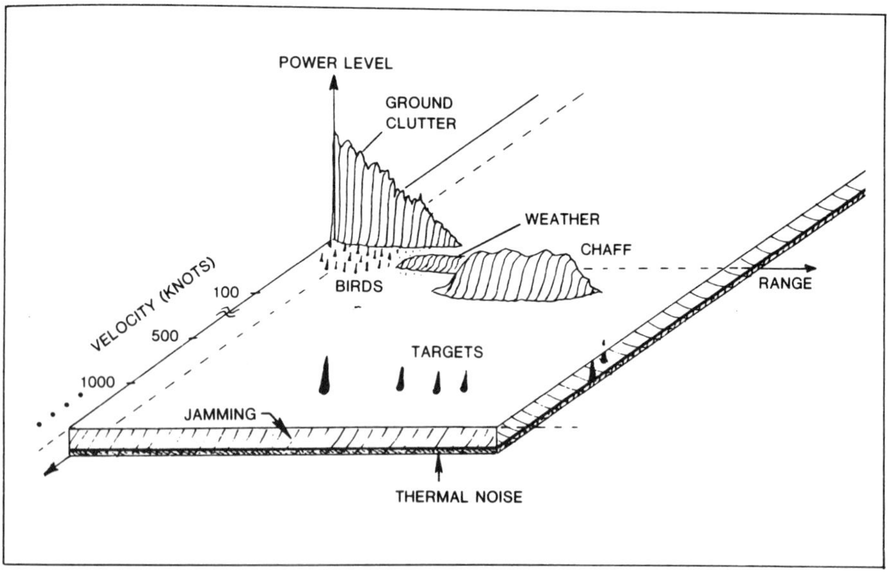

Figure 1.17 Alternative form of environmental diagram.

supplement the doppler filtering processing. The general objectives of these techniques are (1) to reduce the clutter intake of the radar (e.g., pulse compression, dual-beam receiver antenna, circular polarization), (2) to normalize the clutter return (e.g., sensitivity time control, instantaneous *automatic gain control* (AGC), and (3) to suppress or censor particular range cells that have heavy clutter returns (e.g., clutter maps, cell-averaging CFAR).

Use of the last of these techniques is a tacit admission that building a practical doppler processor which suppresses all clutter returns may be impossible. The philosophy of dealing with these clutter "poke throughs" is then to densensitize the radar in those resolution cells, where processing the clutter is impossible, so that an inordinately large number of false alarms do not occur.

1.3 CLUTTER MODELS [1]

A clutter model assumption is implicit in the design or analysis of an MTI or PD radar. The function of the clutter model is to represent the random clutter process in such a manner as to determine system performance. Ideally, the model accurately reflects the clutter process in a realistic manner. Practically, the clutter database is rarely adequate to achieve this objective, and the art of modeling the clutter involves a compromise between accurate representation and analytic convenience.

1.3.1 The Gaussian or Rayleigh Clutter Model [1,13]

The simplest clutter model is the Gaussian model, whereby the random clutter process is represented as a multidimensional Gaussian process. This model has been employed almost exclusively in MTI and PD radar analyses. An attractive feature of this model is that the clutter process is completely specified by its mean value and covariance function. The Gaussian model assumption leads to an optimal doppler processor that is linear.

The Gaussian or Rayleigh envelope clutter model is applicable to distributed clutter returned from a spatially continuous distribution of scatterers wherein no subset of scatterers predominates. This type of clutter is usually associated with weather clutter, chaff, sea clutter observed with low-resolution radar (pulsewidth $> 0.5\ \mu$s), or with a high-resolution radar at high grazing angles ($\phi > 5°$) and land clutter observed from high grazing angles ($\phi > 5°$) over undeveloped terrain.

In the Gaussian model, the clutter signals in the predetection portions of the radar receiver can be represented as the random process $c_t = x_t \cos\omega_c t - y_t \sin\omega_c t$. The term "random process" connotes that the value of c_t does not depend in a completely definite way on the independent time variable t as will be the case for a causal process. The clutter signal information is carried by the jointly normal, zero-mean, identically spectrally distributed, low-pass, independent quadrature processes, x_t

and y_t. A significant property of the c_t process is that any linear operation (such as doppler filtering) on the signal (c_t) does not disturb the Gaussian nature of the distribution, but rather modifies its covariance matrix. Quadrature detection is often used to extract the slowly varying low-pass (the time variation is consistent with the radar's processing bandwidth) quadrature processes, x_t and y_t, which are then jointly processed. When doppler filtering is performed in an MTI or PD radar on the quadrature x_t and y_t processes, the act is generally termed *coherent* (predetection) doppler processing.

The voltage envelope of c_t is given by the random process $v_t = \sqrt{x_t^2 + y_t^2}$, which represents the signal obtained after linear envelope detection. The first *probability density function* (pdf) of v_t describes the probability of finding the envelope-detected voltage v_t in the range $(v, v + dv)$ at any specific time. The envelope-detected voltage, v_t, has a Rayleigh distributed first pdf given by $p(v) = v \cdot \exp(-v^2/2\sigma^2)/\sigma^2$ for $v \geq 0$, where σ is related to the clutter power by the relation $P_c = 2\sigma^2$. For this reason, the clutter model is often termed the "Rayleigh clutter model." However, note that the equivalent *radar cross section* (RCS) of the clutter does not follow a Rayleigh pdf, but rather an exponential pdf, as is described later.

1.3.2 The Radar Cross Section of Clutter

The radar cross section, σ, of an object is defined as the area intercepting that amount of power which, when scattered isotropically, produces an echo equal to that received from the object [14]. For example, a sphere with radius a that is large with respect to a wavelength ($a > \lambda$) scatters isotropically and its mean RCS is given by $\bar{\sigma} = \pi a^2$. The power (P_R) received from a target at range R can then be computed using the standard radar equation given by $P_R = P_t G_t^2 \lambda^2 \bar{\sigma}/(4\pi)^3 R^4$, where P_t is the transmitted power, G_t is the antenna's gain, and λ is the radar's wavelength. Note from this relationship that the mean echo power received by the radar is directly proportional to the mean RCS ($\bar{\sigma}$).

The RCS of extended targets is generally a sensitive function of orientation and wavelength and hence exhibits a range of values. For this situation, an appropriate way to describe the target's RCS is by using a probability density function $p(\sigma)$, the average value of which is given by $\bar{\sigma}$. Through examination of the standard radar equation, one can deduce that the power received by the radar from an extended target exhibits a probability distribution identical to that of its RCS and hence $p(P_R) = p(\sigma)$. Because power is proportional to the voltage squared, the relationship $p(v^2) = p(\sigma)$ is established. Hence, by applying the transformation $\sigma = v^2$ to the Rayleigh voltage distribution for clutter, we can determine that the first probability distribution for the RCS of the Gaussian clutter model is given by the exponential distribution $p(\sigma) = \exp(-\sigma/\bar{\sigma})/\bar{\sigma}$, where $\sigma \geq 0$, and $\bar{\sigma}$ is the clutter's mean RCS. Note that this distribution is completely determined by its mean value $\bar{\sigma}$ and also that the clutter's RCS (σ) is a statistical quantity that varies over a wide dynamic range.

1.3.2.1 Normalized Clutter Backscattering Coefficient

The return from surface or volume clutter is caused by reflections from all scatterers contained within the radar's resolution cell (see Sections 1.1.2 and 1.1.3). Conventional use of the concept of the equivalent RCS replaces the large amount of elemental clutter scatterers contained within a radar resolution cell with a single equivalent scatterer, which returns the same amount of power to the radar and exhibits the same statistical characteristics.

To calculate the RCS of this equivalent scatterer, a normalized parameter called the *backscattering coefficient per unit area,* $\sigma^{\circ}(m^2/m^2)$ is used for surface clutter and the *backscattering coefficient per unit volume,* $\eta(m^2/m^3)$, applies for volumetric clutter. The equivalent clutter RCS is then obtained by multiplying the appropriate backscattering coefficient by either the surface area or volume contained within the radar's resolution cell.

The effective RCS for surface clutter is then given by $\sigma = \sigma^{\circ} A_c$ where A_c is the surface area of the clutter's resolution cell (see Figure 1.3). Note that the clutter's RCS (σ) is proportional to σ° so that both variates have identical statistical properties. In general, the conventional way to specify the backscattering coefficient is by its mean value $\bar{\sigma}^{\circ}$, which is extensively tabulated. For low grazing angles ($\psi < 10^{\circ}$) and narrow azimuthal beamwidths, the resulting mean clutter power received at the radar can then be found by substituting the clutter's RCS value $\bar{\sigma} = (c\tau/2) \cdot \sec\psi \cdot R \cdot \theta \cdot \bar{\sigma}^{\circ}$ into the standard radar equation, where τ is the radar's effective pulse width, c is the velocity of light, R is the slant range to the clutter patch, θ is the radar's 3-dB azimuth beamwidth in radians, and $\bar{\sigma}^{\circ}$ is the tabulated clutter's mean reflectivity per unit surface area.

Figure 1.18 shows the projection of the clutter patch's surface area onto a plane perpendicular to the radar *line of sight* (LOS). This projected incidence area (A_i) determines how much transmitted energy will be intercepted by the surface area (A_c) and is given by the relation $A_i = (c\tau/2) \cdot \tan\psi \cdot R \cdot \theta$ [15]. Note that A_i generally conforms to the usual concept of the capture area of a radar target and, through multiplication by a proportionality factor γ, can be transformed into the effective clutter RCS. The resulting RCS is given by $\sigma = (c\tau/2) \tan\psi \cdot R \cdot \theta \cdot \gamma$ [15]. Comparing this to the alternative representation provides the relationship $\sigma^{\circ} = \gamma \sin\psi$. This representation is sometimes called the *constant Gamma model* for terrain backscatter [16], and its advantage over using the σ° representation is that the parameter γ is independent of the grazing angle. Typical surface backscattering coefficients for both σ° and γ are given in Table 1.1 for 10° grazing angle and an X-band radar [4].

When evaluating the effective RCS of surface clutter (σ), the two-way propagation 3-dB azimuth beamwidth is generally more accurate than the one-way beamwidth (θ) [11,17]. However, most practical data have been tabulated by using the one-way beamwidth [14]. Also, the difference between the two methods is on the order of 1.5 dB, which is generally less than the uncertainty in the actual backscatter data.

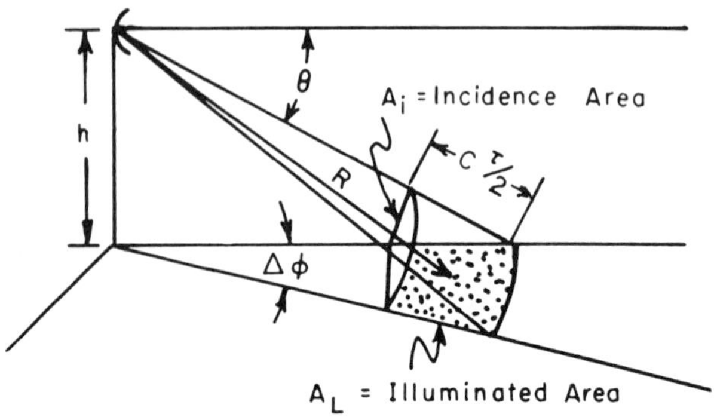

Figure 1.18 Surface clutter area geometry.

Table 1.1
Typical Backscattering Coefficients

Terrain	*Coefficients* (dB)		
	σ^0	γ	
Smooth Water	−53	−45.4	$\psi = 10^0$
Desert	−20	−12.4	$f = 10$ GHz
Wooded Area	−15	−7.4	
Cities	−7	0.6	

The effective RCS of weather or chaff clutter is found through multiplying the radar's volumetric resolution cell (recall Figure 1.7) by the volumetric backscatter coefficient $\eta(\mathrm{m}^2/\mathrm{m}^3)$. The volumetric backscatter coefficient can be approximated over the microwave region for weather clutter as $\eta = 7 \cdot f^4 \cdot r^{1.6} \times 10^{-12} \mathrm{m}^2/\mathrm{m}^3$, where f is the radar frequency in GHz, and r is the rainfall rate in mm/h [18]. The fourth-power variation with frequency generally makes this type of clutter significant at frequencies above L-band for most radars. The radar's volume resolution cell can be found through multiplying the area subtended by the radar's circular antenna beam pattern (possibly elliptical for fan beams) at a particular range by the range equivalent to the radar's effective transmitted pulsewidth. The effective RCS of weather clutter (assuming a Gaussian shaped antenna beam) is given by $\sigma = \pi\eta\theta_A\phi_E R^2 c\tau/16$, where η is the backscattering coefficient in $\mathrm{m}^2/\mathrm{m}^3$, θ_A is the azimuth beamwidth in radians, ϕ_E is the elevation beamwidth in radians, R is the range in meters to the resolution cell, τ is the radar effective pulsewidth in seconds, and c is the velocity of light in meters per second [18]. Note that the two-way antenna response is used in the relationship for the effective RCS (σ).

The ratio of the clutter power to the target power in a resolution cell is given by $P_c/P_s = \bar{\sigma}/\bar{\sigma}_t$, where $\bar{\sigma}$ is the mean clutter effective RCS, and $\bar{\sigma}_t$ is the target's mean RCS. Generally the clutter power exceeds the target power and MTI or PD processing is required to allow target detection. The improvement factor (I) defines the average increase in signal-to-clutter power ratio due to doppler filtering and is given by $I = (P_s/P_c)_o/(P_s/P_c)_i$ [1]. The signal-to-clutter power at the output of the doppler filter can then be found as $(P_s/P_c)_o = \bar{\sigma}_t I/\bar{\sigma}$. Solving this relationship for the doppler filtering improvement factor results in $I = (P_s/P_c)_o/(\bar{\sigma}_t/\bar{\sigma})$, where the detection performance desired from the MTI or PD radar (e.g., $P_d = 0.9$, $P_{fa} = 10^{-6}$) determines the required output signal-to-clutter performance (e.g., $P_s/P_c =$ 13.2 dB).

1.3.3 The Clutter Spectrum and Autocorrelation Function

A primary feature of clutter, which makes it distinguishable from receiver noise, is that clutter is highly correlated from one radar pulse to the next, whereas receiver noise is usually independent on a pulse-to-pulse basis. This property is reflected in the clutter's psd or equivalent autocorrelation function.

The first probability density function (see Section 1.3.1) gives the distribution of the dynamic range or amplitude fluctuation of the clutter and the second pdf, $p(v_1, v_2; \tau)$, which gives the differential probability of measuring v_t in the range $(v_1, v_1 + dv_1)$ at time t_1 and v_t in the range $(v_2, v_2 + dv_2)$ at time t_2, describe the rate of clutter fluctuation. As we are assuming a stationary process (statistics are invariant with time), time enters into consideration only as the difference $\tau = t_1 - t_2$. As τ approaches zero, we would expect the values v_1 and v_2 to coalesce and, as τ becomes infinite, the values would become independent of one another, with each being determined by its respective first pdf. This action is described by a parameter called the *correlation coefficient* (ρ), which is as such if $\tau = 0$, indicating perfect correlation, and zero as τ increases to infinity, indicating no correlation.

The autocorrelation function $R(\tau)$ is obtained by finding the expected value or average of the product of values of the clutter process v_t at two times $(t, t + \tau)$, $R(\tau) = E(v_t v_{t+\tau})$. When $\tau = 0$, this second-order moment function assumes its maximum value equal to the mean squared value (power) of v_t. For a stationary random process, the correlation coefficient ρ can be expressed in terms of the autocorrelation function as $\rho = [R(\tau) - m^2]/\sigma^2$, where m is the mean, and σ^2 is the variance of the process. The function $M(\tau) = R(\tau) - m^2$, called the *covariance function*, is often used by statisticians to describe a random process. In pulsed radar systems, the covariance function is defined only at pairs of instants (e.g., t_1, t_2; t_1, t_3) for which the signals are defined. The values of the covariance function can then be arranged in covariance matrix form:

$$\mathbf{M}_x = \sigma_x^2 \begin{bmatrix} 1 & \rho_{12} & \rho_{13} & \cdots & \rho_1 \\ \rho_{21} & \ddots & & & \rho_{2n} \\ \vdots & & \ddots & & \vdots \\ \rho_{n1} & \cdots & \cdots & \cdots & 1 \end{bmatrix}$$

The covariance matrix is extensively used in MTI analyses to determine the effect of doppler filtering on the random clutter process.

For random processes, another useful concept is to decompose signal power into elemental frequency components, this decomposition being described by the psd function. Frequency generally implies a rate at which the signal changes. When the signal has high-frequency components, it is expected to exhibit variations in the time domain, which will be reflected in the autocorrelation function. Because the autocorrelation function is a time-domain signal, it can be Fourier transformed to obtain a frequency-domain signal. In accordance with the Wiener-Khintchine theorem, the Fourier transform of the autocorrelation function of a random process results in the psd of that process. The psd of various clutters has been measured, and is a fundamental parameter used in the analyses of MTI and PD radars.

The psd associated with the randomly fluctuating portion of ground clutter RCS (discrete clutter targets are excluded) is generally given by the Gaussian shaped spectrum $S_p(f) = P_c \exp(-f^2/2\sigma_f^2)/\sqrt{2\pi\sigma_f^2}$, where P_c is the clutter power and σ_f is the standard deviation of the clutter spectrum, which is related to the rms velocity clutter spread (σ_v) by $\sigma_f = 2\sigma_v/\lambda$ [1]. The autocorrelation function, which is found by the Fourier transform of the psd, is given by $R_p(\tau) = P_c \exp(-\tau^2/2\sigma_\tau^2)$, where $\sigma_\tau = 1/2\pi\sigma_f$. The normalized autocorrelation coefficient is given by $\rho_p = R_p(\tau)/P_c = \exp(-\tau^2/2\sigma_\tau^2)$. Typical power spectral densities for various types of clutter at 1 GHz are shown in Figure 1.19.

In coherent MTI processing, the psd and autocorrelation function of the quadrature random processes x_t and y_t are of interest. We can show that $\rho_p = \rho_x^2 = \rho_y^2$, which relates the normalized autocorrelation function of the clutter power envelope (proportional to the clutter RCS) and the quadrature x_t and y_t processes [13]. Thus, the psd of the power envelope can be found by convolving the power spectral density of the coherent quadrature components, resulting in $S_p(f) = S_x(f) \cdot S_x(f)$. Hence, for Gaussian shaped spectra, all power spectral densities and autocorrelation functions are Gaussian shaped, and the standard deviation of the power envelope can be found from those of the coherent quadrature components by using $\sigma_p^2 = 2\sigma_x^2 = 2\sigma_y^2$. When working with experimental or published data, care must be exercised to ascertain whether the standard deviations cited refer to the coherent (σ_x) or noncoherent (σ_p) standard deviations.

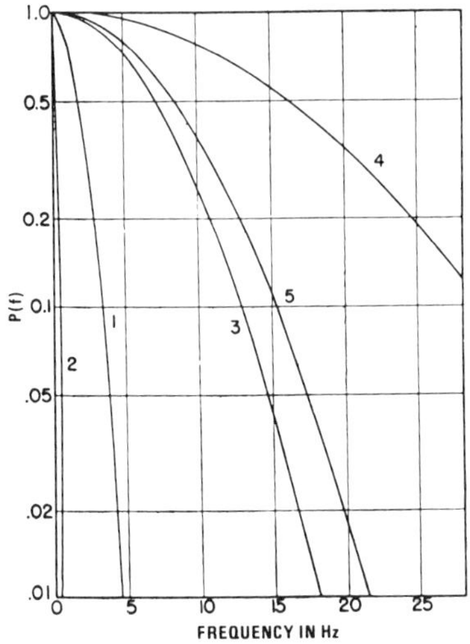

Target	a	Curve in Figure 5-8
Heavily wooded hills, 20-mph wind blowing	2.3×10^{17}	1
Sparsely wooded hills, calm day	3.9×10^{19}	2
Sea echo, windy day	1.4×10^{16}	3
Rain clouds	2.8×10^{15}	4
Window "jamming"	1.0×10^{16}	5

[a]From E.J. Barlow (1949).

Figure 1.19 Power spectral densities of various types of clutter.

The environmental diagram (refer to Figure 1.15) gives some idea of the spectral spread and mean frequency of various types of clutter. In applications involving sea clutter, weather clutter, chaff, and angel clutter, we need to model clutter that has relative radial motion with respect to the radar. This is accomplished by modifying the Gaussian-shaped ground clutter psd by the mean doppler shift frequency ($\bar{f}_d$), resulting in $S_p(f) = P_c \exp[-(f - \bar{f}_d)^2/2\sigma_f^2]$.

Experimental evidence [14,19] suggests that the clutter spectra may have significantly longer tails than indicated by a Gaussian shaped spectrum. A model suggested for ground clutter [1,20] measured with an X-band radar postulates a spectral density shape given by $S(f) = k/(1 + |f/f_c|^3)$. As shown in Figure 1.20, the cubic dependence results in a significantly greater magnitude at high frequencies than the Gaussian-shaped spectrum shown for comparison purposes in the diagram.

Recent measurements employing a wide dynamic range receiver indicate that the clutter spectrum from wind-blown trees can be closely approximated by using an exponentially shaped spectrum down to the -60 dB power level [21]. Typical measured spectra under both calm and windy conditions are depicted in Figure 1.21. The implication of these measurements is that the previously measured spectral data may have been distorted by receiver nonlinearities that caused wider spectral spreads at low power levels (see Section 3.2.5 and Figure 3.20).

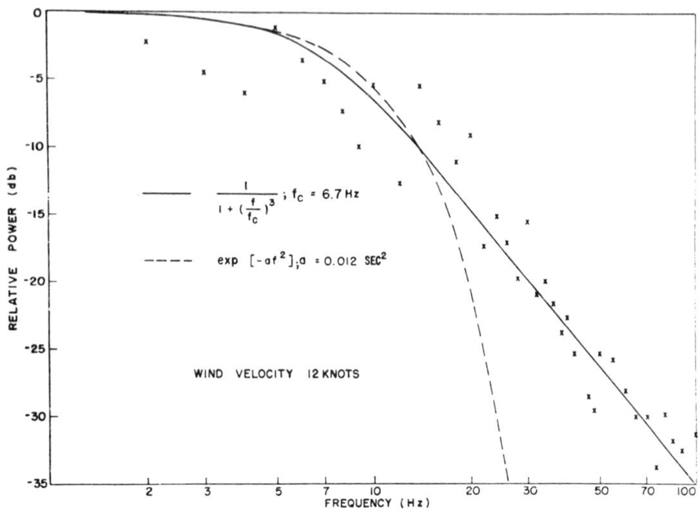

Figure 1.20 Ground clutter power spectral density obtained with an X-band radar.

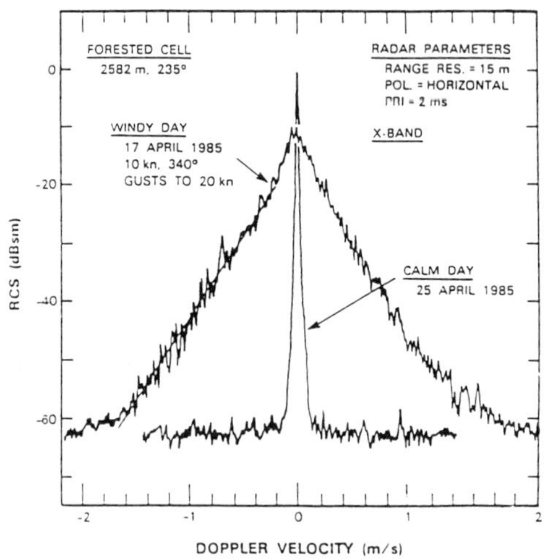

Figure 1.21 Power spectra of X-band radar returns from wind-blown trees.

1.3.3.1 Adaptive Clutter Filtering by Using Spectral Estimation

When only ground clutter is present, we are able to employ doppler filters with fixed rejection bands, which are designed to reject the maximum expected clutter spread. However, when moving clutter (such as sea, weather, or chaff) is present together with zero doppler frequency ground clutter, a more sophisticated filtering approach is required. One such approach uses an adaptive MTI, which adjusts its frequency response in accordance with the estimated clutter spectrum [8,12]. Such approaches rely on the tendency of targets to be spatially concentrated, whereby they are located in single resolution cells, whereas clutter tends to be spatially diffused and exists in multiple resolution cells about the target. This allows the clutter spectrum to be estimated in cells containing only the clutter return and free of any target returns.

The problem of estimating the clutter spectrum in a conventional scanning surveillance radar is compounded by the limited number (on the order of 20 to 50) of data samples provided as the radar beam scans through a radar resolution cell. One method of handling this problem is the *maximum entropy method* (MEM) of spectral estimation. By using the clutter returns from adjacent resolution cells and the MEM algorithm, developing reasonable clutter spectral estimates is possible for use in adaptive MTI systems [12,22].

The MEM algorithm is a modern spectrum analysis technique based on the parametric approach to spectral estimation. In general, the parametric approach involves three steps [23]. First, an appropriate parametric time-series model is selected to represent the expected clutter process. The model must embody information that allows generation of a random process which approximates the data to be measured. Second, an estimate of the parameters involved is on the basis of the measured data. Then, the estimated parameters are inserted into the theoretical psd expression appropriate for that model.

The parametric model generally selected for the random Gaussian clutter process is the *autoregressive* (AR) *time-series model,* which is appropriate if spectra with sharp peaks but no deep nulls, are involved [22,23]. The AR process is formed by exciting a linear all-pole filter with white Gaussian noise samples. The block diagram of an AR parametric model of order p is shown in Figure 1.22. We can show (Wold-Cramér decomposition theorem) that this model can be used to represent an arbitrary, stationary, Gaussian random process, but the order (number of parameters to be estimated) of the model is not necessarily minimized. Also, if the model is not closely matched, the order may be large [23].

The psd of the AR model is given by $S(f) = P_c T / |1 + \Sigma_{k=1}^{p} a_k \exp(-j2\pi fkT)|^2$, where P_c is the clutter power, T is the time between samples, p is the order of the AR process, and a_k denotes the parameters to be estimated [24]. The MEM algorithm provides a method for estimating the unknown parameters of the AR model. The original MEM algorithm developed by Burg used a rationale that extrapolated the time-series autocorrelation function such that it would have maximum entropy or

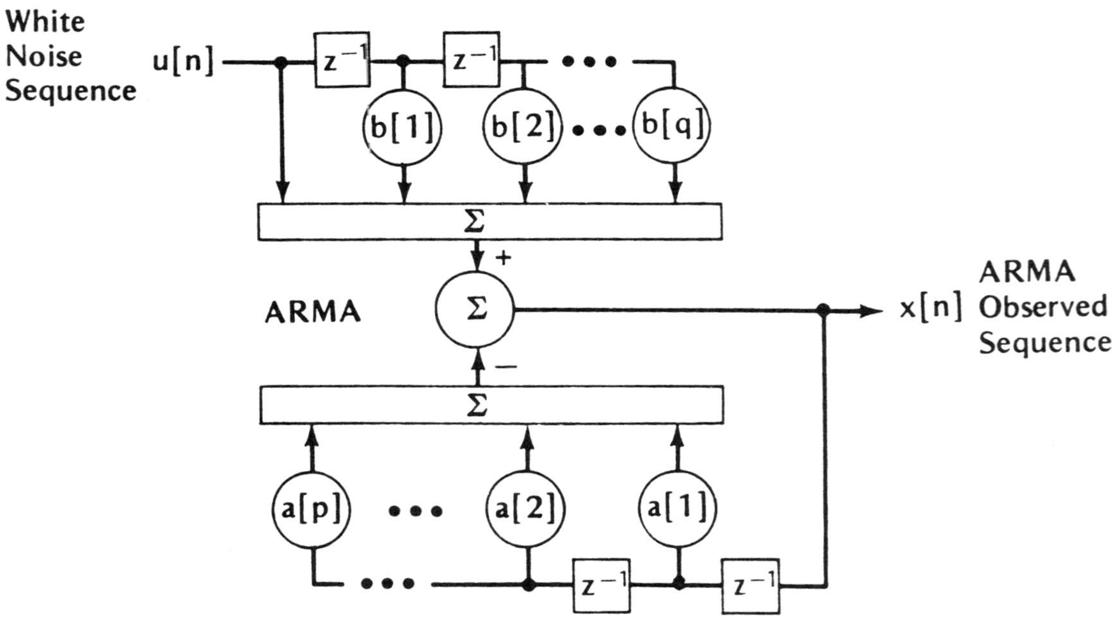

Figure 1.22 Autoregressive filter mode: of order p.

randomness. This results in the flattest spectrum of all possible spectra that have sample autocorrelation functions identical to those of the measured data [24]. Further, the MEM algorithm has been shown to provide a best least-squares fit of the AR model output to the measured data [25].

The usefulness of the MEM approach to spectral analysis depends on a proper choice of the model order (p). Too small an order results in a highly smoothed estimate, obviating the resolution advantages of the MEM, whereas an excessive order introduces spurious detail into the spectrum [24,25]. Experiments with clutter data suggest that a relatively low order of two or three may be adequate [22].

There are several other anomalies with the MEM algorithm. Unlike conventional Fourier spectral estimates, the peak amplitudes in AR spectral estimates are not linearly proportional to the power when the input process consists of a sinusoid in noise. Further, when clutter is estimated in the presence of receiver noise, the AR model is no longer valid, and a serious degradation occurs for low clutter-to-noise power ratios [24].

1.3.4 Non-Gaussian Clutter Models

Experimental measurements indicate that when land or sea clutter is viewed with a high-resolution radar (pulsewidth $< 0.5\ \mu s$) or at low grazing angles ($\phi < 5°$), a larger dynamic range is observed than will be predicted by a Gaussian (Rayleigh envelope) clutter model. Figure 1.23 shows a series of photographs taken with a medium-resolution ($\theta_{az} = 1.3°$, $\tau = 2\ \mu s$) ground surveillance radar viewing mountainous terrain clutter, where the clutter dynamic range exceeds 70 dB [5]. These photographs were taken by inserting a variable attenuator into the receiver channel to reduce the amplitude of the clutter with respect to receiver noise.

The wide dynamic range of clutter presents several problems to the MTI or PD radar. First, any nonlinearity inserted into the receiver to compress the dynamic range is known to spread the spectrum of the clutter. This significantly reduces the detection performance of linear doppler filter processors, which are designed on the basis of the optimum doppler filter theory discussed in Section 1.1.4.2. Second, the optimum doppler filter theory, which leads to a linear processor and is based upon a Gaussian clutter model, may no longer be applicable. These questions lead to the need for a non-Gaussian clutter model, which allows prediction of the detection performance of MTI and PD radars in clutter of wide dynamic range. Regardless of the specific model selected, the inherent detection performance of an MTI or PD radar apparently will be degraded in clutter of wide dynamic range, as compared to that which can be achieved with the Gaussian clutter model.

The determination of a non-Gaussian clutter model is a difficult problem that is the subject of much current research. From a theoretical viewpoint, we need to specify the complete multidimensional pdf, $p(x_1, x_2, \ldots, x_n; t_1 + \tau, t_2 + \tau, \ldots, t_n$

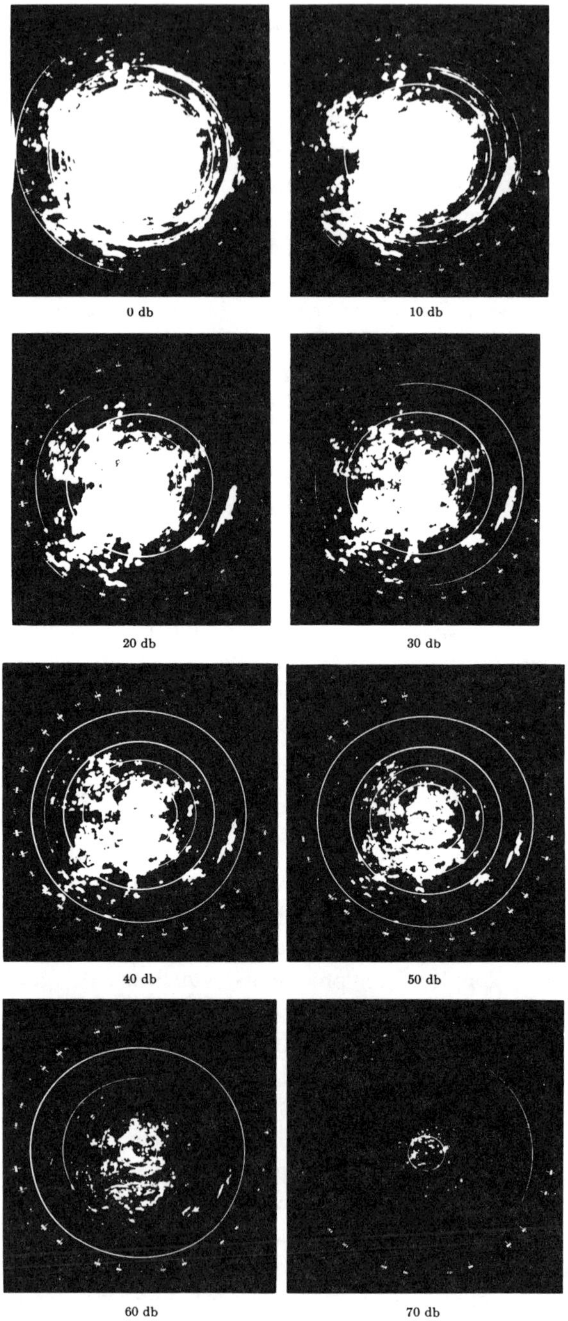

Figure 1.23 Wide dynamic range clutter returns over mountainous terrain.

$+ \tau$), which completely defines a stationary random process. For a stationary Guassian random process, this information is embodied in its mean (amplitude) and covariance (or equivalent psd) functions, which form the basis of most experimental measurements. The specification of non-Guassian processes requires high-order moments (beyond the second-order covariance function), which are not generally available from experimental data. This situation has resulted in a number of approaches toward specifying non-Gaussian clutter models, which result in incomplete, although useful, specification of the non-Gaussian clutter process.

The first approach to modeling a non-Gaussian clutter process involves a specification of the first pdf, $p(\sigma^\circ)$, which describes the variation of the clutter's backscattering coefficient per unit area $\sigma^\circ(\mathrm{m}^2/\mathrm{m}^2)$. Recall that σ° is proportional to the effective clutter RCS ($A = \sigma^\circ A_c$) as well as the power ($P_c = k\sigma$) returned from the clutter cell so that all variates have the same statistical properties. The variation of σ° may be measured on a spatial basis from cell to cell with the understanding that the clutter is homogeneous in the sense that the terrain over which the statistics are measured is the same for each cell used to form the statistics. The probability distribution formed for σ° on an ensemble basis can be used as the probability distribution for independent samples from a single cell, provided that the random clutter process is *ergodic* (i.e., ensemble and time statistics are identical) [26].

Figure 1.24 depicts the first pdf for Rocky Mountain ground clutter taken with an S-band radar having a resolution cell defined by a 2 μs pulsewidth and a 1.5° beamwidth [5,15]. The amplitude probability distribution for the backscatter coefficient σ° is plotted on log-Rayleigh coordinates and has the property that a Weibull probability distribution plots as a straight line. The close fit of the straight line to the data points in Figure 1.24 indicates that the Weibull probability distribution provides a close match to the experimental data, and may represent a good model for this type of clutter.

Note that in Figure 1.24 the data only extend to the 0.99 probability point, whereas detection calculations involving false alarm probabilities usually require values several orders of magnitude beyond this point. Thus illustrated is a typical problem in clutter measurements in that rarely are enough independent data points available to define the shape of the probability distribution in the tail region.

Consider estimating the tail probability clutter distribution given by $Q(y) = 1 - F(y)$, where $F(y)$ is the clutter's cumulative probability distribution function. Usually, the observed data points are formed into a histogram by counting the number of observations that occur above a preset threshold point, Y. The estimate for the cumulative tail pdf at level Y is then $\hat{Q}(y) = 1/N \, \Sigma_{i=1}^{N} D_Y(y)$, where N is the total number of samples, and $D_Y(y)$ has value one if the data point exceeds Y and zero otherwise. The expected value of the estimate for $Q(y)$ is $E[\hat{Q}(y)] = Q(y)$, whereas the standard deviation is $\mathrm{SD}[\hat{Q}(y)] = [Q(y)/N]^{1/2}$. The first relationship indicates that $\hat{Q}(y)$ is an unbiased estimate for $Q(y)$, and the second specifies how many data points are necessary to make an estimate of $Q(y)$ with high precision. A precise estimate is achieved when its standard deviation is much smaller than its mean resulting

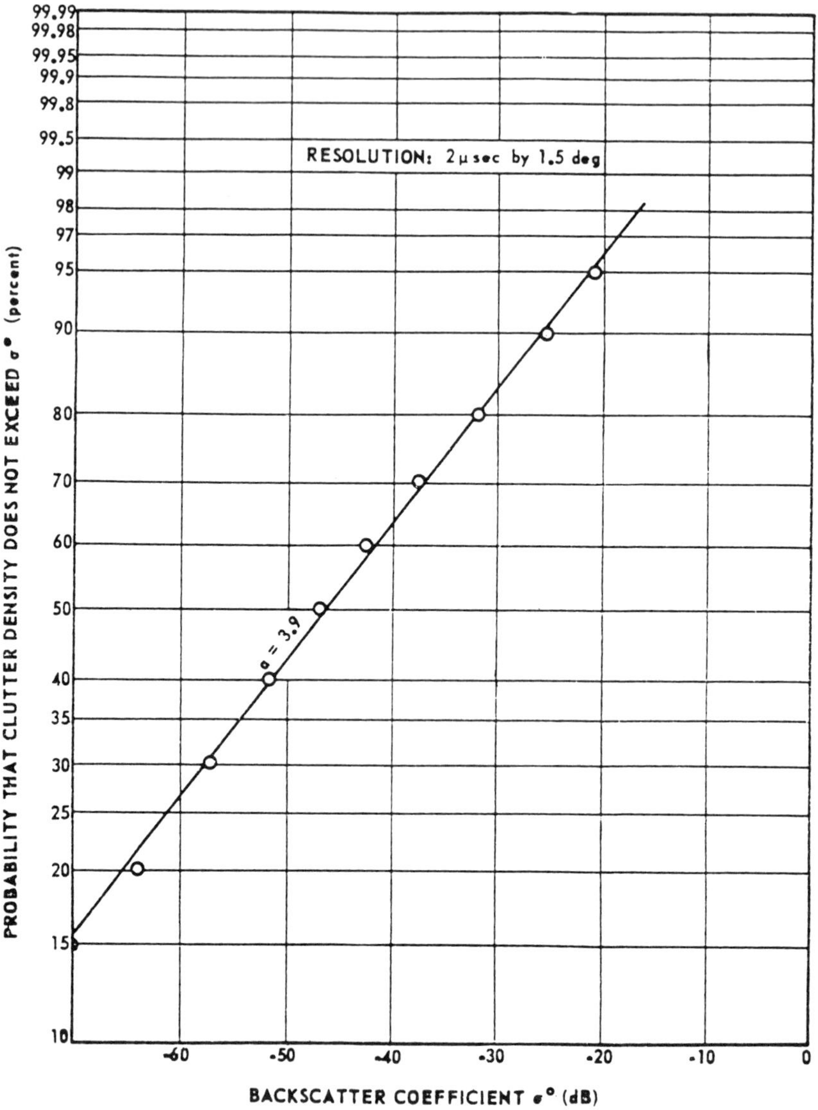

Figure 1.24 Spatial distribution for Rocky Mountain ground clutter with an S-band radar.

in $NQ(y) \gg 0$ [27]. Thus, to make a precise estimate of the tail distribution of the clutter's probability distribution for $Q(y) = 10^{-4}$ would require on the order of 10^5 independent data points, illustrating the difficulty of estimating such a distribution.

Several probability distributions have been empirically fitted to clutter data. The most popular are the log-normal, Weibull, and K distributions. Each of these non-Gaussian models is intended to represent the effective RCS distribution (first probability distribution of σ°) of the long-term average clutter for the measured clutter data in a single clutter resolution cell. These distributions are useful for calculating detection probabilities of fast-scanning radars, where multiple independent clutter samples are obtained on a scan-to-scan basis [28–32]. However, care must be exercised when using these distributions within a single radar scan for a conventional search radar. This relates to the nonstationary behavior of non-Gaussian clutter.

One method which has been used to relate these non-Gaussian probability distributions to conventional radar detection calculations is as follows. From physical reasoning there are usually a sufficient number of clutter scatterers (the minimum number is on the order of five scatterers) within a radar resolution cell to result in a clutter return having a Gaussian (Rayleigh envelope) distribution. The Gaussian distribution has an associated psd that determines its second-order statistics, which are useful for doppler filtering detection calculations. Further, the Gaussian distribution is assumed to have a variable mean $\bar{\sigma}^\circ$, which is a random variable selected from the non-Gaussian first probability spatial distribution of the measured clutter data. The temporal situation is assumed to be represented by two correlation functions: a short-term fast correlation function (orders of 10 ms), applicable over a single radar scan associated with the Gaussian model's power spectral density, and a longer-term correlation function associated with the variable mean $\bar{\sigma}^\circ$ for which there is complete correlation within the radar's dwell on the clutter resolution cell and significant decorrelation over the period within a radar scan (usually on the order of 5 to 10 s).

The detection probability can be found by using this model. First, compute the probability of detection for the Gaussian clutter model as a function of the variable backscatter coefficient $P_d(s|\sigma^\circ)$, and the average over the specified non-Gaussian first pdf, $p(\sigma^\circ)$, which results in $\bar{P}_d(s) = \int_0^\infty P_d(s|\sigma^\circ)p(\sigma^\circ)\,d\sigma^\circ$ [33]. Note that this is the same procedure as that used to compute the probability of detection for the Swerling target models, except that in the preceding relationship s represents the signal-to-interference (receiver noise plus clutter power residuals after doppler filtering) power ratio and σ° represents the clutter RCS probability density rather than that of the actual target [34–36]. Extension to Swerling fluctuating targets is easily made by averaging $\bar{P}_d(s)$ over the expected target distribution (Rayleigh for Swerling 1 and 2 targets) [34].

The cumulative probability distribution functions representing the effective RCS of the clutter for the Rayleigh, log-normal, Weibull, and K-distributed clutter models are given in Table 1.2. Note that the Weibull and log-normal distributed RCS (σ) are normalized to their median values (σ_m), whereas the Rayleigh and K-distributed

Table 1.2
Probability Distributions for Effective Clutter RCS

Clutter Model	*Probability Density, p(σ)*	*Probability Distribution Clutter RCS, P(σ)*
Rayleigh	$\exp(-\sigma/\bar{\sigma})/\bar{\sigma}$	$1 - \exp(-\sigma/\bar{\sigma})$
Weibull	$\beta \cdot \ln 2(\sigma/\sigma_m)^{\beta-1} \exp[-2(\sigma/\sigma_m)^\beta]/\sigma_m$	$1 - \exp[-\ln 2(\sigma/\sigma_m)^\beta]$
Log-Normal	$\exp[-\ln^2(\sigma/\sigma_m)/2\sigma_p^2]/\sqrt{2\pi}\,\sigma_p\sigma$	$1/2[1 + \mathrm{erf}(\ln(\sigma/\sigma_m)/\sqrt{2}\sigma_p)]$
K-Distribution	$2\nu(\nu\sigma/\bar{\sigma})^{\nu-1/2}K_{\nu-1}[2(\nu\sigma/\bar{\sigma})^{1/2}]/\Gamma(\nu)$	$1 - 2(\nu\sigma/\bar{\sigma})^\nu K_\nu[2(\nu\sigma/\bar{\sigma})^{1/2}]/\Gamma(\nu)$

$\bar{\sigma}$ = Mean Clutter RCS; σ_m = Median Clutter RCS

RCS are normalized to the mean values ($\bar{\sigma}$). The Rayleigh distribution model is part of the Weibull family of distributions and can be obtained from the Weibull distribution by setting $\beta = 1$.

The presentation of the probability distributions for the effective clutter RCS given in Table 1.2 is facilitated by plotting the distribution under consideration on special probability graph paper, which results in a straight-line plot on a dB scale. The Weibull distribution can be expressed as $\sigma_{dB} = \sigma_{mdB} + 1.592/\beta + (10/\beta)\log[\ln(1/(1 - P(\sigma))]$ which plots on log-Rayleigh probability graph paper as a straight line. The straight line can be determined by connecting the 0.5 probability point $\sigma_{dB} = \sigma_{mdB}$ with the 0.9999 probability point given by $\sigma_{dB} = \sigma_{mdB} + 11.235/\beta$. Values of the parameter β for the Weibull distribution range from about $\beta = 0.25$ for mountainous land clutter to $\beta = 1$ (Rayleigh clutter) for low-resolution, high-grazing-angle land and sea clutter [1].

For log-normal clutter the RCS distribution plots as a straight line on normal probability graph paper using a dB scale. The straight line is determined by connecting the 0.5 probability point $\sigma_{dB} = \sigma_{mdB}$ with the 0.9999 probability point given by $\sigma_{dB} = \sigma_{mdB} + 16.153\sigma_p$. Values of the parameter σ_p range from approximately $\sigma_p = 0.7$ for rain clutter to over $\sigma_p = 2$ for certain types of land clutter [1].

A parameter used to characterize non-Gaussian clutter is the mean-to-median ratio. The higher this ratio the longer are the tails of the probability distribution. The mean-to-median ratio ($\rho = \bar{\sigma}/\sigma_m$) is given by $\rho = \Gamma(1 + 1/\beta)/(\ln 2)^{1/\beta}$ for the Weibull distribution and $\rho = \exp(\sigma_p^2/2)$ for the log-normal distribution. No simple analytic expression appears to be available for the *K* distribution. Figure 1.25 depicts the mean-to-median ratio for the Weibull, log-normal, and *K*-distributions. Note the mean-to-median ratio for the Rayleigh ($\beta = 1$) distribution is given by $\bar{\sigma}/\sigma_m = 1/\ln 2 = 1.442$.

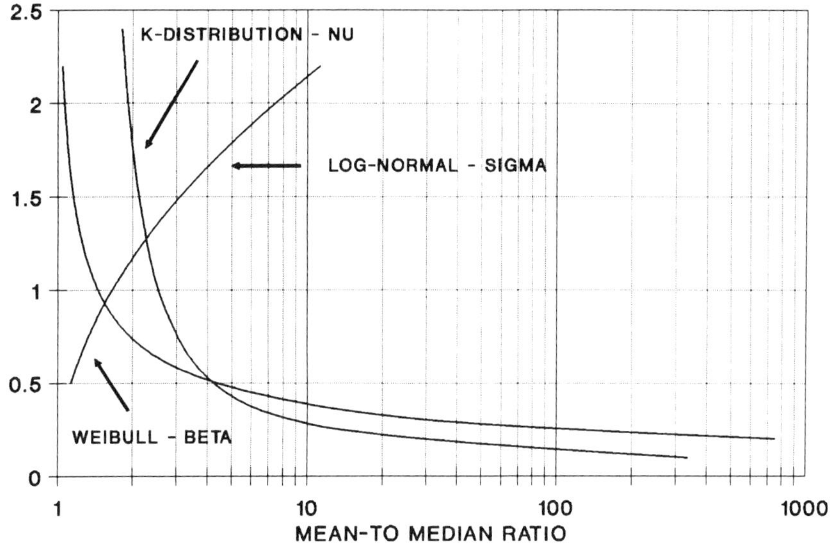

Figure 1.25 Mean-to-median ratios for the Weibull, log-normal, and *K*-distributions.

Figure 1.26 depicts the Rayleigh, Weibull ($\beta = 0.5$), log-normal ($\sigma_p = 1.689$), and *K* distributions ($v = 0.5$) plotted on log-Rayleigh probability graph paper for a mean-to-median ratio of $\rho = 4.163$, all normalized to their median values. The Rayleigh and Weibull distributions plot as straight lines. The *K* distribution in the tail region is very similar to the Weibull distribution, and the log-normal distribution has the largest tail values.

The *K* distribution was developed to represent non-Gaussian sea clutter data, and is claimed to model both the spatial and temporal correlation characteristics of the clutter [37]. The compound form of the *K*-distributed sea clutter model identifies two components of clutter fluctuations. The first component is a spatially varying mean level associated with the sea swell structure. This component has a long correlation time (on the order of seconds) and is unaffected by frequency agility. The second component (called the *speckle* component) is due to the return from multiple scatterers within the radar's resolution cell, and exhibits the general characteristics of Gaussian (Rayleigh envelope) clutter. This component has a relatively short correlation time (on the order of 10 ms) and can be decorrelated by using frequency agility.

The *K* distribution is derived by assuming that the fast component consists of a Rayleigh distributed variate, conditioned on its mean value, which is chi distributed. The chi distribution is an empirical fit to measured sea clutter data [38]. Thus, the

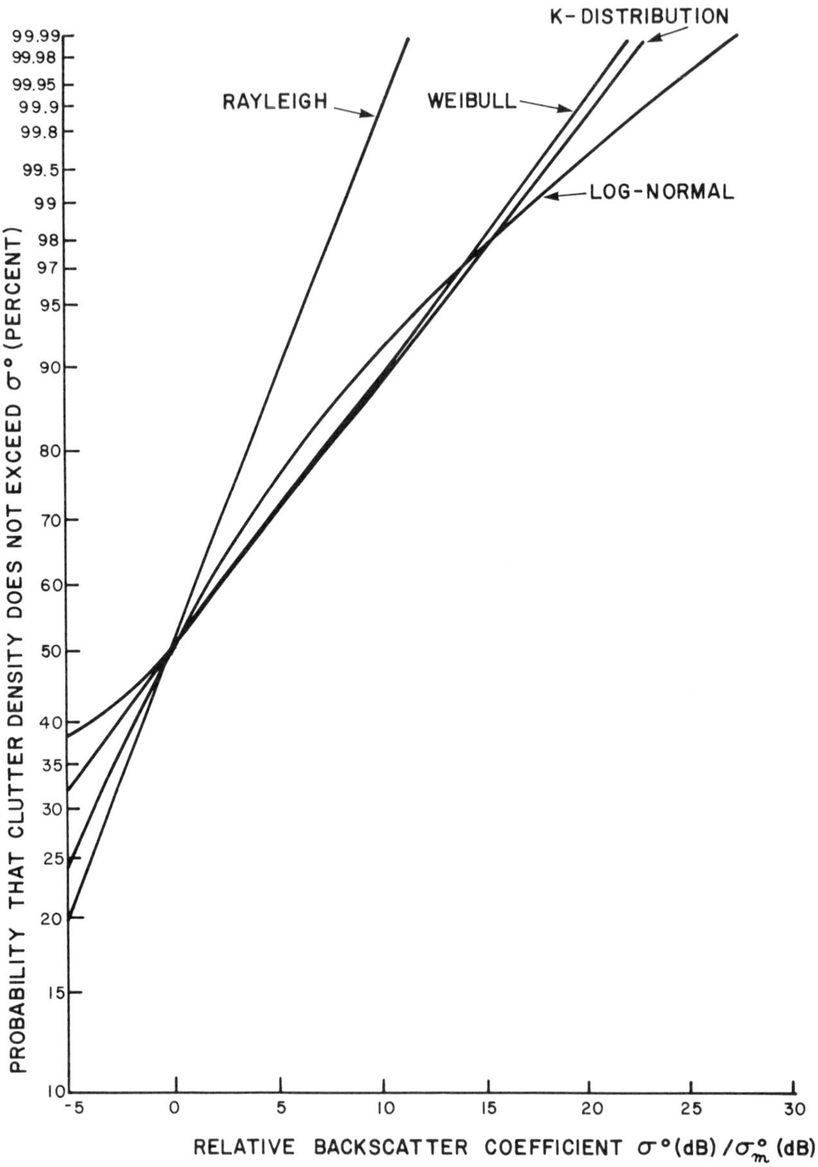

Figure 1.26 Rayleigh, Weibull, log-normal, and *K*-distributed probability distributions.

K-distributed model, like the Weibull and log-normal models, is based upon an empirical fit to measured data, but has the virtue of some physical justification. However, the analytical properties of the *K* distribution are not as well developed as those of the Weibull or log-normal distributions.

1.4 THE MOVING TARGET DETECTOR

The moving target detector or MTD is a digital signal processor specifically designed to allow low-PRF two-dimensional (2-D) doppler *air surveillance radar* (ASR) to function in a multifaceted clutter environment. The processor is designed to provide highly reliable detection of small, low-flying aircraft in the simultaneous presence of ground clutter, precipitation clutter, and angel clutter (for example, refer to the environmental diagram in Figure 1.15).

The original MTD-I processor was developed for the ASR-9 radar, which is a medium-range (60 nmi) 2-D air surveillance terminal radar that operates at most major airports. ASR-9 operates at S-band (2.7–2.9 GHz) with a pulsewidth of 1.05 μs, a 1.3° azimuth beamwidth, an antenna rotation rate of 12.5 rev/min, a PRF of 1200 pulses/s, and a peak power of 1200 kW. Typical performance in ground clutter is depicted by Figure 1.27, where the left photograph shows the normal video return and the right MTD photograph shows a controlled aircraft track over moderate-intensity ground clutter. Figure 1.28 indicates analogous performance of the MTD amid strong weather clutter. The controlled target shown in both MTD photographs was a small, single-engine, general aviation aircraft (RCS $\approx$ 1 m^2). The tracks depicted in the MTD photographs show that no detections are lost due to clutter effects and false alarms are negligible. Further, the tracks were maintained while the aircraft flew

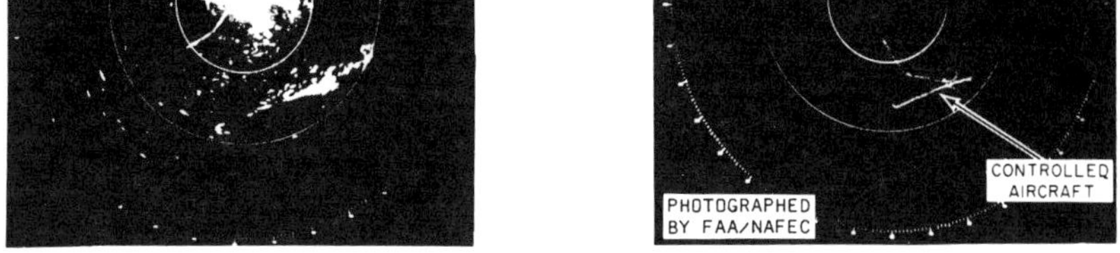

Figure 1.27 Performance of the MTD in ground clutter.

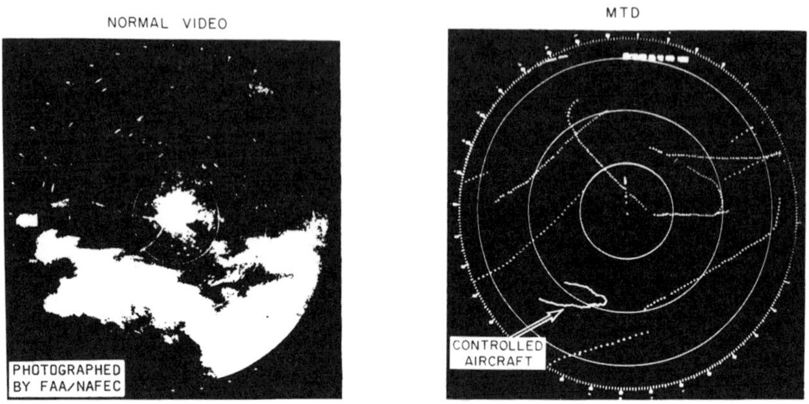

Figure 1.28 Performance of the MTD in precipitation clutter.

tangentially to the radar, which illustrates the *superclutter visibility* capability of the MTD processor.

A block diagram of an early version of the MTD processor (MTD-I) is depicted in Figure 1.29 [36]. The basic elements of the processor are (1) a coherent MTI filter for rejecting ground clutter; (2) an FFT doppler filter bank, which provides coherent integration, used in the suppression of weather clutter and chaff; (3) a digital clutter map, which allows tangential targets to be detected over ground clutter, also used to suppress large clutter false alarms; (4) CFAR processing, which suppresses false alarms; and (5) block PRF staggering, which eliminates MTI blind speeds, *second-time-around clutter,* stationary targets.

The basic idea of the MTD processor is to break the radar coverage into a large number of elemental range-azimuth-velocity (doppler) cells and to select those containing aircraft targets on the basis of the instantaneous radial velocity, motion, and extent of the real targets as compared to the characteristics of extraneous targets including clutter. The ASR-9 MTD processor contains 365,000 range-azimuth cells (1/16 nmi by 3/4 degree), each containing eight velocity cells (150 Hz doppler bandwidth) for a total of 2,920,000 cells.

The MTD processor shown in Figure 1.29 separates the input signal into two channels, depending on the return's radial velocity. The upper channel contains an MTI that suppresses ground clutter by 45 to 50 dB, and a doppler filter bank that accepts targets and weather clutter returns, the radial velocity of which is within the MTI's passband. The lower channel accepts target and ground clutter returns having radial velocities that fall within the lowest doppler filter's passband (±75 Hz). The output of this lower channel is used to generate a ground clutter map by averaging eight consecutive radar scans of each range-azimuth cell.

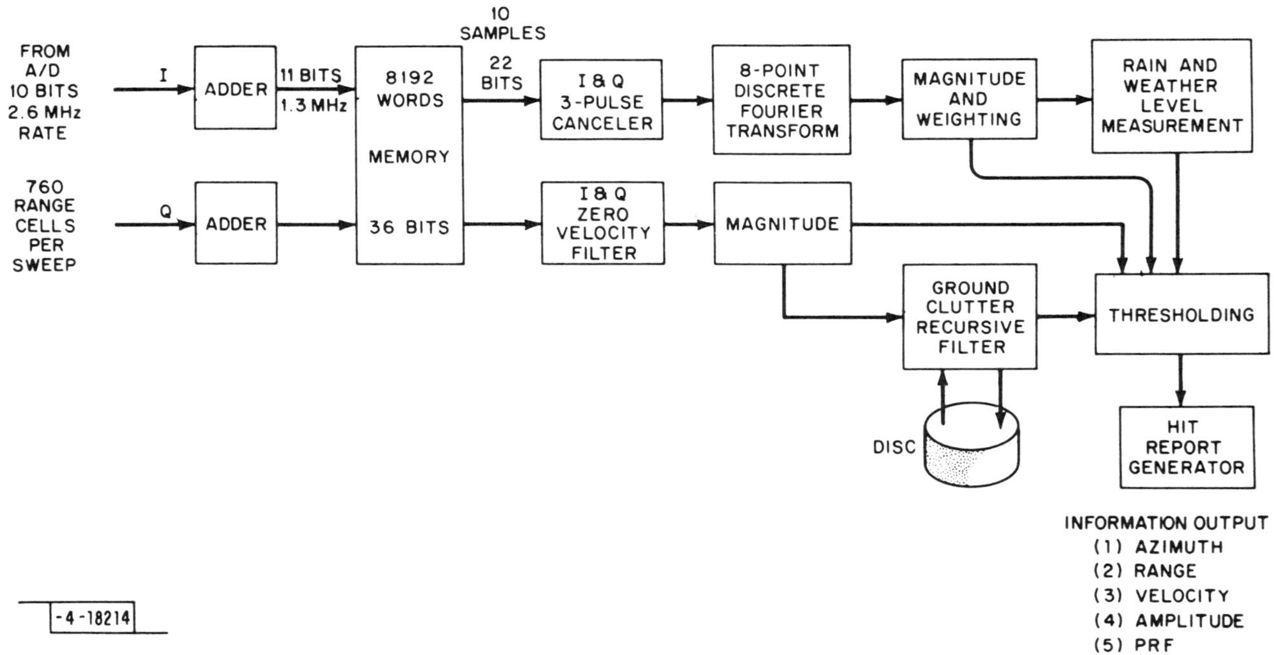

Figure 1.29 Moving target detector (MTD-I) block diagram.

The function of the upper channel is to suppress both ground and weather clutter while passing all moving targets, except those that fall into the low-radial-velocity class associated with the lower channel. The ground clutter is suppressed by the MTI action, whereas the weather clutter, which tends to fall within the first three doppler filters, is suppressed by a mean-level CFAR circuit. The mean-level CFAR sets the threshold in each doppler cell by averaging the returns from the same doppler cell in sixteen range cells, eight on either side of the cell of interest. Weather returns, which are generally distributed throughout this one-mile-plus range extent, fill the doppler cells of interest, thereby creating a high threshold that suppresses the return in this combination of range-doppler cells.

Several complications arise in low-PRF ambiguous-velocity designs of this type. First, all targets having radial doppler velocities that fall in a region about the multiples of the first PRF line (1200 Hz, which corresponds to 125 knots, 250 knots, *et cetera*) are suppressed by the MTI. This is the *blind-speed effect* of the MTI. Second, all targets having radial velocities that are ambiguous with respect to doppler filter cells occupied by weather clutter are suppressed. This target suppression is due to the *folding effect* of a low-PRF radar, whereby the doppler returns of all targets are folded into the region between zero frequency and the first PRF line.

The blind-speed problem and suppression of targets by weather clutter (recall Figure 1.9) is alleviated by using a block-to-block PRF stagger, where a block in the MTD processor of Figure 1.29 corresponds to ten consecutive pulses. The need to maintain the same PRF (and transmitter frequency) for ten pulses can be understood by considering that the three-pulse MTI canceller must be filled before its output can be applied to the eight-point FFT doppler filter bank. The ASR-9 MTD uses a block stagger of about 20 percent between returns obscured by weather clutter or by the MTI blind-speed effect to translate returns into adjacent doppler filters so that they can be detected. The significant effect of this procedure is that 20 pulses must be processed before a target detection can be made.

A particular feature of the MTD processor that distinguishes it from earlier designs is its management of false alarms, which is accomplished by using adaptive CFAR thresholding of each range-doppler cell. Two types of mean levels are formed in the MTD processor. First, a clutter map is formed using a fading-memory recursive filter that operates on the output of the doppler filter centered at zero frequency. The recursive filter has a bandwidth that corresponds to approximately eight radar scans (32 s). The second type of mean level is formed from the output of each nonzero velocity doppler filter averaged in a moving window detector whose width is 16 range cells centered about the doppler cell of interest.

The mean-level clutter map is used to form the thresholds for the zero velocity doppler filter shown in the lower channel of Figure 1.29 and the eight doppler filters of the upper channel that contain residual clutter components folded about the PRF lines. This would provide superclutter visibility whereby large cross-sectional tangential targets (zero radial velocity) could be detected over clutter that otherwise might have fallen into the MTI's blind-speed zone, and thus been eliminated. The moving window mean-level outputs for doppler filters 2 through 7 are used to form the threshold for their respective doppler filters. This threshold is used to provide a CFAR for weather clutter, which can have a mean radial velocity of up to 60 knots and a spectral spread of 25 to 30 knots. In addition, a combination of both mean-level filters is used to form the threshold for the doppler filter adjacent to the zero velocity filter to allow for ground clutter spillover.

In addition to the MTI, doppler filtering, and CFAR processing, a number of other censoring techniques are used in the MTD processor. Pulse type of interference might be received from other radars, which would not be eliminated due to the averaging effect of the CFAR processor. To guard against this condition, the magnitude of any one return is checked against the average of the return over the coherent processing interval (eight to ten pulses). If any pulse exceeds the average by five times, all information from the range-azimuth cell is discarded. Also, saturation of the A/D converter on any pulse causes the information over the whole coherent processing interval to be deleted.

The multiplicity of range-doppler-azimuth cells in the MTD processor increases the possibility of a large target causing multiple range, doppler, and azimuth cell reports. Reports in adjacent cells that appear to originate from a single target can be grouped together to ease the problem of track initiation in the radar data processing subsystem. Also, the target amplitude and doppler reports on a target can be used to eliminate small cross-section and low-speed radar angel (birds and insects) returns before these target reports enter the radar data processing tracking system.

The classic doppler radar design problem addressed by the MTD is the detection of moving target return in the presence of both ground and precipitation clutter (see Section 1.1.3). The MTD-1 handles ground clutter through the combination of a coherent double canceller MTI followed by an eight-point FFT doppler filter bank where the sidelobes of the doppler filter bank further attenuate the residual clutter from the MTI canceller. This cascade can be shown to approximate the linear optimum transversal filter for this situation, but it is inefficient in the sense that this same improvement factor performance can be achieved by using a lower order transversal filter (the optimal transversal filter would use eight *versus* ten pulses for the MTD-I) [39].

The realization of quasioptimum performance of the doppler processor requires linear operation to mitigate the deleterious effects of limiting that spreads the clutter spectrum, thereby causing an improvement factor reduction (see Section 3.2.5 and Figure 3.23). The linear dynamic range of the MTD-I is approximately 50 to 60 dB due to the use of a 10-bit A/D converter in the digital processor. When clutter returns exceed this limit (due primarily to large discrete ground clutter returns), one possibility is to use the clutter map to control a linear attenuator in these limit cells to maintain linear operation.

The suppression of precipitation clutter in the MTD-I is accomplished by generating a doppler filter bank in each range-azimuth angle cell that divides the spectrum between PRF lines into n doppler cells each with bandwidth $f_B = f_r/n$, where n is the number of applied pulses used to form the filter (i.e., eight pulses) and f_r is the radar's PRF. This situation is depicted in Figure 1.30 which shows how the MTD resolution cells are spread through the radar's range, azimuth angle, and doppler coverage. When precipitation clutter (or chaff return) occurs, it generally will be confined to one or two adjacent doppler cells within each doppler filter bank of those located within the range-angle extent of the clutter cloud. The return in each doppler cell containing the clutter can then be attenuated in accordance with a normalizing CFAR signal formed by examining like doppler cells contained in contiguous range cells about the test cell's range (i.e., ± 1 nmi) at the same azimuth angle. This leaves the other doppler cells, which are free of precipitation clutter returns, available to detect moving targets, thus providing subweather clutter visibility. Also, superclutter visibility is provided in those doppler cells containing weather clutter that have target returns sufficiently stronger than the precipitation clutter return.

Although precipitation returns are generally unambiguous in doppler (i.e., $|f_d| \leq 55$ knots), aircraft target returns are not. This causes a situation where target returns are lost due to spectrum folding into doppler cells suppressed by precipitation clutter return (recall Section 1.1.3 and Figure 1.9). The solution to this problem employed by the MTD is to use a block stagger where two coherent processing intervals (CPIs) are consecutively transmitted at PRFs that differ by about 20%.

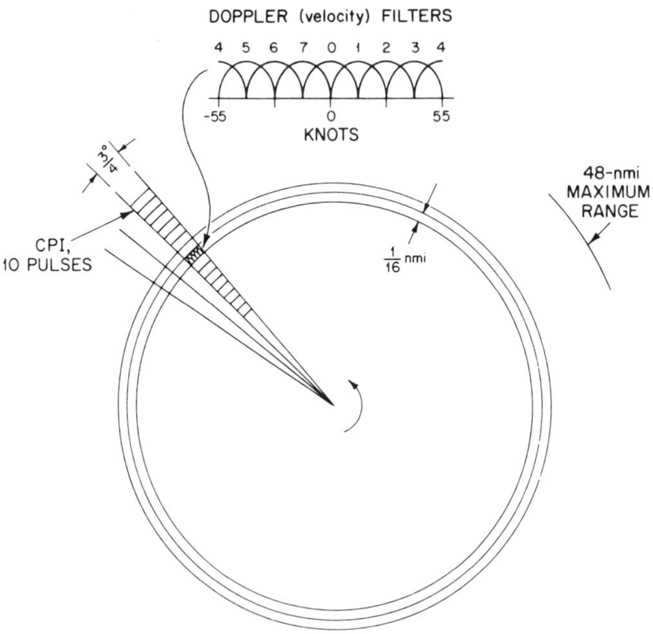

Figure 1.30 MTD resolution cells.

Thus, if the target is suppressed on one PRF, it will be folded into another nonsuppressed doppler filter on the second PRF and thus detected. The block stagger used in the MTD also suppresses blind speeds. These are associated with the low-PRF employed and multiple-time-around clutter returns that occur when the radar is operated under conditions of anomalous propagation or in regions containing reflections from mountainous terrain and urban areas, so that the radar waves intercept ground targets at distances greater than that associated with the first interpulse period.

A critical issue involved with the MTD type of processors that use doppler filter banks to suppress precipitation clutter concerns the sidelobes of the doppler filters. If the sidelobes are too high, strong precipitation clutter spectral components located outside the passband of any particular doppler filter will cause the filter to suppress targets due to the CFAR action that follows the filter. In the MTD-I, the doppler filters consisted of a cascade of three separate filters: a three-pulse canceller for ground clutter suppression, an eight-point FFT for precipitation clutter suppression, and post-FFT frequency-domain weighting to reduce the filter's sidelobe level. Further, the double canceller used ten bits of signal information and coefficient precision (to provide approximately 60 dB of ground and discrete clutter rejection) whereas the FFT carried a lesser number of bits (approximately three or four bits) because this was all that would be required to form the filters. The resulting doppler filter response is depicted in Figure 1.31 for two of the filters. The filter closest to the ground clutter response exhibits poor sidelobes, which are approximately −5 dB below the principal response. This is due in part to the use of a double canceller in front of the FFT, which attenuates the main responses. Also, the small number of bits employed in the FFT further limits the achievable sidelobe level.

In the MTD-II processor shown in Figure 1.32, the doppler filtering is made more efficient, and the sidelobe levels of the filters are reduced [40]. The filters are synthesized by the cascade of a single canceller and eight 7-pulse complex transversal filters. The single canceller suppresses the zero-velocity ground clutter and carries the full dynamic range of the signal information (ten bits). The transversal filters are matched to the target spectrum and shaped to provide low sidelobes. These filters

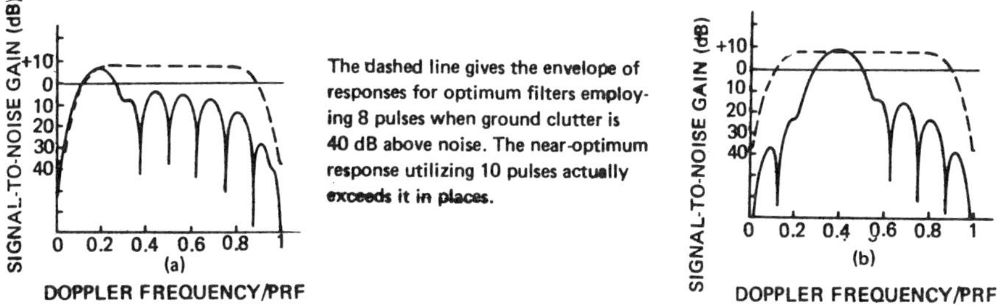

Figure 1.31 Doppler filter response of the MTD-I processor.

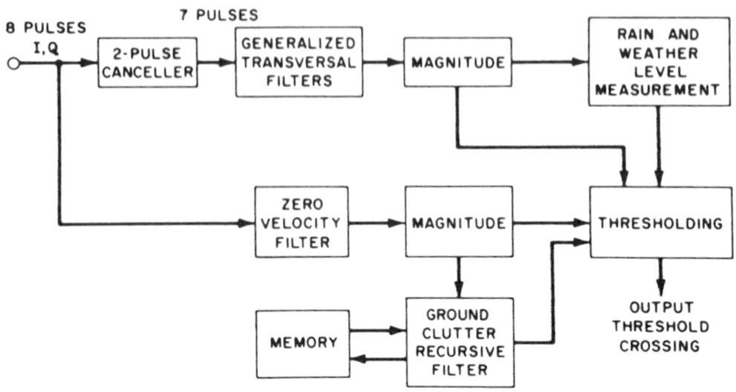

Figure 1.32 The MTD-II processor.

are formed using five-bit weighting coefficients. The improved sidelobe response of the MTD-II is depicted in Figure 1.33.

Note that the efficiency of the doppler portion of the MTD processor was improved in the MTD-II, where eight-pulse transversal filters were used to provide essentially the same result as with the ten-pulse filters used in the MTD-I. This reduces the number of pulses required to implement an MTD-type processor from 20 to 16 pulses (i.e., for two CPI bursts). This improvement results because, although

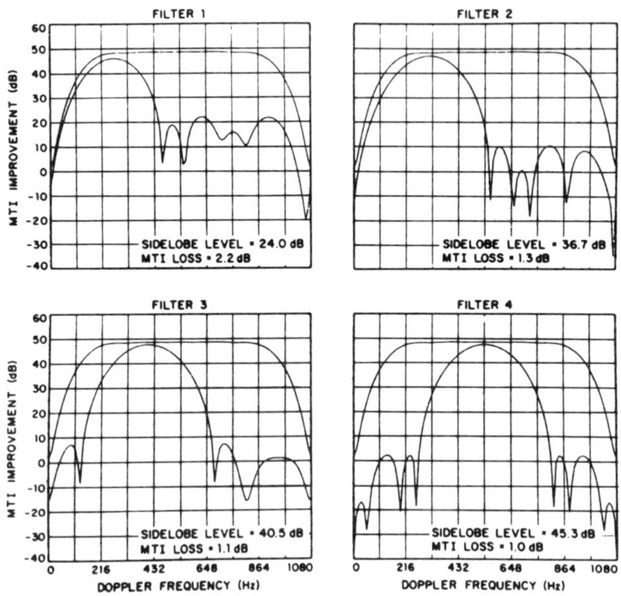

Figure 1.33 MTI improvement factors for the MTD-II processor.

the first two pulses into the MTI are useful in suppressing ground clutter, they are essentially wasted in the formation of the doppler filter bank.

The number of bits required for the coefficient of an MTD type of processor is open to some controversy. If a pure transversal filter configuration (no MTI) is used, each filter in the bank (e.g., eight filters) must be designed to suppress zero-velocity clutter (large discrete and ground terrain clutter) and may require up to 12-bit precision for coefficient weightings [41]. If an MTI prefilter is used, then the transversal filters may require 8–9-bit precision in the filter coefficients to achieve minimum sidelobes, which are of the order of −40 dB below the main response [42]. The sidelobes are also affected by imbalances in the amplitude response of the in-phase (*I*) and quadrature (*Q*) channels. For example, a 4% imbalance would raise the sidelobe level the order of 6 dB [41].

Another factor that reduces the sidelobes of transversal type doppler filters occurs when multiple-time-around clutter is processed. For example, when processing an *n*-pulse burst with low sidelobe weighting, the second-time-around clutter echoes will arrive with an extra time delay so that the first pulse will receive the second weighting and the last pulse will be lost. The erroneous weightings raise the sidelobe level as shown in Figure 1.34 for an eight-pulse processor with a Dolph-Chebyschev type of weighting [43].

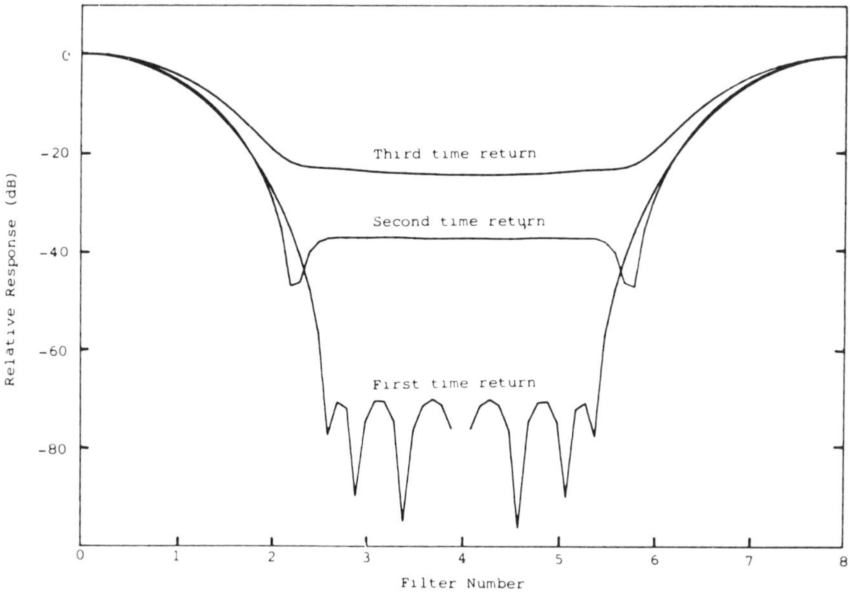

(a) Response of 8-pulse processor.

Figure 1.34 Response of a transversal filter with Dolph-Chebyschev weighting to ambiguous clutter.

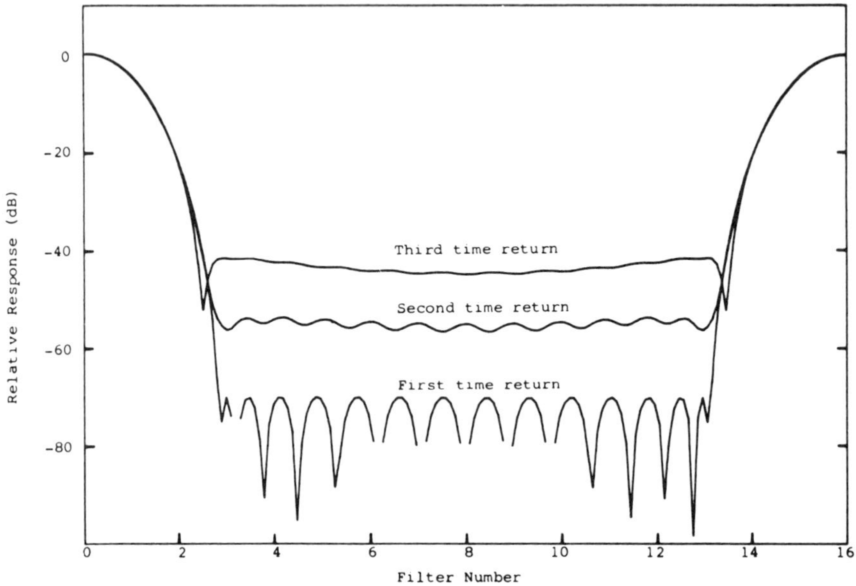

(b) Response of 16-pulse processor.

Figure 1.34 continued.

A significant disadvantage of the MTD processor is the relatively large number of pulses that must be transmitted to achieve the desired performance level. This restricts its use to 2-D surveillance radars or 3-D radars that use the stacked beam approach. Further, the need for a relatively long coherent burst transmission at a stable PRF (i.e., eight to ten pulses) makes the MTD susceptible to electronic countermeasures (ECM) [36]. A responsive jammer could measure the frequency of the first transmitted pulse and then center the jammer to spot jam the following pulses. Also the requirement for a stable PRF precludes the use of pulse-to-pulse jitter, which is one of the most effective techniques against deception and camouflage jammers that rely on anticipating the radar transmitter pulse.

1.4.1 Clutter Maps

One innovation employed in the MTD-type of processor is the use of a clutter map for false-alarm control against a spatially nonstationary variable clutter background (see Section 1.3.4). This is contrasted with a conventional cell-averaging CFAR that provides good performance in a spatially stationary background but poor performance when the sample cells surrounding the target cell have nonuniform statistics [44].

The conventional cell-averaging CFAR can be modified by using a split-gate or by employing nonparametric detection techniques to improve its performance in nonstationary interference. However, when the conditions are appropriate for using a clutter map CFAR, this type of CFAR control will generally outperform (have less CFAR loss) the conventional and derivative forms of cell-averaging CFAR processors.

The basic difference between the conventional cell-averaging and the clutter-map types of CFAR concerns the manner in which the samples are obtained to estimate the background interference level and the method used to censor the target return from the background estimate. In the conventional CFAR, the background estimate is made within a single radar scan by examining statistically independent range (and possibly doppler) cells contiguous to the test cell, which is excluded from the estimate. In the clutter-map approach, the background in the test cell is examined on a scan-to-scan basis where target returns are excluded through doppler discrimination (the MTD uses the zero-velocity doppler filter to estimate the mean clutter level) or through a target detection test [45]. Implicit in the clutter map CFAR method is that independent clutter samples are available on a scan-to-scan basis. This occurs naturally in most conventional surveillance radars, where the scan frame time is on the order of 5–10 s, but the clutter decorrelation time due to scan modulation may be on the order of 100–500 ms .

The clutter-map type of CFAR processors generally use a first-order recursive filter, as shown in Figure 1.35, to estimate the mean clutter level. This type of filter requires only one memory storage cell for each clutter map resolution cell. As there can be many clutter map resolution cells (e.g., 365,000 range-azimuth cells in the ASR-9 MTD), the total memory could be excessive if a moving-window type of averager (e.g., with 16 samples) were used as in the conventional CFAR processor. In the recursive filter type of estimator, a fraction α of the present sample (e.g., α

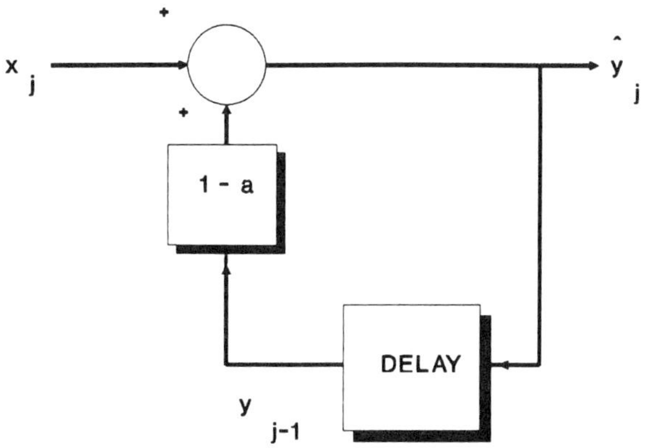

Figure 1.35 Clutter map recursive filter.

= 1/8 in the MTD) is added to $1 - \alpha$ of the previous estimate to form the new estimate. This produces an exponential type of smoothing action, where the effect of each sample fades at an exponential rate.

The state equation for the first-order recursive filter used in the clutter map CFAR is given by $\hat{y}_j = (1 - \alpha)\hat{y}_{j-1} + \alpha x_j$, where $\hat{y}_j$ is the jth output of the power estimate, x_j is the jth sample of the interference (clutter and noise) power, and α is the filter's gain coefficient. The expected value of the power estimate is given by $E(\hat{y}_i) = P_c[1 - (1 - \alpha)^j]$, which approaches $E(y) = P_c$ for large values of j. Note that the value of clutter power (P_c) in each cell is proportional to the backscatter coefficient (σ°), which is assumed to be a random variable in the spatially nonstationary clutter model described in Section 1.3.4. Thus, the clutter map CFAR estimates the backscattering coefficient (σ°) on a cell-to-cell basis and hence accommodates both spatially homogeneous and nonhomogeneous clutter distributions.

The variance of the recursive filter's estimate of the clutter power is given by $\text{var}(\hat{y}_j) = \alpha^2\sigma_x^2[1 - (1 - \alpha)^{2j}]/[1 - (1 - \alpha)^2]$, where σ_x^2 is the variance of the clutter power random variable. For large values of j, this approaches $\text{var}(\hat{y}) = \alpha\sigma_x^2/2 - \alpha$, which illustrates the smoothing effect of the filter. Because the variance is reduced by the factor $n' = (2 - \alpha)/\alpha$, this is equivalent to the use of n' cells in a conventional cell-averaging CFAR [44].

One use of the clutter-map type of CFAR is to provide superclutter visibility in those clutter cells where the target return significantly exceeds the clutter return. In the MTD processor, this allows strong tangential targets (those with small radial velocities) to be detected over weak ground clutter or tangentially moving precipitation returns. This is particularly advantageous in the formation and maintenance of ground tracks that occur in subsequent data processing.

A simplified analysis of the CFAR loss associated with the clutter-map CFAR processor shown in Figure 1.35 can be made for the situation where a Swerling 1 type fluctuating target is to be detected in an unknown power Rayleigh clutter background, where the clutter is completely correlated within a single radar scan [44]. The probability of detection for a square-law detector and a fixed threshold (y_b) is given by $P_{d_1} = \exp[-y_b/(1 + \bar{\gamma})]$, where $\bar{\gamma} = P_s/P_c$ is the mean signal-to-clutter power ratio. The threshold is estimated by the clutter map such that $y_b = C\hat{y}_j/P_c$, where C is a fixed threshold offset, $\hat{y}_j$ is the clutter power estimate from the recursive filter, and P_c is the clutter power. The value for P_d can then be written as $P_d = \exp[-C\hat{y}_j/P_c(1 + \bar{\gamma})]$, which is a function of the estimated clutter power (y_j). If we average over the clutter power estimate, this results in $\bar{P}_{d1} = \int_0^\infty \exp[-C\hat{y}_j/P_c(1 + \bar{\gamma})]P(\hat{y})\,d\hat{y}$, which can be recognized as the characteristic function $M[S = C/P_c(1 + \bar{\gamma})]$ of $\hat{y}$ evaluated at the argument of the exponential. The false alarm probability can be evaluated from this relationship evaluated at $\gamma = 0$ (i.e., $P_{fa} = M(S = C/P_c)$.

The recursive filter of Figure 1.35 provides an estimate of the clutter power formed from the current and previous samples as $\hat{y}_j = \alpha\Sigma_{l=0}^{\infty}(1 = \alpha)^l x_{j-l}$. The probability density of each x clutter sample is given by $p(x) = \exp(-x/P_c)/P_c$ so that the characteristic function of $\hat{y}$ equals $M(s) = \Pi_{l=0}^{\infty} [1 + \alpha(1 - \alpha)^l P_c S]^{-1}$, which upon substitution into the probability of detection relationship becomes $\bar{P}_{d_1} = \Pi_{l=0}^{\infty} [1 + C\alpha(1 - \alpha)^l/(1 + \bar{\gamma})]^{-1}$. This relationship evaluated at $\gamma = 0$ results in $P_{fa} = \Pi_{l=0}^{\infty} [1 + C\alpha(1 - \alpha)^l]^{-1}$.

The CFAR loss can be evaluated by first solving the P_{fa} relationship for C and then substituting this and an appropriate value of $\gamma = \bar{P}_s/P_c$ into the $\bar{P}_{d_1}$ relationship to provide a particular probability of detection. The known power (non-CFAR) value of γ_k can then be calculated from $\gamma_k = \ln P_{fa}/\ln \bar{P}_{d_1} - 1$. The CFAR loss in dB is then found from $L = 10 \log(\gamma - \gamma_k)$. This is plotted in Figure 1.36 as a function of the recursive filter gain α for $P_d = 0.9$ and $P_{fa} = 10^{-6}$.

Figure 1.36 shows that small CFAR loss results from using a small filter gain coefficient (α). This occurs because decreasing α corresponds to using a longer data window to estimate the clutter power. Conversely, map settling time is reduced as α is increased as is shown in Figure 1.37 [44]. Thus, a compromise between low CFAR loss and adequate settling time is required. A filter gain of $\alpha = .125$ is used in the MTD, which represents a compromise between these two factors.

Settling time becomes important for clutter transients caused by either changing or moving clutter. One example of this situation occurs in the MTD processor, where

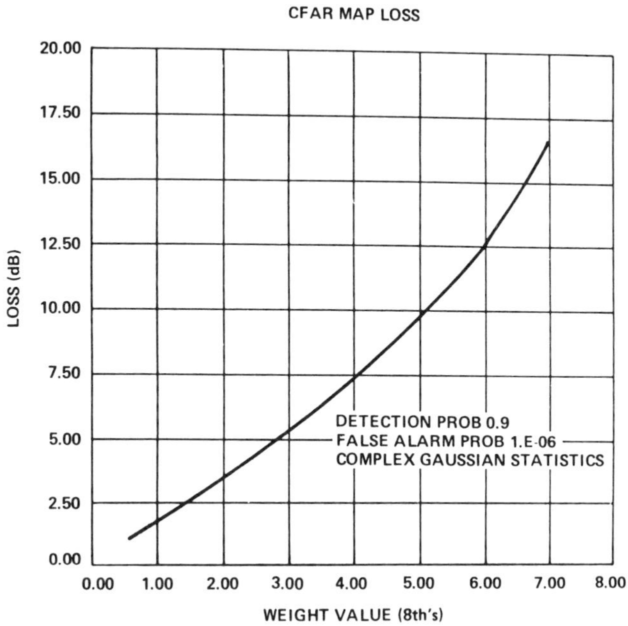

Figure 1.36 Clutter map CFAR loss.

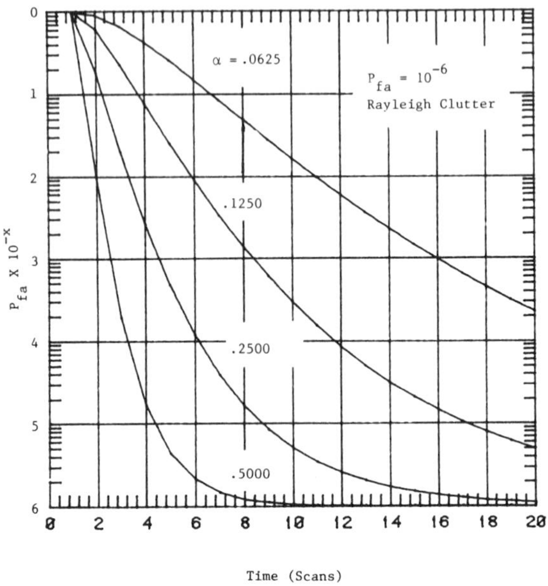

Figure 1.37 Clutter map settling time as a function of clutter map gain.

relatively large zero-velocity returns have been observed to occur from weather fronts moving rapidly through the radar's coverage. The clutter map then reacts to the changing clutter power by adapting its threshold to the new clutter condition. However, a false alarm will be reported until the map reaches its steady state condition. In the MTD processor, this situation was handled in the data processing by not allowing tracks to be initiated on zero-velocity targets.

A digital clutter map obtained with the MTD processor is depicted in Figure 1.38 [46]. The MTD uses 10-bit storage words to preserve the large dynamic range of the clutter signal. The ground clutter patch indicated in the right-hand corner was used for subclutter visibility tests using the MTD. The expected subclutter visibility of the MTD (in the doppler filter channels) is approximately 41 dB, provided the clutter return is within the linear dynamic range of the receiver-processor, which is on the order of 45 dB measured at the A/D converter. To accommodate ground clutter returns that exceed the linear dynamic range, the MTD raises the detection thresholds of the doppler filter channels in direct proportion to the measured clutter level. This in general has the effect of censoring those cells that contain clutter returns beyond the subclutter visibility capability of the doppler processor. Figure 1.39 depicts a situation that might occur in spotty ground clutter where clutter cells containing severe clutter returns are censored from the output. The probability of the radar data

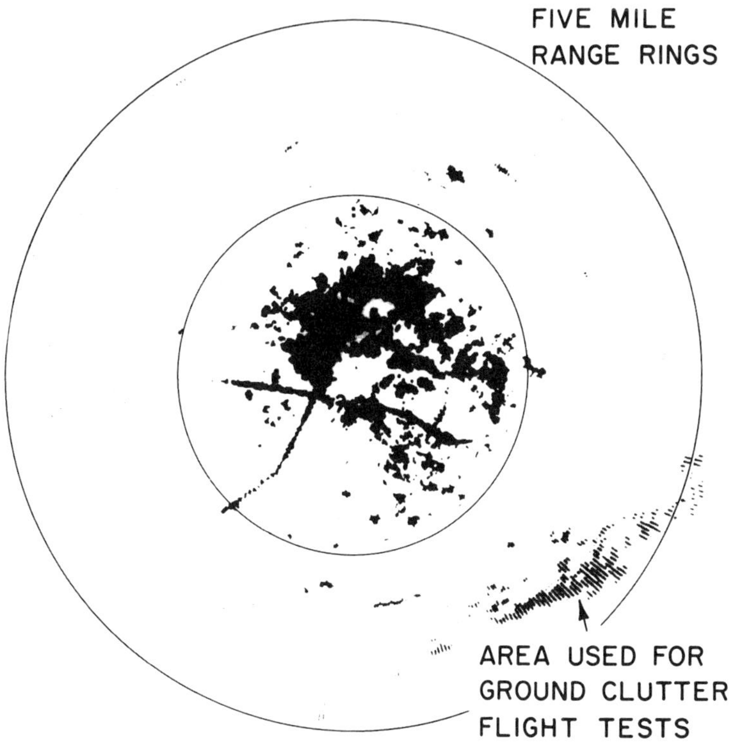

Figure 1.38 Digital clutter map of ground clutter.

processor tracker losing track is generally low in this situation. This is an example of interclutter visibility that generally improves as the radar's resolution is increased.

The minimum dimension of a clutter cell used in a clutter map is the radar's surface resolution cell (the radar range corresponding to the effective pulsewidth and the azimuth beamwidth). If a stationary target exists within this cell, then it will be suppressed by the CFAR action as its return will become part of the estimated clutter return. A slowly moving target also will be partially suppressed if its return is contained within the clutter cell for several data samples. The minimum map-free velocity is a measure of this effect, and is defined as the minimum target velocity that will cause the target not to be suppressed by the clutter map's CFAR action. In the range dimension, this velocity is given by $v_{\mathrm{mfr}} = (c\tau_e/2)/T_s$, where τ_e is the radar's effective pulsewidth, and T_s is the interscan period. For the ASR-9 ($\tau_e = 1.05\ \mu$s, $\theta_e = 1.3°$, $T_s = 4.8$ s) the minimum map-free velocity in the range dimension is then a radial velocity of about 65 knots. In the azimuth dimension, the relationship equals $v_{\mathrm{mfr}} = (\theta R)/T_s$, where θ is the azimuth beamwidth, R is the range to the clutter cell, and

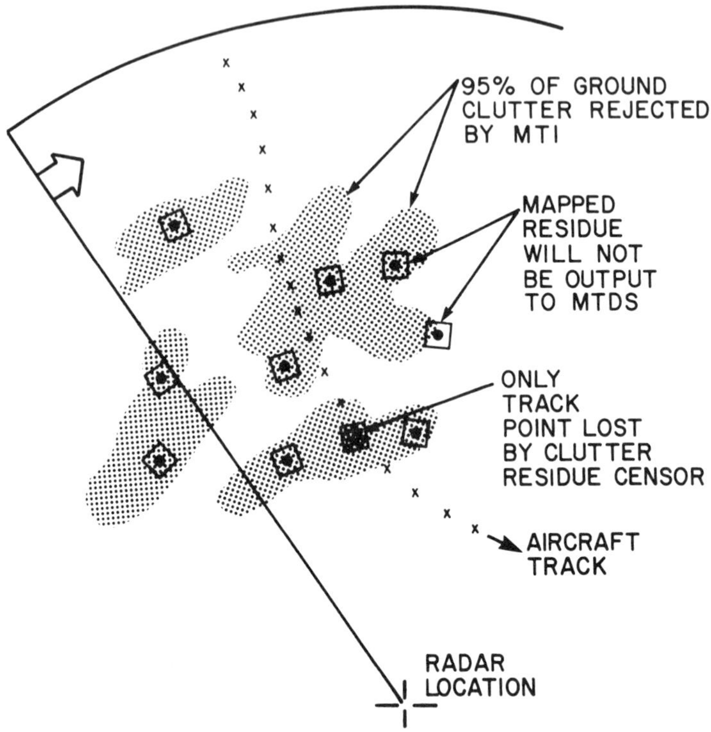

Figure 1.39 Aircraft tracking using MTD processing.

T_s is the interscan period. For the ASR-9 at 10 nmi, the minimum map-free tangential velocity becomes 170 knots.

REFERENCES

1. D.C. Schleher, ed., *MTI Radar,* Artech House, Norwood, MA, 1978.
2. D. Povejsil, R. Raven, and P. Waterman, *Airborne Radar,* D. Van Nostrand, Princeton, NJ, 1961.
3. D.K. Barton, ed., *CW and Doppler Radar,* in *Radars,* Vol. 7, Artech House, Norwood, MA, 1978.
4. G. Stimson, "Introduction to Airborne Radar," Hughes Aircraft Company, El Segundo, CA, 1983.
5. M. Skolnik, ed., *Radar Handbook,* 1st Ed., Chapters 17 and 18, "MTI Radar" and "Airborne MTI," McGraw-Hill, New York, 1970.
6. A. Rihaczek, *Principles of High Resolution Radar,* McGraw-Hill, New York, 1969.

7. D.K. Barton ed., *Radar Resolution and Multipath Effects,* in *Radars,* Vol. 4, Artech House, Norwood, MA, 1975.
8. S. Haykin, Radar Signal Processing, *IEEE ASSP Magazine,* April 1985.
9. M.I. Skolnik, ed., *Radar Handbook,* 1st Ed., Chapter 19, "Pulsed Doppler Radar," McGraw-Hill, New York, 1970.
10. G. Andrews, "Optimal Radar Doppler Processors," NRL Rep. 7727, May 1974.
11. F. Nathanson, *Radar Design Principles,* McGraw-Hill, New York, 1969.
12. V.G. Hansen, Signal Processing Applications to Radar, *Electronic Progress,* Vol. 28, No. 1, 1987.
13. D. Kerr, *Propagation of Short Radio Waves,* McGraw-Hill, New York, 1951.
14. M.W. Long, *Radar Reflectivity of Land and Sea,* Artech House, Norwood, MA, 1983.
15. D.K. Barton, ed., *Radar Clutter,* in *Radars,* Vol. 5, Artech, Norwood, MA, 1975.
16. D.K. Barton, "Land Clutter Models for Radar Design and Analysis," *Proc. IEEE,* Vol. 73, No. 2, February 1985.
17. D.K. Barton, *Radar System Analysis,* Artech House, Norwood, MA, 1976.
18. M.I. Skolnik, *Introduction to Radar Systems,* 2nd Ed., McGraw-Hill, New York, 1980.
19. J. Eaves and E. Reedy, eds., *Principles of Modern Radar,* Van Nostrand Reinhold, New York, 1987.
20. W. Fishbein, S. Graveline, and O. Rittenback, "Clutter Attenuation Analysis," Tech. Rep. ECOM-2808, U.S. Army Electronics Command, Fort Monmouth, NJ, March 1967.
21. J. Billingsley, and J. Larrabee, "Measured Spectral Extent of L- and X-Band Radar Reflections from Wind-Blown Trees," MIT Lincoln Laboratory Project Rep. CMT-57, Lexington, MA, February 1987.
22. S. Haykin, "Maximum-Entropy Spectral Analysis of Radar Clutter," *Proc. IEEE,* Vol. 70, No. 9, September 1982.
23. L. Marple, *Digital Spectral Analysis with Applications,* Prentice-Hall, Englewood Cliffs, NJ, 1987.
24. S. Kay and L. Marple, "Spectrum Analysis—A Modern Perspective," *Proc. IEEE,* Vol. 69, No. 11, November 1981.
25. T. Ulrych and T. Bishop, "Maximum Entropy Spectral Analysis and Autoregressive Decomposition," *Rev. Geophysics Space Physics,* Vol. 13, February 1975.
26. A. Papoulis, *Probability, Random Variables and Stochastic Processes,* McGraw-Hill, New York, 1965.
27. R. Mitchell, "Importance Sampling Applied to Simulation of False Alarm Statistics," *IEEE Trans.,* Vol. AES-17, No. 1, January 1981.
28. D.C. Schleher, ed., *Automatic Detection and Radar Data Processing,* Artech House, Norwood, MA, 1980.
29. G. Trunk, "Non-Rayleigh Sea Clutter: Properties and Detection of Targets," NRL Rep. 7986, June 1976.
30. D.C. Schleher, "Radar Detection in Log-Normal Clutter," *IEEE Int. Radar Conf.,* Washington, DC, April 1975.
31. D.C. Schleher, "Radar Detection in Weibull Clutter," *IEEE Trans.,* Vol. AES-12, No. 6, November 1976.
32. D.C. Schleher, "Harbor Surveillance Radar Detection Performance," *IEEE J. Oceanic Engineering,* Vol. OE-2, No. 4, October 1977.
33. D.C. Schleher, "MTI Detection Performance in Rayleigh and Log-Normal Clutter," *IEEE Int. Radar Conf.,* Washington, DC, April 1980.
34. P. Swerling, "Probability of Detection for Fluctuating Targets," *IEEE Trans.,* Vol. IT-6, April 1960.
35. J. DiFranco and W. Rubin, *Radar Detection,* Artech House, Norwood, MA, 1980.

36. D.C. Schleher, *Introduction to Electronic Warfare,* Artech House, Norwood, MA, 1986.
37. K. Ward and S. Watts, "Radar Sea Clutter," *Microwave J.*, Vol. 28, No. 6, June 1985.
38. S. Watts, "Radar Detection Prediction in Sea Clutter Using the Compound K-Distribution Model," *IEE Proc.*, Vol. 132, Pt. F, No. 7, December 1985.
39. D.C. Schleher, "Performance Comparison of MTI and Coherent Doppler Processors," *IEE Int. Radar Conf.*, London, November 1982.
40. R. O'Donnell, C. Muehe, "Automated Tracking for Aircraft Surveillance Radar Systems," *IEEE Trans.*, Vol. AES-15, No. 4, July 1979.
41. J. Taylor, "Sacrifices in Radar Clutter Suppression Due to Compromises in Implemention of Digital Doppler Filters," *IEE Int. Radar Conf.*, London, November 1982.
42. J. Taylor, E. D'Addio, and G. Galati, "Effects of Finite Word Length on FIR Filters for MTD Processing," *IEE Proc.*, Vol. 130, Pt. F, No. 6, October 1983.
43. D.K. Barton, *Modern Radar System Analysis,* Artech House, Norwood, MA, 1988.
44. E. Khoury and J. Hoyle, "Clutter Maps: Design and Peformance," *Nat. Radar Conf.*, Atlanta, March 1984.
45. R. Nitzberg, "Clutter Map CFAR Analysis," *IEEE Trans.*, Vol. AES-22, No. 4, July 1986.
46. L. Cartledge and R. O'Donnell, "Description and Performance Evaluation of the Moving Target Detector," Rep. No. FAA-RD-76-190, MIT Lincoln Laboratory, Lexington, MA, March 1977.

Chapter 2
Doppler Radar Waveform Design and Filtering

In Chapter 1, we discussed the theory of optimum doppler filter design for detection of a doppler frequency shifted coherent pulse train. The MTI solution was identified as appropriate when there was no *a priori* knowledge about the doppler shift of the target return, and the pulsed doppler design resulted when it was assumed that the doppler shift of the target return was precisely known. In practice, the pulsed doppler design is implemented by filling the expected region of target dopplers with a suitable number of point-optimized complex weighted linear transversal doppler filters. Although the structure of the doppler filters depends on the assumptions associated with the velocity (doppler) characteristics of the expected targets, the detailed filter parameters depend on the power spectral density characteristics of the clutter, which, in turn, depends on the radar's waveform design.

The primary consideration in doppler radar waveform design is concerned with the range and doppler ambiguities associated with the particular waveform under consideration. One way to visualize this situation is through the use of the radar ambiguity function. An idealized form of the ambiguity surface for a coherent pulse train with a large number of pulses is shown in Figure 2.1. Note that range ambiguities occur at multiples of the pulse repetition intervals given by $T = 1/f_r$, where f_r is the radar's pulse repetition frequency. Doppler (velocity) ambiguities are also related to the radar's PRF and are given by $v_t = \lambda f_r/2$, where λ is the radar's wavelength. Thus the radar's PRF (f_r) determines both range and doppler ambiguities and is the critical parameter in a doppler radar's waveform design.

The radar ambiguity function provides the response of a matched filter receiver to a point scatterer whose range and doppler are centered about the target. Consider first a fixed based radar where ground clutter occurs at ranges randomly distributed throughout the range delay space and at zero doppler. By superimposing the radar ambiguity responses onto the target and clutter we can see that the clutter responses will be distributed throughout the range dimension; but if the target possesses a differential doppler with respect to the clutter, it can be separated in the doppler domain by appropriate filtering. If target range is to be measured unambiguously,

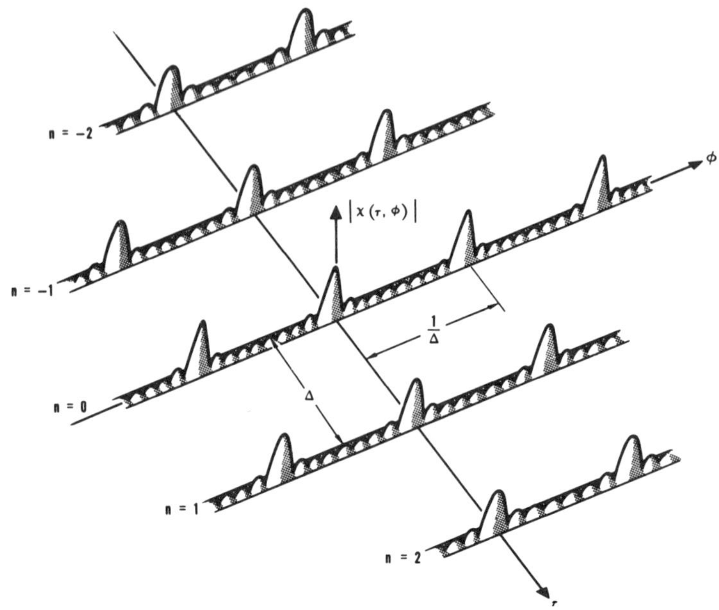

Figure 2.1 Idealized radar ambiguity surface for a coherent pulse train with a large number of pulses.

then the multiple responses in the range domain should be separated by greater than the range delay associated with the maximum expected target range. This results in the low-PRF radar waveform design appropriate for an MTI radar and a fixed based radar operating against ground clutter. Note that if the doppler associated with the target exceeds the radar's PRF then the doppler generally can be separated by doppler filtering but the radar's doppler response will be ambiguous. When the target's doppler responses occur at multiples of the radar's PRF then the targets cannot be separated by doppler filtering, resulting in the so-called blind speeds associated with low-PRF doppler radar design. In this case, we must jitter or stagger the radar's PRF to achieve target detection.

Next consider the case of an airborne radar operating against ground clutter. The aircraft motion will cause the return from the clutter scatterers to have a doppler response that depends on the relative closing velocity between the aircraft and the particular scatterer. The magnitude of the response will depend on the two-way antenna pattern in the direction of the scatterer. Hence, in effect, the aircraft motion spreads the doppler return of the clutter throughout the doppler region corresponding to the aircraft's velocity ($f_d = \pm 2v_r/\lambda$). If a low-PRF design is used with ambiguous doppler response, then sidelobe clutter will generally appear within the same doppler region as the target. Therefore, the low-PRF airborne radar clutter improvement factor generally cannot exceed the magnitude of the radar antenna's sidelobe two-way radiation pattern as compared to the magnitude of the antenna's peak response in its mainlobe.

To circumvent this difficulty, the radar's PRF can be increased to greater than $f_d = 2v_r/\lambda$ plus the maximum doppler frequency of the target. Then the target's response is unambiguous in doppler, and closing targets can be separated from the clutter by using appropriate doppler filtering. This type of radar waveform design results in a high-PRF radar appropriate for a PD radar. Note that this type of radar generally is highly ambiguous in range, which requires PRF coding to resolve the range ambiguities.

However, another significant difference between the high- and low-PRF airborne radar designs affects the strength of the clutter return. In the low-PRF radar design, only clutter returns occurring within the same range resolution cell as the target compete with the target. In a high-PRF radar design, clutter returns from shorter ambiguous ranges compete with the target return in a range resolution cell. Because the shorter range clutter returns are much stronger than those at the target, the high-PRF radar must provide a much greater clutter improvement factor than the low-PRF design for equal detection performance. Another consequence of this effect is that detection performance in high-PRF radars is limited against tail-aspect targets because these targets produce doppler returns that must compete against range ambiguous sidelobe clutter doppler returns.

A compromise airborne radar waveform design uses a medium-PRF waveform that is ambiguous in both the range and doppler domains. This type of radar waveform can provide performance against closing targets that is somewhat less than that provided by a high-PRF design and performance against tail-aspect targets that is somewhat less than that provided by a low-PRF design. The medium-PRF airborne radar design therefore is useful when all-aspect coverage is required and the tail-aspect limitations of high-PRF designs or the closing aspect limitations of low-PRF designs cannot be tolerated.

The medium-PRF airborne radar waveform finds application in situations where the radar must look down in heavy ground clutter against tail-aspect targets. These targets appear in the sidelobe clutter region of the range-doppler clutter surface shown in Figure 2.2 for the high-PRF waveform. Hence any doppler filter placed in this region collects both target and clutter returns as contrasted to the noise-limited region that offers good sensitivity for high-radial-speed closing targets. The heavy sidelobe clutter shown in Figure 2.2 for the high-PRF design is emphasized by the collection of close-in clutter due to the highly range-ambiguous nature of this type of waveform. The medium-PRF range-doppler clutter surface also depicted in Figure 2.2 shows a reduction of the sidelobe clutter's magnitude due to spreading of the range ambiguities. However, the targets now must generally compete with sidelobe clutter, which occurs at all ranges and radial velocities, thereby placing a premium on the low-sidelobe antenna design.

Medium- and high-PRF waveform designs are also useful in surface based radars for detecting low-altitude targets in heavy ground clutter or rejecting sky clutter (weather or chaff). In low-PRF radars weather or chaff returns often are aliased into the first doppler region, where they overlap fundamental and aliased target doppler returns.

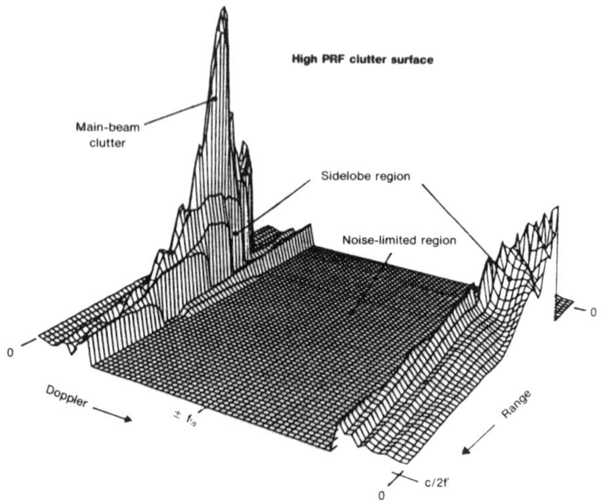

2. High-PRF radar signals *offer good sensitivity to approaching targets moving at high speed. This range-Doppler diagram shows noise-limited performance in which ground echo does not compete with such targets.*

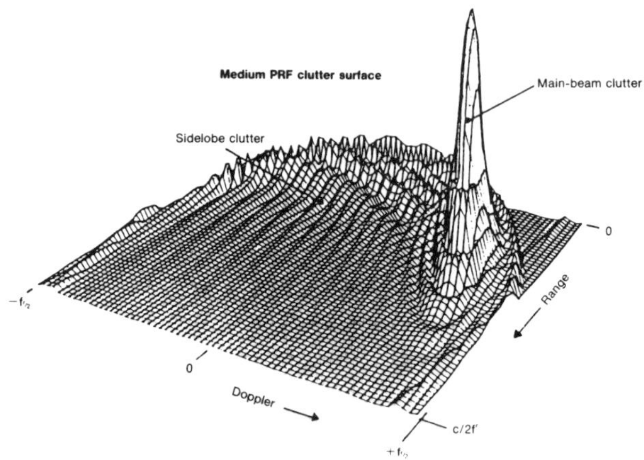

3. Medium-PRF signals *are subject to sidelobe clutter at all ranges, but careful antenna design can reduce this limitation to satisfactory levels. Medium PRF can be a good compromise between high- and low-PRF signals.*

Figure 2.2 High- and medium-PRF airborne radar range-doppler clutter surface.

2.1 DEFINITION OF PRF CLASSES [1]

The preceding discussion points to the desirability of classifying doppler radars on the basis of their PRFs. What has become common practice, particularly in the airborne radar field, is to classify radars as low-, medium-, and high-PRF radars. The generally accepted definitions are as follows [1]:

- In a low-PRF radar the maximum range the radar is designed to handle falls in the first range zone given approximately in nautical miles by $R_u = 80/f_r$, where f_r is the radar's PRF in kHz. In the absence of a return from beyond this zone, range is unambiguous;
- In a high-PRF radar the observed doppler frequencies of all significant targets are unambiguous, but the range generally is highly ambiguous;
- In a medium-PRF radar, both range and doppler frequency are ambiguous.

The blind speeds ($f_{d\,\text{blind}}$) for the radar waveform occur when the return doppler corresponds to multiples ($n = 1, 2, \ldots$) of the radar's PRF ($f_{d\,\text{blind}} = nf_r$). The corresponding blind speeds are given by $v_{d\,\text{blind}} = n\lambda f_r/2$. The unambiguous range is given by $R_u = c/2f_r$, where f_r is the radar's PRF. Using these equations, unambiguous velocity (first blind speed) can be plotted against unambiguous range and PRF, which is depicted in Figure 2.3.

As an example, consider the situation for a fixed based S-band (3 GHz) air surveillance radar with a maximum range of 50 nmi that searches for targets with radial velocities in the ±300 knot region. Reference to Figure 2.3 indicates that setting the first blind speed at 300 knots results in a PRF of approximately 3100 Hz, which provides a maximum unambiguous range of approximately 26 nmi. Alternatively setting the PRF at 1600 Hz results in an unambiguous range of about 50 nmi but reduces the first blind speed to about 155 knots. Therefore unambiguous range and velocity cannot be achieved at the specified frequency for the example given. However, reducing the transmitter frequency to L-band (1 GHz) would allow unambiguous operation in both range and doppler.

In the air surveillance radar example, the S-band version operating at a PRF of 1600 Hz would be classified as a low-PRF radar. The same radar operating at a PRF of 3100 Hz would be classified as a medium-PRF radar. The L-band version operating at a PRF of 1600 Hz would generally be classified as a low-PRF radar even though it provides unambiguous velocity information. In general, the PRF class does not depend on the numerical value of the radar's PRF in a precise way but rather is strongly influenced by the radar's operating frequency and the maximum expected range of both target and clutter returns and their relative velocities with respect to the radar.

Airborne radars for high performance aircraft generally are required to perform a number of functions that require the use of several PRF classes. An X-band multimode radar might use a low-PRF mode, typically from 250 to 4000 Hz; PRFs in the medium

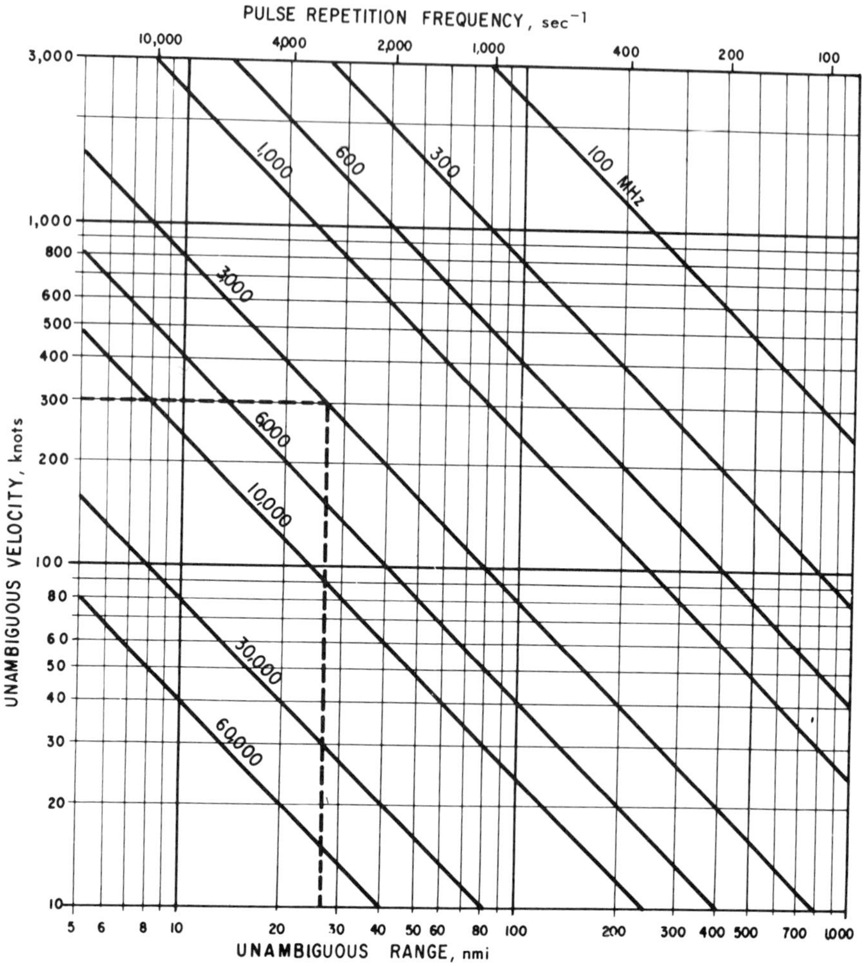

Figure 2.3 Unambiguous velocity *versus* unambiguous range.

category, from 10 to 20 kHz; and PRFs in the high category, from 100 to 300 kHz [1]. The three categories of PRF are typically interlaced as the radar scans through its coverage volume.

2.2 SIGNAL PROCESSING AND COMPARATIVE SYSTEM CONSIDERATIONS

Each PRF choice has advantages and limitations that relate to a number of complex factors. The range and doppler ambiguities of the radar waveform generally determine the radar's response to target and clutter returns. This determines the radar's signal processing configuration and, ultimately, the radar's overall performance capability.

2.2.1 Low-PRF Radar

Consider the low-PRF airborne radar whose ground-clutter doppler power spectral density for the antenna pointing along the ground track is depicted in Figure 2.4 [2]. The generally ambiguous doppler responses of this type of design cause target and clutter returns to be folded into the region between the reference doppler (doppler return from the ground corresponding to the spatial direction of the antenna's mainlobe peak radiation pattern) and the first blind speed. The objective of the doppler comb filter is to eliminate all the clutter returns while retaining all the target returns in this region. This becomes increasingly difficult as the number of clutter and target returns that are aliased into this region increases, causing possible overlapping clutter and target returns. Note that clutter sidelobe returns are translated into the filter's passband due to aircraft motion and hence cannot generally be eliminated by doppler filtering. The width of the central clutter lobe along the ground track generally is set by antenna scanning modulation induced onto each clutter scatterer illuminated by the antenna's mainlobe radiation pattern.

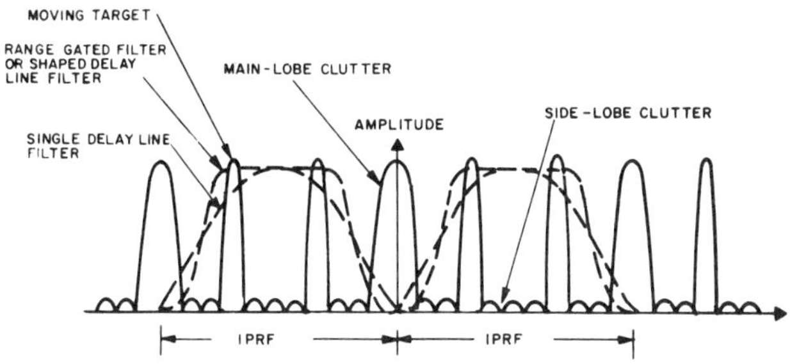

Figure 2.4 MTI filter and clutter spectrum (antenna pointing along the ground track).

In Figure 2.5, the ground clutter doppler power spectral density is depicted for the antenna pointing away from the aircraft's ground track. The width of the central clutter lobe is now set by the differential doppler responses caused by the different closing velocities for the velocity vectors to each clutter scatterer and that to the doppler reference vector. In particular the effective width of the main lobe can be shown to be given by $\sigma_p = kv_{ac} \sin\theta/D_{az}$, where v_{ac} is the aircraft's ground track velocity, θ is the antenna's azimuth pointing angle relative to the aircraft's ground track, and D_{az} is the antenna's aperture width in the azimuth dimension [2]. Note that, for high-performance aircraft, the doppler region available for target detection is severely restricted.

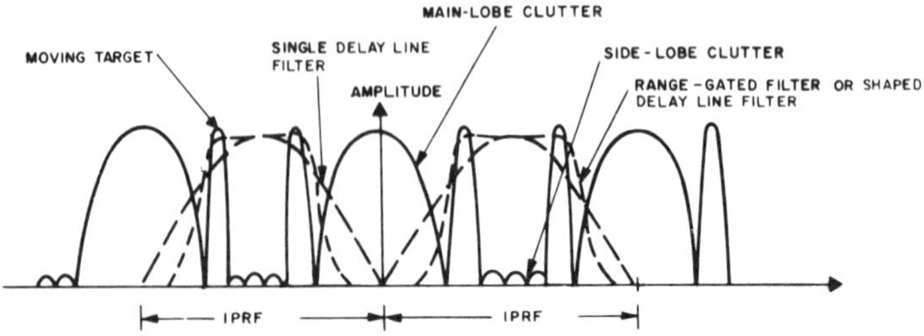

Figure 2.5 MTI filter and clutter spectrum (antenna pointing away from the ground track).

At low PRFs the target's doppler frequency may be many times the PRF, so the target is about as likely to appear at any one point, within a span of frequencies equal to the PRF, as at any other. Therefore, the probability of a target being in the blind zones at any one time is roughly equal to the ratio of the width of the rejection notches to the PRF.

Figure 2.6 shows two applications of low-PRF airborne radar that are designed to have equal ratios of doppler-clear to blind regions. The first is typical of an *airborne early warning* (AEW) radar (e.g., the E-2C radar), where the speed of the radar bearing aircraft is modest and a large antenna can be mounted in a rotodome above the aircraft. For the example depicted (v_{ac} = 300 knots, θ = 90°, D_{az} = 20 feet) the doppler-clear region will be at least as wide as the blind region at a PRF as low as 200 Hz. Typical transmitter frequencies for low-PRF AEW radars are UHF (420–450 MHz) or L-band (1215–1400 MHz). The first blind velocity for a 200 Hz PRF would be about 135 knots at UHF and 45 knots for L-band.

The second application shown in Figure 2.6 is that of an airborne interceptor radar in which the antenna size is limited and radar speeds may be high. For low-PRF designs, the two recourses open are to limit the look angle (±30° in the example) or to use the highest possible PRF. Typical transmitter frequencies for this type of radar are X-band (8.5–10.7 GHz) and Ku-band (15.7–17.7 GHz). The first blind velocity for a 4000 Hz PRF would be about 117 knots at X-band and 71 knots at Ku-band.

Low-PRF airborne radar operation has several advantages. The unambiguous range capability allows (1) direct measurement of the target's range; (2) *sensitivity time control* (STC) in the receiver to equalize returns as a function of range, thereby reducing the overall dynamic range; and (3) minimizing clutter intake by range gating to just those in the target's range resolution cell.

Limitations on low-PRF airborne radars involve (1) poor look-down capability where excessive target returns are rejected along with mainlobe clutter; (2) doppler ambiguities generally too severe to be resolved; and (3) ground moving targets indistinguishable from airborne moving targets. These limitations are generally too

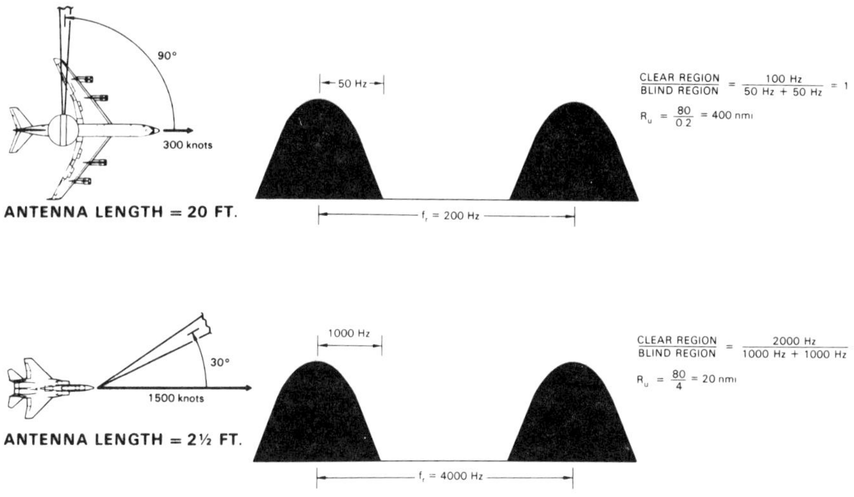

Figure 2.6 Ground clutter spectra for two applications of low-PRF airborne radar.

severe for airborne interceptor radar, forcing use of other PRF classes. Low-PRF waveforms are most useful in AEW radars and for surface based radars that do not have the severe clutter intake of look-down airborne interceptor radars.

A block diagram of a typical coherent low-PRF radar digital signal processor is depicted in Figure 2.7. The receiver output is fed to a synchronous detector that converts it to in-phase and quadrature video signals. The frequency of the reference signal supplied to the detector is such as to place the central line of mainlobe clutter at zero frequency. An A/D converter samples the video signals at intervals matching the width of the transmitted pulses. The output of the converter therefore is a stream of numbers representing the I and Q components of the returns from successive range increments. The numbers are sorted by range increment into separate range bins.

An MTI comb filter clutter canceller in each I and Q channel is used to reduce the mainlobe clutter component. To reduce the mainlobe clutter residue in the output of the canceller as well as minimize the amount of noise and simultaneously received sidelobe clutter with which a target must compete, the output of each clutter canceller is generally integrated in a bank of doppler filters. At the end of every filter integration time, the magnitude of each desired filter's output is detected. If the integration time is less than the length of the time-on-target for the radar beam, then *postdetection integration* (PDI) may be provided.

A threshold detector is set at a predetermined amount, higher than the average of the output of the corresponding filters for several range bins on either side. This controls the probability of false alarms while providing adequate sensitivity for detecting

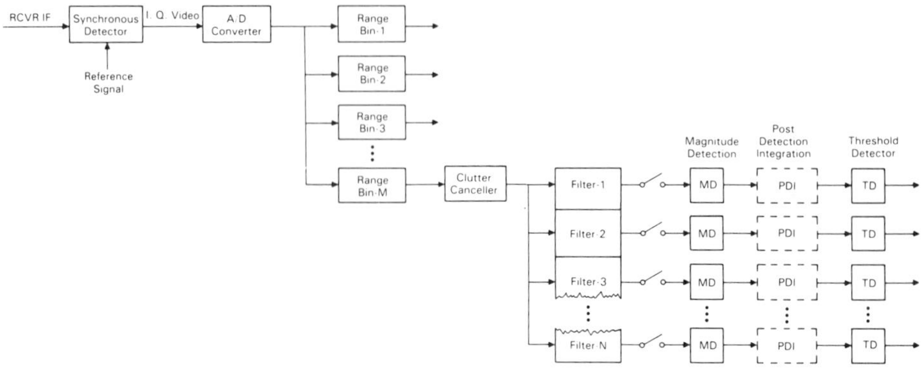

Figure 2.7 Signal processor for a low-PRF radar.

targets. A frequency offset is required to adjust the synchronous detectors reference frequency to the changing center frequency (due to changing antenna look angles) of the ground clutter spectrum. This usually is accomplished to a first order using aircraft navigation data coupled with the antenna's known look angle. Fine tracking uses a clutter spectral tracking servo that adjusts the reference frequency so that mainlobe clutter is centered within the canceller's rejection notch.

Figure 2.8 depicts a single-channel low-PRF signal processor employed in older or less sophisticated modern radars. This type of processor uses only the *I* channel from the synchronous detector, resulting in an average 3 dB loss in signal energy and the introduction of blind phases (doppler phases causing zero output). Elimination of the *Q* channel, doppler filters, and magnitude detectors considerably simplifies the processing hardware.

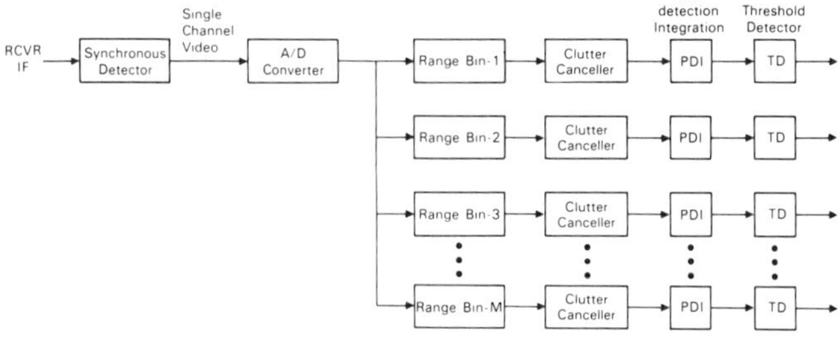

Figure 2.8 Single-channel low-PRF signal processor.

2.2.2 High-PRF Radar

A typical power spectral density for an airborne high-PRF radar is depicted in Figure 2.9. All mainlobe and sidelobe doppler returns from ground clutter are located in the region $f_d = \pm 2v_r/\lambda$, where v_r is the velocity of the aircraft carrying the radar. The absence of clutter returns in the region between the band of central-line sidelobe clutter frequencies and the first repetition of this band opens up a region that is clutter free. In this region, which corresponds to targets closing on the airborne radar at a speed greater than the radar's ground speed, detection can be made against a noisy background using a matched-filter signal processor. From an operational viewpoint long detection ranges are desired for these approaching targets.

The high-PRF radar allows the average transmitter power to be maximized by increasing the PRF. This further enhances the detection capability against closing

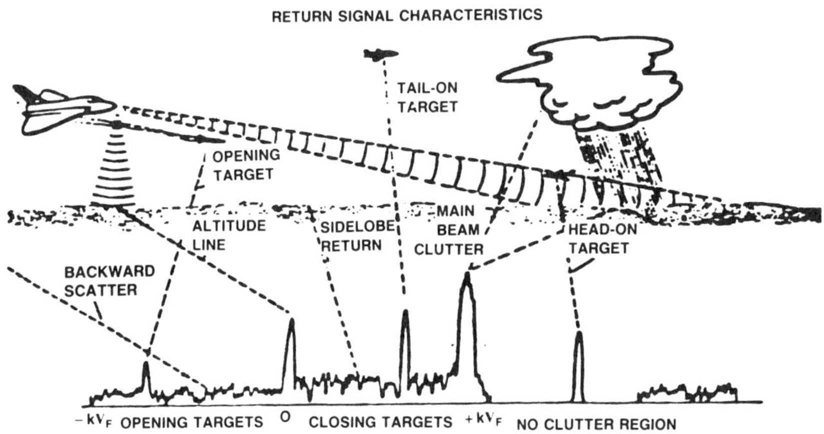

Figure 2.9 High-PRF radar clutter and target spectra.

targets. At X-band typical PRFs for airborne radars are on the order of 100 to 300 kHz with duty factors of up to 50%. To achieve the same amount of average power with a low-PRF radar, subject to a transmitter peak power limitation, requires increasing the radar's pulse width and employing large amounts of pulse compression to provide the degree of range resolution essential to low-PRF radar operation.

The principal limitation of high-PRF airborne radar operation is that detection performance may be degraded by sidelobe and backlobe clutter when operating against low closing-rate and opening-rate tail-aspect targets. As we can see from Figure 2.9, these targets must compete with sidelobe and backlobe clutter with spectral responses that overlap the target's spectral response. At the high PRFs typically employed in modern high-performance aircraft, the highly ambiguous range response causes sidelobe clutter from virtually all ranges to fold within the range interval corresponding to the interpulse period between successive radar transmissions. Consequently, the sidelobe clutter can be rejected only by resolving the return in doppler frequency. Much of the sidelobe clutter falling within the same resolvable frequency increment as a target's return will have been reflected from much shorter ranges and so will be quite strong. Thus, unless the target is at a short range or has a very large RCS, its return may be lost in the clutter.

Another disadvantage of the high-PRF radar waveform is the necessity to resolve the target's range ambiguities to obtain the target's range. This can be accomplished by either switching the radar PRFs or by using a frequency modulated transmitter. In either case, the losses incurred in resolving ambiguities usually cause a significant reduction in overall detection range compared to that calculated on a power aperture basis. Because range measurement reduces detection range, where maximum detection range is desired, a special velocity search mode may be provided in which range is not measured.

High-PRF radars, when operating at high duty factors experience significant eclipsing losses that occur due to returns being received while the radar is transmitting and the receiver blanked. This may be minimized by switching PRFs or providing multiple-range gates. However, switching PRFs reduces detection range in the doppler clear region.

A typical high-PRF airborne radar analog signal processor is depicted in Figure 2.10. The receiver output is first applied to a bandpass filter that passes only the central spectral band, thereby converting its output into a continous signal, and all subsequent processing is handled as in a *continuous wave* (CW) radar. The output of the central band filter is passed through a filter that removes the return signals having zero doppler frequency, consisting of any transmitter spillover and altitude return, composed of specular returns from the terrain directly under the aircraft. Along with them, the echoes of any zero-closing-rate targets will unavoidably be rejected.

The entire doppler spectrum is then shifted in frequency to center the mainlobe clutter in the rejection notch of the *mainlobe clutter* (MLC) filter. The doppler spectrum output of the MLC filter is then multiplexed into two subbands. The first subband spans the doppler clutter clear region, and the second covers the sidelobe and backlobe clutter regions. By applying automatic gain control separately in each subband, strong return in one band is prevented from desensitizing the other bands.

Signal levels equalized, the subbands are recombined and applied to a bank of doppler filters. The output of each doppler filter is then detected and compared against a detection threshold. The doppler filter output builds up for a time $T_i = 1/B_D$, where B_D is the doppler filter's bandwidth. When T_i is less than T_D, the time on target for the radar antenna, then the detector outputs are added up in postdetection integrators before comparison to separate threshold detectors.

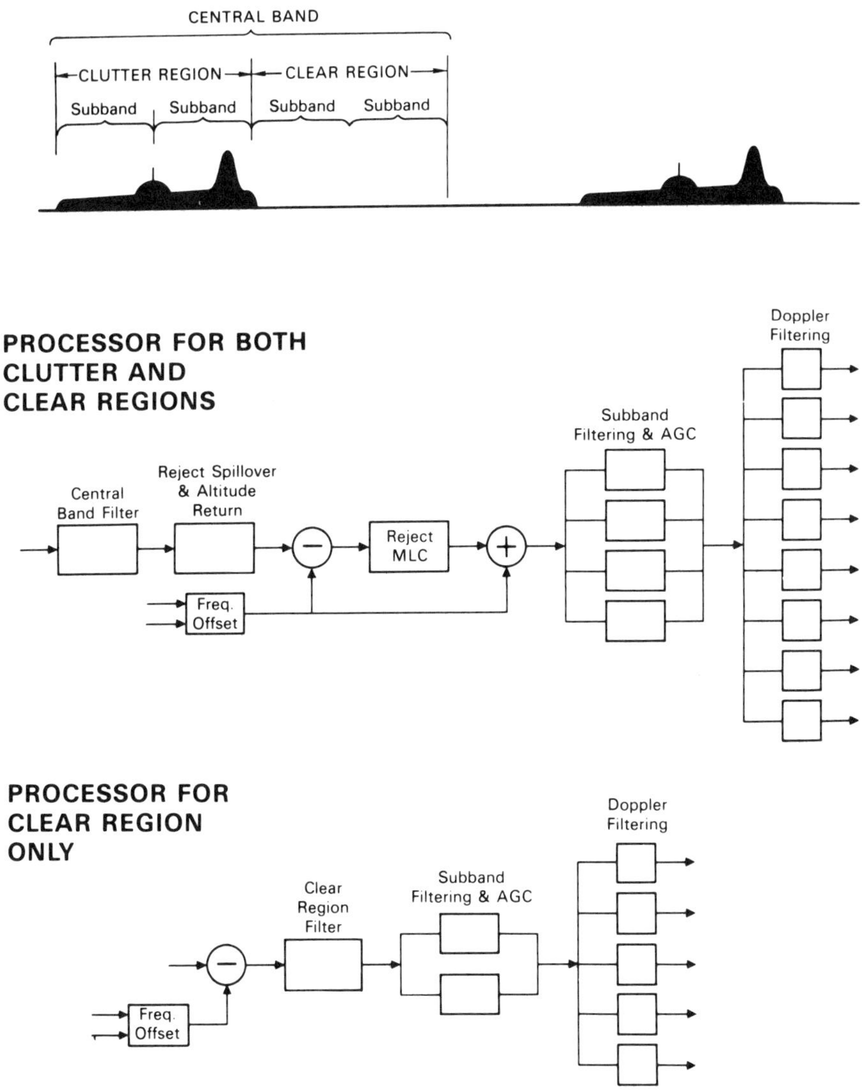

Figure 2.10 Signal processor for a high-PRF radar.

2.2.3 Medium-PRF Radar [3–5]

Medium-PRF airborne radar was conceived as a solution to the problems of detecting tail-aspect targets in the presence of both mainlobe and strong sidelobe clutter, thereby providing good overall coverage from all aspects. For modest operating ranges, the PRF can be set high enough to provide adequate separation between the periodic repetitions of the mainlobe clutter spectrum without incurring particularly severe range ambiguities. Mainlobe clutter can then be rejected through doppler filtering, and individual targets can be extracted from sidelobe clutter using a combination of doppler filtering and range resolution discrimination. Also ground moving targets, being similar in characteristic to mainlobe clutter, can be rejected using a medium-PRF design.

The key feature of medium-PRF airborne radar operation is its ability to detect tail-aspect moving targets in a sidelobe clutter background. Increasing the PRF over low-PRF operation opens the clear region between mainlobe clutter responses that occur at multiples of the radar's PRF (e.g., recall Figure 2.6). This makes it easier to construct practical doppler filters to both reject mainlobe clutter and bracket the target's response, thereby rejecting that sidelobe clutter not in the target's doppler resolution cell (e.g., $B_D = 1/T_d$, where T_d is the time on target).

However, in the process of increasing the PRF, range ambiguities are created that allow close in mainlobe and sidelobe clutter to enter the range resolution cell containing the target. Mainlobe clutter can be rejected through the use of a comb filter (e.g., MTI canceller) just as for a low-PRF design except that the rejection capability must be increased due to the ambiguous clutter. Sidelobe clutter must be rejected through increased doppler resolution (e.g., bandpass filtering), which serves the dual function of attenuating the clutter outside the target's doppler resolution cell while minimizing the unavoidable sidelobe clutter intake within the doppler filter containing the target. A prime consideration in the design of medium-PRF radar involves the minimization of the unavoidable sidelobe clutter intake that ultimately limits the radar's performance.

The clutter intake into a medium-PRF airborne radar is regulated through sorting by both range and doppler frequency. Sorting by range is accomplished using range gates that isolate the returns received from relatively narrow strips of ground at constant range. Because of the range ambiguities, the return passed by each gate will come from several strips of ground. In Figure 2.11, the concentric circles represent the projection of the radar's ambiguous ranges onto the ground through which the clutter return entering a single range gate must pass.

Sorting of sidelobe clutter by doppler frequency is accomplished by applying the output of each range gate to a bank of doppler filters. They will isolate the returns received from strips of ground lying between lines of constant angle relative to the radar's velocity, which are called *isodops* (see Section 1.1.4.1). Because of doppler ambiguities, any one filter will pass the returns from several strips. In Figure 2.11, the narrow strips between the hyperbolic lines represent the strips of ground providing clutter returns to a single doppler filter.

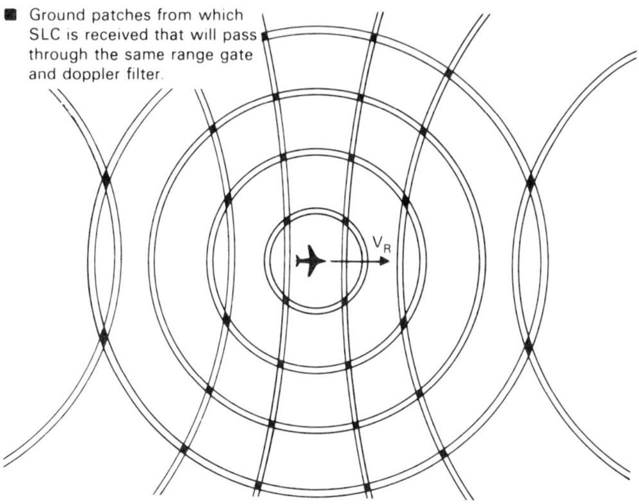

Figure 2.11 Ground patches providing sidelobe clutter returns in medium-PRF airborne radar.

The shaded areas in Figure 2.11 depict those ground patches contributing returns to the output of a single range-gated doppler filter. Note that each doppler filter receives only a portion of the total sidelobe return passed by a single range gate. Increasing range and doppler resolution reduces the shaded areas in Figure 2.11, thereby minimizing the unavoidable clutter intake that competes with target returns in each doppler filter.

The level of sidelobe clutter in a doppler filter for low-, medium-, and high-PRF designs of an X-band radar are depicted in Figure 2.12 [5]. In the medium-PRF design, as most of the sidelobe clutter comes from relatively short ranges, the background of clutter against which targets must be detected is generally stronger than the background noise falling in the passband of a doppler filter. Therefore, detection performance is limited by the target-to-sidelobe clutter ratio rather than the signal-to-noise ratio. As the target's range is continuously increased, a point eventually is reached where targets become lost in the clutter. However, before this range is reached, there are alternating regions where the target return exceeds the clutter return and those where the target is obscured by the clutter, which are called *range blind zones* (e.g., blind speeds in low-PRF radar). The positions of the range blind zones shift with changes in the PRF. Fortunately, the shift is such that the same PRF switching used to reduce doppler blind zones are also effective in reducing range blind zones. Therefore, by applying multiple PRFs in the medium-PRF design, smoothing the detection performance for blind range zones is possible in the same manner as doppler blind zones are reduced in a low-PRF radar design.

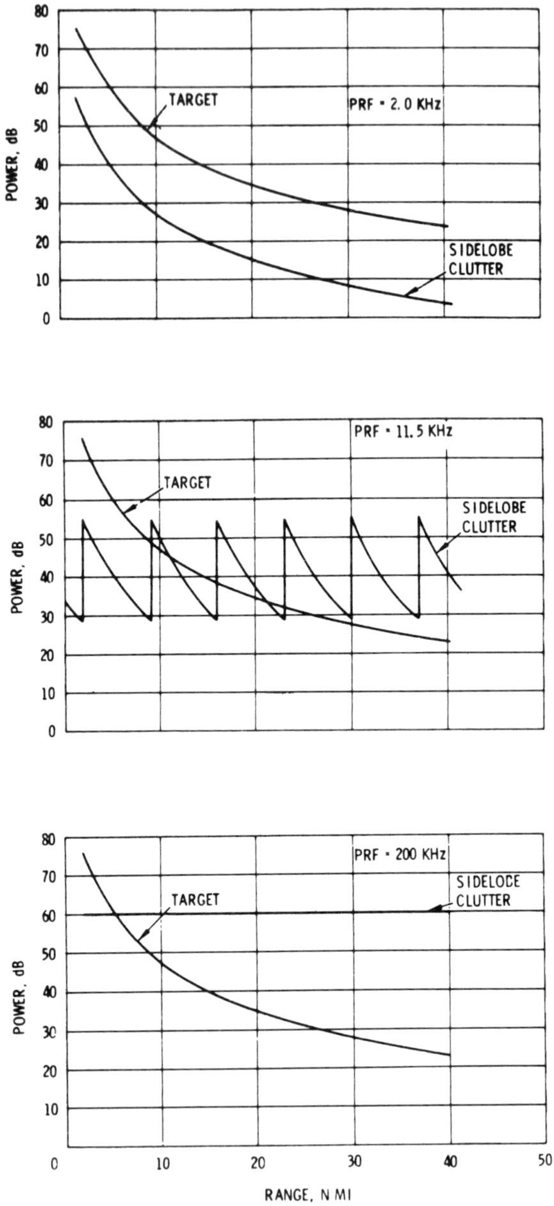

Figure 2.12 Sidelobe clutter and target responses for low-, medium-, and high-PRF airborne radar.

From the preceding discussion, the key design issue for medium-PRF airborne radars clearly is the minimization of sidelobe clutter. Reducing the antenna's sidelobes is the most fundamental action that can be taken in this regard. Recent developments in this area have produced ultralow-sidelobe designs (i.e., relative one-way sidelobe responses −40 dB or more below the mainbeam response), but these are difficult to achieve in typical high-performance airborne installations due to installation and radome considerations.

Another action that can reduce sidelobe clutter is to increase the radar's range resolution by narrowing the radar's effective pulse width and correspondingly narrowing the radar's range gates. Narrowing the radar's effective pulse width reduces the sidelobe clutter level in the same proportion by which the pulse width is narrowed. Additional range gates are required to provide full range coverage and more doppler filters must be formed, as a separate bank of doppler filters is required for each range gate.

One way to narrow the effective radar pulse width without reducing the average transmitted power is pulse compression. A common practice is to maximize the average transmitter power without incurring an unacceptable loss due to eclipsing and then use enough pulse compression to achieve the desired range resolution. An alternative approach that minimizes eclipsing is to transmit very narrow pulses of higher peak power.

The sidelobe clutter with which a target must compete can be further reduced by narrowing the passbands of the doppler filters. To retain all the power in the target return, the filters must be wide enough to pass the target's spectrum. The target spectrum generally is inversely proportional to the radar's time on target (T_d) plus an allowance for internal target motion and acceleration effects.

An important form of sidelobe clutter in medium-PRF airborne radars results from large RCS targets on the ground (e.g., buildings, water towers, *et cetera*), which are particularly prevalent in urban or mountainous areas. Also included in this class of clutter are unwanted moving ground vehicular traffic that, because of the range and doppler ambiguities, conflict with the detection of airborne targets.

Although an airborne radar is vulnerable to such targets when operating at low-PRF, it is more vulnerable at medium PRFs due to the severe range ambiguities in this type of operation. One way to reduce these unwanted targets is to provide the radar with a guard channel that consists of a separate receiver processor whose input is supplied by a small horn antenna mounted on the main radar antenna. The width of the horn's mainlobe is sufficient to encompass the entire region illuminated by the radar antenna's principal sidelobes, and the gain of the horn's mainlobe is greater than any of the main antenna's sidelobes. Any detectable target in the radar antenna's sidelobes will produce a stronger output from the guard receiver than from the main receiver. Conversely, any detectable target in the mainlobe will produce a stronger signal than in the guard receiver. Consequently, by comparing the output of the two receivers and inhibiting the output of the main receiver when that of the guard receiver is stronger, we can eliminate the unwanted targets in the radar's sidelobe region.

At medium PRFs, signal processing is handled very much as at low PRFs with three main exceptions. First, because range ambiguities preclude the use of sensitivity time control, additional automatic gain control is needed to avoid saturation of the A/D converter. Second, the passband of the doppler filters employed must be made considerably narrower to exclude the excess sidelobe clutter that builds up due to doppler ambiguities. Third, additional processing is required to resolve range and doppler ambiguities.

The block diagram of a typical medium-PRF airborne radar digital signal processor is depicted in Figure 2.13 [1]. The first step in processing the receiver output is to translate the doppler spectrum to in-phase and quadrature channel baseband signals for digital sampling. The shift must be dynamically controlled to account for changes in radar velocity and antenna look angle. The *I* and *Q* output are sampled at intervals on the order of the transmitted pulsewidth and digitized.

However, to reduce the dynamic range required of the A/D converter, automatic gain control is provided ahead of the converter. To accomplish this, the converter's output is monitored and a continuously updated profile of the output over the course of the interpulse period is stored. On the basis of this profile, a gain control signal is produced and applied to the amplifiers ahead of the A/D converter. By reducing the gain when the mainlobe clutter and strong close-in sidelobe clutter are being received, the control signal keeps the converter from being saturated while maintaining the input to the converter well above the local noise level when a weaker return is being received. Because the control signal is derived after the return has been digitized, this is called *digital automatic gain control* (DAGC). The use of DAGC in medium-PRF radar causes target suppression, in the same way that STC would, for targets with apparent ranges that place them in strong clutter regions.

The output of the A/D converter is then sorted into range bins, after which mainlobe clutter is largely eliminated by using a clutter canceller. The clutter canceller

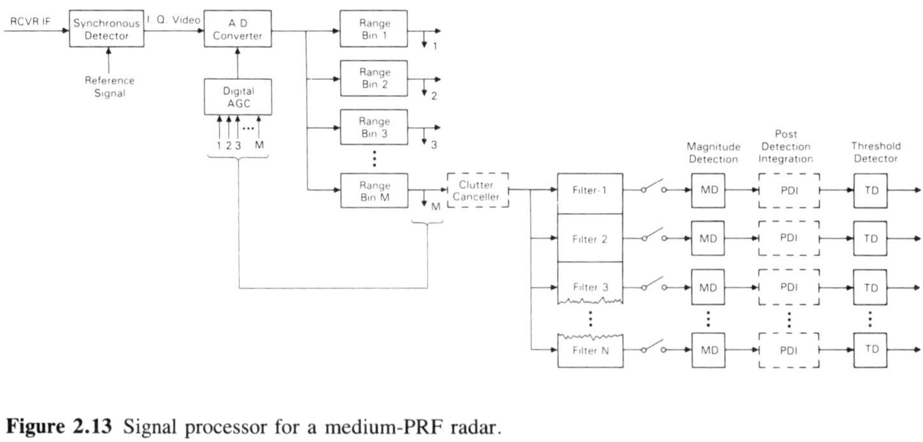

Figure 2.13 Signal processor for a medium-PRF radar.

serves the dual purpose of reducing both the mainlobe clutter and the dynamic range into the doppler filter bank. The return in each clutter cancelled range bin is then applied to a doppler filter bank as depicted in Figure 2.14. Note that the first and last few doppler filter output are discarded to eliminate ground moving targets.

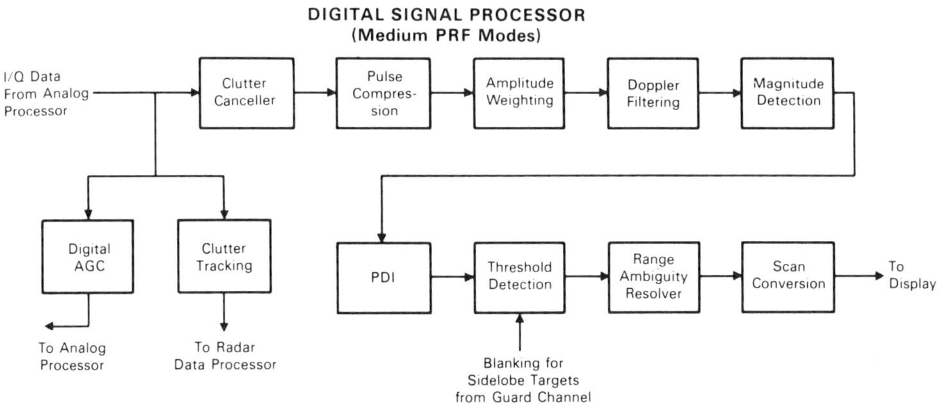

Figure 2.14 Medium-PRF radar clutter canceller and doppler filter bank.

The output (I and Q) of each doppler filter is then magnitude detected at the end of each processing integration period. The integration period (T_i) is inversely proportional to the doppler filter's bandwidth ($T_i = 1/B_D$). If T_i is less than the time on target (T_d) then each magnitude detected doppler filter output is subsequently integrated in a postdetection integrator. The integrated output is then applied to a threshold detector, which is adaptively set to keep the probability of clutter producing false alarms acceptively low. The adaptive threshold is generally set by averaging range and doppler bins on either side of the one in question.

The apparent target range and doppler is determined by observing which range-gated doppler bin provides the output. Because both range and doppler are ambiguous, PRF switching is required to resolve these ambiguities.

2.3 DOPPLER FILTERING

The heart of any MTI or PD radar is the doppler filter that extracts the desired doppler shifted target from the interfering clutter background. The basic filter associated with an MTI radar functions to attenuate those doppler frequency regions associated with any clutter return while passing all other doppler spectral regions. Because the spectral region associated with ground clutter tends to be concentrated about the PRF spectral lines (e.g., recall Figure 2.4), the periodic MTI filter response assumes a comblike shape and hence is called a *comb filter*.

The pulsed doppler filter response, on the other hand, is generally matched to the target's doppler shifted spectral response. Because the magnitude of the target's doppler frequency shift relative to the radar can appear anywhere in the region bounded by the doppler frequency corresponding to the sum of the maximum radar platform speed (v_r) and target speed (e.g., $v_m = \pm |v_r + v_t|$), the target spectral region must be filled with a bank of doppler filters covering all possible expected target doppler shifts. Hence, the PD filter response generally consists of a bank of doppler filters that fill most of the region between the transmitted waveform's PRF spectral lines, as shown in Figure 2.15.

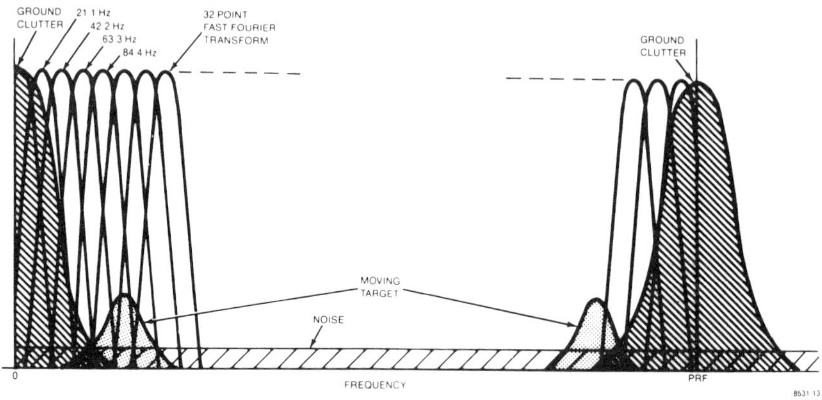

Figure 2.15 Pulse doppler filter response.

The MTI and PD filtering responses are not mutually exclusive and many systems employ a cascade of the two types of responses. In a cascade, the MTI filter is generally used first to eliminate much of the clutter before it reaches the bandpass doppler filters. This reduces the requirements on the skirt selectivity of the many bandpass doppler filters, which otherwise would have to both eliminate the clutter and match the target response. In a digital design, which might typically use the FFT algorithm, this simplifies the filter arithmetic that would carry the full dynamic range of the clutter (e.g., 10–12 bits) to where only the target's dynamic range need be carried (e.g., 3–4 bits).

2.3.1 Coherent and Noncoherent MTI

The ground clutter return from a surface-based low-PRF radar was described in Section 1.1.2 and depicted in Figure 1.3. This return consists of the vector sum from the reflections of many elemental scatterers that combine to form a waveform at the carrier frequency of the radar, whose amplitude slowly fluctuates due to both the internal scatterer motion and induced radar antenna scanning motion. The waveform associated with this return can be visualized through the use of a phasor diagram as shown in Figure 2.16.

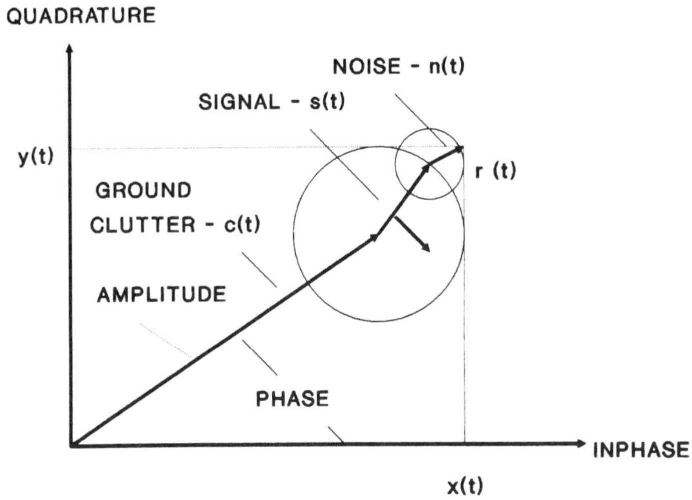

Figure 2.16 Phasor diagram of doppler shifted target in ground clutter and receiver noise.

The phasor diagram shows the instantaneous composite signal (r_t) due to ground clutter, target, and receiver noise components with respect to an inphase reference I_t (x-axis of phasor diagram) and a quadrature reference Q_t (y-axis of phasor diagram), where both references are synchronized to the carrier frequency of the radar (ω_c) but displaced in phase by 90°. In this diagram, the clutter signal (c_t) slowly fluctuates in both phase (ϕ_c) and amplitude at a rate corresponding to its spectral spread. The target vector rotates about the clutter vector at a rate that corresponds to its doppler frequency shift (ω_d) with respect to the clutter. The noise component fluctuates rapidly both in amplitude and phase about its mean value such that its position varies randomly each time the phasor diagram is sampled.

In the phasor diagram, the magnitude of the vector r_t represents the envelope of the return signal. This envelope is processed in an noncoherent MTI (see Figure 2.17) to provide detection of the doppler shifted target. When no target is present, the envelope fluctuates slowly due to (1) internal clutter motion, (2) antenna scanning modulation, and (3) platform motion; the envelope fluctuates rapidly due to the receiver noise component. However, when the target is present, the envelope varies at a rate corresponding to the target's doppler shift. Targets can be detected by using a clutter

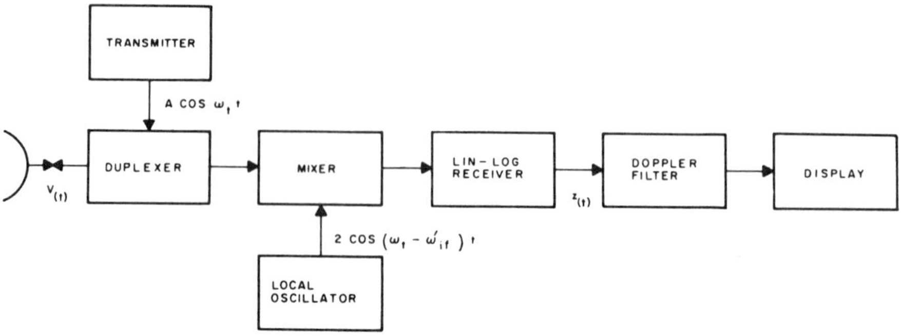

Figure 2.17 Block diagram of noncoherent MTI.

canceller. The simplest form of this is the single delay line canceller depicted in Figure 2.18, which functions by subtracting the envelope signal returned from one previous interpulse period ($T = 1/f_r$) from the present signal. The output of the single delay line canceller thus responds to the change in the envelope signal from successive radar transmitted pulses. When a doppler shifted target is present, this fluctuation is emphasized; when no target is present, the fluctuation is minimized. The fluctuation will generally be maximum when the phase shift of the target in an interpulse period ($\Delta\phi = 2\pi v_d T$) equals π radians. This occurs when the doppler frequency shift of the target is at one-half of the radar PRF ($f_d = f_r/2$), which is sometimes called an *optimum speed target*.

Note that if the target vector in the phasor diagram of Figure 2.16 is orthogonal to the clutter vector, there is no output of the MTI filters for an optimum speed target. This is known as a *blind phase*, which is inherent in single channel MTI designs. Also note that the MTI filter's power gain response, when averaged over all target velocities, provides equal response to both signal and noise components and hence does not alter the signal-to-noise ratio.

The noncoherent MTI just described is also called a *clutter referenced MTI*, because clutter must be present for this type of MTI to function. To circumvent this

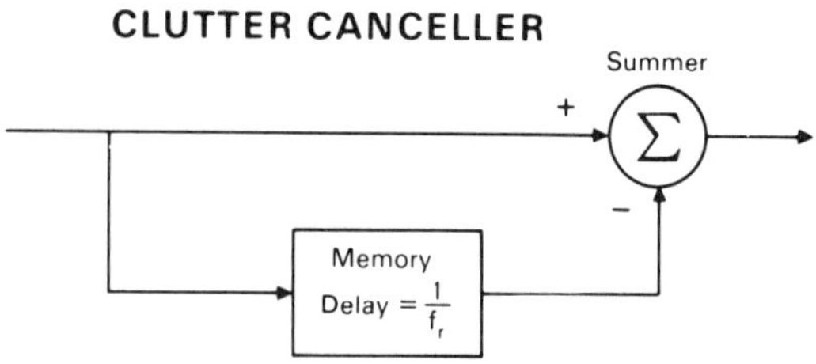

Figure 2.18 Single delay line canceller.

problem, the output of the noncoherent MTI can be switched between MTI and normal video, with the MTI selected only when there is sufficient clutter for it to operate. The performance of the noncoherent MTI will also be limited by the use of a relatively noisy reference (the clutter itself) as compared to the stable internal reference used in coherent MTI.

A coherent MTI utilizes the phase information (ϕ_r) illustrated in the phasor diagram of Figure 2.16. When only clutter is present, the phase of the vector r_t with respect to the in-phase reference I_t is primarily pertubated by the receiver noise vector. When a target is present, the phase varies at a rate determined primarily by the doppler frequency of the target. The bipolar video signal produced by phase detecting the return signal against the coherent in-phase reference signal (I_t) is depicted in Figure 2.19 for a single interpulse period and also for several interpulse periods superimposed on the same sweep. When a moving target is present, a butterfly type of pattern is generated; otherwise, the trace is relatively stable except for the receiver noise output component.

Examination of Figure 2.19 indicates that the same type of delay line canceller used in noncoherent MTI systems will also be effective in processing the phase detected bipolar video output of a coherent MTI system. The basic action of the delay line

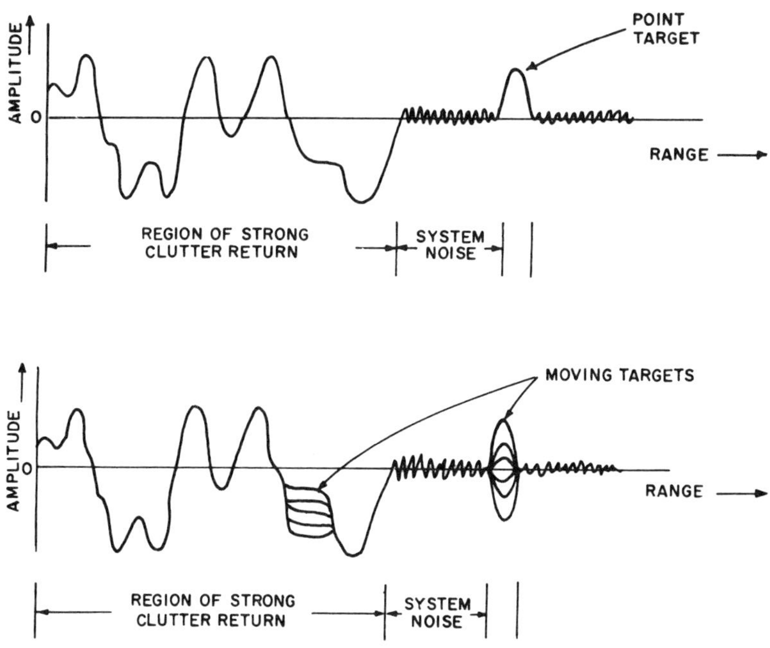

Figure 2.19 Bipolar video from a coherent MTI system.

canceller in this case is to suppress those phase detected signals due to clutter that remain stable from one pulse period to the next and accentuate those target signals that fluctuate from one interpulse period to the next. This action is equally effective whether processing envelope detected amplitude data or phase detected bipolar video.

Note that the coherent MTI does not require clutter to operate, which is a distinct advantage relative to noncoherent MTI. Nonetheless, the blind phase problem will remain when a single reference channel is used. The solution is to add a second quadrature channel using reference vector Q_t, the phase of which is shifted 90° with respect to the first reference vector. From Figure 2.16, we can see that when the phase of the doppler shifted target can produce a blind phase in the first channel, its output will be maximum in the second quadrature channel.

The use of a single-channel coherent MTI, as depicted in Figure 2.8, results in a sensitivity loss as compared to fully coherent MTI systems that process both in-phase and quadrature channels. The most severe loss occurs for nonfluctuating targets with high detection probabilities when no postdetection integration is used. For example, the relative detection loss has been evaluated as 13.2 dB for a $P_d = 0.9$ with a $P_{fa} = 10^{-6}$. The loss is reduced to 2.75 dB for a $P_d = 0.5$ with a $P_{fa} = 10^{-6}$ and overall the relative loss is generally insensitive to the false alarm probability. The relative loss (as compared to that for a nonfluctuating target) is reduced for Swerling 1 fluctuating targets with high detection probabilities ($L = 10.66$ dB, $P_d = 0.9$, $P_{fa} = 10^{-6}$), but increased for moderate detection probabilities ($L = 4.35$, $P_d = 0.5$, $P_{fa} = 10^{-6}$) [6].

The cause of the sensitivity loss in a single-channel coherent MTI is due to a modification of the detection statistics. The relative loss is reduced in a noisy background by using postdetection integration [7,8]. The relative loss with postdetection integration can be related to a reduction in the number of effective pulses integrated by a factor of two, which causes a loss in integrated signal-to-noise ratio (S/N) of 1.5 to 2.5 dB [7]. However, the analyses apparently have not considered the results of residual clutter components, which can result in degraded performance with postdetection integration [9].

The single-channel coherent MTI was used extensively in analog processors due to its simplicity. Most modern MTIs utilize digital processing, which allows use of quadrature-channel MTI systems as depicted in Figure 2.7.

2.3.1.1 Coherence

Generally, coherence implies a consistency, or continuity, in the phase of a signal's carrier frequency from one pulse to the next. Coherent MTI operation requires that the phase of the receiving reference signal into the MTI's phase detector be consistent with the radar's transmitted signal. A common way of achieving this coherence in a coherent MTI or PD radar is depicted in the block diagram shown in Figure 2.20.

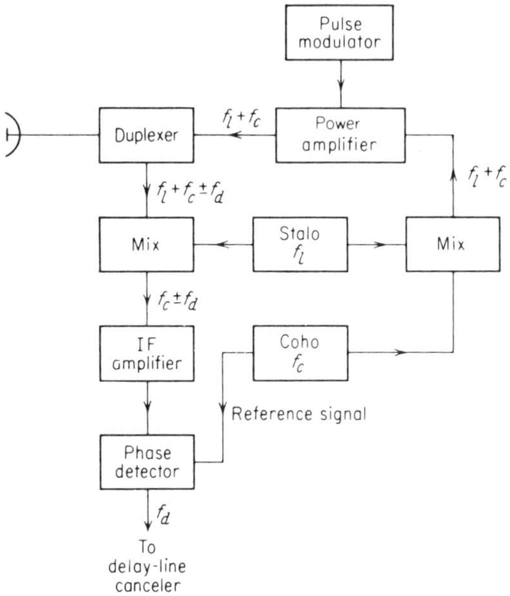

Figure 2.20 Block diagram of an MTI radar with a master oscillator power amplifier.

The system shown in Figure 2.20 is called the *master oscillator power amplifer* (MOPA) approach, because the basic coherent reference signal is developed as a CW signal within the radar. This signal is then amplified in a pulsed power amplifier such as a Klystron, traveling wave tube, or solid state amplifier and radiated through the radar antenna.

The MOPA system generally uses two stable oscillators to develop the coherent signals needed in the radar receiver and transmitter. A stable *coherent oscillator* (COHO) develops a signal at the radar's intermediate frequency (e.g., 60 MHz) that is used as the coherent reference signal in the MTI's phase detector. Another *stable local oscillator* (STALO) signal is used as the receiver's local oscillator to heterodyne the received RF signal to a lower intermediate frequency signal, which is compared against the COHO in the MTI's phase detector.

The transmitter signal is formed by mixing the COHO (f_c) and STALO (f_s) signals resulting in a transmitter signal at $f_t = f_s + f_c$. The doppler shifted return signal ($f_r = f_s + f_c \pm f_d$) is then mixed with the stalo resulting in a signal at $f_c \pm f_d$, which is compared with the COHO in the phase detector. The net effect of this process is an addition and subtraction of the STALO signal in the signal path so that the absolute value of the phase (or frequency) of the STALO is cancelled provided it does not change over the processing time of the MTI.

The MOPA approach can be viewed as providing coherence by cutting a stream of pulses out of a continuous wave signal. An alternate method of achieving coherence, called the *coherent-on-receive approach*, is depicted in Figure 2.21. This approach is useful in situations that employ a power oscillator (e.g., magnetron) instead of a power amplifier as the radar transmitter. In a pulsed power oscillator, the phase of the RF from one pulse to the next is not coherent. However, by locking the phase of the COHO onto each transmitted pulse, we can achieve quasicoherence during each phase immediately following an interpulse period. Note that multiple-time-around range-ambiguous clutter and targets are not coherent when using this basic scheme as they are with the MOPA approach. This can result in large range-ambiguous clutter returns (e.g., mountains, cities), that are not cancelled by this type of MTI system.

In the coherent-on-receive block diagram shown in Figure 2.21, a portion of the transmitted signal is mixed with the STALO output to produce a sample pulse at IF, which is the phase coherent with the transmitted output. The IF pulse is applied to the COHO and causes the phase of the COHO to lock onto that of the transmitter pulse. Operation of the system is then functionally equivalent to that of the MOPA system shown in Figure 2.20. Performance in this type of system is generally less than that of the MOPA system due to the difficulty in achieving a high degree of coherence with a locked COHO and the previously described multiple-time-around clutter cancellation problem.

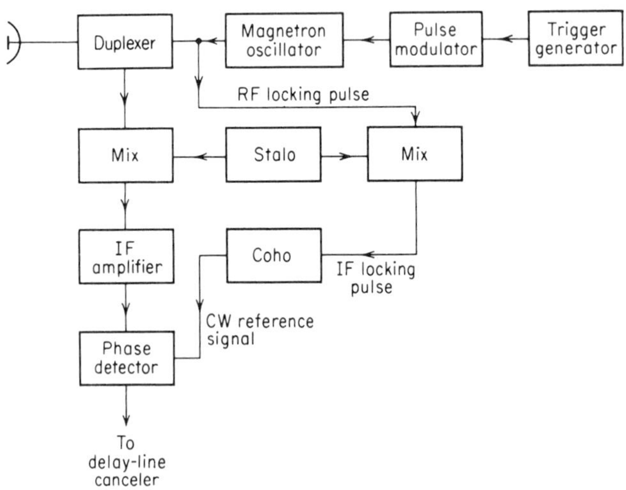

Figure 2.21 Block diagram of a coherent-on-receive MTI radar.

2.3.2 Doppler Filter Characteristics in the Frequency Domain

The single delay line canceller depicted in Figure 2.18 is an example of an elemental doppler filter, whose basic action can best be visualized by considering its operation in the time domain. Consider that the target vector signal depicted in Figure 2.16 is applied to the single delay line canceller. This vector rotates at the doppler frequency shifted rate of the target and, when its phase is sampled at the radar's PRF, provides an unchanged output for rotations of $2n\pi$, where $n = 0, 1, 2, 3, \ldots$ When this unchanged vector is applied to the delay line canceller, a null output results (due to the differencing of successive samples). This nulled output is denoted as the *blind speeds* of the MTI doppler filter. Examination of the single delay line canceller's characteristics shows that these occur where the target's doppler frequency shift is either zero or a multiple of the radar's PRF ($f_d = nf_r$, $n = 0, 1, 2, 3, \ldots$). When an optimum-speed target is applied to the single delay line canceller (i.e., $f_d = nf_r/2$, $n = 1, 3, 5, \ldots$), then the output is maximum or twice that of the input target vector. Further consideration of various doppler shifted target vectors applied to the single delay line canceller results in the well-known sinusoidal voltage amplitude characteristic ($v_c = |2 \sin \pi f_d/f_r|$) as depicted in Figure 2.22.

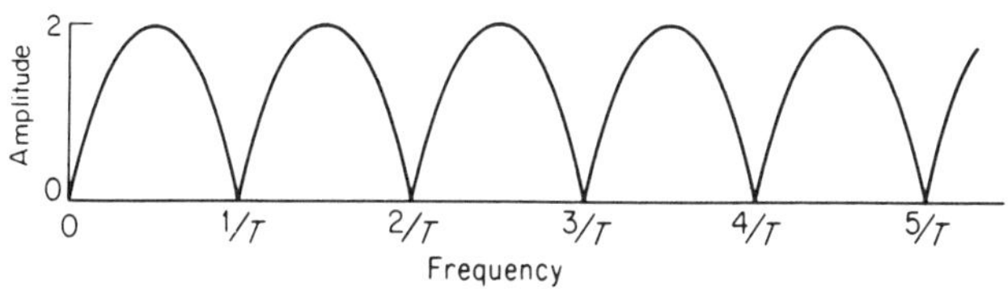

Figure 2.22 Single delay line canceller frequency response.

Thus, the single delay line canceller's amplitude characteristic has a periodic comb filter structure that repeats at multiples of the radar's PRF and is a function of the target's doppler shift with respect to the radar. When viewed as a doppler filter, the rejection notches are concentrated at doppler frequency shifts, which occur at frequencies corresponding to multiples of the radar's PRF (i.e., $f_d = nf_r$, $n = 0, 1, 2, 3, \ldots$), whereas the acceptance passband generally occurs at frequencies between the rejector notches with a maximum response for optimum speed targets (i.e., $f_d = nf_r/2$, $n = 1, 3, 5, \ldots$).

The single delay line canceller is a linear system and can be analyzed using classical Fourier transform theory [10]. The impulse response of the single canceller is given by $h(t) = \delta(t) - \delta(t - T)$. The frequency response is found by Fourier transforming the impulse response as $H(\omega) = 1 - \exp(-j\omega T) = 2j \exp(-j\omega T/2) \sin(\omega T/2)$. The magnitude of the frequency response is then $|H(\omega)| = |2 \sin \omega T/2|$.

The rejection notch of the single delay line canceller is not always wide enough to provide adequate clutter cancellation. The notch width can be increased by using a cascade of two single cancellers, which is called a *double canceller*. The frequency response of this configuration is the product of the response of the two single cancellers, which is given by $|H(\omega)| = |4 \sin^2(\omega T/2)|$. The increased notch width is depicted in Figure 2.23, which shows both the frequency and impulse response of both single and double delay line cancellers.

Another way of characterizing the double delay line canceller is through its impulse response, which is given by $h(t) = \delta(t) - 2\delta(t - T) + \delta(t - 2T)$. In this form, it can be seen that the double delay line canceller processes three pulses using a tapped delay line (i.e., taps at nT, $n = 0, 1, 2$) with weights $w = (1, -2, 1)$. This general type of filter, depicted in Figure 2.24, is called a *transversal* or *nonrecursive doppler filter*. In its most general form, complex weights (phase and amplitude) are employed (see Chapter 5); however, in a basic MTI canceller, only real weights are used. For MTI use, one way of achieving the effect of complex weights is to use transversal filters with real weights in both the I and Q channels. Further, as a note of interest, the basic cascade of two single cancellers, rather than a transversal filter, was usually preferred in analog MTI applications, as any amplitude imbalance in the first canceller was compensated for in the second canceller. This is no longer a problem because digital implementations are generally used in most modern MTIs.

Cascades of single delay line cancellers produce an amplitude frequency response given by $|H(\omega)| = |2 \sin(\omega T/2)|^n$, where n is the number of single delay line cancellers. We can show [2] that the impulse response of such a cascade uses weights with binomial coefficients given by $w_k = (-1)^k\binom{n}{k} = (-1)^k n!/k!\,(n - k)!$, where k is the weight of the kth tap and n is the number of delay sections. The normalized power transfer function ($|H(\omega)|^2$) of an MTI having the indicated number of cancellers is shown in Figure 2.25. MTIs with binomial coefficients are known to produce close to optimum MTI improvement factors. Note that the MTI's acceptance response becomes progressively narrower as the number of delay line cancellers is increased. In practice, generally no more than three delay sections are used because the excessive apparent clutter improvement factor (not supported by MTI stability considerations) lowers the target response due to the narrowing of the MTI filter's acceptance band.

An alternative approach to the transversal filters is the recursive MTI filter whose canonical form is depicted in Figure 2.26. This type of doppler filter employs both feedforward and feedback paths and is sometimes called a *shaped delay line canceller* [2]. The type of frequency response produced by this type of MTI canceller is depicted in Figure 2.23. The ability to shape the passband is an attribute of this MTI filter, but its poor transient response is a definite disadvantage. The poor transient response results from the feedback paths in this type of filter, which allow interferring pulses to recirculate through the filter for many interpulse periods (e.g., recall Figure 2.23).

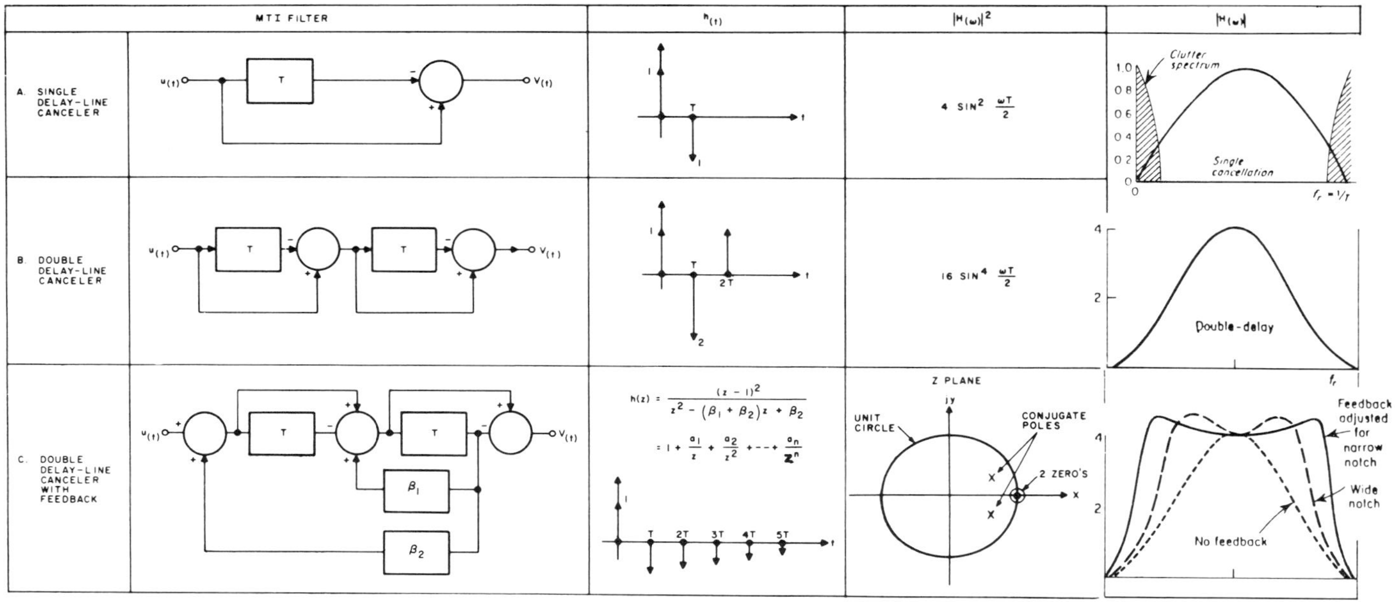

Figure 2.23 Frequency and impulse response of delay line cancellers.

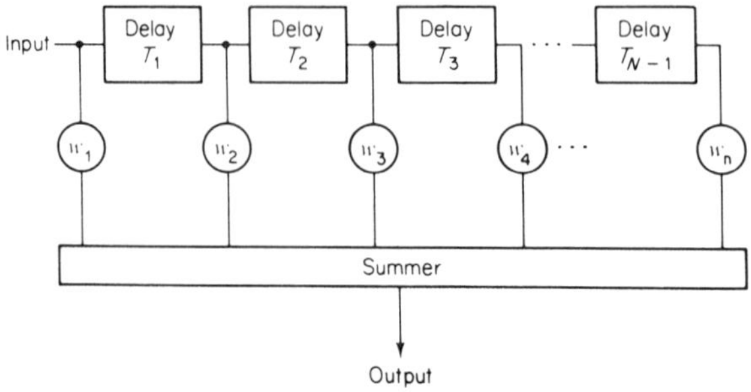

Figure 2.24 General transversal filter.

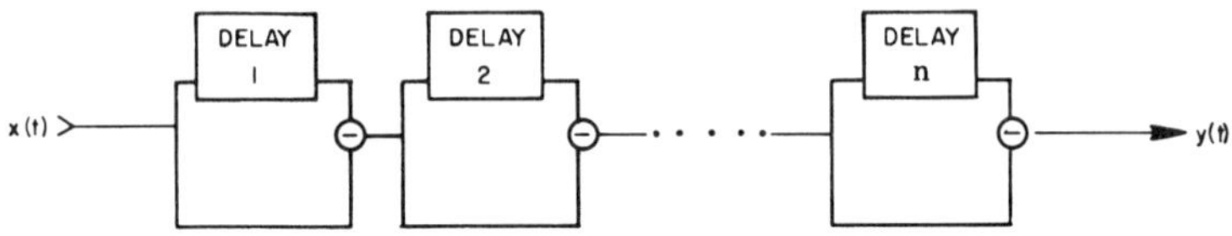

Fig. 2 — An n-stage, or n-canceler, MTI

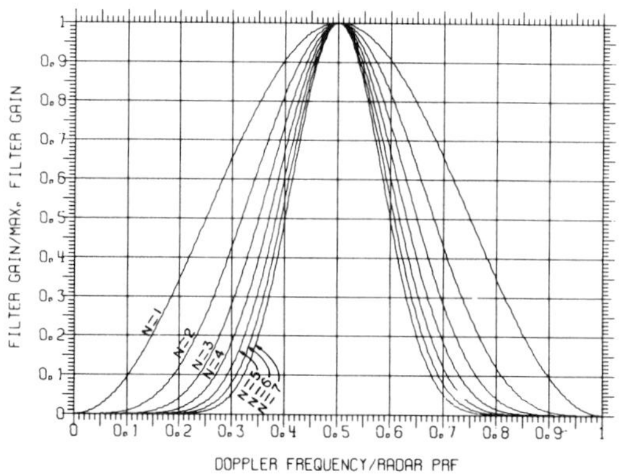

Figure 2.25 Normalized power transfer function for an MTI employing binomial coefficients.

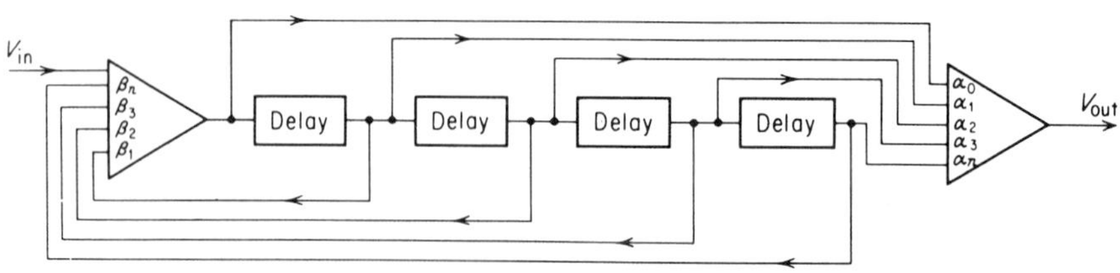

Figure 2.26 Canonical form of MTI comb filter.

Recursive MTI doppler delay line filters can be synthesized from classical filter design theory by using the z-transform (see Appendix C) [11]. The z-transformation is given by $z = \exp(pT)$ where $p = j\omega$ and T is the radar's interpulse interval. The z-transform of a unit impulse delayed in time $\delta(t - nT)$ is given by z^{-n}. Hence the z-transform for the impulse response of a single delay line canceller would be given by $H(z) = 1 - 1/z = (z - 1)/z$. Note that for nonrecursive filters, the z-transform of the impulse response is given by $H(z) = \Sigma_{k=0}^{n} w_k z^{-k}$, where k is the tap number, n is the number of delay lines, and w_k are the tap weights.

The amplitude frequency response of an MTI filter can be characterized by its pole-zero pattern. In conventional filter design, the pole-zero pattern is plotted in the p-plane, where $p = j\omega$ is the variable utilized in the Laplace transform of the filter's impulse response. The transformation utilized in the z-transform [$z = \exp(pT)$] maps the left-hand p-plane into the unit circle in the z-plane as depicted in Figure 2.27. The transformation is effective because it allows one period of the filter's transfer characteristic to be examined by a single rotation about the unit circle in the z-plane.

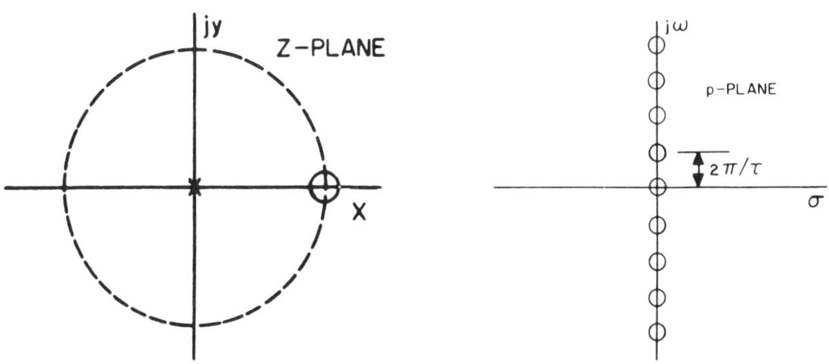

Figure 2.27 Pole-zero pattern in p- and z-planes for a single delay line canceller.

The single delay line canceller has a zero at $z = 1$ and a pole at $z = 0$, as depicted in Figure 2.27. The amplitude response for any point on the unit circle, located at an angular position of Ω radians ($f_d = \Omega f_r/2\pi$), can be found using the following procedure [2]. First, form vectors from the point on the unit circle to all zeros and poles. Then, the amplitude function is given by the product of all zero vectors divided by the product of all pole vectors. For the single canceller pole-zero pattern shown in Figure 2.27, the amplitude response is given by the length of the vector from the zero to the point on the unit circle. Simple geometry shows this to be equal to

$$|H(\omega)| = \sqrt{2(1 - \cos \omega T)} = 2 \sin(\omega T/2)$$

However, the advantage of the z-transform method stems from the availability of a simple transformation that allows poles and zeros in the p-plane (obtained from classical network theory) to be transformed into the z-plane [11]. Thus standard Butterworth, Chebyschev, Bessel, or ellipical filter characteristics can be easily translated into shaped delay line filter designs. When a standard low-pass filter design is to be translated into a high-pass MTI canceller design the appropriate bilinear transformation that transforms the poles and zero in the p-plane (low-pass filters p-plane) to the MTI filter's z-plane is given by $z = (p - \Omega)/(p + \Omega)$, where Ω is a scale adjusting factor related to the filter's acceptance bandwidth given by $\Omega = \omega_{co} \tan(\omega_{co}T/2)$ [2].

An example of a double shaped canceller design is given in Figure 2.23. The transfer function in the z-plane for this type of design is given by $H(z) = (z - 1)^2/z^2 - (\beta_1 + \beta_2)z + \beta_2$, and its response is shown in Figure 2.23. This type of design can be used to provide the varying filter shapes that are useful in situations with variable clutter spectral spreads such as might be found in radars employing variable antenna scan rates or airborne radars that scan away from the ground track.

2.3.3 Digital Doppler Filters

In most modern radars, signal processing is accomplished digitally, reflecting both the capability and the economy associated with modern digital components. All the doppler filtering functions described in the previous sections can be accomplished digitally. Although the theoretical capabilities of analog and digital signal processing are equivalent, the detailed implementation is significantly different. Basically, in digital signal processing, the input waveform is converted to a set of numbers that then are manipulated in accordance with a set of algorithms to accomplish the desired target detection function.

A block diagram of the basic conversion from the input analog waveform into digital numbers is shown in Figure 2.28. The first part of this process is a conversion of the radar's input waveform into lowpass I and Q video signals. This conversion is accomplished by using synchronous detectors (i.e., phase detectors) that multiply the input IF signal with quadrature reference signals coherently related to the radar transmitting signal. The purpose of this conversion is to translate the signal to baseband where a minimal sampling rate preserves the information content of the input waveform (see Appendix A, quadrature decomposition theorem).

The next step in the digital conversion process is to sample the I and Q video waveforms. Successive samples are taken at intervals corresponding to the radar's range resolution cell (i.e., the radar range corresponding to the effective transmitter pulsewidth) throughout the range coverage of the radar. This range gating process results in a sampling rate equal to the PRF times the number of range gates. Each sample thus represents the return of a single pulse from a given range increment—

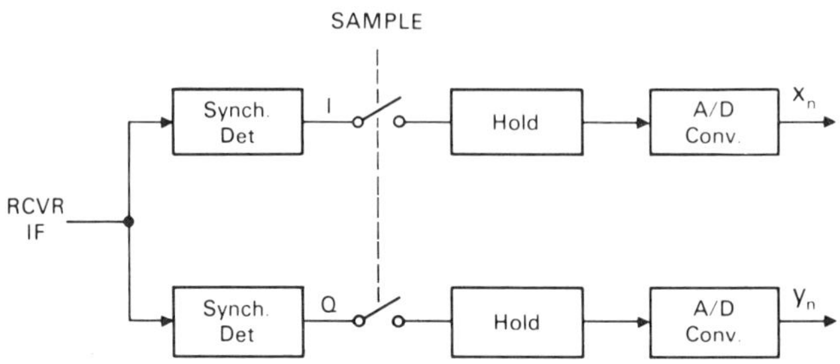

Figure 2.28 Block diagram of digital conversion in digital signal processing.

or, if range is ambiguous, from a number of range increments separated by the unambiguous range (R_u).

The sampling is generally followed by a holding circuit. Separate circuits are provided from the outputs of both the I and Q phase detectors. The overall sample-and-hold circuits sense the instantaneous values of the waveform at precisely the right time and hold the sample long enough for them to be converted to digital numbers, then dump them, and the process repeats.

The *analog-to-digital* (A/D) converter is the device that translates the sample-and-hold voltage output into a digital number. A/D converters generally work by comparing the voltage of each sample with a succession of progressively higher voltages of precisely known values. When the closest of these is found, the converter produces a binary number equal to the known voltage. Separate A/D converters are required for both the I and Q samples.

The output of the A/D converters due to one transmitter pulse are then a set of numbers in signal space $\mathbf{s} = (x_1, x_2, \ldots, x_m; y_1, y_2, \ldots, y_m)$, where m is the range gate number and the pair (x_n, y_n), $n = 1, 2, \ldots, m$, represents the return in a particular range gate. The function of the signal processor is to take the sets of numbers from successive radar transmitter pulse returns $(\mathbf{s}_1, \mathbf{s}_2, \ldots, \mathbf{s}_l)$ and manipulate them to perform the desired target detection. The general procedure is to form a decision function $D(z)$, where $z = f(\mathbf{s}_1, \mathbf{s}_2, \ldots, \mathbf{s}_l)$, which is compared against a criterion (e.g., threshold) to decide upon the presence or absence of a target.

A key function performed within the signal processor is doppler filtering. It is generally performed by examining the set of returns within a particular range gate so that the number set into the doppler filter is given by $\mathbf{s} = (x_{n1}, x_{n2}, \ldots, x_{nl}, y_{n1}, y_{n2}, \ldots, y_{nl})$, where n is the range gate number and l is the number of pulses processed. This range-gated doppler filter then has the capacity to produce l complex (I and Q) doppler filters for each range gate resulting in an overall number of nl complex doppler filters. When MTI doppler filters are involved, only one complex filter is produced for each range gate, resulting in the formation of n doppler filters.

Doppler filters are generally formed in a particular range gate by taking the weighted sum of the l quadrature returns ($x_{n1}, x_{n2}, \ldots, x_{nl}; y_{n1}, y_{n2}, \ldots, y_{nl}$). In simple signal processors, the weighted sums are stored in shift registers as depicted in Figure 2.29. The shift register has one storage position for each range increment to be processed. Each time a new product is formed by multiplying the input sample by an appropriate weight, the stored sums are shifted one positon to the right. The new product is thus added to the stored sum of the previous products resulting in a vector output equal to $\Sigma_{i=1}^{l} w_i x_{ni}$ for the situation depicted in Figure 2.29. When all return pulses have been processed, the weighted sum output of the shift register is then used to find the magnitude of the vectors in the I and Q channels.

A block diagram of a digital implementation of a single delay line canceller is depicted in Figure 2.30. The vector outputs of the shift register using the weights (1, −1) are equal to $(\mathbf{I}, \mathbf{Q}) = x_{n1} - x_{n2}, y_{n1} - y_{n2})$. Note that the weights in both the I and Q channels are equal, and no cross coupling between the I and Q channels is involved (i.e., the complex weights are real). The output of the single canceller is formed by finding the magnitude of the I and Q vectors, which is given by $z = \sqrt{|I|^2 + |Q|^2}$. Because this is diffficult to accomplish digitally, an approximation algorithm is frequently employed.

One algorithm [1,12] used to find the magnitude of the vector sum of the I and Q channels is as follows. First, subtract the magnitude of the I vector from the magnitude of the Q vector to determine which is smaller. For example, assume $|Q|$ is the smaller vector. Then divide $|Q|$ by 2, which is simply accomplished in binary arithmetic by shifting the number to the right by one binary place. Then add the result to the larger quantity resulting in $z = |I| + |Q|/2$ for the example. The error of the approximation in terms of the S/N ratio to achieve a specific probability of detection varies with the value of the phase angle between the vector and in-phase component, but generally is less than a fraction of a decibel [12].

2.3.3.1 Digital Doppler Filter Banks

The most general form of a digital doppler filter of the transversal type is depicted in Figure 2.31. This uses two channels: a real channel and an immaginary channel, each employing a shift register. For each target return from the same range increment, the filter receives an in-phase (x_n) and a quadrature (y_n) input. These inputs are multiplied by both real (w^{R}) and imaginary (w^{I}) weights.

The analysis of transversal filters is best handled by expressing the pertinent quantities in complex form (see Appendix A). In this notation, a signal return from a range increment is represented by using its in-phase (x_n) and quadrature (y_n) parts as $\mathbf{s}_n = x_n + jy_n$. The transversal filter weights applied to this signal element are given by $\mathbf{w}_n = w_n^{\mathrm{R}} + jw_n^{\mathrm{I}}$. The output of the transversal filter is then given by $\mathbf{c}_n = r_n + ji_n = (w_n^{\mathrm{R}} + jw_n^{\mathrm{I}})(x_n + jy_n)$. The real part of this output is therefore given by $r_n = w_n^{\mathrm{R}} x_n - w_n^{\mathrm{I}} y_n$ and the imaginary part equals $i_n = w_n^{\mathrm{I}} x_n + w_n^{\mathrm{R}} y_n$.

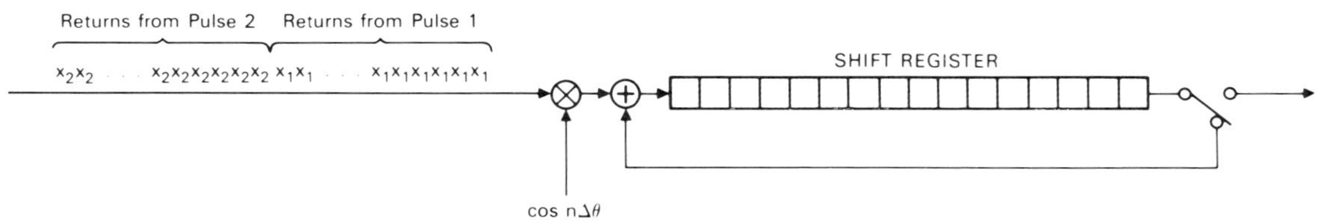

Figure 2.29 Formation of digital filters using shift registers.

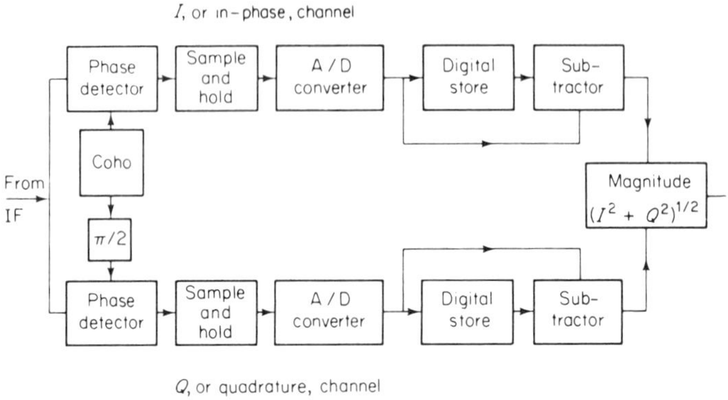

Figure 2.30 Block diagram of a digital single canceller MTI.

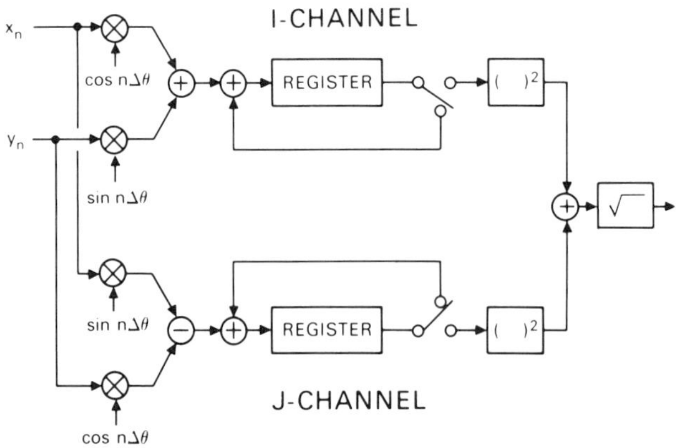

Figure 2.31 General digital transversal doppler filter.

The basic action of the doppler filter can be understood from the phasor diagram depicted in Figure 2.32. This diagram shows the target signal return that rotates at the doppler shift rate (f_d) superimposed upon two sets of coordinates displaced by an angle θ_m. The first set of coordinates represents the in-phase and quadrature reference signals (recall Figure 2.16). The target signal return has an in-phase component of x_n and a quadrature component of y_n in this coordinate system. The filter formation is accomplished through the projection of target inputs onto the second coordinate system, which represents the filter's incremental output. If the filter is tuned to the target's doppler, then the second coordinate system rotates in synchrony with the target vector. Otherwise, if tuned to a different doppler frequency, the second set of

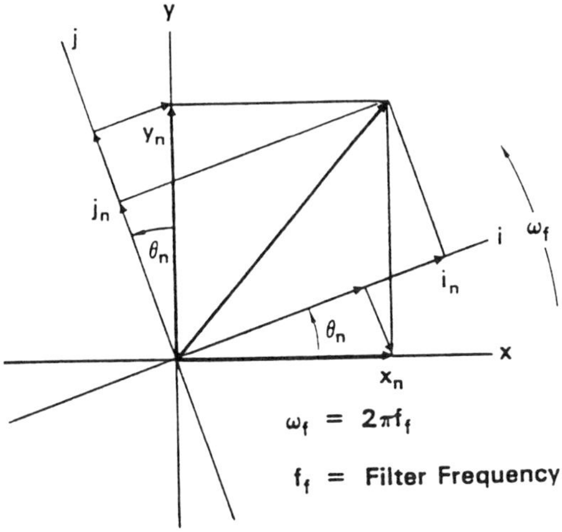

Figure 2.32 Phasor diagram of a coherent integrating digital doppler filter.

coordinates rotates at a rate different than the target vector. The overall filter output is the sum of all projections that result from the l returns processed by the filter. This sum is maximized when the filter coordinate system rotates at the same rate as the target doppler. When this occurs, the net effect is a coherent integration of the target return (this is sometimes called coherent integration of doppler echoes).

The algorithm that defines the computations required to form a bank of digital doppler filters can be deduced from Figure 2.32. The projection of the in-phase and quadrature components of the successive samples of the input waveform (x_n and y_n) onto the rotating coordinates (i and j) are given by $r_n = x_n \cos\theta_n + y_n \sin\theta_n$ and $i_n = y_n \cos\theta_n - x_n \sin\theta_n$. Comparison of these projections to the complex representation of a transversal filter allows the evaluation of the real weight as $w_n^{\mathrm{R}} = \cos\theta_n$ and the imaginary weight as $w_n^{\mathrm{I}} = -\sin\theta_n$. Alternatively, the complex weight can be expressed as $\mathbf{w}_n = \exp(-j\theta_n)$. This indicates that the doppler filter bank's characteristics are a function of only the rotational position of the parameter θ_n.

Next, the initial position and rotational rate of the filter coordinate system are determined by setting $\theta_n = (p - 1)\Delta\theta_n$, where $p = 1, 2, 3, \ldots, l$ is the pulse number and $\Delta\theta_n$ is the incremental angle through which the coordinate system rotates when the filter is tuned to doppler frequency f_f. By causing the filter and signal cordinates to coincide for $n = 1$, the filter's weight is normalized so that $\mathbf{w}_{n1} = 1(w_{n1}^{\mathrm{R}} = 1, w_{n1}^{\mathrm{I}} = 0)$.

The doppler filter bank for an eight-pulse processor is depicted in Figure 2.33. This filter bank is formed by finding the magnitude of the weighted sums of the real and imaginary incremental projections onto the filter's coordinate system. Note the

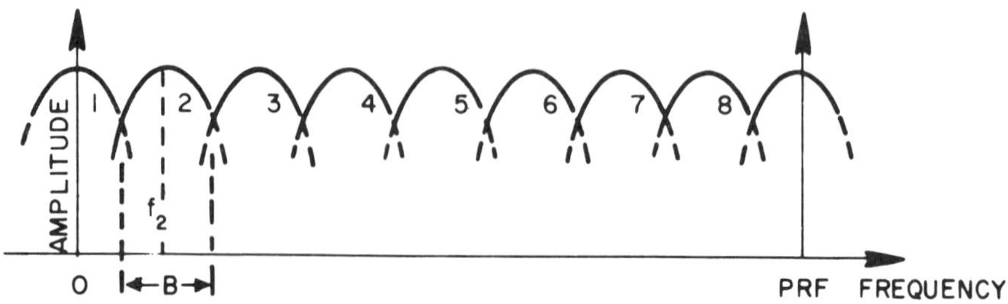

Figure 2.33 Coherent doppler filter bank.

periodic nature of the doppler filter bank whereby the filters recur periodically with a period equal to the radar's PRF. This requires that the filter's coordinate system rotate through an angle of 2π radians in the time required to process l pulses (e.g., $l = 8$). In addition, filter tuning requires that $\Delta\theta_n = 2\pi f_f/\text{PRF}$, where f_f is the doppler frequency to which the filter is tuned. The relationship for the incremental angle rotation of the filter's coordinate system can then be expresed as $\Delta\theta_n = 2\pi(k - 1)/l$, where $k = 1, 2, \ldots, l$ is the filter number. This results in $\theta_n = 2\pi(p - 1)(k - 1)/l$ where p is the pulse number, k is the filter number, and l is the total number of pulses processed. Substituting this into the weight relationship results in $\mathbf{w}_n = \exp[-j2\pi(p - 1)(k - 1)/l]$ as the complex weight factor that must be applied to input sample p to form the kth filter of the doppler filter bank.

For example, when two samples ($l = 2$) are processed, the first filter ($k = 1$) weights become (1, 1), which implies coherent integration of fixed target echoes, and the second filter ($k = 2$) has weights of (1, −1), which provides for the coherent integration of an optimum speed target. For a four-pulse sample, the weights for the corresponding four filters become (1, 1, 1, 1), (j, 1, $-j$, −1), (−1, 1, −1, 1), and ($-j$, 1, j, −1). Note that for more than two pulse samples, the weights generally become complex, which is necessary to coherently integrate doppler shifted target returns.

2.3.3.2 The Discrete Fourier Transform

In the previous sections, doppler filtering has been described as the algorithmic process of projecting the radar's input return signal (x_n, y_n) onto a rotating filter coordinate system (i, j) and adding up the projections. The rate at which the coordinates rotate relative to the x and y axes is the frequency to which the filter is tuned.

Alternatively, doppler filtering can be viewed from a spectral analysis viewpoint, which involves performing a *discrete Fourier transform* (DFT) on the signal. The algorithms for both procedures are identical. The Fourier transform approach has the advantage that the computational procedures using Fourier analysis are well developed with many fast algorithms, the most notable of which is the *fast Fourier transform* (FFT). The FFT provides a dramatic reduction in the immense volume of computing otherwise required to form a large filter bank with the DFT.

The discrete Fourier transform is given by $F(p, k) = \Sigma_{p=1}^{l} s_n(p)w^{(p-1)(k-1)}$, where $s_n(p) = x_n + jy_n$ is the complex signal at sample time $p = 1, 2, \ldots, l$, l is the number of sample points, $k = 1, 2, \ldots, l$ is the output transform point, and $w = \exp(-j2\pi/l)$ is the complex weight kernel (incremental angle rotation of filter coordinates). The equivalent normalized transfer function of the DFT can be shown [2] equal to $|H(\omega)| = |\sin[\pi(lf/f_r - k + 1)]/l \sin[\pi/l(lf/f_r - k + 1)]|$. The transfer function is depicted in Figure 2.34 for both a rectangular window (uniform input weighting) and an amplitude taper that provides substantially reduced sidelobes.

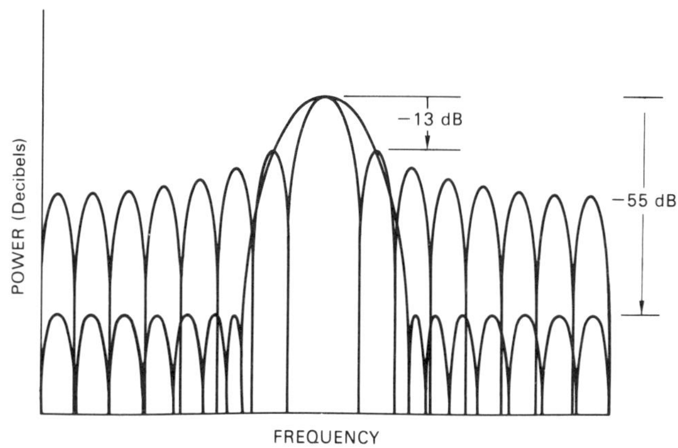

Figure 2.34 Transfer function of a discrete Fourier transform with regular and weighted inputs.

A method for providing a reduced filter sidelobe structure is depicted in Figure 2.35. This process, called *amplitude weighting*, is applied to the digitized video (after A/D conversion) before it is supplied to the doppler filters. Following each transmission, the I and Q receiving signals (x_n, y_n) returned from each range increment are multiplied by a weighting coefficient. The coefficient is changed from one pulse to the next according to a prescribed pattern (e.g., Hamming or Hann (i.e, hanning) weighting), which is repeated for each train of pulses to be integrated. Amplitude tapers can be selected to reduce sidelobes to very low levels (e.g., −35 to −60 dB); however, in this process, the filter's mainlobe response is significantly widened (e.g., Hamming weighting; −43 dB sidelobes, 1.30 3 dB bandwidth expansion). Alternatively, post-DFT weighting can reduce filter sidelobes [2].

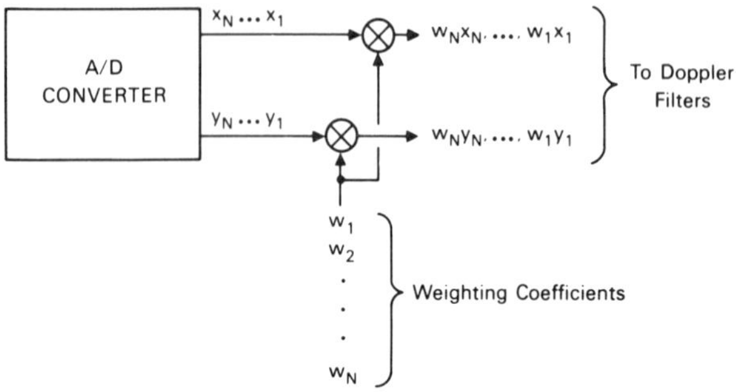

Figure 2.35 Reducing doppler filter sidelobes through amplitude weighting.

2.3.3.3 The Fast Fourier Transform

To obtain an idea of the computations required to form a doppler filter bank using DFT, consider the block diagram of the transversal filter implementation depicted in Figure 2.31. First, each in-phase and quadrature sample must be multiplied by real weights and then summed, resulting in eight mathematical operations per sample. To form l filters, $8l$ operations per complex input sample are required. Because l pulses are required to completely form each filter, the overall requirement for each range cell is then $8l^2$ operations. The number of filter banks (N_B) that must be formed is equal to the number of range cells (i.e., one range-gated doppler filter bank per range resolution cell). Because the samples occur at the radar's PRF rate, the total computation rate is given by $R_c = 8lN_B$ PRF operations per second. For example, if we assume for an airborne radar that a 16-sample DFT filter bank is to be formed using 200 range gates with a radar PRF of 8 kHz, the resulting computation rate is approximately 205 MOPS. This rate represents a formidable processing requirement that would severely stress the current digital array signal processors [3].

This example points to a requirement for substantially reducing the computational requirements involved in forming digital doppler filter banks. This requirement is satisfied by the fast Fourier transform, an algorithm that evaluates the DFT in a computationally efficient manner. Basically, the FFT exploits the periodic structure of the DFT weighting coefficients to significantly reduce the number of arithmetic operations required to form the DFT.

The first step in forming a computationally efficient doppler filter bank is to require the filter bank to span a range of frequencies equal to the sampling frequency (i.e., the PRF). The next step is to require that the number of filters in the bank be equal to an integer raised to some power, where the most popular choice is the radix-2 (i.e., l = 2, 4, 8, ..., 2^m); both steps result in a highly symmetrical distribution of the filter coefficients about the x- and y-axes (recall Figure 2.32), which, in turn, minimizes the number of multiplications required in the algorithm. The final step, which is critical in achieving high efficiency, is to add or subtract all inputs having a particular coefficient by the coefficients before multiplication instead of afterward. For this, all samples (x_n, y_n) that must be integrated to form the entire filter bank must be received before the multiplication operation can be accomplished. In other words, the algorithm cannot be performed on-line but must be accomplished in block fashion after all l samples have been received. The algorithm that accomplishes these steps, called the *FFT*, reduces the number of computations from the order of l^2 required for the DFT to the order of $l \log_2(l)/2$. The dramatic reduction in the number of computations achieved by using the FFT as opposed to direct evaluation of the DFT is depicted in Figure 2.36.

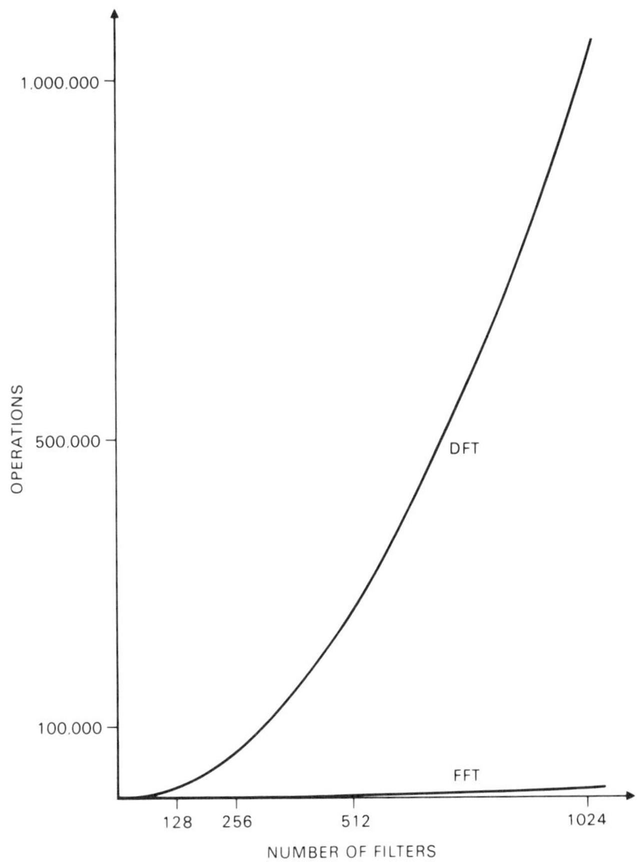

Figure 2.36 Operations required by using FFT and DFT for formation of large filter banks.

The DFT can be conveniently written in matrix form as $[F(k)] = [w^{(p-1)(k-1)}][s_{n(p)}]$, where $w = \exp(-j2\pi/l)$ and $k = 1, 2, \ldots, l$ is the output transform point, and $p = 1, 2, \ldots, l$ is the input sample point. In this representation, $[w^{(p-1)(k-1)}]$ is an l by l square matrix. Using the Cooley-Tukey algorithm, this matrix can be factored into $m = \log_2 l$ square matrices of dimension l by l [14]. The unique property of this factored form is that it allows formation of the FFT by using a cascade of repetitive operations where each operation is known as a *butterfly*. A flow graph of the conventional method for computing the DFT by using on the order of l^2 operations and the radix-2 FFT method by using $l \log_2(l)/2$ butterfly operations is depicted in Figure 2.37.

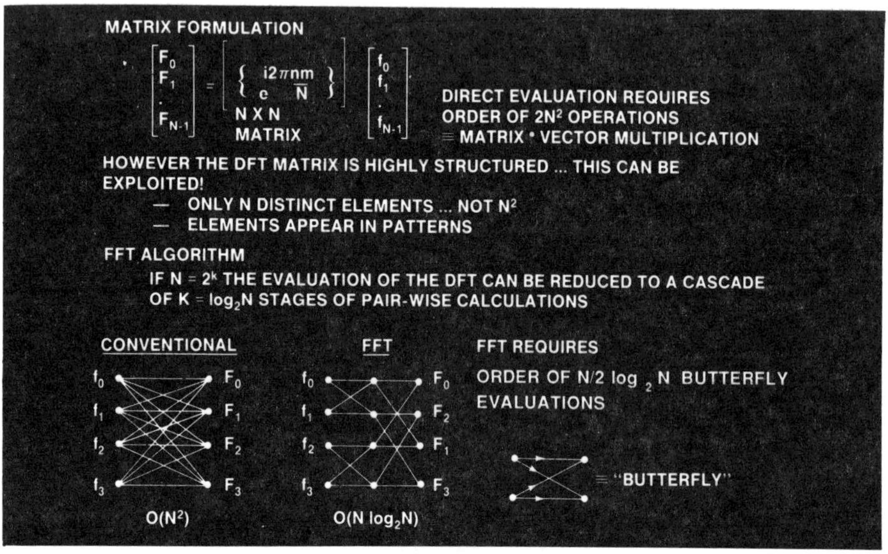

Figure 2.37 Flow graph of FFT using butterfly stages.

The butterfly operation in the radix-2 FFT is performed on a pair of samples in the mth stage to produce a corresponding pair of samples in the $m + 1$ stage. For an l-point FFT with $R_m = 2^m$, the butterfly operation can be expressed as the pair:

$$u_i(m + 1) = u_i(m) + w^r v_{i+l/Rm}(m)$$

and

$$v_{i+l/Rm}(m + 1) = u_i(m) - w^r v_{i+l/Rm}(m)$$

where u_i and $v_{i+l/Rm}$ are complex numbers representing the ith FFT data sample pair of the mth stage and r is an integer that is a function of the index i and stage m [15]. Because the input and output samples of the butterfly operation have the same indices, the computation can be executed in place by writing the output results over the input data. Thus, in theory, the FFT may be implemented with only l complex storage locations, plus auxiliary storage registers to support the butterfly computation.

In the butterfly operation, the complex weight value w^r can be determined using the bit-reversal method [14]. This is done by writing $i - 1$ as a t-bit binary number, scaling this number by $t - m$ bits to the right, filling the newly opened bit positions on the left by zeros, and reversing the order of the bits to obtain the value for r. For example, in the four-point FFT depicted in Figure 2.37, consider the second stage and third data sample (i.e., $m = 2$, $t = 2$). The binary number index is then given by 10, which, when scaled $t - m = 0$ bit to the right, results in the number 10. When bit-reversed, this results in 01, which yields $r = 1$ (i.e., w^1). The remaining weights for the four-point FFT can be found as w^0 ($i = 1, 2$; $m = 1$ and $i = 1$; $m = 2$) and w^1 ($i = 3$; $m = 2$), which allows the complete transform to be found. Note that in the representation given in Figure 2.37, the weights were determined from a tree-graph construction (rather than a butterfly representation) but due to symmetry $w^2 = -w^0$ and $w^3 = -w^1$, which reconciles the two representations [14].

The bit-reversal process necessary to determine the butterfly weights is simply implemented in hardware form by using a counter that counts a bit-reversed notation [15]. For an eight-point FFT, a conventional 3-bit counter yields the sucessive integers 0, 1, 2, 3, 4, 5, 6, 7. If the counter bit positions are reversed, then the resulting counter output is 0, 4, 2, 6, 1, 5, 3, 7, which gives a one-to-one correspondence between the natural order sequence and the bit-reversed order sequence.

An analysis of the butterfly relationships shows that each butterfly can be accomplished with four real multiplications and six real additions (a total of ten real operations). The butterfly operation can be considered a fundamental kernel associated with the FFT algorithm. The complexity of a signal processor that performs an FFT operation can be specified in terms of an equivalent number of butterfly operations (i.e., $l \log_2(l)/2$), and the computation time to perform a butterfly operation becomes a basic figure of merit for a radar signal processor [2].

Many of the FFT algorithms were developed to calculate the DFT by using a high-level language on a general purpose mainframe or a minicomputer. The emphasis in these algorithms was generally on minimizing the execution time, and hence the cost, of performing the DFT. The approach taken was to minimize the total number of operations performed by the arithmetic unit with particular emphasis on the total number of multiply operations, as this function represented the major limitation on processing speed. Consequently, most FFT algorithms are designed simply to minimize the total number of multiplications required to calculate the DFT, often for an increase in the number of additions, memory access, and control complexity. Increased control complexity is not usually a major penalty in software-based systems as, in simplistic terms, the control part of an FFT algorithm implemented on a high-level machine comes for free; however, in hardware-based processors, this is not the case.

The hardware designer must consider control complexity as well as try to optimize the processor arithmetic unit, as the control structure usually represents the major portion of the hardware system cost. As a result, DFTs implemented on general purpose computers are usually calculated by using highly composite data block sizes, where the arithmetic complexity can be reduced by decomposing the DFT into a large number of nested small-radix transforms, whereas hardware-based DFTs often use minimum block partitioning to simplify their control structure and microcode generation [16].

Emphasis in FFT hardware used for radar doppler filtering applications is on throughput rate (rate at which data can be absorbed) rather than algorithm processing time. This leads to pipeline structures that allow increased amounts of data to be processed in a given algorithm processing time. The pipeline structure is analogous to a long pipe through which water is pumped, where water enters continuously rather than waiting for one gallon to exit the pipe before another enters. The pipeline structure thus effectively trades the ability to absorb data at a higher rate for processing time delay.

For example, consider the flow of data and computations through the eight-point radix-2 FFT depicted in Figure 2.38. The radix-2 configuration breaks the FFT into $l \log_2(l)/2 = 12$ nested butterfly operations, which are grouped into three stages with each providing four butterfly operations. Data pairs separated by a distance of $l/2 = 4$ are fetched from the input storage buffer memory and operated on by the butterfly operations with the results returned to the original memory location. The process is repeated $l/2 = 4$ times until all l data points have been operated upon. The first pass, therefore, consists of $l/2$ butterfly operations performed on the data with a spacing of $l/2$. On the second pass, the process is repeated with a spacing of $l/4$. Successive passes use spacings that decrease by one-half until the spacing is such that adjacent data points are used. At this point, the desired spectrum has been computed and is resident in memory.

A pipeline FFT processor can be formed from the configuration depicted in Figure 2.38 by inserting storage registers between computational elements; this allows data for the second transform to be entered as soon as the immediate results of the first transform have been moved to the second stage. A double buffer memory would seem to be required: one would accommodate new data entering the computational element, and the second would hold the data block being worked on. However, by using a special "shuffle" memory [17], we can process the data blocks through the pipelined FFT using only a single buffer memory equal to the block length of the FFT being processed (e.g., an eight-point FFT would require only eight words of memory). Note also that the rotational vector weight (w^r) is a function of the data displacement number and hence varies in a predictable way from one computational stage to the next.

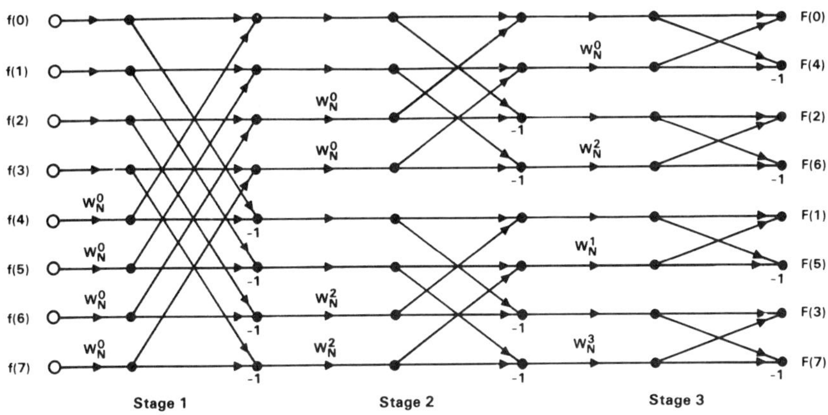

Figure 2.38 Flow of data through an eight-point radix-2 FFT.

The structure of the FFT algorithm is such that each of the $\log_2 l = M$ computational stages can be realized with a basic computational module, the general form of which is shown in Figure 2.39. The basic computational module is assembled to form a pipeline-FFT processor by combining M modules as depicted in Figure 2.40. Each module contains an arithmetic unit that computes the butterfly operation $[u(m + 1) = u(m) + w^r v(m); v(m + 1) = u(m) - w^r v(m)]$ and a shift register (delay line) whose storage capacity is 2^{M-m}, where M is the total number of stages and m is the stage number. For example, the first stage ($m = 1$) of an eight-point FFT ($M = 3$) would use a shift register whose capacity is $2^2 = 4$ words; the second stage, two words; and the third stage, one word capacity.

Each pipeline FFT module alternately transfers blocks of data samples into the delay line and the arithmetic unit [18]. Initially, $l/2$ data samples are loaded into

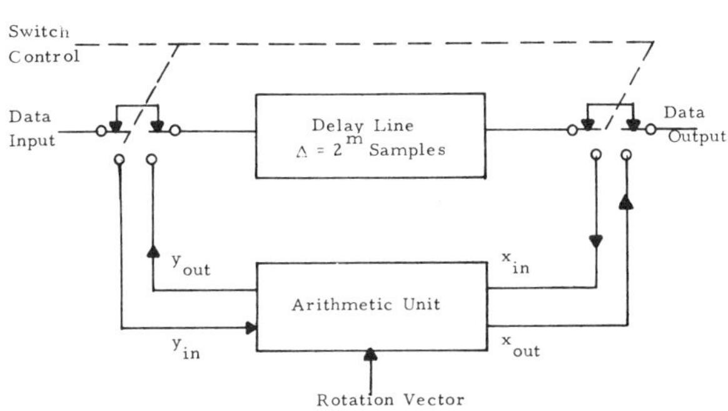

Figure 2.39 Pipeline FFT module.

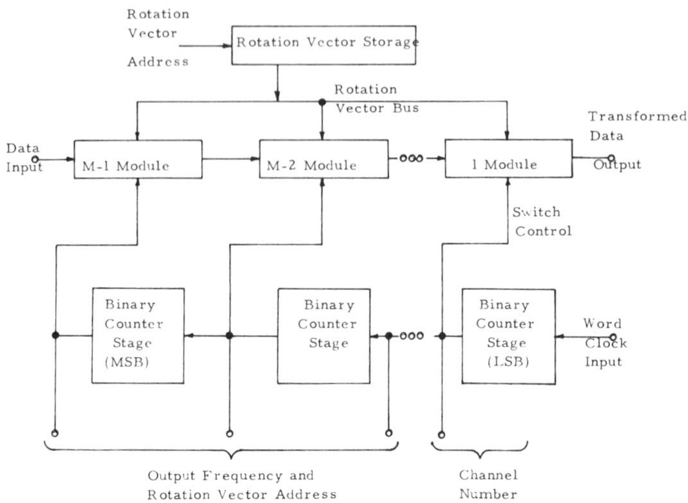

Figure 2.40 Pipeline FFT processor.

the delay line of the first-stage FFT module. Then, the first-stage arithmetic unit obtains a rotational weight vector (from a read-only memory) and begins its operation. The next $l/2$ data samples in conjunction with the first block stored in the delay line are loaded into the first FFT module to produce two successive complex butterfly outputs. One output is immediately transferred to the next stage, and the other output is sent to the delay line. Thus, in the interim period when the delay line is filled with fresh input data, the contents of the line containing the results of processing the earlier blocks are transferred to the next stage.

In the pipeline FFT processor shown in Figure 2.40, the rotational weight vector storage consists of a table of roots of unity (sines and cosines) shared by all M modules. In general, $l/2$ different rotational weight vectors must be read to process one block of l samples. If they are produced in the order required for the earlier stages, they may be obtained by strobing the memory storage in the proper time sequence. Thus, the sine and cosine weights can be obtained by providing a register for each arithmetic unit, all driven from a common bus. Note that $M = \log_2 l$ arithmetic units and $l - 1$ complex data points of storage are needed to implement the pipeline FFT processor depicted in Figure 2.40. Also, the first transform output is obtained immediately after the last data sample in the block of l is received.

The control mechanism in the pipeline FFT processor uses a binary counter driven in synchrony with the data stream. This counter generates a signal that designates the processing interval and switch position for each stage (see Figure 2.39). In general, when the counter output is supplying the output for the $k + 1$ stage, then the lower k bits of the counter, in normal order, give the channel number of the data currently being produced, whereas the upper $M - k$ bits, taken in bit-reversed order, give the frequency currently being produced. In addition, the counter supplies the address of the complex weights (sine and cosine) needed in the current computation of the $k + 1$ stage. Thus the control mechanism contains all the information needed to carry out the spectral analysis as well as to descramble the output data. In many applications, descrambling the output data is not necessary because the precise order of the output frequency samples is always known [17].

The pipeline FFT processor just described represents only one of several types of FFT processing hardware architectures that can be used. The types available, which range from the least hardware–lowest throughput to the most complex hardware–highest throughput, are called the (1) sequential, (2) cascade (pipeline FFT), (3) parallel-iterative, and (4) array processors [19].

The sequential FFT processor performs a single butterfly operation on $l \log_2 l/2$ pairs of data in a serial manner. The same memory can be used to store the input data, the intermediate results, and the resulting Fourier coefficients. Because the processor operates on batches of data, a real-time environment would dictate that a buffer memory precede or be incorporated into the data memory. The sequential FFT processor can be characterized as having (1) one arithmetic unit; (2) $l \log_2 l/2$ butterfly operations performed sequentially; and (3) an execution time of $Bl \log_2 l/2$ where B is the time to perform the butterfly operation.

The cascade FFT processor introduces parallelism by using a separate arithmetic unit for each iteration stage, thereby increasing the throughput over the sequential processor by the factor $\log_2 l$. In the flow diagram of Figure 2.38, the arithmetic unit associated with each stage of the algorithm processes $l/2 = 4$ pairs of data. The buffering required for processing a continuous stream of data is incorporated directly into the processor organization in the form of a delay line (shift register). The cascade FFT processor can be characterized as having (1) $M = \log_2 l$ arithmetic units and iterations performed in parallel, (2) $l/2$ operations performed sequentially, (3) an execution time of $B \cdot l/2$ seconds, and (4) buffering incorporated within the processor in the form of time delays.

The parallel iterative FFT processor increases throughput by introducing parallelism into each processing stage. The processor performs the $l/2$ butterfly operations within each stage in parallel, and the $\log_2 l$ iterative operations are performed sequentially. The parallel iterative processor can be characterized as having (1) $l/2$ arithmetic units and operations performed in parallel, (2) $M = \log_2 l$ iterations performed sequentially, and (3) an execution time of $B \log_2 l$ seconds. In practice, due to cost constraints, this architecture is often combined with that of the sequential processor so that only the degree of parallelism actually needed to produce the required throughput is implemented.

The array processor is configured such that all $l \log_2 l/2$ butterfly operations are performed in parallel. This is accomplished by simultaneously pipelining $\log_2 l$ sets of data through the processor, which results in an execution time equal to the that required to perform a basic butterfly operation. The FFT-array processor can be characterized as having (1) $l/2 \log_2 l$ arithmetic units and operations performed in parallel, and (2) an execution time of B seconds.

As an example, a 64-point FFT is to be performed using a butterfly unit that has an execution time of 0.5 μs ($l = 64$, $n = 6$, $B = 0.5$ μs). Table 2.1 compares the various processing architectures, indicating the increasing data rate associated with an increased degree of parallel operation.

At present, FFT hardware emphasizes optimal use of standard digital integrated circuit chips to generate butterfly modules. Complete FFT processors can then be formed by using the butterfly modules in combination with appropriate digital control circuitry. One such hybrid unit, which provides a 4- to 16,384-point FFT, utilizes a radix-4 butterfly module that performs a basic butterfly operation in 100 ns [20].

Further increases in FFT hardware capacity involve utilizing *very large scale integration* (VLSI) circuit techniques to form a number of butterflies on a single chip. For example, one recent hardware survey projected that 16 butterflies could be implemented on a single 50 mm^2 chip [13]. A resulting problem with this approach is that the FFT butterfly interconnections lead to long wires between butterflies [21]. Not only do these wires consume a lot of area but require high-power semiconductors to drive this capacitive load at useful speeds. Thus VLSI designs are not necessarily well matched to the implementation of the conventional FFT butterfly algorithm.

One of the most computationally efficient algorithms for the DFT is the *Winograd Fourier transform* (WFT) algorithm, which minimizes the number of multiplication operations required for a given l-point FFT [15]. The WFT reduces the number of multiplications to the order of l to compute an l-point transform, and the number of additional operations is comparable to that required by the conventional FFT. This algorithm is, therefore, most effective when the multiplication operation involved in the implementation is relatively slow (e.g., minicomputers and microcomputers). Of

Table 2.1
FFT Hardware Architecture Data Rate to Compute a 64-Point FFT

Architecture	*Butterfly Units*	*Execution Time (μs)*	*Data Rate (MHz)*
Sequential	1	96	0.67
Cascade	6	16	4
Parallel Iterative	32	3	21.3
Array	192	0.5	128

$B = 0.5$ s

theoretical note, the WFT provides a provable lower bound on spectrum analyzer complexity and shows that the conventional radix-2 FFT is no more than an order of magnitude slower than the most efficient algorithm.

In the WFT, the weight matrix $D_l = [w^{(p-1)(k-1)}]$ is factored into three matrices $D_l = S_l C_l T_l$, where T_l is an $J \times l$ (row × column) incidence matrix (i.e., has elements whose values are 0, 1, and -1), C_l is an $J \times J$ diagonal matrix (i.e., has values only on the diagonal), and S_l is an $l \times J$ incidence matrix. Winograd has developed formulations for small-l point transforms (e.g., l = 2, 3, 4, 5, 6, 7, 8, 9, and 16), whereby the magnitude of J is the order of l. Because C_l is a diagonal matrix with elements that can be shown to be either real or imaginary, the total number of multiplications involved in the algorithm is then the order of l, which corresponds to the number of points in the transform [22].

A number of small-l WFTs can be combined into a larger WFT by using the *prime factor algorithm* (PFA) [15]. The PFA involves mapping the long-sequence WFT into a multidimensional sequence, the size of the individual dimensions being those of the small-l WFT algorithms [21]. The implementation of the PFA to generate a long WFT involves a reordering of the input data before processing and output data after processing. For this reordering, l must be factorable into $l_1, l_2, \ldots, l_s$ (e.g., $l = \Pi_{i=1}^{n} l_{si}$), where the l_s are those associated with the small-l WFT and additionally must be mutually prime (i.e., the largest common divisor equals one). For example, to form a 12-point WFT, small-l WFTs with $l_1 = 4$ and $l_2 = 3$ would be used.

A hardware implementation of the WFT involves the following three steps. The l reordered complex input data samples are multiplied by either $+1$, -1, or 0 in accordance with the $J \times l$ incidence matrix T yielding J complex values. Next, these J complex values are multiplied by a coefficient that is either purely real or purely imaginary as determined by the $J \times J$ diagonal matrix C_l again yielding J complex numbers (note that only two real multiplications are required for each complex value). Finally, the J complex numbers out of the second stage are multiplied by either $+1$, -1, or 0 in accordance with the $l \times J$ incidence matrix S_l and reordered to give the l complex output terms of the WFT.

A hardware design of a WFT processor using bit-level systolic arrays has been recently described that minimizes the number of connections involved in a VLSI design [21]. The systolic array architecture utilizes nearest-neighbor interconnections, and the efficiency of the WFT provides a potentially high throughput rate. An estimate of the throughput rate for a 16-bit word size VLSI design with a cell computation rate of 20 MHz is 1.42×10^8 complex words per second or a complete new 240-point WFT every 1.7 μs.

REFERENCES

1. G. Stimson, Introduction to Airborne Radar, Hughes, El Segundo, CA, 1983.
2. D.C. Schleher, ed., *MTI Radar*, Artech House, Norwood, MA, 1978.

3. F. Williams and M. Radant, "Airborne Radar and the Three PRFs," *Microwave Journal,* July 1983.
4. J. Roulston, "Design Decisions Guide Airborne Radar," *Microwaves and RF,* March 1985.
5. E. Aronoff and N. Greenblatt, "Medium PRF Radar Design and Performance," *20th Tri-Service Radar Symp.,* 1974.
6. F. Nathanson and P. Luke, "Loss from Approximations to Square-Law Detectors in Quadrature Systems with Postdetection Integration," *IEEE Trans.,* Vol. AES-8, No. 1, January 1972.
7. V. Vannicola, "Fluctuation Loss and Diversity Gain for In-Phase Systems with Post-Detection Integration," *IEEE Trans.,* Vol. AES-9, No. 2, March 1973.
8. M. Weiss and I. Gertner, "Loss in Single-Channel MTI with Post-Detection Integration," *IEEE Trans.,* Vol. AES-18, No. 2, March 1982.
9. D.C. Schleher, "Performance Comparison of MTI and Coherent Doppler Processors," IEEE Int. Radar Conf., London, November 1982.
10. A. Papoulis, *The Fourier Integral and Its Applications,* McGraw-Hill, New York, 1962.
11. W. White and A. Ruvin, "Recent Advances in the Synthesis of Comb Filters," *IRE Nat. Conv. Record,* 1957.
12. A. Filip, "A Baker's Dozen Magnitude Approximations and Their Detection Statistics," *IEEE Trans.,* Vol. AES-12, No. 1, January 1976.
13. J. Mavor and P. Grant, "Operating Principles and Recent Developments in Analogue and Digital Signal Processing Hardware," *IEE Proc.,* Vol. 134, Pt. F, No. 4, July 1987.
14. E. Brigham and R. Morrow, "The Fast Fourier Transform," *IEEE Spectrum,* December 1967.
15. H. Nussbaumer, *Fast Fourier Transform and Convolution Algorithms,* Springer-Verlag, New York, 1981.
16. T. Curtis and J. Wickenden, "Hardware-Based Fourier Transforms: Algorithms and Architectures," *IEE Proc.,* Vol. 130, Pt. F, No. 5, August 1983.
17. H. Groginsky and G. Works, "A Pipeline Fast Fourier Transform," *IEEE Trans.,* Vol. C-19, November 1970.
18. E. Brookner, ed., *Radar Technology,* Artech House, Norwood, MA, 1977.
19. G. Bergland, "Fast Fourier Transform Hardware Implementations—An Overview," *IEEE Trans. Audio and Electroacoustics,* Vol. 15, June 1969.
20. E. Swartzlander, "Spectrum Analysis and the Unconventional Butterfly," *Defense Science and Engineering,* October 1985.
21. J. Ward, J. McCanny, and J. McWhirter, "Bit-Level Systolic Array Implementation of the Winograd Fourier Transform Algorithim," *IEE Proc.,* Vol. 132, Pt. F, No. 6, October 1985.
22. H. Silverman, "An Introduction to Programming the Winograd Fourier Transform Algorithm (WFTA)," *IEEE Trans.,* Vol. ASSP-23, April 1977.

Chapter 3
Doppler Radar Performance Measures

A basic consideration for any radar is the range at which the radar can detect the target. This obviously depends on the parameters of the radar (e.g., transmitter power, antenna gain, *et cetera*) and on the reflection characteristics of the target (e.g., radar cross section). However, a fundamental limitation is that the target must be detected in an interference background that, as a minimum, consists of the everpresent receiver thermal noise. Because receiver noise is a random process, it must be described in terms of its statistical properties. Therefore, radar detection is a statistical problem, and when specified, the range of a radar must be stated in a statistical manner to be meaningful.

Marcum and Swerling [1,2] were among the first to develop a consistent statistical theory of radar detection against a noisy receiving background. The two statistical parameters they used to qualify the statement of a radar's detection range were the probability of detection (e.g., $P_d = 0.9$) and the probability of false alarm (e.g., $P_{fa} = 10^{-6}$). The Marcum and Swerling (M-S) theory is documented by extensive sets of curves and computer generated tabulations that list the required signal-to-noise ratio to provide a specified P_d and P_{fa} for a given number of integrated pulses and Swerling target types (i.e., fluctuating and fixed) [3].

Calculation of the detection performance of a practical radar in a noisy background can be determined from the M-S theory by introducing the concept of detection loss. Detection loss refers to the additional S/N required to achieve a specified performance level relative to the M-S curves under conditions that deviate from the idealized ones embodied in the derivation of the standard curves. Many types of detection losses have been identified, and when these are considered, very accurate predictions of radar performance can be obtained that provide good agreement with actual test results [4].

The probability of detection analysis against a receiver noise background developed by Marcum and Swerling is not directly applicable to doppler-type radars, which must operate in an interference background that contains clutter as well as receiver noise. The basic reason for this is that, as receiver noise (i.e., white noise) is uncorrelated from pulse-to-pulse, the corresponding analysis can be based on well-known statistical techniques for integrating independent statistics. This is not the case for doppler radars that must operate against clutter, which is generally highly correlated from pulse to pulse (the cancellation of fixed echoes depends on this). Even if the clutter were not so correlated, independence would be destroyed by the pulse-combining action of the doppler filters. Consequently, the corresponding analysis for doppler radars is more involved and therefore less highly developed than the M-S analysis.

The present state of detection analysis for doppler radars in a clutter and noise background follows. First, a general theory for determining the probability of detection for a doppler radar employing an optimum doppler processor (i.e., transversal filter), as shown in Figure 3.1, is available and is discussed in Chapter 5 (see Section 5.1) [5]. This theory is useful because it establishes an upper bound on the detection probability of a doppler radar. The theory also allows the detection probabilities to be determined for doppler radars employing nonoptimum doppler processors, which follow the general structure of a coherent linear doppler filter followed by an envelope detector (see Section 5.1). For example, the analysis allows the determination of the detection probabilities of a coherent MTI followed by an envelope detector and an MTI-coherent integrator (FFT coherent filter bank) cascade, followed by an envelope detector, as depicted in Figure 3.2.

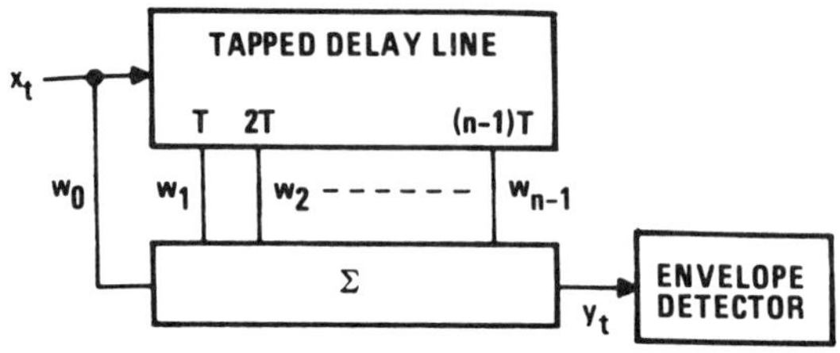

Figure 3.1 Optimum transversal doppler filter processor.

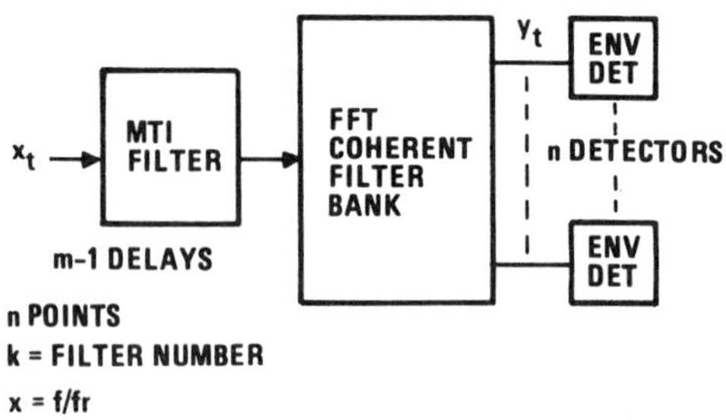

Figure 3.2 Cascaded MTI and coherent integrator.

The problem of computing the probability of detection for a coherent MTI (or MTI-coherent integrator cascade), followed by a postdetection noncoherent integrator, as depicted in Figure 3.3, is difficult but susceptible to an analysis that decorrelates the correlated MTI output by using an orthogonal transformation to determine detection thresholds (see Section 5.3) [6,7]. To determine the probability of detection, we must expand the signal plus interference probability density into a Gram-Charlier series [5]. The tedious nature of this theory makes it more adaptable to the calculation of special situations rather than the generation of generalized results, such as those available from the M-S analysis.

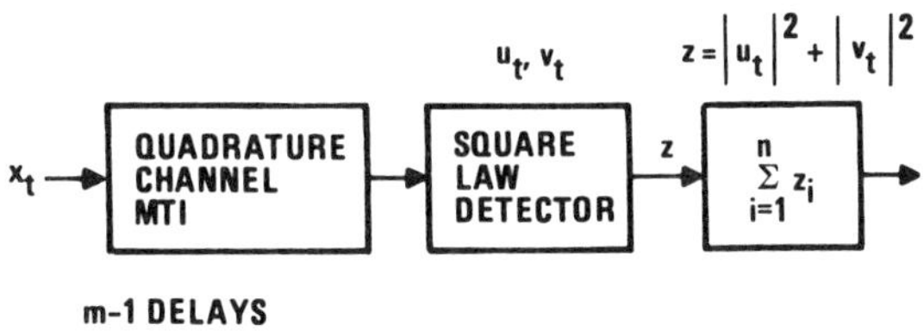

Figure 3.3 Cascaded MTI and noncoherent integrator.

This theory has been applied to two special situations. First, it has been used to determine loss when a target in receiver noise is passed through an MTI and then noncoherently integrated (see Section 5.3) [7]. Second, the theory has been applied to determine the detection loss when a target in clutter and receiver noise is passed through an MTI and then noncoherently integrated [5]. Note that the first case is a subset of the second case and also that the detection loss associated with the clutter plus noise situation is a function of the MTI's capability to suppress the clutter. A typical curve depicting the detection loss for a coherent single canceller MTI followed by a video integrator is given in Figure 3.4.

The next level of complexity in doppler radar probability of detection analysis involves a noncoherent MTI where the MTI filtering occurs after envelope detection. Analytical techniques are available that would allow probability of detection calculations for this type of MTI by using a square-law envelope detector [8]. However, the complexities of this technique are such that it has been generally applied to second-moment (power) analyses rather than to determine detection probabilities [9,10].

In practice, most doppler radar processing techniques are compared on the basis of their ability to enhance the ratio of the target signal power to the interferring clutter power background. To facilitate this comparison, several figures of merit have been defined by the IEEE standards and in other references. The primary figure of merit employed is called the *improvement factor,* which generally encompasses the other criteria. The improvement factor criterion was generally developed by Emerson (who called it the *reference gain*) and accounts for both the attenuation of the clutter and

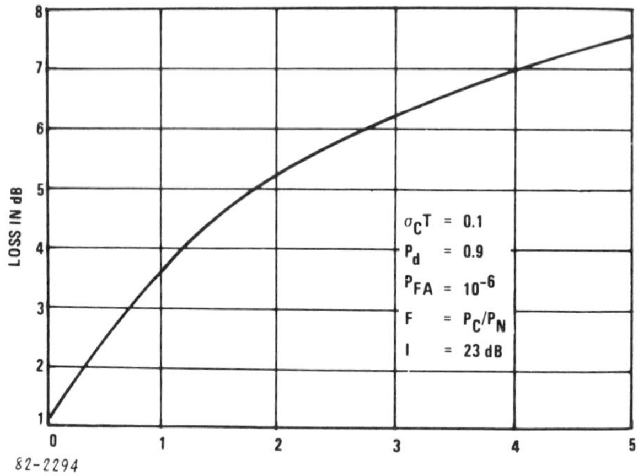

Figure 3.4 Video integration loss with a single canceller MTI.

enhancement of the target response by the doppler filter [10]. Because the improvement factor criterion was initially developed for use with MTI systems, it assumes that the target's velocity is uniformly distributed over all possible values and ignores the effect of receiver noise on the system performance. We can modify this criterion to apply to both receiver noise and clutter, and, in certain situations, to relate to probability of detection analysis.

3.1 THE MTI IMPROVEMENT FACTOR (I_c)

The MTI improvement factor is defined as "the signal-to-clutter ratio at the output of the MTI system divided by the signal-to-clutter ratio at the input of the system, averaged uniformly over all target radial velocities of interest," which can be expressed as

$$I_c = \overline{(S/C)_o}/\overline{(S/C)_i} \tag{3.1}$$

The improvement factor can be rewritten as the product

$$I_c = (\overline{S_o/S_i})(C_i/C_o)$$

where the first term $\bar{G} = (\overline{S_o/S_i})$ is called the *MTI gain* (the average MTI response to targets) and the second CA = (C_i/C_o) is called the *clutter attenuation*. In this form, we can see that the improvement factor embodies both clutter attenuation and target enhancement effects in its definition. As defined, the improvement factor is independent of the target velocity and depends only on the weights of the MTI, the power spectrum of the clutter, and the number of pulses processed.

Another related MTI performance criterion is called the *subclutter visibility* (SCV): "the ratio by which the target echo power may be weaker than the coincident clutter echo power and still be detected with specified detection and false-alarm probabilities. Target and clutter powers are measured on a single pulse return and all target radial velocities are assumed equally likely." The SCV is a measure of an MTI's system's ability to detect a moving-target signal superimposed on clutter signals.

A subclutter visibility of 40 dB implies that a moving target can be detected in the presence of clutter, although the clutter echo power is 10,000 times the target echo. Two radars with the same subclutter visibility might not have the same ability to detect targets in clutter if the clutter resolution cell of one is greater than the other, causing it to accept a greater clutter power. Thus a radar with a 10 μs pulse length and 10° beamwidth would generally need 20 dB more subclutter visibility than a radar with a 1 μs pulse width and 1° beamwidth for equal performance in a specified cluttered environment.

The subclutter visibility of a radar is generally less than the improvement factor associated with that radar. For example, assume that a 1 m^2 target was detected with $P_d = 0.9$ and $P_{fa} = 10^{-6}$ in a clutter-free environment. The radar would then require a subclutter visibility of 20 dB to detect the same target in an environment where the clutter background had an equivalent RCS of 100 m^2. However, an improvement factor of 20 dB in the radar's doppler processor would produce a *signal-to-clutter* (S/C) ratio of only one at its output, which is insufficient to produce the specified P_d. This situation is further complicated if the radar employs postdetection integration because the integration gain will be greater against noise than against clutter, which generally is highly correlated.

In the case of a doppler processor followed by an envelope detector, the subclutter visibility can be directly related to the processor's improvement factor. A generalized relation (see Section 5.1.1) for the probability of detection of a doppler processor is given by

$$P_d = Q\{[2P_s I_f / P_n(1 + F)]^{1/2}, \quad [2 \ln(1/P_{fa})]^{1/2}\} \tag{3.2}$$

where $F = P_c/P_n$ is the clutter-to-noise power ratio, P_s and P_n are the signal and noise powers, I_f is the interference improvement factor, P_{fa} is the probability of false alarm, and $Q(\cdot, \cdot)$ is Marcum's Q function [1]. In the preceding relationship, the interference improvement factor is given by

$$I_f = \mathbf{W}^T \mathbf{M}_s \mathbf{W}^* / \mathbf{W}^T \mathbf{M}_x \mathbf{W}^* \tag{3.3}$$

where $\mathbf{M}_s$ and $\mathbf{M}_x$ are the signal and interference covariance matrices, and $\mathbf{W}^T$ (transpose) and $\mathbf{W}^*$ (complex conjugate) form the complex weight vector. Note that the doppler processor functions to improve both the signal-to-clutter and signal-to-noise ratios. This accounts for the use of I_f (the interference improvement factor) rather than I_c (the MTI improvement factor), which operates just on the clutter alone.

The interference improvement factor (I_f) can be expressed in terms of the MTI clutter improvement factor (I_c) as

$$I_f = [(1 + F)/F] \cdot I_c/(1 + I_c/F) \tag{3.4}$$

Thus, as F (the clutter-to-noise power ratio) becomes large, I_f tends to approach I_c. Also, for a very large I_c, the maximum improvement against interference approaches $F + 1$, indicating a maximum improvement relating to the clutter-to-noise power ratio.

The general improvement factor relationship (I_f) can be specialized to the MTI improvement factor (I_c) case by setting $\mathbf{M}_s = \mathbf{I}$ (all target velocities are equally likely) and $\mathbf{M}_x = \mathbf{M}_c$, where $\mathbf{I}$ is the identity matrix and $\mathbf{M}_c$ is the clutter's covariance matrix. The expression for the probability of detection then becomes

$$P_d = Q\{[2P_s/P_n(1 + F/I_c)]^{1/2}, \quad [2\ln(1/P_{fa})]^{1/2}\} \tag{3.5}$$

which demonstrates that as the MTI improvement factor becomes large, the detection performance approaches that of the noise-limited case (i.e., the MTI provides no improvement in the S/N).

The MTI improvement factor for the coherent MTI case can be evaluated as

$$I_c = \sum_{j=0}^{n-1} w_j^2 \Bigg/ \sum_{j=0}^{n-1} \sum_{k=0}^{n-1} w_j w_k \rho_c(j - k) \tag{3.6}$$

where

$$\bar{G} = \sum_{j=0}^{n-1} w_j^2$$

is the MTI gain, w_j is the MTI's real weights, $\rho_c(\cdot)$ is the clutter's correlation coefficient, and n is the number of pulses processed by the MTI. For a Gaussian-shaped spectrum, the correlation coefficient is given by

$$\rho_c(i - j) = \exp[-(i - j)^2\Omega^2/2] = \rho_c(1)^{(i-j)^2} \tag{3.7}$$

where $\Omega = \sigma_c T$, σ_c is the standard deviation of the clutter spectrum in rad/s, and T is the radar's interpulse period.

Explicit evaluation of the MTI improvement factor requires a specification of the weights used in the MTI. Optimum weights can be determined by using Emerson's procedure (see Section 5.1.2) [10]. However, in practice, binomial weights (the binomial coefficients with alternating signs) are generally employed, as they provide performance approaching the optimum (for $n = 2$ pulses the binomial weights are optimum), and the values for the optimum and binomial weights are very nearly equal for narrow clutter spectra (i.e., $\sigma_c T \rightarrow 0$).

Table 3.1 lists the MTI gain and binomial weights for values of $n = 2$ to 8 pulses processed by the MTI. The binomial weight improvement factor for N delay sections ($N = n - 1$) is given by

$$I_c = \frac{1}{1 - 2\dfrac{N}{N+1}\rho_c(1) + 2\dfrac{N}{N+1}\dfrac{N-1}{N+2}\rho_c(2) + \cdots} \tag{3.8}$$

which can be approximated as $I_c = 2^{n-1}/(\sigma_c T)^{2(n-1)}(n - 1)!$ for $\sigma_c T \ll 1$ [9]. The MTI improvement factor for binomial weights is depicted in Figure 3.5, and a comparison of the improvement factor for optimum and binomial weights is given in Figure 3.6. Note that the optimized weights provide the most advantage when (1) more than two pulses are processed, (2) the clutter spectrum is narrow, and (3) the filter weights are accurate. (This follows from the small numerical difference for low $\sigma_c T$ between the optimum and binomial weights and relatively large difference in improvement factor.)

An alternative method for computing the MTI improvement factor in the spectral domain is given by

$$I_c = |\overline{H(\omega)}|^2 \int_0^{2\pi f_r} S_i(\omega)\, d\omega \Big/ \int_0^{2\pi f_r} |H(\omega)|^2 S_i(\omega)\, d\omega \tag{3.9}$$

Table 3.1
MTI Power Gain

n	$\overline{G} = \Sigma_{j=0}^{n-1} w_j$	*Binomial Weights*
2	2	1, −1
3	6	1, −2, 1
4	20	1, −3, 3, −1
5	70	1, −4, 6, −4, 1
6	252	1, −5, 10, −10, 5, −1
7	923	1, −6, 15, −20, 15, −6, 1
8	3432	1, −7, 21, −35, 35, −21, 7, −1

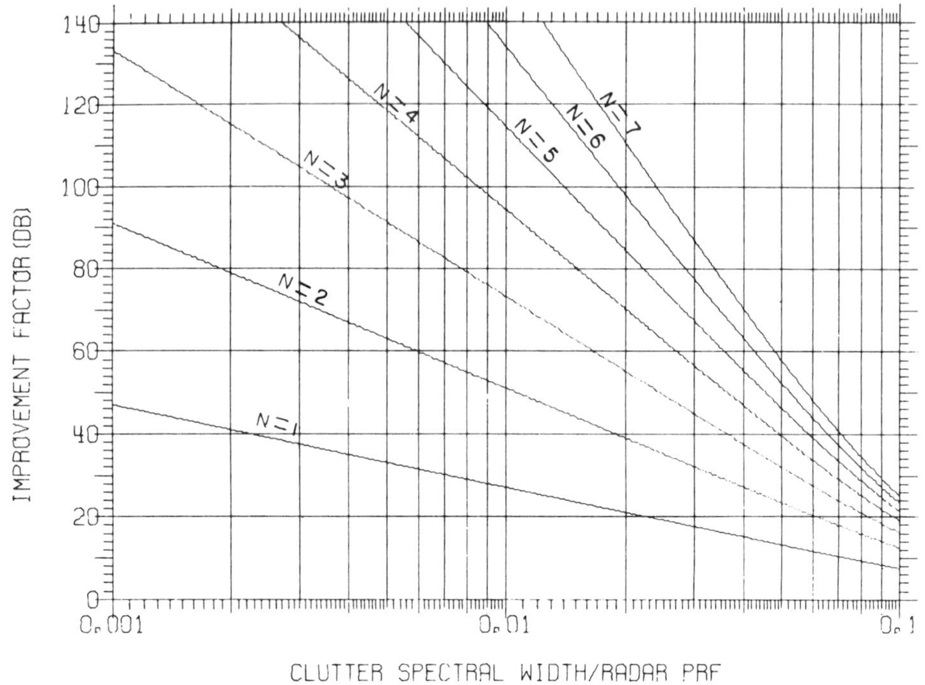

Figure 3.5 MTI improvement factor for binomial weights (n = 2 to 8).

where $S_i(\omega)$ is the normalized clutter power spectral density [e.g., $S_i(\omega) = \exp(-\omega^2/2\sigma_c^2)/\sqrt{2\pi\sigma_c^2}$], f_r is the radar's PRF, and $|H(\omega)|^2$ is the MTI filter's power transfer function (e.g., $|H(\omega)|^2 = [2\sin(\omega T/2)]^{2(n-1)}$ for an MTI with binomial weights). Relationship (3.9) can be extended to compute the interference (clutter plus noise) improvement factor (I_f) by redefining $S_i(\omega)$ to be the clutter-plus-noise power spectral density and evaluating $|H(\omega)|^2$ at the frequencies of interest (see Section 5.2 for evaluation of the cascaded MTI and coherent integrator).

The MTI improvement factor of conventional surface-based search radars often is limited by the antenna scanning modulation induced onto the clutter return. The antenna scanning modulation causes a spreading of the clutter's spectrum, which has a maximum frequency that corresponds to the doppler shift associated with the tip speed of the antenna's aperture. For analytic convenience, the antenna scanning clutter spectrum is customarily approximated by a Gaussian-shaped function, although this implies clutter frequency components greater than those associated with the antenna's tip speed. The MTI improvement factor due to antenna scanning modulation is best

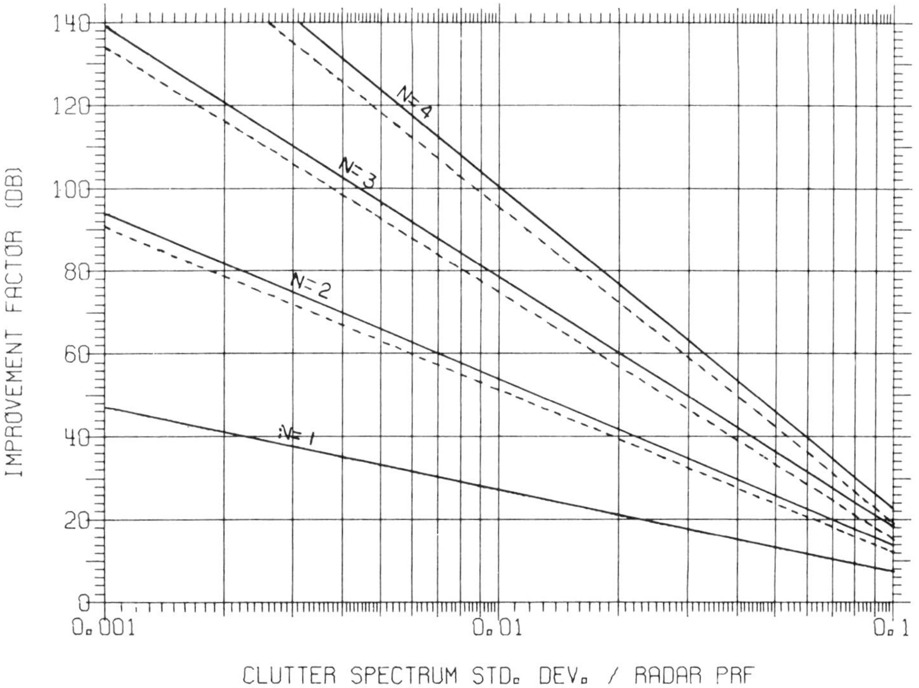

Figure 3.6 MTI improvement factor for optimum and binomial weights (n = 2 to 5).

calculated using the spectral analysis method, illustrated in the following text for a single canceller type of MTI.

The one-way antenna pattern is assumed to be given by

$$G(\theta) = G_o \exp(-2.776\theta^2/\theta_B^2) \tag{3.10}$$

where θ_B is the antenna's one-way 3 dB beamwidth. The scanning action of the antenna induces a Gaussian-shaped voltage modulation onto each clutter scatterer, which is given by $v(t) = v_o \exp(2.776t^2/t_o^2)$, where the time on target is given by $t_o = n_B T$ and n_B is the number of hits per 3 dB antenna beamwidth, while T is the radar's interpulse period. Next, the voltage modulation is Fourier transformed to find the voltage spectrum and then squared to find the clutter's power spectrum, resulting in

$$S(\omega) = P_o \exp[-(\omega t_o)^2/2 \cdot 2.776]$$

This can then be substituted into the relationship for the MTI improvement factor by using

$$|H(\omega)|^2 = [2 \sin(\omega T/2)]^2 \tag{3.11}$$

for the single canceller MTI. However, because the clutter spectrum is Gaussian shaped, its standard deviation can be determined as $\sigma_c = 1.666/t_o$. Substitution of $t_o = n_B T$ results in $\sigma_c T = 1.666/n_B$, which can be used to find the limitation to the MTI improvement factor caused by antenna scanning as $I_c = 2/(\sigma_c T)^2 = n_B^2/1.388$. In a similar manner, the MTI improvement factor under an antenna scanning limitation for a double canceller can be determined as $I_c = n_B^4/3.853$ and a triple canceller as $I_c = n_B^6/16$. These values are plotted in Figure 3.7.

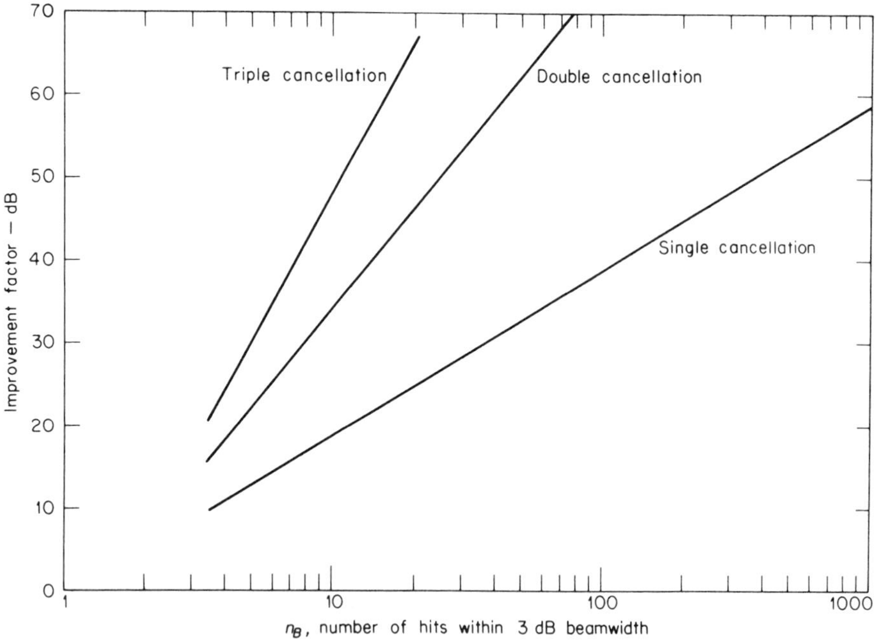

Figure 3.7 MTI improvement factor limitation due to antenna scanning.

3.1.1 Noncoherent MTI Improvement Factor

The MTI improvement factor for a noncoherent MTI (recall Figure 2.17) can be found by using a procedure similar to that used for a coherent MTI. The basic difference between coherent and noncoherent MTI is that the former employs synchronous (or phase) detection prior to MTI filtering, whereas the latter uses envelope detection. The synchronous detection used in the coherent MTI is a linear heterodyning operation that preserves the RF spectral shapes of the various target, clutter, and noise spectra. However, the envelope detection used in the noncoherent MTI widens the video target and clutter spectra, which generally results in lower performance than available from a coherent MTI.

The noise-suppressed video spectrums for coherent MTI and noncoherent MTI employing a square-law envelope detector are depicted in Figure 3.8. The essential action of the square-law envelope detector is to cause the coherent MTI spectrum to be convolved with itself [9,11]. For Gaussian-shaped spectra, the resulting noncoherent MTI clutter spectrum ($C \times C$) has a variance given by $\sigma_{nc}^2 = 2\sigma_c^2$ whereas the target spectrum ($C \times T$) is $\sigma_{tnc}^2 = \sigma_t^2 + \sigma_c^2$, where σ_c^2 is the variance of the coherent spectrum and σ_t^2 is the variance of the coherent target spectrum. Note that the convolution action also causes the noncoherent MTI to have an additional target sideband, which is not present in a coherent MTI.

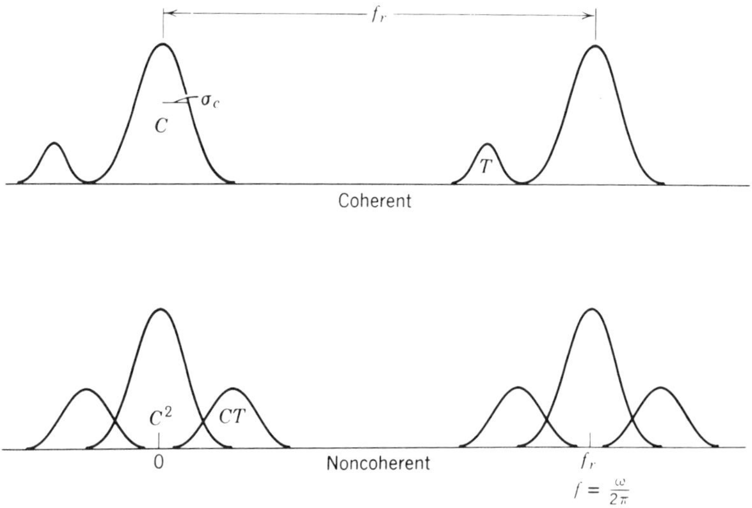

Figure 3.8 Video spectra of coherent and nocoherent MTIs.

The MTI improvement factor for a noncoherent MTI with binomial weights can be deduced from the expression for the MTI improvement factor of a coherent MTI by multiplying by a factor of two (due to the additional sideband) and using the noncoherent MTI spectral variance ($\sigma_{nc}^2 = 2\sigma_c^2$) as

$$I_{nc} = 2 \cdot 2^{n-1}/(\sqrt{2}\sigma_c T)^{2(n-1)}(n-1)! = 2/(\sigma_c T)^{2(n-1)}(n-1)! \quad (3.12)$$

for $\sigma_c T \ll 1$. The resulting MTI improvement factor for both coherent and noncoherent MTIs with binomial weights is tabulated in Table 3.2 (σ_c^2 is the clutter variance and T is radar's interpulse period). The resulting scanning limitations (n_B = number of pulses per 3 dB beamwidth) are listed in Table 3.3. Note that the improvement factors and scanning limitation are equal for a single MTI canceller ($n = 2$), whereas the coherent MTI provides progressively greater performance as the order (number of delay lines employed) of the MTI increases. Alternatively, a general expression for the noncoherent MTI improvement factor can be shown to be given by

$$I_{nc} = 2\sum_{j=0}^{n-1} w_j^2 \Bigg/ \sum_{j=0}^{n-1}\sum_{k=0}^{n-1} w_j w_k \rho_c^2(j-k) \quad (3.13)$$

Table 3.2
Coherent and Noncoherent MTI Improvement Factors for Binomial Weights ($\sigma_c T \ll 1$)

n	*Coherent*	*Noncoherent*
2	$2/(\sigma_c T)^2$	$2/(\sigma_c T)^2$
3	$2/(\sigma_c T)^4$	$1/(\sigma_c T)^4$
4	$4/3(\sigma_c T)^6$	$1/3(\sigma_c T)^6$

Table 3.3
MTI Improvement Factor Scanning Limitation for Coherent and Noncoherent MTIs with Binomial Weights ($\sigma_c T \ll 1$)

n	*Coherent*	*Noncoherent*
2	$n_B^2/1.388$	$n_B^2/1.388$
3	$n_B^4/3.853$	$n_B^4/7.715$
4	$n_B^6/16$	$n_B^6/64.3$

where w denotes the MTI weights and $\rho_c(\cdot)$ is the clutter's correlation coefficient (the ρ_c^2 term results from spectral convolution in the frequency domain that translates to multiplication in the time domain) [10].

The identical improvement factor provided by the single canceller ($n = 2$) for both the coherent and noncoherent MTI might suggest that the detection performance of the two types is equivalent. However, the detection performance of the coherent MTI is known to be optimum (see Section 5.1.2); hence, the performance of the suboptimal noncoherent MTI must be lower than that of the coherent MTI. The explanation for this apparent inconsistency is that the improvement factor criterion is based on a second moment analysis that compares only signal-to-clutter power, ignoring the shape of the probability density functions involved under the alternate hypotheses. Hence, the translation of an MTI's improvement factor into a probability of detection is different for noncoherent and coherent MTIs, precluding a direct comparison of the two types on this basis. Unfortunately, the non-Gaussian nature of the probability distributions in the noncoherent MTI do not allow an analytical solution that represents the probability of detection in terms of the improvement factor.

These results for the MTI improvement factor of a noncoherent MTI primarily apply when a square-law envelope detector (the square of the envelope) is employed. An analysis of a noncoherent MTI employing a linear envelope detector indicates that the MTI improvement factor is reduced from that provided by a square-law envelope detector [12]. We can show that the clutter power residue from a single canceller noncoherent MTI using a linear envelope detector is given by $C_o = 4\sigma^2 [1 - F(\rho)]$, where $2\sigma^2$ is the input clutter power before envelope detection; $F(\rho) = (\pi/4)\,{}_2F_1(-1/2, -1/2, 1, \rho^2)$, where ${}_2F_1(\cdot, \cdot, \cdot, \cdot)$ is the Gaussian hypergeometric function; and ρ is the clutter correlation coefficient [12]. This generalizes for higher order MTIs so that the clutter power residue is given by $C_o = \mathbf{W}^T\mathbf{M}_c'\mathbf{W}^*$, where $\mathbf{M}_c'$ is the coherent MTI's covariance matrix ($\mathbf{M}_c$) with ρ replaced by $F(\rho)$. Use of this relationship in combination with a Monte Carlo simulation to determine the MTI's target responses provides the curves depicted in Figure 3.9 which show the noncoherent MTI's improvement factors for square-law and linear envelope detectors [12].

The curves depicted in Figure 3.9 show a deterioration of MTI improvement factor for noncoherent MTIs that use a linear envelope detector instead of a square-law envelope detector. The decrease in performance varies with the clutter's spectral spread and the order (number of delay lines) of the MTI filter employed. This is significant for noncoherent MTIs because many practical detectors operate in the square-law region for small signals and the linear region for large applied signals. As clutter signals are usually much larger than receiver noise, most practical detectors employed in noncoherent MTIs probably will work in the linear region.

(a)

(b)

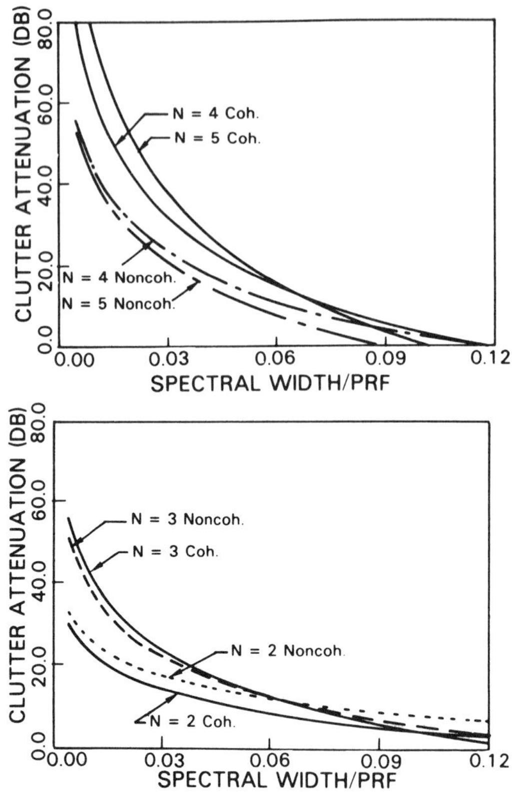

Figure 3.9 MTI improvement factors for nocoherent MTIs using linear and square-law envelope detectors.

3.1.2 The Cancellation Ratio

The *cancellation ratio* (CR) is a commonly used yardstick for measuring the instability noise generated within a doppler radar and its effect on overall system performance. The CR is defined as that amount of attenuation creditable to the doppler processor for truly fixed target returns when the latent amplifications (e.g., MTI gain) or losses have been normalized. The CR sets an upper bound on the improvement factor achievable by a particular doppler radar design.

The CR is sometimes determined by presenting a completely correlated signal to the doppler processor and measuring the power residue in the output. The input power is then divided by this residue, and the ratio, after accounting for the latent gain or loss, is the cancellation ratio. This measurement gives the CR for the doppler processor, but does not account for any noise introduced by the transmitter or receiver.

A more complete measurement involves searchlighting the radar onto a truly fixed target; but such a target is hard to find and great care must be exercised in performing the experiment so that the results will be meaningful. Many times by judiciously combining both the direct and indirect measurements, we can determine the CR with reasonable accuracy.

There are many possible causes of system instability in both the transmitter and receiver. These instabilities are due to the inevitable imperfections in installed and operating equipment that give rise to system instability noise. The most prominent sources of system instability noise are

1. pulse-time jitter,
2. pulsewidth jitter,
3. pulse amplitude jitter,
4. pulse distortion,
5. receiver phase and gain variations,
6. frequency modulation within the transmitted pulse,
7. transmitter, STALO, or COHO frequency drift,
8. COHO locking instability,
9. gating noise,
10. quantization noise present in digital systems.

Note that pulse-time jitter does not generally apply to doppler radars using digital processors where the system clock is started from the detected RF envelope of the transmitted pulse. Also, frequency modulation within the pulse and COHO locking instability generally apply to coherent-on-receive MTI systems (see Figure 2.21) and not MOPA systems (see Figure 2.20). In pulsed oscillator systems, the transmitter often exhibits frequency modulation (i.e., phase run-out) within the pulse, causing the COHO to lock to a frequency which differs from that of the leading edge of the transmitted pulse. Generally, system instabilities are related to any mechanism that induces pulse-to-pulse (i.e., relative to the radar's interpulse interval) variations onto fixed target signals presented to the doppler processor.

The CR limitation due to various peak pulse-to-pulse instabilities is listed in Table 3.4 [13]. In a pulse compression system, the compressed pulse width ($\tau_e = 1/B$) is reduced by the factor $\beta = B \cdot \tau$ from the uncompressed pulsewidth (τ), where B is the chirp bandwidth. The power residue in a pulse compression system for pulse timing and pulsewidth jitter is thus increased by this factor (β), which results in the CR being reduced by the time bandwidth factor $B\tau$. For simple pulsed systems, the time-bandwidth factor equals unity ($B\tau = 1$), which should be substituted into the expressions for the cancellation ratios for these types of systems. For phase-coded pulse compression systems (e.g., Barker code), the effective pulsewidth is the

Table 3.4
Cancellation Ratio Limitation Due to System Instabilities

Pulse-to-Pulse Instability	*Cancellation Ratio*
Transmitter Frequency	$(\pi \Delta f \tau)^{-2}$
STALO or COHO Frequency	$(2\pi \Delta f T_R)^{-2}$
Transmitter Phase Shift	$(\Delta\phi)^{-2}$
COHO Locking	$(\Delta\phi)^{-2}$
Pulse Timing	$\tau^2/[(\Delta t)^2 2B\tau]$
Pulsewidth	$\tau^2/[(\Delta\tau)^2 B\tau]$
Pulse Amplitude	$(A/\Delta A)^2$
Quantization Noise	$(3/4)(2^n - 1)^2$

uncompressed pulsewidth divided by the number of elements in the code (e.g., $\tau_e = \tau/13$ for a thirteen-element Barker code) and hence the number of elements of the code is substituted for $B\tau$ (e.g., $B\tau = 13$) in these types of systems. Other factors used in Table 3.4 are Δf = interpulse frequency change, T_r = transmission time to and from the target, $\Delta\phi$ = interpulse phase change, and n = number of bits utilized in the A/D converter. For random fluctuations, the peak values utilized in Table 3.4 can be replaced by their rms value to compute the averge CR limitation (e.g., $\sigma_\phi = \sqrt{2}\Delta\phi$, $\sigma_\tau = \sqrt{2}\Delta\tau$, *et cetera*) [10].

Dopper radars using digital signal processing are susceptible to quantization errors that occur in the system's analog-to-digital (A/D) converter. The A/D converter converts each sample to a binary number by comparing its voltage with a succession of progressively higher voltages of precisely known value. When the closest of these is found, the converter displays a binary number equal to the known voltage. Figure 3.10 illustrates the effect of applying a linearly changing voltage to the A/D converter whereby the output consists of the sum of two signals: a quantized replica of the original analog signal; and a triangular error wave having a peak amplitude equal to half the value of the *least significant digit* (LSD) or quantization step. For the uniformly distributed error depicted in Figure 3.10, the rms value of the quantization error voltage is given by $\text{LSD}/\sqrt{12}$.

In a coherent MTI, the A/D converter samples the bipolar video output of the MTI's synchronous detector (recall Figure 2.28). For an n-bit A/D converter, the LSD is given by $2v_p/(2^n - 1)$, where v_p is the peak signal level available from the MTI's synchronous detector. The rms value of the quantization noise is then given by $\sigma_q = 2v_p/(2^n - 1)\sqrt{12}$. If we assume that quantization noise behaves like thermal noise, the pulse-to-pulse power variation is given by $E(n_1 - n_2)^2 = 2\sigma_q^2$ resulting in

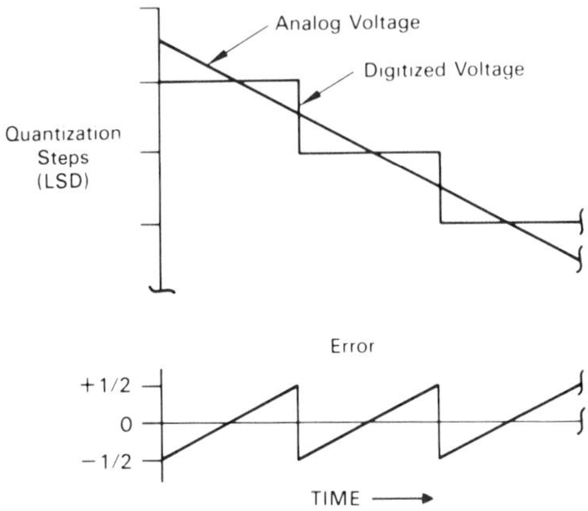

Figure 3.10 Quantization error for analog-to-digital converter.

an rms variation equal to $\sqrt{2}\sigma_q$. The fixed target signal out of the bipolar phase detector (synchronous detector) can be represented as $v_p \cos\phi$, where ϕ is the uniformly distributed phase difference between the target return signal and the MTI's coherent reference. The average fixed signal power from the synchronous detector is given by $E(v_p \cos\phi)^2 = v_p^2/2$ resulting in a rms value of $v_p/\sqrt{2}$. The resulting cancellation ratio can then be derived [13] as the fixed signal power divided by the residue quantization noise power, which provides

$$\mathrm{CR} = (v_p/\sqrt{2})^2/[2\sqrt{2}v_p/(2^n - 1)\sqrt{12}]^2 = 0.75(2^n - 1)^2 \qquad (3.13)$$

The quantization noise CR limitation is tabulated in Table 3.5 as a function of the number of bits (n) employed in the A/D converter.

From an analytical viewpoint, a convenient approach is to break the fixed target returns into two parts: an unperturbed part (s_t) that is not affected by system instabilities; and a perturbed part (n_t) caused by the system instabilities. The cancellation ratio is then given by

$$\mathrm{CR} = \int S_s(\omega)\, d\omega \Big/ \int \sum_{n=1}^{N} S_n(\omega)\, d\omega \qquad (3.14)$$

where $S_s(\omega)$ is the spectrum of the unperturbed part, $S_n(\omega)$ is the spectrum of each noiselike additive independent perturbation element, and N is the number of perturbation sources. Interchanging integration and summation results in $\mathrm{CR} = [1/\mathrm{CR}_1 + 1/\mathrm{CR}_2 + 1/\mathrm{CR}_n]^{-1}$, where CR_n is the cancellation ratio for each individual effect. This

Table 3.5
Cancellation Ratio Limitation for Quantization Noise

Number of Bits *(n)*	*Cancellation Ratio* *Limitation (dB)*
4	22.3
5	28.6
6	34.7
7	40.8
8	46.9
9	52.9
10	59.0
11	65.0
12	71.0

expression allows the overall CR to be determined from the CRs of each individual effect [9].

The effect of the CR on the overall improvement factor can be derived by assuming that the system instabilities (n_t) modulate the return signal. Because the instabilities are independent of the clutter returns, the *autocorrelation function* (ACF) of the received signal can be written as the product $R_{s+n}(\tau)\rho_c(\tau)$, where $R_{s+n}(\tau)$ is the ACF of the sum of the perturbed and unperturbed signal components and $\rho_c(\tau)$ is the normalized ACF of the clutter (see Section 3.1). The actual improvement factor is then given by

$$(I_{\text{actual}})^{-1} = \sum_{j=0}^{n-1}\sum_{k=0}^{n-1} w_j w_k \rho_c(j-k) R_{s+n}(j-k)/R_{s+n}(0) \sum_{j=0}^{n-1} w_j^2 \tag{3.15}$$

Substituting $R_{s+n}(j-k) = R_s(j-k) + R_n(j-k)$ and recognizing that $R_s(j-k)$ has a steady value $R_s(0)$ and $R_n(j-k)$ is a delta function with value $R_n(0)$ only at $j-k$ results in

$$(I_{\text{actual}})^{-1} = [R_s(0)/R_s(0) + R_n(0)][I_{\text{ideal}}^{-1} + R_N(0)/R_s(0)] \tag{3.16}$$

Further, recognizing that the $CR = R_s(0)/R_n(0)$ results in

$$(I_{\text{actual}})^{-1} = [\text{CR}/(1+\text{CR})](I_{\text{ideal}}^{-1} + \text{CR}^{-1})$$

which simplifies to

$$I_{\text{actual}}^{-1} = I_{\text{ideal}}^{-1} + \text{CR}^{-1}$$

for sufficiently large CR (15 dB or more). This relationship, which is plotted in Figure 3.11, indicates the effect of the CR on the ideal improvement factor (I_{ideal}), which is computed on the assumption that the instabilities have no effect on the clutter returns (i.e., CR $\rightarrow \infty$) [9].

As an example, consider a 3 GHz radar with a 100 nmi range that transmits a 2 μs pulse. We want to impose a 50 dB (316:1 voltage ratio) CR limitation on each instability component [13]. If a pulse oscillator is used, its peak pulse-to-pulse frequency change must be less than $\Delta f = 1/316\pi\tau = 504$ Hz, which is a stability of 2 parts in 10^7, and the COHO locking must be within $\Delta\phi = 1/316 = 3.16 \times 10^{-3} = 0.18°$. If a MOPA type of transmitter is used, the power amplifier must have a pulse-to-pulse phase change of less than $\Delta\phi = 1/316 = 0.18°$. The STALO or COHO pulse-to-pulse

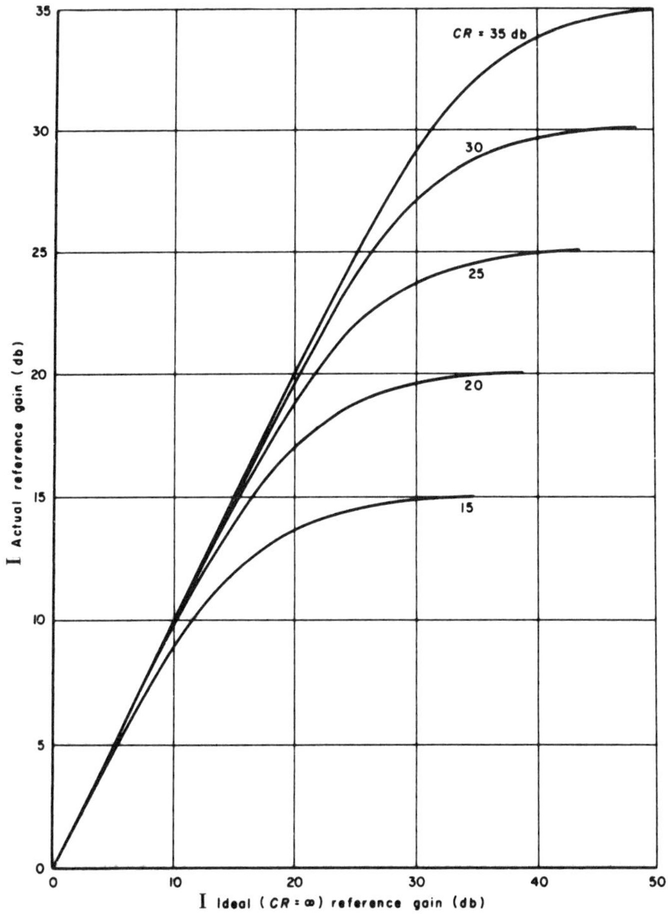

Figure 3.11 Improvement factor degradation due to system instability noise.

frequency change to provide performance at a 100 nmi range ($T_r = 12.4 \times 100 \times 10^{-6}$ s) must be less than $\Delta f = 1/632\pi T_r = 0.4$ Hz. This is short-term stability for the STALO of one part in 10^{10} and one part in 10^8 for the COHO assuming a 30 MHz COHO frequency. The pulse-timing jitter must be less than $\Delta t = \tau/316\sqrt{2} = 4.5 \times 10^{-9}$ s, the pulsewidth jitter less than $\Delta\tau = \tau/316 = 6 \times 10^{-9}$ s , and the pulse-to-pulse amplitude jitter less than $\Delta A/A = 1/316$ or 0.3%. For a digital processor, the A/D converter must use more than 9 bits.

3.2 CLUTTER EFFECTS ON DOPPLER RADAR PERFORMANCE

In the previous section, we discussed the general performance measures associated with doppler radars and how they would affect system performance. The improvement factor was identified as the primary performance measure utilized to evaluate doppler radars. For fully coherent systems followed by an envelope detector, the improvement factor is related directly to the probability of detection; otherwise, the improvement factor can be used as a measure to compare various alternative approaches. The improvement factor was shown to be composed of two components: one due to system instabilities, characterized as the cancellation ratio, and the other due to the effect of the doppler filtering on the clutter signals, termed the *ideal improvement factor* (I_{ideal}). This section deals primarily with the determination of I_{ideal}, which is concerned with spreading the clutter spectrum due to factors induced by the radar as well as those intrinsic to the clutter.

The characteristic fluctuation of any particular patch of clutter return arises from the combined influence of several statistically independent factors. First among these is the relative motion of the reflecting scatterers constituting the clutter mass. This "internal motion"—the extent of which generally depends on wind velocity and such diverse factors as sea state, turbulence, and individual scatterer mobility—accounts for the basic randomness of the return.

In addition to the internal motion fluctuations, the echo intensity induced by the scanning motion of the antenna with respect to the individual clutter scatterers also will vary. The severity of this scanning noise depends on the ratio of the scanning rate to the antenna's beamwidth, which determines the time that the antenna illuminates a particular scatterer. As this dwell time is reduced, the scanning noise becomes more severe. The scanning modulation clutter spectrum is truncated at a doppler frequency that corresponds to the tip speed of a mechanically scanning radar antenna. For a low-PRF ground-based surveillance radar, the ground clutter's spectral width is determined primarily by the scanning modulation component. The scanning modulation component can be significantly reduced by "step-scanning" the antenna, which is accomplished by freezing the antenna scanning motion (e.g., generally using an electronic scan) over the processing time of the doppler processor.

A third factor affecting the clutter statistics is the so-called platform motion noise. This applies to airborne or shipborne installations, where the radar platform undergoes a translational motion with respect to the clutter mass. Motion noise is caused by the differential radial velocities (referenced to the radial velocity along the antenna's boresight) induced onto various clutter scatterers illuminated by the radar's antenna beam. This results in an increased spread of the clutter power spectrum, the extent of which depends directly on the planar translation velocity of the antenna and inversely on the antenna size. Thus, for large platform velocities and small antennas pointed away from their ground track, the motion noise can be severe; and it is the most important single factor affecting MTI performance in high-speed *airborne intercept* (AI) radars. As with scanning modulation noise, platform motion noise can be reduced by freezing the antenna motion over the processing time of the doppler processor. Under certain circumstances this can be accomplished by electronically varying the phase center of the antenna (e.g., displaced phase center antenna), causing the antenna to appear stationary in space.

Another factor that tends to spread the clutter spectrum is the transmitter frequency instability. The first effect, which occurs during pulse-to-pulse transmitter frequency changes, is a partial decorrelation of the clutter returns similar to that which occurs by using frequency agility. This causes a pulse-to-pulse variation in the clutter signal at the output of the MTI's phase detector not cancelled in the doppler processor. The second effect is from the noise sidebands (called *phase* or *frequency noise*) in every practical transmitter. These sidebands, which occur in the vicinity of the transmitter carrier frequency, are reflected by each clutter scatterer. The net result is that the clutter return of other spreading effects (i.e., internal motion, scanning, platform motion) is convolved with a noisy carrier rather than a line spectra in forming the total clutter spectrum. This effect is most noticeable in high-PRF doppler radars due to the folding of the many ambiguous clutter cells that compete with the target, thereby increasing the required doppler processor improvement factor.

A general objective in doppler radar processor design is to maintain a linear response over the full dynamic range of the clutter and design the processor with an improvement factor sufficient to reduce the clutter residue to the receiver noise level. In many cases, the dynamic range of the clutter is very large (e.g., see Figure 1.23 where the clutter dynamic range exceeds 60 dB) and requires large improvement factors. One solution to this problem employed in older MTIs is to use a limiter in the IF portion of the receiver. The nonlinear action of the limiter generally spreads the clutter spectrum.

The additional clutter spectral components generated by the limiter are caused in part by the sharp discontinuities in the clutter envelope as the clutter reaches the limit level. This is depicted in Figure 3.12, which shows the voltage response of a single point scatterer illuminated by a scanning radar for limiting and nonlimiting situations. The response for the nonlimiting situation provides a relatively smooth signal, which can be successively cancelled (the figure shows single, double, and triple cancellers) in the MTI processor to provide a high degree of improvement factor. When a signal having a peak that exceeds the limit level by 20 dB is applied to the MTI canceller, the resulting residues through the cancellers are much higher

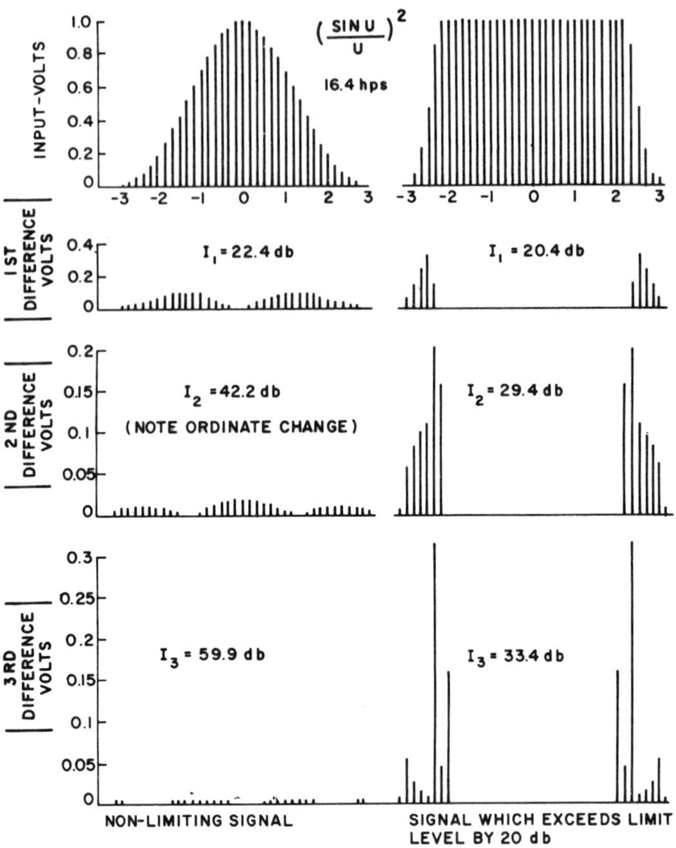

Figure 3.12 Improvement factor limitation caused by a limiter.

than in the nonlimited case, resulting in much lower improvement factors. In fact, in this case, additional stages of cancellation beyond a single canceller provide marginal additional improvement factor.

In low-PRF MTI radar designs, the doppler frequency response is generally highly ambiguous (see Figure 2.3), resulting in blind speeds at doppler frequencies corresponding to multiples of the radar's pulse repetition frequency. One way to resolve the blind speeds is to stagger the radar's PRF on a pulse-to-pulse basis. For example, if the periods of the staggered waveform have the relationship $n_1/T_1 = n_2/T_2 = \ldots n_N/T_N$, where $n_1, n_2, \ldots, n_N$ are integers, and if v_B is equal to the first blind speed of a nonstaggered waveform with a constant period equal to $T_{av} = (T_1 + T_2 + \ldots + T_N)$, the first blind speed is $v_1 = v_B(n_1 + n_2 + \ldots + n_N)/N$. However, when a staggered PRF is used, the improvement factor for a scanning radar is limited in proportion to the square of the number of hits per beamwdith (n_B). The scanning limitation is given by $I = [2.4n_B/(\gamma - 1)]^2$, where γ is the PRI stagger ratio [13].

3.2.1 Internal Motion Clutter Spectrum

The internal motion clutter spectrum is generally assumed to be Gaussian-shaped: $S(f) = P_c \exp(-f^2/2\sigma_f^2)/\sqrt{2\pi}\sigma_f$, where σ_f is the standard deviation of the clutter spectrum in Hz and P_c is the clutter power. The corresponding ACF is given by $\rho(\tau) = P_c \exp(-\tau^2/2\sigma_\tau^2)$, where $\sigma_\tau = 1/2\pi\sigma_f$. This can be related to the wind velocity through the relation $\sigma_f = 2\sigma_v/\lambda$, where σ_v is the standard deviation of the particle velocities about the mean velocity of the clutter mass. The improvement factor for a binary canceller is then given by

$$I_c = (\lambda f_r/\pi\sigma_v)^{2(n-1)}/[2^{3(n-1)} \cdot (n - 1)!] \quad (3.17)$$

where λ is the radar wavelength (m/s), f_r is the pulse repetition frequency (Hz), σ_v is the rms velocity spread (m/s), and n is the number of pulses processed. Table 3.6 gives some typical values measured for various clutter sources [9], and the improvement factor is plotted in Figure 3.13 for a double canceller.

An empirical relationship for the coherent spectral width (before envelope detection) of sea clutter based on a curve fit to experimental data is given by $\sigma_v = .051v_w$, where v_w is the wind velocity in knots and σ_v(m/s) is the rms velocity spread [14]. This relation must be multiplied by $\sqrt{2}$ for the noncoherent spectral width (after envelope detection). A similar empirical relationship for wooded terrain type of land clutter is given by $\sigma_v = 0.0033v_w^{1.26}$, where v_w is the wind speed in knots and σ_v (m/s) is the rms velocity spread [9,14].

Table 3.6
Apparent rms Velocity Spreads of Clutter Internal Motion

Description of Clutter	σ_v (m/s)
Sparsely Wooded Hills, Calm Day	0.005
Heavily Wooded Hills, 20 knot wind	0.3
Sea Clutter, Windy Day	0.9
Chaff	1
Rain Clouds	2

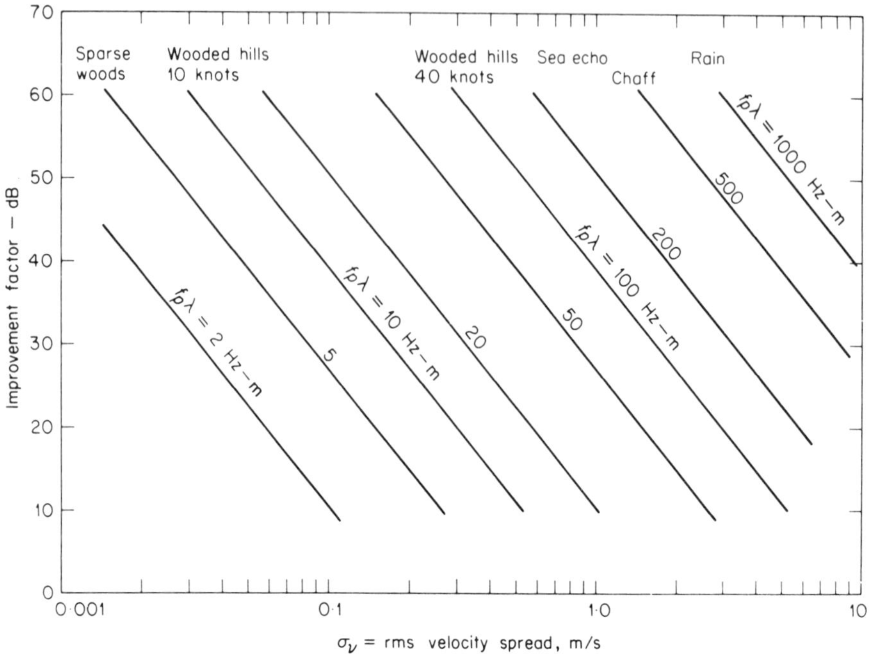

Figure 3.13 Improvement factor for internal motion limitation.

MTI systems that search for slowly moving ground targets (e.g., combat surveillance radars) employ slowly scanning or starring antennas. In this type of system, internal motion clutter usually determines the overall clutter spectrum shape. Experiments have indicated that the spectrum measured with this type of radar have a longer tail than would be predicted with a Gaussian-shaped spectrum. The measured spectrum has been matched by using a cubic dependence as $S(f) = P_o/(1 + |f/f_c|^3)$, where $P_o = 3\sqrt{3}P_c/4\pi f_c$, P_c is the clutter power, $f_c = 1.33 \exp(.1356 v_w)$, and v_w is the wind velocity in knots [9,15]. The measured power spectrum fitted with a cubic spectrum and a reference Gaussian-shaped spectrum is depicted in Figure 1.20. More recent experimental data (see Figure 1.21) indicates that an exponential-shaped spectrum may be appropriate.

The internal motion spectrum with a cubic dependence has significantly greater magnitude at high frequencies than the normally assumed Gaussian-shaped spectrum. When this effect dominates in determining the clutter shape, the MTI improvement factor will be poorer than that predicted by using a Gaussian-shaped spectrum. An example of a range-gated filter combat surveillance MTI used in a 20-knot wind resulted in a 7.5 dB degradation (i.e., from 28.5 dB to 21 dB) of the improvement factor for the cubic spectrum as compared to the Gaussian-shaped spectrum [9,15].

The decorrelating effect of wind-driven particle motion discussed earlier describes the effect of particle motion relative to the center of gravity of the clutter mass. For ground clutter, the center of gravity of the clutter mass will be fixed, regardless of the wind velocity. However, for airborne clutter, such as rain clouds and chaff, the center of gravity will drift at the wind velocity (v_w) with respect to the ground. If the velocity is steady, this drift will produce a doppler shift of the entire clutter spectrum away from the central ground-clutter frequency by an amount $f_w = 2v_w/\lambda$. The spectrum for Gaussian-shaped clutter is then given by $S(f) = P_c \exp[-(f - f_w)^2/2\sigma_f^2)/\sqrt{2\pi\sigma_f^2}$. This mean velocity for rain and chaff, as well as for sea clutter, must be taken into account when designing the MTI processor.

3.2.2 Antenna Scanning Clutter Spectrum

As the radar antenna scans by a point target, it induces an amplitude modulation onto the target return that has a shape corresponding to the antenna's two-way pattern, as depicted in Figure 3.14. If the antenna has a Gaussian-shaped pattern with standard deviation σ_θ, the voltage spectrum will be Gaussian-shaped with standard deviation $\sigma_w = \alpha/\sigma_\theta$, where α is the angular scan rate. If the clutter patch is assumed to be composed of many individual point scatterers, the resulting spectrum will be the sum of the spectra of the individual scatterers and have the same shape as that of the target.

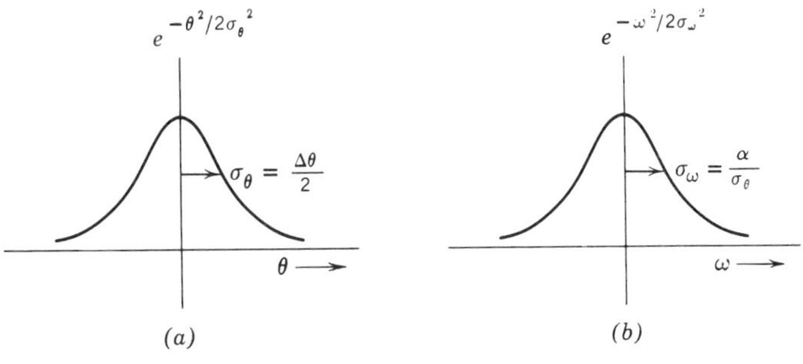

Figure 3.14 Gaussian antenna pattern and resultant Gaussian voltage spectrum.

A Gaussian one-way power antenna pattern is given by $G(\theta) = G_o \exp(-2.776\theta^2/\theta_B^2)$, where θ_B is the antenna beamwidth (in degrees). The standard deviation of this pattern is given by $\sigma_\theta = \theta_B/2.356$, which translates to a voltage spectrum standard deviation $\sigma_w = 2.356\alpha/\theta_B$. The power spectrum (voltage squared) has a standard deviation that is reduced by the factor $\sqrt{2}$ resulting in $\sigma_c = 1.666/t_o$, where $t_o = \theta_B/\alpha$ is the time-on-target and α (degrees per second) is the scan rate. Alternately, the time-on-target can be expressed as $t_o = n_B T$ ($\sigma_c T = 1.666/n_B$), where n_B is the number of hits per 3 dB antenna beamwidth and T is the interpulse period of the radar. The net result is that the Gaussian-shaped scanning modulation clutter spectrum is given by

$$S(f) = P_c \exp(-f^2/2\sigma_f^2)/\sqrt{2\pi}\sigma_f \tag{3.18}$$

where P_c is the clutter power and $\sigma_f = 1.666\alpha/2\pi\theta_B = 0.833\alpha/\pi\theta_B$ is the standard deviation of the power spectrum. The improvement factor for a binomial canceller is given by

$$I_c = 2^{n-1}(f_r\theta_B/1.666\alpha)^{2(n-1)}/(n-1)! \tag{3.19}$$

or alternatively as

$$I_c = 2^{n-1}(n_B/1.666)^{2(n-1)}/(n-1)! \tag{3.20}$$

where f_r is the pulse repetition frequency, θ_B is the antenna's beamwidth, α is the antenna's scan rate, n_B is the number of hits per 3 dB beamwidth, and n is the number of pulses processed by the MTI canceller.

The assumed Gaussian-shaped antenna pattern generally produces greater high-frequency components in the power spectral density than those associated with a real antenna pattern. This may be significant in analyzing high-performance MTIs, the improvement factor of which is limited by scanning modulation. To understand this effect, consider a uniformly illuminated antenna aperture scanning at constant speed that induces an amplitude modulation onto the target given by

$$v_t = V_m \sin^2(a_t/2)/(a_t/2)^2$$

where $a = 5.6\alpha/\theta_B$, α is the scan rate (rad/s), and θ_B is the antenna's beamwidth [9]. Fourier transforming and squaring to find the scanning modulation power spectrum results in

$$S(\omega) = V_m^2(1 - \omega/a)^2$$

for $\omega \leq a$ and zero for $\omega > a$. Thus, the spectrum is truncated at a frequency given by $f = 0.88\alpha/\theta_B$. Now, the 3 dB beamwidth of a rectangular aperture is given by $\theta_B = 0.88\lambda/D$, where D is the antenna's aperture width [16], resulting in $f = \alpha D/\lambda$, which is precisely the doppler frequency corresponding to the tip velocity of the antenna's aperture.

This analysis can be simplified by initially considering the antenna's illumination function rather than the antenna's beam pattern [9]. The power spectral density is then obtained by convolving and squaring the illumination function and scaling the frequency so that the truncation occurs at the tip velocity, given by $f = \alpha D/\lambda$. The voltage spectrum of the scanning clutter spectrum is obtained by using several standard illumination functions, as depicted in Figure 3.15.

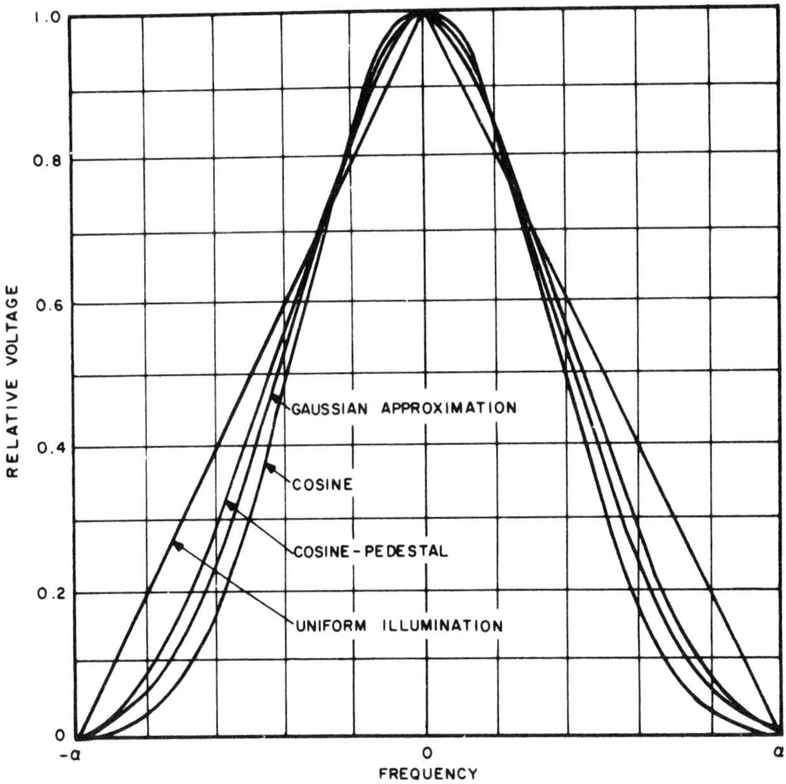

Figure 3.15 Voltage spectrum due to scanning antenna for various illumination functions.

3.2.3 Platform Motion Clutter Spectrum

If the radar platform is in motion, such as when the radar is installed on an aircraft or naval vessel, the radar motion induces a different velocity onto each clutter scatterer located within its field of view. In the low-PRF MTI case, the general clutter geometry is as depicted in Figure 3.16. The mean clutter velocity to the center of the ground clutter patch can be found by resolving the aircraft's doppler frequency $f_{ac} = 2v_{ac}/\lambda$ into a radial translational component given by $f_c = f_{ac} \cos\phi_0 \cos\theta_0$, where v_{ac} is the

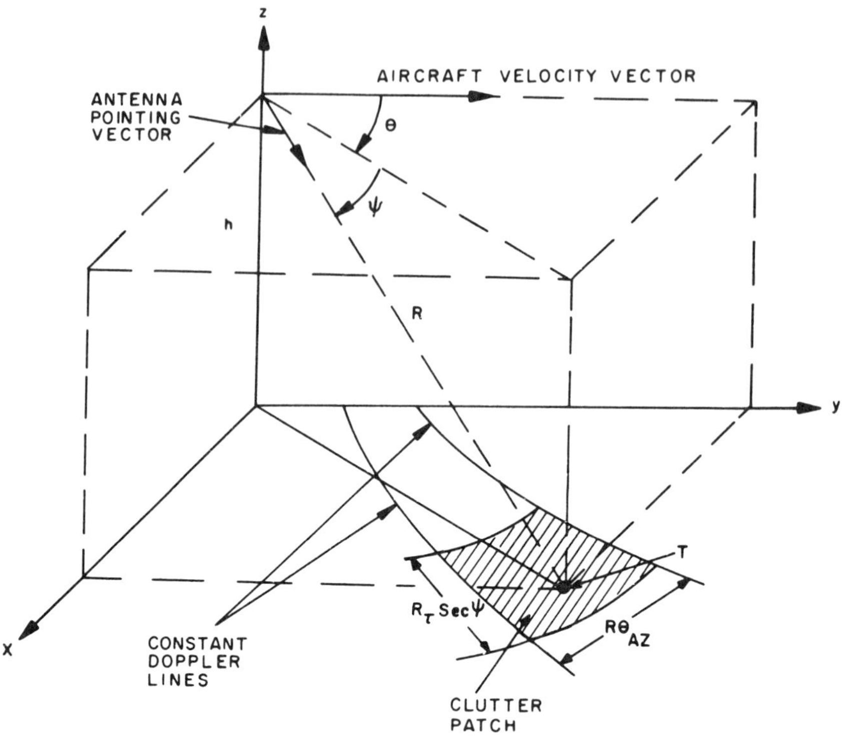

Figure 3.16 Airborne clutter geometry for MTI radar.

platform velocity, λ is the radar wavelength, θ_0 is the antenna's azimuth angle, and ϕ_0 is the antenna's depression angle. This frequency (f_c) represents the center or mean frequency of the ground clutter patch spectrum due to platform motion. The MTI must generally adjust its rejection notch to correspond to this mean frequency. One method for accomplishing this adjustment is a clutter-locking system called TACCAR (time-averaged clutter coherent airborne radar), which adapts the MTI frequency null of the MTI to the mean clutter frequency using a voltage-controlled oscillator (VCO) in a phase-locked loop (PLL). However, because the adjustment frequency is a function of both the azimuth angle and range to the clutter patch, compensation is never perfect; and this must be taken into account in analyzing this type of MTI (see Section 3.2.7). Note also that in a noncoherent MTI, the average component is eliminated by the envelope detector, which is an advantage for this type of MTI on a moving platform.

Examination of Figure 3.16 reveals that the platform motion spectrum will have a spectral width determined by the variations in both azimuth and depression angles.

For small depression angles, the mainlobe rms width is generally determined by the radar's azimuth beamwidth (θ_B), which, in turn, is proportional to the antenna's aperture width (e.g., $\theta_B = 0.88\lambda/D$). The spectral width due to the antenna's azimuth beamwidth is found by subtracting the radial doppler frequency components, which occur at the half-power edges of the azimuth beam as

$$\Delta f_0 = f_{ac} \cos\phi_0[\cos(\theta_0 + \theta_B/2) - \cos(\theta_0 - \theta_B/2)] \qquad (3.21)$$

For small azimuth beamwidths, this results in $\Delta f_\theta = f_{ac}\theta_B \cos\phi_0 \sin\theta_0$, where f_{ac} is the platform's doppler frequency, θ_B is the antenna's azimuth beamwidth (rad), ϕ_0 is the depression angle (assumed small), and θ_0 is the azimuth angle. For example, an X-band airborne radar ($f = 10$ GHz) moving at 300 knots with a 3° azimuth beamwidth would have a spectral width due to platform motion equal to 500 Hz at an azimuth angle of 90° and 250 Hz for an azimuth angle of 30°.

The preceding method for computing the doppler spectral width due to the finite azimuth beamwidth breaks down at $\theta_0 = 0$, where it predicts a zero spectral width. The spread at $\theta_0 = 0$ can be found by considering the difference between the mean doppler frequency (f_c) and the edge of the azimuth beam ($\theta_B/2$) as

$$\Delta f_\theta = f_{ac} \cos\phi_0[1 - \cos(\theta_B/2)] = f_{ac} \cos\phi_0\theta_B^2/8$$

for small azimuth beamwidths. For the previous example ($v = 300$ knots, $\theta_B = 3°$, $f = 10$ GHz), the spectral width at $\theta_0 = 0$ becomes a negligible 3 Hz.

A larger effect is due to the change of depression angle over the clutter patch when the depression angle is greater than zero. The magnitude of this effect can be derived in a way analogous to the method used in deriving the azimuth effect as $\Delta f_\phi = f_{ac}\Delta\phi \cos\theta_0 \sin\phi_0$, where $\Delta\phi$ is the depression angle spread over the clutter patch caused by the finite radar pulsewidth (see Figure 3.16). The value for $\Delta\phi = x/R$ ($x = R_\tau \tan\phi_0$ and $h = R \sin\phi_0$) can be found as $\Delta\phi = R_\tau \sin^3\phi_0/h \cos\phi_0$ (rad), where R_τ is the range equivalent to the radar's pulsewidth (e.g., 1 μs equals 500 ft), ϕ_0 is the depression angle, and h is the platform height. The resulting doppler spectral width due to the depression angle effect then becomes $\Delta f_\theta = f_{ac} \cos\theta_0 R_\tau \sin^3\phi_0/h \cos\phi_0$. At $\theta_0 = 0$ for the previous example ($v = 300$ knots, $f = 10$ GHz) with a pulsewidth of 1 μs and an altitude of 10,000 ft, the doppler spread becomes 72 Hz at a depression angle of 30° and only 2.6 Hz at a depression angle of 10°.

The azimuth and depression angle effects can be combined to form an overall spectral width (corresponding to antenna half-power points) relationship as

$$\Delta f = f_{ac}\{\theta_B \cos\phi_0 \sin\theta_0 + [(\theta_B^2 \cos\phi_0 \cos\theta_0)/8] + [(R_\tau \sin^3\phi_0 \cos\theta_0)/(h \cos\phi_0)]\}$$

where θ_B is the azimuth beamwidth, θ_0 is the azimuth angle, ϕ_0 is the depression angle, R_τ is the range equivalent of the radar's pulsewidth, and h is the platform altitude. For small depression angles, the relationship $\Delta f = f_{ac}\theta_B \sin\theta_0 = 1.76 v_{ac} \sin\theta_0/D$ is often appropriate, indicating that the spectrum spread is independent of frequency, whereas aircraft speed and antenna aperture size dictate the maximum clutter spread due to aircraft motion.

Another way of looking at platform motion noise is that the aircraft motion induces a form of scanning motion onto the clutter scatterers. This is illustrated in Figure 3.17, where to a moving observer (located on the aircraft) the clutter mass will appear to rotate about the point, 0, at an angular rate $\omega = v_{ac} \sin\theta_0/R$. Consequently, the clutter scatterer at p will exhibit a relative radial velocity for small beamwidths approximately equal to $v_p = \omega R = v_{ac}\theta \sin\theta_0$.

Each clutter scatterer within the range resolution cell defining the clutter patch will be illuminated by the one-way voltage antenna pattern. The return will then be weighted by the antenna's two-way voltage pattern or by the one-way antenna power pattern $G(\theta)$. The platform motion induces a doppler spectral distribution that is a function of the angular displacement from the antenna sighting direction (i.e., $f = 2v\theta \sin\theta_0/\lambda$). The resulting power spectral density is given by $G^2(\theta = \lambda f/2v \sin\theta_0)$ and has a shape corresponding to the two-way power antenna pattern. This relationship indicates that the two-way power antenna sidelobes are important because they weight frequency components that appear in the passband of the MTI (see Figure 2.5), which sets an upper limit on the performance of an airborne MTI.

For a Gaussian-shaped one-way power antenna pattern given by

$$G(\theta) = G_0 \exp(-2.776\theta^2/\theta_B^2)$$

the power spectral density is given by

$$S(f) = P_0 \exp[-5.552 f^2/(f_{ac}\theta_B \sin\theta_0 \cos\phi_0)^2]$$

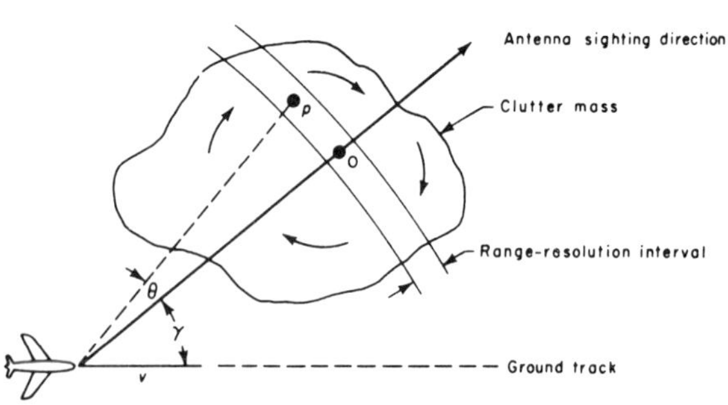

Figure 3.17 Relative rotation of clutter mass due to platform motion.

The platform motion power spectral density can then be written as

$$S(f) = P_c \exp(-f^2/2\sigma_{pm}^2)/\sqrt{2\pi}\sigma_{pm} \tag{3.22}$$

where

$$\sigma_{pm} = 0.3 f_{ac}\theta_B \sin\theta_0 \cos\phi_0 = 0.528 v_{ac} \sin\theta_0 \cos\phi_0/D$$

The improvement factor for an n-pulse coherent MTI binomial canceller with platform motion is given by

$$I_c = 2^{n-1}/1.885(f_{ac}T\theta_B \sin\theta_0 \cos\phi_0)^{2(n-1)}(n-1)!$$

which for a single canceller equals

$$I_c = 1/1.777(f_{ac}T\theta_B \sin\theta_0 \cos\phi_0)^2$$

and for a double canceller equals

$$I_c = 1/6.313(f_{ac}T\theta_B \sin\theta_0 \cos\phi_0)^4$$

For the previous example (f = 10 GHz, v_{ac} = 300 knots, θ_B = 3°, θ_0 = 30°, ϕ_0 = 0) at a PRF of 5000 Hz, the improvement factor for a single canceller would be 23.5 dB and 44 dB for a double canceller.

Because the antenna sidelobes essentially cover 360°, the doppler shifts received through the sidelobes range from $-2v_r/\lambda$ to $2v_r/\lambda$, where v_r is the platform velocity. In a low-PRF airborne radar, these sidelobe returns are aliased through the entire MTI ambiguous frequency band. As a result, these sidelobe returns are not cancelled, and the MTI improvement factor is limited by the integrated mainlobe-to-sidelobe ratio. This is the ratio of the integrated energy entering the system through the mainlobe to that entering through the sidelobes for the two-way antenna pattern. Thus, the integrated mainlobe-to-sidelobe ratio is a quality factor for the antenna used in this type of application [9].

Antennas are frequently specified on the basis of the amount of power radiated in their sidelobes compared to that radiated in their mainlobe (e.g., −10 dBi implies 10% power in the sidelobes). Ordinary antennas have sidelobes with 0 to −5 dBi rms sidelobes, low-sidelobe antennas have −5 to −20 dBi rms sidelobes, and ultralow-sidelobe antennas have rms sidelobes below −20 dBi. For example, a low-sidelobe antenna would provide an improvement factor limit from 10 to 40 dB in a low-PRF airborne MTI. Ultralow-sidelobe antennas are required to provide an improvement factor limit due to this effect of the order of 60 dB, which is required in high-performance AEW applications [9].

3.2.4 Transmitter Instability Spectrum Spreading

Pulse-to-pulse transmitter frequency instability causes two primary effects in a low-PRF coherent MTI system. First, for a fixed target, fluctuations occur at the output of the MTI's synchronous detector that are not cancelled by the MTI processor. This limitation is included in the cancellation ratio determination for the MTI (see Table 3.4). Note also that this effect is not a factor for a noncoherent MTI. The second effect is to cause a spectral spreading of the clutter spectrum due to the decorrelating effect of the pulse-to-pulse frequency changes on the clutter return, which is equivalent to the action of frequency agility on a distributed target.

To examine this effect, the clutter patch of Figure 3.18 is sliced in range at intervals separated by a radar wavelength (a 1 μs pulse at 10 GHz contains 10,000 slices). When the scatterers located in each slice are illuminated by the radar, they return in-phase signals. If the radar's transmitter frequency changes by an amount sufficient to cause a change of one in the number of wavelengths within one radar pulsewidth, the slices return a signal that contains all possible phases. Hence, the signal returns at the two frequencies are completely decorrelated.

We can show that the normalized autocorrelation function of the clutter return power for a rectangular pulse, which describes the correlation between return power P_1 at frequency f and return power P_2 for the same set of scatterers at frequency $f + \Delta f$, is given by $\rho = (\sin \pi\tau\Delta f / \pi\tau\Delta f)^2$, where τ is the radar's pulsewidth and Δf is the pulse-to-pulse frequency change [9]. The power spectral density can be found as $S(f) = P_0(1 - f/\Delta f)$, $|f| \leq \Delta f$. Matching a Gaussian spectrum at the half-power points results in $S(f) = P_c \exp(-f^2/2\sigma_i^2)/\sqrt{2\pi}\,\sigma_i$, where $\sigma_i = 0.424\Delta f$.

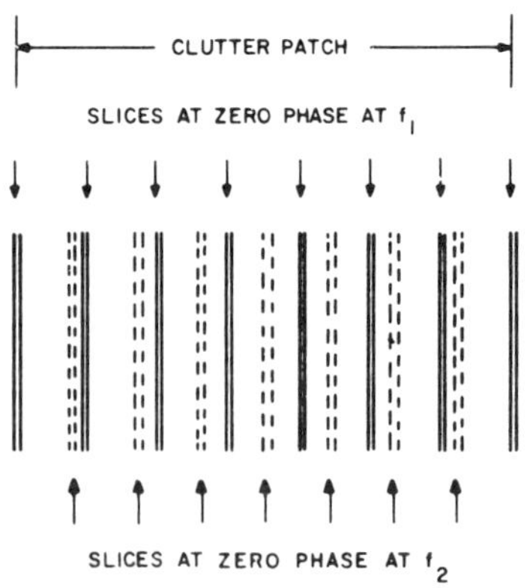

Figure 3.18 Phase of a clutter slice for frequencies f_1 and f_2.

3.2.5 Clutter Limiting Spectral Spreading

Modern radars generally attempt to maintain sufficient dynamic range to pass clutter peaks without saturation. When this is not possible (e.g., in a digital MTI processor because of limited A/D converter range), an IF limiter may be employed to keep the clutter's dynamic range within that of the MTI's receiver-processor. When an IF limiter is employed (as was prevalent in older MTI radars), the nonlinear action of the limiter spreads the clutter spectrum, causing a reduction in the MTI's improvement factor. Note that distortion of the clutter statistics by the limiter causes the optimum linear MTI processor dictated by the clutter's Gaussian statistics to be nonoptimum and theoretically a nonlinear processor provides greater improvement in limited clutter than the generally employed linear MTI processor.

A convenient way to analyze the spectral spreading effect of clutter limiting is to employ a soft limiter (which has an error function characteristic) because this provides an analytic expression for the limiter's autocorrelation function (and power spectral density) [9,17]. A block diagram of the IF limiter configuration is depicted in Figure 3.19, where the function of the bandpass filter is to eliminate harmonics of the IF caused by the nonlinear action of the limiter.

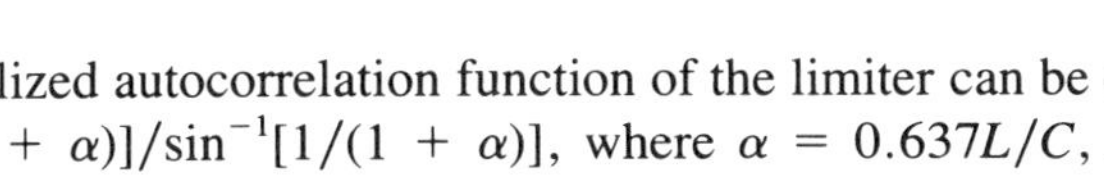

Figure 3.19 Block diagram of limiter and coherent MTI.

The normalized autocorrelation function of the limiter can be expressed as $\rho_l(\tau) = \sin^{-1}[\rho(\tau)/(1 + \alpha)]/\sin^{-1}[1/(1 + \alpha)]$, where $\alpha = 0.637L/C$, L is the clutter's power saturation point, C is the clutter power, and $\rho(\tau)$ is the normalized ACF of the clutter [9,17,18]. For hard limiting, $\alpha = 0$ and the clutter's autocorrelation function reduces to $\rho_l(\tau) = (2/\pi)\sin^{-1}[\rho(\tau)]$. The IF autocorrelation function of clutter with a Gaussian-shaped spectrum can be expressed as $\rho(\tau) = \exp[-(\sigma_c\tau)^2/2]\cos\omega_0\tau$, where σ_c(rad/s) is the standard deviation of the clutter spread, τ is the time separation between samples, and ω_0 is the IF radian frequency.

Expanding the arcsine function into a power series:

$$\sin^{-1}(x) = \sum_{p=0}^{\infty} a_p x^{2p+1}$$

allows the limiter ACF to be expressed as a power series given by

$$\rho_l(\tau) = k \sum_{p=0}^{\infty} C_p \exp[-(2p + 1)(\sigma_c\tau)^2/2]$$

where

$$C_p = a_p b_p/(1 + \alpha)^{2p+1}, \quad k = 1 \Big/ \sum_{p=0}^{\infty} C_p$$

and

$$a_p = 1 \cdot 3 \cdot \ldots (2p - 1)/2 \cdot 4 \cdot \ldots (2p)(2p + 1)$$

where $b_p = \binom{2p+1}{p}/2^{2p}$. The factor b_p results from expanding $\cos^n \omega_0 \tau$ and retaining only the $\cos\omega_0\tau$ terms that pass through the bandpass filter; for example, $\cos^3 x = (\cos 3x + 3 \cos x)/4 \rightarrow 3 \cos x/4$. The corresponding spectrum is a sum of Gaussian-shaped spectral densities given by $S(\omega) = P_0 \Sigma C_p \exp[-\omega^2/(2p + 1)2\sigma_c^2]$, where each term has a progressively larger frequency spread with standard deviation given by $\sqrt{2p + 1}\sigma_c$, but is diminished in amplitude by the coefficient of the power series (C_p).

The spectrum for limited Gaussian-shaped clutter is depicted in Figure 3.20 for various ratios of clutter power to limit power (C/L). The curve is plotted in terms of the prelimited 3 dB video bandwidth of the clutter spectrum, which is given by $B_3 = \sqrt{(2 \ln 2/\pi^2)}\sigma_c$. Note that, as the ratio of clutter power to limit level increases, the clutter spectrum is increasingly spread until hard limiting is achieved. The increased clutter spectral spread will cause additional clutter components in the MTI's passband, which, in turn, will cause a reduction in the MTI's improvement factor.

The improvement factor for limited Gaussian-shaped clutter can be found by substituting the limited normalized ACF into the improvement factor expressions for a binomial canceller given in Section 3.1; that is,

$$I_1 = 1/1 - \rho_l(T)$$

$$I_2 = 3/[(3 - 4\rho_l(T) + \rho_l(2T)]$$

$$I_3 = 10/[10 - 15\rho_l(T) + 6\rho_l(2T) - \rho_l(3T)]$$

For example, the single canceller improvement factor is given by

$$I_1 = \sum_{p=0}^{\infty} C_p \Big/ \sum_{p=0}^{\infty} C_p\{1 - \exp[-(2p + 1)(\sigma_c T)^2/2]\}$$

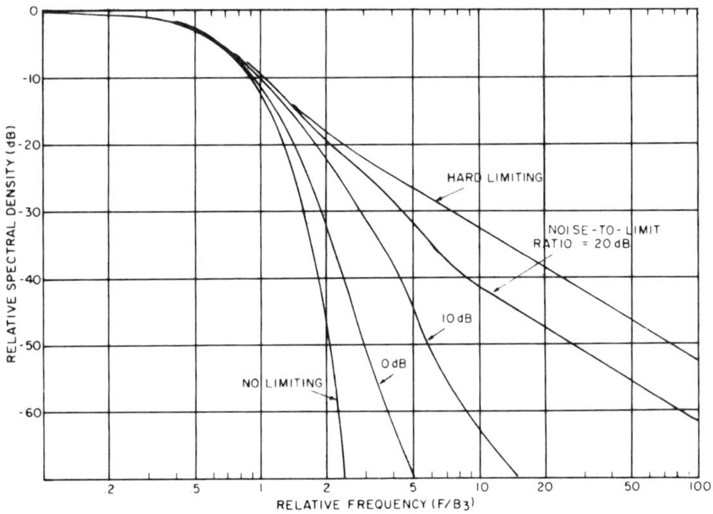

Figure 3.20 Power spectral density for limited Gaussian-shaped clutter.

and

$$I_2 = \sum_{p=0}^{\infty} 3C_p \Big/ \sum_{p=0}^{\infty} C_p\{3 - 4\exp[-(2p+1)(\sigma_c T)^2/2] + \exp[-2(2p+1)(\sigma_c T)^2]\}$$

is the double canceller improvement factor.

As an example, consider a single canceller with hard limiting. The improvement factor is given by

$$I_1 = \sum_{p=0}^{\infty} a_p b_p \Big/ \sum_{p=0}^{\infty} a_p b_p\{1 - \exp[-(2p+1)(\sigma_c T)^2/2]\}$$

which for sufficiently small $\sigma_c T$ can be written as

$$I_1 = [2/(\sigma_c T)^2]\left[\sum_{p=0}^{\infty} a_p b_p \Big/ \sum_{p=0}^{\infty} a_p b_p (2p+1)\right]$$

The first term can be recognized as the improvement factor without limiting, and the second factor can be identified as an asymptotic loss associated with the hard limiter. In this case, $\Sigma_{p=0}^{\infty} a_p b_p = 1.273$ and $\Sigma_{p=0}^{\infty} a_p b_p (2p + 1) = 6.722$, resulting in a loss of 7.2 dB. The hard limited improvement factor for $n = 100$ pulses would then be the nonlimited improvement factor $I_1 = 10 \log(100^2/1.388) = 38.6$ dB diminished by the limiter loss (7.2 dB), resulting in $I_{l1} = 31.4$ dB.

Figure 3.21 depicts the improvement factor achieved with various levels of limiting for single, double, and triple cancellers [17]. In practice, a limiter will be set at a level that equalizes the clutter residuals to the receiver noise level. A typical situation is shown in Figure 3.22, which compares improvement factors for single, double, and triple cancellers for clutter 20 dB above the limit level (a situation common to most clutter areas) and for hard limiting. Note that very little advantage in improvement factor is gained by using higher order cancellers (i.e., triple and greater) as hard limiting is approached. In this situation, for typical MTI search radars employing 10 to 20 hits per scan, practical improvement factors of 20–25 dB with hard limiting appear achievable using a double canceller. The result is that targets are suppressed

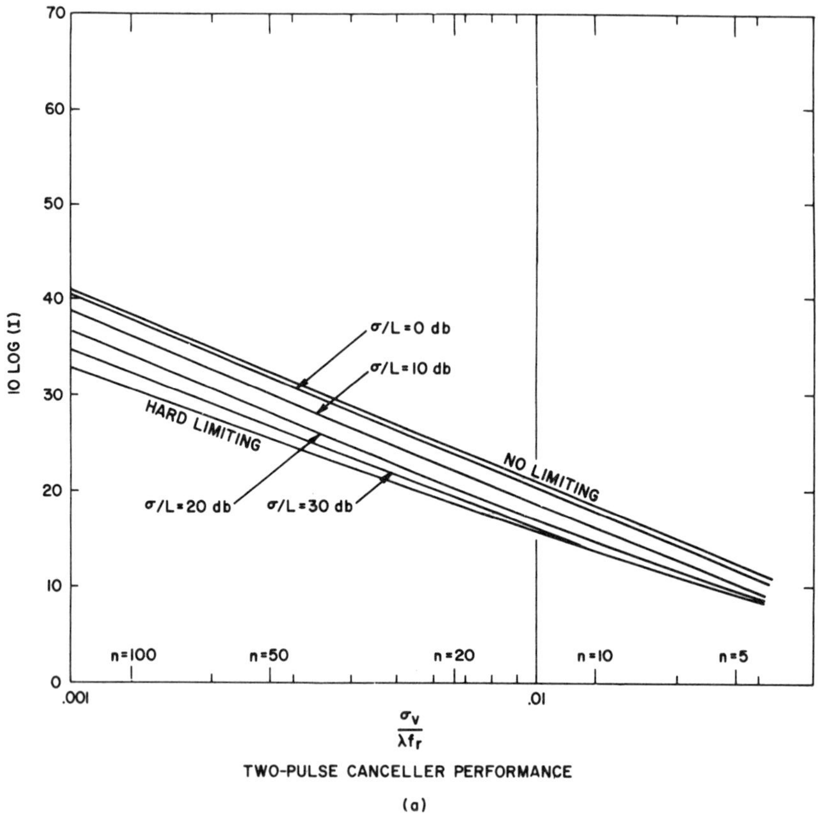

Figure 3.21 MTI improvement factor for single, double, and triple cancellers for various clutter power to limit levels.

in those resolution cells containing strong clutter, which require heavy limiting. In other resolution cells, having weaker clutter (which is less than 20 dB above the limit level), targets will frequently be detected. This property, called *interclutter visibility,* points to the desirability of using moderately small spatial resolution cells in the radar to take advantage of the spotty nature of land clutter. This mode of operation allows MTI radars with limiting to achieve reasonable functional performance in many practical situations.

Figure 3.23 compares the improvement factor available from an ASR-7 radar for single and double cancellers with and without limiting [9,19]. An improvement factor of 46 dB is required to detect a 1 m^2 target (small target) at 15 nmi in moderately heavy ground clutter. As is obvious from the curve, this degree of performance is not available from an MTI that uses limiting to control the clutter's dynamic range.

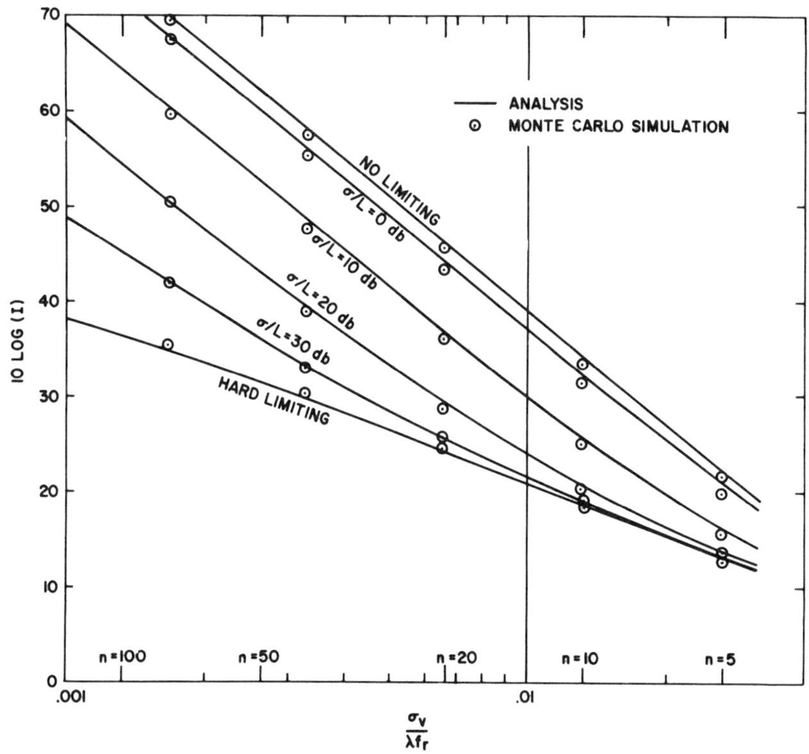

THREE-PULSE CANCELLER PERFORMANCE

(b)

Figure 3.21 continued.

The solution applied in the ASR-9 radar is to use a ground clutter map that measures the average clutter strength in each radar resolution cell. The measured clutter strength is then used to set a threshold that controls the false alarm rate to the specified value while allowing the MTI processor to operate in a linear mode. This provides the maximum MTI capability available from the radar and, in addition, provides superclutter visibility in those cells containing targets stronger than the clutter background.

The ground clutter map is a form of CFAR that provides operation in ground clutter (rather than receiver noise) and allows for the spotty nature of ground clutter where appreciable variations occur from one resolution cell to the next. The ground clutter map threshold approach tends to censor those resolution cells containing ground clutter with a dynamic range that exceeds the improvement factor capability of the MTI radar. Note that the ground clutter map approach is most appropriate to clutter

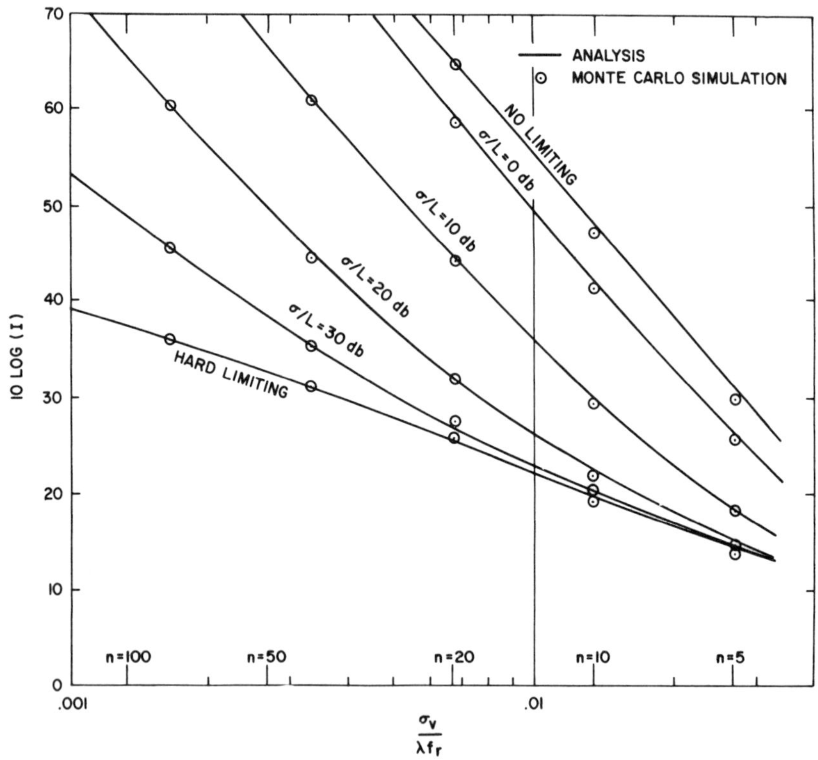

FOUR-PULSE CANCELLER PERFORMANCE

(c)

Figure 3.21 continued.

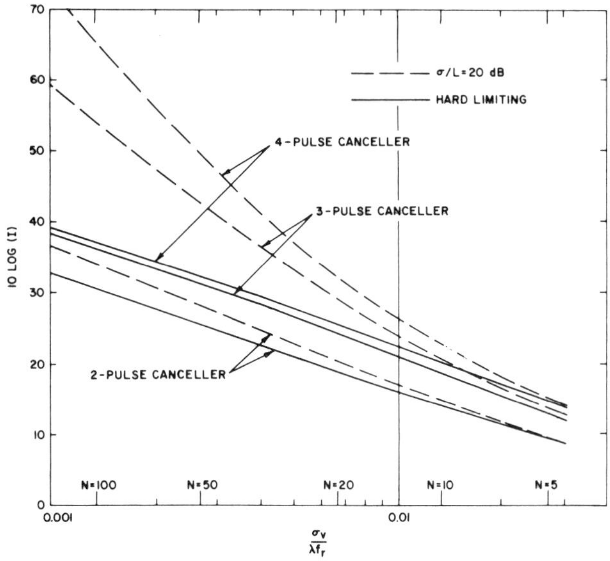

Figure 3.22 Comparison of single, double, and triple cancellers in limiting clutter.

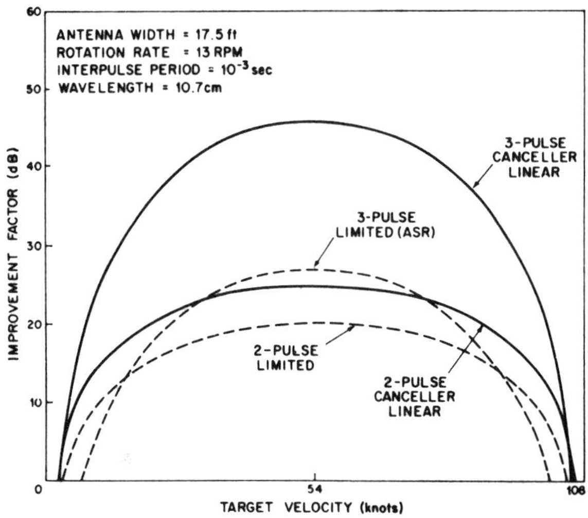

Figure 3.23 Improvement factors available with limiting and linear operation from an airport surveillance radar.

sources having a character that changes slowly. It is ideal for ground clutter from a stationary radar, but the applicability of this approach becomes questionable against moving clutter (e.g., weather or chaff) or from a moving platform (e.g., airborne radar).

3.2.6 Improvement Factor Restriction Caused by PRF Staggering

PRF staggering is used in low-PRF MTI radars to detect targets at blind speeds associated with targets having doppler velocities ($f_d = 2v/\lambda$) that fall into the stopband of the MTI filter (recall Figures 2.4 and 2.5). The periodic nature of the MTI comb filter creates blind speeds when f_d equals or is a harmonic of the PRF. Staggering the interval between pulses makes possible the alteration of the MTI filter's passband shape and elimination of blind speeds below some very high speed.

Figure 3.24 depicts a common method used to generate a staggered PRF: a short (relative to the average interpulse period) delay line is switched in and out of

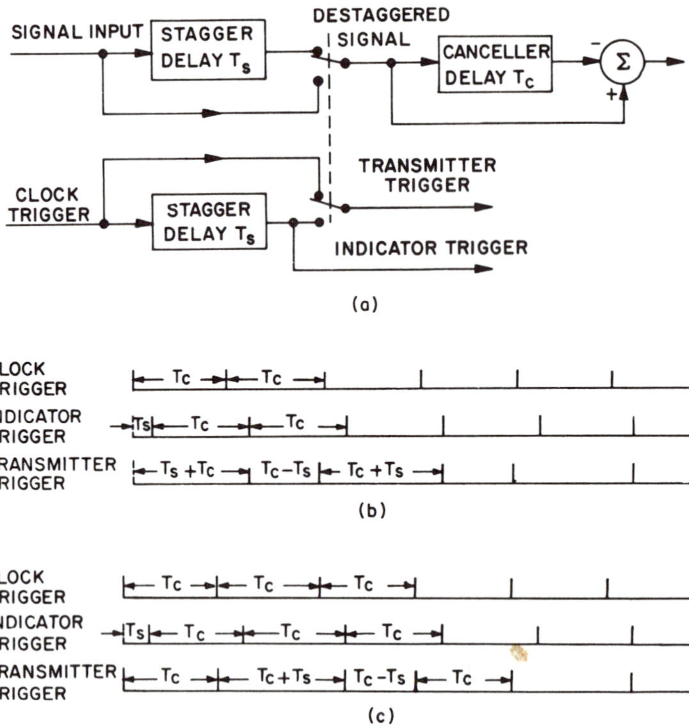

Figure 3.24 Block diagram and waveforms for two- and three-period staggered PRFs.

the signal path on alternate pulses. The transmitter trigger is also put through the same delay line on alternate pulses, causing the transmitter to be pulsed at the staggered rate. The stagger ratio is given by $\gamma = T_c + T_s/T_c - T_s$, where T_c is the average interpulse period and T_s is the short differential period. Note that the destaggered signal comes out at a constant rate and thus can be applied to a conventional canceller (e.g., single, double, or triple) with no modification to the canceller.

The PRF staggering modifies the passband shape of the MTI filter (relative to the nonstaggered shape), which causes a reduction in the MTI's improvement factor. To consider this effect, the two interpulse periods (see Figure 3.24) are written as $T_1 = (1 + e)T$ and $T_2 = (1 - e)T$. Applying this to a double canceller MTI results in an impulse response given by $h(t) = \delta[t + T(1 - e)] - 2\delta(t) + \delta[t - T(1 + e)]$. This can be Fourier transformed to obtain the MTI's voltge transfer function as $H(\omega) = \exp[j\omega T(1 - e)] - 2 + \exp[-j\omega T(1 + e)]$. The staggered power transfer function is then given by

$$|H(\omega)|^2 = 4[1 - 2\cos(\omega Te)\cos(\omega T) + \cos^2(\omega T)] \tag{3.23}$$

which, for $e = 0$, reduces to the standard double canceller response given by

$$|H(\omega)|^2 = 16\sin^4(\omega T/2)$$

which is periodic $\omega = 2\pi/T$.

The staggered power transfer function is periodic for that value of ω that $\cos(\omega T)$ and $\cos(\omega Te)$ are both $+1$ or -1. For $e \ll 1$, the approximate depth of the first null relative to the peak is then given by $|H(2\pi/T)|^2/|H_{\max}|^2 = 8[1 - \cos(2\pi e)]/16 = (\pi e)^2$, or, equivalently $[\pi(\gamma - 1)]^2/4$, where γ is the stagger ratio.

The double canceller improvement factor with PRF staggering can be found by first substituting the approximation ($e \ll 1$) $\cos(\omega\tau e) = 1 - (\omega\tau e)^2/2$ into the power transfer function, resulting in $|H(\omega)|^2 = 16\sin^4(\omega T)/2 + 4(\omega Te)^2$. Next, the residual clutter power after MTI filtering for the second term is found from $\int_{-\infty}^{\infty}|H(\omega)|^2 S(\omega)\, d\omega$ as $4(\sigma_c Te)^2$, where $S(\omega)$ is the Gaussian-shaped clutter distribution. The improvement factor is given by (see Section 3.1)

$$I = |H(\omega)|^2 \int_{-\infty}^{\infty} S(\omega)\, d\omega \Big/ \int_{-\infty}^{\infty} |H(\omega)|^2 S(\omega)\, d\omega$$

This results in $I_s = I_2/[1 + I_2(\sigma_c Te)^2/4]$, which for $I_2 \gg 1$ results in $I_s = 4/(\sigma_c Te)$. Substituting $e = (\gamma - 1)/2$ and $n_B = \sigma_c T/1.666$ yields $I_s = (2.4 n_B/\gamma - 1)^2$ as the MTI improvement factor limitation for PRF staggering in scanning clutter, where n_B is the number of hits per 3 dB beamwidth and γ is the stagger ratio. This relationship is depicted in Figure 3.25. For internal motion clutter, $\sigma_c T = 4\pi\sigma_v/\lambda$ (see Section

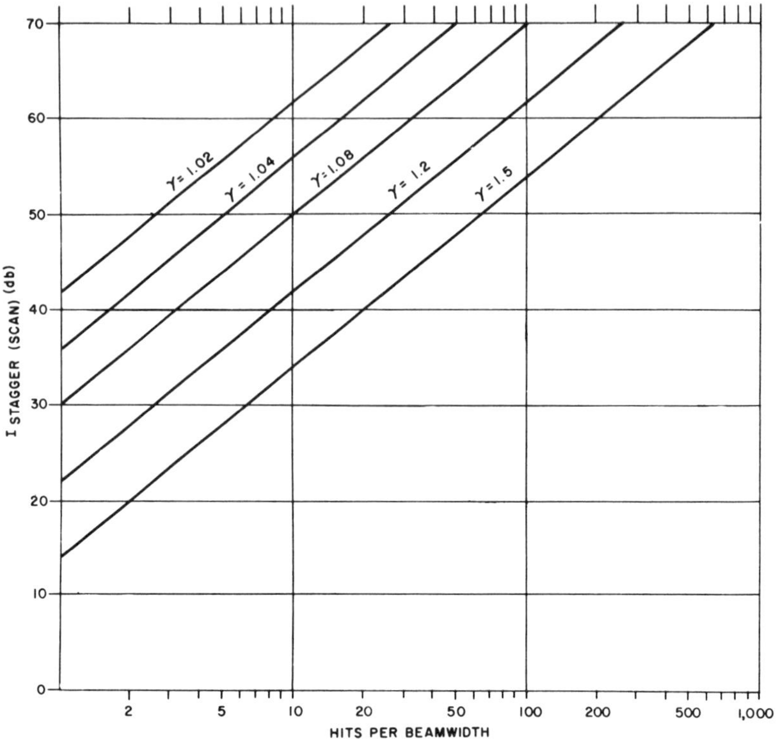

Figure 3.25 MTI improvement factor restriction due to PRF staggering and scanning.

3.2.1) and the PRF staggering limitation becomes $I_s = (0.32/\gamma - 1)^2(\lambda f_r/\sigma_v)^2$, where f_r is the pulse repetition frequency and σ_v is the rms velocity spread of the clutter (see Section 3.2.1). This relationship is depicted in Figure 3.26. The MTI improvement factor limitations due to PRF staggering tend to apply independently of the type of canceller employed following the staggering network.

Figure 3.27 illustrates the frequency response of an MTI used in a long-range air traffic control radar with a four-period stagger with ratio 25:30:27:31 [13]. The first blind speed is (25 + 30 + 27 + 31)/4 = 28.25 times that of a constant PRF waveform with the same average period. A disadvantage of the staggered PRF is its inability to cancel second-time-around and higher clutter echoes. Such clutter does not appear at the same range from pulse to pulse and thus produces uncancelled residues.

The problem of uncancelled multiple-time-around echoes in PRF staggering can be avoided by using a coherent PRF-diversity MTI, where the pulse repetition interval (PRI) is held constant until echoes from the longest range clutter are received

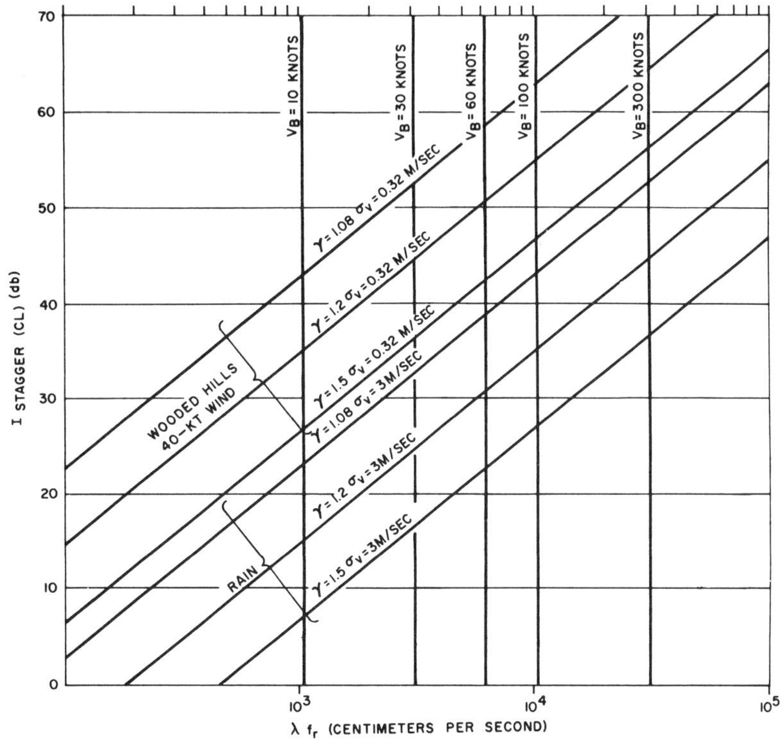

Figure 3.26 MTI improvement factor restriction due to PRF staggering and internal clutter motion.

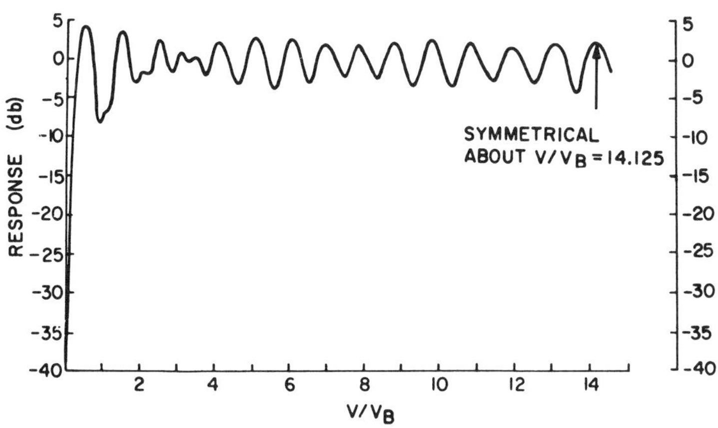

Figure 3.27 Frequency response of a four-period PRF stagger (25:30:27:31).

and processed through the coherent MTI processor. The PRI is then changed on a burst-to-burst basis. After each constant-PRF burst is batch processed in the MTI canceller, the output must be blanked until the first of the longest-range echoes from the next burst has passed the cancellers. Generally, each burst is processed coherently, followed by envelope detection, and then noncoherently integrated.

3.2.7 Improvement Factor Restriction Due to Motion Compensation Losses—TACCAR and DPCA

When a coherent MTI or PD radar is operated from a moving platform (e.g., airborne or naval radar), a doppler shift is induced onto the clutter by the radar motion. This situation is depicted in Figure 3.16 for an airborne radar in which the center of the ground clutter patch has an induced doppler shift equal to $f_c = f_{ac} \cos\phi_0 \cos\theta_0$, where $f_{ac} = 2v_{ac}/\lambda$ is the doppler due to the platform velocity, ϕ_0 is the antenna's depression angle, and θ_0 is the antenna's azimuth angle. In general, the mean doppler shift to each clutter patch will be different (as either θ_0 or ϕ_0 will be different). Therefore, the clutter rejection notch needed to cancel clutter for an airborne or naval radar cannot be fixed but must vary. This problem is more difficult for an airborne radar than a shipborne radar because of the higher speeds and the greater range of depression angles, which results in a greater variation of the clutter spectrum.

Several methods are available to compensate the coherent doppler processor for the mean doppler shift induced by platform motion. One method, which is most appropriate against ground clutter for a high-PRF doppler radar, uses the platform's inertial navigation system, which in conjunction with knowledge of the antenna's pointing direction provides sufficient information to appropriately shift the doppler reference in the radar's processor (recall Figure 2.10). However, most doppler processors employ an adaptive type of platform motion compensation that adjusts the rejection null of the frequency response of the doppler processor to the mean doppler frequency of the clutter spectrum. One implementation of this method adjusts the frequency of the coherent oscillator to compensate for the shift in the clutter doppler frequency. This may be accomplished by mixing the output of the COHO with a signal from a tunable voltage controlled oscillator to compensate for the shift in the clutter doppler frequency [9,20]. Another implementation inserts a phase shifter in one branch of an IF vector MTI canceller (I and Q channels) and adjusts its phase to shift the null of the frequency response (a phase shift in one branch corresponds to a frequency shift $2\pi f_d = \theta/T$, where T is the PRI) [21]. However, regardless of the motion compensation employed, there is generally an error in the adjustment of the doppler processor null that restricts the improvement factor for these types of systems.

Figure 3.28 depicts the situation where the null of the doppler filter is adjusted to a frequency that differs from the mean of the clutter spectrum. Figure 3.29 shows

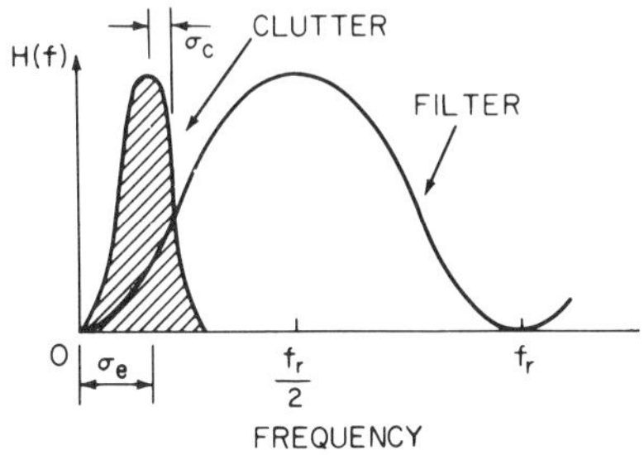

Figure 3.28 Effect of doppler offset error on an MTI clutter filter.

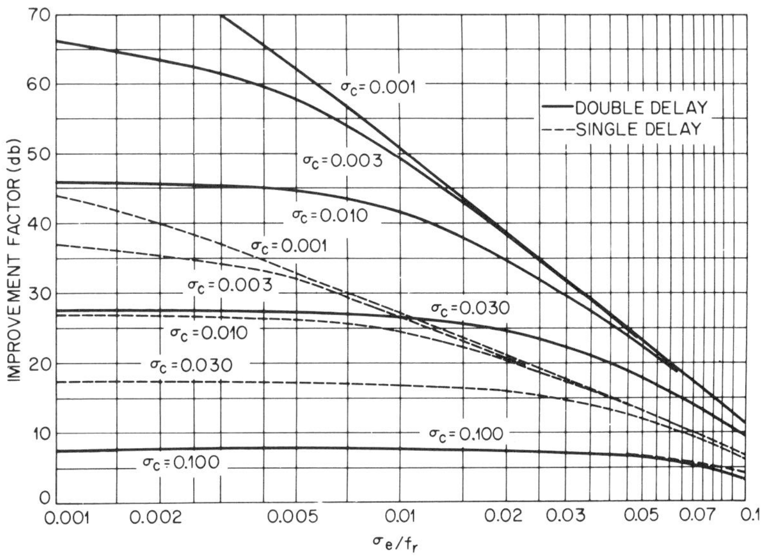

Figure 3.29 Improvement factor *versus* normalized doppler offset as a function of clutter spectrum width.

the resulting improvement factor for a single and double canceller as a function of the error frequency-to-PRF ratio (f_e/f_r) for a Gaussian-shaped clutter spectrum with various clutter spectrum standard deviations (σ_c). The improvement factor degrades as the clutter spectrum extends into more of the filter's passband, which occurs as the frequency offset (f_e) becomes a larger fraction of the PRF (i.e., f_e/f_r increases) and also as the clutter spectral width becomes a larger fraction of the PRF (i.e., σ_c/f_r increases). In an airborne radar, the clutter spread (σ_c) due to platform motion is minimal when the radar's antenna points along the aircraft's flight path (see Section

3.2.3), whereas the offset error (f_e) is generally maximal under this condition, resulting in partially counterbalancing effects.

Time average clutter coherent airborne radar (TACCAR) is a motion compensation technique used widely in both airborne and shipborne radars. A block diagram of one implementation of this technique is shown in Figure 3.30 [20]. The processing estimates the average doppler shift of the returns from all objects illuminated by the radar's antenna and within some range sampling interval. This average doppler is removed by shifting the relative difference frequency between the transmitter and the local oscillator.

The basic mechanism employed in TACCAR is a gated phase-locked loop, which is used to correct for the average doppler shift during a selected interval of time as shown in Figure 3.30. For TACCAR, the phase-locked loop is gated on at a time interval that corresponds to the selected range interval, which contains the clutter to be used as the reference. The integrator and sample-and-hold network provides an estimate of the average doppler within this selected range interval. This estimate is repeated during each pulse repetition interval. The time constant of the TACCAR phase-locked loop must be large compared to the interpulse period to maintain the pulse-to-pulse coherency required for successful doppler processing.

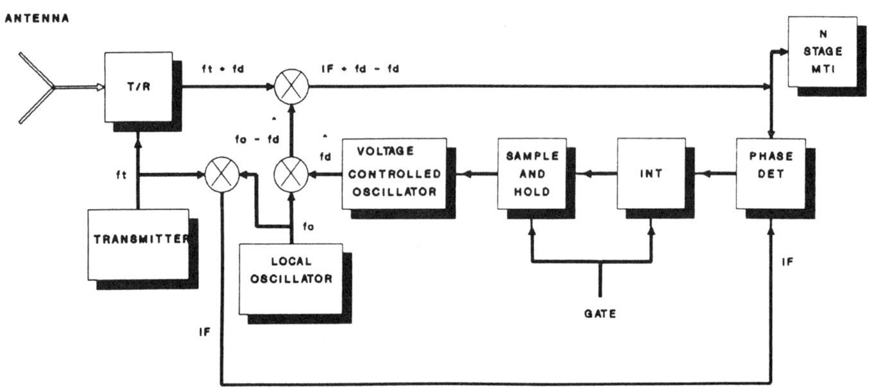

Figure 3.30 TACCAR block diagram.

When TACCAR samples the clutter return from a particular range interval, the doppler processor is compensated for the averge doppler frequency within this interval. Because the depression angle generally varies over the samples interval, this average doppler frequency estimate is generally in error for every clutter patch. A performance evaluation of TACCAR must therefore consider the variation of the frequency error (f_e) of this estimate with range.

In the design of the usual TACCAR loop, a single sample range interval is selected, over which the average doppler frequency ($\bar{f}_d$) is developed. Because this estimate does not provide an optimum correction at all ranges, the problem is to select a range interval that will minimize the peak error over this interval. Defining the range extremes that bracket the estimation interval as R_{min} and R_{max}, which have corresponding depression angles of ϕ_{min} and ϕ_{max}, results in a clutter doppler frequency estimate of

$$\hat{f}_d = f_{ac}\cos\theta_a(\cos\phi_{max} + \cos\phi_{min})/2$$

where f_{ac} is the radar's doppler frequency and θ_a is the antenna's azimuth pointing angle. The TACCAR range error is then given by $f_e = \hat{f}_d - \bar{f}_d$, which is equal to $f_e = f_{ac}\cos\theta_a\,[\cos\phi_R - (\cos\phi_{max} + \cos\phi_{min})/2]$, where ϕ_R is the depression angle to each particular clutter cell. The peak error occurs at $\phi_R = \phi_{max}$ (see Figure 3.31), which results in a normalized peak error of $\hat{e}_n = f_e/f_{ac}\cos\theta_a = (\cos\phi_{max} - \cos\phi_{min})/2$. The normalized TACCAR range error is depicted in Figure 3.32 as a function of the range normalized to the horizon range (R_{max}). In Figure 3.32, $R_{max} = \sqrt{h^2 + 2hr}$ with $R_{min} = R_{max}/10$, where h is the aircraft's height and r equals the 4/3 radius of the earth, approximately 4600 nmi.

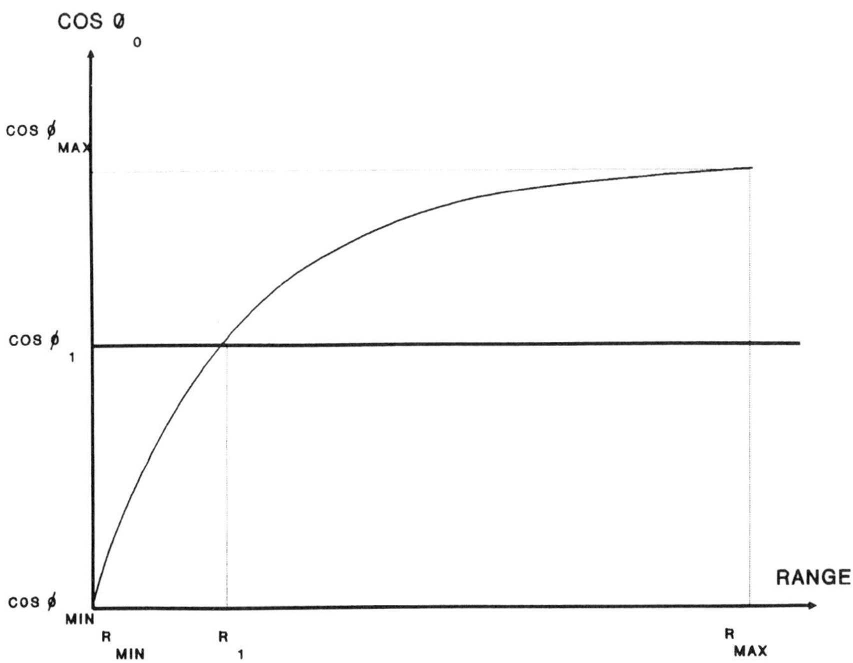

Figure 3.31 TACCAR range interval estimates.

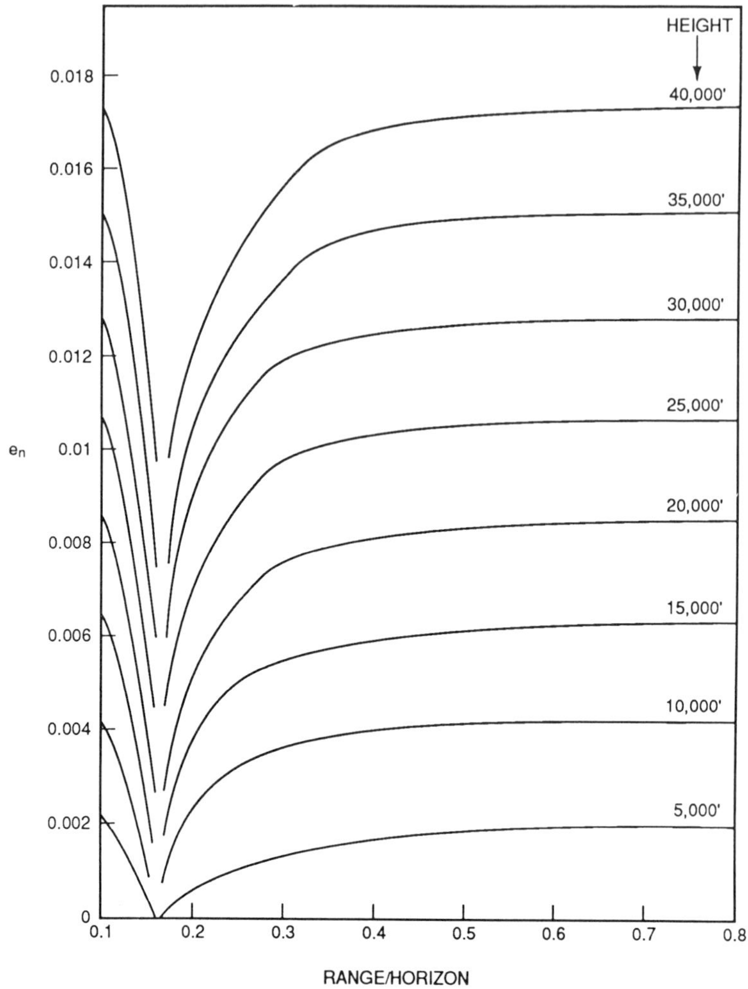

Figure 3.32 Normalized TACCAR error.

The TACCAR loss can be derived as

$$L_n = (f_e/\sigma_c)^{2N} + \sum_{k=1}^{N} \binom{2N}{2k} (f_e/\sigma_c)^{2(N-k)} 1 \cdot 3 \ldots 2k - 1)/(1 \cdot 3 \ldots N - 1) \quad (3.24)$$

where N is the number of delay sections, f_e is the frequency offset, and σ_c is the rms clutter width [20]. This function is plotted in Figure 3.33 for a single, double,

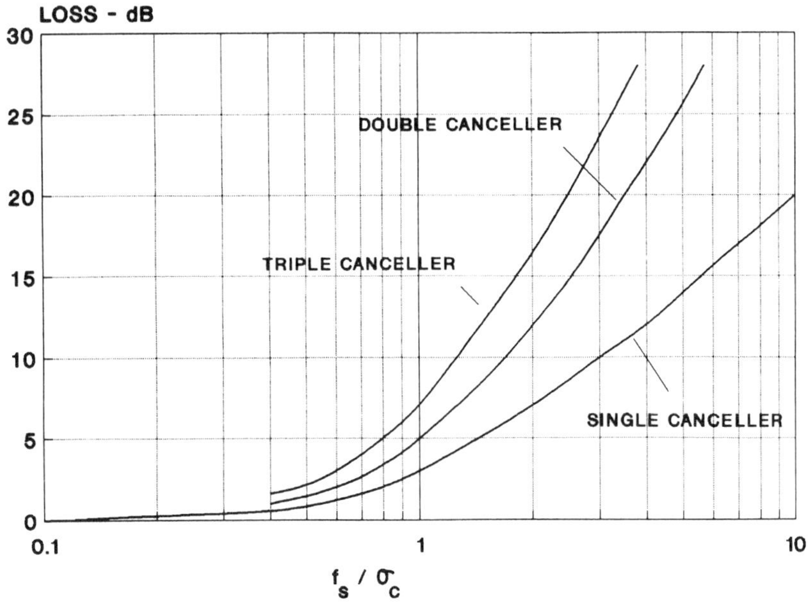

Figure 3.33 TACCAR losses for single, double, and triple MTI cancellers.

and triple canceller. The improvement factor is then given by $I_N = I/L_N$, where I is the improvement factor as derived by assuming perfect motion compensation.

To illustrate the effect of TACCAR losses on the improvement factor, consider an airborne early warning radar with the following parameters: f = 425 MHz, α = 12 rev/min, D = 25 ft, f_r = 300 pulses/s and v_{ac} = 300 knots [9,22]. The AEW radar is to operate over land with 25-knot surface winds at an altitude of 25,000 ft. The AEW radar uses a coherent double canceller type of MTI with single-loop TACCAR motion compensation.

Consider the situation when the radar antenna points along the aircraft's ground track ($\theta_a = 0°$). Internal clutter motion causes an rms velocity spread $\sigma_w = .0033v_w^{1.261} = 0.2$ m/s, which results in an rms clutter frequency spread of $\sigma_i = 0.57$ Hz. The scanning clutter spread is given by $\sigma_s = 0.833\alpha/\pi\theta_B = 4.12$ Hz($\alpha = 72°$/s, $\theta_B = 4.63°$). The combined rms clutter spread is then $\sigma_f = \sqrt{\sigma_c^2 + \sigma_s^2} = 4.16$ Hz, which results in an improvement factor $I_2 = 2/(\sigma_c T)^4$ of 45.4 dB with perfect TACCAR motion compensation. Note that the frequency spread due to platform motion is negligible when the antenna points along the aircraft's ground track ($\sigma_{pm} \rightarrow 0$).

From Figure 3.32, the normalized TACCAR range error (e_n) is on the order of 0.01 over most of the radar's range coverage. This results in a frequency offset error $f_e = e_n f_{ac} \cos\theta_a = .01 \cdot 438 = 4.38$ Hz when the radar antenna points along the aircraft's ground track. The ratio $f_e/\sigma_c = 4.38/4.16 = 1.05$, which, from Figure

3.33, results in a TACCAR improvement factor loss for the double canceller (N = 2) of L_2 = 3 dB. The overall improvement factor is then given by I_N = 45.4 − 3 = 41.4 dB.

As the radar antenna points away from the aircraft's ground track, the frequency spread due to platform motion (see Section 3.2.3) tends to dominate, reaching its maximum value at $\theta_a = 90°$. For example, at $\theta_a = 90°$, the spectrum spread due to platform motion is $\sigma_{pm} = 0.528\ v_{ac} \sin\theta_a/D = 10.7$ Hz, whereas, at $\theta_a = 30°$, the spread becomes σ_{pm} = 5.35 Hz. By combining with the other spreading factors, the total spread ($\sigma_f = \sqrt{\sigma_c^2 + \sigma_s^2 + \sigma_{pm}^2}$) becomes σ_f = 11.5 Hz at $\theta_a = 90°$, resulting in an improvement factor of I_2 = 28 dB. At $\theta_a = 30°$, the total spread becomes σ_f = 6.8 Hz, the resulting improvement factor is I_2 = 37 dB, the TACCAR range error is f_e = 3.8 Hz, the TACCAR loss is L_2 = 1 dB, and the resulting overall improvement factor thus is I_N = 37 − 1 = 36 dB.

The preceding example illustrates that only modest improvement factors are available from conventional AEW radars of this type. More advanced systems require improvement factors on the order of 60 dB. To accomplish this, several improvements are necessary:

1. an integrated two-way mainlobe-to-sidelobe antenna power ratio of 60 dB,
2. system stability factors of 10^{-9},
3. use of triple instead of double cancellers,
4. use of MTI-coherent integrator combinations (16-pulse),
5. TACCAR correction at three points in range,
6. improved platform motion cancellation using *displaced phase center antennas* (DPCA) [9, 22].

The affect of the platform motion on the antenna aperture can be resolved into a normal component ($v_n = v_{ac} \cos\theta_0 \cos\phi_0$) and a tangential or rotational component ($v_t = v_{ac} \sin\theta_0 \cos\phi_0$). The normal component causes a shift in the mean frequency of the clutter spectrum that is corrected over a particular range interval by using a TACCAR loop as discussed earlier. The rotational component causes a widening of the clutter spectrum, which, as the example illustrated, limits the MTI improvement factor as the antenna points away from the aircraft's ground track. This effect can be corrected by using an antenna array that electronically displaces the phase center of the antenna by an amount equal and opposite to the physical displacement of the antenna caused by platform motion. The displaced phase center antenna technique in conjunction with TACCAR thus provides complete motion compensation for an airborne MTI.

The rotational doppler shift caused by platform motion is given by $f_d = 2v_{ac}\theta \sin\theta_0 \cos\phi_0/\lambda$ (see Section 3.2.3 and Figure 3.17). The differential phase shift due to this component over an interpulse period (T) is given by $\eta = \int_0^T 2\pi f_d\, dt \approx 2\pi f_d T$ or $\eta = 4\pi v_{ac} T\theta \sin\theta_0 \cos\phi_0/\lambda$, where v_{ac} is the aircraft's velocity, $T = 1/f_r$ is the radar's interpulse period, θ_0 is the antenna's azimuth angle, ϕ_0 is the antenna's depression angle, λ is the radar's wavelength, and θ is the azimuth angular displacement from boresight. A vector diagram (see Figures 2.16 and 2.32) showing the phase rotation of the clutter vector due to the platform motion rotational doppler component is depicted in Figure 3.34(a). Figure 3.34(b) shows how the phase advance can be compensated with the addition of two quadrature vectors (e_1 and e_2). The quadrature vectors are given by $e_1 = jx_1 \tan\eta/2$ and $e_2 = -jx_2 \tan\eta/2$, where x_1 and x_2 are the clutter returns from successive transmitted pulses and η is the phase rotation. If the one-way power antenna pattern is $G(\theta)$, the amplitudes x_1 and x_2 are modulated by the two-way pattern $G^2(\theta)$, resulting in $e_1 = jG^2(\theta) \tan\eta/2$ and $e_2 = jG^2(\theta) \tan\eta/2 = \Delta(\theta)$. These vectors can be realized by transmitting the pattern $G(\theta)$ and receiving pattern $G(\theta)$ plus an additional pattern $\Delta(\theta)$.

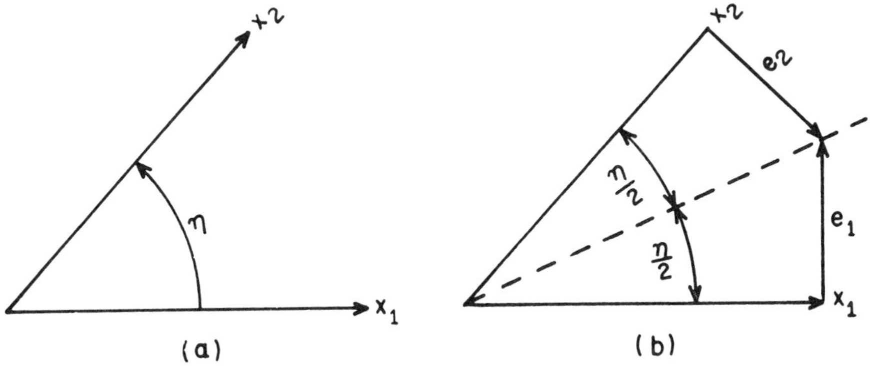

Figure 3.34 Vectorial representation of DPCA platform motion compensation.

Figure 3.35 depicts the method used to apply DPCA compensation to the first section of a multistage MTI canceller [23,24]. The returns received through the $\Delta(\theta)$ pattern are subtracted from the returns from the first pulse received through pattern $G(\theta)$ and added to the returns from the second pulse. The $\Delta(\theta)$ pattern is approximated by the difference pattern of a conventional amplitude monopulse antenna. Because the antenna pattern defined by $\Delta(\theta)$ generally must be approximated, the error in this pattern results in an imperfect DPCA correction and consequently in degraded MTI performance. Also, DPCA corrects only one canceller of a multiple-stage MTI, resulting in less than complete correction when double and higher order cancellers are employed.

The power transfer function for the canceller with DPCA correction depicted in Figure 3.35 is given by

$$|H_\eta(f)|^2 = (1 + \tan^2 \eta/2)[2 \sin\pi(f - f_d)T]^2[2 \sin\pi fT]^{2N-2} \qquad (3.25)$$

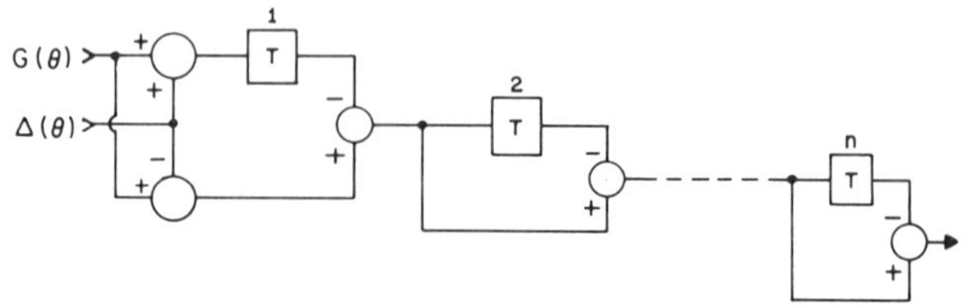

Figure 3.35 Displaced phase center antenna correction applied to the first stage of an N-stage MTI canceller.

where η is the interpulse phase advance, f_d is the rotational component of the platform motion doppler shift, T is the interpulse period, and $N = n - 1$ is the number of MTI cancelled stages [23]. This equation demonstrates that the null of the first canceller is shifted over the rotational component of the platform motion doppler and the other canceller stages are not affected by the DPCA correction. The improvement factor with DPCA correction can be derived, assuming a Gaussian-shaped clutter spectrum, as

$$I_N = 2^N(1 \cdot 3 \ldots 2N - 1)\sqrt{2\pi}\,\sigma_c \int_{-\theta_0}^{\theta_0} G^4(\theta)\,d\theta/n! \int_{-\theta_0}^{\theta_0} \cdot G^4(\theta)(1 + \tan^2\pi f_d T)H_n(\theta)\,d\theta \quad (3.26)$$

where the integration is over the mainbeam (θ_0) of the antenna pattern [23]. The improvement factors derived from this equation for a $(\sin x)/x$ antenna pattern with double and triple cancellers is depicted in Figures 3.36 and 3.37 as a function of the fraction of the antenna aperture displaced per interpulse period ($x = v_p T \cos\phi_0 \sin\theta_a / D$). The dashed curves represent the MTI improvement factor with platform motion compensated by DPCA and the solid curves illustrate the platform motion limitation.

Comparison of the MTI improvement factor for a single canceller in conjunction with a $(\sin x)/x$ antenna pattern shows that if the fraction of antenna aperture displaced per interpulse period is less than $x = 1/10$, the MTI improvement factor almost perfectly compensates for the rotational platform motion [23]. Therefore, in this case, the improvement factor for surface clutter generally will be determined by the antenna scanning and internal motion limitations (see Sections 3.2.1 and 3.2.2). Antenna scanning modulation of the clutter return can also be compensated by using an implementation similar to that employed for DPCA [24]. Scanning motion compensation adds one-half the interpulse differential amplitude caused by scanning the two-way power antenna pattern, $\Delta G^2(\theta)$, to one pulse and then subtracts it from the next pulse, using the same implementation depicted for DPCA (see Figure 3.35).

Figures 3.3 and 3.37 illustrate that DPCA compensation of double and triple cancellers is not perfect, but there is appreciable correction, especially for ground

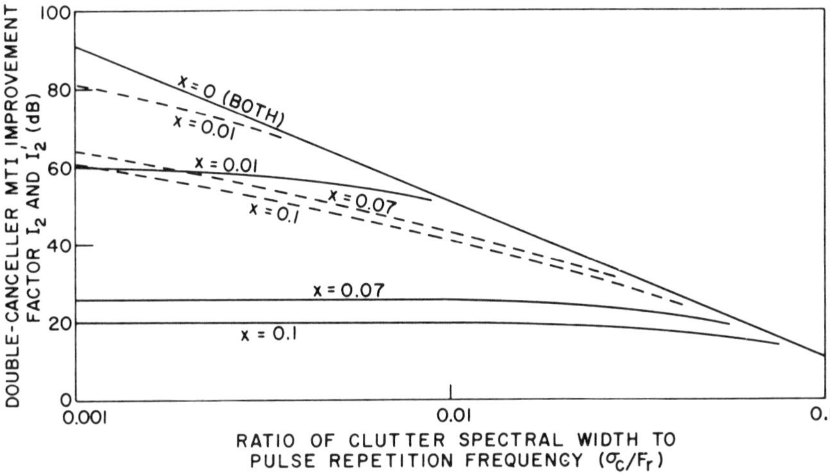

Figure 3.36 Comparison of the MTI improvement factor for a double canceller with and without DPCA correction.

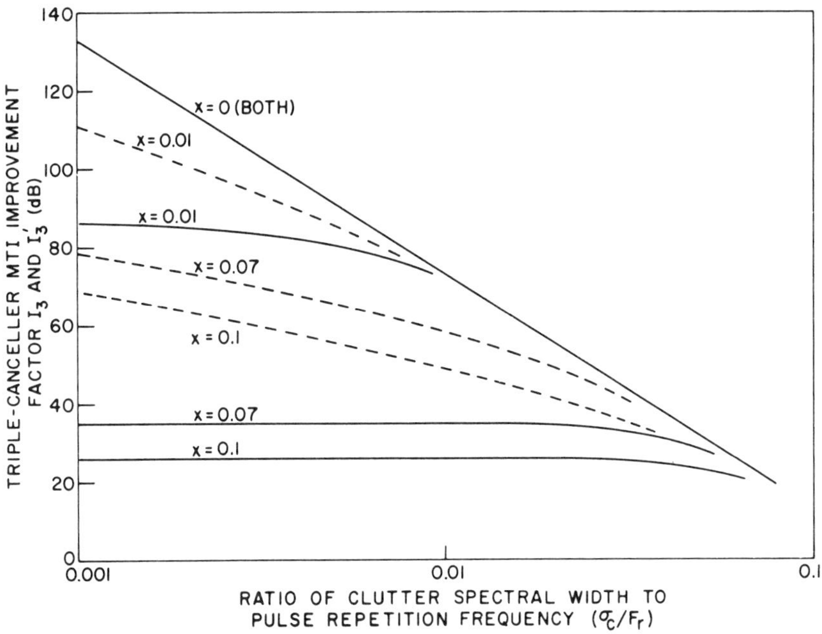

Figure 3.37 Comparison of the MTI improvement factor for a triple canceller with and without DPCA correction.

clutter, where the intrinsic spectral widths are small. These figures also illustrate the additional improvement that can be achieved by a more complete compensation. Perfect compensation for motion is represented by the curves labeled $x = 0$, and comparing these curves with the dashed curves indicates the additional improvement that ideal motion compensation may provide.

3.2.8 Combining Improvement Factor Effects

The previous sections each derived an improvement factor limitation on the assumption that all other effects were absent. In practice, many of the effects will be simultaneously present and the composite improvement factor must consider the interaction of all such effects.

The basic assumption in combining the individual effects is that all are statistically independent and each can be represented as a separate modulation on a correlated carrier. When this condition exists, then the received clutter voltage is given by $v_t = a_t b_t c_t d_t \ldots \cos\omega t$, which results in a demodulated autocorrelation function, given by $\rho_c(\tau) = \rho_a(\tau)\rho_b(\tau)\rho_c(\tau) \ldots$. The power spectral density is found by Fourier transforming the ACF as $S_c(\omega) = S_a(\omega) * S_b(\omega) * S_c(\omega) \ldots$ where $*$ indicates the convolution procedure $S_1(\omega) * S_2(\omega) = \int_{-\infty}^{\infty} S_1(\omega)S_2(\omega - y)\, dy/2\pi$. The procedure is particularly convenient when the individual effects can be represented by a Gaussian-shaped power spectral density $S(\omega) = P_c \exp(-\omega^2/2\sigma_c^2)/\sqrt{2\pi}\sigma_c$, where σ_c is the standard deviation of the clutter spectrum (rad/s) related to the 3 dB bandwidth (ω_3 – rad/s) as $\sigma_c = \omega_3/\sqrt{\ln 4} = 0.849\omega_3$. Note that when synchronous detection is involved, the 3 dB video bandwidth (B_3) is given by $B_3 = \sqrt{\ln 4}\sigma_c/\pi = (\sigma_c/2)/1.334$.

When individual Gaussian-shaped power spectral densities are convolved, the resulting clutter power spectral density is also Gaussian-shaped, but has a variance σ_c^2, which is the sum of the variances of each of the individual effects (i.e., $\sigma_c^2 = \sigma_1^2 + \sigma_2^2 + \ldots \sigma_n^2$). For example, if the situation involves clutter spreading due to internal motion (σ_i), scanning modulation (σ_s), platform motion (σ_{pm}), and transmitter instability (σ_x), the variance of the Gaussian-shaped clutter spectrum is $\sigma_c^2 = \sigma_i^2 + \sigma_s^2 + \sigma_{pm}^2 + \sigma_x^2$. If noncoherent MTI using square-law detection were involved, the clutter variance would be given by $\sigma_{nc}^2 = 2\sigma_c^2$, although the spectrum would still be Gaussian shaped. In a coherent MTI, the target spectrum also is spread by antenna scanning modulation as with the clutter spectrum and often is represented by a Gaussian-shaped spectral density. With a noncoherent MTI, the predetection target spectrum is convolved with the clutter spectrum to form the video target spectrum (i.e., $\sigma_t^2 = 2\sigma_s^2 + \sigma_i^2 + \sigma_{pm}^2 + \sigma_x^2$). The video coherent and noncoherent MTI target and clutter spectra for both fixed and airborne radars is tabulated in Table 3.7.

The improvement factor can be found by considering the composite clutter spectrum with its associated composite variance. Alternatively, the improvement factor

Table 3.7
Video Target and Clutter Spectrum Variances for Fixed and Airborne Radars

Coherent MTI	*Noncoherent MTI*
Fixed:	
Target σ_s^2	$2\sigma_s^2 + \sigma_x^2 + \sigma_i^2$
Clutter $\sigma_s^2 + \sigma_x^2 + \sigma_i^2$	$2\sigma_s^2 + 2\sigma_x^2 + 2\sigma_i^2$
Airborne:	
Target σ_s^2	$2\sigma_s^2 + \sigma_x^2 + \sigma_i^2 + \sigma_{pm}^2$
Clutter $\sigma_s^2 + \sigma_x^2 + \sigma_i^2 + \sigma_{pm}^2$	$2\sigma_s^2 + 2\sigma_x^2 + 2\sigma_i^2 + 2\sigma_{pm}^2$
Variances:	
σ_i^2 = Internal motion	σ_s^2 = Scanning modulation
σ_x^2 = Transmitter instability	σ_{pm}^2 = Platform motion

can be found by computing the individual improvement factor for each additive noiselike effect and combining these as $I^{-1} = I_1^{-1} + I_2^{-1} + \ldots I_N^{-1}$. This relationship can be developed by considering $I = \overline{G}\text{CA}$, where $\overline{G}$ is the MTI gain and CA is the clutter attenuation. The clutter attenuation is given by

$$\text{CA} = \int_0^\infty \sum_{L=1}^{N} S_n(\omega)\, d\omega \Big/ \int_0^\infty \sum_{L=1}^{N} S_N(\omega)|H(\omega)|^2\, d\omega \tag{3.27}$$

where $S_N(\omega)$ is the individual effect clutter spectral density and $H(\omega)$ is the MTI filter response. If we interchange summation and integration, and note that the numerator always equals the clutter power, we have a relationship for combining the clutter attenuation for the individual effects (CA_N) as $(\text{CA})^{-1} = (\text{CA}_1)^{-1} + (\text{CA}_2)^{-1} + \ldots (\text{CA}_N)^{-1}$. Multiplication of the clutter attenuation by the MTI gain then provides the relationship for the improvement factor:

$$(\overline{G} \cdot \text{CA})^{-1} = (\overline{G} \cdot \text{CA}_1)^{-1} + (\overline{G} \cdot \text{CA}_2)^{-1} + \ldots (\overline{G} \cdot \text{CA}_N)^{-1}$$

or

$$I^{-1} = I_1^{-1} + I_2^{-1} + \ldots I_N^{-1}$$

When additive noiselike disturbances (i.e., system instabilities) are combined with multiplicative effects (i.e., clutter spreading), these effects must be considered separately before finding the overall improvement factor. For example, consider primary instability effects due to transmitter frequency change, STALO frequency change,

COHO frequency change, pulse timing jitter, pulsewidth jitter, and pulse amplitude change, where each individual instability causes an improvement factor limitation of 50 dB. The composite improvement factor due to system instabilities would then be equal to $I_s = I_{\mathrm{IND}}/6$ or 42.2 dB.

Now, assume that the improvement factor limitation due to clutter spectral spreading is caused by antenna scanning and internal clutter motion effects, where the limitation due to each effect is 43 dB when a double canceller MTI is used. The improvement factor for a double binomial canceller using Gaussian-shaped clutter is given by $I_c = 2/(\sigma_{c1}T)^4$ and results in $\sigma_{c1}T = 0.1$ rad for the example. The improvement factor considering both effects is then given by $I_c = 2/(\sigma_{c1}^2 + \sigma_{c2}^2)^2T^4$, which reduces to $I_c = 1/2(\sigma_{c1}T)^4$ or 37 dB. The resulting overall improvement factor is hence given by $I = (I_c^{-1} + I_s^{-1})^{-1}$ or 35.9 dB for the example.

3.3 OSCILLATOR EFFECTS ON DOPPLER RADAR PERFORMANCE

The performance of a coherent doppler radar system often is limited by phase noise generated in the STALO or master oscillator. In the master oscillator power amplifier system (see Figures 1.10 and 2.20), the same STALO is used as the reference oscillator in the receiver and, with an IF offset, as the transmitted signal. When phase noise is present on the STALO, the resulting transmitted signal contains noise sidebands that extend into the frequency region occupied by doppler shifted targets. When this noisy transmitted signal is reflected from clutter scatterers, it induces a noise background that is sometimes characterized as "transmitted clutter." This noise is both a function of the strength of the clutter scatterer returns and the level of sideband phase noise measured with respect to the carrier amplitude. The situation is further complicated when the return signal is mixed with the noisy STALO in the radar receiver to produce an IF signal that is subsequently synchronously detected by using the offset oscillator. The phase noise in the STALO is thus correlated between transmission and reception with the degree of correlation depending on the range of the scatterers with respect to the radar. This correlation produces a cancellation effect for short-range clutter scatterers (which pose the most severe problem) while multiplying the noise power by a factor of four for the most unfavorably positioned scatterers.

STALO phase noise, although important for low-PRF MTI radars, becomes the critical performance factor for high-PRF PD radars, because in high-PRF PD radar the highly ambiguous range response causes close-in clutter returns to compete with long-range target returns. The resulting clutter attenuation required for a high-PRF PD system is then on the order of 80–100 dB as compared to the 50–60 dB required for low-PRF doppler systems.

Figure 3.38 depicts the block diagram of a typical doppler radar transmitter-receiver system. If the phase noise associated with the microwave reference oscillator (STALO) is given by ϕ_t, then the transmitted signal (T_o) can be represented as v_t =

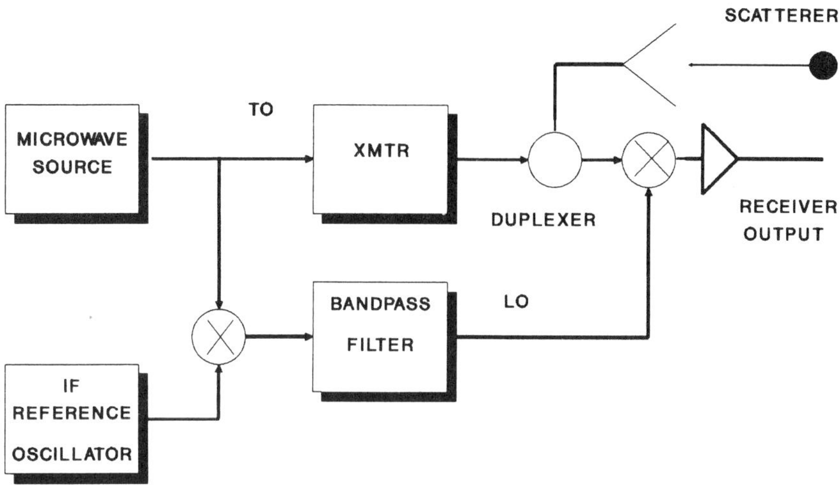

Figure 3.38 Block diagram of common oscillator doppler radar transmitter-receiver.

$\cos(\omega_t t + \phi_t)$. The local oscillator is given by $v = \cos(\omega t + \phi_t)$, where the phase noise from the IF offset oscillator is assumed to be negligible compared to that from the microwave STALO due to the large frequency ratio between the two sources. The return from a clutter scatterer at range R is then given by $v_c = \cos[\omega_t(t - t_d) + \phi_t - t_d]$, where $t_d = 2R/c$ is the two-way scatterer return time delay. The receiver output is obtained by mixing the clutter signal with the local oscillator signal as $v_r = \cos(\omega_{\text{IF}} t - \omega_t t_d + \phi_c)$, where $\phi_c = \phi_{t-t_d} - \phi_t$. If the phase noise is assumed to be given by $\phi_t = \Delta\phi \sin\omega_m t$, where ω_m is the noise radian frequency, the resulting output of the coherent doppler radar synchronous phase detector can be represented as $v_p = \Delta\phi[\sin\omega_m(t - t_d) - \sin\omega_m t]$, which reduces to $v_p = \Delta\phi 2 \sin(\omega_m t_d/2) \cos(\omega_m t + \phi_0)$, where the voltage correlation factor $2 \sin(\omega_m t_d/2)$ directly multiplies the phase noise voltage spectral density, $S_\phi(\omega)$. The psd seen by the doppler processor due to phase noise is therefore given by $S_p(\omega) = 4 \sin^2(\omega_m t_d/2) S_\phi(\omega)$, where $S_\phi(\omega)$ is the master oscillator (STALO) phase noise power spectral density, and t_d is a time delay, which depends on the range to the clutter scatterer. The correlation factor is plotted in Figure 3.39 for $t_d = 25$ μs, where a peak (+6 dB) is reached at $f_m = 1/2t_d$ (20 kHz in Figure 3.39) followed by a null at $f_m = 1/t_d$ (40 kHz in Figure 3.39).

In a low-PRF coherent MTI using a binomial canceller, the clutter attenuation can be expressed as

$$\text{CA} = 8 \int_0^\infty \sin^2(\pi f t_d) \sin^{2N}(\pi f T) S_\phi(f)\, df$$

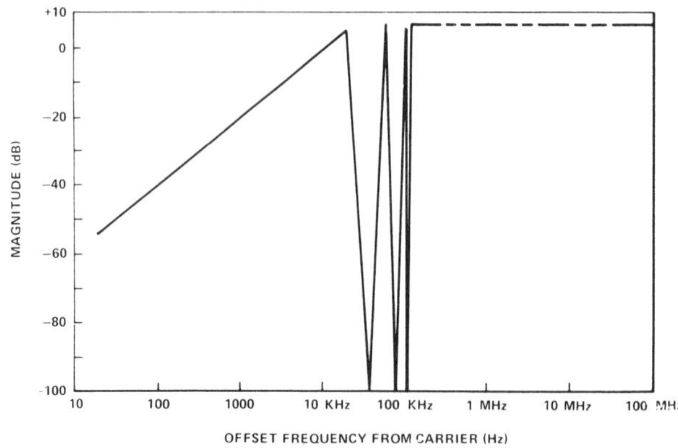

Figure 3.39 Correlation factor for master oscillator phase noise (t_d = 25 μs).

where N is the number of canceller stages ($n = N + 1$ pulses processed), t_d is the two-way range delay, T is the radar's interpulse interval, and $S_\phi(f)$ is the master oscillator phase noise power spectrum normalized to the carrier power level. Figure 3.40 depicts several permissible phase noise master oscillator spectra ($S_\phi(f)$) for a single canceller MTI that provide the indicated level of clutter attenuation at approximately a 20 nmi range (t_d = 250 μs) with a radar PRF of 1000 pulses/s (T = 1000 μs). The oscillator phase noise spectrum is composed of three distinct components [25]: flicker noise, the phase noise spectrum of which is proportional to $1/f^3$ at small values of f_m; external additive oscillator noise, the phase spectrum of which is "white" or flat with frequency; and internal oscillator noise, the phase spectrum of which is proportional to $1/f^2$.

Several measured master oscillator phase noise spectral densities are shown superimposed on the permissible phase noise spectra in Figure 3.41. Generally, cavity stabilized oscillators (e.g., dielectric resonant oscillators, electrostatically focused klystrons, cavity stabilized klystrons) have less phase noise than multiplied (e.g., varactor multiplied, step recovery) crystal oscillators, which generally have a low fundamental frequency, under 100 MHz. Also, in general, lower fundamental frequency master oscillators have less phase noise than higher fundamental frequency oscillators, resulting in lower frequency MTI radars having greater performance potential than higher frequency MTI radars.

To understand the situation for high-PRF doppler radars, we must first consider the clutter attenuation requirement for such radars. The requirements are different for surface-based and airborne PD radars due to both the geometrical and platform-motion aspects of these radars. The clutter return for surface-based high-PRF radars is similar to that for a CW radar except that the CW clutter return is multiplied by the duty factor ($D_u = \tau/T$, where τ = pulsewidth, and T = interpulse period) of the radar. This produces a large mainbeam close-in clutter return that is a function of

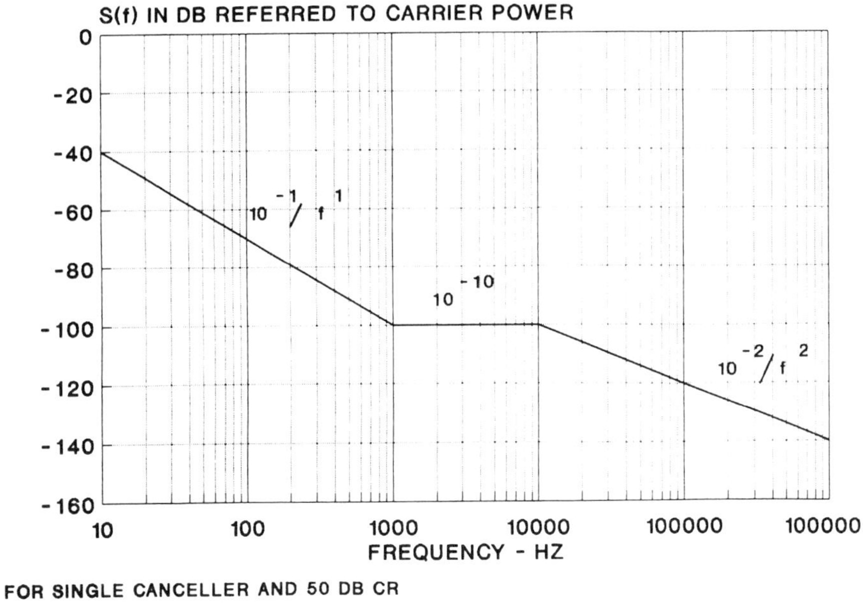

Figure 3.40 Permissible master oscillator phase noise spectra (t_d = 250 μs).

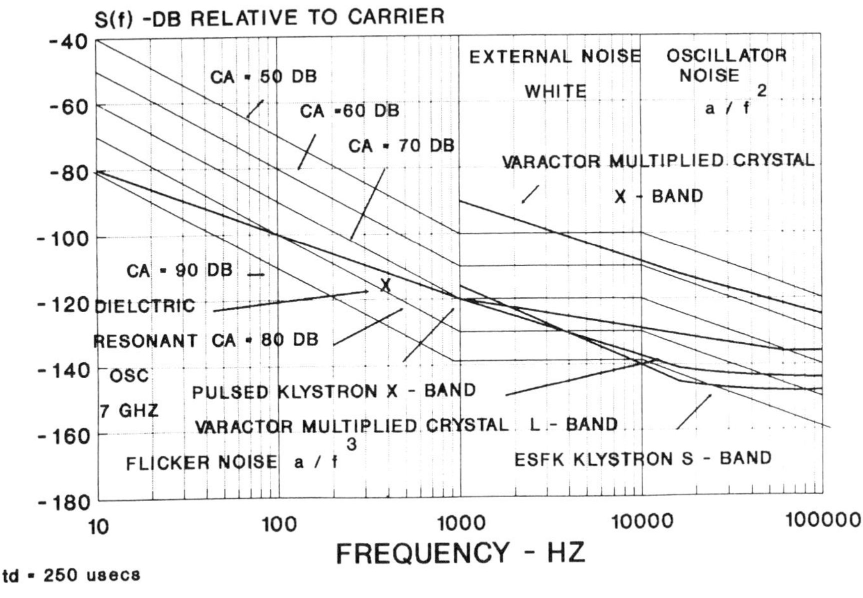

Figure 3.41 Permissible master oscillator phase noise spectral densities (t_d = 250 μs).

the radar's elevation antenna pattern with respect to the clutter scatterers in close proximity to the radar.

The clutter return for an airborne high-PRF radar is a function of the platform's speed and altitude as well as the antenna's radiation pattern. The mainbeam clutter return is modulated by the transmitter's phase noise into the frequency band covered by the doppler filter bank, where it then competes with the doppler frequencies returned from targets. The mainbeam clutter return for an airborne radar is less than that of a surface-based radar because the minimum range returning ground clutter always must be greater than the altitude of the aircraft. However, the mainbeam intercepts the clutter patch at an increased grazing angle, which has higher reflectivity than the lower grazing angles of surface-based radars.

The clutter power returned to a surface-based CW radar can be expressed as $C = (P_t G^2 \lambda^2 \theta_a/(4\pi)^3 L) \cdot \int_0^\infty (\sigma^\circ F^4/r^3)\, dr$, where P_t is the transmitter power, θ_a is the azimuth beamwidth (rad), σ° is the clutter backscatter coefficient, F^4 is the two-way power antenna pattern propagation factor, L is the transmission and beam shape loss, and r is the range to the clutter scatterer [26]. If a Gaussian-shaped antenna beam is assumed, $F^4 = \exp(-5.54\phi^2/\phi_e^2)$, where ϕ_e is the 3 dB elevation beamwidth. Further, using the small angle approximation $\phi = h/r$, where h is the height of the antenna's phase center above the surface, and assuming that σ° is independent of grazing angle results in $\int_0^\infty (\sigma^\circ \exp(-5.54h^2/\phi_e r^2)/r^3)\, dr = \sigma^\circ \phi_e/(3.33h)^2$. The CW clutter power can then be expressed as

$$C = P_T G^2 \lambda^2 \theta_a \phi_e^2 \sigma^\circ/(4\pi)^3 (3.33h)^2 L \tag{3.28}$$

Further simplifications can be made by substituting $G = 4\pi/(\theta_a \phi_e) L_n$ and $\theta_a = 1.5\lambda/D_a$, where D_a is the azimuth antenna aperture and $L_n = 1.2$ is the antenna pattern loss experienced when aperture weighting is used to provide low antenna sidelobes. The resulting CW clutter power is given by $C = P_t \sigma^\circ \lambda D_a/400\ h^2$, where P_t is the transmitter power, σ° is the clutter backscatter coefficient, λ is the radar wavelength, D_a is the antenna's azimuth width, h is the antenna's phase center height above the surface, and a beam shape loss of $L_B = 1.33$ is assumed. The high-PRF PD clutter average power return can then be determined as $C = P_t d_u \sigma^\circ \lambda D_a/400\ h^2$, where d_u is the PD radar's duty cycle. If the clutter power returned from the transmitter carrier (center line of spectrum) is desired, $C = P_t d_u^2 \sigma^\circ \lambda D_a/400\ h^2$.

A further refinement of the PD radar clutter return equation can be obtained by using the constant reflectivity (γ) model in which the backscattering coefficient is given by $\sigma^\circ = \gamma/\sin\phi$, (see Section 1.3.2.1). If we use the approximation $\sin\phi = h/r$, the integral in the clutter power equation is

$$\int_0^\infty (\sigma^0 F^4/r^3)\, dr = \int_0^\infty (\gamma h \exp(-5.54h^2/\phi_e^2 r^2)/r^4)\, dr$$

which equals $\gamma\phi_e^3/29.4h^2$. Substituting this into the CW clutter power equation results in

$$C = P_t G^2 \lambda^2 \theta_a \phi_e^3 \gamma / 29.4 (4\pi)^3 L h^2$$

which can be further simplified ($\phi_e = 1.5\lambda/D_e$, where D_e is the antenna's elevation aperture), resulting in

$$C = P_t d_u \lambda^2 \gamma D_a / 533 D_e h^2$$

where the terms are as previously defined.

As an example, assume an X-band radar with a circular aperture ($D_a = D_e$) operating at a height $h = 3$ m above a ground surface with $\gamma = 0.1$. For a transmitted power of 10 kW and a duty cycle $d_u = 0.1$ ($P_{\text{ave}} = 1$ kW), the average clutter return power is $C = 1.9 \times 10^{-5}$ W, whereas the clutter power associated with the central line is 1.9×10^{-6} W [27]. If a doppler filter bank using filters with a doppler bandwidth of 1 kHz ($B_D = 1$ kHz, $t_0 = 10^{-3}$ s) were employed and if the total clutter energy were concentrated in the filter, then it would equal 1.9×10^{-8} W-s. The noise energy level assuming a noise temperature $T_s = 3000$ K would be 4×10^{-20} W/Hz. The clutter-to-noise ratio would then be $C/N = 4.75 \times 10^{11} = 117$ dB. Hence, the clutter attenuation requirement (to reduce the clutter's rms level to the noise level) for this example would be on the order of 120 dB, thus establishing a stringent requirement on the master oscillator phase noise stability and the doppler filter sidelobe capability. Note that the correlation factor must be taken into account when establishing the master oscillator's allowable phase noise spectrum.

In the airborne case, the ground clutter return for a high-PRF PD radar is spread over a velocity range equal to $\pm 2v_{ac}/\lambda$ as shown in Figure 3.42 (see Section 1.1.4). The prominent features of the clutter return are (1) the mainlobe clutter return, which is most severe when the radar is in a look-down mode and occurs at a mean frequency equal to $f_{ML} = 2v_{ac} \cos\theta_a \cos\phi_e/\lambda$; (2) the altitude line clutter, which represents the specularlike reflections received from the scatterers directly under the aircraft returned through the radar's sidelobes at a mean frequency $f = 0$; and (3) the sidelobe and backlobe clutter returns, which range in frequency from $f_d = 2v_{ac}/\lambda$ for the forward directed sidelobes to $f_d = -2v_a/\lambda$ for the backward directed backlobes. Also shown in Figure 3.42 are the frequency placements of the doppler filter bank: in the clutter-free region to detect closing aircraft or missile targets; in the forward clutter sidelobe region to detect chased targets or for doppler beam sharpening of ground returns.

Figure 3.43 depicts the key elements of the search processing generally employed in a high-PRF PD airborne radar. Mainlobe and altitude-line clutter rejection notch filters are used to bring those strong clutter returns to a manageable level. This is followed by a bank of contiguous doppler filters that are matched to the target return spectrum. The notch filtering is generally accomplished by using analog filters that

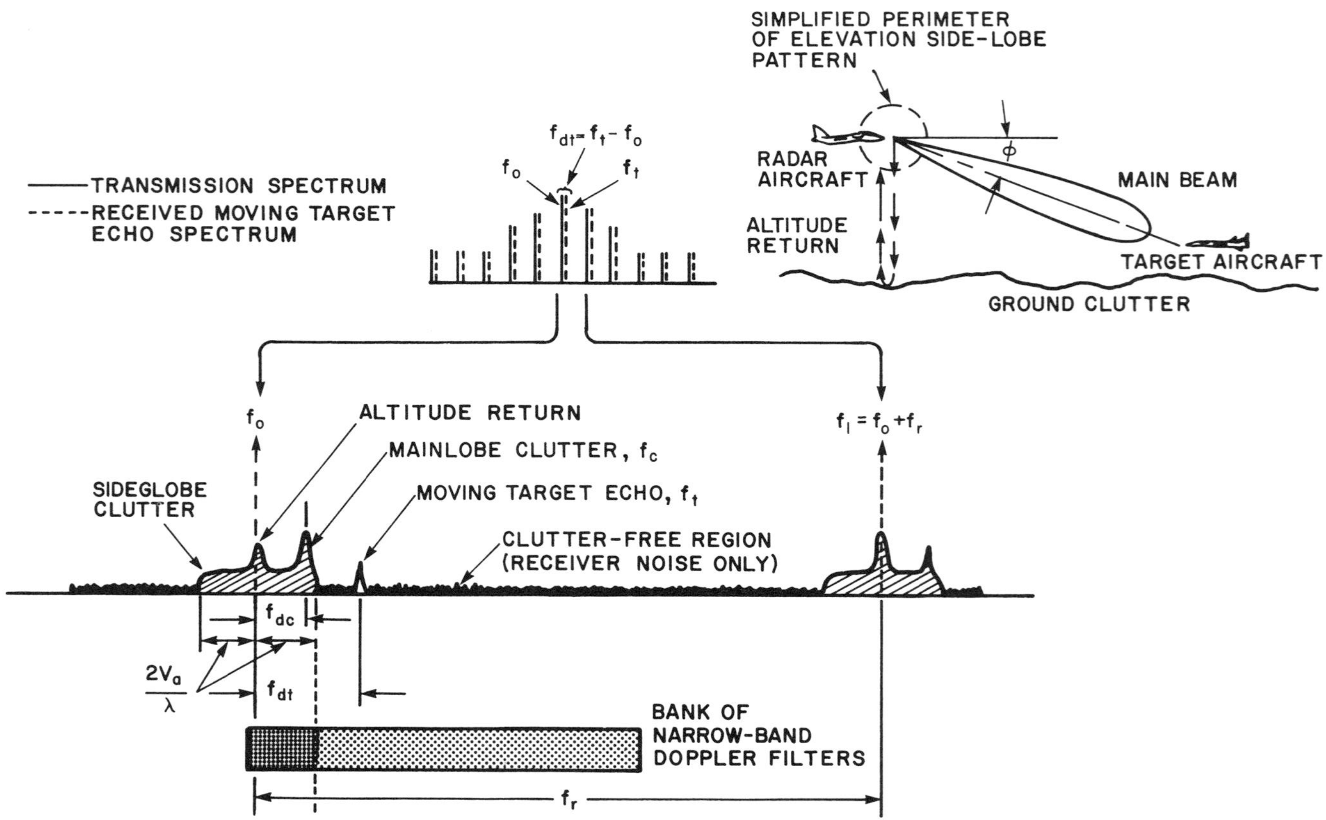

Figure 3.42 Clutter return and doppler filter placement for a high-PRF PD radar.

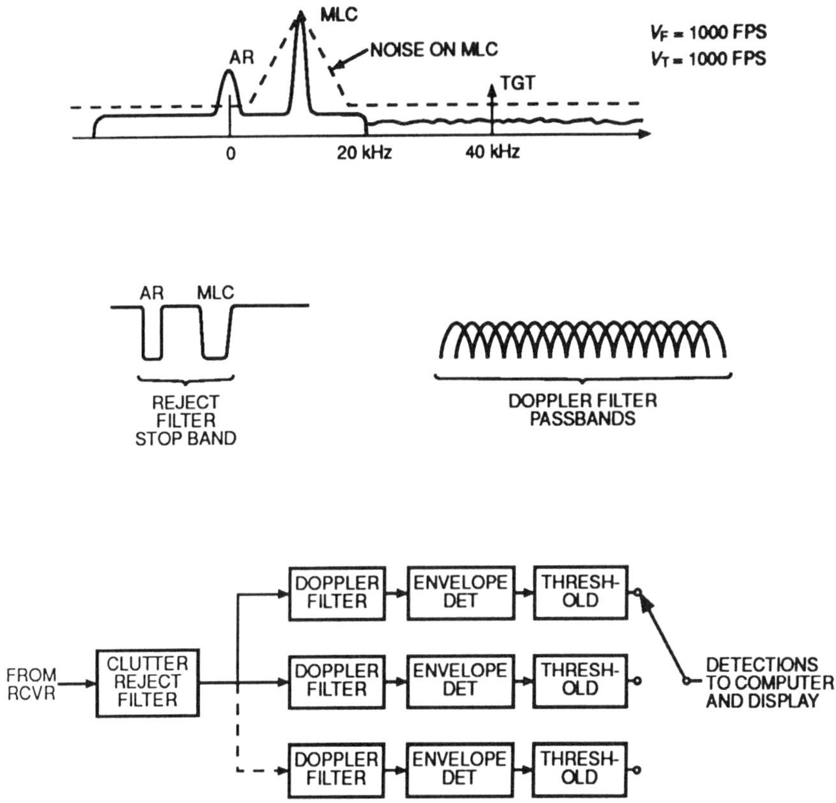

Figure 3.43 High-PRF PD search processing.

serve the dual purpose of reducing both the dynamic range and sidelobe requirements of the doppler filter bank. The doppler filter bank frequently is implemented through a digital FFT processing.

Figure 3.43 also depicts the effects of transmitter phase noise, which spreads the clutter return spectrum into the frequency region occupied by the doppler filter bank. This effect can be visualized by considering the transmitted spectrum as a band of frequencies about each individual line spectrum (due to phase noise), rather than as a single line spectrum. When this spectrum is convolved with each individual clutter scatterer (to form the total clutter return), the effect is to modulate each frequency component of the pure clutter return spectrum (no phase noise) by the phase noise spectrum. Because the phase noise spectrum has considerable frequency extent, it translates components of the mainlobe clutter return from the notch rejection filter into the frequency region occupied by the doppler filter bank. The objective is then to hold the phase noise spectrum to a level with respect to the carrier level such that the translated clutter spectrum has an rms value in each doppler filter less than the receiver noise level. When this objective cannot be accomplished (generally in the doppler filters closest to the mainlobe clutter frequency), an adaptive CFAR is needed to desensitize the doppler filter to this clutter leakage.

To consider the effect of mainlobe clutter leakage, we need to determine the magnitude of the mainlobe clutter return. Assuming that the airborne radar uses a pencil beam, the approximate equivalent clutter radar cross section is given by $\sigma_c = \sigma^\circ \pi(\theta R)^2/4$, where θ is the two-way power antenna beamwidth ($\theta = \theta_{3\ \mathrm{dB}}/\sqrt{2}$). Substituting this into the radar equation results in

$$C_{ML} = P_t d_u G^2 \theta_3^2 \sigma^\circ \lambda^2 / (\sqrt{2}\ 16\pi)^2 R^2 L$$

where d_u is the radar's duty factor, θ_3 is the antenna's one-way power 3 dB beamwidth, R is the range to the MLC patch, and L is a system loss term. Substituting $G = \pi^2/\theta_3^2 = (\pi D/\lambda)^2$, where D is the circular antenna aperture diameter, results in $C_{ML} = P_t d_u \sigma^\circ D^2/50R^2 L$ [27].

Assuming the same parameters as for the surface PD radar (i.e., $P_t = 10$ kW, $d_u = 0.1$, $\sigma^\circ = 0.1$) with $D = 0.7$ m, $L = 1$, and $R = 10$ km, results in an average mainlobe clutter power return equal to $C_{ML} = 10^{-8}$ W or a return associated with the central spectral line of 10^{-9} W. The clutter energy concentrated in a 1 kHz doppler filter is 10^{-11} W-s and the receiver noise energy ($T_s = 3000$ K) is $N = 4 \times 10^{-20}$ W/Hz, resulting in a $C/N = 2.5 \times 10^8 = +84$ dB, requiring a clutter attenuation on the order of 80 dB. This results in a less stringent requirement on oscillator phase noise for the airborne case than the surface radar case. A typical master oscillator phase noise spectrum is shown in Figure 3.44.

As an additional example, consider an X-band airborne intercept high-PRF PD radar. The parameters are $P_t = 2.5$ kW, $d_u = 0.5$, $\theta_3 = 2.2°$, $NF = 5$ dB, $B_d = 200$ Hz, $G = 37$ dB, $L = 3.9$ dB, and $\sigma^\circ = 0.15$. The mainlobe clutter return in the central line at $R = 100$ nmi then equals $C_{ML} = -78$ dBm (the central line average power is $d_u^2 P_t$). The noise power in the 200 Hz doppler bandwidth equals $P_N = (-174 - 3 + 5 + 23) = -149$ dBm, where the noise power in the doppler filter is reduced by the duty factor ($d_u = -3$ dB) because the receiver is gated on only half the time. The resulting C/N at 100 nmi is $+71$ dB, which implies that the master oscillator phase noise spectrum must be on the order of 80 dB below the carrier power level over the frequency region covering the doppler filter bank. At 100 nmi, the central line power returned from a 5 m^2 target is about -149 dBm, so that $R_0 = 100$ nmi (R_0 is the range where $S/N = 1$). The mainlobe clutter and target power levels for this example are plotted in Figure 3.45.

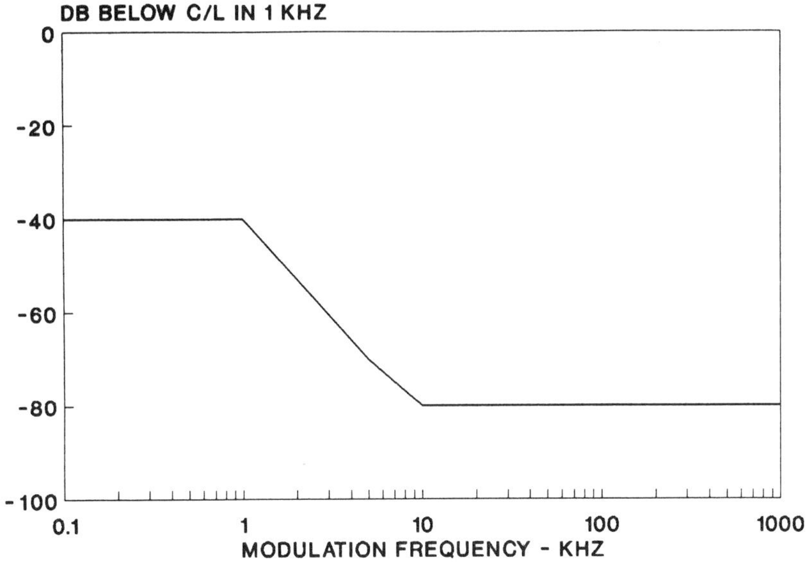

Figure 3.44 Typical high-PRF PD phase noise spectrum.

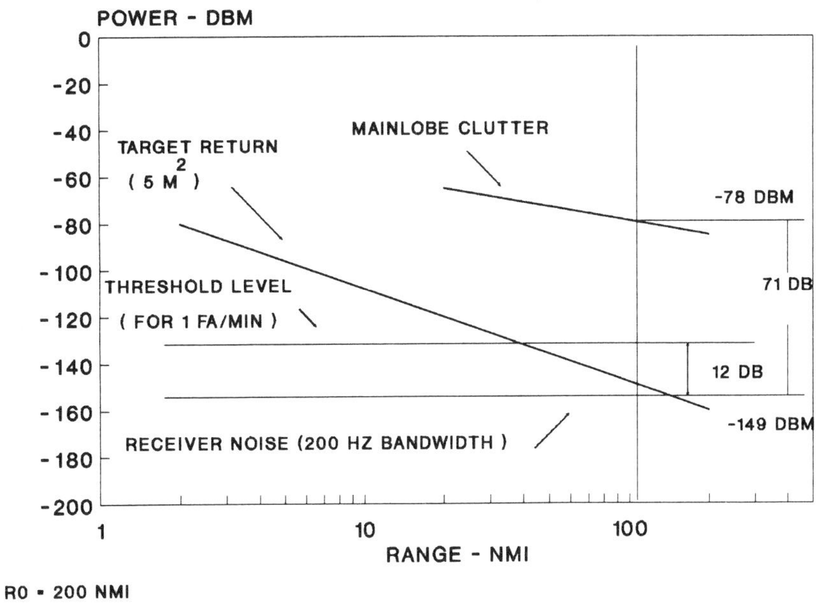

Figure 3.45 Airborne intercept PD radar mainlobe clutter and noise levels.

REFERENCES

1. J. Marcum, "A Statistical Theory of Target Detection by Pulsed Radar," *IRE Trans.*, Vol. IT-6, No. 2, April 1960.
2. P. Swerling, "Probability of Detection for Fluctuating Targets," *IRE Trans.*, Vol. IT-6, No. 2, April 1960.
3. D.C. Schleher, *Introduction to Electronic Warfare,* Artech House, Norwood, MA, 1986.
4. D.K. Barton, ed., *The Radar Equation, Radars,* Vol. 2, Artech House, Norwood, MA, 1974.
5. D.C. Schleher, "Performance Comparison of MTI and Coherent Doppler Processors," IEEE Int. Radar Conf., London, November 1982.
6. I. Kanter, "A Generalization of the Detection Theory of Swerling," IEEE EASCON Conf., Washington, DC, September 1974.
7. G. Dillard and J. Richard, "Performance of an MTI Followed by an Incoherent Integrator for Nonfluctuating Signals," IEEE Int. Radar Conf., Washington, DC, April 1980.
8. R. Emerson, "First Probability Densities for Receivers with Square Law Detectors," J. Applied Physics, Vol. 24, No. 9, September 1953.
9. D.C. Schleher, ed., *MTI Radar,* Artech House, Norwood, MA, 1978.
10. R. Emerson, "Some Pulsed Doppler MTI and AMTI Techniques," the Rand Corporation, R-274, March 1954.
11. R. Berkowitz, *Modern Radar,* John Wiley and Sons, New York, 1965.
12. F. Kretschmer, F. Lin, and B. Lewis, "A Comparison of Noncoherent and Coherent MTI Improvement Factors," *IEEE Trans.*, Vol. AES-19, No. 3, May 1983.
13. M.I. Skolnik, ed., *Radar Handbook,* 1st Ed., McGraw-Hill, New York, 1970, Chapter 17, MTI Radar.
14. F. Nathanson, *Radar Design Principles,* McGraw-Hill, New York, 1969.
15. W. Fishbein, S. Graveline, and O. Rittenbach, "Clutter Attenuation Analysis," ECOM-2808, U.S. Army Electronics Command, Fort Monmouth, NJ, March 1967.
16. G. Stimson, *Introduction to Airborne Radar,* Hughes Aircraft Company, El Segundo, CA, 1983.
17. H. Ward and W. Shrader, "MTI Performance Degradation Caused by Limiting," *IEEE EASCON Rec.*, Washington, DC, September 1968.
18. H. Ward, "The Effect of Bandpass Limiting on Noise with a Gaussian Spectrum," *Proc. IEEE,* November 1969.
19. C. Muehe, L. Cartledge, W. Durry, E. Hofstetter, M. Labitt, P. McCorison, and V. Sferrino, "New Techniques Applied to Air-Traffic Control Radars," *Proc. IEEE,* Vol. 62, No. 6, June 1974.
20. G. Andrews, "Airborne Radar Motion Compensation Techniques, Evaluation of TACCAR," NRL Rep. NRL-7407, Washington, DC, April 1972.
21. M.I. Skolnik, *Introduction to Radar Systems,* 2nd Ed., McGraw-Hill, New York, 1980.
22. T. Ap Rhys and G. Andrews, "AEW Radar Antennas," *Antennas for Avionics,* AGARD Conference Preprint No. 139, 1974.
23. G. Andrews, "Airborne Radar Motion Compensation Techniques, Evaluation of DPCA," NRL Rep. 7426, Washington, DC, July 1972.
24. M.I. Skolnik, ed., *Radar Handbook,* 1st Ed., McGraw-Hill, New York, 1970, Chapter 18, Airborne MTI.
25. R. Raven, "Requirements on Master Oscillators for Coherent Radar," *Proc. IEEE,* Vol. 54, No. 2, February 1966.
26. D.K. Barton, ed., *CW and Doppler Radar, Radars,* Vol. 7, Artech House, Norwood, MA, 1974.
27. D.K. Barton, *Modern Radar System Analysis,* Artech House, Norwood, MA, 1988.

Chapter 4
Clutter Properties and Data

Radar clutter is defined as unwanted echoes, typically from the ground, sea, rain or other precipitation, chaff, birds, insects, and aurora. The designer of a doppler radar is generally interested in discriminating against these unwanted returns, and hence we must understand their properties to provide an effective radar design.

Distributed clutter is defined as radar signals backscattered from a spatially continuous distribution of scatterers with no subset of scatterers predominating. This type of clutter is generally observed from sea return, precipitation return, and returns from most undeveloped land terrain (excluding mountainous terrain). When the radar's wavelength is very small compared to the spatial resolving power of the radar, a reasonable approach is to model distributed clutter as the summation of phase-independent returns from a large number of small scatterers. This leads to representing the clutter as a random process having the statistical properties of Gaussian noise. Hence, the expected (mean) clutter power received from any incremental resolution element (surface or volume region) depends solely on the physical area of that region and a proportionality constant, σ°, which represents the local clutter cross section per unit area, or η, which represents the local clutter cross section per unit volume (see Section 1.3.2.1). The significance of this result is that, if σ° (or η) is specified at each point of the clutter surface, we can completely determine the temporal, spectral, and statistical properties of the clutter input to any radar. Consequently, any description that permits σ° (or η) to be realistically specified over the expected clutter surface can be regarded as a suitable model for distributed clutter [1].

The clutter backscattering coefficient σ° for distributed surface clutter is determined by a number of factors, which include the radar's wavelength, the transmitted and received polarizations, the grazing angle of the incident wave to the surface, and the type of surface or terrain. The grazing angle to the terrain surface can never be known exactly, and hence must be specified in terms of the mean surface. This will cause σ° to fluctuate for a fixed grazing angle due to local variations in the surface slope. Similarly, to isolate and identify every conceivable environment is not possible, resulting in the use of broad classifications for the surface type (e.g., desert, farmland, sea state number). Within each broad environmental type, σ° will vary further because of the large variety of possible subclasses contained within each classification. As is apparent from this discussion, σ° is not given by a precise number for each surface classification, but is a random variable characterized by a probability distribution. In Section 1.3.1, we described the Gaussian or Rayleigh clutter model, which generally represents the statistical properties of σ° for distributed clutter. With this model, the complete first-order statistical properties of σ° are determined by the specification of $\bar{\sigma}^\circ$, the mean value of σ°.

A considerable amount of experimental data is available on the measurement of $\bar{\sigma}^\circ$ for various surfaces. Unfortunately, these data vary widely, which generally reflects the complex nature of the surface and the difficulty in extracting the precise condition under which the data was obtained. This has led to the use of highly averaged values for $\bar{\sigma}^\circ$ that represent the combination of data obtained through many experiments under supposedly similar conditions, but with wide variability in the data [2].

The concept of distributed clutter represents a consistent theory whereby a physical model results in an analytical model (i.e., the Gaussian random process), which can then be used to design and analyze the performance of a doppler radar. With this model, the clutter's amplitude probability density function, its correlation with time and spectral density, and its variation over time and location is completely specified (see Section 1.3). Further, the form of the clutter process is independent of the radar's parameters (e.g., pulsewidth and antenna pattern) and homogeneous over the complete clutter field. With these significant advantages, the distributed clutter model is not surprisingly used as the basis for most clutter descriptions and the objective of most experimental programs is to measure the parameters (e.g., $\bar{\sigma}^\circ$, amplitude probability density function, correlation function) associated with this model.

Unfortunately, distributed clutter very rarely exists in nature. Generally, clutter is not stationary; it varies both temporally and spatially. Clutter is sensitive to the parameters of the radar used to induce it and cannot in general be represented by a stationary random process. There is much confusion between its short- and long-term temporal properties and its spatial distribution.

To illustrate some of these points, consider the situation for land clutter, which represents one of the most stringent environments for a doppler radar (see Figures 1.15 and 1.16). Figure 4.1 depicts how land clutter appears on a ground-based search radar with several azimuth beamwidths (1.3° and 6°) and pulsewidths (2 μs and 10 μs) [3]. The spatial variation of the clutter is apparent, as is its sensitivity to the radar's parameters. In addition, the dynamic range of land clutter generally varies over a much wider dynamic range than is predicted by the Rayleigh clutter distribution (e.g., 20 dB from the 5% to the 99% probability points for Rayleigh clutter) [4]. Because of this, many investigators have used the log-normal or the Weibull amplitude probability distributions to describe land and sea clutter with wide dynamic ranges (see Section 1.3.4).

One problem in representing land clutter as distributed clutter is the presence of the many discrete scatterers over developed land terrain. Discrete clutter, sometimes

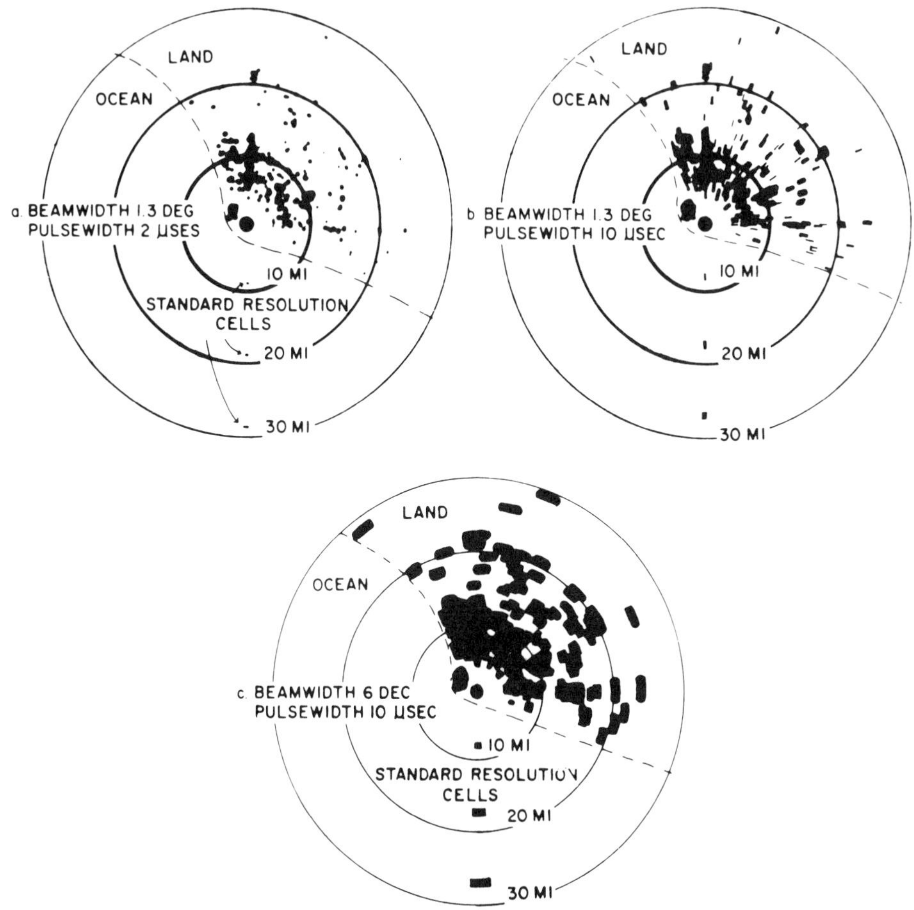

Figure 4.1 Land clutter return for several azimuth beamwidths and pulsewidths.

called *specular* or *point clutter,* is backscattered from physically isolated dominant reflecting objects on the terrain. This clutter includes radar signals backscattered from both developed areas (e.g., cities and urban areas, where the discrete scatterers occur in clusters) and physically isolated structures (e.g., water towers and silos). Radar returns from mountainous terrain may also be regarded as a form of discrete clutter in which the dominant reflectors are the scattering facets of the mountain slopes.

Discrete clutter signals [1,5] are distinguished from distributed clutter in that they generally exhibit larger amplitudes (e.g., the RCS for an individual scatterer might range from 10^2 to 10^6 m^2), fluctuate more slowly, and cannot be described by Gaussian statistics. Experimental evidence shows that discrete scatterers exist in sufficient numbers over land to have a significant effect on radar performance. In general, land clutter can be viewed as a composite between distributed and discrete clutter. The uncertainty of the discrete field of scatterers over land terrain is generally the mechanism that imparts a wide dynamic range to land clutter and its large spatial variability. Note that most experimental clutter measurements over developed land include composite discrete and distributed clutter, which cannot be separated into component parts, and hence measurement statistics of parameters such as $\bar{\sigma}^\circ$ and the probability distribution of σ° represent combined statistical data.

Other anomalies occur when land clutter is viewed at low grazing angles such as with ground-based radar. For relatively flat terrain at low grazing angles, the clutter reflectivity can be viewed as the product of two factors $\sigma^\circ F_c^4$ [6]. The first is the distributed clutter backscatter coefficient (σ°), and the second represents the two-way propagation factor for the radar, clutter, and path geometry. For lower grazing angles, below the critical angle, the propagation factor (F_c^4) produces a significant effect and hence must be evaluated to determine the clutter reflectivity. For higher grazing angles, F_c^4 generally approaches a value of one. The wide dynamic range observed for low grazing angles of both land and sea clutter can be predicted by the dependence of the propagation factor on the fourth power of the clutter source height. A difference of 1 or 2 m of height can result in a change in tens of decibels in observed reflectivity [6].

Backscatter from clutter returns can be either coherent or noncoherent [4]. If the phase of a wave is constant or varies in a deterministic manner, it is called *coherent*. If the phase of a wave is random and uniformly distributed over a 2π interval of phase, it is called *noncoherent*. The clutter cross section can be expressed as

$$\sigma = \left| \sum_{i=1}^{n} \sqrt{\sigma_i}\, e^{j\phi_i} \right|^2$$

where σ_i is the individual scatterer RCS and ϕ_i is its phase. The phase factor $e^{j\phi_i}$ includes phase change on reflection and phase delay due to the range of the radar to the scatterer. As an example, if all scatterers were of equal strength ($\sigma_i = \sigma$), for noncoherent scattering $\bar{\sigma}_c = \Sigma_{i=1}^n \sigma = n\sigma$, whereas for coherent scattering with all scatterers in phase $\bar{\sigma} = |\Sigma_{i=1}^n \sqrt{\sigma}|^2 = |n\sqrt{\sigma}|^2 = n^2\sigma$. Hence, if there were ten $\sigma = 1\ m^2$ scatterers in a clutter patch, the RCS could vary from 10 m^2 to 100 m^2 depending on the type of scattering involved.

In general, the scattering properties of clutter return are dependent on the incident polarization and the properties of the scattering surface. The rectangular scattering matrix can be expressed in terms of the clutter RCS as

$$S_c = \begin{bmatrix} (\sigma_{HH})^{1/2}\, e^{j\phi_{HH}} & (\sigma_{HV})^{1/2}\, e^{j\phi_{HV}} \\ (\sigma_{VH})^{1/2}\, e^{j\phi_{VH}} & (\sigma_{VV})^{1/2}\, e^{j\phi_{VV}} \end{bmatrix} \tag{4.1}$$

where the subscripts denote horizontal (H) and vertical (V) polarization on transmitting and receiving, respectively. For a monostatic radar $(\sigma_{HV})^{1/2} = (\sigma_{VH})^{1/2}$ and $\phi_{HV} = \phi_{VH}$, resulting in a symmetrical matrix. Scattering surfaces many times depolarize the incident wave. For example, measurements on deciduous or pine trees showed ratios of $\bar{\sigma}_{HH}/\bar{\sigma}_{HV}$ and $\bar{\sigma}_{VV}/\bar{\sigma}_{VH}$ in the range of 3 to 10 dB [4].

4.1 SURFACE CLUTTER

Clutter return backscattered from the land or sea is caused by the various scatterers located within the radar resolution cell of the radar. The scatterers are characterized by the normalized parameter $\sigma°$, which represents the radar cross section per unit area of surface. Thus, the radar cross section can be expressed as

$$\sigma_c = \sigma° A_c \tag{4.2}$$

where $\sigma°(\text{m}^2/\text{m}^2)$ is the normalized clutter reflectivity per unit area and $A_c(\text{m}^2)$ is the illuminated surface area within the radar resolution cell. The clutter power returned from the clutter patch with RCS of σ_c can be expressed as

$$P_c = \frac{P_t G_t^2 \lambda^2 F_c^4 \sigma_c}{(4\pi)^3 R_c^4 L_c} \tag{4.3}$$

where P_t is the transmitter power, G_t is the antenna gain, λ is the radar wavelength, L_c is the clutter transmission and beamshape losses, F_c^4 is the clutter pattern propagation factor, and R_c is the range to the clutter patch. For range-ambiguous radars, the clutter RCS must be weighted by the range and pattern propagation factor to the various clutter patches resulting in

$$P_c = \frac{P_t G_t^2 \lambda^2}{(4\pi)^3 L_c} \sum_{i=1}^{n} \frac{\sigma_{ci} F_{ci}^4}{R_{ci}^4} \tag{4.4}$$

where σ_{ci}, F_{ci}^4, and R_{ci}^4 are the respective values associated with each of the ambiguous clutter elements. Note that, in a range-ambiguous radar (e.g., a PD radar), the clutter element closest to the radar generally contributes the most incremental clutter power. Also, the pattern propagation factor is most significant for low grazing angles and often can be represented as the two-way power antenna radiation pattern (e.g., $F_c^4 = \exp(-5.54\phi^2/\phi_{\text{el}}^2)$ for a Gaussian-shaped beam) as described in Section 3.3.

The surface clutter area (A) illuminated by the radar can be represented by the two cases depicted in Figure 4.2 [7]. The first case (beamwidth-limited) applies for large grazing angles where the range associated with the pulsewidth ($R_\tau = c\tau/2$) is

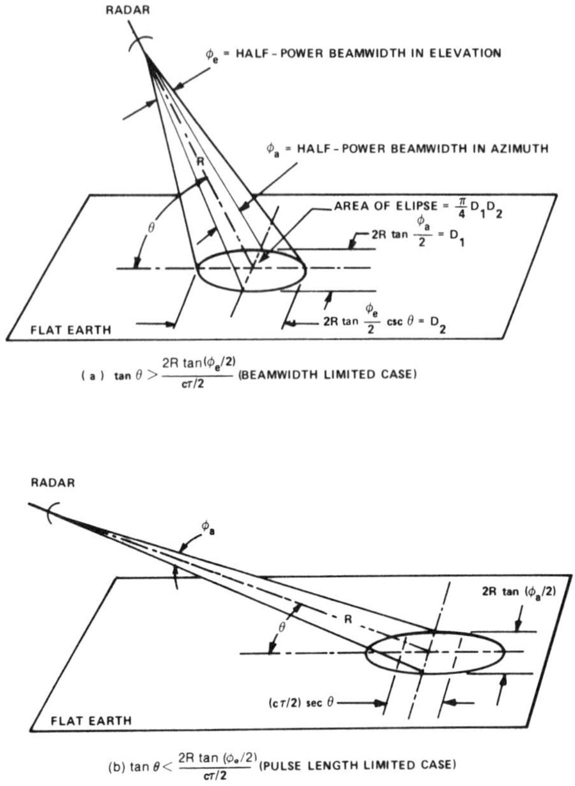

Figure 4.2 Surface clutter area illuminated by a pulsed radar.

large compared to the projected length in the radial direction of the beam. The second, more usual case (pulse-length-limited) applies for low and medium grazing angles where the pulse length derived range is less than the projected radial beam length. For small two-way power azimuth (θ_{az}) and elevation (ϕ_{el}) beamwidths less than 10° the clutter area can be expressed for the beam-limited case as

$$A_c = \frac{\pi R^2}{4}\,\theta_{az}\phi_{el}\csc\psi; \quad \left(\tan\psi > \frac{\phi_{el}R}{R_\tau}\right) \tag{4.5}$$

and for the pulsewidth-limited case as

$$A_c = R\theta_{az}\frac{c\tau}{2}\sec\psi; \quad \left(\tan\psi < \frac{\phi_{el}R}{R_\tau}\right) \tag{4.6}$$

Equation 4.6 generally applies to surface-based radars, where the clutter patch is simply the projection of the area formed by the azimuth beam dimension at the clutter's range ($R\theta_{az}$) and the projected range associated with the radar pulsewidth ($c\tau/2 \cdot \sec\psi$). Also note that the one-way power beamwidths for θ_{az} and ϕ_{el} are often used in place of the more accurate two-way power antenna beamwidths. The resulting difference in equations (4.5) and (4.6) is less than 1.5 dB, which generally is insignificant when compared to the uncertainty in the associated clutter data parameters used to compute the effective clutter RCS.

As an example, consider an S-band surface-based radar (i.e., ASR-9 with τ = 1.05 μs and θ_{az} 1.3°), which views land clutter at a range of 10 nmi with a corresponding grazing angle of $\psi = 1°$. The pulse-limited clutter area is then found from equation (4.6) as $A_c = 6.6 \times 10^4$ m^2. If the clutter consisted of wooded terrain, the backscatter coefficient would be approximately $\sigma° = -30$ dB, so that the clutter RCS ($\sigma_c = \sigma° A_c$) would be $\sigma_c = 66$ m^2. Over desert terrain, $\sigma° = -45$ dB, so that the clutter RCS would be $\sigma_c = 2$ m^2. For sea clutter (i.e., sea state 3), the backscatter coefficient is further reduced to $\sigma° = -50$ dB at 1° grazing angle. The RCS for sea clutter with these parameters would then be $\sigma_c = 0.66$ m^2.

As another example, consider an X-band low-PRF airborne radar operating at an altitude of 20,000 ft and looking down on a low-altitude target at 10 nmi slant range. The pertinent parameters of the radar are a pulsewidth of 0.5 μs and an azimuth beamwidth of 3°. First, the grazing angle to the ground clutter patch is given by

$$\psi = \sin^{-1}\left[\frac{h}{R}\left(1 + \frac{h}{2r_e}\right) - \frac{R}{2r_e}\right] \approx \sin^{-1}\left(\frac{h}{R}\right) \tag{4.7}$$

where h is the antenna's height, R is the slant range, and r_e = 4587 nmi is the 4/3 earth radius. The grazing angle can be calculated as 19°. Substituting into (4.6) results in $A_c = 7.7 \times 10^4$ m^2. For wooded terrain at $\psi = 19°$, $\sigma° = -20$ dB. The clutter RCS is then equal to $\sigma_c = 770$ m^2.

In the preceding examples, if we assume that both the target and clutter return are correlated completely over the period that the antenna dwells on the target and the clutter power is much greater than the noise power, the signal-to-clutter power ratio can be expressed as

$$\frac{S}{C} = \frac{P_s}{P_c} = \frac{\sigma_t}{\sigma° R\theta_{az}(c\tau/2)\sec\psi} \tag{4.8}$$

The maximum clutter limited range is then given by

$$R_{max} = \frac{\sigma_t}{(S/C)_{min}\sigma°\theta_{az}(c\tau/2)\sec\psi} \tag{4.9}$$

Note that the clutter-limited range is generally independent of the transmitted power. Also, equation (4.9) can be used for partially correlated targets and clutter if $(S/C)_{\min}$ is defined as the minimum signal to clutter required for the actual correlation. Note that the substitution of the constant-γ model into (4.9), at angles above the critical angle, leads to clutter that is constant with range, depending instead on radar antenna height.

As a further example, consider the case where the radar observes surface clutter at near vertical incidence. The value of σ° will depend on the surface reflectivity and any quasispecular reflections from generally randomly tilted surface facets. For a 10° antenna beam, a value for σ° on the order of +17 dB has been predicted [8]. Experimental data taken at 3.8 GHz indicate values of σ° over water on the order of +20 dB, whereas for land the values varied from about 0 dB to +10 dB depending on the terrain involved [4]. The clutter power can be calculated by substituting the value for the clutter area given by (4.5) into (4.3) and substituting $G = \pi^2/\theta_{\text{az}}\phi_{\text{el}}$, resulting in

$$P_c = \frac{\pi P_t A_e \sigma^\circ}{64R^2 \sin\psi} \tag{4.10}$$

where $A_e = G_t\lambda^2/4\pi$ is the antenna's capture area, R is the range to the surface, and ψ is the near-vertical incidence grazing angle. The clutter power thus varies inversely as the square of the range to the surface.

In an airborne radar, the strong reflection from near-vertical incidence results in "altitude line" clutter (see Figure 2.9), the doppler frequency spectrum of which is centered about zero frequency. The clutter power can generally be computed from (4.10) as modified to account for reception of the clutter return through the antenna's sidelobe structure (i.e., $P_{\text{al}} = P_c/G_{SL}^2$, where G_{SL} is the ratio of the peak antenna gain to the sidelobe gain in the direction of the clutter). This equation applies in the beam-limited case, where the altitude or beamwidth is such that the trailing edge of the pulse does not reach the ground before the leading edge passes out of the antenna beam. Otherwise, the pulse-limited case applies, where the clutter return is now inversely proportional to the third power of range, resulting in a weaker clutter return.

From a radar scattering viewpoint, the range of possible grazing angles can be divided into three distinct regions: the near grazing incidence region, the plateau region, and the near-vertical incidence region, as depicted in Figure 4.3 [4]. Within each of these regions, the dependence of σ° on the grazing angle and wavelength can be characterized to some extent. However, the boundaries of the three regions change with wavelength, surface conditions, and polarization.

In the near grazing incidence region, σ° varies rapidly with the grazing angle. It generally decreases rapidly with a decreasing grazing angle and longer wavelengths. When the surface in a scattering sense is rough, σ° is consistently larger for all

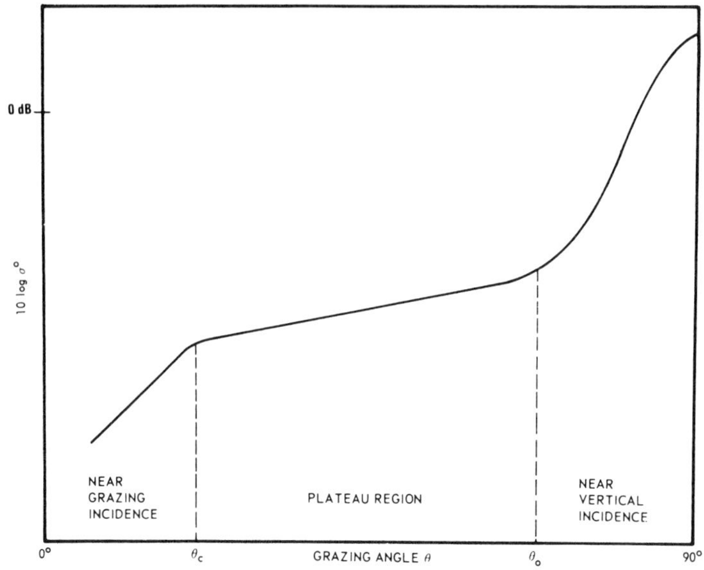

Figure 4.3 General dependence of the radar backscattering coefficient σ° on grazing angle for surface clutter.

grazing angles and frequencies compared to a smooth surface. Generally, roughness can be defined (i.e., Rayleigh criterion) as

$$\sigma_h \sin\psi \geq \lambda/8 \tag{4.11}$$

where σ_h is the rms height of the surface irregularities, ψ is the grazing angle, and λ is the radar's wavelength. Thus, roughness in the scattering sense depends on both surface irregularities and the frequency and grazing angle of the incident radiation. For example, at a grazing angle of 1° and a frequency of 2 GHz ($\lambda = 0.15$ m) a surface with irregularities on the order of $\Delta h \geq 1.07$ m will appear rough. Generally, rough surfaces tend to backscatter more of the incident radiation than smooth surfaces. In the limit, a perfectly smooth surface at less than near-vertical incidence will forward-scatter all incident radiation and σ° will then be zero.

The interference concept is frequently used to explain certain properties of surface clutter observed in the low-grazing-angle region [9]. This concept postulates that the return from each scatterer can be modeled as the combined return from a direct reflection from the scatterer and an indirect reflection scattered from the underlying clutter surface. The clutter backscatter is then modified by the pattern propagation factor F_c^4 that applies for multipath phenomena and generates a lobe pattern caused by destructive and constructive interference [4]. The critical angle, ψ_{c1} (rad), which defines this region, is approximately equal to the angle for the first maximum of the lobe pattern given by

$$\psi_{c1} = \sin^{-1}(\lambda/4\pi\sigma_h) \tag{4.12}$$

where σ_h is the rms deviation of the surface from the mean surface. Equation (4.12) generally applies only for sea clutter illuminated with horizontally polarized radiation (i.e., when the reflection coefficient is $\Gamma = -1$). However, the general form of the equation (i.e., $\psi_{c1} = \sin^{-1}(\lambda/k\sigma_h)$, with k a constant depending on polarization and surface type) could be expected to apply for other situations [10]. The backscatter coefficient for grazing angles less than the critical angle then equals $\sigma^\circ = kF_c^4\sigma_p^\circ$, where F_c^4 is the pattern propagation factor in the region below the first interference lobe maximum:

$$F_c^4 = 16\sin^4(2\pi\sigma_h \sin\psi/\lambda) \tag{4.13}$$

and σ_p° is the value of σ° at the critical angle. Equation (4.13) predicts that the backscattering coefficient σ° for surface clutter will vary approximately proportional to ψ^4 and λ^{-4} below the critical angle (ψ_c). Some experimental evidence confirms this behavior in accordance with interference theory [4].

Interference theory can also be used to explain the difference in σ° for sea clutter between horizontal and vertical polarization, the difference lying in their respective reflection coefficients. For horizontal polarization, the reflection coefficient for perfectly conducting surfaces (i.e., sea clutter) has a magnitude of one and a phase shift of 180° (i.e., $\Gamma = -1$). This results in destructive interference at grazing angles below the critical angle (ψ_{c_1}) given in (4.12). However, for vertical polarization the situation is quite different. The magnitude and phase of the reflection coefficient each vary with both grazing angle and frequency. The grazing angle, for which the phase is $\pi/2$ and the magnitude is a minimum, is known as *Brewster's angle*. Above Brewster's angle, the reflection phase shift becomes small (i.e., $\phi \rightarrow 0$); hence, destructive interference can occur only well below Brewster's angle. Therefore, the critical angle for vertical polarization depends on the grazing angle at which the frequency-dependent Brewster's angle occurs and generally is independent of the surface roughness.

The shape of the backscattering coefficient curve (σ°) for horizontal and vertical polarization is depicted in Figure 4.4 for UHF. In this frequency range, Brewster's angle is expected to be on the order of 3° to 4°, whereas the critical angle for horizontal polarization and a moderate sea (σ_h = 4 ft) is on the order of ψ_c = 9°. Hence, the backscattering coefficient (σ°) would be expected to be generally higher for vertical polarization than for horizontal polarization at low frequencies. As the radar's frequency is increased, Brewster's angle increases and the critical angle decreases to become approximately equal (i.e., ψ_c = 5°) at 1 GHz. For further increases in frequency,

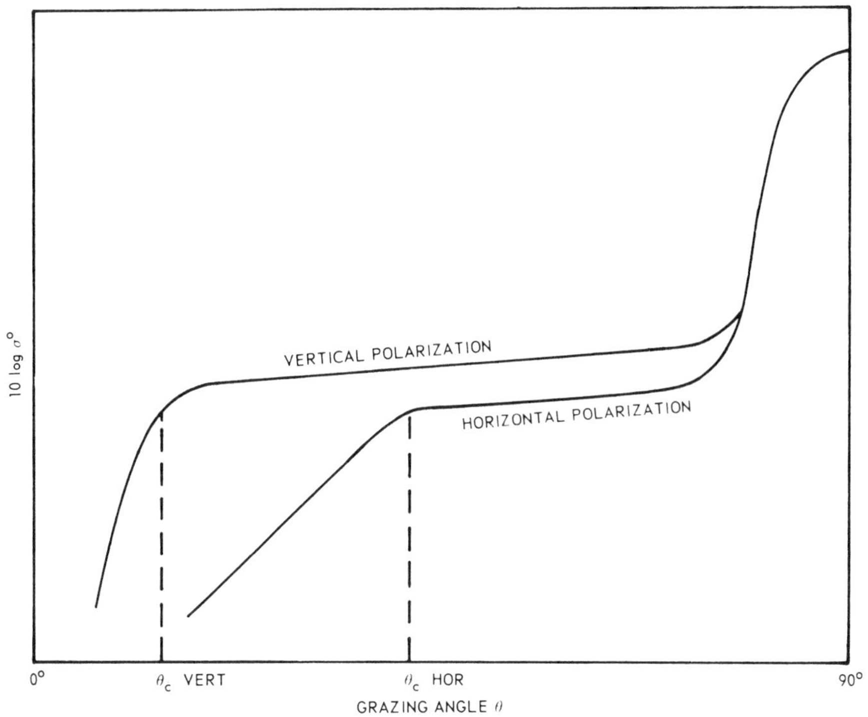

Source: Long et al. (1965).

Figure 4.4 General appearance of radar backscattering coefficient σ° at UHF.

Brewster's angle becomes much greater than the critical angle so that the backscattering coefficients (σ°) at near-grazing angle incidence would be expected to coalesce for horizontal and vertical polarization. For example, at 10 GHz, Brewster's angle is 8° and the critical angle is usually less than 1° [4].

In general, the surface clutter return for vertical polarization tends to be greater than for horizontal polarization. The return for vertical polarization tends to be stronger because the phase lag of the reflection coefficient is such that more energy exists at points close to the surface. As λ decreases or the scatterers are raised higher above the surface (corresponding to a rougher surface) more of the interference pattern tends to be averaged and its value at points close to the surface becomes less significant. For very small λ or a very rough surface, the phase of the reflection coefficient becomes unimportant and σ° depends only on the average field strength (i.e., the magnitude of the reflection coefficient). Finally, at very small grazing angles, only those scatterers with exceptional height will be illuminated causing quasispecular reflections [9]. For sea clutter, this results in spiky returns that are characterized by non-Gaussian probability distributions (see Section 1.3.4).

In the plateau region, the change of σ° with grazing angle is slow and generally can be approximated by the constant gamma model [4], given by

$$\sigma^\circ = \gamma \sin\psi \tag{4.14}$$

where γ is a constant and ψ is the grazing angle. In this region, most surfaces generally appear rough in a scattering sense (see equation (4.11)) and the scattering is incoherent. The backscattering coefficient (σ°) changes slowly with wavelength (i.e., σ° is proportional to λ^{-1} in the microwave region), whereas its magnitude generally increases with the surface roughness. The transition angle between the plateau and near-vertical incidence region is on the order of 60°, but this angle varies with the surface roughness whereas the transition is gradual rather than sharp. The amplitude probability distribution of the surface clutter in the plateau region is generally characterized by Rayleigh statistics.

At near-vertical incidence, the backscatter is larger when the surface is smooth than when it is rough. This is in contrast to the plateau region, where the opposite condition applies. Clutter returns in this region become significant for airborne radars that experience strong returns from terrain directly underneath the aircraft. This type of return generally enters the radar through its antenna sidelobes and has zero doppler frequency. The clutter return from near vertical incidence is strongest over smooth water or land (e.g., asphalt or concrete roads) and weakest over rough terrain (e.g., sea clutter with high sea states and wooded or cultivated land) [4].

4.1.1 Land Clutter

Land clutter is one of the most difficult types of clutter to characterize because it generally lacks a homogeneous structure. The large variation of natural and human-made objects over most terrain makes this clutter difficult to classify into type or obtain consistent experimental data. Under apparently similar conditions, experimental measurements obtained by using an airborne radar will differ from those of a ground radar. One explanation is that airborne measurements tend to smooth the large spatial variations prevalent in land clutter data (thereby producing a spatial average), whereas fixed ground measurements tend to accentuate shadowing and geometric anomalies, which are then averaged over time to produce smoothed statistical clutter data [7]. The result of this uncertainty in land clutter data is that most characterizations are composed of highly smoothed data that must be used with caution in the design of doppler radars.

The general approach followed in describing land clutter is to state its properties in terms of an average backscatter coefficient ($\bar{\sigma}^\circ$), an amplitude probability distribution, and a power spectral density. However, because of the spatial variability of land clutter, we also need to specify its spatial distribution. One measure of the spatial distribution is to consider σ° as a random variable, the probability distribution of which represents values of σ° that may be found in independent spatial cells [11]. The specification as to what may constitute an independent spatial cell is an open question. One study of this effect, based upon high-resolution radar data, considered triangular correlations consisting of a short correlation distance (i.e., no correlation after 50 ft) and a long correlation distance (i.e., no correlation after 1000–2000 ft), which was a function of the type of terrain involved [1].

Some general characteristics of land clutter may be deduced from the general properties of surface clutter. The properties of smooth terrain (see equation (4.11)), such as pavement, desert, or very flat land, follow the general characteristics of surface clutter illustrated in Figure 4.3, which depicts the three regional variations of σ° with grazing angle. Within this class is an increase of σ° with frequency usually no faster than linear with the radiation frequency. In the interference region, the reflection coefficient of terrain is tempered by the dielectric properties of the surface, which reduces its magnitude (i.e., $\rho < 1$). Because this factor determines the degree of destructive interference caused by the reflected wave, land clutter could be expected to exhibit a slower variation of σ° with grazing angle (ψ) in this region and also be less sensitive to polarization changes (e.g., horizontal *versus* vertical) than for a perfectly reflecting surface.

As the surface roughness of the terrain increases, the various regions tend to merge into a single region. In this region, the diffusion law applies and the backscattering coefficient (σ°) exhibits a dependence on grazing angle given by the constant-γ relationship ($\sigma^\circ = \gamma \sin\psi$). However, σ° does not tend to zero as the grazing angle approaches zero, but rather approaches some finite value that apparently depends on the frequency and terrain [12]. The backscattering coefficient exhibits some variation with polarization, but no systematic effects are definable [7]. Frequency variations are, at most, a linear increase of σ° with frequency [13], but the experimental data defining this variation has such a large variance that it is difficult to define a systematic change with frequency or state that any difference really exists [1]. In urban areas, certain building surfaces may be more effective flat-plate reflectors when the surface irregularities are small compared to the radiation wavelength, which might cause an increase in σ° for a decrease in frequency [10].

Examples of rough terrain are mountains, cities, and urban areas with a significant number of human-made structures. However, other terrains might be considered rough (in the scattering sense) if their irregularities would become significant when compared to the wavelength of the incident radiation (i.e., $\sigma_h \sin\psi \geq \lambda/8$). For example, at X-band and a grazing angle of 30°, a surface would be considered rough if $\sigma_h \geq \lambda/4 = 0.75$ cm. At this frequency, most land surfaces would appear rough, and the rough terrain model characteristics would apply.

For surface-based radar, land clutter is generally returned from scatterers with a grazing angle of less than 1°. Very little experimental data is available in this region to determine the magnitude and extent of land clutter. An analytical model developed for this region (called the *horizon and transhorizon region*) has been used to determine the magnitude of clutter returns at X-band for both level and hilly terrain [6]. The essence of this model is to express the backscatter coefficient as the product of the two-way pattern propagation factor (F_c^4) and the constant-γ backscatter coefficient ($\sigma^\circ = \gamma \sin\psi$). The X-band clutter reflectivity predicted over level terrain depicted in Figure 4.5 indicates that due to diffraction low clutter levels may be returned from beyond the geometrical horizon. These returns may be significant in overpowered or sensitive radars designed to detect small RCS targets. A similar curve for clutter reflectivity over rolling or hilly terrain is depicted in Figure 4.6. Note that the clutter is extended to greater ranges and the decline beyond the geometrical horizon is less steep. Figure 4.7 depicts the effect of frequency on the clutter reflectivity at low grazing angles for both level and hilly terrain. The effect at UHF is most pronounced where the increased scatterer height (from σ_h = 1 m to σ_h = 10 m) can increase the clutter strength by the order of 60 dB. At these low grazing angles, the high sensitivity to terrain height irregularities will cause large variations in clutter reflectivity, and the clutter's amplitude probability distribution will generally be described by non-Gaussian statistics.

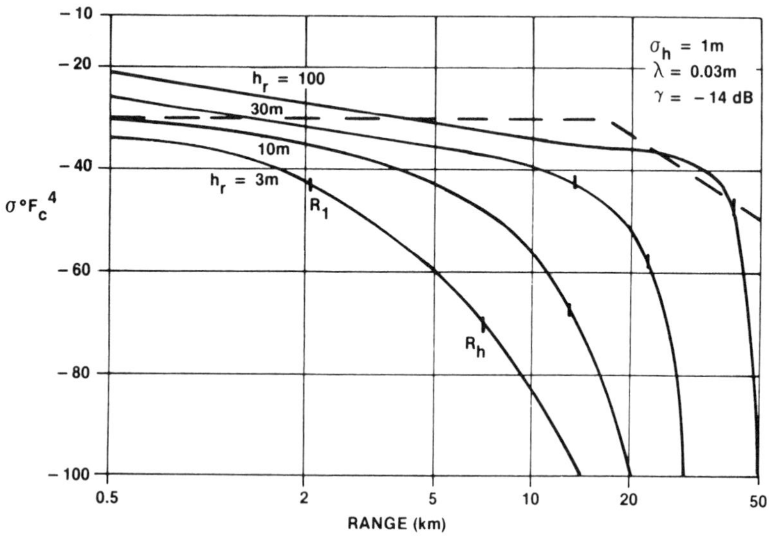

Figure 4.5 Land clutter backscatter coefficient for low grazing angles at X-band over level terrain.

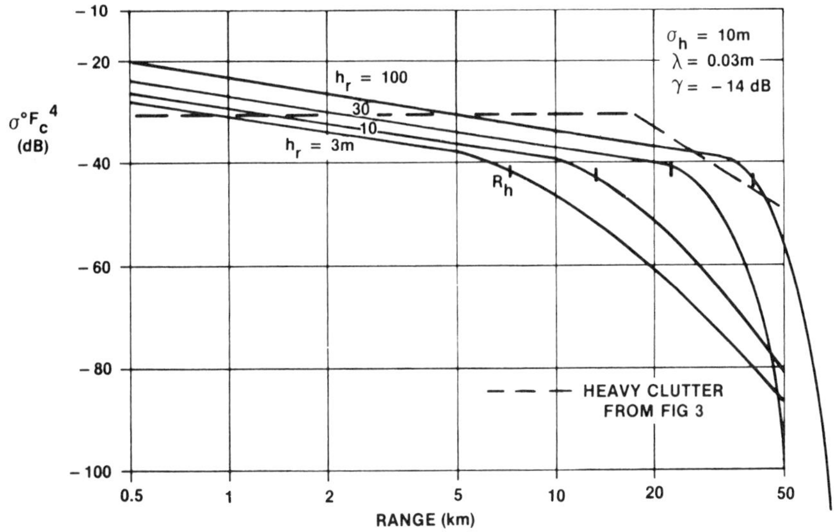

Figure 4.6 Land clutter backscatter coefficient for low grazing angles at X-band over rolling or hilly terrain.

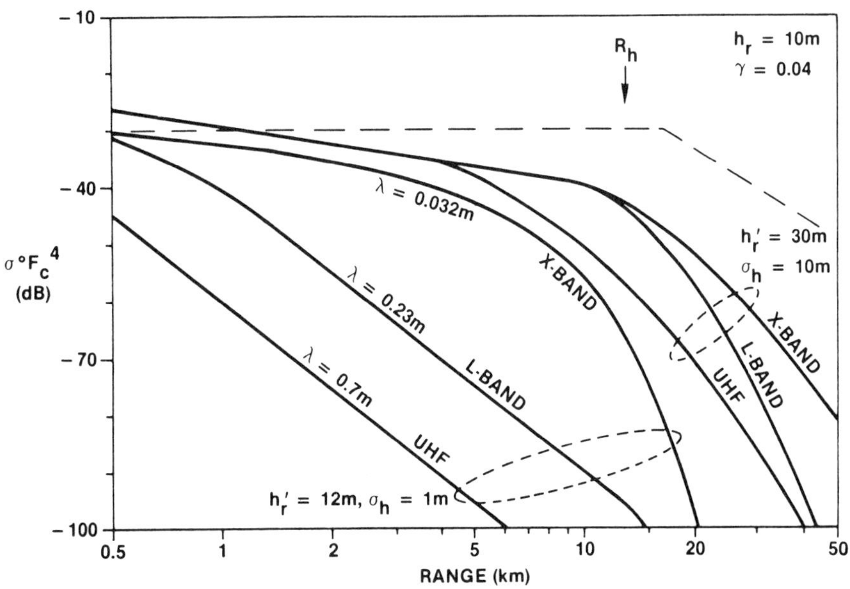

Figure 4.7 Land clutter backscatter coefficient for low grazing angles as a function of frequency.

4.1.1.1 Average σ° for Various Terrains

The backscatter coefficient (σ°) for land clutter is difficult to specify because of the wide variety of terrain types and spatial variability of this parameter. Nevertheless, σ° is a useful parameter for describing extended target areas because the radar cross section is normalized with respect to the surface area illuminated by the radar. The wide variability of the data makes the use of an average value of σ° an appropriate parameter, which will again be designated $\bar{\sigma}^\circ$. For airborne measurement data, $\bar{\sigma}^\circ$ naturally becomes a spatial average. For ground-derived data, $\bar{\sigma}^\circ$ becomes a temporal average for high-resolution data and a composite (both temporal and spatial) average for low-resolution data. Further, many clutter bases represent heavily averaged data that include measurements from many experimental programs that are normalized and then averaged.

In general, $\bar{\sigma}^\circ$ depends on grazing angle, frequency, polarization, surface roughness, and the type of terrain involved; polarization effects are not pronounced. The wind does not generally affect $\bar{\sigma}^\circ$, but does affect the spectra of some types of land clutter. Reflectivity values for land clutter are generally higher than any other form of radar clutter.

Different terrain types are categorized into generic classes to systematize the data into a useable form. Note that the reflection characteristics vary greatly within each terrain type. A general classification of types is as follows:

1. relatively flat land, deserts;
2. rural rolling countryside, farmland;
3. forests, heavy woods, jungle;
4. mountainous regions;
5. urban areas with a high density of artificial structures [7].

Over the years, a number of major efforts have been conducted to collect and provide a definitive model for the backscattering coefficient over land. We shall describe and discuss one effort, associated with the Overland Radar Program, and give several other data summaries. Note that most summaries use a similar database, but differ in their interpretation of the data or extrapolations to fill voids in them.

Figure 4.8 depicts the average backscattering coefficient ($\bar{\sigma}^\circ$) for five terrains associated with distributed land clutter. The curves represent an average of about 24 different experimental programs that measured σ° associated with the types of terrain listed. The data utilized span a frequency range from UHF to X-band, involve both horizontal and vertical polarizations, with grazing angles from 1/4° to 70° [1, 14, 15]. The data, with the exception of those concerned with desert terrain, represent an average of all frequencies and polarizations associated with a particular type of terrain. The rationale for this approach is that the data revealed no systematic trend with frequency or polarization and hence represented fluctuations about a mean value.

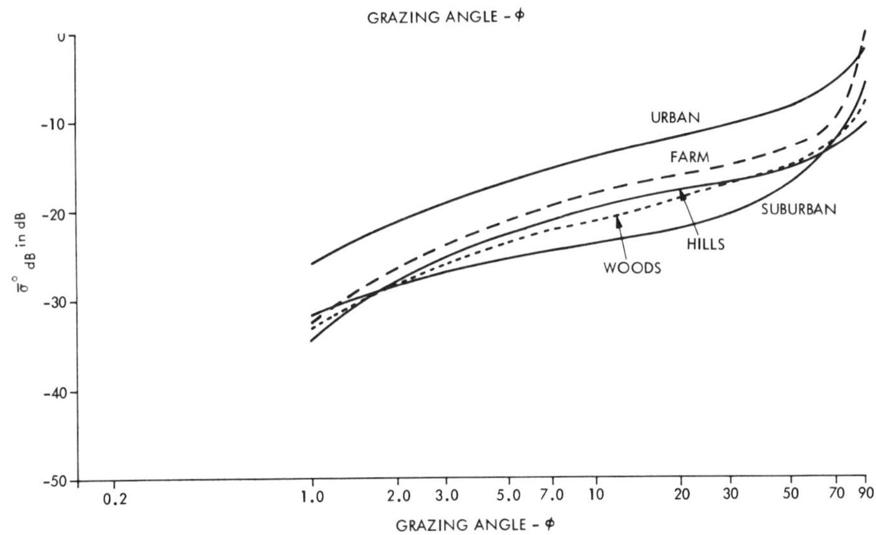

Figure 4.8 Mean backscattering coefficient $\bar{\sigma}^\circ$ dB for five terrains.

The backscatter coefficient data illustrated in Figure 4.8 extend down to a grazing angle of 1° and up to 90°. However, the data utilized provided high statistical confidence only over a range of 10–90°. The data below 10° were extrapolated from the data in the 10–90° range by using a constant-γ reflectivity model (i.e., $\sigma^\circ = \gamma \sin\psi$). The data points in this region (i.e., 1–10°), although sparse, tended to confirm the theoretical model used to extend the higher angle data into this region [1].

The curves for the backscattering coefficient presented in Figure 4.8 represent the average for many data points. These curves are intended to represent the statistical spatial average for $\bar{\sigma}^\circ$ within the spatially independent resolution patch that will be given later in Table 4.4 for the various terrains listed. Further, the clutter is assumed to be log-normally distributed within each independent elemental clutter patch. The random variable σ°_{dB} (σ° plotted on a dB scale) for each elemental clutter patch will then be normally distributed. Further, σ°_{dB} for the radar's surface clutter cell (formed by a number of elemental clutter patches) will also be normally distributed. The standard deviation for each elemental clutter patch of the backscattering coefficient σ°_{dB} is depicted in Figure 4.9 for farmland clutter and in Figure 4.10 for woodland clutter (see also Figure 4.52 for other terrains).

The surface clutter resolution cell will, in general, be larger than the resolution patch that will be given in Table 4.3, resulting in a number (n) of independent clutter patches located within the overall surface clutter resolution cell. The spatial amplitude distribution of σ°_{dB} will then be normally distributed with standard deviation $s_2 = \Delta\sigma^\circ_{\text{dB}} = \Sigma_{i=1}^{n}(\Delta\sigma^\circ_{\text{dB}i})^2/n$. For example, if nine independent cells ($A_c = 10^6$ ft^2) were

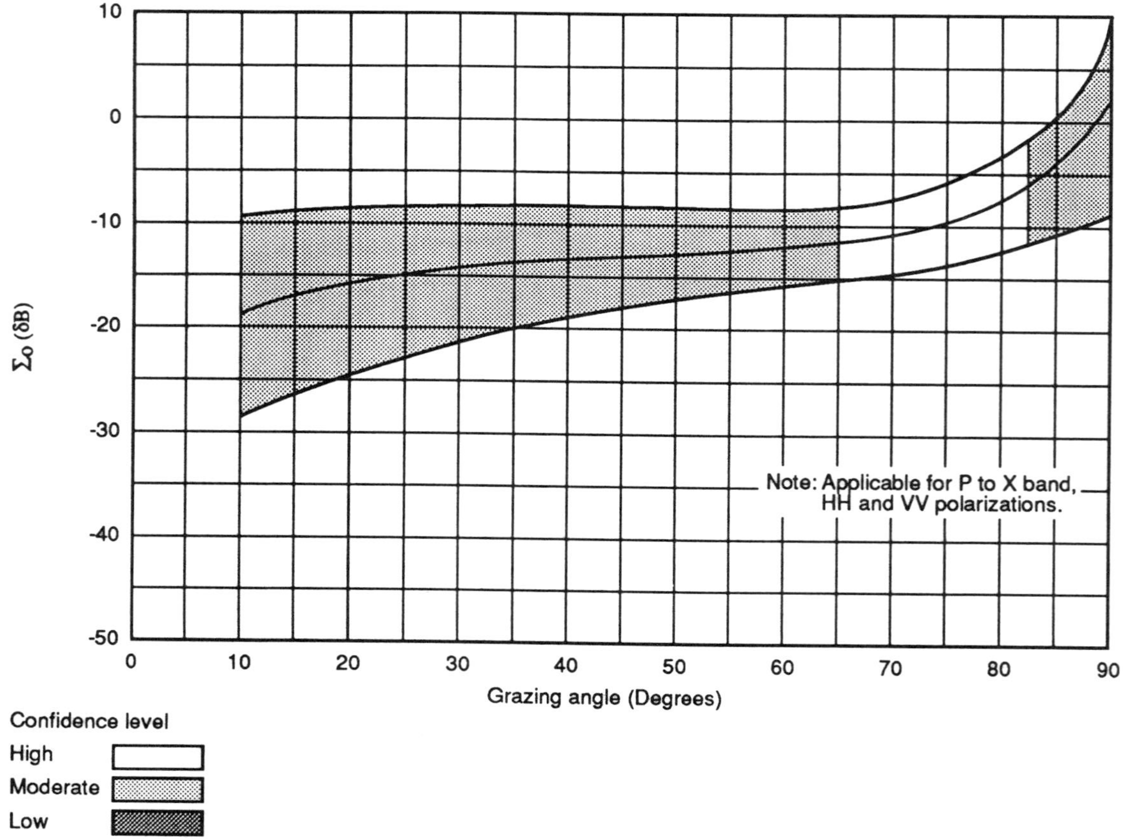

Figure 4.9 Range of backscatter coefficient for farmland clutter.

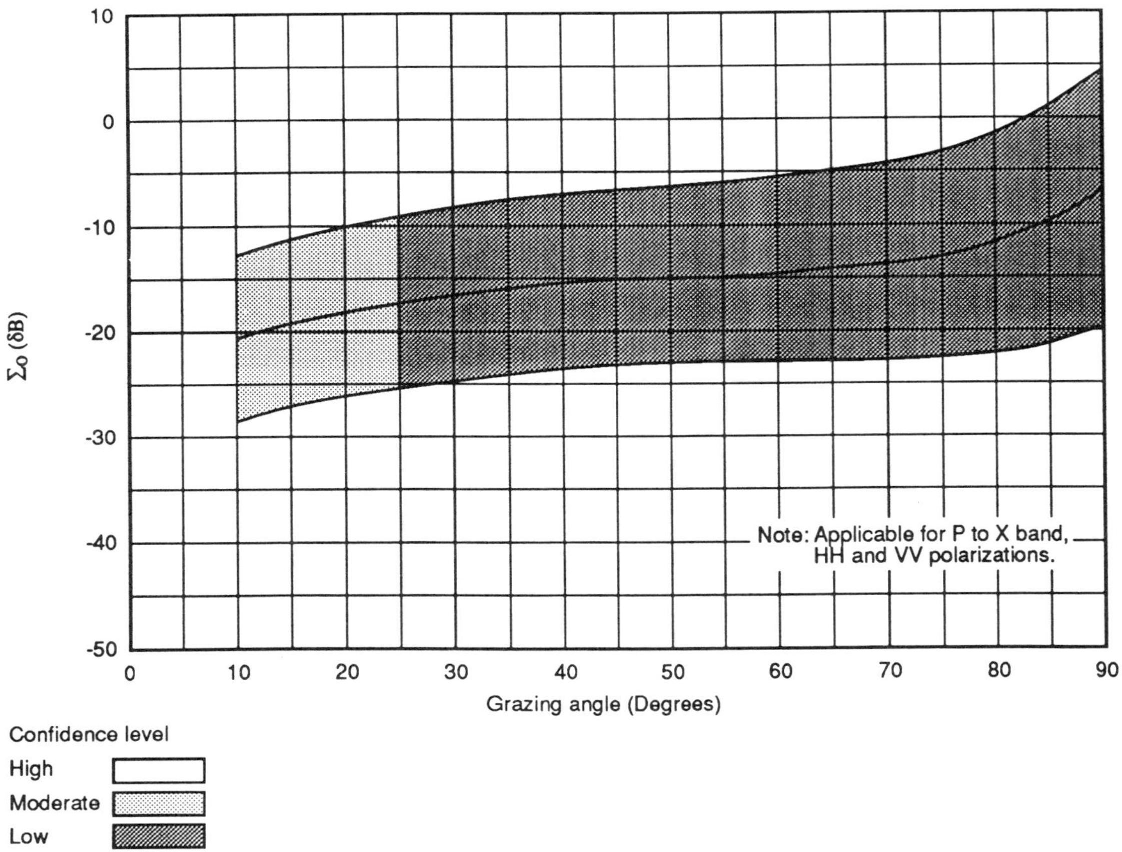

Figure 4.10 Range of backscatter coefficient for woodland clutter.

located within the surface clutter resolution cell for woodland clutter ($\bar{\sigma}^{\circ}_{dB} = -20$ dB and $s_2 = 7.5$ dB at 10°), then the standard deviation of the clutter patch would be 2.5 dB. Thus (see Figure 4.10), at the 84th percentile (the one σ point), the value of σ°_{dB} would be less than -17.5 dB, and at the 99th percentile (the two σ point), the value of σ°_{dB} would be less than -15 dB. Note that the temporal variation (e.g., Rayleigh distribution with mean σ°_{dB}), which is important for dynamic range considerations in doppler radars, would be superimposed upon the spatial amplitude distribution.

Figure 4.11 depicts the average backscatter coefficient ($\bar{\sigma}^{\circ}$) associated with desert terrain. These curves show that $\bar{\sigma}^{\circ}$ increases with increasing frequency. The curve at X-band was derived from high fidelity data available from 10–70° grazing angle. Available data at other frequencies indicated a frequency dependence, but were not of sufficient fidelity to allow a complete determination of $\bar{\sigma}^{\circ}$ *versus* grazing angle. Therefore, the X-band curve was modified by the factor 40–10 log$f$(MHz)dB to obtain the curves at S-band and UHF. These curves were extended into the grazing angle region below 10° by using a constant-γ model ($\sigma^{\circ} = \gamma \sin\psi$). The available data at near-vertical incidence showed no frequency dependence, and hence the curves at all frequencies were linearily interpolated from their value at 70° into a single value at 90°, which was determined from the X-band data [1].

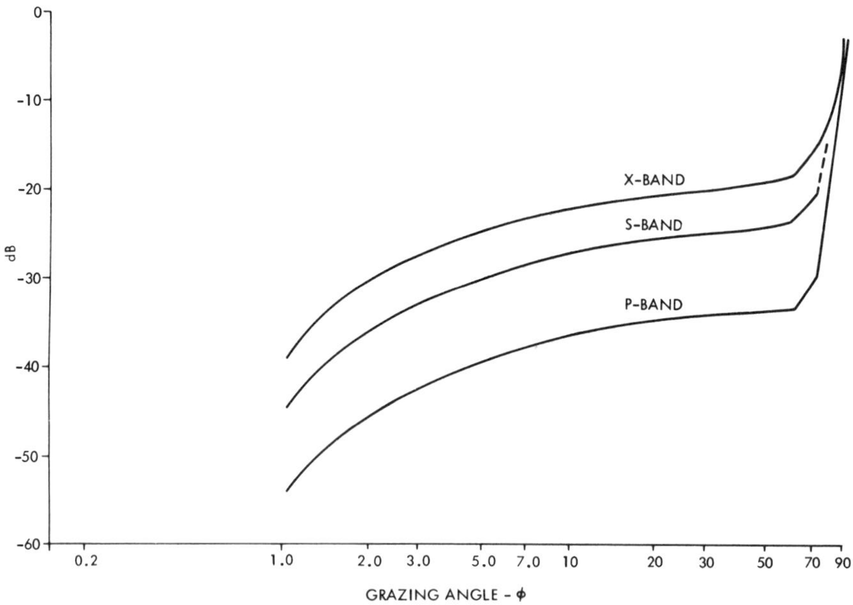

Figure 4.11 Mean backscatter coefficient $\bar{\sigma}^{\circ}$ dB for desert terrain.

The confidence interval for desert terrain that contains 90% of the data points is depicted in Figure 4.12 for X-band data. These X-band data should be scaled by the factor 40 − 10 logf_o(MHz) to provide confidence intervals for other frequency bands. The data analyzed showed no polarization dependence for desert terrain, and hence the curves apply for both horizontal and vertical polarization.

Another set of data that characterizes the mean backscattering coefficient ($\bar{\sigma}^\circ$) at S-band (λ = 0.1 m) is depicted in Figure 4.13 [13]. These data include curves for mountainous terrain and smooth pavement not provided in the Overland Radar Program data. In the diffuse region where the $\bar{\sigma}^\circ$ (dB) curve is virtually a straight line, the data may be scaled to other microwave frequencies by using the following relation:

$$\sigma^\circ = \sigma^\circ(3\text{ GHz})\cdot f(\text{GHz})/3 \tag{4.15}$$

Further, the critical angle (ψ_c) can be scaled to other microwave frequencies by way of the relation:

$$\psi_c = \psi_c(3\text{ GHz})\cdot 3/f(\text{GHz}) \tag{4.16}$$

The data given in Figure 4.13 indicate a generally smaller value of $\bar{\sigma}^\circ$ for city and farmland terrain than given by the Overland Radar Program data. The curves exhibit the three grazing-angle regions characteristic of surface clutter. At high grazing angles, there is a quasispecular region for which the reflectivity is very high and is higher for smooth than rough terrain. In the intermediate-angle diffuse scattering region, the grazing angle dependence is given by a constant-γ model ($\sigma^\circ = \gamma \sin\psi$). At very low grazing angles, there is an interference region for which the grazing angle dependence is much steeper than in the diffuse region. In the quasispecular region at near-vertical incidence grazing angles where the curves for $\bar{\sigma}^\circ$ rise steeply, there is essentially no frequency dependence.

The linear frequency dependence of land clutter postulated for the model given in Figure 4.13 is generally considered to apply throughout the microwave region if the surface is rough relative to a wavelength (see (4.11)). If the same variation were applied throughout the millimeter-wave (MMW) region, very large values of $\bar{\sigma}^\circ$ would be expected. Limited data [16], depicted in Figures 4.14, 4.15, and 4.16 for backscatter from wooded terrain, indicate that the linear variation does not apply for these higher frequencies. However, the data show that the average backscatter coefficient ($\bar{\sigma}^\circ$) is generally larger in the MMW region, but with an apparent anomaly at low grazing angles, where the frequency variation seems to reverse itself.

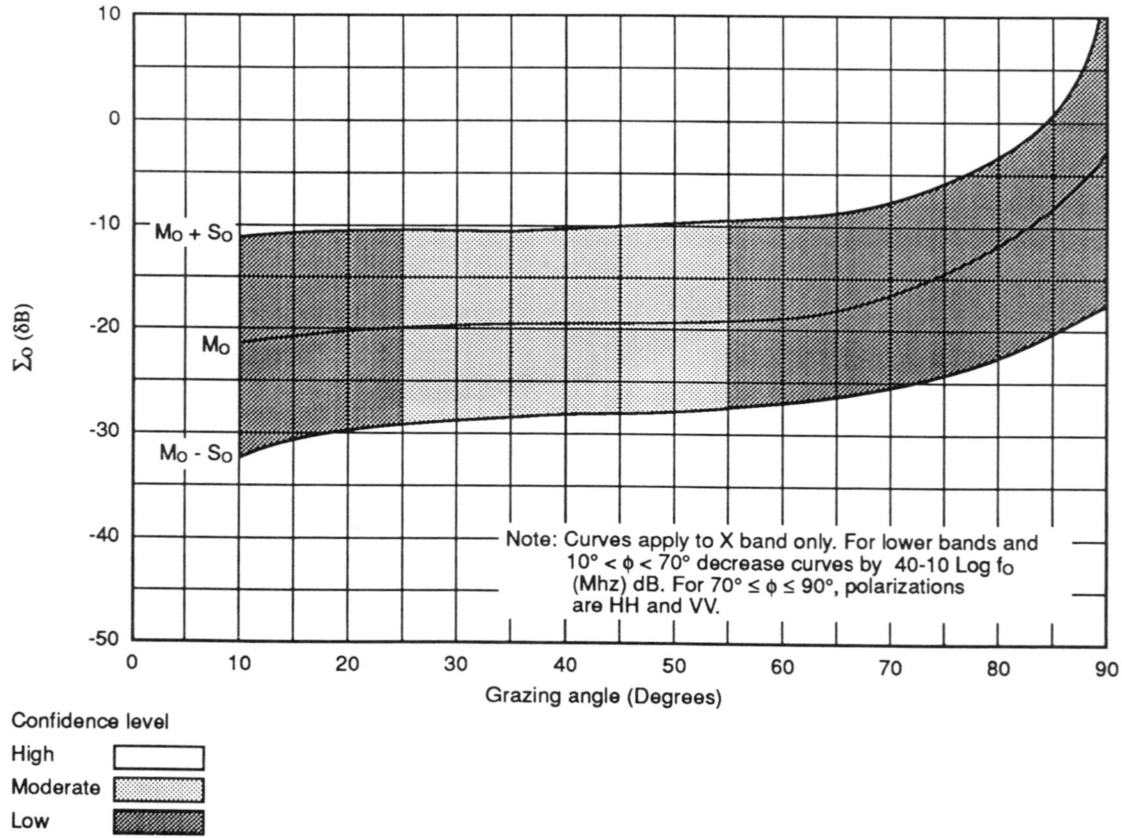

Figure 4.12 Range of backscatter coefficient for desert terrain at X-band.

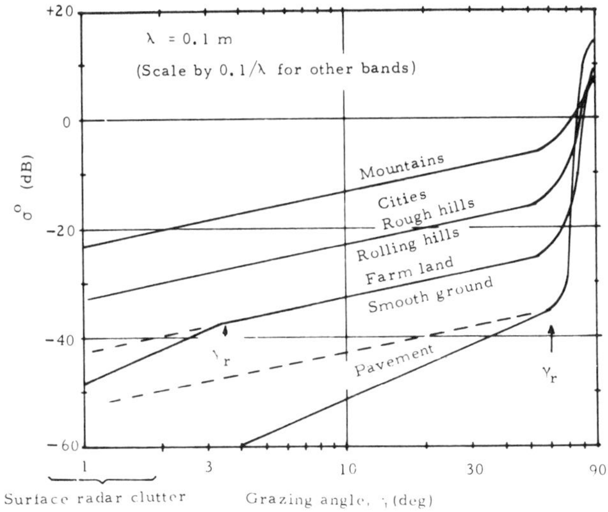

Figure 4.13 Land clutter reflectivity data at S-band.

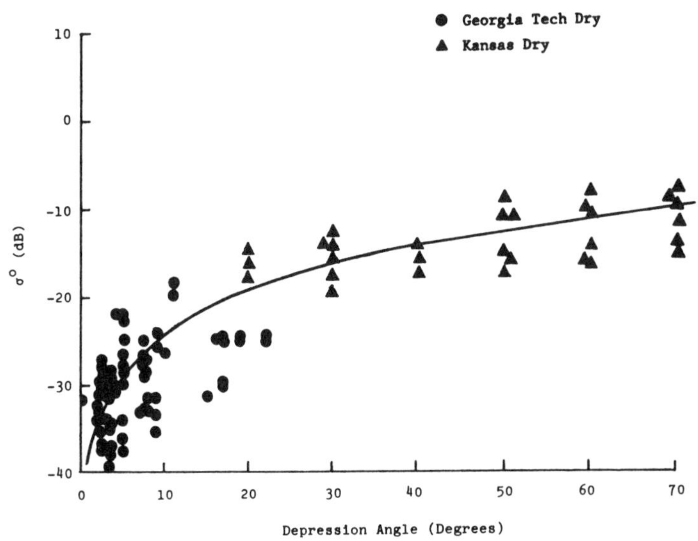

Figure 4.14 Backscatter coefficient for wooded terrain at X-band.

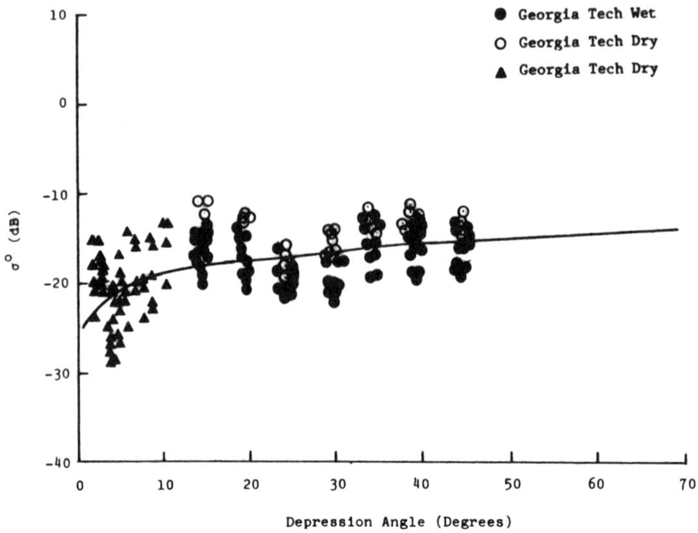

Figure 4.15 Backscatter coefficient for wooded terrain at 35 GHz.

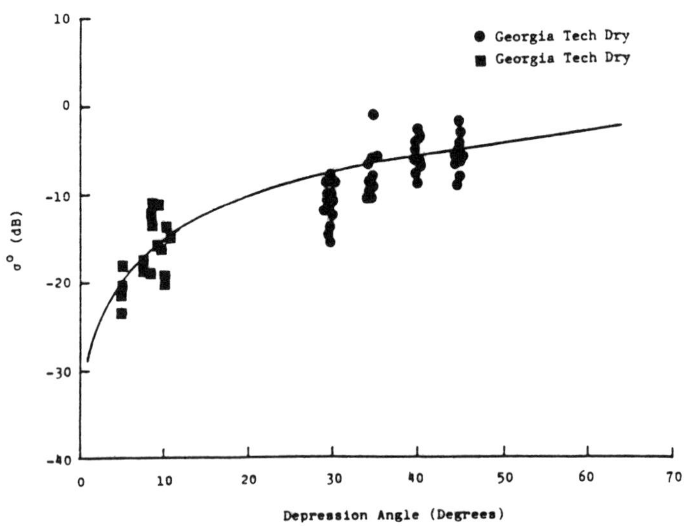

Figure 4.16 Backscatter coefficient for wooded terrain at 95 GHz.

4.1.1.2 Amplitude Probability Distribution Function

Two types of amplitude statistics are useful in describing land clutter: temporal and spatial amplitude distributions. Temporal amplitude statistics refer to fluctuations over time in the amplitude of the clutter return due to the relative motion of the individual scatterers. The relative scatterer motion can be caused by either the physical movement of the radar platform or antenna scanning, which induces apparent scatterer motion due to the changing aspect angle between the scatterers and the radar. This temporal fluctuation can be described in terms of an amplitude probability distribution function, where the fluctuation rate determines the frequency spectrum of the clutter. The temporal amplitude statistics give information on the percentage of time during which the returns have a given range of values, and the frequency spectrum provides information on how rapidly the values change.

If a ground-based radar illuminates a particular clutter patch, any clutter amplitude fluctuation observed will be due to intrinsic temporal fluctuation of the scatterers due to wind effects. Now, if the radar is pointed at another land clutter patch containing scatterers unrelated to the first patch, a different temporal amplitude distribution will generally be observed. This effect, which is prevalent for land clutter, is due to significant physical variations in the clutter field from point to point. This variability results from the inhomogeneity of the clutter and can be described in terms of a spatial probability distribution that defines the underlying statistics for each point in the clutter field.

When land clutter consists primarily of a large number of randomly distributed scatterers, approximately equal in size, the amplitude variation of the clutter (v_c) can be represented by a Rayleigh probability density function:

$$p(v_c) = v_c \exp[-v_c^2/2\sigma^\circ]/\sigma^\circ; \quad v_c \geq 0 \tag{4.17}$$

where the average clutter power $P_c = 2\sigma^2$. Because the instantaneous clutter power is given by $P = v^2 = k\sigma^\circ$ the clutter's backscatter coefficient (σ°) can be represented by the exponential probability density function given by (see Section 1.3.1)

$$p(\sigma^\circ) = \exp(-\sigma^\circ/\bar{\sigma}^\circ)/\bar{\sigma}^\circ \tag{4.18}$$

where $\bar{\sigma}^\circ$ is the mean clutter backscattering coefficient.

Rayleigh statistics are generally considered to apply for distributed land clutter viewed at high grazing angles ($\psi > 5°$) with a medium- or low-resolution radar (pulsewidth > 0.5 μs) [14]. When discrete clutter (e.g., buildings or other artificial structures) are present in addition to distributed Rayleigh clutter, the clutter's Rician probability density function can be written as [14]

$$p(P) = (1 + m^2) \exp(-m^2 - P(1 + m^2)/\bar{P}_c) I_o(2m\sqrt{(1 + m^2)P/\bar{P}_c})/\bar{P}_c \tag{4.19}$$

where $m^2 = S^2/P_o$ is the ratio of the discrete (S^2) to the distributed (P_o) clutter power and $I_o(\cdot)$ is the modified Bessel function. In this case, the total clutter power $\bar{P}_c = S^2 + P_o$ is proportional to the clutter's mean backscattering coefficient $\bar{\sigma}°$.

Data related to the spatial variation of land clutter are relatively scarce. One report that summarized data from mountainous terrain at S-band, wooded terrain at L-band, and both wooded and cultivated terrain at X-band recommended the use of the Weibull probability distribution to represent the spatial statistics of land clutter [11]. Figure 4.17 depicts X-band data for cultivated land, which show an increasing dynamic range with decreasing grazing angle. These data are fitted with a Weibull

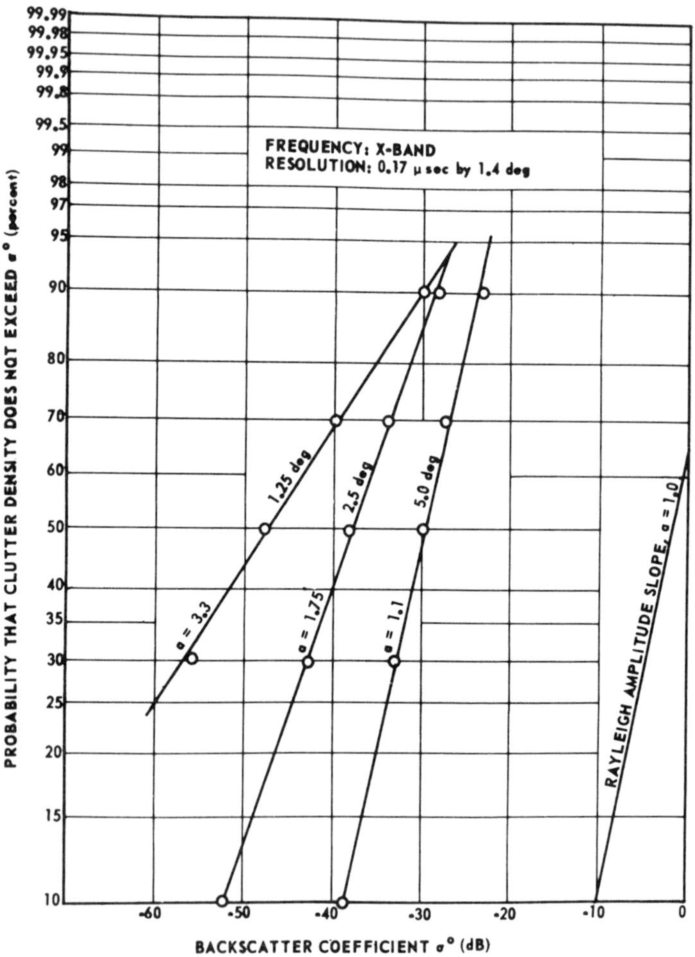

Figure 4.17 Land clutter spatial distribution for cultivated land at X-band.

probability distribution function (see Section 1.3.4), given by

$$P(\sigma^\circ) = 1 - \exp[-\ln 2(\sigma^2/\sigma^\circ_m)^\beta] \qquad (4.20)$$

where $\beta = \alpha/2$ is the Weibull parameter and σ°_m is the median clutter backscattering coefficient that can be related to the mean backscattering coefficient ($\bar{\sigma}^\circ$) as

$$\bar{\sigma}^\circ = \sigma^\circ_m \Gamma(1 + 1/\beta)/(\ln 2)^{1/\beta} \qquad (4.21)$$

Table 4.1 lists some values of β and σ°_m that apply to several different types of terrain [11]. Note that the Weibull distribution reduces to the exponential probability distribution (Rayleigh amplitude) for $\beta = 1$.

Table 4.1
Weibull Parameters for Land Clutter Distributions

Terrain	*Frequency Band*	*Grazing Angle (°)*	*Resolution (μs x°)*	σ°_m (dB)	β
Rocky Mountains	S	—	2.0 × 1.5	−46.3	.256
Forest	X	0.7	.17 × 1.4	−42.5	.253
Forest	X	0.7	.17 × 1.4	−36.4	.266
Cultivated Land	X	1.25	.17 × 1.4	−47.8	.303
Cultivated Land	X	2.5	.17 × 1.4	−38	.571
Cultivated Land	X	5.0	.17 × 1.4	−29.8	.909
Cultivated Land	X	1.25	.17 × 1.4	−46.5	.352
Cultivated Land	X	2.5	.17 × 1.4	−37.8	.429
Cultivated Land	X	5.0	.17 × 1.4	−25.3	1.0

Another distribution used to represent the spatial distribution of land clutter is the log-normal distribution (see Section 1.3.4) [1]. The log-normal probability distribution function for the clutter's backscatter coefficient (σ°) is given by

$$P(\sigma^\circ) = 1 + \text{erf}\,[\ln(\sigma^\circ/\sigma^\circ_m)/\sqrt{2}\sigma_p]/2 \qquad (4.22)$$

where σ_p is the standard deviation of the underlying normal distribution and σ°_m is the median backscatter coefficient. The mean backscatter coefficient is given by

$$\bar{\sigma}^\circ = \rho\sigma^\circ_m \qquad (4.23)$$

where ρ is the mean-to-median ratio given by

$$\rho = \exp(\sigma_p^2/2) \tag{4.24}$$

Table 4.2 lists some parameteric values for land clutter log-normal distributions [14].

A study [17] to measure the worst-case amplitude statistics experienced by a minimally attended L-band air search radar, deployed in Alaska, determined that the distribution functions range from contaminated normal [18] to log-normal. The measurements were taken with depression angles from the radar to the clutter that varied from $-0.8°$ to $1.2°$ with pulsewidths of 0.5, 1, 3, and 6 μs. The worst-case clutter statistics obtained assuming a log-normal distribution for mountains, hills, and valleys are given in Table 4.3. These values were used in the design specification for the ARSR-4 air route surveillance radar [19].

Table 4.2
Log-Normal Parameters for Land Clutter Distributions

Terrain	*Frequency Band*	*Grazing Angle (°)*	σ_p
Discrete	S	low	3.916
Distributed	S	low	1.380
Various	UHF-K_a	10–70	.728–1.584

Table 4.3
Worst-Case Land Clutter Model at Low Grazing Angles—Alaskan Environment

			$\sigma°$ *Values*	
Terrain	*Vegetation*	*Height (ft)*	*Median*	*84th Percentile*
Mountains	Bare	4000	−20 dB	−14 dB
Wooded Hills	Small Trees, Grass	1000–4000	−19 dB	−12 dB
Valleys, Muskeg	Heavily Wooded	1000	−20 dB	−13 dB
Composite (15–100 nmi)		—	−23.5 dB	−17.5 dB

The spatial distributions are of interest for scanning radars where the returns from several points on the ground are generally processed before a detection decision is made. Of interest is the depth of correlation associated with each clutter cell before adjacent cells can be considered decorrelated. An autocorrelation analysis of side-looking airborne radar (SLAR) data provided that depicted in Figure 4.18, which indicate the distance require to decorrelate various types of terrain [15].

The decorrelation function in Figure 4.18 can be approximated by two correlation functions with associated decorrelation distances. The first decorrelation distance (d_1) provides a rapid decorrelation to about the 0.4 level, and the second decorrelation distance (d_2) is much longer and represents a long-term correlation. The fast decorrelation distance was found to be on the order of 50 ft for all terrain types; the values for the long decorrelation distances are given in Table 4.4. Note that the value of the long decorrelation distance is significantly larger for more or less homogeneous terrains such as farms and woodland than for urban or mountainous terrains, which have sharp facets in the terrain structure.

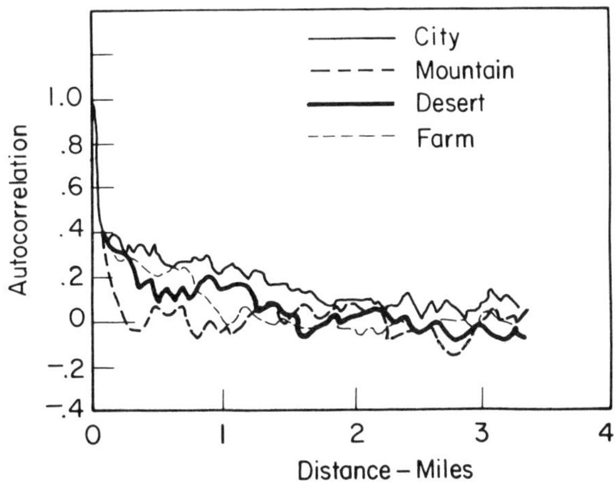

Figure 4.18 Spatial decorrelation distances for various types of land clutter.

In general, land clutter displays non-Gaussian amplitude statistics (i.e., wide clutter dynamic ranges) when viewed at low grazing angles or with a high-resolution radar. The low-grazing-angle effect, which is important for surface-based radars, is generally attributed to shadowing effects in which significant portions of the clutter may be obscured from the radar, depending on its aspect with respect to the radar. As the aspect changes, different parts of the clutter send strong reflections back to the radar, causing large amplitude fluctuations. The same sort of effect generally occurs for mountainous or city clutter, except that this effect can also occur for high grazing angles and low-resolution radar. Non-Gaussian amplitude statistics for high-resolution radar are generally caused by the radar resolving individual clutter scatterers with returns that vary widely in amplitude. Figure 4.19 depicts several non-Gaussian land clutter distributions obtained for surface-based radars at low grazing angles, showing the wide dynamic range associated with this type of clutter.

Table 4.4
Decorrelation Distances for Land Clutter

Terrain Type	*Long Decorrelation Distance (ft)*
Woodland	500–1000
Desert	1000–2000
Farmland	1000
Mountains, Hills	500
Urban, Suburban	500

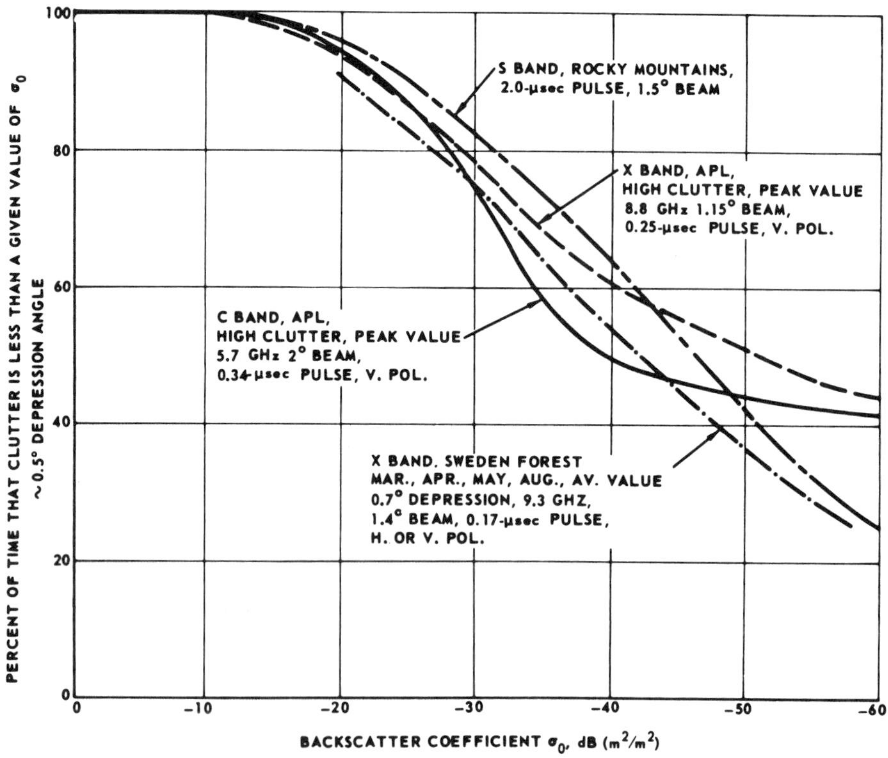

Figure 4.19 Land clutter backscatter distributions from surface radars at low grazing angles.

4.1.1.3 Land Clutter Spectra

The intrinsic spectrum of land clutter results from the relative motion of wind-driven scatterers (e.g., tree branches, leaves, vegetation). As the wind increases, the motion increases and the spectrum broadens. In general, the intrinsic spectral width is almost directly proportional to the transmitted frequency throughout the microwave region. Note that further spectral spreading of the clutter occurs due to antenna scanning, platform motion, transmitter instability, and clutter-limiting effects. These effects are covered in Sections 3.2.2 to 3.2.5.

The intrinsic spectral width for wooded land clutter is depicted in Figure 4.20 as a function of wind speed (see also Table 3.6). The standard deviation of the spectrum expressed in velocity units (σ_v – m/s) given in the figure is intended to

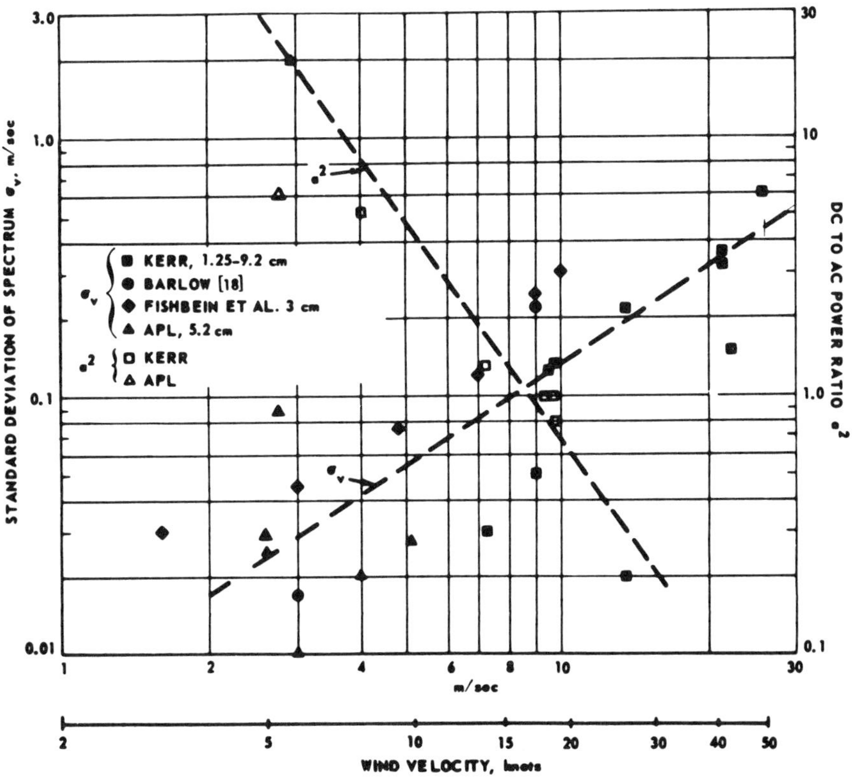

Figure 4.20 Spectral width for wooded terrain as a function of wind speed.

be used in a Gaussian-shaped spectrum. The Gaussian-shaped spectrum is given by

$$S(f) = P_c \exp(-f^2/2\sigma_f^2)/\sqrt{2\pi}\sigma_f \quad (4.25)$$

where $\sigma_f = 2\sigma_v/\lambda$ is the standard deviation of the spectrum and P_c is the clutter power. Also shown is the ratio of the fluctuating distributed clutter component to the fixed discrete clutter component (e.g., see (4.19); $m^2 = S^2/P_o$). Equation (4.25) applies for the spectrum of the fluctuating component, whereas the discrete clutter component is represented by an impulse function $\delta(f)$ at zero frequency.

Experimental data indicates that the Gaussian-shaped spectrum given by (4.25) may underestimate the higher-frequency components of the land clutter spectrum. A cubic equation has been suggested as a better fit to experimental *X*-band data than the Gaussian-shaped spectrum [20,21]. Data at other frequencies suggested that a more general relationship for the clutter spectrum, given by

$$S(f) = k/[1 + (f/f_c)^n] \quad (4.26)$$

is appropriate. Here, f_c is the corner frequency proportional to the wind speed and n is an exponent that varies with the radar's transmitted frequency [17,22]. Parameters for this relationship with 6–15 mi/hr wind speed are given in Table 4.5.

Table 4.5
Power Spectrum Function Parameters for Land Clutter at Various Frequencies

	Frequency			
Parameter	*9.5 GHz*	*16 GHz*	*35 GHz*	*95 GHz*
n	3	3	2.5	2
f_c(Hz)	9	16	21	35

Other measurements [1] indicate that the spectrum can be approximated by an exponentially shaped spectrum. Typical power-spectra measured under both calm and windy conditions are depicted in Figure 1.21.

4.1.2 Sea Clutter

Sea clutter depends on radar parameters such as frequency, polarization, grazing angle, pulsewidth, and azimuth beamwidth. With the exception of near-vertical incidence, the reflectivity of sea clutter would be negligible were it not for wind-generated waves and capillaries (small waves caused by wind ripples superimposed on main waves). The properties of sea clutter are thus related to the sea-surface configuration, the most prominent feature of which are sea waves. The sea condition is generally described in terms of sea state, which relates to the height of the waves.

The sea-surface condition is related to the interaction of two somewhat independent disturbances. Distant winds cause swell, a fairly regular surface of nearly sinusoidal waves. Local winds cause steeper, less regular waves that occur in a random manner over the sea surface. Local winds blowing for an appreciable period of time produce a fully arisen sea, which achieves a quasisteady-state condition with recognizable upwind, crosswind, and downwind characteristics. The conventional description of sea state relates it to wind speed and assumes a fully arisen condition. In general, backscatter from the sea depends on sea state (i.e., wave height and wind speed) and the direction of the waves relative to that of the radar beam.

The reflectivity characteristics of the sea are generally related to sea state number in the same manner as land clutter reflectivity is related to terrain type. The sea states listed in Table 4.6 use the Douglas scale, which relates the sea condition to the wave height [4]. Because the wave height is instrumental in establishing the backscattering properties of sea clutter, the Douglas scale is an appropriate measure.

Table 4.6
Douglas Sea-State Levels

Sea State	*Description*	*Wave Height* ($h_{1/3}$ *ft*)	*Approximate Wind Speeds (knots)*
1	Smooth	0–1	0–6
2	Slight	1–3	6–12
3	Moderate	3–5	12–15
4	Rough	5–8	15–20 (white caps)
5	Very Rough	8–12	20–25
6	High	12–20	25–30 (gale)
7	Very High	20–40	30–50

The statistical wave height parameter used in the Douglas scale $h_{1/3}$ denotes the average crest-to-trough sea height of the highest one-third of all the waves. If the wave heights are assumed to have a Gaussian amplitude distribution with standard deviation σ_h, we can deduce that $h_{1/3} = 4\sigma_h$ and $h_{av} = 2.6\sigma_h$ [10].

The theory of sea clutter attributes part of the reflection to the existence of capillary waves (ripples) generated by local wind gusts of short duration. The capillary waves, which have wavelengths on the order of an inch, are superimposed on the larger wave structure and, because of their small size, assume greater importance as the frequency of the radar is increased. In this theory, the capillary waves are assumed to act approximately like flat-plate reflectors of varying size and orientation. This capillary wave structure apparently causes the interference effect at low grazing angles that results in σ° being sensitive to the radar's frequency and polarization. The capillary waves, in effect, become scatterers above the average sea surface that reflect both a direct and an indirect wave (off the sea surface), resulting in destructive interference at low grazing angles [4].

Sea clutter is generally described by its backscatter coefficient (σ°), amplitude probability distribution, and power spectral density. This type of clutter differs from land clutter in its generally homogeneous structure, its average velocity proportional to wind velocity, which affects the central frequency of its spectrum, and the effects caused by the different dielectric properties of the sea as compared to land terrain, which affect the magnitude and phase of the reflection coefficient. The nonzero doppler shift associated with sea clutter requires using motion compensation techniques in MTI radars.

The variation of average backscattering coefficient ($\bar{\sigma}^\circ$) with grazing angle (ψ) for sea clutter follows the three scattering regions of general surface clutter, depicted in Figure 4.3. For microwaves, the only strong frequency dependence is associated with the interference effect, which determines the location of the critical angle (ψ_c) and makes σ° a strong function of λ for angles less than ψ_c. For a given sea state, ψ_c is proportional to the wavelength (i.e., $\psi_c = \sin^{-1}\lambda/4\pi\sigma_h$). For horizontal polarization, the curve of σ° *versus* ψ for low frequencies is expected to be similar to that for microwaves, except that the critical angle will be larger. This creates an expected variation, as depicted in Figure 4.21, in which the low-grazing-angle region (i.e., ψ^4 dependence; see (4.13)) is extended and the plateau region is foreshortened as the frequency is lowered into the UHF, VHF, and HF regions [4].

For vertical polarization, σ° would not be expected to vary as ψ^4 in the low-grazing-angle region because the magnitude of the reflection coefficient is generally less than one and the phase angle is not 180°. The curve for vertical polarization is expected to resemble that for horizontal polarization while differing in several important respects. First, σ°_{VV} will generally be larger than σ°_{HH} for most grazing angles. Second, the critical angle ψ_c will not be proportional to wavelength. Third, the functional dependence of σ° on ψ_c in the low-grazing-angle region will not be proportional to ψ^4.

For vertical polarization, destructive interference can occur only at grazing angles below Brewster's angle, where the reflection coefficient approaches $\Gamma = -1$. At higher frequencies (e.g., X-band), the critical angle generally occurs at grazing angles less than Brewster's angle. For example, at X-band frequencies the critical angle might be 1° whereas Brewster's angle would be about 8° [4]. Thus, as the grazing angle is decreased at high microwave frequencies, destructive interference would generally be expected to occur at about the same angle for both horizontal and vertical polarizations. This condition persists until the frequency approaches about 1 GHz,

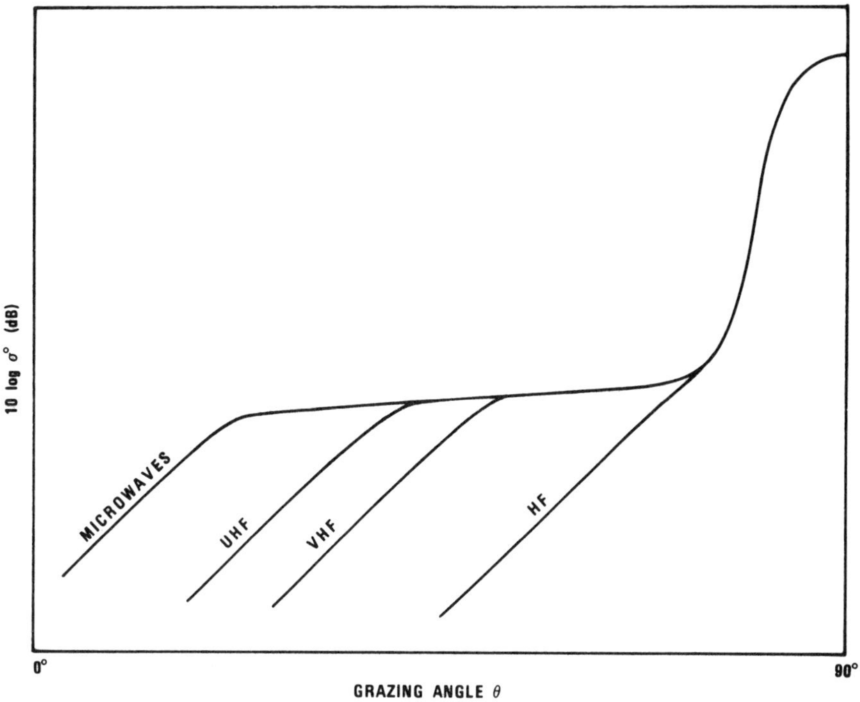

Source: Adapted from Long et al. (1965).

Figure 4.21 Backscatter coefficient σ° as a function of grazing angle for horizontal polarization at various frequencies.

where Brewster's angle and the critical grazing angle are approximately equal. As the frequency is further decreased, Brewster's angle is less than the critical grazing angle and the shape of the σ° curve for vertical polarization will become significantly different from that for horizontal polarization.

The situation for vertical polarization and low frequencies (UHF) is depicted in Figure 4.4. For low frequencies, as the grazing angle decreases, destructive interference will not be expected until the grazing angle becomes very small (on the order of 1° at 30 MHz). Therefore, the plateau region of the σ°_{VV} curve is expected to extend to much smaller angles than for σ°_{HH}. Further, the critical angle for low frequencies and vertical polarization will be generally independent of surface irregularities. The critical angle will be determined primarily by the Brewster angle effect, which is dependent on frequency. As the grazing angle approaches zero, σ°_{VV} should approach σ°_{HH} because the vertical polarized reflection coefficient approaches unity with a phase shift of 180° [4].

The backscatter from the sea depends on the direction of the wind relative to the direction of the radar antenna beam. Backscattering is generally higher when the radar beam looks into the wind than when looking downwind or crosswind. There might be a variation as great as 5 to 10 dB in σ° as the antenna scans 360° in azimuth. The increased backscatter into the wind is apparently due to the asymmetrical nature of sea waves, which tend to be steeper on the upwind side.

A large number of radars that operate over water are horizontally polarized to take advantage of the apparently lower values of σ° with this polarization. However, when a horizontally polarized short-pulse radar ($\tau \leq 0.5$ μs) is used at low grazing angles (typical for surface-based radars), the clutter backscatter return becomes spatially nonuniform. The clutter return then contains occasional spiky quasispecular echoes that can often persist for several seconds. The random spiky returns have a radar cross section at X-band that is on the order of 1 m^2 for an individual spike. To resolve individual spikes requires a radar pulsewidth on the order of tens of nanoseconds. The spikes are apparently associated with breaking waves that form whitecaps. Sea spikes are also observed with vertically polarized short-pulse radars, but the effect is much less pronounced than for horizontal polarization.

Thus, as the resolution cell size of the radar is reduced, the nature of the sea clutter changes. The effect of sea spikes on the radar performance can be profound, causing a large increase in false alarms, or a reduction in sensitivity to desired targets as the threshold is increased to reduce the false alarms. The resolution of a radar operating in sea clutter is therefore an important design parameter to be optimized for maximum detection performance.

4.1.2.1 Average σ° for Various Sea States

Because sea clutter is a fluctuating quantity, experimental σ° data must be evaluated statistically. The usual statistical measure used to quantify σ° is the time average, which we will specify as $\bar{\sigma}^\circ$. In some cases [4], the median (value of σ° for which the probability distribution function has a value of 50%) is used to specify the clutter's backscattering coefficients (σ°) average value. If the probability distribution of the fluctuations is known, we are able to relate the median to the mean value. For example, for a Rayleigh distribution, the mean-to-median ratio is given by $10 \log(\bar{\sigma}^\circ/\sigma^\circ_m) = 1.6$ dB (see Section 1.3.4). We should note that the correct value to use in the radar range equation for σ° is the mean value, and standard detection curves (computed for thermal noise interference) can be used to specify signal-to-clutter ratios, provided the sea clutter amplitude follows a Rayleigh probability density function [10].

Most compilations of $\bar{\sigma}^\circ$ show wide variations due to the problems of defining sea conditions, the experimental difficulties of sea clutter measurement, and the many factors on which sea clutter backscatter depends. One approach [7] is to average the available data over many of the variables, which provides an indication of trends and can be used for initial system design. A compilation obtained by averaging and extending data on $\bar{\sigma}^{\circ}$ acquired from a large number of experiments for grazing angles between 0.1° and 30° is given in Tables 4.7 to 4.12 [7,23]. These data represent (1) a time average over several tens of milliseconds; (2) an average of upwind, crosswind, and downwind values; and (3) radar pulsewidths in the 0.5–10 μs region, where Rayleigh statistics apply and the resolution does not allow resolving the individual wave structure. Note that for the small grazing angles tabulated (i.e., $\psi = 0.1°$ and 0.3°), the depression angle of the radar antenna beam may be different due to earth curvature (see equation 4.7).

Sea state 3 data given in Tables 4.7 to 4.12 for the mean backscattering coefficient $\bar{\sigma}^{\circ}$ as a function of grazing angle for horizontal polarization are plotted in Figure 4.22. A similar curve for vertical polarization is depicted in Figure 4.23. These curves employ some smoothing of the data points to provide continuous curves. The sea state 3 data were chosen as representative of typical or average sea conditions [10].

Table 4.7
Normalized Mean Sea Backscatter Coefficient σ° for a Grazing Angle of 0.1°

		Reflection Coefficient, dB, *Below 1* m^2/m^2 *at Indicated Carrier Frequency*, GHz						
Sea State	*Polarization*	*UHF 0.5*	*L 1.25*	*S 3.0*	*C 5.6*	*X 9.3*	*Ku 17*	*Ka 35*
0	V	94	87	80	75	67	61	55
	H	103	96	89	84	76	70	64
1	V	89	82	75	70	62	56	50
	H	96	89	82	77	69	63	57
2	V	92	83	74	65	56	47	37
	H	96	87	78	69	59	50	51
3	V	86	76	67	57	47	38	28
	H	90	80	71	61	51	42	32
4	V	71	65	59	53	47	41	35
	H	72	66	60	54	48	42	36
5	V	71	63	56	48	41	34	26
	H	71	63	56	48	41	34	26

Note: Monostatic radar, 0.5–10 μs pulse.

Table 4.8
Normalized Mean Sea Backscatter Coefficient σ° for a Grazing Angle of 0.3°

		Reflection Coefficient, dB, *Below 1* m²/m² *at Indicated Carrier Frequency,* GHz						
Sea State	*Polarization*	*UHF 0.5*	*L 1.25*	*S 3.0*	*C 5.6*	*X 9.3*	*Ku 17*	*Ka 35*
0	V	84	78	73	67	61	55	49
	H	94	89	83	78	72	67	61
1	V	76	71	66	61	56	51	46
	H	89	83	76	70	63	57	50
2	V	77	70	64	57	50	44	37
	H	84	76	69	61	53	46	38
3	V	70	64	57	50	44	37	30
	H	76	69	61	53	45	38	30
4	V	62	57	53	48	43	38	33
	H	62	57	52	47	42	37	31
5	V	58	53	48	43	38	34	29
	H	62	56	49	42	36	29	23

Note: Monostatic radar, 0.5–10 μs pulse.

Table 4.9
Normalized Mean Sea Backscatter Coefficient σ° for a Grazing Angle of 1.0°

		Reflection Coefficient, dB, *Below 1* m²/m² *at Indicated Carrier Frequency,* GHz						
Sea State	*Polarization*	*UHF 0.5*	*L 1.25*	*S 3.0*	*C 5.6*	*X 9.3*	*Ku 17*	*Ka 35*
0	V	75	69	64	59	53	48	43
	H	85	80	75	70	65	60	55
1	V	68	64	59	55	50	45	41
	H	81	74	66	59	52	45	38
2	V	62	58	53	49	45	41	37
	H	74	67	60	54	47	40	33
3	V	57	53	49	45	41	37	38
	H	67	61	55	49	42	36	30

Table 4.9 cont'd.

Sea State	Polarization	*Reflection Coefficient*, dB, *Below 1* m²/m² *at Indicated Carrier Frequency*, GHz						
		UHF *0.5*	*L* *1.25*	S *3.0*	C *5.6*	*X* *9.3*	*Ku* *17*	*Ka* *35*
4	V	51	47	44	40	37	34	30
	H	53	49	45	41	36	32	28
5	V	43	41	39	36	34	32	30
	H	49	52	46	39	33	26	20

Note: Monostatic radar, 0.5–10 μs pulse.

Table 4.10
Normalized Mean Sea Backscatter Coefficient σ° for a Grazing Angle of 3.0°

Sea State	Polarization	*Reflection Coefficient*, dB, *Below 1* m²/m² *at Indicated Carrier Frequency*, GHz						
		UHF *0.5*	*L* *1.25*	*S* *3.0*	*C* *5.6*	*X* *9.3*	*Ku* *17*	*Ka* *35*
0	V	64	61	59	57	54	52	50
	H	76	72	67	63	59	54	50
1	V	58	55	52	49	46	43	40
	H	68	63	59	54	50	45	41
2	V	55	52	49	46	42	39	36
	H	63	59	54	50	45	41	36
3	V	45	43	41	40	38	36	35
	H	58	54	50	45	41	37	33
4	V	39	38	37	36	34	33	32
	H	51	47	44	41	37	34	31
5	V	38	37	35	34	32	31	29
	H	50	45	41	36	32	27	23

Note: Monostatic radar, 0.5–10 μs pulse.

Table 4.11
Normalized Mean Sea Backscatter Coefficient σ° for a Grazing Angle of 10°

		Reflection Coefficient, dB, *Below 1* m²/m² *at Indicated Carrier Frequency*, GHz						
Sea State	*Polarization*	*UHF 0.5*	*L 1.25*	*S 3.0*	*C 5.6*	*X 9.3*	*Ku 17*	*Ka 35*
0	V	48	47	46	46	45	45	44
	H	62	59	57	54	52	49	47
1	V	46	45	44	43	42	41	39
	H	58	56	53	50	48	45	42
2	V	38	38	37	36	36	35	34
	H	55	52	49	46	44	41	38
3	V	34	34	33	33	32	32	32
	H	51	48	44	41	37	34	30
4	V	31	31	30	30	30	30	30
	H	47	44	41	37	34	31	28
5	V	27	27	27	27	27	26	26
	H	44	41	38	35	31	28	25

Note: Monostatic radar, 0.5–10 μs pulse.

Table 4.12
Normalized Mean Sea Backscatter Coefficient σ° for a Grazing Angle of 30°

		Reflection Coefficient, dB, *Below 1* m²/m² *at Indicated Carrier Frequency*, GHz						
Sea State	*Polarization*	*UHF 0.5*	*L 1.25*	*S 3.0*	*C 5.6*	*X 9.3*	*Ku 17*	*Ka 35*
0	V	43	43	42	42	41	41	40
	H	47	48	48	49	50	50	51
1	V	39	38	38	38	38	37	37
	H	45	45	45	45	45	45	45
2	V	34	33	33	32	32	31	31
	H	42	41	41	40	40	39	39

Table 4.12 cont'd.

Sea State	Polarization	*Reflection Coefficient*, dB, *Below 1* m^2/m^2 *at Indicated Carrier Frequency*, GHz						
		UHF 0.5	*L 1.25*	*S 3.0*	*C 5.6*	*X 9.3*	*Ku 17*	*Ka 35*
3	V	29	29	28	28	28	27	27
	H	40	39	38	36	35	33	32
4	V	29	28	27	26	24	23	22
	H	39	38	36	35	33	32	31
5	V	21	21	20	20	20	20	19
	H	39	35	32	29	26	23	20

Note: Monostatic radar, 0.5–10 μs pulse.

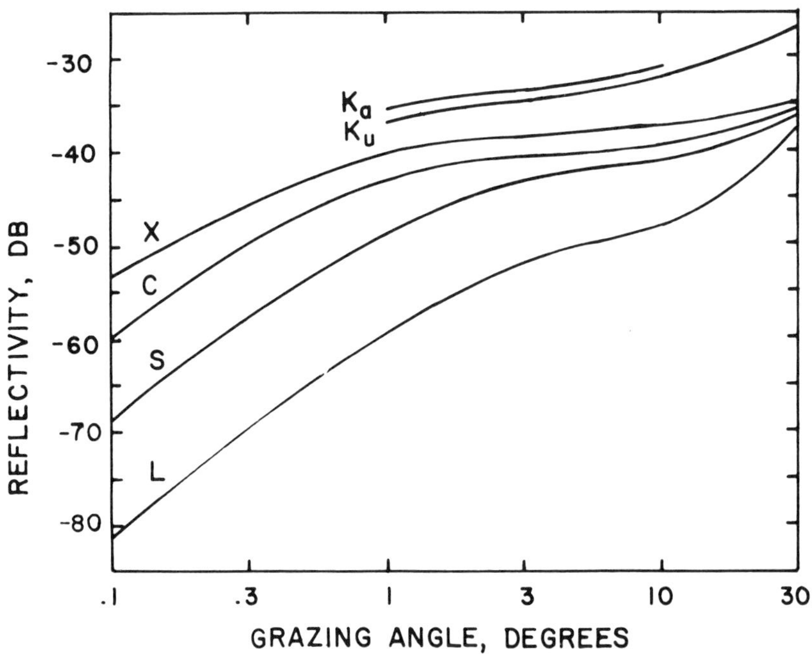

Figure 4.22 Mean backscattering coefficient $\bar{\sigma}^\circ$ as a function of grazing angle for horizontal polarization of various frequencies (sea state 3).

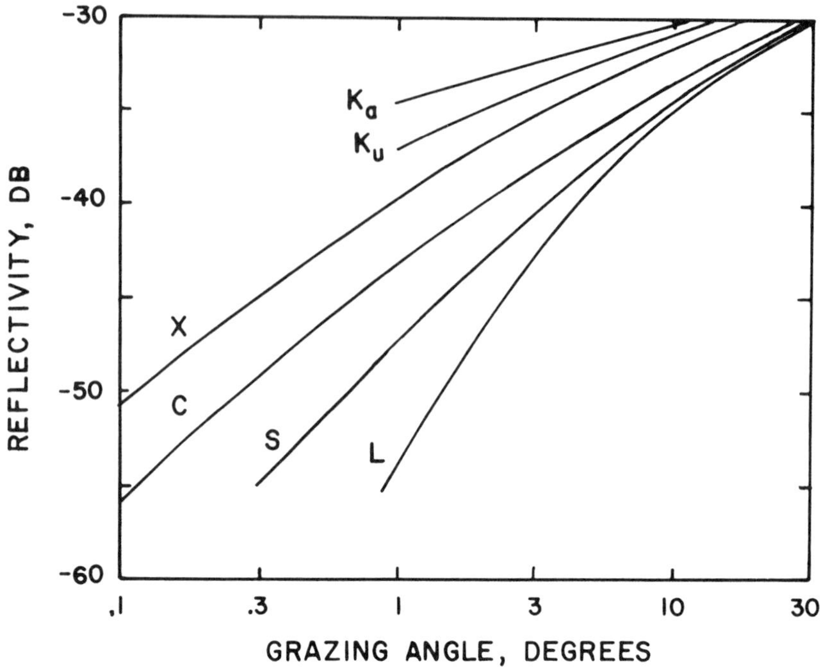

Figure 4.23 Mean backscattering coefficient $\bar{\sigma}^{\circ}$ as a function of grazing angle for vertical polarization at various frequencies (sea state 3).

The Naval Research Laboratory (NRL) has developed a four-frequency airborne radar system capable of making surface clutter measurements at frequencies ranging from UHF (428 MHz) to X-band (8910 MHz) [24]. The coherent pulsed radar is capable of transmitting a sequence of four frequencies (8910, 4455, 1228, and 428 MHz) alternately on horizontal and vertical polarization. Pulsewidths range from 0.1 to 2 μs, allowing almost simultaneous measurement of surface clutter over a wide frequency range with both horizontal and vertical polarization. The advantage of this measurement system is that it removes the surface condition variability, yet allows a parametric study of frequency and polarization sensitivity.

Tables 4.13 and 4.14 show sea clutter backscatter data taken under rough sea conditions (i.e., wave height = 16.4 ft, approximately SS-6), which compares sea clutter return for a wide range of frequency, polarization, and relative wind direction. Note that the calibration accuracy for these data is ±2 dB, and the backscattering data are given in median values rather than mean values (i.e., 1.6 dB should be added to the median values to provide mean values).

Table 4.13 provides rough sea backscatter data *versus* grazing angle for vertically and horizontally polarized transmissions and receptions (i.e., σ_{VV} and σ_{HH}), different

Table 4.13
Median Values of the Normalized Radar Cross Section $\sigma°$ for Directly Polarized Signals Measured at X, C, L, and P Bands (from [25])

Depression Angle (degrees)	*Wind Direction*[b]	$\sigma°$ *(dB)*[a]							
		X_{VV}	X_{HH}	C_{VV}	C_{HH}	L_{VV}	L_{HH}	P_{VV}	P_{HH}
5	U	−35.5	−37.5	−34.5	−39.5	−40.5	−45.0	−41.0	−47.0
	D	−38.5	−43.0	−40.0	−44.0	−43.0	—	−43.0	—
	C	−37.0	−40.5	−39.5	−41.5	−44.0	−48.5	−42.5	—
10	U	−31.5	−35.0	−28.5	−38.5	−36.0	−48.0	−36.0	−46.5
	D	−33.5	−42.5	−31.5	−43.5	−34.5	−51.0	−36.0	−49.0
	C	−34.5	−39.5	−33.5	−41.5	−35.5	−48.5	−36.0	−49.0
20	U	−27.0	−33.5	−25.0	−35.0	−31.0	−41.5	−30.0	−47.5
	D	−29.5	−38.5	−28.0	−40.0	−29.0	−42.5	−28.5	−46.5
	C	−30.0	−38.5	−30.0	−40.0	−30.5	−43.5	−30.5	−45.5
30	U	−25.5	−30.5	−22.0	−32.5	−28.0	−37.5	−28.5	−37.0
	D	−27.0	−35.5	−23.5	−36.5	−26.0	−37.5	−28.0	−36.5
	C	−29.0	−35.5	−25.5	−35.5	−26.5	−36.5	−28.0	−37.5
45	U	−20.5	−24.0	−19.0	−27.0	−27.0	−33.0	−25.5	−31.0
	D	−23.0	−28.5	−21.5	−31.0	−25.0	−32.0	−24.5	−30.5
	C	−25.5	−29.0	−23.5	−30.0	−27.0	−34.5	−25.5	−32.5
60	U	−17.5	−18.5	−14.5	−19.0	−20.5	−22.0	−23.0	−25.5
	D	−18.5	−20.0	−16.0	−21.0	−19.0	−21.5	−22.0	−24.5
	C	−19.5	−20.5	−17.5	−21.0	−22.0	−24.0	−21.0	−24.0
75	—	−9.5	−10.0	−7.0	−9.0	—	—	−13.0	−15.0
90	—	−1.0	−1.0	+2.0	+1.0	—	—	+0.5	−2.0

[a]February 20, 1969: Wind velocity, 29 knots; wave height, 16.4 ft.
[b]U = upwind, D = downwind, and C = crosswind.

wind direction (upwind, downwind, and crosswind) and various frequencies betwen UHF and X-band. The data indicate no apparent backscatter frequency sensitivity for vertical polarization, whereas the frequency sensitivity for horizontal polarization exhibits a λ^{-1} to $\lambda^{-1/4}$ dependence. As expected, vertical polarization backscatter generally exceeds horizontally polarized backscatter, and upwind backscatter generally exceeds that of crosswind or downwind. Another effect noticed with these sets of experimental measurements was a saturation effect, whereby the backscatter tends to have an upper limit as a function of grazing angle [24]. This upper limit is attributed to the tilting effect of the composite wave surface (i.e., capillary waves superimposed on sea surface).

Table 4.14
Median Values of the Normalized Radar Cross Section σ° for Cross-Polarized Signals Measured at X, C, L, and P Bands (from [25])

Depression Angle (degrees)	*Wind Direction*[b]	σ° (dB)[a]							
		X_{VH}	X_{HV}	C_{VH}	C_{HV}	L_{VH}	L_{HV}	P_{VH}	P_{HV}
5	U	−44.0	−47.5	−42.0	—	−47.0	—	−47.0	—
	D	−46.0	−51.0	−43.0	—	−48.5	—	−48.0	—
	C	−44.5	−47.5	−43.5	—	—	—	−47.5	—
10	U	−42.5	−43.5	−41.0	−43.0	−48.5	−50.0	−45.5	−51.0
	D	−46.0	−46.5	−44.5	−46.0	−48.0	−48.5	−46.0	−51.0
	C	−45.5	−47.0	−44.0	−47.0	−47.5	−48.5	−46.5	−51.5
20	U	−39.0	−40.0	−38.0	−39.0	−42.5	−45.0	−42.0	−46.0
	D	−42.0	−43.5	−40.5	−40.5	−42.0	−41.5	−42.0	−45.0
	C	−43.0	−43.5	−41.0	−43.0	−43.5	−45.0	−42.0	−45.0
30	U	−38.0	−39.0	−36.5	−36.0	−41.0	−41.5	−40.0	−43.5
	D	−40.0	−41.0	−38.0	−38.0	−39.5	−39.0	−40.5	−44.0
	C	−40.0	−40.5	−38.0	−38.5	−39.0	−39.5	−40.5	−43.5
45	U	−36.0	−35.0	−34.5	−33.5	−41.0	−40.5	−40.0	−42.5
	D	−38.0	−37.5	−36.0	−36.0	−38.5	−38.0	−38.5	−41.5
	C	−37.5	−37.0	−36.5	−37.0	−41.5	−42.0	−40.0	−43.0
60	U	−35.0	−33.5	—	—	—	—	−38.5	−40.5
	D	−35.0	−34.0	—	—	—	—	−37.5	−39.5
	C	−34.5	−33.5	—	—	—	—	−37.0	−39.5

[a]February 20, 1969: Wind velocity, 29 knots; wave height, 16.4 ft.
[b]U = upwind, D = downwind, and C = crosswind.

Table 4.14 lists cross-polarization backscatter values (i.e., σ_{VH} and σ_{HV}), where transmission is on one polarization (e.g., vertical) and reception is on the orthogonal polarization (e.g., horizontal). The data show agreement between values for σ_{HV} and σ_{VH}, indicating a symmetrical scattering matrix. The values for the cross-polarized components generally range on the order of 10 dB below the vertically polarized component. However, for horizontally polarized backscatter, the values range from on the order of 10 dB above the cross-polarized component at high frequencies (X-band) to about that of the cross-polarized component at low frequencies (L-band and UHF) and low grazing angles ($\psi < 20°$).

4.1.2.2 Amplitude Probability Distribution Functions for Sea Clutter

The amplitude of sea clutter fluctuates in a statistical manner that can be characterized in terms of a probability density function. When the return from sea clutter can be modeled as resulting from a number of independent random scatterers with no one scattering predominating, the envelope detected amplitude fluctuation voltage can be described by the Rayleigh amplitude probability density function given by

$$p(v) = \frac{v}{\sigma^2}\exp\left(-\frac{v^2}{2\sigma^2}\right) \tag{4.27}$$

which is a function of only the clutter power given by $P_c = 2\sigma^2$ (see also equation (4.18) for the exponential power probability density function). The Rayleigh probability density function applies to sea clutter when the resolution cell (proportional to the pulsewidth and azimuth beamwidth) or area illuminated by the radar is relatively large. Experimental data show that this distribution does not apply for a high-resolution radar that illuminates only a small area of the sea.

In general, the Rayleigh distribution applies for sea clutter when the size of the area of ocean illuminated instantaneously by the radar is large compared to the wave structure. In a typical sea, as has been estimated, if the area has linear dimensions of about 250 ft, a sufficient number of scatterers (i.e., five to ten) contributes to the backscatter to justify the use of Rayleigh statistics. This corresponds to a pulsewidth of 0.5 μs or larger and an azimuth beamwidth of 2.5° at 1 nmi or 0.25° at 10 nmi.

One measure used to quantify the departure from Rayleigh statistics is the ratio of the standard deviation to the mean value of σ°, which equals one for Rayleigh distributed clutter (this follows from (4.18), where $E(\sigma^\circ) = \bar{\sigma}^\circ$ and $E[(\sigma^\circ)^2] = 2(\bar{\sigma}^\circ)^2$). Figure 4.24 depicts this measure when applied to sea return from both C-band and X-band surface radars covering pulse lengths between 0.008 and 3 μs with an azimuth beamwidth of 0.9° to 3° [7]. These data indicate an increasing deviation from Rayleigh statistics as the pulse-width is decreased, especially on horizontal polarization. The increased probability of high amplitudes is generally attributed to the presence of spikes or specular reflections in the sea clutter return.

Sea clutter may also be non-Rayleigh at very low grazing angles, even when the resolution cell is large enough to result in Rayleigh clutter at higher grazing angles. This is due to a shadowing effect that results in fewer significant scattering elements in a given area of ocean. This effect also was found to be more pronounced for horizontal than vertical polarization [10].

Figure 4.25 depicts the cumulative probability distribution measured for sea clutter by using a short pulse (0.4 μs) *X*-band surface-based radar with a narrow (0.8°) beamwidth [4]. These data are plotted on a normal probability scale against a decibel scale, where a straight line represents a log-normally distributed probability

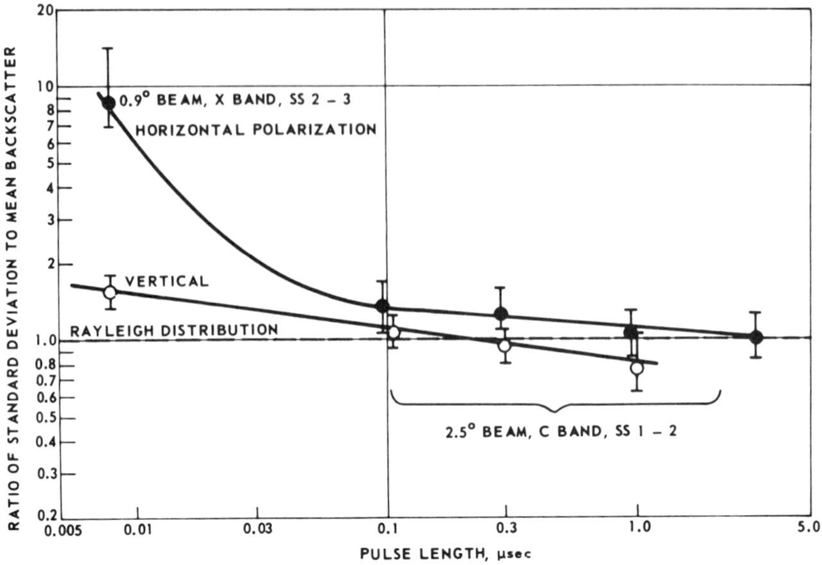

Figure 4.24 Distribution of the standard deviation to mean for short-pulse sea backscatter σ°.

function (see (4.22)). That the measured sea clutter data can be approximated in the tail region by a log-normal distribution (i.e., a straight line approximation) is apparent from inspection of the data shown on the curve. Also apparent is the longer tail of the distribution associated with horizontal as compared to vertical polarization. Measured standard deviations were 7 ± 0.6 dB for horizontal polarization and 4 ± 0.6 dB for vertical polarization [4]. A reference Rayleigh distribution is also shown in Figure 4.25, which allows determining the increased dynamic range associated with the log-normal clutter at the higher cumulative probability values.

Figure 4.26 depicts experimental high-resolution (pulsewidth = 20 ns, beamwidth = 0.5°) sea clutter data taken at a grazing angle of 4.7° for vertical polarization [18]. The amplitude probability data generally fall between the Rayleigh and log-normal cumulative probability distribution functions shown for comparison on the curves. This situation is typical for most high-resolution radars, where the Rayleigh distribution tends to underestimate the amplitude distribution. As shown in the curve, two effects are associated with increasing sea roughness (i.e., increasing sea state). First, the tails of the distribution generally increase with increasing sea state. In addition, the tails tend to approach a log-normal distribution as the sea becomes rougher. Both effects combine to increase the dynamic range of the sea clutter, which require the threshold detector to be set at a higher value than if the clutter were Rayleigh distributed to avoid excessive false alarms.

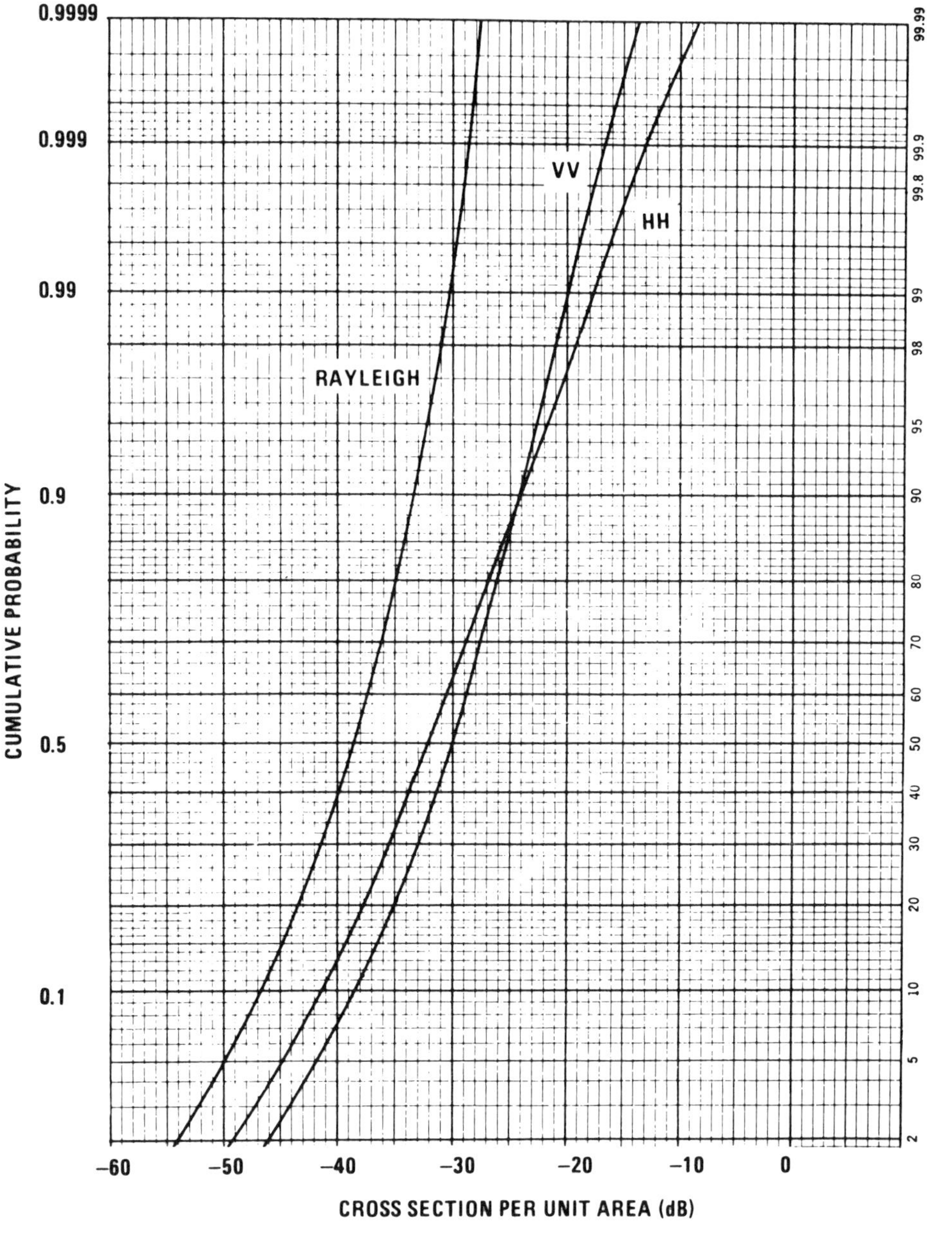

Figure 4.25 Cumulative amplitude probability distribution function for an X-band sea backscatter coefficient.

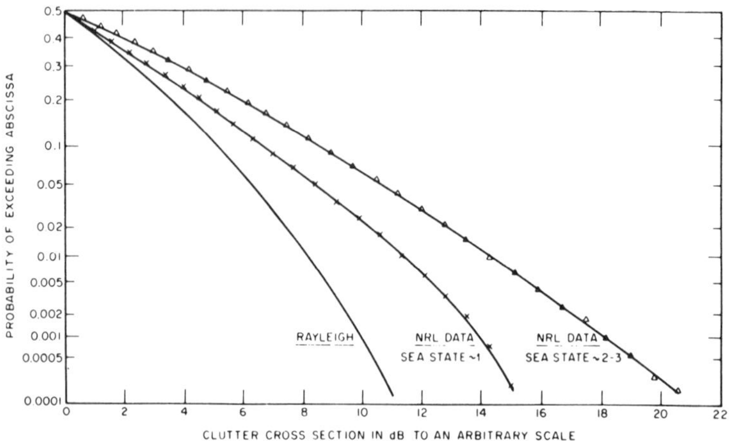

Figure 4.26 Cumulative probability distribution function for vertically polarized, high-resolution sea clutter data.

The log-normal amplitude voltage probability density function representing envelope-detected sea clutter is given by

$$p(v) = \frac{\exp[-\ln^2(v/v_m)/2\sigma_v^2]}{\sqrt{2\pi}\sigma_v v}; \quad v > 0 \tag{4.28}$$

where σ_v is the standard deviation of the underlying normal distribution and v_m is the median value of the log-normal distribution. Transforming this distribution to represent the distribution for clutter power ($P = v^2$), which is proportional to the backscatter coefficient (see Section 1.3.1), results in

$$p(\sigma^\circ) = \frac{\exp[-\ln^2(\sigma^\circ/\sigma_m^\circ)/2\sigma_p^2]}{\sqrt{2\pi}\sigma_p\sigma^\circ} \tag{4.29}$$

where σ_m° is the median value of σ° and $\sigma_p = 2\sigma_v$ is the standard deviation of the underlying normal generating distribution (i.e., standard deviation of $\ln \sigma^\circ$). The mean-to-median ratio $\rho = \bar{\sigma}^\circ/\sigma_m^\circ$ is frequently used to characterize the tails of a probability distribution (see Section 1.3.4). The higher this ratio, the longer are the tails of the distribution. For the log-normal distribution, the mean-to-median ratio is given by (4.24) as $\rho = \exp(\sigma_p^2/2)$. Figure 4.27 depicts values of ρ expressed on a decibel scale (i.e., $\rho_{dB} = 10 \log \rho$) as a function of grazing angle and sea state [14]. These data were obtained experimentally by using a Ku-band radar transmitting a short pulse (pulsewidth = 0.1 μs). The values given for ρ allow various log-normal fits

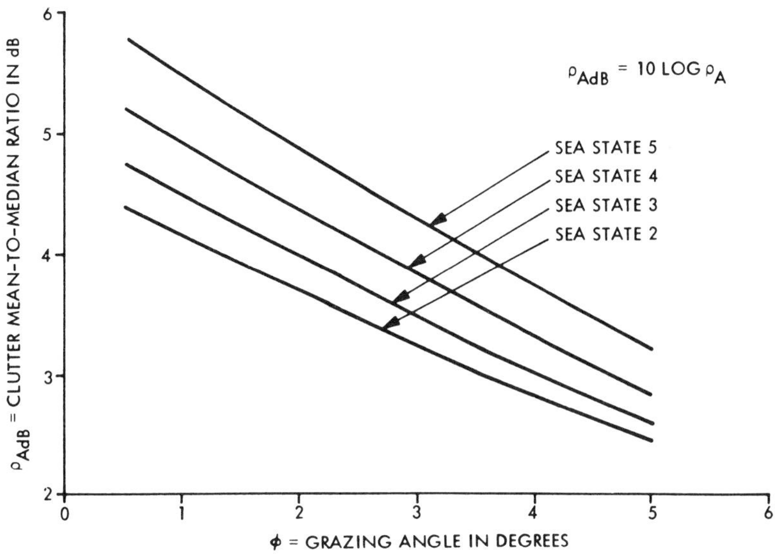

Figure 4.27 Sea clutter mean-to-median ratios for log-normally distributed sea clutter for various sea states and grazing angles.

to be synthesized for characterizing sea clutter at various grazing angles and sea states.

Because the log-normal distribution generally overestimates the real clutter's amplitude probability distribution (see Figure 4.26), several other distributions have been proposed for a closer match. The Weibull distribution, which was originally developed for land clutter, has also been used to model sea clutter data [26]. This distribution generally provides a fit closer to sea clutter data than the log-normal distribution, especially in the tail region, and has the added advantage that the Rayleigh distribution is a member of the Weibull family of distributions. Figure 4.28 depicts amplitude probability distributions plotted for sea clutter data (SS-3, Ku-band, horizontal polarization, 0.1 μs pulsewidth, into the wind) using a log-Rayleigh *versus* dB scale in which a Weibull distribution plots as a straight line (see Section 1.3.4). The data show a fit close to a Weibull distribution with longer tails at the lower grazing angles. Note that the distribution approaches a Rayleigh clutter distribution at a grazing angle $\psi = 30°$.

The Weibull probability density function for the clutter backscatter (σ°) is given by

$$p(\sigma^\circ) = \beta \cdot \ln 2(\sigma^\circ/\sigma^\circ_n)^{\beta-1} \exp[-\ln 2(\sigma^\circ/\sigma^\circ_m)^\beta]/\sigma^\circ_m \tag{4.30}$$

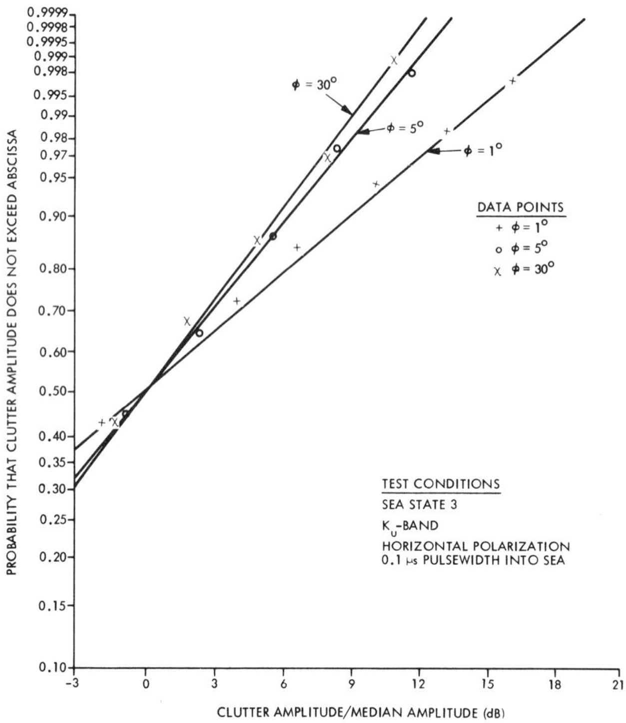

Figure 4.28 Weibull distribution for sea clutter at several grazing angles.

where σ_m° is the median value of the clutter and β is a parameter relating to the skewness of the distribution [14]. The mean-to-median ratio $(\bar{\sigma}^\circ/\sigma_m^\circ)$ for Weibull clutter is given by

$$\rho = \Gamma(1 + 1/\beta)/(\ln 2)^{1/\beta} \tag{4.31}$$

Note that for $\beta = 1$ the Weibull distribution reverts to the exponential density function $p(\sigma^\circ) = \exp(-\sigma^\circ/\bar{\sigma}^\circ)/\bar{\sigma}^\circ$, which characterizes Rayleigh clutter. Table 4.15 lists some values obtained experimentally for the Weibull skewness parameter β [14].

The *K*-distribution sea clutter model (see Section 1.3.4) was developed to provide a mathematical description for the composite model sea surface; the model theorizes that sea clutter is due to the compound effect of two fluctuating components. The speckle component due to capillary waves is modeled as a classical Rayleigh-distributed clutter (see Section 1.3.1). This component has a relatively short decorrelation time (on the order of 10 ms) that can be further decorrelated by using frequency agility (but this is not appropriate for an MTI radar). The other component is associated

Table 4.15
Weibull Skewness Parameter for Several Grazing Angles and Sea States

Sea State	*Frequency Band*	*Beamwidth (degrees)*	ψ *(degrees)*	*Pulsewidth* (μs)	β
SS-1	*X*	0.5	4.7	0.02	0.726
SS-3	Ku	5	1–30	0.1	0.58–0.8915

with the sea swell structure of a fully arisen sea, which has a long correlation time (on the order of seconds) and is not affected by frequency agility. This component is modeled with a chi-distribution obtained by using an empirical fit to experimental sea clutter data [27].

The *K*-distribution is derived by assuming that the mean value of the Rayleigh distributed component is a random variable with chi-distributed probability distribution. The resulting voltage amplitude probability density function is found from the relation

$$p(v) = \int_0^\infty p(v|y)p(y)\,dy \tag{4.32}$$

where $p(v|y)$ is the Rayleigh probability density conditioned on its mean value y and $p(y)$ is a chi probability density function. The mean value (y) of the Rayleigh probability density function (see (4.27)) is equal to $y = \sqrt{\pi/2}\,\sigma$ so that the Rayleigh probability density function conditioned on its mean level y is given by

$$p(v|y) = \frac{\pi v}{2y^2}\exp\left(-\frac{\pi v^2}{4y^2}\right); \quad v \geq 0 \tag{4.33}$$

The chi probability density function that represents the fluctuation of the mean level is given by

$$p(y) = \frac{2b^{2\nu}}{\Gamma(\nu)} y^{2\nu-1}\exp(-b^2y^2); \quad y \geq 0 \tag{4.34}$$

The resulting envelope detected voltage amplitude density function of the sea clutter can then be evaluated from (4.32) as

$$p(v) = 4c(cv)^\nu K_{\nu-1}(2cv)/\Gamma(\nu) \tag{4.35}$$

where $c = \sqrt{\pi/4}\, b$ is a scale parameter, ν is a shape parameter, and $K_{\nu-1}(\cdot)$ is the modified Bessel function of order $\nu - 1$. The nth moments of $p(v)$ are given by

$$E(v^n) = \Gamma(\nu + n/2)\,\Gamma(1 + n/2)/c^n\,\Gamma(\nu) \tag{4.36}$$

so that the clutter power is given by $P_c = E(v^2) = \nu/c^2$. If we use the transformation $\sigma^\circ = kv^2$ in (4.35) and substitute for the clutter mean value $\bar{\sigma}^\circ = \nu/c^2$, we obtain the sea clutter's backscattering coefficient (σ°) probability density function:

$$p(\sigma^\circ) = 2\nu(\nu\sigma^\circ/\bar{\sigma}^\circ)^{\nu-1/2}\, K_{\nu-1}(2\sqrt{\nu\sigma^\circ/\bar{\sigma}^\circ})/\Gamma(\nu)\,\bar{\sigma}^\circ \tag{4.37}$$

The probability distribution function for K-distribution clutter can then be found as

$$P(\sigma^\circ) = 1 - 2(\nu\sigma^\circ/\bar{\sigma}^\circ)^{\nu}\, K_{\nu}\,(2\sqrt{\nu\sigma^\circ/\bar{\sigma}^\circ})/\Gamma(\nu) \tag{4.38}$$

where $\bar{\sigma}^\circ$ is the mean backscattering coefficient and ν is a shape parameter that determines the dynamic range of the clutter.

The K-distribution backscatter coefficient (σ°) mth order moments can be deduced from (4.36) as

$$E[(\sigma^\circ)^m] = \frac{\Gamma(\nu + m)\Gamma(m + 1)}{c^{2m}\,\Gamma(\nu)} \tag{4.39}$$

which in normalized form ($M_m = \overline{(\sigma^\circ)^m}/(\bar{\sigma}^\circ)^m$) becomes equal to [28]

$$M_m = \frac{\Gamma(\nu + m)\Gamma(m + 1)}{\nu^m\,\Gamma(\nu)} \tag{4.40}$$

The second-order moment $M_2 = 2(\nu + 1)/\nu$ then becomes a measure of the dispersion. Figure 4.29 depicts sea clutter data fitted with a K-distribution. Run 2 on the illustration denotes SS-1 to -2, 10–15 knots upwind, 1.5° grazing angle, circular polarization, a 270 ns pulsewidth, and $M_2 = 2.97$. Run 18 denotes SS-3, 18 knots downwind, with 1° grazing angle, vertical polarization, and $M_2 = 2.88$. Figure 4.30 depicts wide dynamic-range sea clutter data fitted with a K-distribution. Run 9 on the figure denotes SS-2, 15 knots upwind, 1° grazing angle, horizontal polarization, a 70 ns pulsewidth, and $M_2 = 5.98$. Both Figures 4.29 and 4.30 have Rayleigh distributions superimposed to illustrate the extent of the tail distribution. In general, sea clutter distributions at grazing angles between 1° to 1.5° measured for pulsewidths of 70 and 270 ns at SS-1 to SS-5 with horizontal, vertical and circular polarizations had values of ν ranging from 0.1 to 3 [28].

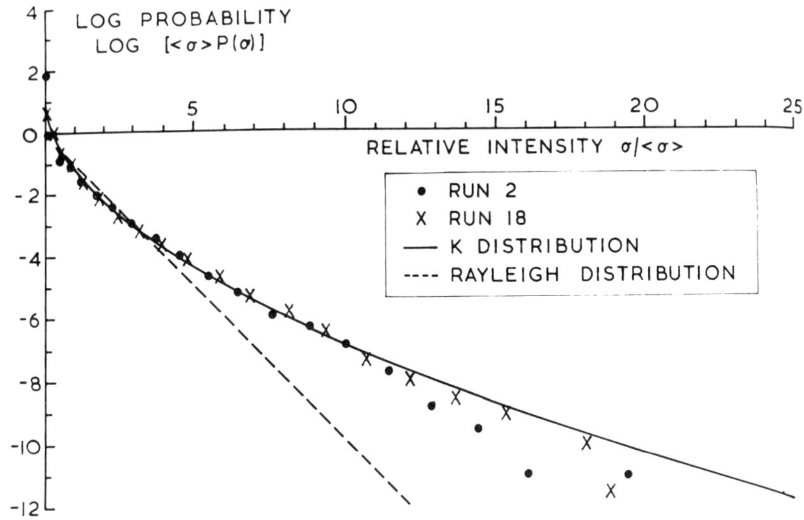

Figure 4.29 *K*-distributed sea clutter for circular and vertical polarization.

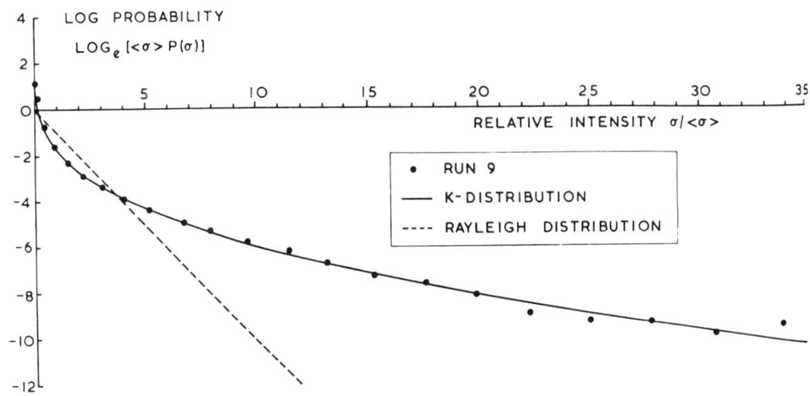

Figure 4.30 *K*-distributed sea clutter for horizontal polarization.

The K-distribution provides a clutter tail distribution very similar to that provided by the Weibull distribution. However, its properties are more difficult to quantify analytically because of the unavailability of formulations for the modified Bessel function, $K_\upsilon(\cdot)$. The value of the K-distribution lies primarily in the area of devising signal processing configurations based on its compound structure (i.e., processing the Rayleigh speckle component). The signal processor can be configured to process the Rayleigh distributed speckle component with the understanding that its mean value fluctuates with a chi distribution. For example, if the probability of detection $P_d(\bar{\sigma}^\circ)$ is evaluated for Rayleigh distributed clutter, then the overall probability of detection is found from

$$P_d = \int_0^\infty P_d(\bar{\sigma}^\circ) p(\sigma^\circ)\, d\sigma^\circ$$

4.1.2.3 Sea Clutter Spectra

The relative motion of the sea surface with respect to the radar causes an intrinsic doppler frequency shift of the return signal from the individual scatterers. Two effects are of interest to designers of MTI and PD radars. The first is the spectral width and shape, which result from the spread of velocity of the elemental scatterers of the sea surface. The second, a mean doppler frequency shift of the entire spectrum, can be either up or down from the central frequency of the sea clutter return. The amount of the shift reflects the average radial velocity of the waves relative to the radar, which is typically a few knots or less. This mean or average doppler shift is given by

$$f_m = \frac{2\upsilon_w}{\lambda} \tag{4.41}$$

where υ_w is the average radial velocity of the waves with respect to the radar. Note that υ_w can be either positive or negative, depending on wind conditions that generate the sea surface conditions. This shift is sometimes important in MTI radars, requiring motion compensation techniques to adjust the MTI filter to coincide with the center frequency of the doppler spectrum (see Section 3.2.7). In other cases, the shift is so small relative to the MTI filter's passband (the MTI filter's passband is approximately equal to a band of doppler frequencies corresponding to the radar's PRF) that it falls within the MTI clutter filter rejection notch, and no motion compensation is required.

The spectrum of sea clutter often is assumed to have a Gaussian shape. There is some question whether this is correct [10], but it is an analytically convenient assumption. With a Gaussian shape, the spectral width is characterized by its standard deviation σ_f, which can be converted to velocity units as

$$\sigma_v = \frac{\lambda}{2}\sigma_f \tag{4.42}$$

The width of the velocity spectrum varies with the agitation of the sea surface, becoming broader as the sea state increases. Expressing the width in velocity units makes the spread independent of frequency, thus allowing data from different experiments to be compared on an equivalent basis.

Figure 4.31 depicts the variation of bandwidth (double-sided half-power spectral width) with sea state (see Table 4.6), which is compiled from many experiments involving horizontal, vertical, and circular polarizations. Note that these data represent the spectrum spread prior to envelope detection (i.e., coherent detection) and involve measurements taken over a frequency range of 220 MHz to 9.5 GHz [7]. The double-sided half-power spectral width can be related to the standard deviation of a Gaussian spectrum by using the following relation:

$$\sigma_v = 0.425\Delta v \tag{4.43}$$

where Δv is the half-power width. The linear relationship for the coherent half-power bandwidth indicated by the dotted line can be expressed as

$$\Delta f_{\text{Hz}} = 3.6 f_{\text{GHz}}\, S \tag{4.44}$$

where f_{GHz} is the radar's frequency in GHz and S is the sea state (see Table 4.5).

Figure 4.32 depicts the variation of bandwidth for envelope detected sea clutter as a function of sea state, where experimental data are taken with both horizontal and vertical polarization [7]. The spectral width of the envelope-detected video can be derived as the convolution of the coherent spectrum, which itself results in

$$\sigma^2_{\text{video}} = 2\sigma^2_{\text{doppler}} \tag{4.45}$$

where $\sigma^2_{\text{doppler}}$ is the coherent spectral variance. This relationship indicates that the half-power width of the video spectrum is greater by the factor $\sqrt{2}$ than the coherent half-power spectral width.

Figure 4.33 depicts the upwind-downwind mean doppler shift for sea clutter as a function of wind speed. The data show an upper limit on the spectral shift on the order of 7 knots and a difference between horizontal and vertical polarized data. The difference is apparently caused by the sensitivity of the vertically polarized return to just wave height, whereas sensitivity of the horizontal polarized return is to both wave height and wind speed [7].

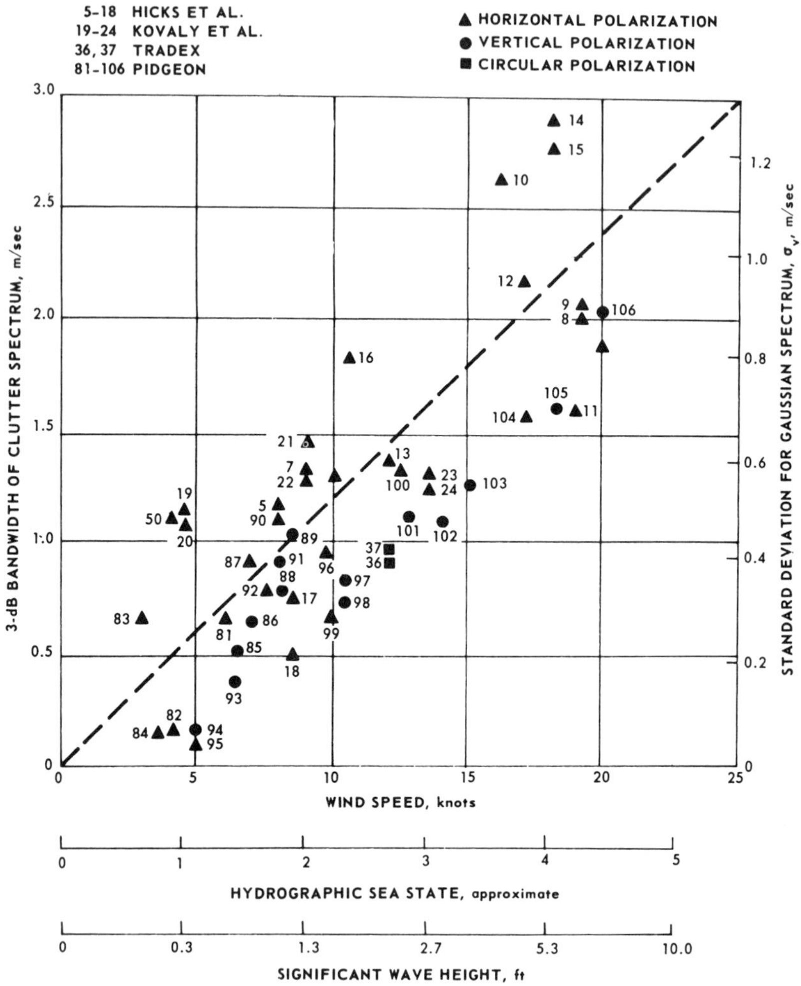

Figure 4.31 Bandwidth of sea clutter spectrum for coherently detected sea clutter.

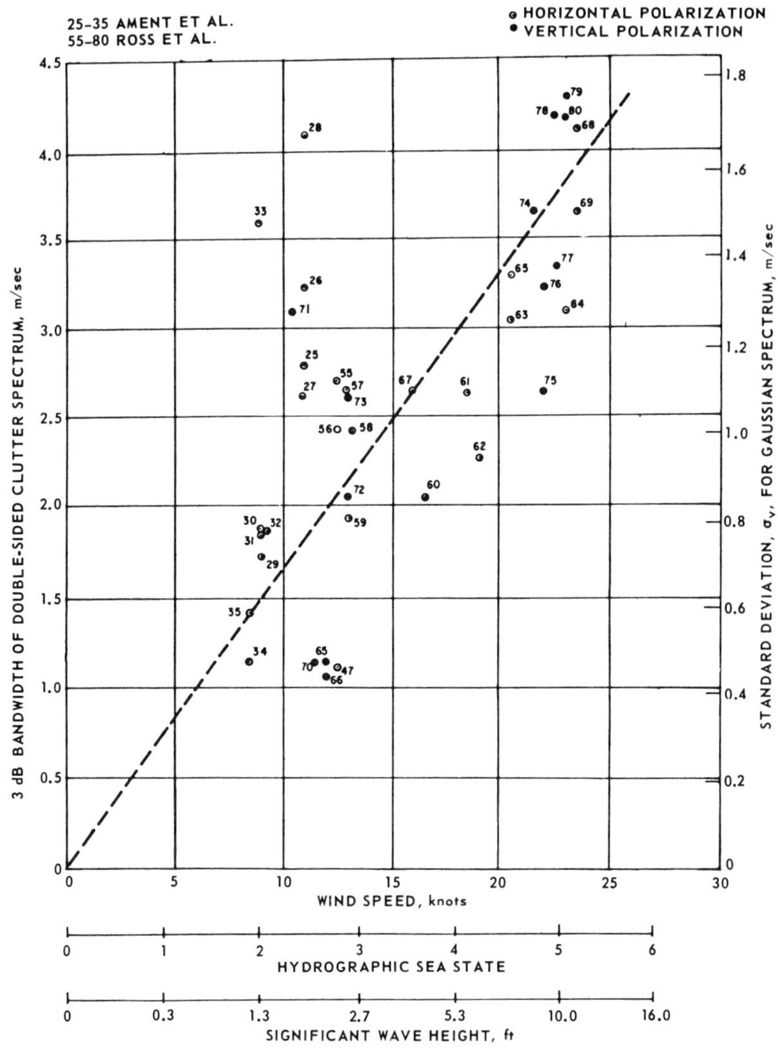

Figure 4.32 Bandwidth of sea clutter spectrum for envelope-detected sea clutter.

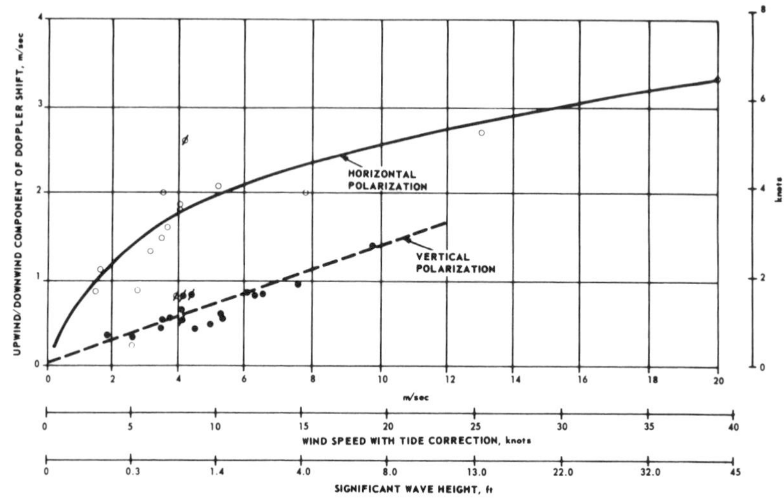

Figure 4.33 Mean doppler shift for sea clutter.

The basic cause for the sea clutter spectra is the relative motion of the individual scattering areas on the surface. This motion is caused by both the large swell waves and the short capillary waves that ride on the swell surface. One theory of the backscattering mechanism attributes it to a Bragg resonance from the capillary wave structure tilted by the large swell waves. This model accounts for both a frequency dependence (the resonant water wavelength $\lambda_r = 2\lambda_w \cos\psi$, where λ_w is the water wavelength) and a polarization dependence [2].

A typical envelope-detected autocorrelation function for sea clutter is depicted in Figure 4.34 [4]. Note that this function consists of a long decorrelation time related to the period of the individual waves passing through the radar clutter patch and a short decorrelation time due to the effect of the wind on the wave structure. The fast fluctuations have a spectral width on the order of 100 Hz, whereas the slow fluctuation spectrum is on the order of 1 Hz [4].

Of significance is the ratio of the power contained in the fast fluctuating component (A) compared to that in the slow fluctuating component (B). Typical ratios for B/A are 0.7, 0.07, and 0.2, respectively, for *HH*, *VV*, and *HV* polarizations [4]. The significant difference of these values for different polarizations indicates a basic difference in the scattering mechanism associated with various polarizations. This also can be observed from the character of the echo return, which appears spiky for *HH* polarization and relatively noiselike for vertical polarization.

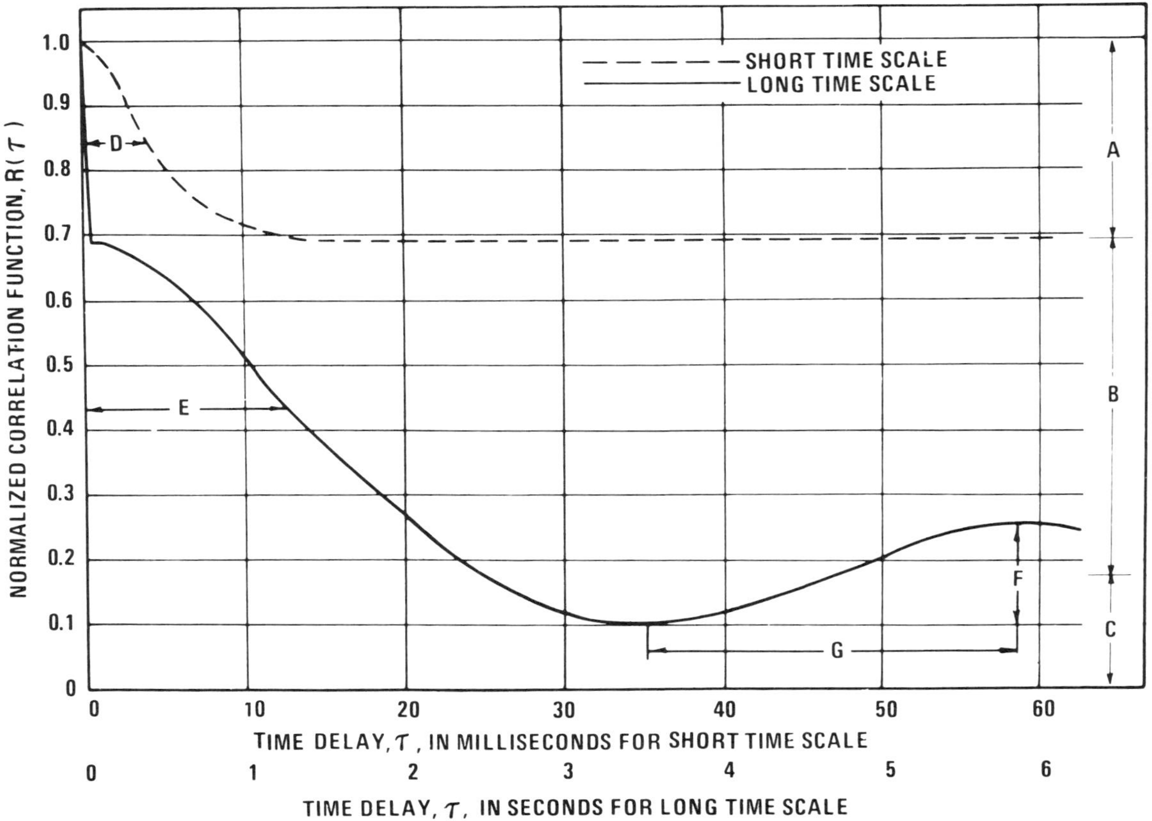

Source: Long (1974).

Figure 4.34 Typical sea clutter autocorrelation function.

4.2 ATMOSPHERIC CLUTTER

A key attribute of radar is its ability to function in all weather at long ranges. However, as the frequency of the radar is increased, the reflection from precipitation generally increases as the fourth power of frequency throughout the microwave region, becoming significant for conventional radars that operate above L-band frequencies. High- or medium-PRF doppler radars generally can separate the precipitation echoes from those of other targets contained within the same resolution cell. This property can be used to discriminate against precipitation clutter or, in the case of doppler weather radar, to analyze the weather return.

The properties of chaff clutter are similar to those of precipitation clutter. However, the reflectivity of chaff, which consists of many lightweight metallic dipoles, generally is uniform over a wide frequency band. This can be achieved by using dipoles of various lengths that are resonant throughout the frequency coverage band of the chaff.

Clutter return backscattered from precipitation or chaff is caused by the various scatterers generally assumed to be uniformly distributed within the radar's volumetric resolution cell. The scatterers are characterized by the normalized parameter η, which represents the radar cross section per unit volume. Thus, the radar cross section of volume-filling clutter can be expressed as

$$\sigma_c = \eta \, V_c \tag{4.46}$$

where η is the volume clutter reflectivity (m^2/m^3) and V_c is the radar's volume resolution cell. The volume resolution cell depicted in Figure 4.35 is given by

$$V_c = \frac{\pi}{8} \, \theta_{az} \, \phi_{el} \, R^2 \, \frac{c\tau_e}{2} \tag{4.47}$$

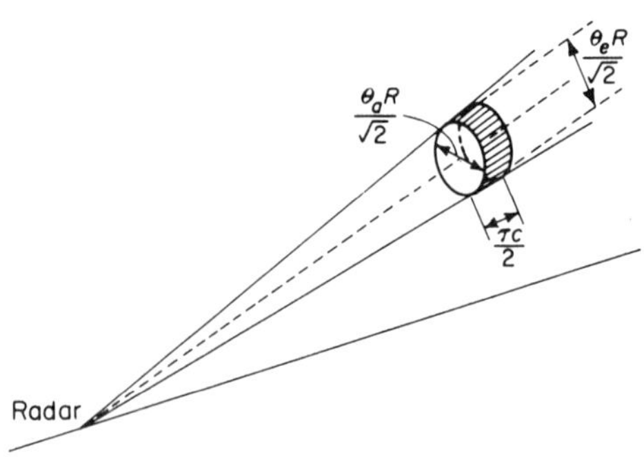

Figure 4.35 Resolution cell for volume clutter.

where θ_{az} is the azimuth beamwidth in radians, ϕ_{el} is the elevation beamwidth in radians, R is the range in meters to the resolution cell, τ_e is the effective radar pulsewidth in seconds, and c is the velocity of light in m/s. The power returned to the radar from the clutter is then equal to

$$P_c = \frac{P_t G_t G_r \lambda^2 \eta \theta_{az} \phi_{el} c \tau_e}{1024 \pi^2 R^2 L} \tag{4.48}$$

where P_t is the transmitted power; G_t and G_r are the respective transmitting and receiving antenna gains; L is a loss term; and the other terms are as previously defined.

Other atmospheric clutter consists of reflections from birds and insects. Reflections from individual birds are generally small but the cumulative effect from migrating birds or flocks of birds is significantly greater. Table 4.16 gives the RCS taken with vertical polarization of some individual birds at UHF, S-band, and X-band frequencies. The increased RCS at S-band is due to a resonant effect from the bird's size. Birds typically fly at speeds from 15 to 40 knots, which can be troublesome for MTIs with low blind speeds.

Table 4.16
Mean Radar Cross Section (dB/m^2) of Birds

	Frequency Band		
Bird	*UHF*	*S*	*X*
Grackle	−42	−26	−28
Sparrow	−57	−29	−38
Pigeon	−30	−21	−28

Returns from single birds at S-band can range from 10^{-4} to 10^{-2} m^2. The return is principally from the body with very little from the wings. For large birds, the body is resonant near L-band and is in the Rayleigh (steady-state) region at UHF. Typically, anywhere from one to several hundred birds may occupy a resolution cell. Although the mean return from a typical flock of birds may be low (i.e., on the order of 10^{-2} m^2), the tail of the distribution has been observed to return up to 10 m^2. Birds usually fly at altitudes less than 7000 ft with air speeds between 15 and 45 knots [14]. Returns from birds viewed by an L-band search radar with a display range of 50 km are depicted in Figure 4.36.

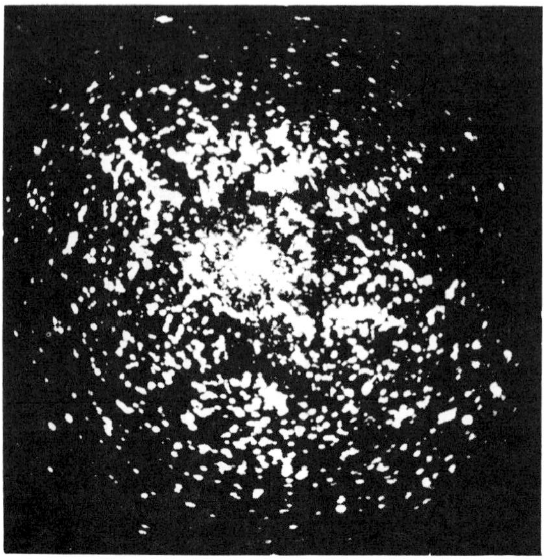

Figure 4.36 Birds viewed from an *L*-band search radar.

The RCS for several different kinds of insects taken with vertical polarization are listed in Table 4.17 [7]. As with birds, their RCS increases rapidly as the resonance region is reached. The RCS of some insects (especially the hawk moth) are large at X-band frequencies.

Swarms of insects have been observed by meteorologists using powerful radars. The swarms may cover large areas and, in general, drift with the winds. The large number of insects within a radar resolution can substantially increase the mean value over that of an individual insect. In addition, the RCS of both birds and insects have been reported to follow a log-normal probability distribution, resulting in large tail returns of up to 10 m^2.

Table 4.17
Typical Mean RCS (dB/m^2) of Individual Insects

	Frequency Band		
Insect	*UHF*	*S*	*X*
Hawk Moth (5 cm)	−54	−30	−18
Worker Bee (1.5 cm)	−52	−37	−28
Dragonfly	−52	−44	−30

4.2.1 Precipitation Clutter

The backscattering properties of rain depend on the transmitted frequency, polarization, and number and size of the raindrops. Because the raindrops are approximately spherical, scattering in the Rayleigh region occurs when the ratio of the raindrops' diameter to the wavelength is small. In this region, the RCS of the raindrops exhibit a fourth power variation with frequency (i.e., $\sigma_c = kf^4$). The Rayleigh scattering theory approximation generally applies for rain below C-band and, except for the heaviest rain, represents a good approximation for X-band. Figure 4.37 depicts the variation of the backscattering coefficient η (m^2/m^3) as a function of the radar frequency and precipitation rate for rain and snow [8]. These curves apply to frequencies up to the order of 100 GHz, where the backscatter becomes relatively independent of frequency, but rather depends on the number of rain scatterers contained in a resolution cell. Figure 4.38 depicts the variation of the backscatter coefficient through the

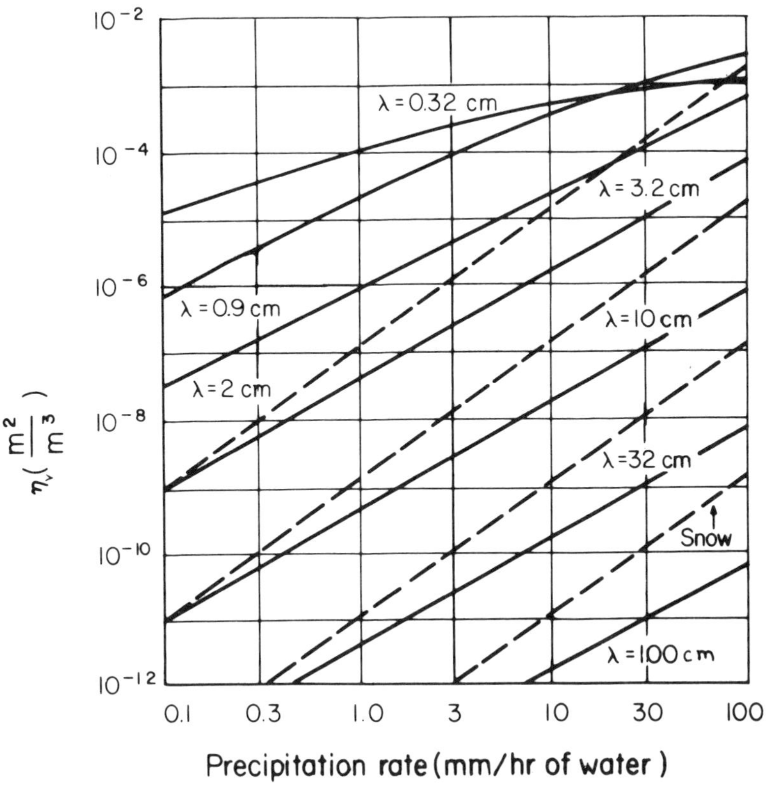

Figure 4.37 Backscattering coefficient η for rain and snow.

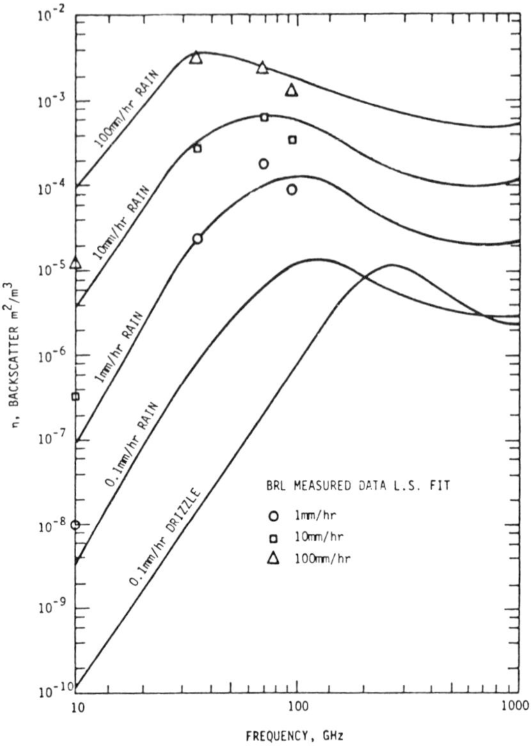

Figure 4.38 Calculated rain backscatter coefficient *versus* frequency.

millimeter-wave frequency region for different rain rate intensities. The curves given in Figures 4.37 and 4.38 apply for linearily polarized radar or for circularly polarized radar using an opposite sense for transmission and reception. When same-sense circular polarization is used, an attenuation of approximately 20 dB occurs in the backscattering coefficient.

The problem of calculating radar performance in precipitation clutter is complicated by the variability of raindrop intensity within a rainstorm, both horizontally and vertically. The vertical extent of rain may be as high as 25,000 ft for thunderstorms, but usually is not above 12,000 ft for steady rain. This can result in a considerable variation in rain reflectivity. In one rainstorm, for example, a 10 dB variation was observed within a range of 1 nmi [7]. More typical variations for moderate rain are a variation of ±1 dB in the mean value over a 10 nmi distance. The approximate extent of various rain rates is tabulated in Table 4.18 [33].

Table 4.18
Extent of Various Rain Rates

Rate	*Diameter (nm)*	*Maximum Altitude (ft)*
Light (0–1 mm/hr)	Unlimited	10,000
Moderate (4 mm/hr)	30–50	10,000–15,000
Heavy (16 mm/hr)	20	20,000
Intense (50 mm/hr)	2	40,000

When precipitation starts as snow or hail at a high altitude and then melts as it drops, a phenomenon is created called the *brightband effect*. This effect results from the increased reflectivity (about 7 dB) of the melting snow or hail at the melting altitude, typically around 11,000 to 12,000 ft in a band of about 1000 ft thickness [7].

4.2.1.1 Amplitude Probability Distribution Function for Rain Clutter

The backscatter from rain clutter results from the cumulative effect of a large number of independent scatterers (i.e., raindrops). Thus, in accordance with the central limit theorem, the rain clutter should be described by a Gaussian distribution. Thus, the received RF clutter voltage has a two-dimensional normal probability density function, and the envelope-detected voltage has a Rayleigh distribution. This has been confirmed by experimental measurement performed at X-band frequencies [29]. The Rayleigh envelope distribution should apply even for high-resolution radars because these contain a large number of individual scatterers within a radar resolution cell.

Other measurements indicate that the envelope-detected amplitude probability density function has longer tails than indicated by a Rayleigh distribution. One set of data taken at 95 GHz with a vertically polarized radar and a rain rate of 5 mm/hr was closely fitted with a log-normal distribution [30]. This distribution applied for rain rates between 2 to 60 mm/hr and overall frequencies measured (i.e., 9.375, 35, 70, and 95 GHz). Typical standard deviations for these measurements with vertical and circular polarization and rain rates from 5 to 20 mm/hr were 7.5 to 8.5 dB at 9.375 GHz, 2.5 to 3.0 dB at 35 GHz, and 2.9 to 3.2 dB at 95 GHz. The standard deviations were most sensitive to frequency change (i.e., 9.375 GHz to 35 GHz) and insensitive to polarization or rain rate change.

Another set of data fitted the return from rain and clouds with a Weibull distribution [31]. The data were obtained with an L-band (1.3 GHz) ARSR radar.

Weibull-shaped parameters between $\beta = 0.68$ and 1 were used to fit the observed data. Unfortunately, the small amount of data samples collected did not permit the fit to be used with high confidence in the tail region (see Section 1.3.4).

4.2.1.2 Precipitation Clutter Spectra

The relative motion of the scatterers (raindrops) within the weather clutter's resolution cell produces a spread of velocities and doppler frequencies, resulting in a frequency spectrum of width related to the spread of velocity. The rain clutter spectrum is spread about some mean value determined by the wind velocity. This is particularly significant in an MTI radar operating in the microwave frequency region (above *L*-band), where the weather clutter reflectivity is sufficient to result in an appreciable clutter return and the first MTI blind speed is such that higher-speed targets are aliased (translated from a multiple-PRF zone) into the same frequency range as the weather clutter. This situation becomes progressively worse as the radar's frequency is increased. A radar processor specifically designed to cope with this situation (i.e., the MTD) is described in Section 1.4. An estimate of the mean velocity extent of rain clutter (e.g., up to about 30 m/s) and its spread (e.g., from 1 m/s at short range to 5 m/s at 200 nmi) was given in the environmental diagram of Figure 1.15 for an air traffic control radar.

In general, four atmospheric effects contribute to the total spectral width of rain clutter [7]. Wind shear involves a change in wind speed with altitude, which results in a distribution of radial velocities over the vertical extent of the beam, as depicted in Figure 4.39. This component is generally the most significant of all the contributing factors in determining the spectrum of weather clutter and most pronounced for radars with broad elevation beams (e.g., cosecant-squared antenna beams). The other components of the weather clutter's spectrum are beam broadening, turbulence, and falling velocity distribution. Beam-broadening is due to the finite width of the radar beam, which causes a spread of radial velocity components when the radar is looking crosswind. Turbulence is caused by the fluctuating currents of the wind, which causes a radial velocity distribution centered about the mean wind velocity. The falling velocity distribution involves a spread in the velocity components among the reflecting particles that constitute the weather clutter located within the radar's resolution cell. For a Gaussian-shaped power spectral density given by $S(f) = P_c \exp(-f^2/2\sigma_f^2)/\sqrt{2\pi\sigma_f^2}$, the frequency spectral variance (σ_f^2) is related to the doppler velocity spectral variance (σ_v) as $\sigma_f = 2\sigma_v/\lambda$, The variance of the doppler velocity spectrum can be represented as the sum of the variances of each contributing factor [7]:

$$\sigma_v^2 = \sigma_{\text{shear}}^2 + \sigma_{\text{beam}}^2 + \sigma_{\text{turb}}^2 + \sigma_{\text{fall}}^2 \tag{4.49}$$

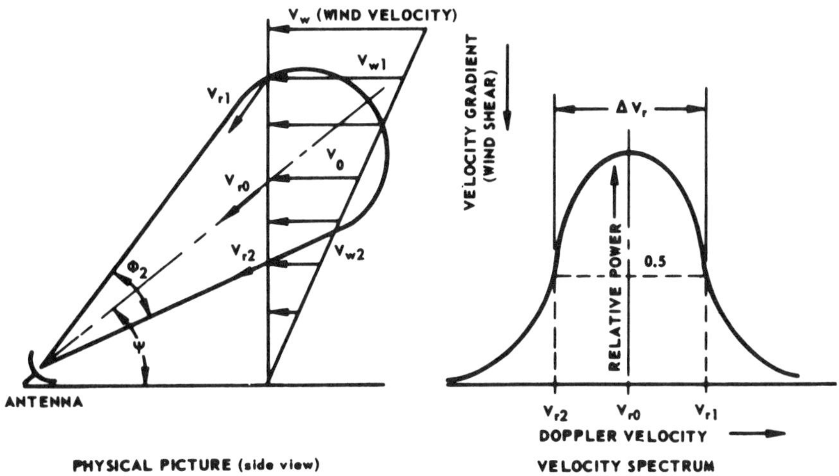

Figure 4.39 Effects of wind shear on the doppler spectrum.

The standard deviation of the wind shear component is given by

$$\sigma_{\text{shear}} = 0.42kR\phi_{\text{el}} \tag{4.50}$$

where k is the wind shear velocity gradient and ϕ_{el} is the two-way, half-power radar's elevation beamwidth. A value of $k = 5.7$ m/(s-km) is the maximum gradient and for an arbitrary azimuth $k = 4$ m/(s-km) is appropriate. Care must be exercised in applying (4.50) to ensure that the elevation beam is filled with precipitation. In practice, σ_{shear} is limited to about 6 m/s for beamwidths of 2.5° or less because the precipitation is limited to some altitude ceiling (e.g., 15,000 to 20,000 ft) [7].

The beam-broadening velocity spectral component has a standard deviation:

$$\sigma_{\text{beam}} = 0.42\, v_o\, \theta_{\text{az}} \sin\beta \tag{4.51}$$

where v_o is the wind velocity at wind center, θ_{az} is the two-way half-power azimuth beamwidth, and β is the azimuth angle relative to the wind velocity at beam center. The beam-broadening component is generally small (e.g., less than 0.5 m/s for a wind speed of 60 knots and a beamwidth of 2°).

The standard deviation of falling velocity for rain is approximately $\sigma_{\text{fall}} = 1$ m/s for the vertical velocity component and is nearly independent of the rain intensity. At elevation angle σ_{el}, the approximate standard deviation is given by [7]

$$\sigma_{\text{fall}} = \sin\phi_{\text{el}} \text{ m/s} \tag{4.52}$$

For small elevation angles, this component is insignificant compared to that due to shear effects.

Fluctuations about the mean value of the spectrum are called *turbulence*. The standard deviation due to turbulence is nearly independent of height and generally varies over a range of 0.5 to 2 m/s with an average value on the order of $\sigma_{turb} = 1$ m/s.

The doppler velocity spectral width is generally a function of the range, elevation and azimuth beamwidths, wind speed at beam center, and relative direction of the wind with respect to the radar beam. Experimental data for various ranges and an elevation beamwidth of 1.4° is depicted in Figure 4.40.

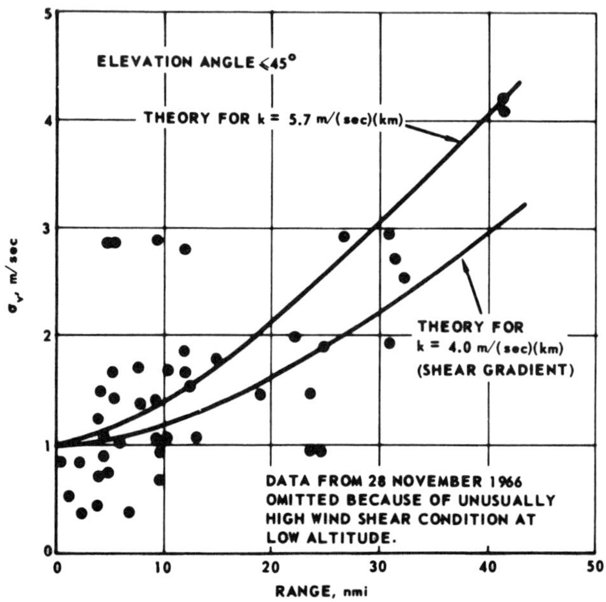

Figure 4.40 Experimental rain clutter standard deviation for a 1.4° elevation beamwidth.

The normalized frequency spectrum for rain clutter at 9.375 GHz with vertical polarization is depicted in Figure 4.41 [32]. These data are fitted with a spectrum of the form $S(f) = P_o/1 + (f/f_c)^n$, where f_c is the corner frequency and n is an exponent equal to either 2 or 3. Measured values for f_c and n for frequencies of 10, 35, 70, and 95 GHz with rain rates of 5 and 100 mm/hr are given in Table 4.19.

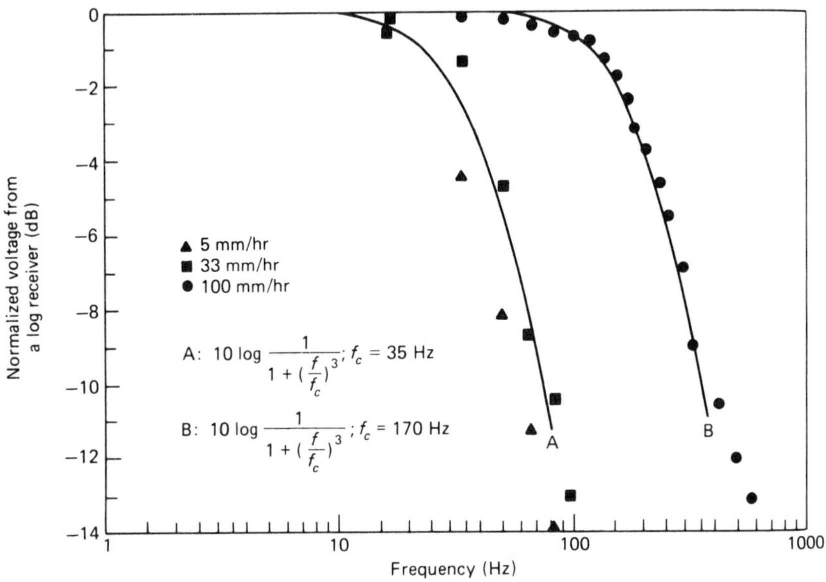

Figure 4.41 Normalized frequency spectrum of rain return at 9.375 GHz and vertical polarization.

Table 4.19
Measured Rain Spectrum Parameters

Frequency (GHz)	*Rate (mm/hr)*	f_c*(Hz)*	*n*
10	5	35	3
	100	70	3
35	5	80	2
	100	120	2
70	5	175	2
	100	500	2
95	5	140	2
	100	500	2

4.2.2 Chaff

Chaff is a form of volumetric radar clutter composed of distributed metalized reflectors, dispensed into the atmosphere to interfere with and confuse radar operation. The chaff usually consists of a large number of dipoles designed to resonate at the frequencies of the radars that they are attempting to confuse. Chaff can be used in either a corridor-laying or self-protection mode. Chaff dispensed from an aircraft at a steady rate over a fairly long period is used to form a corridor that conceals following aircraft. Self-protection involves launching relatively small quantities of chaff in controlled bursts to cause a weapon-associated tracking radar to point at the chaff rather than at the protected vehicle. The design of a chaff dispensing and deployment system is quite different for each of the two modes of operation [33]. Types of chaff distribution are depicted in Figure 4.42.

An elemental piece of chaff can be considered a dipole antenna whose output terminals are short-circuited. When the chaff dipole is excited with RF energy, the shortest length just long enough to permit an electrical charge to propagate from one end to the other and back during an RF cycle is on the order of one-half the wavelength of the incident energy. When the chaff is cut for this resonance condition (typically, 0.46 to 0.48λ), the RCS of the elemental chaff dipole averaged over all possible

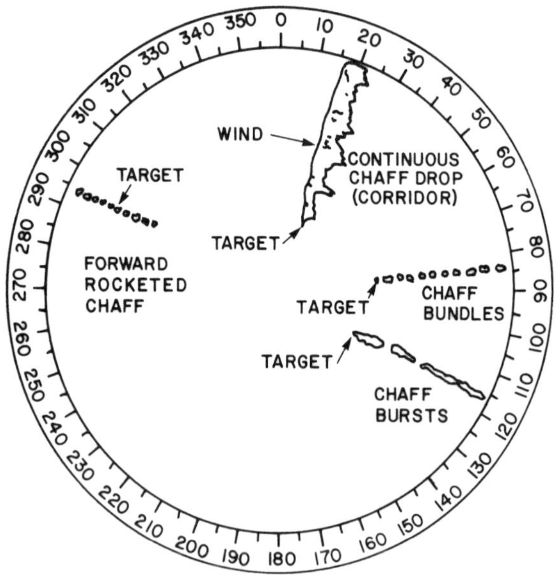

Figure 4.42 Types of chaff distribution.

orientations is given by

$$\sigma_c = 0.15\lambda^2 \tag{4.53}$$

where λ is the radar wavelength. Resonance can occur at harmonic frequencies (shorter wavelengths) as shown in Figure 4.43, but have lower absolute RCS due to the dependence on wavelength of the relationship given by (4.53) [33].

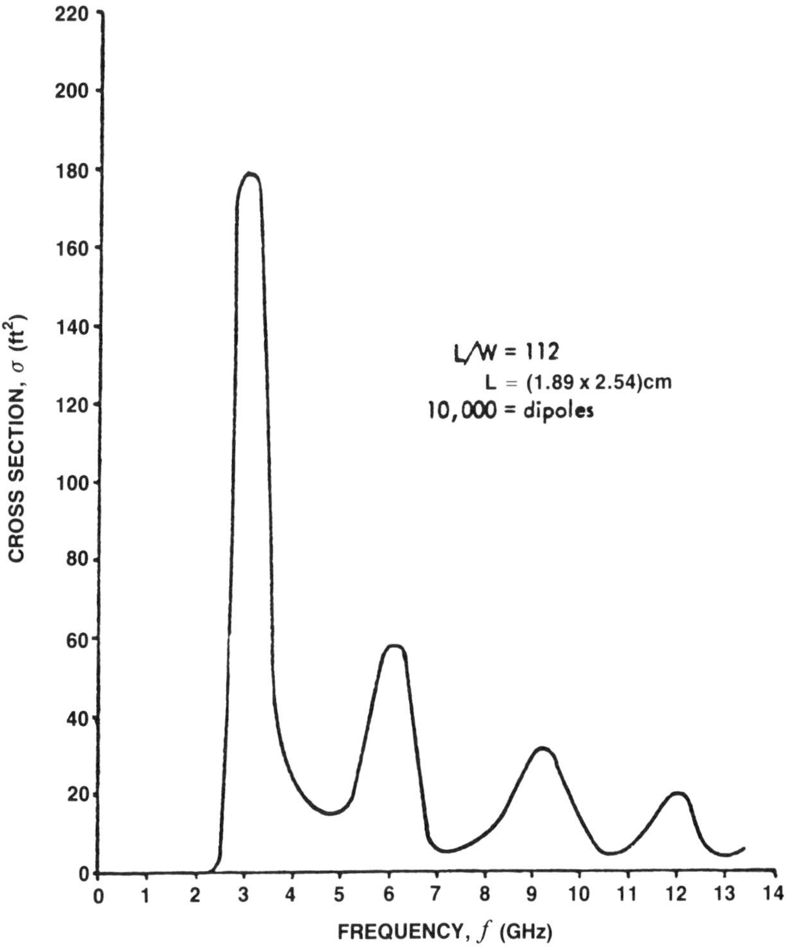

Figure 4.43 Frequency response of chaff dipoles radar cross section.

If a number of elemental chaff dipoles are combined to form a chaff cloud with spacing between elements on the order of two wavelengths or greater, the average RCS of the cloud is given by

$$\sigma_c = 0.15N\lambda^2 \tag{4.54}$$

where N is the number of effective chaff dipoles. The bandwidth of an elemental dipole is a function of the length-to-diameter (L/d) ratio for circular dipoles and the length-to-width ($4L/w$) ratio for rectangular dipoles. The chaff bandwidth ranges from 10 to 25% of the resonant frequency for L/d of 100–10,000, where the higher bandwidths are associated with the lower ratios. A typical design provides a bandwidth approximately equal to 15% of the center frequency for an L/d of 1000. Figure 4.44 shows the RCS *versus* frequency for a representative design of a broadband chaff package covering the range from 2 to 12 GHz, where various elements are stagger-tuned to cover the whole band of interest. Note that the number of dipoles associated with the higher frequency chaff elements are substantially greater than those associated with the lower frequency chaff elements. Also, the RCS response at any particular

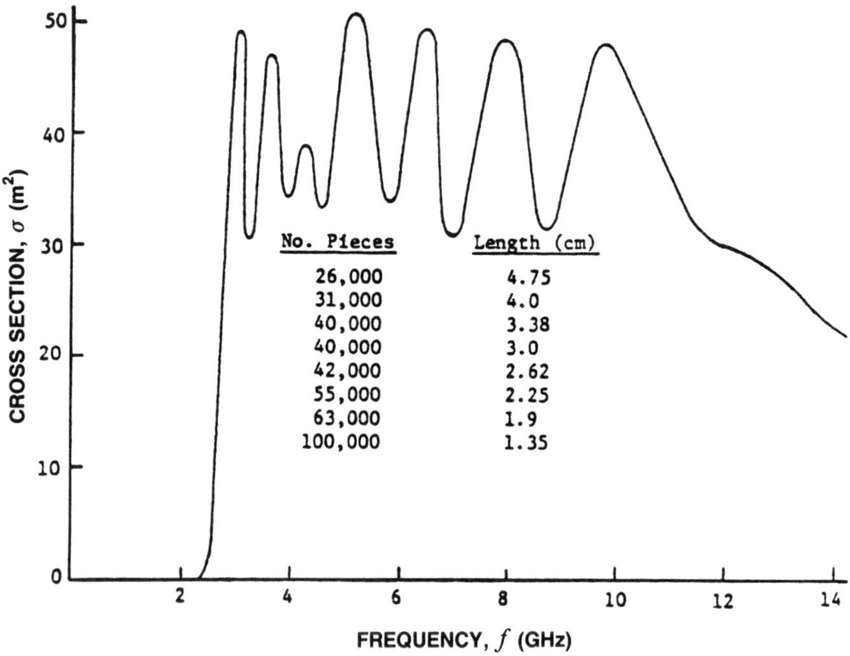

Figure 4.44 Radar cross section of multiband chaff package.

frequency is the sum of the responses of all chaff dipoles resonant at any value lower than the design frequency. Thus, the RCS associated with a particular group of elements cut to the same length (e.g., typically $L = 0.48\lambda$) is generally lower than the actual RCS at this frequency, except for the group with the longest cut lengths. Note that the weight of 25 × 50 μm aluminum foil chaff to achieve the multiband RCS of 40 to 50 m^2 depicted in Figure 4.44 is approximately 0.1 pound [33].

The reflectivity of chaff depends on the deployment doctrine (i.e., corridor or self-protection missions) employed and whether the chaff has bloomed fully or is in a transient state [33]. In the corridor saturation mission, a large amount of chaff is dispensed throughout the corridor such that any aircraft entering the corridor are shielded from radar detection and tracking. Although the primary function of the chaff corridor is to prevent radar acquisition or tracking of aircraft within the corridor, a secondary function is to provide protection for the chaff-deploying aircraft against radar-directed weapons [34].

For effectively denying information regarding strike composition and timing, a chaff corridor should originate outside the radar detection range and be uninterrupted throughout its length. The chaff RCS in the radar resolution cells should be sufficient to prevent detection of the shielded aircraft, taking into consideration any chaff discrimination capabilities of the radar. The chaff RCS initially deployed by the chaff-dispensing aircraft must be greater than the chaff RCS subsequently required for corridor use by the penetrating aircraft. This is necessary to allow for chaff cloud growth and chaff diffusion across the resolution cell boundaries. How much greater the initial chaff RCS should be compared to the chaff RCS per resolution cell at time of use would depend on several factors, such as the chaff-dispensing system, the RCS of the aircraft to be shielded within the corridor, the corridor growth and chaff fall characteristics, the length and use time required of the corridor, and the resolution cell's volume.

As an example, consider the ASR-9 radar ($\lambda = 0.1$ m, $\theta_{az} = 1.3°$, $\phi_{el} = 4.8°$, $\tau = 1.05$ μs) that at a range of 50 nmi has a volume resolution cell $V = 10^9$ m^3 (see equation (4.47)). Assume that we want to generate a chaff corridor with an RCS of 50 m^2 within this volume cell. The chaff reflectivity will then equal $\eta = 5 \times 10^{-8}$ m^2/m^3. The number of chaff dipoles contained within the radar's resolution cell can be found from (4.54) as $N = 3.33 \times 10^4$ dipoles. The density of dipoles within the resolution cell is then 3.33×10^{-5} dipoles/m or 1 dipole per 31 m^3. The most efficient chaff is made by aluminizing cylindrical glass dipoles. The diameter of each chaff dipole is 25 μm and its density is 2550 kg/m^3 [33]. The volume of each dipole is given by $V = 0.48\ \pi d^2\lambda/4 = 0.377 d^2\lambda$. The weight of each dipole is thus $w_1 = 961 d^2\lambda$ kg. The total weight of the aluminized glass dipoles to provide an RCS = 50 m^2 within the radar's volume resolution cell at a wavelength $\lambda = 0.1$ m is then $w = Nw_1/e$, where e is an efficiency factor that allows for breakage, mistuning, mismatching, and other chaff inefficiencies. For a chaff efficiency of 50%, the total weight of chaff to fill a resolution cell would hence equal 4×10^{-3} kg.

The preceding analysis can be expressed as a relationship of the chaff's RCS with respect to the weight of chaff required to produce a given RCS at a specified frequency. This relationship is given by

$$\sigma_c = k w_{\text{lbs}} / f_{\text{GHz}} \tag{4.55}$$

where w_{lbs} is the weight of the chaff in pounds, f_{GHz} is the radar's frequency in GHz, and k is a constant that depends on the properties of the type of chaff utilized. The value of k ranges from 17,000 for aluminized glass chaff to on the order of 3000 for aluminum foil chaff [33].

4.2.2.1 Amplitude Probability Distribution Function for Chaff

The large numbers of randomly moving dipoles that make up the chaff cloud ideally satisfy the conditions for Gaussian-distributed clutter (see Section 1.3.1). In fact, experimental analysis of chaff echo returns show full agreement with the Gaussian clutter theory [29]. This means that the chaff's envelope-detected amplitude probability density function follows a Rayleigh distribution, $p(v) = v \exp(-v^2/2\sigma^2)/\sigma^2$; $P_c = 2\sigma^2$, whereas the RCS for chaff has an exponential probability density function, $p(\eta) = \exp(-\eta/\bar{\eta})/\bar{\eta}$.

Note that, although this distribution is valid when the chaff has reached its steady-state conditions (i.e., the rate of change of $\bar{\eta}$ with time is small), the probability distribution changes significantly when the chaff cloud is growing in a transient state. The preceding reflects the difficulty of modeling a nonstationary process (the growing chaff cloud) by a stationary Gaussian random process.

4.2.2.2 Chaff Clutter Spectra

The main *electronic counter-countermeasures* (ECCM) technique used to combat chaff is doppler processing. In an MTI radar, the techniques used in the MTD processor (see Section 1.4) are applicable. These techniques generally allow the rejection of both ground clutter and chaff (or precipitation) clutter, although these two types of clutter occupy different parts of the frequency spectrum. High-PRF PD radar is considered to be an even better method for eliminating chaff because its unambiguous doppler response allows a clutter rejection filter to be set above the velocity response of the chaff. However, care must be exercised in radars operated in extended clutter with range ambiguities (e.g., high- and medium-PRF PD) because the spectral response at each range-ambiguous clutter cell will be different (both center frequency and spread), and the total clutter spectrum is the sum of the individual clutter spectra.

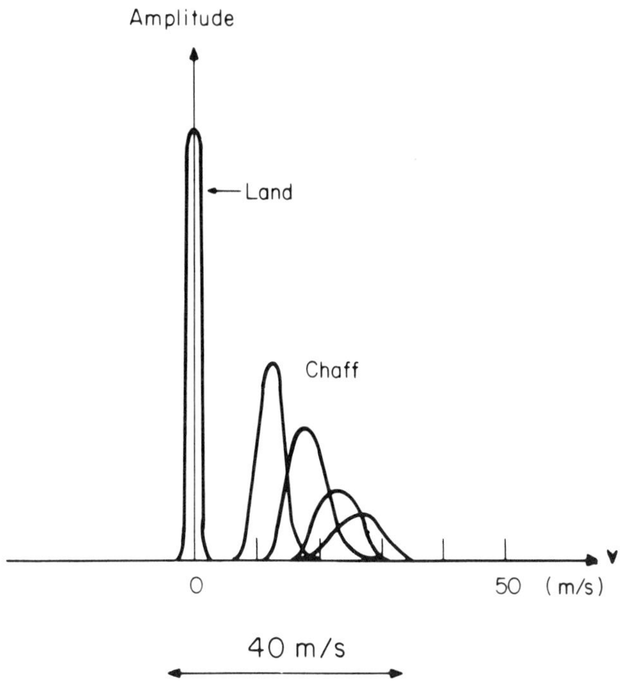

Figure 4.45 Spectrum of range ambiguous chaff clutter.

This is illustrated in Figure 4.45 for four separate range ambiguities, where the chaff clutter spectrum is spread over a significantly wider range than that associated with an individual range cell [8].

The voltage autocorrelation function for chaff obtained from experimental data is depicted in Figure 4.46 [29]. The curve shows an autocorrelation different from one at $\tau = 0$ due to the effect of receiver noise. Also, the curve approaches $\rho = 0$ very slowly, which is indicative of a slow secular variation of the average clutter power. This variation has been attributed to the change of orientation of the chaff dipoles as they fall, which causes an accompanying change in the chaff's RCS. When these anomalies are removed from the experimental autocorrelation function, we find that it can be well approximated by a Gaussian-shaped function. Thus, the power spectral density will also be approximated by a Gaussian-shaped function.

Chaff dipoles have high aerodynamic drag that causes their velocity to drop to that of the local wind conditions a few seconds after being dispensed. This means

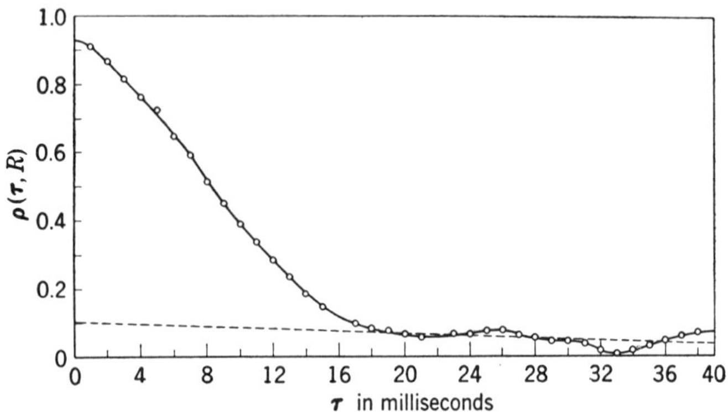

Figure 4.46 Voltage autocorrelation of chaff at X-band.

that the horizontal velocity of the chaff will be determined primarily by the wind speed in the radar resolution cell containing the chaff. Its vertical velocity (falling rate) depends on the dipole dimensions and material, with a falling velocity of 0.3 m/s attributed to aluminized glass at sea level.

The width of the doppler spectra for chaff can be divided into four components in a manner similar to that for precipitation clutter (see (4.49)). The components due to wind shear and beam broadening due to the finite widths of the beam should be identical to those for precipitation (see (4.50) and (4.51)). The value for turbulence is on the order of $\sigma_{\text{turb}} = 1$ m/s below 12,000 ft altitude and 0.7 m/s above that altitude. The final component of the chaff spectra is the doppler spread due to the variation in falling velocity of the dipoles. Because chaff dipoles fall more slowly than raindrops, this component is estimated at $\sigma_{\text{fall}} = 0.45 \sin\phi_{\text{el}}$ m/s [7].

As with precipitation, the spectra of chaff clutter is generally determined by the wind shear component, which is given by $\sigma_{\text{shear}} = 0.42\ kR\phi_{\text{el}}$, where k is the velocity gradient (4 m/s − km is suggested for arbitrary azimuth angles), R is the slant range to the clutter resolution cell in km, and ϕ_{el} is the two-way elevation antenna beamwidth (rad). However, in contrast with precipitation, chaff altitude is not generally limited to 20,000 ft, and spectral spreads greater than 6 m/s may be observed when chaff is dispensed above this altitude. Spectral widths for this component are depicted in Figure 4.47 [7]. Note that observed spectral widths for chaff are apparently not affected by the age of the chaff (assuming the chaff has bloomed) and, when normalized to wavelength, are independent of frequency (i.e., σ_v does not depend on frequency) [29].

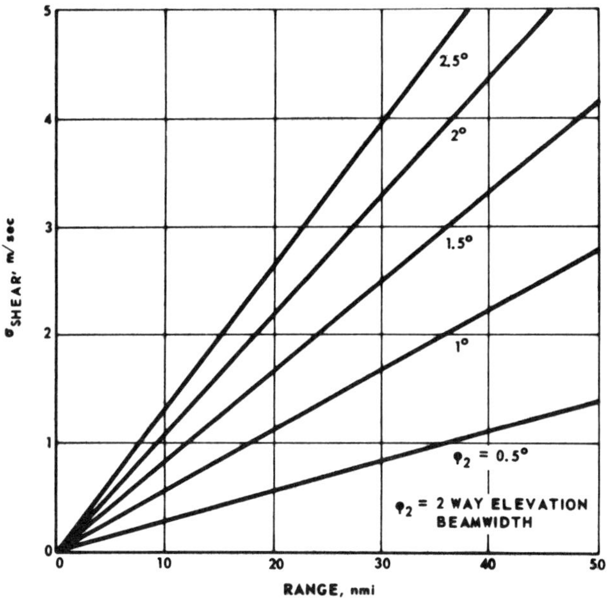

Figure 4.47 Spectral width of chaff due to wind shear (average over 360°).

4.3 APPROXIMATIONS TO CLUTTER AMPLITUDE STATISTICS

Determination of the performance of an MTI or PD radar in clutter generally requires the representation of the clutter as a random process. The Gaussian clutter model (see Section 1.3.1) is an example of a physical model developed into an analytic formulation that allows meaningful representation of the clutter's properties. In this case, the clutter process is completely specified by its mean value and covariance function (i.e., spectral characteristics). However, the mean value, which is proportional to the clutter's average backscattering coefficient ($\bar{\sigma}^\circ$ or $\bar{\eta}$), is a function of many variables (e.g., terrain type, grazing angle, frequency, polarization). Therefore, we want analytic parametric expressions that allow determining the clutter's mean as a function of the variables associated with a particular representation. The parametric representation of the clutter's backscattering coefficient is usually obtained by using an approximation procedure (e.g., a least mean square fit) to a set of experimental data.

When the clutter cannot be described by using a Gaussian clutter model, then a complete representation requires the specification of the clutter's multidimensional probability density function (see Section 1.3.4). Experimental data generally is not available to accomplish this, and a compromise is to specify the first probability density function of the clutter's backscattering coefficient, $p(\sigma^{\circ})$, and the clutter's covariance function. This involves an empirical fit, where one criterion applied is that the resulting amplitude probability density function have analytic properties that allow determining the detection performance in clutter. The most popular non-Gaussian amplitude probability density functions employed are the log-normal, Weibull, and *K*-distributions.

4.3.1 Approximate Average Backscatter Coefficient—$\bar{\sigma}^{\circ}$

The following models are provided for the average backscattering coefficient of rain, land, and sea clutter. These models generally are based on averaged experimental data obtained from a number of sources that is then combined and fitted with a parametric model.

4.3.1.1 Parametric Rain Clutter Model

Based on an analytic model, the backscatter coefficient per unit volume of rain clutter as a function of frequency and rainfall rate can be determined as

$$\bar{\eta} = 7 f_{\text{GHz}}^{4}\, r^{1.6} \times 10^{-12}\ \text{m}^2/\text{m}^3 \tag{4.56}$$

where f_{GHz} is the radar frequency in GHz and r is the rainfall rate in mm/hr [2].

An empirical fit to rain clutter data in the high microwave and millimeter-wave region provides the following parametric relationship for the backscattering coefficient of precipitation clutter:

$$\bar{\eta} = A r^{B}\ \text{m}^2/\text{m}^3 \tag{4.57}$$

where r is the rain rate in mm/hr and A and B are as given in Table 4.20 [30].

4.3.1.2 Parametric Land Clutter Model

A parametric model that provides an empirical fit to a large sample of experimental land clutter data for the average backscatter coefficient (σ°) is given by

$$\bar{\sigma}^{\circ} = A(\psi + C)^{B} \exp\left[-D \Big/ \left(1 + \frac{0.1\,\sigma_h}{\lambda}\right)\right] \tag{4.58}$$

where ψ is the grazing angle (rad), σ_h is the standard deviation of the surface (cm),

Table 4.20
Rain Backscatter Model Coefficient

Frequency (GHz)	*A*	*B*
9.4	1.3×10^{-8}	1.6
35	1.2×10^{-6}	1.6
70	4.2×10^{-5}	1.1
95	1.5×10^{-5}	1.0

λ is the radar wavelength (cm), and A, B, C, and D are empirically derived constants [21]. The model provides $\bar{\sigma}^\circ$ for various types of land clutter including soil, grass, trees, sand, rocks, urban areas, and snow-covered ground as a function of gazing angle, surface roughness, and frequency. The values for the parameters A, B, C, and D are given in Table 4.21. Note that $\bar{\sigma}^\circ$ variations with surface roughness occur in the model for only soil or sand ($\sigma_h \approx 0.2$ to 2.5 cm) at frequencies below 15 GHz. The parameters for trees have been adjusted to match available data at X-band [35] and 95 GHz [16]. In addition, parameters for soil or sand, grass, and crops have been adjusted to be consistent with the model at other frequencies. Figure 4.48 compares the model for crops at X-band against data obtained in various experimental programs.

4.3.1.3 Parametric Sea Clutter Model

A number of the unique characteristics of sea clutter are illustrated in a sea clutter model developed for use with airborne radars. The backscattering coefficient (σ°) for sea clutter is dependent on frequency, grazing angle, sea and wind aspect, sea state and wind speed, and polarization [14].

The variation with polarization is the most difficult to quantify. For moderate to low sea states, horizontal polarization has yielded consistently lower sea returns than vertical polarization in the upwind and downwind directions. However, in crosswind, vertical polarization may give a lower return. In rough seas, the differences between the two linear polarizations may diminish or reverse at large grazing angles. A disadvantage of horizontal polarization is the appearance of highly correlated "sea spikes," especially upwind. Because of this uncertainty, a model for horizontal polarization is developed first and then modified for vertical polarization.

The general shape of the curve for backscattering coefficient (σ°) as a function of grazing angle is shown in Figure 4.3. The three major regions of dependence are the small-angle region, the plateau region, and the large-angle region.

Table 4.21
Land Clutter Model Constants

Constant	*Frequency*	*Soil, Sand*	*Grass*	*Tall Grass, Crops*	*Trees*	*Urban*	*Wet Snow*	*Dry Snow*
A	3	0.0045	0.0071	0.0071	0.0028	0.362	—	—
	5	0.0096	0.015	0.015	0.0047	0.779	—	—
	10	0.025	0.039	0.039	0.0095	2.0	0.0246	0.195
	15	0.05	0.079	0.079	0.019	2.0	—	—
	35	—	0.125	0.301	0.036	—	0.195	2.45
	95	—	—	—	0.046	—	1.138	3.6
B	3	0.83	1.5	1.5	0.64	1.8	—	—
	5	0.83	1.5	1.5	0.64	1.8	—	—
	10	0.83	1.5	1.5	0.64	1.8	1.7	1.7
	15	0.83	1.5	1.5	0.64	1.8	—	—
	35	—	1.5	1.5	0.64	—	1.7	1.7
	95	—	1.5	1.5	1.1	—	0.83	0.83
C	3	0.0013	0.012	0.012	0.002	0.015	—	—
	5	0.0013	0.012	0.012	0.002	0.015	—	—
	10	0.0013	0.012	0.012	0.002	0.015	0.0016	0.0016
	15	0.0013	0.012	0.012	0.002	0.015	—	—
	35	—	0.012	0.012	0.012	—	0.008	0.0016
	95	—	0.012	0.012	0.012	—	0.008	0.0016
D	3	2.3	0.0	0.0	0.0	0.0	—	—
	5	2.3	0.0	0.0	0.0	0.0	—	—
	10	2.3	0.0	0.0	0.0	0.0	0.0	0.0
	15	2.3	0.0	0.0	0.0	0.0	—	—
	35	—	0.0	0.0	0.0	—	0.0	0.0
	95	—	0.0	0.0	0.0	—	0.0	0.0

The critical angle (ϕ_c), where the curves for the plateau and small-angle regions intersect, is given by

$$\phi_c = \frac{\lambda}{2.5h_w} \tag{4.59}$$

where ϕ_c = critical angle in radians, λ = wavelength in feet, and h_w = peak-to-trough wave height in feet of the highest 10% of the waves.

An approximate relationship for h_w (in feet) is given by

$$h_w = 1.05 \exp(0.495s) \text{ (ft)} \tag{4.60}$$

where s is the Douglas sea state.

Ohio State Data (X Band)

GRAPH	TERRAIN	POL.
A	4' Oats	VV
B	3' Soybeans	VV
C	12" Wheat stubble	VV
D	Disked field	VV
E	Plowed field	HH
F	1/2" Green wheat	VV

Other Data (X Band)

GRAPH	AGENCY	TERRAIN	POL.
G	Goodyear	Arizona Irrigated farmland	HH
H	Goodyear	Arizona Dry grassland	HH
J	NRL	New Mexico Pecan orchards	HH
K	NRL, Grant & Yoolee	Green grass Sandy Loam	VV

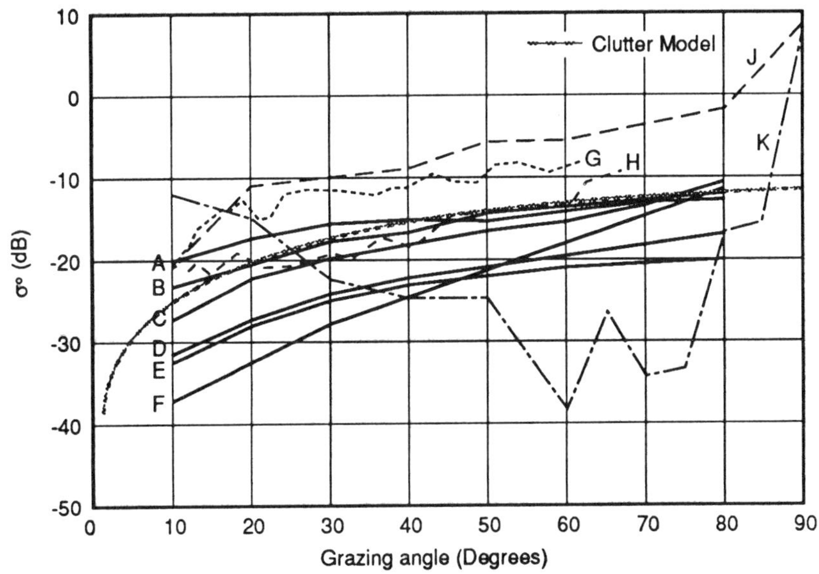

FARMLAND CLUTTER CROSS SECTION MEASUREMENTS OF VARIOUS INVESTIGATORS

Figure 4.48 Land clutter model at X-band for crops *versus* experimentally determined farmland clutter.

The variation of σ° in the plateau region is taken as proportional to $\phi^{0.4}$ and the variation in the small-angle region is $\phi^{5.0}$.

The aspect dependence is modeled as

$$\sigma^{\circ}_{\mathrm{dB}}(\beta) = \sigma^{\circ}_{1\mathrm{dB}} - 4 + 4\cos\beta \qquad (4.61)$$

where β = angle between antenna boresight and upwind direction and $\sigma^{\circ}_{1\mathrm{dB}}$ is the backscatter coefficient for upwind direction. A frequency dependence of $\lambda^{-1/2}$ was chosen as best representing the data for the plateau region.

A calibration point is needed to quantify the model. This was taken as $\lambda = 3$ cm, $\phi = 10°$, $s = 4$, and $\sigma^\circ = -42$ dB. The value of σ_p° in the plateau region is then

$$\sigma_p^\circ = \frac{k\phi^{0.4}\exp(0.792s)}{\lambda^{1/2}} \tag{4.62}$$

where s = Douglas sea state, ϕ = grazing angle in degrees, λ = wavelength in cm, and $k = 1.83 \times 10^{-6}$. The value of σ_s° in the small-angle region is given by

$$\sigma_s^\circ = \frac{\phi^5\exp(3.069s)}{c\lambda^{5.1}} \tag{4.63}$$

where $c = 1.17 \times 10^5$.

The value of σ_L° in the large-angle region was not modeled, but the transition grazing angle appears to lie in the 30° to 60° region.

Analytical expressions were not obtained for vertical polarization. Grazing angle and frequency dependence were observed to be the same as for horizontal polarization. The dependence on sea state is very complicated.

4.4 ANALYTICAL CLUTTER REPRESENTATIONS

Analytical clutter models are useful in the design and analysis of MTI and PD radars. These models attempt to represent the random clutter in such a manner that system performance can be determined analytically. Many of the characteristics of these models are discussed elsewhere in this text (e.g., see Section 1.3). This section collects the many relationships involved with the various models and presents them in a consistent manner.

The simplest clutter model is the Gaussian model that represents the random clutter process as a multidimensional Gaussian process. The model has been employed almost exclusively in MTI analyses. An attractive feature of this model is that the clutter process is specified completely by its mean value and covariance function. The assumption of the Gaussian model leads to an optimal MTI processor that is linear.

The Gaussian clutter assumption generally leads to an optimistic evaluation of MTI performance. At present, none of the analytical techniques available allow determining the performance of MTI systems in the presence of non-Gaussian clutter. The principal method of attacking this problem is through a Monte Carlo computer simulation, in which the actual MTI system is exercised by applying statistical samples of the postulated clutter and target processes.

4.4.1 The Gaussian Process Representation

Distributed clutter that is returned from a spatially continuous distribution of scatterers with no subset of scatterers predominating can generally be represented by a Gaussian process. Such clutter is usually associated with weather clutter, chaff, sea clutter observed with a low-resolution radar (pulsewidth $\tau > 0.5\ \mu s$), or with a high-resolution radar at high grazing angles ($\phi > 5°$) and land clutter observed from high grazing angles ($\phi > 5°$) over undeveloped terrain.

The random clutter process is represented at the clutter carrier frequency (f_c) as

$$c_t = x_t \cos\omega_c t - y_t \sin\omega_c t \tag{4.64}$$

where x_t and y_t are zero mean, identically distributed, low-pass independent normal processes with variance σ^2. The clutter process c_t is strict sense stationary and hence its characteristics are independent of the time origin.

The voltage envelope of c_t is given by

$$v_t = \sqrt{x_t^2 + y_t^2} \tag{4.65}$$

and has a Rayleigh first probability density function:

$$p_v(v) = \frac{v}{\sigma^2} \exp(-v^2/2\sigma^2); \quad v \geq 0 \tag{4.66}$$

independent of time due to the stationarity, and is a function of only the clutter power:

$$P_c = 2\sigma^2 \tag{4.67}$$

The power amplitude distribution function can be found by applying the transformation $P = v^2$:

$$P_p(P) = \frac{1}{P_c} \exp(-P/P_c) \tag{4.68}$$

The clutter cross-sectional area (backscatter coefficient $\times$ intercepted area) is proportional to the power envelope, resulting in an exponential probability density function for the clutter cross section (σ):

$$p_p(\sigma) = \frac{1}{\bar{\sigma}} \exp(-\sigma/\bar{\sigma}) \tag{4.69}$$

Here, $\bar{\sigma}$ is the mean clutter cross section, given by

$$\bar{\sigma} = \bar{\sigma}^{\circ} A_c \tag{4.70}$$

where $\bar{\sigma}^{\circ}$ is the mean backscattering coefficient and A_c is the area intercepted by the radar cell.

The power spectral density associated with the randomly fluctuating portion of the clutter cross section is usually assumed to be Gaussian shaped in that it is the result of several independent effects. The power spectral density of the power envelope is given by

$$S_p(f) = \frac{P_c}{\sqrt{2\pi\sigma_f^2}} \exp(-f^2/2\sigma_f^2) \tag{4.71}$$

where P_c is the clutter power and σ_f is the standard deviation of the spectrum. The standard deviation of the spectrum (σ_f) is related to the rms velocity clutter spread (σ_v) by

$$\sigma_f = \frac{2\sigma_v}{\lambda} \tag{4.72}$$

The autocorrelation function associated with the power envelope is given by

$$R_p(\tau) = P_c \exp(-\tau^2/2\sigma_\tau^2) \tag{4.73}$$

where

$$\sigma_\tau = 1/2\,\pi\sigma_f \tag{4.74}$$

In many applications, to work with the normalized autocorrelation function is convenient:

$$\rho_p(\tau) = R_p(\tau)/P_c = \exp(-\tau^2/2\sigma_\tau^2) \tag{4.75}$$

Of interest in MTI work and other coherent processing systems are the power spectral density and autocorrelation function of x_t and y_t. The relationship between the normalized autocorrelation functions of P_t and x_t is known [29] to be given by

$$\rho_p(\tau) = \rho_x^2(\tau) = \rho_y^2(\tau) \tag{4.76}$$

Thus, the spectral density of the power envelope can be found from the power spectral density of the coherent quadrature components by

$$S_p(\omega) = S_x(\omega) * S_x(\omega) \tag{4.77}$$

where * represents convolution. Hence, for a Gaussian-shaped spectrum, all power spectral densities and autocorrelations are Gaussian shaped, and the standard deviation of the power envelope can be found from those of the coherent quadrature components by using

$$\sigma_p^2 = 2\sigma_x^2 = 2\sigma_y^2 \tag{4.78}$$

To summarize, for MTI work, the power spectral density or autocorrelation functions are given by (4.71), (4.73), and (4.75), using the appropriate standard deviation given by (4.78). When working with experimental or published data, care must be exercised to ascertain whether the standard deviations referenced are the coherent (σ_x) or noncoherent (σ_p) standard deviations.

Some MTI applications require modeling clutter that has relative motion with respect to the radar. This is accomplished through modifying (4.71) by the mean clutter doppler shift ($\bar{f}_d$), resulting in

$$S_p(f) = \frac{P_c}{\sqrt{2\pi\sigma_f^2}} \exp[-(f - f_d)^2/2\sigma_f^2] \tag{4.79}$$

When the clutter process (4.64) is sampled by a pulsed radar, the probability density of the samples is given by a multidimensional Gaussian distribution:

$$p_{x_1 \dots x_n}(x_1 \dots x_n) = \frac{\exp(-\mathbf{x}^T\mathbf{M}_x^{-1}\mathbf{x}/2)}{(2\pi)^{n/2}|\mathbf{M}_x|^{1/2}} \tag{4.80}$$

where $\mathbf{x} = (x_1, x_2, \dots, x_n)^T$, and the covariance matrix is given by

$$\mathbf{M}_x = \sigma_x^2 \begin{bmatrix} 1 & \rho_{12} & \rho_{13} & \rho_{1n} \\ \rho_{21} & \ddots & & \rho_{2n} \\ & & \ddots & \\ \rho_{n_1} & \cdots & \cdots & 1 \end{bmatrix} \tag{4.81}$$

where

$$\rho_{ij} = \exp[-(i - j)^2 \Omega^2/2]; \quad \Omega = 2\pi\sigma_f T \tag{4.82}$$

for a Gaussian-shaped distribution.

If the clutter is passed through a linear MTI, the Gaussian form of the distribution is preserved. The probability density function at the output of the MTI can be found by using (4.80) with the covariance matrix modified by the effect of the MTI filter on the clutter process.

For analytical convenience, the clutter process is sometimes assumed to be a Gaussian Markov process. This process has an exponential normalized function given by

$$\rho_{\alpha\beta} = \exp(-\gamma|\alpha - \beta|T) \tag{4.83}$$

and a power spectral density given by

$$S(\omega) = 2\gamma P_c/(\gamma^2 + \omega^2) \tag{4.84}$$

The inverse covariance matrix ($\mathbf{M}_x^{-1}$) has a particularly simple form:

$$\mathbf{M}_x^{-1} = \frac{1}{1 - \rho_{12}^2} \begin{bmatrix} 1 & -\rho_{12} & 0 & \cdots & & 0 \\ -\rho_{21} & 1 + \rho_{12}^2 & -\rho_{12} & \cdots & 0 & 0 \\ & & & \ddots & & \\ 0 & -\rho_{21} & 1 + \rho_{12}^2 & \cdots & 0 & 0 \\ 0 & 0 & 0 & \cdots & 1 + \rho_{12}^2 & -\rho_{12} \\ 0 & 0 & 0 & \cdots & -\rho_{21} & 1 \end{bmatrix} \tag{4.85}$$

4.4.2 The Rician Clutter Representation

This model is similar to the Gaussian process representation except that a predominant steady scatterer (S) is added to the distributed clutter. The first probability density function corresponding to (4.68) becomes

$$p_p(P) = \frac{1 + m^2}{\bar{P}} \mathrm{e}^{-m^2} \mathrm{e}^{-[P/\bar{P}(1+m^2)]} I_0(2m\sqrt{(1 + m^2)P/\bar{P}}) \tag{4.86}$$

where the ratio of the steady (S^2) to distributed (P_0) power ratio is given by

$$m^2 = S^2/P_0 \tag{4.87}$$

and the total power ($\bar{P}$) proportional to the cross-sectional area is given by

$$\bar{P} = S^2 + P_0 \tag{4.88}$$

For MTI work, the clutter process at the carrier frequency is obtained by modifying (4.64) to

$$c_t = (S + x_t)\cos\omega_c t - y_t \sin\omega_c t \tag{4.89}$$

The autocorrelation function for the in-phase (I) process and Gaussian-shaped distributed clutter is

$$R(\tau) = P_0[m^2 + \exp(-\tau^2/2\sigma_\tau^2)] \tag{4.90}$$

The resulting power spectral density is

$$S(f) = m^2 P_0 \delta(f) + P_0 \exp(-f^2/2\sigma_f^2)/\sqrt{2\pi\sigma_f^2} \tag{4.91}$$

which illustrates that the MTI radars operating in this clutter must be designed to provide a null at zero frequency.

4.4.3 The Log-Normal Clutter Representation

The log-normal clutter representation has been used to model high-resolution (pulsewidth $\tau < .5$ μs) sea-clutter data, where the sea clutter is observed at grazing angles less than 5°. In addition, ground clutter observed at low grazing angles has also been characterized by using the log-normal model.

The log-normal distribution is highly skewed, resulting in a relatively high probability of large clutter amplitudes. False alarm probabilities are controlled by raising the detection thresholds, thereby causing a reduced probability of detection (P_d) as compared to that achieved in Rayleigh clutter.

The log-normal model generally tends to overestimate the dynamic range of the real clutter distribution, whereas the Rayleigh model tends to underestimate the dynamic range. Thus, analyses with the log-normal model produce conservative results, whereas optimistic results are produced with the Rayleigh model. A good design procedure is to design for the range of distributions between Rayleigh and log-normal.

The log-normal sea clutter model is associated with sea spikes, a quasispecular reflection from wave crests during the formation of white caps. The log-normal model for ground clutter seems to be associated with large directive scatterers that, depending on viewing aspect, produce bright ridges and shadowing, resulting in a large clutter dynamic range.

The log-normal character of sea clutter is shown in Figure 4.49 for data taken with a harbor surveillance radar [36]. As is apparent, the data above the median can be closely approximated by straight lines, indicating the close fit provided by the log-normal model.

The log-normal character of ground clutter data is apparent for clutter returned from cities as shown in Figure 4.50. These data were synthesized from high-resolution synthetic aperture radar data [15].

The experimental log-normal data depicted in Figures 4.49 and 4.50 represent long-time averages of the clutter cross section in a resolution cell. From an MTI

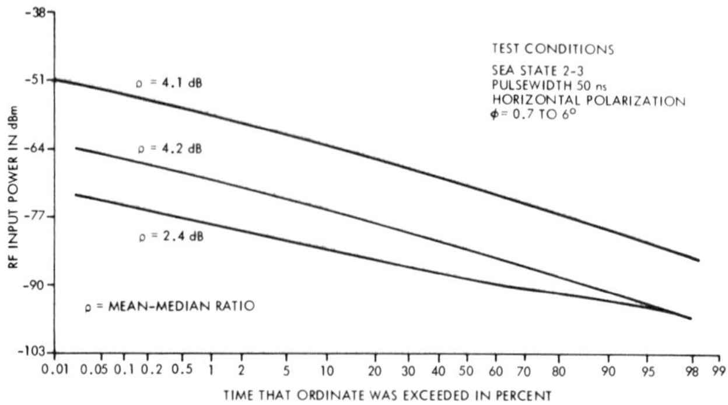

Figure 4.49 Harbor surveillance radar, sea clutter data.

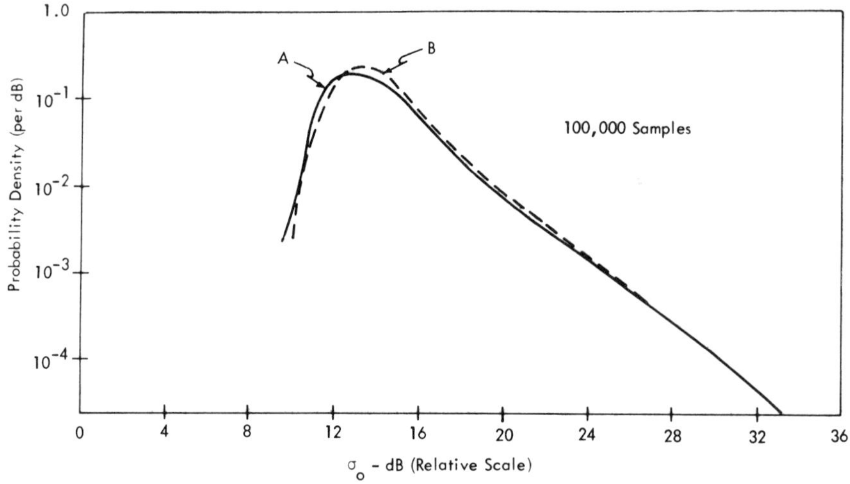

Figure 4.50 Probability density function of two cities.

standpoint, the short-time characteristics of the clutter are important, and two alternative interpretations of the data may be conjectured.

The first interpretation represents the random clutter process as a wide-sense stationary log-normal process. This is obtained analytically by passing the multidimensional Gaussian process represented by (4.80) through an exponential nonlinearity. To achieve the desired stationarity, the x_t and y_t processes of (4.64) must be uncorrelated. This interpretation has the advantage that it is a straightforward extension of the Gaussian clutter model, and many of the properties of this log-normal model can be deduced from corresponding properties of the Gaussian model.

The second interpretation models the clutter as a time-varying process having parameters that are stationary within a radar observation period over a specified spatial correlation distance. The conditional probability density function $p_v(v|\sigma^\circ)$ is Rayleigh or Rician over the spatial correlation distance associated with each spatial correlation block that varies statistically in accordance with a log-normal distribution. Parameters of this model are discussed later in Section 4.5.1 on the IIT clutter model.

Both models lead to the same approximate result if uncorrelated (either temporally or spatially) samples are taken from the distribution. This situation occurs in fast-scanning radars that sample the clutter on a scan-to-scan basis. For MTI systems, the two interpretations can lead to different results, but no guidelines are currently available to indicate a clear choice.

By using the first interpretation, the log-normal clutter model is formed by expressing (4.64) as

$$c_t = v_t \cos(\omega_c t + \phi_c) \tag{4.92}$$

where v_t is the envelope of (4.65) and ϕ_c is a uniformly distributed random variable. If the quadrature components $x_t = v_t \cos\phi_c$ and $y_t = v_t \sin\phi_c$ are zero mean, uncorrelated, and have identical band-limited spectral densities, c_t is a wide-sense stationary process. The first probability density function of the envelope is given by

$$p_v(v) = \frac{1}{\sqrt{2\pi}\sigma_v v} \exp\left[-\frac{(\ln v/v_m)^2}{2\sigma_v^2}\right]; \quad v \geq 0 \tag{4.93}$$

where σ_v is the standard deviation of the underlying normal distribution and v_m is the median value of the log-normal distribution.

The power envelope density function (proportional to the clutter cross-sectional area) is found by applying the transformation $A = v^2$ to (4.93) resulting in

$$p_p(\sigma) = \frac{1}{\sqrt{2\pi}\sigma_p \sigma} \exp\left[-\frac{(\ln\sigma/\sigma_m)^2}{2\sigma_p^2}\right]; \quad \sigma \geq 0 \tag{4.94}$$

where $\sigma_p = 2\sigma_v$, and σ_m is the median value of the clutter cross section. The mean value of the clutter cross section is given by

$$\bar{\sigma} = \rho_A \sigma_m \tag{4.95}$$

where ρ_A is the mean-to-median ratio:

$$\rho_A = \exp(\sigma_p^2/2) \tag{4.96}$$

The distribution function for the cross-sectional area is given by

$$P_p(\sigma) = \frac{1}{2}\left[1 + \operatorname{erf}\left(\frac{1}{\sqrt{2}\sigma_p} \ln \frac{\sigma}{\sigma_m}\right)\right] \tag{4.97}$$

On log-probability paper, this function plots as a straight line, as depicted in Figure 4.51 ($A_{\text{dB}} = 10 \log\sigma$). The straight line is easily determined for either the voltage envelope or cross-sectional area using the relationships provided in Figure 4.51 for the 0.5 and 0.9999 cumulative probability points. The mean or median cross-sectional area is determined from a knowledge of the clutter's backscattering coefficient.

The characteristics of log-normal clutter are often given in terms of its underlying normal distribution. We can see from Figure 4.51 that the log-normal clutter distribution may be expressed as

$$p_N\,(A_{\text{dB}}) = \frac{1}{\sqrt{2\pi}\sigma_{\text{dB}}} \exp\left[-\frac{A_{\text{dB}}^2}{2\sigma_{\text{dB}}^2}\right] \tag{4.98}$$

where

$$A_{\text{dB}} = 10 \log_{10}\sigma/\sigma_m \tag{4.99}$$

Transforming (4.98) by using (4.99) results in the log-normal distribution of (4.94) with

$$\sigma_p = 0.2303\ \sigma_{\text{dB}} \tag{4.100}$$

Equation (4.100) can be used to relate the log-normal clutter parameters in natural units (σ_p) to those in dB units (σ_{dB}).

Table 4.22 gives some reported values of σ_p for various types of clutter. For MTI work, the correlation properties of the sample log-normal clutter model are important. A complete characterization of the log-normal model can be obtained by passing the multidimensional Gaussian distribution of (4.80) through an exponential nonlinearity. This is discussed in [37] and involves specification of the mean and covariance matrices associated with the multidimensional Gaussian distribution. The elements of the Gaussian covariance matrix (m_{xij}) are related to the elements of the log-normal covariance matrix, $m_{\ln(ij)}$, by

$$m_{xij} = \ln\left(1 + \frac{m_{\ln(ij)}}{\bar{x}_i\,\bar{x}_j}\right) \tag{4.101}$$

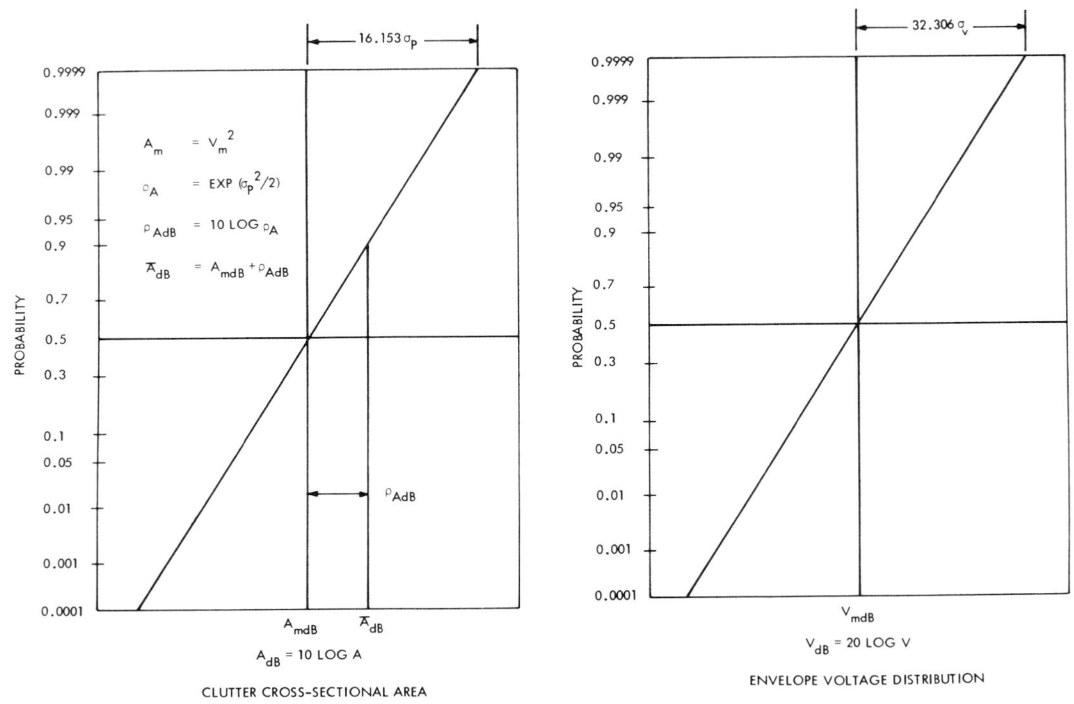

Figure 4.51 Log-normal clutter model.

Table 4.22
Log-Normal Clutter Parameters

Terrain or Sea State	*Band Frequency*	ϕ	σ_p
Sea State 2–3	X	4.7°	1.382
Sea State 3	Ku	1–5°	1.440–1.960
Sea State 4	X	0.24°	1.548
Sea State 5[a]	Ku	0.50°	1.634
Land Clutter (discrete)	S	low	3.916
Land Clutter (distributed)	S	low	1.380
Land Clutter[a]	P to Ka	10°–70°	.728–2.584
Rain Clutter	X and 95 GHz	—	.680
Radar Angels	—	—	1.352–1.620

[a]Extrapolated values.

where $\bar{x}_{i,j}$ denotes the elements of the mean vectors of the log-normal clutter. Several relationships useful in converting from Gaussian (x_t) to log-normal (z_t) variables through an exponential nonlinearity, $A \exp(\cdot)$ are

$$E(z_t) = A \exp\left[\frac{R_x(0)}{2}\right] \tag{4.102}$$

$$R_z(\tau) = A^2 \exp[R_x(0) + R_x(\tau)] \tag{4.103}$$

$$C_z(\tau) = A^2 \exp[R_x(0)]\{\exp[R_x(\tau)] - 1\} \tag{4.104}$$

where $R_x(\tau)$ is the ACF of the Gaussian variable, and $R_z(\tau)$ is the ACF of the log-normal variable, and $C_z(\tau)$ is the covariance function of the log-normal variable.

4.4.4 The Weibull Clutter Representation

The Weibull clutter representation has properties that lie between the Rayleigh (a member of the Weibull family) and log-normal clutter models. The Weibull model has been used to model both land and sea clutter and offers the potential to accurately represent the real clutter distribution over a much wider range of conditions than either the log-normal or Rayleigh model.

Figure 4.28 showed experimental sea clutter distributions plotted on log-Rayleigh probability paper, and Figure 4.17 presented experimental land clutter data. The close fits provided by the straight lines indicate that the data can be closely modeled by Weibull statistics. In addition, the decreasing skewness with increasing grazing angle indicates that the distribution is approaching a Rayleigh distribution.

The two clutter model interpretations that apply to the log-normal model also apply to the Weibull clutter model. For MTI systems, the two interpretations can lead to different results, but no guidelines are currently available to indicate a clear choice.

In a Weibull clutter model, the first probability distribution of the voltage output of an envelope detector normalized to the median v_m is given by

$$p_v(R) = \alpha \cdot \ln 2 \cdot R^{-i\alpha - 1} \exp(-\ln 2 \cdot R^{\alpha}), \quad R > 0 \tag{4.105}$$

where $R = v/v_m$ and α is a parameter that relates to the skewness of the distribution. This is related to the amplitude probability density function of the clutter cross section σ by the transformation $\sigma = v^2$, resulting in

$$p_A(\sigma_c) = \beta \cdot \ln 2 \cdot \sigma_c^{\beta - 1} \exp(-\ln 2 \cdot \sigma_c^{\beta}], \quad \sigma_c > 0 \tag{4.106}$$

where $\sigma_c = \sigma/\sigma_m$, $\beta = \alpha/2$, and σ_m is the median of the clutter cross-sectional area. The mean of the clutter cross section is given by

$$\bar{\sigma} = \sigma_m \, \Gamma(1 + 1/\beta)/(\ln 2)^{1/\beta} \tag{4.107}$$

The distribution function representing the clutter cross-sectional area on a dB scale is given by

$$\sigma_{\text{dB}} = \sigma_{m_{\text{dB}}} + 1.592/\beta + (10/\beta) \log\{\ln[1/(1 - P_A(\sigma))]\} \tag{4.108}$$

This function plots as a straight line on log-Rayleigh probability paper. The straight line is easily determined by plotting the 0.5 probability point as

$$\sigma_{\text{dB}} = \sigma_{m_{\text{dB}}} \tag{4.109}$$

and the 0.9999 probability point as

$$\sigma_{\text{dB}} = \sigma_{m_{\text{dB}}} + 11.235/\beta \tag{4.110}$$

The Weibull clutter distribution is sometimes defined in terms of a slope parameter (a), given by

$$a = 1/\beta \tag{4.111}$$

Some reported values of the Weibull model parameters are given in Table 4.23.

Table 4.23
Weibull Clutter Parameters

Terrain or Sea State	*Band Frequency*	*Beamwidth (degrees)*	ϕ *(degrees)*	*Pulsewidth* (μs)	β
Rocky Mountains	S	1.5	—	2	0.256
Wooded Hills	L	1.7	≈0.5	3	0.313
Forest	X	1.4	0.7	0.17	0.253–0.266
Cultivated Land	X	1.4	0.7–5	0.17	0.303–1
Sea State 1	X	0.5	4.7	0.02	0.726
Sea State 3	Ku	5	1–30	0.1	0.58–0.8915

The correlation properties of the sample Weibull clutter model are important for MTI work. The correlation matrix of the Weibull distribution can be related to the correlation matrix of a Gaussian distribution by the following set of transformations [38]. A Rayleigh vector (**v**) is formed from two zero-mean independent Gaussian vectors **x** and **y** by

$$\mathbf{v}_i = \sqrt{\mathbf{x}_i^2 + \mathbf{y}_i^2} \tag{4.112}$$

The Rayleigh vector is transformed into a Weibull vector (**R**) by

$$\mathbf{r}_i = \mathbf{v}_i^{2/\alpha} \tag{4.113}$$

Elements of the correlation matrix (m_{xij}) of the Gaussian vectors are related to elements of the correlation matrix of the Weibull vector (s_{ij}) by

$$\begin{aligned} s_{ij} = &\{\Gamma^2(1 + 1/\alpha)/[\Gamma(1 + 2/\alpha) - \Gamma^2(1 + 1/\alpha)]\} \\ &[{}_2F_1(-1/\alpha, -1/\alpha; m_{xij}^2) - 1] \end{aligned} \tag{4.114}$$

where ${}_2F_1(\cdot)$ is the hypergeometric function.

4.4.5 The *K*-Distribution Clutter Representation

A representation of *K*-distribution clutter has been used to model the envelope-detected return from both land and sea clutter [39]. The properties of the *K*-distribution are similar to those of the Weibull distribution (see Section 4.1.2.2) with values in the tail region ranging from that of the Rayleigh distribution (a member of the *K*-distribution family, as is the Weibull) to approach that of the log-normal distribution. The *K*-

distribution representation is derived from a composite structure that assumes that the clutter can be represented as a Gaussian process with a varying mean. The mean is further assumed to fluctuate slowly so that its value is fixed over the processing time in the radar signal processor. This allows analyzing the system performance by using Gaussian (Rayleigh envelope) statistics, which are then averaged over the ensemble of mean values to determine the average system performance in a manner analogous to that used to determine the detection performance for Swerling fluctuating targets. Thus, although the ensemble probability distribution of the clutter's backscatter coefficient is represented by a *K*-distribution, the actual analysis is generally performed on the compound distributions that consist of the Rayleigh envelope for the fast fluctuating component and the chi-square distribution for the slowly fluctuating mean.

The Rayleigh probability density function of the envelope-detected *K*-distribution voltage conditioned on its mean value ($y = \sqrt{\pi/2}\,\sigma$) is given by

$$p(v|y) = \frac{\pi v}{2y^2} \exp\left(-\frac{\pi v^2}{4y^2}\right); \quad v \geq 0 \tag{4.115}$$

where $\sqrt{2}\sigma$ is the standard deviation of the clutter power associated with this component. The chi probability density function that represents the fluctuation of the mean level is given by

$$p(y) = \frac{2b^{2\nu}}{\Gamma(\nu)} y^{2\nu-1} \exp(-b^2y^2); \quad y \geq 0 \tag{4.116}$$

where ν is a shape parameter and b is a scale parameter. The resulting *K*-distributed probability density function of the envelope-detected clutter is then given by

$$p(v) = 4c(cv)^{\nu} K_{\nu-1}(2cv)/\Gamma(\nu) \tag{4.117}$$

where $c = \sqrt{\pi/4}\, b$ is a scale parameter, ν is a shape parameter, and $K_{\nu-1}(\cdot)$ is the modified Bessel function of order $\nu - 1$. The nth moments of $p(v)$ are given by

$$E(v^n) = \Gamma(\nu + n/2)\,\Gamma(1 + n/2)/c^n\,\Gamma(\nu) \tag{4.118}$$

so that the clutter power is given by $P_c = E(v^2) = \nu/c^2$. The probability density function for the clutter power can be derived by using the transformation $p = v^2$ as

$$p(p) = 2c^{\nu+1} p^{\nu-1/2} K_{\nu-1}(2c\sqrt{p})/\Gamma(\nu) \tag{4.119}$$

The cumulative probability distribution function for the clutter's power that is proportional to the clutter's backscattering coefficient (σ°) is then given by

$$P(p) = 1 - 2c^{\nu} p^{\nu/2} K_{\nu}(2c\sqrt{p})/\Gamma(\nu) \tag{4.120}$$

For most clutter situations, the value of the shape parameter (ν) varies over the range $0.1 < \nu < \infty$, where the smaller values (i.e., $\nu \rightarrow 0.1$) correspond to clutter with high tails, and as ν approaches ∞, the clutter distribution approaches a Rayleigh distribution. Values of ν between 0.1 and 3 have been experimentally determined to approximate high-resolution (τ = 70–270 ns) sea clutter data at low grazing angles (ϕ = 1°–1.5°) [28]. We also can show that the moments of K-distributed clutter range from that of the Rayleigh distribution to approach those of the log-normal distribution [39].

4.5 CLUTTER MODEL ENVIRONMENTS

This section describes two clutter model environments developed to represent those seen by an airborne search radar (thè IIT model) and a ground-based search radar.

4.5.1 The IIT Radar Clutter Model [15]

Representing one of the more comprehensive clutter models, the IIT model was developed for use with overland airborne radars on the basis of extensive data analysis. The model provides for the spatial distribution of the clutter and does not explicitly address intrinsic or induced variations in the clutter return.

The IIT model treats radar ground clutter as consisting of two additive components: distributed clutter and discrete clutter. The former is indicative of returns from natural terrains; the latter corresponds to returns from physically isolated and dominant artificial scatterers.

The complete model is intended for use in computer simulations and is far too complex for manual calculations. In the following section, a general description of the model is given to provide insight into clutter modeling and a basis for approximations that may be used for simple calculations. The reader is referred to [15] for a complete description of the model.

The distributed clutter backscattering cross section per unit area (σ^0) is a function of frequency, polarization, grazing angle, terrain, and environmental factors. The first four factors are deterministic in nature, whereas environmental factors are random and account for variations in grazing angle.

Figures 4.8 and 4.11 depicted average backscattering coefficients for various terrains as a function of grazing angle. The data given in Figure 4.8 were independent of frequency and polarization because uncertainties in the database would be larger than variations due to these parameters. For desert terrain, the variation with frequency was significant and hence should be included in the model.

The instantaneous clutter power returned to the radar is proportional to the effective clutter cross-sectional area, the distribution of which is given by

$$p_p(\sigma) = \frac{1}{\bar{\sigma}} \exp(-\sigma/\bar{\sigma}) \tag{4.121}$$

The factor $\bar{\sigma}$ is a random variable having a value that is a function of the spatial properties of the clutter distribution, given by

$$\bar{\sigma} = \bar{\sigma}^{\circ}(x, y)A_c \tag{4.122}$$

where $\bar{\sigma}^{\circ}(x, y)$ is the spatially dependent clutter backscatter coefficient and A_c is the area intercepted by the radar resolution cell.

For any fixed set of spatial parameters (x_0, y_0), the value of the backscatter coefficient $\sigma^{\circ}(x_0, y_0)$ is selected from a log-normal distribution defined by

$$p_{N(\sigma^{\circ}\text{dB})} = \frac{1}{\sqrt{2\pi}s_0} \exp\left[-\frac{1}{2}\left(\frac{\sigma^{\circ}_{\text{dB}} - \bar{\sigma}^{\circ}_{\text{dB}}}{s_0}\right)^2\right] \tag{4.123}$$

where

$$\sigma^{\circ}_{\text{dB}} = 10 \log_{10}\sigma^{\circ} \tag{4.124}$$

and s_0 is the standard deviation of $\sigma^{\circ}_{\text{dB}}$. The term $\bar{\sigma}^{\circ}$ is given in Figures 4.8 and 4.11 above, whereas s_0 is given by

$$s_0 = \sqrt{s_1^2 + s_2^2} \tag{4.125}$$

where s_1 and s_2 are defined in Figures 4.52 and 4.53. Both $\bar{\sigma}^{\circ}_{\text{dB}}$ and s_0 are functions of the grazing angle and type of terrain.

As an intermediate step in the development of the model, consider that the terrain is divided into a checkerboard kind of pattern. Each cell of the checkerboard

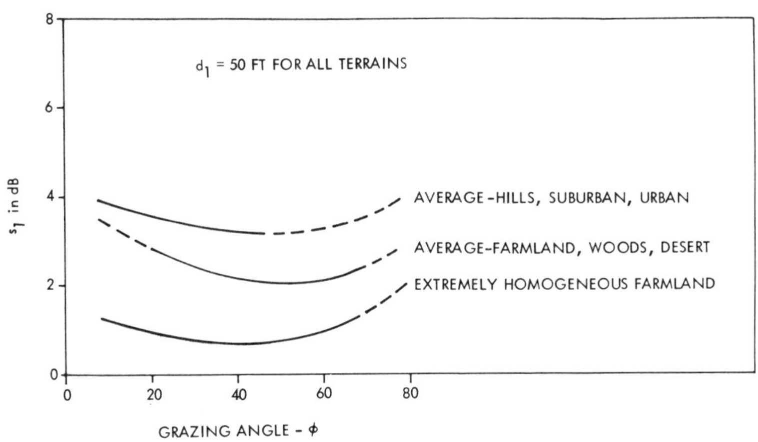

Figure 4.52 Standard deviation of short-range spatial correlation for various terrains.

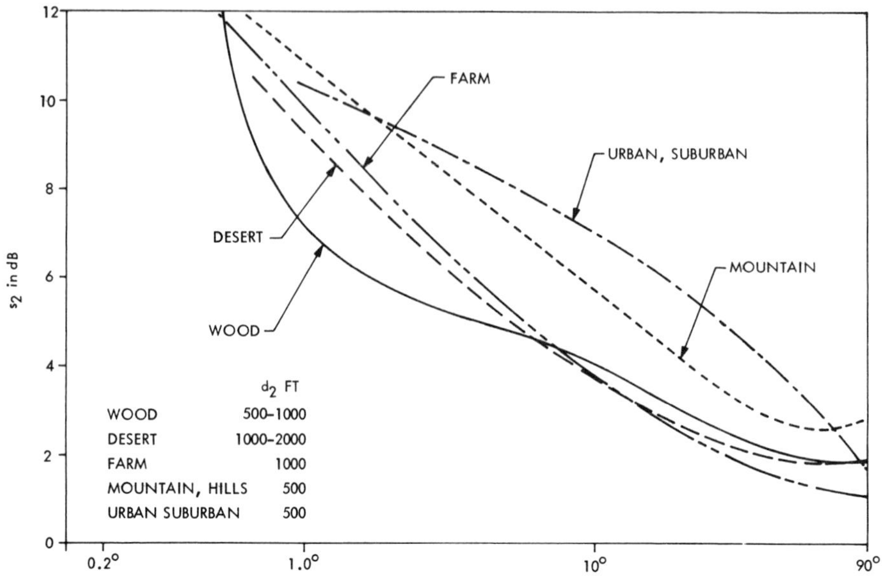

Figure 4.53 Standard deviation of long-range spatial correlation.

is defined by a correlation distance (d) over which the value of $\sigma^\circ(x_0, y_0)$ applies. The average value of the backscatter coefficient, $\bar{\sigma}^\circ(x, y)$, to be used in (4.122) is found by integrating over those cells of the checkerboard contained within the window formed by the radar's resolution cell. For high accuracy, the radar's two-way antenna pattern is used as an amplitude weighting for one dimension of the window.

As the radar antenna is scanned, or when the radar is on a moving platform, new cells of the checkerboard enter the radar resolution window and old cells are discarded. Thus, the radar induces a temporal variation in the clutter statistics that accounts for the correlation properties of the clutter.

The next refinement of the model adds a triangular correlation function across each cell of the checkerboard. This spatial ACF is given by

$$r(\Delta x, \Delta y) = \frac{\exp[(.2303\, s_0)^2 \rho_{12}^2(\Delta x, \Delta y)] - 1}{\exp[(.2303 s_0)^2] - 1} \tag{4.126}$$

where

$$\rho_{12}^2(\Delta x, \Delta y) = \rho(\Delta x)\rho(\Delta y) \tag{4.127}$$

$$\rho(\Delta z) = 1 - \left|\frac{\Delta z}{d}\right|; \quad |\Delta z = \sqrt{(\Delta x)^2 + (\Delta y)^2}| \leq d = 0; \quad |\Delta z| > d \tag{4.128}$$

and d is the correlation distance.

For computer simulation, the model at any point in the terrain field can be generated by

$$\sigma^{\circ}_{\mathrm{dB}}\,(x,\,y) = \bar{\sigma}^{\circ}_{\mathrm{dB}} + s_0 U_0(x,\,y) \tag{4.129}$$

where $U_0\,(x,\,y)$ is a correlated, zero-mean, unit variance, sample function of a two-dimensional Gaussian random process. Standard computer programs are available for generating $U_0\,(x,\,y)$. For this model, values of $d = d_2$ from Figure 4.54 should be used.

Analysis of *synthetic aperture radar* (SAR) data [15] indicates that a further refinement of the ACF of the model is required for accurately depicting the clutter return. The revised ACF depicted in Figure 4.54 is a two-segment approximation to an exponential function. The first correlation distance (d_1) provides a rapid decorrelation,

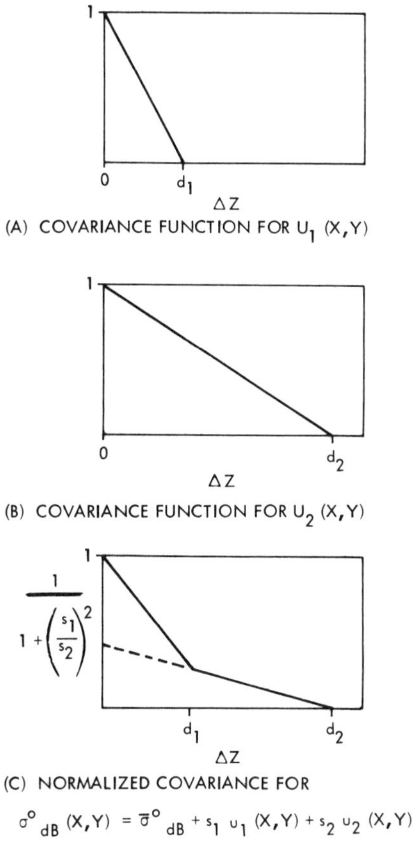

Figure 4.54 Covariance function for long- and short-range spatial covariance.

and the second correlation distance (d_2) is much longer and represents a long-term correlation. The value of d_2 is much longer for homogeneous terrain such as farms and woodland than for urban or mountainous terrain where sharp changes occur.

For computer simulation, (4.129) is modified so that

$$\sigma^{o}_{dB}(x, y) = \bar{\sigma}^{o}_{dB} + s_1 U_1(x, y) + s_2 U_2(x, y) \tag{4.130}$$

where s_1 and s_2 are given in Figures 4.52 and 4.53, and $U_1(x, y)$ and $U_2(x, y)$ are sample functions from correlated, zero-mean, unit variance, two-dimensional Gaussian processes.

The amplitude probability distribution of discrete clutter is given by a log-normal distribution defined as

$$p_N(A_{dB}) = \frac{1}{\sqrt{2\pi}\, s_d} \exp\left[-\frac{1}{2} \left(\frac{A_{dB} - m_d}{s_d} \right)^2 \right] \tag{4.131}$$

where

$$A_{dB} = 10 \log A_d (\text{dB/m}^2) \tag{4.132}$$

s_d is the standard deviation of A_{dB}, and m_d is the median of A_d with respect to a cross section of 1 m^2. The value of s_d is approximately equal to 5 dB. The value of m_d, given in Figure 4.55, ranges from 10^2 to 10^6 m^2 with a triangular probability distribution.

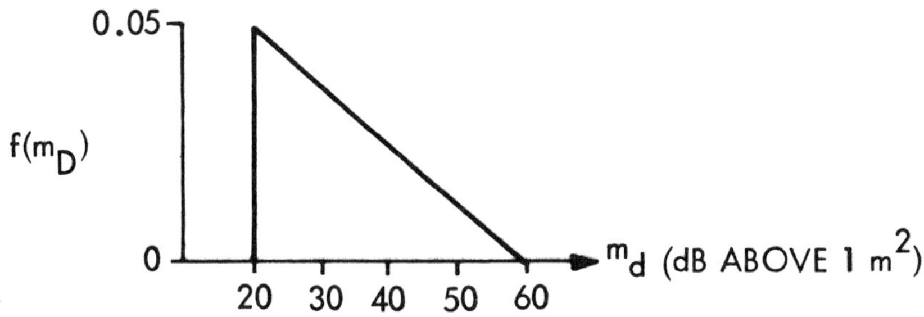

Figure 4.55 Probability distribution law for discrete clutter.

4.5.2 A Search Radar Model Environment

Several environmental models have been postulated that apply to a ground-based air search radar in a benign environment [5, 41]. These models generally consist of an environment that generates a collection of radar returns corresponding to those most commonly observed on the display of a ground-based air search radar. The environment comprises land and sea clutter, precipitation clutter, and spurious targets such as birds and ground vehicles that contaminate the radar display.

The model environment allows various competing radar designs to be compared against a standard reference. The environment is such that there are points in the coverage where desired targets cannot be detected and others where false alarms are indicated. False alarms are generally considered to be a far more serious problem because they tend to overload the data processing function of the radar (i.e., radar display or automatic detection equipment), thus limiting its effectiveness. For this reason, false alarm elimination is generally a high priority.

A model environment for an air search radar is listed in Table 4.24 [5]. This model is intended to stress the radar's signal processor rather than to be representative to a typical environment. In the model, land clutter is represented by point surface

Table 4.24
An Air Search Radar Model Environment

Return	*Cross Section*		*Azimuth (°)*	*Range (km)*	*Radial Velocity (m/s)*	*Height (m)*
	Number	*σ of each (m^2)*				
Land:			0–180	0–40	0	(surface)
Fixed points	450	10^4				
uniformly	900	10^3				
distributed in	4500	10^2				
area	6800	10				
Sea (Sea State 4)	$\sigma° = 3 \times 10^{-5}$	(m^2/m^2)	180–360	0–20	2–4	(surface)
Rain (4 mm/hr):	*Frequency*	η (m^2/m^3)	−25–25	20–140	0–30	500–2000
	1300	2×10^{-10}				
	3000	5×10^{-9}				
	5000	5×10^{-8}				
	10000	5×10^{-7}				
	Number of birds	*σ of each (m^2)*				
Birds			0–360	0–50	15	200
	2000	10^{-2}				
	Number of cars	*σ of each (m^2)*				
Cars			0–180	0–40	15	(surface)
	50	1				

Interference: Four identical radars at a distance $R_{max}/2$ from the radar under analysis;
There is a clear line of sight to each;
PRFs differ slightly, whereas center frequencies are displaced by 2, 4, 6, and 8% of f_0.

reflectors, which vary in amplitude from 10 to 10^4 m^2. The points are spread randomly over a sector from 0° to 180° in azimuth and 40 km in range. If the radar is completely surrounded by a land mass throughout its operating range, then the coverage would be extended to 360° in azimuth with the same reflector density.

An alternate model [41] represents land clutter in a distributed manner, where the RCS per resolution cell depends on the size of the cell given by the half-power azimuth beamwidth (θ_{az}) in degrees multiplied by the pulse length (τ) in seconds. With θ_{az} in the range 0.4° to 4° and τ in the range 0.1 to 10 μs, the probability distribution is log normal and has the parameters for linear polarization given in Table 4.25. In each case, the distribution is truncated at 50 dB m^2. Circular polarization reduces the RCS by 3 dB, and the clutter return is independent of frequency. The velocity spectrum is Gaussian in shape with zero-mean velocity and a standard deviation of 0.3 m/s.

Table 4.25
Model Parameters for Log-Normal Ground Clutter

$\theta_{az}\tau$ (°-s)	*Median* (dB/m^2)	*Standard Deviation (dB)*	*Number of Resolution Cells Affected by Clutter*
10^{-5}	0	20	5×10^3
10^{-6}	−15	20.1	5×10^4
10^{-7}	−30	20.8	5×10^5

In the model, a second 180° azimuth sector is devoted to sea clutter and extends to the radar horizon (antenna height is 20 m) at approximately 20 km. A sea state 4 condition is postulated with a spectral spread corresponding to radial velocities of 2 to 4 m/s.

Rain falling at a rate of 4 mm/hr in a 30° sector from 20 to 140 km is used to represent a storm in the radar's coverage. Spectrum spreads corresponding to radial velocities from 0 to 30 m/s (depending upon the wind direction) are used with the highest rain densities occurring at altitudes from 500 to 2000 m. Note that the backscattering coefficient (η) is a strong function of the radar transmitted frequency so that different values are given for the various radar bands.

Alternatively [41], the dimensions of the radar coverage volume affected by the rain as a function of precipitation rate are given in Table 4.26. The duration is the maximum expected in any one place in a year. The backscattering coefficient for rain is given by $\bar{\eta} = 7 \times 10^{-48} \cdot f^4 \cdot r^{1.6}$ m^2/m^3, where f is the frequency in Hz and r is the rain rate in mm/hr. The velocity spectrum of the rain is given by [1 +

Table 4.26
Precipitation Volume as a Function of Rain Rate

Precipitation Rate (mm/hr)	*Diameter (km)*	*Ceiling (km)*	*Maximum Duration (hr)*
2	300	4	13
4	45	4	5
8	35	8	1.8
16	20	8	0.6
32	8	8	0.2
64	1	8	0.06

$(7.3 \cdot 10^{-3} \cdot k \cdot R \cdot \phi_{el})]^{1/2}$ m/s, where k is the wind shear across the vertical beamwidth in m/s-m, R is the range in m, and ϕ_{el} is the vertical two-way half-power beamwidth in degrees. The mean velocity of the precipitation is determined by the wind speed, which varies with height, as tabulated in Table 4.27.

Birds are a source of moving clutter for low-PRF air search radar. The model uses an arbitrary number of 2000 birds uniformly distributed in all azimuths over a range from 0 to 50 km. The small RCS of each bird is 0.01 m^2, so only sensitive radars that search for small targets will be bothered by this source of clutter. A random radial velocity distributed between ± 15 m/s with a maximum height of 200 m is assumed for each bird.

The distribution of the number of birds expected within a particular distance from the radar has been modeled by using ornithological data [42]. The density of birds is a random variable that depends on many factors, including the migratory pattern of the bird flocks. The model for the bird distributions is given in Table 4.28.

Table 4.27
Wind Speed Variation with Height

	Wind Speed Exceeded (% time)		
Height (km)	*50% (m/s)*	*10% (m/s)*	*1% (m/s)*
Surface	5	9	13.5
2	11.5	21	31
4	13.5	24.5	36
6	18.5	34	50
8	23.5	43	63

Table 4.28
Bird Distribution Model

Observation Percentile (%)	*Bird Density (per mi²)*
36.2	None
50	3
90	48
99	400

Note that the number of birds used in the alternative search radar model (i.e., 2000) approximately corresponds to the 50th percentile of the bird distribution model.

Moving ground vehicles such as cars are another source of radar returns that may bother a low-PRF air search radar. Fifty cars are specified, which are uniformly distributed throughout the land clutter region and have a random uniformly distributed ±15 m/s radial velocity. Return from cars as viewed by an L-band search radar using MTI and a display range of 50 km is depicted in Figure 4.56.

Interference is specified as that which would exist if radiation from identical radars with a small variation of parameters were received by the radar under evaluation.

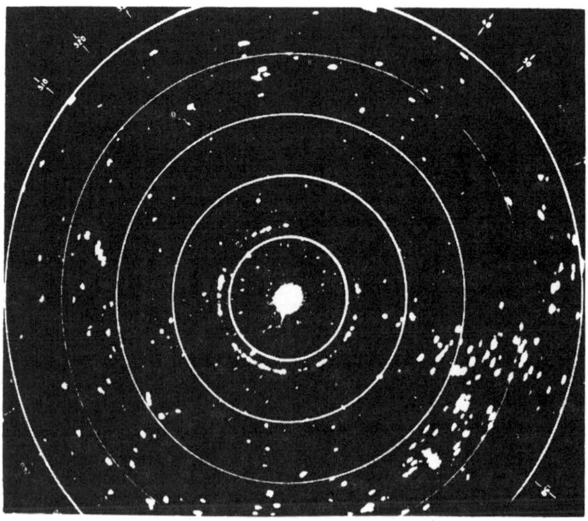

Figure 4.56 Cars viewed by an L-band search radar using MTI and a display range of 50 km.

For this purpose, four interfering radars are specified, each located within the line of sight at a distance equal to half the maximum range. PRFs differ slightly, whereas center frequencies are displaced 2, 4, 6, and 8% from the carrier frequency. This type of interference produces so-called rabbits, which move at a rate determined by the difference between the PRF of the radar under evaluation and that of the interfering radar. Return from an interfering radar and land clutter, as viewed on an L-band search radar with 50 km range, is depicted in Figure 4.57.

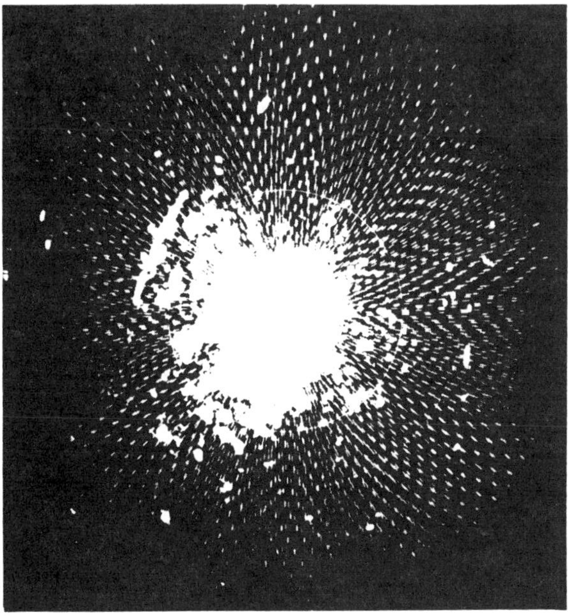

Figure 4.57 Interference and land clutter viewed on an L-band search radar.

REFERENCES

1. L. Greenstein, A. Brindley, and R. Carlson, "A Comprehensive Ground Clutter Model for Airborne Radar," Overland Radar Technology Program Rep., IIT Research Institute, Chicago, September 1969.
2. M.I. Skolnik, *Introduction to Radar Systems*, 2nd Ed., McGraw-Hill, New York, 1980.
3. D.K. Barton and W. Shrader, "Interclutter Visibility in MTI Systems," *IEEE EASCON Record*, Washington, DC, 1969.
4. M.W. Long, *Radar Reflectivity of Land and Sea*, Artech House, Norwood, MA, 1983.
5. H. Ward, "A Model Environment for Search Radar Evaluation," *IEEE EASCON Record*, Washington, DC, 1971.

6. D.K. Barton, "Land Clutter Models for Radar Design and Analysis," *Proc. IEEE*, Vol. 73, No. 2, February 1985.
7. F. Nathanson, *Radar Design Principles*, 1st Ed., McGraw-Hill, New York, 1969.
8. D.K. Barton, *Modern Radar System Analysis*, Artech House, Norwood, MA, 1988.
9. N. Durlach, "Influence of the Earth's Surface on Radar," MIT Lincoln Laboratory, Technical Rep. 373, January 1965, AD 627635.
10. L.V. Blake, *Radar Range-Performance Analysis*, Artech House, Norwood, MA, 1986.
11. R. Boothe, "The Weibull Distribution Applied to Ground Clutter Backscatter Coefficient," Rep. No. RE-TR-69-15, U.S. Army Missile Command, Huntsville, AL, June 1969.
12. R. Hayes and F. Dyer, "Land Clutter Characteristics for Computer Modeling of Fire Control Radar Systems," EES/GIT Project A-1485, Georgia Institute of Technology, Atlanta, AD 912490, May 1973.
13. D.K. Barton, ed., *Radar Clutter*, *Radars*, Vol. 5, Artech House, Norwood, MA, 1975.
14. D.C. Schleher, ed., *MTI Radar*, Artech House, Norwood, MA, 1978.
15. S. Kazel, J. Patten, J. Dabkowski, J. Pipkin, and A. Brindley, *Extensions to the ORT Clutter Model*, IIT Research Institute, Chicago, June 1971.
16. N.C. Currie and S. Zehner, "Millimeter Wave Land Clutter Model," *IEE Int. Radar Conf.*, London, October 1982.
17. W. Simkins, V. Vannicola, and J. Ryan, "Seek Igloo Radar Clutter Study," RADC-TR-338, Griffiss AFB, NY, October 1977, ADA047897.
18. G. Trunk and S. George, "Detection of Targets in Non-Gaussian Sea Clutter," *IEEE Trans.*, Vol. AES-6, No. 5, September 1970.
19. R. Lay, J. Taylor, and G. Brunins, "ARSR-4: Unique Solutions to Long-Recognized Radar Problems," *IEEE Int. Radar Conf.*, Washington, DC, May 1990.
20. W. Fishbein, S. Graveline, and O. Rittenback, "Clutter Attenuation Analysis," Tech. Rep. ECOM-2808, U.S. Army Electronics Command, Fort Monmouth, NJ, March 1967.
21. J. Eaves and E. Reedy, *Principles of Modern Radar*, Van Nostrand Reinhold, New York, 1987.
22. N.C. Currie, F. Dyer, and R. Hayes, "Radar Land Clutter Measurements at 9.5, 16, 35 and 95 GHz," Tech. Rep. No. 3 DAA 25-73-0256, Georgia Institute of Technology, Atlanta, March 1975.
23. G. Petts, "Radar Systems," in *Handbook of Electronic System Design*, C. Harper, ed., McGraw-Hill, New York, 1980.
24. N. Guinard and J. Daley, "An Experimental Study of a Sea Clutter Model," *Proc. IEEE*, Vol. 58, No. 4, April 1970.
25. J. Daley, W. Davis, and N. Mills, "Radar Sea Returns in High Sea States," Naval Research Laboratory Report 7142, September 25, 1970.
26. D.C. Schleher, "Radar Detection in Weibull Clutter," *IEEE Trans.*, Vol. AES-12, No. 6, November 1976.
27. S. Watts, "Radar Detection Prediction in Sea Clutter Using the Compound K-Distributed Model," *IEE Proc.*, Vol. 132, Pt. F, No. 7, December 1985.
28. E. Jakeman and P. Pusey, "Statistics of Non-Rayleigh Sea Echo," *IEE Int. Radar Conf.*, London, November 1977.
29. D. Kerr, *Propagation of Short Radar Waves*, McGraw-Hill, New York, 1951.
30. N.C. Currie, F. Dyer, and R. Hayes, "Some Properties of Radar Returns from Rain at 9.375, 35, 70 and 95 GHz," *IEEE Int. Radar Conf.*, Washington, DC, 1975.
31. M. Sekine, T. Musha, Y. Tomita, T. Hagisawa, T. Irabu, and E. Kiuchi, "On Weibull-Distributed Weather Clutter," *IEEE Trans.*, Vol. AES-15, No. 6, November 1979.
32. N.C. Currie, F. Dyer, and R. Hayes, "Analysis of Radar Rain Return at Frequencies of 9.5, 35, 70 and 95 GHz," Tech. Rep. No. 2, Army Contract DAA 25-73-0256, Georgia Tech, Atlanta, February 1975.

33. D.C. Schleher, *Introduction to Electronic Warfare*, Artech House, Norwood, MA, 1986.
34. P. Mitchell and R. Short, "How to Plan a Chaff Corridor," *International Countermeasures Handbook*, EW Communications, Los Altos, CA, 1979.
35. M. Katzin, D. Ringwalt, T. Weaver, "Evaluation of Airborne Overland Radar Techniques and Testing," Vol. II, *Clutter Data Appendixes*, RTD-TR-65, Air Force Systems Command, Wright Patterson AFB, Ohio, September 1965.
36. D.C. Schleher, "Harbor Surveillance Radar Detection Performance," *IEEE J. Oceanic Engineering*, Vol. OE-2, No. 4, October 1977.
37. P. Peebles, "The Generation of Correlated Log-Normal Clutter for Radar Simulations," *IEEE Trans.*, Vol. AES-7, No. 6, November 1971.
38. W. Szajnowski, "The Generation of Correlated Weibull Clutter for Signal Detection Problems," *IEEE Trans.*, Vol. AES-13, No. 5, September 1977.
39. R. Fante, "Detection of Multiscatter Targets in K-Distributed Clutter," *IEEE Trans.*, Vol. AP-32, No. 12, December 1984.
40. E. Jakeman, P. Pusey, "A Model for Non-Rayleigh Sea Echo," *IEEE Trans.*, Vol. AP-24, No. 6, November 1976.
41. A. Edgar, E. Dodsworth, and M. Warden, "The Design of a Modern Surveillance Radar," *IEE Int. Radar Conf.*, London, October 1973.
42. G. Pollon, "Distributions of Radar Angels," *IEEE Trans.*, Vol. AES-8, No. 6, November 1972.

Chapter 5
Theory of Optimum Radar Doppler Processing

Most surface and airborne radars operate in an environment where the clutter return obscures targets of interest. If the target is moving relative to the clutter, we can filter out the undesired clutter return by exploiting the differential doppler shift produced by the target relative to clutter radial motion. Doppler signal processors are used to extract the doppler-shifted target from the background interference, which typically consists of radar clutter plus receiver noise.

In the design of doppler signal processors, we may be initially tempted to ignore the receiver noise component of the background interference and concentrate on the interfering clutter. This leads to a design that attenuates the portion of the doppler spectral region of interest, which contains the clutter, and passes those signals that lie outside the attenuation region. This type of doppler processor, called a *moving target indicator,* provides a comb filter type of response that ideally has a flat passband in the region of expected targets and a stopband in the region of heavy clutter returns.

The MTI is effective in improving the system signal-to-clutter ratio, but has no capability in improving the system signal-to-noise ratio. This outcome limits detection performance of the basic MTI to the single-pulse performance determined by using the standard Marcum-Swerling radar detection analysis [1]. A natural extension of the MTI doppler processor is the addition of an integrator to improve its detection performance through enhancing the S/N.

The integrator can be implemented either coherently or incoherently (after detection). A bank of coherent integration filters is conveniently formed by using the FFT algorithm. The cascaded MTI-coherent integrator bank doppler processor is usually called a *pulsed doppler processor,* which means that the processor's filter response is matched to the target spectrum. The term *MTI processor* means that the processor's doppler response is open to all possible target doppler frequencies.

A processor that uses considerably less hardware than the PD processor is the cascaded MTI-incoherent integrator. This processor is implemented by envelope detection of the MTI output, followed by a video integrator. However, the MTI correlates receiver noise that, when incoherently integrated, causes a detection loss

[2–5]. The loss increases as the number (m) of pulses processed by the MTI increases [4]. This loss is compounded by any residual clutter passed by the MTI, which, in the extreme, can completely negate any improvement provided by the video integration [5].

As is apparent from this discussion, detection performance of the MTI-coherent integrator doppler processor is superior to that of the cascaded MTI-incoherent integrator processor. Also, a careful examination of any MTI integrator cascade using batch processing shows that only one of the m pulses employed in the MTI is effective in building up the processor's S/N, which implies an inherent detection loss [5]. Hence, none of the previously described doppler processors provide the optimal performance, which is the motivation for development of a general theory of optimal doppler processors.

The fundamental problem to be considered is the detection of n coherently related, doppler shifted target pulses embedded in a background of Gaussian-distributed interference, consisting of clutter and receiver noise. The initial RF phase of the coherent pulse train returned from the target is uncertain and hence will be randomly distributed with respect to the phase of the coherently transmitted waveform.

If all the parameters associated with the signal and interference were known, we could apply the optimal radar signal processing techniques described in Appendix B to the fundamental problem. The resulting optimal processor would be a complex-weight (amplitude and phase processing) transversal filter, followed by an envelope detector. The processor synthesis problem is then to find the optimal weights of the transversal filter. Because the optimal weights (equation (B.36) in Appendix B) involve the assumed signal distribution, separate solutions would be natural in accordance with the *a priori* assumption of the doppler region occupied by the target.

Direct application of the theory of Appendix B results in a processor that is optimum at a single doppler frequency. This will be called the *point doppler processor*. Because the target doppler shift is not known in most applications, a bank of transversal filters can be used to cover the doppler frequency band of interest. Each filter is then optimum at its center frequency, but mismatched otherwise. In this chapter, we will explore the question of straddling losses that occur between the optimal point responses of this type of doppler processor. The cascaded MTI-coherent integrator compares most directly with the point doppler processor, using less hardware, but with only one of the m pulses that forms the MTI being effective in building up the S/N.

The theory of Appendix B can also be applied to develop a doppler processor that is optimum for a signal with a doppler shift that has uniform probability for any value within the detection region. This criterion results in an optimal quadrature channel MTI doppler processor [6–9]. The characteristics of the optimum MTI will be derived in this chapter as well and its performance is compared with conventional MTI cancellers derived by using binomial weights.

The doppler point and MTI processors each represent an extreme in their assumptions relative to the doppler signal distribution. A third type of optimization criterion would be to divide the doppler region of interest into equal intervals and provide a separate optimal processor for each region. We shall call this the *equal interval processor* and compare its performance with that of the point doppler processor.

5.1 OPTIMUM RADAR DOPPLER PROCESSORS

The structure of the optimal radar doppler processor consists of a complex-weight transversal filter, followed by an envelope detector, as depicted in Figure 5.1. This structure was derived in Appendix B by using statistical detection theory and, as shown, can be derived equivalently by maximizing the signal-to-interference ratio. The linear nature of the solution (i.e., a transversal filter) is a result of the Gaussian distribution of the interference and the envelope detector is a consequence of the initially unknown phase of the coherent signal pulse train. The objective of the synthesis procedure for optimal radar doppler processors is to choose the complex weights of the transversal filter so that the output signal-to-interference ratio is maximized for a specified doppler input signal distribution.

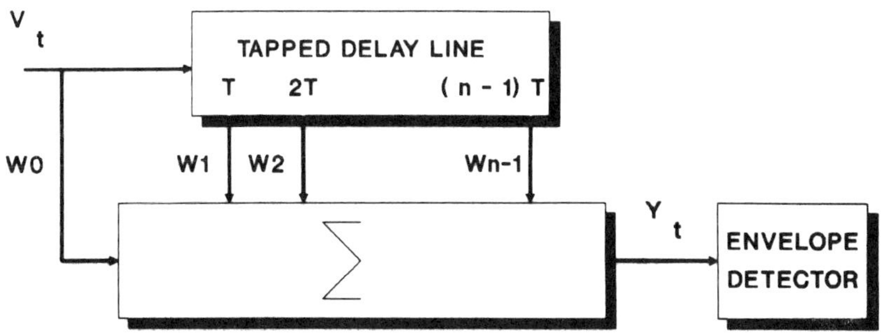

Figure 5.1 Optimum transversal filter processor.

In our study of optimal radar doppler signal processors, we shall first develop some general properties of transversal filters including performance definitions. Then, we shall develop formulas for the performance and complex weights of point, equal-interval, and MTI doppler processors.

5.1.1 Transversal Filters

The complex output of the transversal filter depicted in Figure 5.2 is given by

$$r = \sum_{i=0}^{n-1} w_i v_i = \mathbf{w}^T \mathbf{v} = \mathbf{v}^T \mathbf{w} \tag{5.1}$$

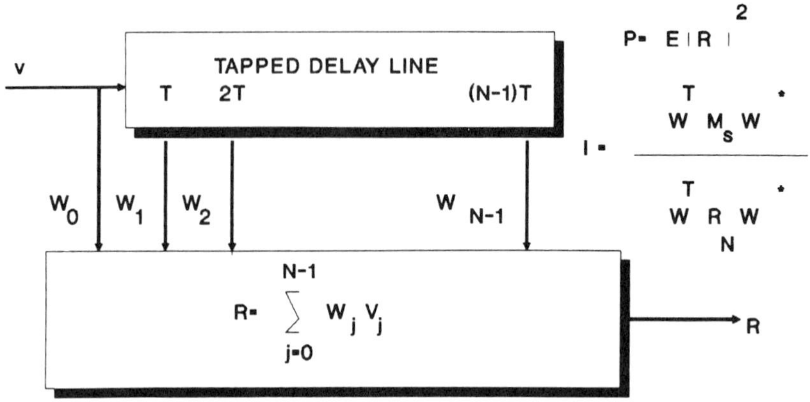

Figure 5.2 Transversal filter.

where $\mathbf{w}^T = (w_0, w_1, \ldots, w_{n-1})$ is the complex weight vector, $\mathbf{v}^T = (v_0, v_1, \ldots, v_{n-1})$ is the complex input signal vector, and the superscript T represents a matrix transpose operation. The output power of the transversal filter is given by

$$P_0 = E|\mathbf{r}|^2 = E(\mathbf{rr}^*) = \mathbf{w}E(\mathbf{vv}^{*T})\mathbf{w}^* \tag{5.2}$$

where the asterisk indicates a complex conjugate operation. When the input is interference the output power is given by

$$P_n = \mathbf{w}^T\mathbf{R}_n\mathbf{w}^* \tag{5.3}$$

where $\mathbf{R}_n$ is the interference's covariance matrix given by

$$\mathbf{R}_n = E(\mathbf{NN}^{*T}) \tag{5.4}$$

and $\mathbf{N}^T = (N_0, N_1, \ldots, N_{n-1})$ is the complex input noise vector. When the input is a complex signal vector given by $\mathbf{s}\ e^{j\phi}$, where ϕ is randomly distributed 0–2π, and $\mathbf{s}^T = (s_0, s_1, \ldots, s_{n-1})$, the output power is given by

$$\mathbf{P}_s = \mathbf{w}^T E(\mathbf{s}e^{j\phi}\mathbf{s}^{*T}e^{-j\phi})\mathbf{w}^* = \mathbf{w}^T\mathbf{M}_s\mathbf{w}^* \tag{5.5}$$

where $\mathbf{M}_s$ is the signal covariance matrix:

$$\mathbf{M}_s = E(\mathbf{s}\ \mathbf{s}^{*T}) \tag{5.6}$$

The improvement factor of a doppler processor is defined as the output S/N divided by the input S/N:

$$I_f = \frac{P_{os}/P_{on}}{P_{is}/P_{in}} = \frac{(S/N)_o}{(S/N)_i} \tag{5.7}$$

The power gain of the processor is given by

$$G_s = P_{os}/P_{is} \tag{5.8}$$

and the interference gain is given by

$$G_n = P_{in}/P_{on} \tag{5.9}$$

so that the improvement factor can be written as

$$I_f = G_s G_n \tag{5.10}$$

The normalized improvement factor of a transversal filter ($P_{is} = P_{in}$) is given by

$$I_f = \frac{\mathbf{w}^T \mathbf{M}_s \mathbf{w}^*}{\mathbf{w}^T \mathbf{R}_n \mathbf{w}^*} \tag{5.11}$$

The performance of the transversal filter is optimized by finding solutions for the weights ($\mathbf{w}$) of (5.11) that maximize the improvement factor I_f. When I_f is maximized, the signal-to-interference ratio at the output of the transversal filter is maximized; in accordance with the theory presented in Appendix B, this results in a maximum P_d when the interference is Gaussian distributed. The relationship between P_d and I_f will be developed later in this section.

The maximum improvement factor occurs at a stationary point in (5.11) for which

$$\mathbf{M}_s \mathbf{w}^* = \gamma \mathbf{R}_n \mathbf{w}^* \tag{5.12}$$

where γ is a scalar. Equation (5.12) can be written as a matrix equation:

$$(\mathbf{R}_n^{-1} \mathbf{M}_s - \gamma \mathbf{I}) \mathbf{w}^* = \mathbf{0} \tag{5.13}$$

where $\mathbf{R}_n^{-1}$ is the matrix inverse of the covariance matrix and $\mathbf{I}$ is the identity matrix. Nontrivial solutions for (5.13) exist only when the determinant:

$$|\mathbf{R}_n^{-1} \mathbf{M}_s - \gamma \mathbf{I}| = 0 \tag{5.14}$$

In general for a doppler processor that processes n pulses there are n solutions for γ, which are called the *eigenvalues* of the matrix $\mathbf{R}_n^{-1}\mathbf{M}_s$. The value of the improvement factor is given by

$$I_f = \gamma_{max} \tag{5.15}$$

where γ_{max} is the largest eigenvalue. Each eigenvalue has an associated eigenvector that can be found from (5.13). The eigenvector associated with γ_{max} allows determining the optimal complex weights ($\mathbf{w}_0$). Equations (5.13) and (5.14) will be used later to find the optimal weights and improvement factors for the point, equal-interval, and MTI doppler processors.

In general, the weights associated with optimal doppler processors are complex. This means that both the phase and amplitude of each signal sample must be appropriately weighted in the transversal filter. Frequently, we want to implement the complex transversal filter by using only real operations. This can be accomplished by using a quadrature-channel processor, where both the real and imaginary components of the weights operate on the signal in each channel. Note that MTI doppler processing requires only real weights, as will be derived in a later section.

Consider the complex transversal filter depicted in Figure 5.3, which is driven by the complex envelope signal $\tilde{v}_t$, has a complex envelope output $\tilde{r}_t$ and complex weights $\tilde{w}$ (see the complex notation described in Appendix A). Recall that the real input signal v_t is related to the complex envelope by

$$v_t = \text{Re}[\tilde{v}_t e^{j\omega t}] = v_{ct} \cos\omega t + v_{st} \sin\omega t \tag{5.16}$$

where v_{ct} and v_{st} are quadrature baseband signal components. The complex envelope in terms of the quadrature components is given by

$$\tilde{v}_t = v_{ct} - jv_{st} \tag{5.17}$$

and the complex weights are given by

$$\tilde{w} = w_R - jw_I \tag{5.18}$$

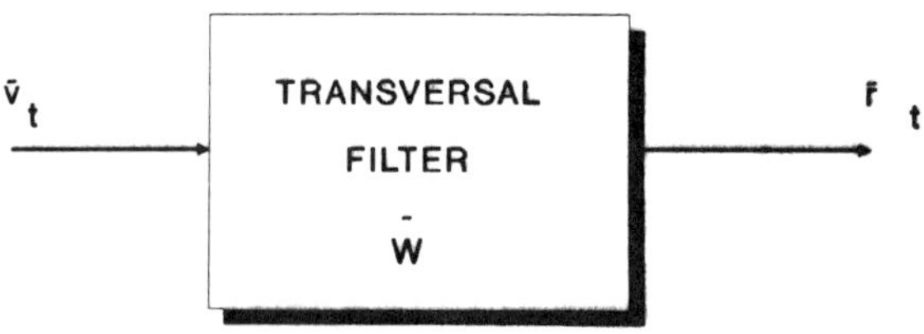

Figure 5.3 Complex transversal filter.

where w_R is the real component and w_I is the imaginary component. The complex envelope of the output is given by [10]

$$\tilde{r} = \tilde{\mathbf{w}}^T \tilde{\mathbf{v}} \tag{5.19}$$

where $\tilde{\mathbf{w}}^T = (\tilde{w}_0, \tilde{w}_1, \ldots, \tilde{w}_{n-1})$ and $\tilde{\mathbf{v}}^T = (\tilde{v}_0, \tilde{v}_1, \ldots, \tilde{v}_{n-1})$ are vector samples of the complex envelope. Substitution of equations (5.17) and (5.18) into (5.19) results in the complex-envelope output of the transversal filter in terms of the quadrature input signals and the real and imaginary weight vectors:

$$\tilde{r} = (\mathbf{w}_R^T \mathbf{v}_c - \mathbf{w}_I^T \mathbf{v}_s) - j(\mathbf{w}_R^T \mathbf{v}_s + \mathbf{w}_I^T \mathbf{v}_c) \tag{5.20}$$

The square-law envelope-detected output of the transversal filter is given by

$$|\tilde{r}|^2 = (w_R^T \mathbf{v}_c - \mathbf{w}_I^T \mathbf{v}_s)^2 + (\mathbf{w}_R^T \mathbf{v}_s + \mathbf{w}_I^T \mathbf{v}_c)^2 \tag{5.21}$$

A block diagram of the complex transversal filter implemented with real components is depicted in Figure 5.4.

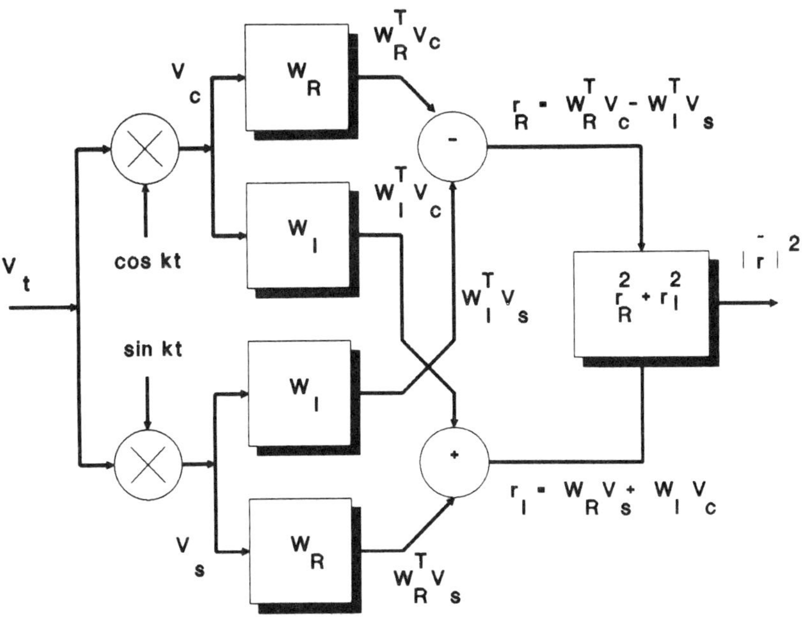

Figure 5.4 Block diagram of complex transversal filter implemented with real operations.

Of considerable interest is the detection performance of transversal filter doppler processors. For maximum utility, the performance should be based on a probability of detection measure, as this will allow comparison with other types of doppler and nondoppler processors in a framework of the Marcum-Swerling radar detection analysis. In addition, the relationship between the doppler processor's improvement factor and its probability of detection should be developed due to the extensive use of the improvement factor to measure the quality of the processor. As previously discussed, the probability of detection is increased as the improvement factor is increased [5].

Consider the real implementation of a complex transversal filter, as depicted in Figure 5.4. The input signal can be represented as

$$v_t = \sqrt{2P_s}\cos[(\omega_c + \omega_d)t + \phi] + x_t \cos\omega_{ct} - y_t \sin\omega_{ct} \tag{5.22}$$

where x_t and y_t include clutter plus noise components, P_s is the signal power, σ_c^2 is the clutter power, σ_n^2 is the receiver noise power, and ω_d is the target's radian doppler frequency. The in-phase and quadrature components of the input are given by

$$i_t = \sqrt{2P_s}\cos(\omega_d t + \phi) + x_t \tag{5.23}$$

$$q_t = \sqrt{2P_s}\sin(\omega_d t + \phi) + y_t \tag{5.24}$$

The quadrature channel outputs of the transversal filter are given by

$$\alpha = \sum_{j=0}^{n-1} w_{\mathrm{R}j} i(t - jT) - \sum_{j=0}^{n-1} w_{\mathrm{I}j} q(t - jT) \tag{5.25}$$

$$\beta = \sum_{j=0}^{n-1} w_{\mathrm{R}j} q(t - jT) + \sum_{j=0}^{n-1} w_{\mathrm{I}j} i(t - jT) \tag{5.26}$$

Now, both α and β are Gaussian-distributed random variables because they result from linear operations on Gaussian-distributed random processes. Therefore, the mean and variance of α and β will specify the joint probability density function $p(\alpha, \beta|s)$. The expected values of α and β can be found as

$$\begin{aligned} s_x = E(\alpha) &= \sum_{j=0}^{n-1} w_{\mathrm{R}j} \sqrt{2P_s}\cos[\omega_d(t - jT) + \phi] \\ &\quad - \sum_{j=0}^{n-1} w_{\mathrm{I}j} \sqrt{2P_s}\sin[\omega_d(t - jT) + \phi] \end{aligned} \tag{5.27}$$

$$\begin{aligned} s_y = E(\beta) &= \sum_{j=0}^{n-1} w_{\mathrm{R}j} \sqrt{2P_s}\sin[\omega_d(t - jT) + \phi] \\ &\quad + \sum_{j=0}^{n-1} w_{\mathrm{I}j} \sqrt{2P_s}\cos[\omega_d(t - jT) + \phi] \end{aligned} \tag{5.28}$$

The variance of α or β is given by

$$\sigma_{\alpha\beta}^2 = \sigma_n^2 \sum_{j=0}^{n-1} (w_{Rj}^2 + w_{Ij}^2) + \sigma_c^2 \sum_{j=0}^{n-1} \sum_{i=0}^{n-1} (w_{Rj} w_{Ri} + w_{Ij} w_{Ii}) \rho(i-j) \quad (5.29)$$

where ρ is the clutter's correlation coefficient, which for a Gaussian-shaped spectrum is

$$\rho(i-j) = \exp[-(i-j)^2 \Omega^2/2] = \rho(1)^{(i-j)^2} \quad (5.30)$$

and

$$\Omega = 2\pi\sigma_f T \quad (5.31)$$

The power out of the transversal filter is given by

$$P_o = E(\tilde{r}\tilde{r}^*) = 2P_s \left[\sum_{j=0}^{n-1} (w_{Rj}^2 + w_{Ij}^2) \right] \quad (5.32)$$

The average power gain from (5.8) can be found as

$$\bar{G}_s = \sum_{j=0}^{n-1} (w_{Rj}^2 + w_{Ij}^2) \quad (5.33)$$

The interference gain (see (5.9)) for only the clutter component of the interference, called *clutter attenuation,* is given by

$$\mathrm{CA} = \sum_{j=0}^{n-1} \sum_{i=0}^{n-1} (w_{Rj} w_{Ri} + w_{Ij} w_{Ii}) \rho(i-j) \quad (5.34)$$

The normalized improvement factor for clutter alone ($2P_s = \sigma_c^2$) using (5.10) can be found as

$$I_{f_c} = \frac{\sum_{j=0}^{n-1} (w_{Rj}^2 + w_{Ij}^2)}{\sum_{j=0}^{n-1} \sum_{i=0}^{n-1} (w_{Rj} w_{Ri} + w_{Ij} w_{Ii}) \rho(i-j)} \quad (5.35)$$

We further define the clutter-to-noise power ratio as

$$F = \frac{\sigma_c^2}{\sigma_n^2} \quad (5.36)$$

and we have

$$\sigma_{\alpha\beta}^2 = \sigma_n^2 \bar{G}_s (1 + F/I_{f_c}) \tag{5.37}$$

Equation (5.37) states that the variance of the Gaussian-distributed interference at the output of the transversal filter is simply the sum of the receiver noise power plus that of the residual clutter power multipled by the power gain of the processor. The residual clutter power can be found by dividing the input clutter power by the clutter improvement factor, I_{f_c}. Because the processor power gain (G_s) multiplies both the signal and interference powers, we would expect this factor to drop out when the signal-to-interference ratio is formed.

An interesting relation is the normalized improvement factor (I_{fc}) for clutter alone as a function of that for the total interference. This relationship can be developed as

$$I_f = \frac{1 + F}{F} \frac{I_{fc}}{1 + I_{fc}/F} \tag{5.38}$$

The overall improvement factor I_f tends to approach that for clutter alone (I_{fc}) as F becomes large. For I_{fc} that is very large compared to F, the overall improvement factor I_f approaches $F + 1$, indicating a maximum improvement relative to the clutter-to-noise power ratio.

The next step in the development of the probability of detection for the complex transversal filter is to determine the correlation between the quadrature channel outputs, α and β. This can be determined as

$$\rho_{\alpha\beta} = E[(\alpha - s_x)(\beta - s_y)] = 0 \tag{5.39}$$

which follows from the basic property of the interference representation $E(x_t y_t) = 0$. Equation (5.39) specifies that the random variables α and β are uncorrelated, and as they are Gaussian, they are independent. Therefore the joint density function can be written as

$$p(\alpha, \beta | s) = \frac{1}{2\pi\sigma_{\alpha\beta}^2} \exp\left[-\frac{(\alpha - s_x)^2 + (\beta - s_y)^2}{2\sigma_{\alpha\beta}^2} \right] \tag{5.40}$$

The envelope (r) is found by using $\alpha = r\cos\theta$ and $\beta = r\sin\theta$ resulting in

$$p(r|s) = \frac{r}{\sigma_{\alpha\beta}^2} \exp\left(-\frac{r^2 + s^2}{2\sigma_{\alpha\beta}^2} \right) I_0\left(\frac{rs}{\sigma_{\alpha\beta}^2} \right) \tag{5.41}$$

The square-law detected output is found by applying the transformation $z = r^2$, resulting in

$$p(z|s) = \frac{1}{2\sigma_{\alpha\beta}^2} \exp\left(-\frac{z + s^2}{2\sigma_{\alpha\beta}^2}\right) I_0\left(\frac{\sqrt{zs}}{\sigma_{\alpha\beta}^2}\right) \tag{5.42}$$

where $s^2 = s_x^2 + s_y^2 = 2\bar{G}_s P_s$.

The detection threshold T_d is found from (5.42) with $s = 0$ to be

$$T_d = 2\sigma_{\alpha\beta}^2 \ln(1/P_{fa}) \tag{5.43}$$

where P_{fa} is the false alarm probability. The probability of detection can be found from (5.40), (5.41), and (5.37) to be

$$P_d = Q\left\{\left[\frac{2P_s}{\sigma_n^2(1 + F/I_{fc})}\right]^{1/2}, [2 \ln(1/P_{fa})]^{1/2}\right\} \tag{5.44}$$

where $Q\{\cdot, \cdot\}$ is Marcum's Q function.

Equation (5.44) relates the complex transversal filter's probability of detection to the improvement factor for clutter alone, given by (5.35). Examination of (5.44) shows that P_d increases as the improvement factor increases. Also, this relation reveals the intuitively satisfying result that the detection probability depends on the ratio of signal to residual interference, where the residual interference is the receiver noise power plus the clutter power divided by the clutter improvement factor.

An alternative solution for the complex transversal filter's probability of detection can be found by solving (5.38) for I_{fc} in terms of I_f and substituting into (5.44):

$$P_d = Q\left\{\left[\frac{2P_s}{\sigma_n^2(1 + F)/I_f}\right]^{1/2}, [2 \ln(1/P_{fa})]^{1/2}\right\} \tag{5.45}$$

This equation illustrates that the complex transversal filter is effective against both clutter and noise. Equation (5.45) is appropriate when the doppler band of interest is limited, thereby improving the S/N, whereas (5.44) is used for MTI transversal filters, for which there is no S/N improvement.

5.1.2 The Point Doppler Processor

The point doppler transversal filter is optimized at a point contained within the desired doppler processing band and is nonoptimal or mismatched at all other points in the doppler band. A doppler processor is formed by using a number of point transversal filters to fill the doppler region of interest. Any number of transversal filters can be used, although in practice the band is filled with n equally spaced filters, where n

is the number of coherent pulses processed. This situation is depicted in Figure 5.5, and an example will be presented in this section.

The complex input signal to the point doppler transversal filter is given by

$$s_t = A\,e^{j\omega_d t} \sum_{i=0}^{n-1} \delta(t - iT) \tag{5.46}$$

where A is the amplitude, ω_d is the radian doppler frequency, n is the number of pulses processed, and T is the radar's interpulse interval. The signal vector is then $\mathbf{s}^T = (s_1, s_2, \ldots, s_n)$, where

$$s_k = A\,e^{j\omega_d(k-1)T}; \quad k = 1, 2, \ldots, n \tag{5.47}$$

The normalized (divided by A^2) signal covariance matrix can be found by using (5.6) as

$$\mathbf{M}_s = \begin{bmatrix} 1 & \ddots & e^{-j\omega_d T} & \cdots & e^{-j\omega_d(n-1)T} \\ & & \ddots & & \vdots \\ e^{j\omega_d T} & & & \ddots & \vdots \\ e^{j\omega_d(n-1)T} & \cdots & & \cdots & 1 \end{bmatrix} \tag{5.48}$$

An element (m_s) of the normalized covariance matrix can be found as

$$m_{s_{kj}} = e^{j\omega_d(k-1)}\,e^{-j\omega_d(j-1)} = e^{j\omega_d(k-j)}; \quad \begin{matrix} k = 1, 2, \ldots, n \\ j = 1, 2, \ldots, n \end{matrix} \tag{5.49}$$

where k is the row number and j is the column number.

The optimal weights can be found as the eigenvector that corresponds to the maximum eigenvalue of $\mathbf{R}_n^{-1}\,\mathbf{M}_s$ (see equation (5.13)). However, solution of (5.13) is not necessary because we have already determined in Appendix B, using statistical detection theory (see (B.36)), that the optimum weights are

$$\mathbf{w}_o = \mathbf{R}_n^{-1}\mathbf{s}^* \tag{5.50}$$

We can test this solution for the optimal weights by confirming that those given by (5.50) provide the maximum improvement factor. Substitution of (5.50) into (5.11) for the improvement factor and use of the solution ($\mathbf{R}_n^{-1}\mathbf{M}_s = \gamma_{\max}\mathbf{I}$) will result in

$$I_f = \frac{\mathbf{w}_o^T\mathbf{M}_s\mathbf{w}_o^*}{\mathbf{w}_o^T\mathbf{R}_n\mathbf{w}_o^*} = \frac{\mathbf{w}_o^T\overbrace{M_s(\mathbf{R}_n^{-1}}^{\mathbf{I}\gamma_{\max}}\mathbf{s}^*)^*}{\mathbf{w}_o^T\underbrace{\mathbf{R}_n(\mathbf{R}_n^{-1}}_{\mathbf{I}}\mathbf{s}^*)^*} = \gamma_{\max} \tag{5.51}$$

Equation (5.51) confirms that the maximum likelihood weights given by (5.50) do, in fact, provide the maximum improvement factor. This is just another demonstration that, for Gaussian interference, the maximum likelihood and the maximum signal-to-interference ratio criteria are equivalent.

In addition to the optimal weights, we need to determine an expression for the improvement factor. An expression can be found by using a number of matrix theorems [5]. First, observe that the rank of the signal covariance matrix $\mathbf{M}_s$ is one because this is the order of the largest nonzero determinant in the matrix. This means that the rank of $\mathbf{R}_n^{-1}\mathbf{M}_s$ is one because the rank of a product of two matrices is equal to or less than the rank of either matrix. Because the number of nonzero eigenvalues is equal to the rank of a matrix, $\mathbf{R}_n^{-1}\mathbf{M}_s$ has only one nonzero eigenvalue that is real (complex eigenvalues are in conjugate pairs). The trace of a matrix is equal to the sum of all its eigenvalues, and because in this case there is only one eigenvalue, it must be equal to the improvement factor; therefore,

$$I_f = \text{Trace}(\mathbf{R}_n^{-1}\mathbf{M}_s) \tag{5.52}$$

Note that the improvement factor given by (5.52) applies only for the point at which the transversal filter is matched.

A general expression for the optimal weights and improvement factor of the point-optimized transversal filter can be derived in terms of the inverse interference covariance matrix $\mathbf{R}_n^{-1}$. Generally, we must invert the interference covariance matrix $\mathbf{R}_n$, which is given by

$$\mathbf{R}_n = \sigma_n^2 \begin{bmatrix} 1+F & F\rho_1 & \cdots\cdots & F\rho_{n-1} \\ F\rho_1 & \ddots & & \vdots \\ \vdots & & \ddots & \vdots \\ \vdots & & \ddots & \vdots \\ F\rho_{n-1} & \cdots & \cdots\cdots & 1+F \end{bmatrix} \tag{5.53}$$

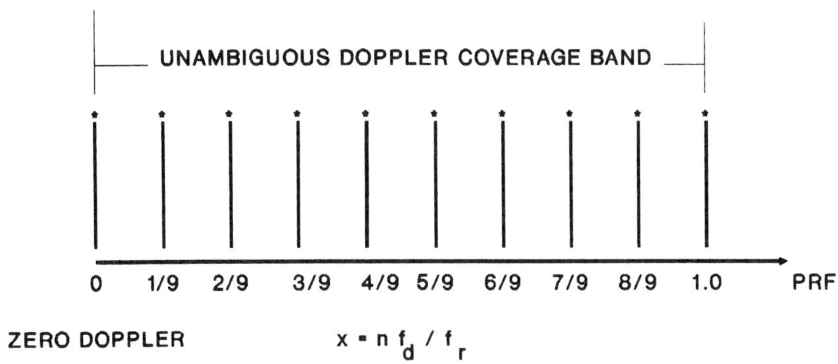

Figure 5.5 Design criterion for a point-optimized doppler processor.

by using an appropriate computer program. However, for a small number of samples ($n \leq 3$), we can obtain an analytic solution.

If the resulting elements of the inverse covariance matrix are given by α_{ij}:

$$\mathbf{R}_n^{-1} = \begin{bmatrix} \alpha_{11} & \alpha_{12} & \cdots & \cdots & \alpha_{1n} \\ \alpha_{12} & \ddots & & & \vdots \\ \vdots & & \ddots & & \vdots \\ \vdots & & & \ddots & \vdots \\ \alpha_{1n} & \cdots & \cdots & \cdots & \alpha_{11} \end{bmatrix} \tag{5.54}$$

then the optimum complex weights given by (5.50) are

$$w_{i-1} = \sum_{j=1}^{n} \alpha_{ij} \cos(j-1)\omega_d T - j \sum_{i=1}^{n} \alpha_{ij} \sin(j-1)\omega_d T \tag{5.55}$$

whereas the improvement factor given by (5.11) is

$$I_f = \frac{\sum_{i=0}^{n-1} |w_i|^2 + 2\sum_{i=0}^{n-2}\sum_{j=i+1}^{n-1} [\mathrm{Re}(w_i^* w_j)\cos(j-i)\omega_d T - \mathrm{Im}(w_i^* w_j)\sin(j-i)\omega_d T]}{\sum_{i=0}^{n-1} |w_i|^2 + 2\sum_{i=0}^{n-2}\sum_{j=i+1}^{n-1} \rho(i-j)\,\mathrm{Re}(w_i^* w_j)} \tag{5.56}$$

where

$$\mathrm{Re}(w^*{}_i w_j) = w_{\mathrm{R}i} w_{\mathrm{R}j} + w_{\mathrm{I}i} w_{\mathrm{I}j} \tag{5.57}$$

$$\mathrm{Im}(w_i^* w_j) = w_{\mathrm{R}i} w_{\mathrm{I}j} - w_{\mathrm{I}i} w_{\mathrm{R}i} \tag{5.58}$$

Equation (5.56) is a general expression that allows computing the improvement factor at any doppler frequency ω_d. However, remember that the weights in (5.56) must be computed at the point doppler frequency where the transversal filter is optimized. Equation (5.52) can be utilized to find the improvement factor at the optimum point:

$$I_{f_o} = \sum_{i=1}^{n} \alpha_{ii} + 2\sum_{i=1}^{n-1}\sum_{j=i+1}^{n} \alpha_{ij}\cos(j-i)\omega_d T \tag{5.59}$$

Figure 5.6 depicts the results of an improvement factor calculation for a nine-pulse point-optimized doppler processor. The curve labeled *optimum envelope* is derived by using (5.59) and represents the maximum improvement factor that can be obtained

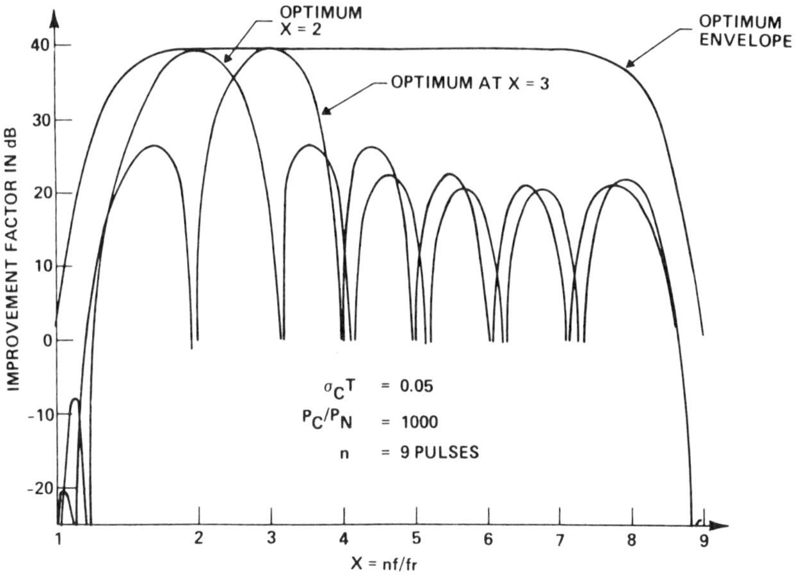

Figure 5.6 Improvement factor for point-optimized transversal filters ($x = 2, 3$).

by an optimized transversal filter centered at that doppler frequency. The improvement factors are provided for two transversal filters obtained by using (5.56). The filters are optimized at doppler frequencies equal to 2/9 and 3/9 of the radar's PRF. The two transversal filters cross over at a point where the improvement factor is approximately -3 dB below the peak. The probability of detection for the transversal filter at each doppler frequency can be found from (5.45) by using the appropriate improvement factor. Note that, when a significant response occurs in two contiguous filters, the joint probabilities of detection must be considered in determining the probability of detection at that point.

Figure 5.6 illustrates the characteristics of transversal filters that provide both S/C and S/N improvement. Note the deep nulls provided in the region about the PRF lines, where the heaviest clutter components are concentrated. Also, the width of the mainlobe is approximately matched to the signal spectrum, which is assumed (in the design of the filter) to be centered about the point where the transversal filter is optimized.

Figure 5.7 illustrates a design optimized about a doppler frequency that is 1/9 the radar's PRF. Note that the filter's response lies within the optimum improvement factor envelope, as expected. However, the conflicting requirements of maximizing the signal response in a region close to the heavy clutter region while minimizing the clutter response have caused a compromise in the target response: the mainlobe width is reduced and the peak improvement factor is reduced as well.

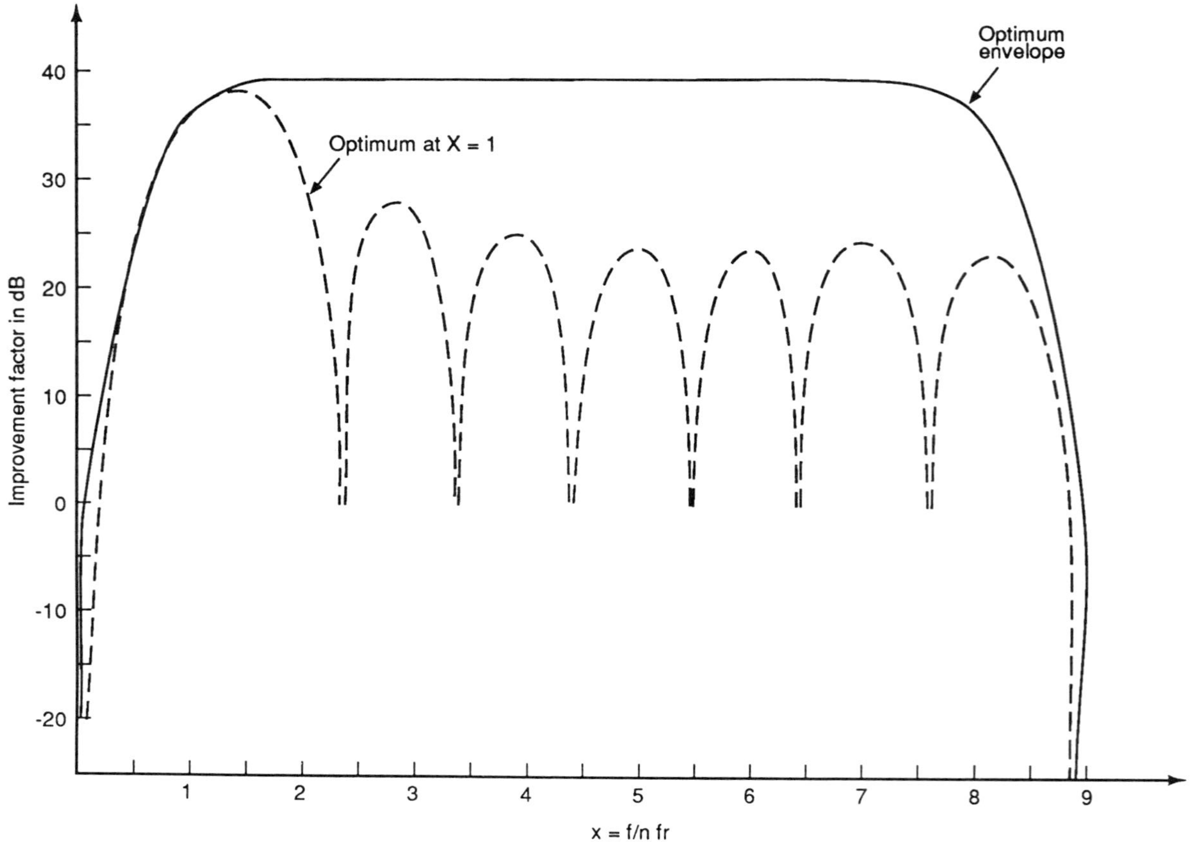

Figure 5.7 Improvement factor for point-optimized transversal filter ($x = 1$).

Figure 5.8 illustrates a design optimized at a doppler frequency that is 4/9 the radar's PRF. Note that the filter's sidelobe response in the vicinity of the heavy clutter region is reduced further from that of the previously illustrated designs (Figures 5.6 and 5.7), which were optimized at points close to the heavy clutter region.

5.1.3 The Optimized MTI Processor

The basic assumption of the MTI processor is that there is no *a priori* knowledge of the doppler shift of the target return. The resulting target doppler probability distribution is given by

$$p(f_d) = \begin{matrix} T, & 0 < f_d < 1/T \\ 0, & \text{otherwise} \end{matrix} \tag{5.60}$$

The signal covariance matrix can be found by averaging the signal covariance matrix for each doppler frequency, (5.49), with respect to the probability distribution of expected doppler frequencies, (5.60). The resulting normalized signal covariance matrix element is given by

$$m_{s_{kj}} = e^{j\pi(k-j)} \frac{\sin \pi(k-j)}{\pi(k-j)} \tag{5.61}$$

where k is the row number and j is the column number. Equation (5.61) for $m_{s_{kj}}$ is equal to 1 when $k = j$ and 0 when $k \neq j$. Therefore, the signal covariance matrix for the MTI case is

$$\mathbf{M}_s = \mathbf{I} \tag{5.62}$$

where **I** is the identity matrix.

The solution for the optimum weights is then given by (5.12), which can be written in the form:

$$(\mathbf{R}_n - \lambda \mathbf{I})\mathbf{w}^* = \mathbf{0} \tag{5.63}$$

where λ are the scalar eigenvalues of $\mathbf{R}_n$. The improvement factor in this case is obtained from (5.11) and is given by

$$I_f = 1/\lambda_{\min} \tag{5.64}$$

where the minimum eigenvalue is chosen to maximize I_f. The optimal weights are then found by solving (5.63), using the minimum eigenvalue $\lambda_{\min}$. To summarize,

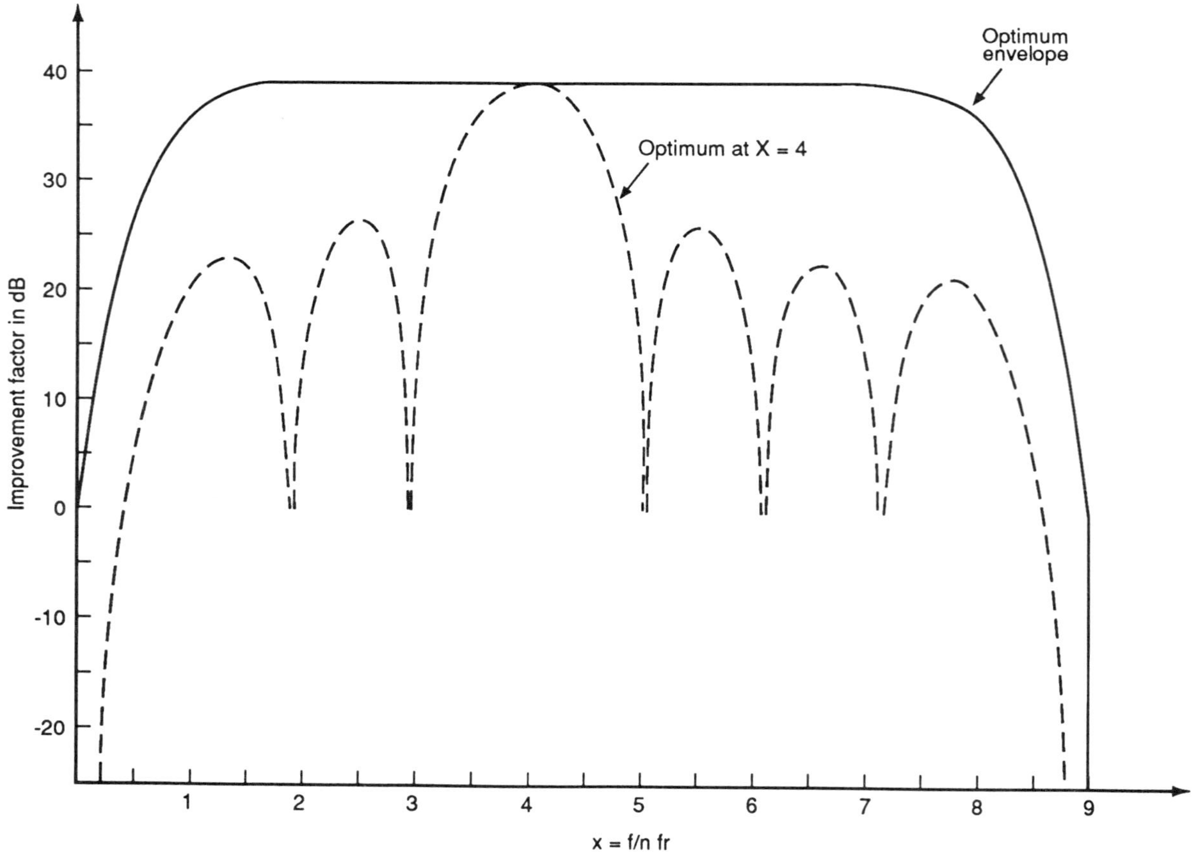

Figure 5.8 Improvement factor for point-optimized transversal filter ($x = 4$).

the optimal weights ($\mathbf{w}_o$) constitute the eigenvector that corresponds to the minimum eigenvalue:

$$(\mathbf{R}_n - \lambda_{\min}\mathbf{I})\mathbf{w}_o^* = \mathbf{0} \tag{5.65}$$

The eigenvalues of $\mathbf{R}_n$ must be real as $\mathbf{R}_n$ is a real symmetric matrix [11]. Furthermore, this means that the optimal weights ($\mathbf{w}_o$) can always be taken as real [12], which simplifies the structure of the transversal filter depicted in Figure 5.4 to the form with only real weights depicted in Figure 5.9. This form is commonly denoted as a *quadrature channel MTI*.

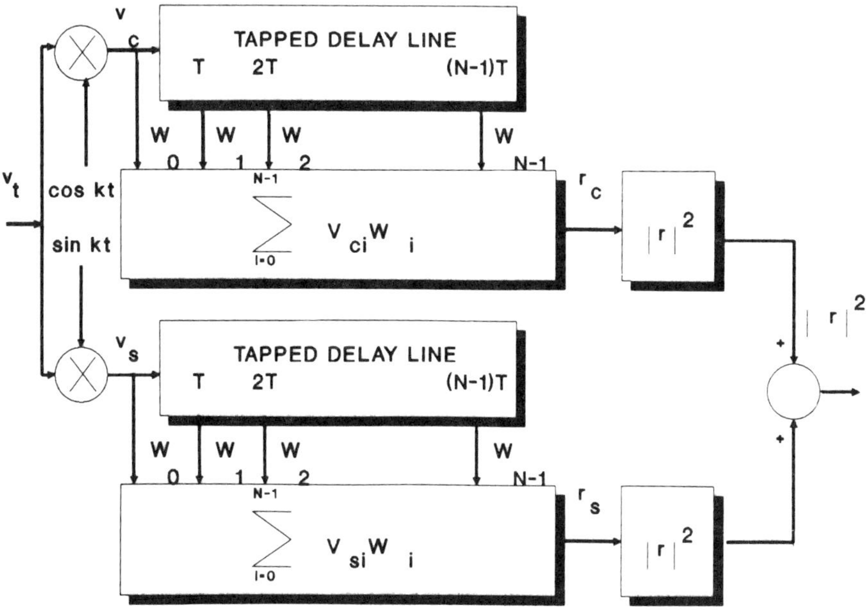

Figure 5.9 Quadrature channel MTI.

The interference covariance matrix $\mathbf{R}_n$ given by (5.53) can be written as the sum of a clutter covariance matrix, $\mathbf{R}_c$, and a receiver noise matrix:

$$\mathbf{R}_n = \mathbf{R}_c + \sigma_n^2\mathbf{I} \tag{5.66}$$

where

$$\mathbf{R}_c = \sigma_c^2 \begin{bmatrix} 1 & \rho_1 & \cdots\cdots & \rho_{n-1} \\ \rho_1 & \ddots & & \vdots \\ \vdots & & \ddots & \vdots \\ \vdots & & \ddots & \vdots \\ \rho_{n-1} & \cdots\cdots & \cdots & 1 \end{bmatrix} \tag{5.67}$$

Substitution of (5.66) into the solution (5.65) results in

$$(\mathbf{R}_c - \lambda_c \mathbf{I})\mathbf{w}_o^* = \mathbf{0} \tag{5.68}$$

where the eigenvalues of $\mathbf{R}_n$ are

$$\lambda_n = \frac{1}{F + 1}(F\lambda_c + 1) \tag{5.69}$$

and λ_c are the eigenvalues of $\mathbf{R}_c$. The improvement factor given by (5.64) can be rewritten in terms of (5.69) as

$$I_f = \frac{F + 1}{F/I_{fc} + 1} \tag{5.70}$$

From (5.70), we can see that the weights which maximize I_{f_c} also maximize I_f. Hence, solving (5.68) for the optimum weights by using the clutter covariance matrix is equivalent to solving (5.38) for the optimum weights by using the interference covariance matrix. Also, as F approaches 0, I_f approaches 1, indicating that the MTI provides no improvement against receiver noise. In addition, as the clutter improvement factor becomes large, the overall improvement factor approaches $F + 1$.

The preceding discussion indicates that the MTI design can be optimized on the basis of only clutter considerations (5.68), without any loss of generality. Therefore, we will use this approach in determining the MTI weights and improvement factors that follow. The approach is consistent with much of the literature on MTI and focuses attention on the main function of the MTI, which is to eliminate clutter.

The improvement factor for the quadrature channel MTI is given by (5.35), specialized to real weights:

$$I_{f_c} = \frac{\sum_{j=0}^{n-1} w_j^2}{\sum_{i=0}^{n-1}\sum_{j=0}^{n-1} w_i w_j \rho(i - j)} \tag{5.71}$$

Specialized formulas for the optimal weights and associated improvement factors can be derived by using (5.63) for $n = 2, 3, 4$ [6,8,9]; these are given in Table 5.1. Beyond $n = 4$, solving (5.68) analytically becomes difficult, and computer solutions thus become appropriate.

Table 5.1
MTI Optimal Weights and Improvement Factors

Pulses	*Weights*	*Improvement Factor*
2	1, −1	$1/(1-\rho_1)$
3	$1; -\dfrac{\rho_2 + \sqrt{\rho_2^2 + 8\rho_1^2}}{2\rho_1}; 1$	$1\Big/\left\{1 - \dfrac{\rho_2}{2}\left[\sqrt{1 + 8\left(\dfrac{\rho_1^2}{\rho_2}\right)} - 1\right]\right\}$
4	$w_1 = -\dfrac{\sqrt{(\rho_1-\rho_3)^2/4 + (\rho_1-\rho_2)^2} + (\rho_1-\rho_3)/2}{\rho_1-\rho_2}$ $1; w_1; -w_1; -1$	$\dfrac{1}{1 - \dfrac{\rho_1+\rho_3}{2} - \sqrt{(\rho_1-\rho_3)^2/4 + (\rho_1-\rho_2)^2}}$

The solutions given by (5.68) for the optimum MTI have a number of undesirable qualities that tend to restrict their practical utility [13]. First, the MTI filter's transfer function becomes increasingly narrow as the number of processed pulses increases, asymptotically approaching an impulse function at the minimum value of the clutter spectrum. This tendency is depicted in Figure 5.10, which illustrates transfer functions for up to ten pulses. The improvement factor also increases with the number of processed pulses. This feature is illustrated in Figure 5.11, which depicts improvement factors for up to 16 processed pulses and various clutter spectral spreads.

Second, to have the sum of the filter weights equal to 0 is desirable, which eliminates fixed, discrete targets. Table 5.1 indicates that this condition is satisfied only when the optimized MTI processes an even number of pulses.

An alternative optimization procedure using quadratic programming [14] allows excess improvement factor to be traded for the flexibility of specifying a desired MTI velocity response [13]. An example is depicted in Figure 5.12, which provides an improvement factor in excess of 30 dB with a significantly widened filter response that is essentially flat in the MTI's passband. The quadratic programming optimization procedure also allows for specifying that the sum of MTI filter weights equal zero regardless of the number of processed pulses.

The improvement factor for binomial weights is given in Figure 5.11 in addition to that for the optimized MTI design. For a narrow clutter spectrum, the performance achieved with binomial weights approaches that for the optimized design. The

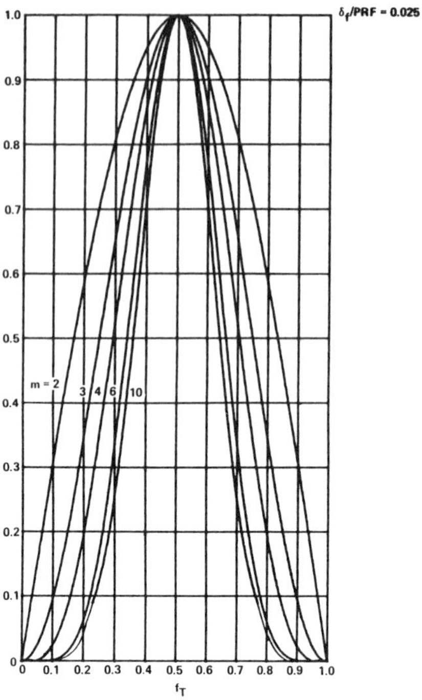

Figure 5.10 Velocity responses for optimum MTI.

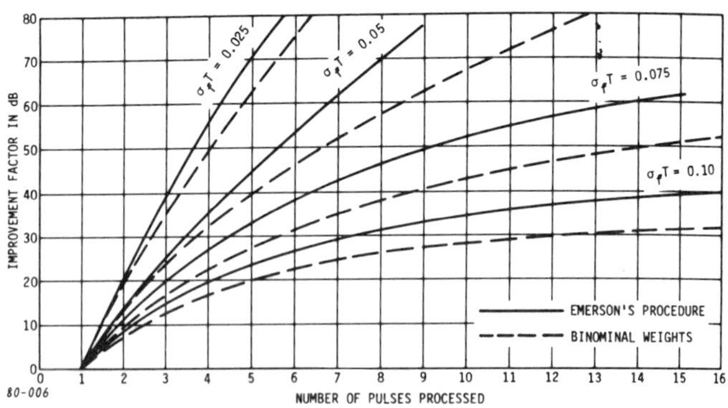

Figure 5.11 Improvement factor for optimum and binomial MTI.

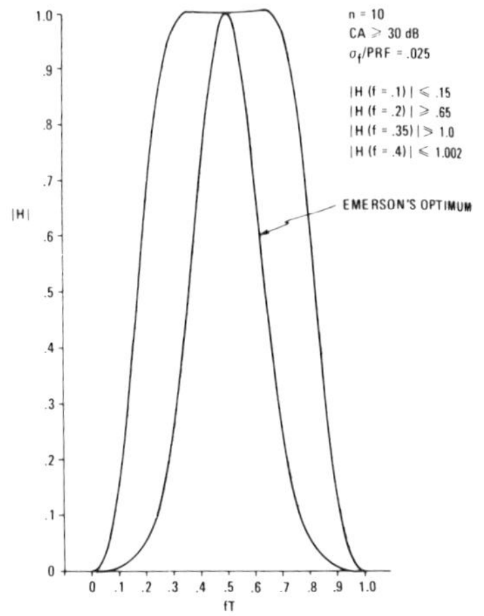

Figure 5.12 MTI velocity response—quadrature programming.

discrepancy becomes greater as the number of processed pulses and the clutter spectrum increase. For a small number of processed pulses ($n \leq 5$), the simplicity of using binomial weights is usually traded for the relatively modest improvement factor gain that can be achieved by using an optimized MTI design.

The binomial weights for an MTI design are given by

$$w_k = (-1)^k \binom{n}{k}, \quad k = 0, 1, \ldots, n - 1 \tag{5.72}$$

The improvement factor is given by

$$I_{fc} = \frac{\sum_{j=0}^{n-1} \binom{n}{j}^2}{\sum_{j=0}^{n-1} \sum_{k=0}^{n-1} (-1)^{k+j} \binom{n}{j} \binom{n}{k} \rho(j-k)} \tag{5.73}$$

which also can be written as [6]

$$I_{fc} = \frac{1}{1 - 2\dfrac{n}{n+1}\rho_1 + 2\dfrac{n}{n+1}\dfrac{n-1}{n-2}\rho_2 + \ldots} \approx \frac{2^n}{n!(\Omega)^{2n}} \tag{5.74}$$

where the approximation applies for $\Omega = 2\pi\sigma_f T \ll 1$. The average power gain is given by

$$\bar{G} = 1 + n^2 + \left[\frac{n(n-1)}{2!}\right]^2 + \left[\frac{n(n-1)(n-2)}{3!}\right]^2 + \ldots$$
$$\approx 2^n\,[1 \cdot 3 \cdot 5 \ldots (2n-1)]/n! \qquad (5.75)$$

The binomial weights represent an MTI design that consists of a cascade of $n - 1$ single MTI cancellers (single delay and subtract MTIs). A popular doppler processor design uses a single or double MTI canceller to eliminate the clutter, followed by a coherent or incoherent integrator to build up the S/N.

5.1.4 The Equal-Interval Doppler Processor

The equal-interval doppler transversal filter optimizes the detection performance within an interval that is a subset of the doppler processing band. A doppler processor is formed by dividing the doppler coverage band into equal intervals and employing an optimized transversal filter in each. This criterion is depicted in Figure 5.13: the doppler band is divided into n intervals, where n is the number of processed pulses.

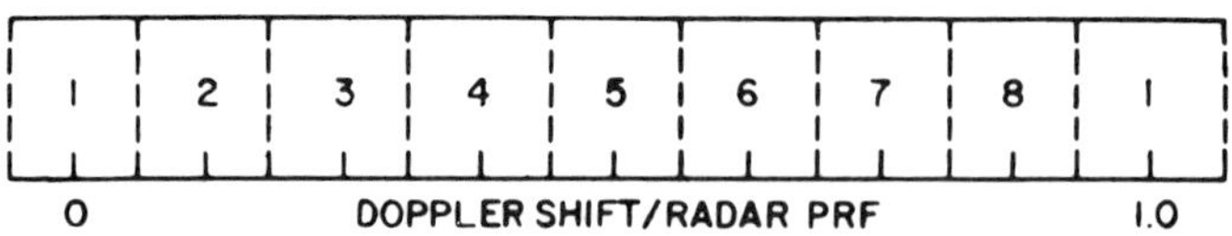

Figure 5.13 Design criterion for equal interval optimized transversal filters.

The equal-interval doppler processor reflects a balanced assumption relative to the distribution of the input signals doppler shift. This processor is optimized over a wider region than the point doppler processor and provides S/N improvement not available with the MTI processor. However, the computation of optimal weights and improvement factor is more difficult than for the other types of processors.

For the equal interval doppler processor, the target doppler probability distribution is given by

$$p(f_d) = \begin{cases} nT, & (2k-1)/2nT \le f_d \le (2k+1)/2nT \\ 0, & \text{otherwise} \end{cases} \qquad (5.76)$$

where $k = 0, 1, 2, \ldots, n-1$ is the filter number. The signal covariance matrix is found by averaging (5.49) for each doppler frequency with respect to the probability distribution of expected doppler frequencies, (5.76). The resulting normalized signal covariance matrix element is given by

$$m_{s_{il}} = e^{j2\pi k(i-l)/n} \frac{\sin[\pi(i-l)/n]}{\pi(i-l)/n} \tag{5.77}$$

where i is the row number, l is the column number, k is the filter number, and n is the number of equal intervals (number of processed pulses).

The optimal weights can be found as the eigenvector corresponding to the maximum eigenvalue of $\mathbf{R}_n^{-1}\mathbf{M}_s$ (see (5.13) and (5.14)). The improvement factor can be found as the value of the maximum eigenvalue ($\gamma_{\max}$) of $\mathbf{R}_n^{-1}\mathbf{M}_s$ (see(5.15)). This solution is more difficult than for the point-optimized processor because $\mathbf{M}_s$ is no longer a singular matrix. Thus, the improvement factor is not equal to the trace of the matrix $\mathbf{R}_n^{-1}\mathbf{M}_s$ (see (5.52)) and, in addition, a simple expression for the weights is not available (see (5.50)). In fact, the solution using (5.14) is somewhat involved, as the matrix $\mathbf{R}_n^{-1}\mathbf{M}_s$ is the product of a real symmetric matrix and a Hermitian matrix, which results in a matrix that has complex elements and apparently no simple properties. Therefore, a solution using the general relations (5.13), (5.14), and (5.15) is most suitable for a computer program that finds eigenvalues and eigenvectors for complex matrices or when a small number of pulses are processed, for which an analytic solution can be found.

We shall proceed in a somewhat intuitive way that provides a simple solution which ought to approach the general solution. The solution for the optimal weights at any point in the doppler domain is given by $\mathbf{w}_o = \mathbf{R}_n^{-1}\mathbf{s}^*$. A quasioptimal solution over an interval in the doppler domain is obtained by using the average of the optimal weights over that interval. This, in turn, can be obtained by averaging the point-optimized weights of (5.50), using the probability distribution of (5.76). The resulting equal-interval-optimized complex weights are given by

$$\bar{w}_{i-1} = \sum_{j=1}^{n} \left[\alpha_{ij} \cos(j-1)\bar{\omega}_d T \frac{\sin(j-1)\pi/n}{(j-1)\pi/n} - j\alpha_{ij} \sin(j-1)\bar{\omega}_d T \frac{\sin(j-1)\pi/n}{(j-1)\pi/n} \right] \tag{5.78}$$

where α_{ij} are the elements of the inverse covariance matrix (see (5.54)), n is the number of equal intervals, $\bar{\omega}_d$ is the doppler radian frequency at the center of the interval ($\bar{\omega}_d T = k/n$), and T is the radar's interpulse interval. Note that (5.78) can also be interpreted as using (5.50) for the optimal point weights with an averaged signal vector $\bar{\mathbf{s}}^*$ substituted for the monochromatic signal $\mathbf{s}^*$. The averaged signal vector is found by averaging $\mathbf{s}^*$, using the probability distribution of (5.76). The corresponding improvement factor is given by (5.56), using the averaged weights of (5.78):

$$I_f = \frac{\sum_{i=0}^{n-1} |\bar{w}_i|^2 + 2\sum_{i=0}^{n-2}\sum_{j=i+1}^{n-1} [\mathrm{Re}(\bar{w}_i^*\bar{w}_j)\cos(j-i)\bar{\omega}_d T - \mathrm{Im}(\bar{w}_i^* w_j)\sin(j-i)\bar{\omega}_d T]}{\sum_{i=0}^{n-1} |\bar{w}_i|^2 + 2\sum_{i=0}^{n-2}\sum_{j=i+1}^{n-1} \rho(i-j)\,\mathrm{Re}(\bar{w}_i^*\bar{w}_j)} \tag{5.79}$$

Figure 5.14 depicts the results of an improvement factor calculation for a nine-pulse equal-interval-optimized doppler processor. The improvement factors are obtained by using (5.79) for two transversal filters optimized about doppler frequencies centered at 2/9 and 3/9 of the radar's PRF. The response of these filters should be compared with that of equivalent filters shown in Figure 5.6, which were optimized by using the point doppler criterion. The principal differences are that the peak response of the interval optimized transversal filter is −1 dB below the point-optimized filter and the interval-optimized filters cross over at 2 dB rather than the 3 dB of the point-optimized filters. The result is that the equal-interval-optimized transversal filter will provide a more uniform response than the point-optimized design for an equal number of distributed filters. Nonetheless, the point-optimized design will have improvement factor peaks throughout the doppler coverage region, exceeding the peak of the interval-optimized filter.

Another advantage claimed for the interval-optimized transversal design is a better response when the filter is optimized about zero doppler shift [15]. The filter response for this condition is depicted in Figure 5.15. Note that this filter also has a response about the PRF line at $x = 9$ because the optimized interval is ambiguous about $\pm$ 1/2 unit around the PRF lines that include both the zero doppler frequency and the first PRF line ($x = 9$).

Figure 5.16 depicts the improvement factor response for transversal filters that are point optimized at $x = 0$ and $x = 1/2$. The design optimized at $x = 0$ has a poor response due to difficulty in coping with the conflicting requirements of a maximum response for a signal at zero doppler frequency and maximum attenuation for clutter centered about the same frequency. Therefore, the interval-optimized design has a better response than the point-optimized design when the design point is a zero doppler shift. However, note that this situation is easily corrected by point optimizing the transversal filter about $x = 1/2$, which, in effect, removes the conflicting requirements.

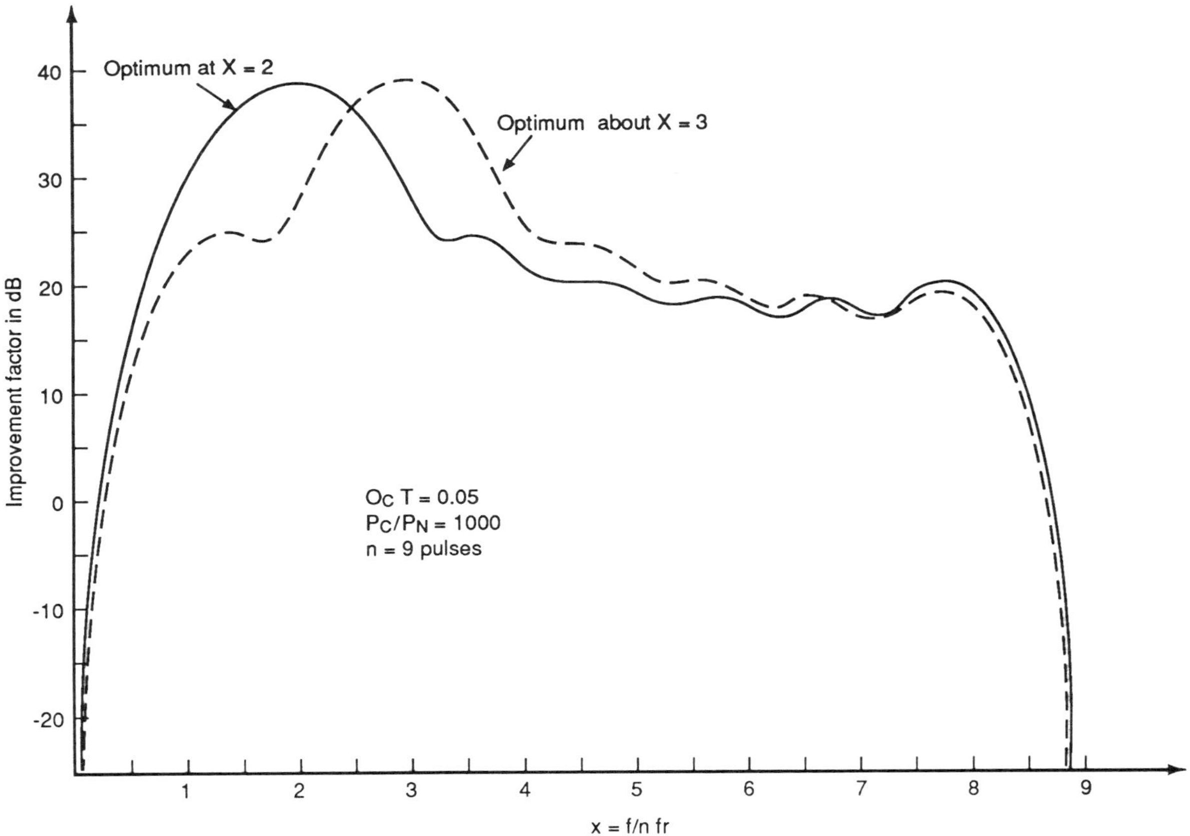

Figure 5.14 Improvement factor for equal interval optimized transversal filters ($x = 2, 3$).

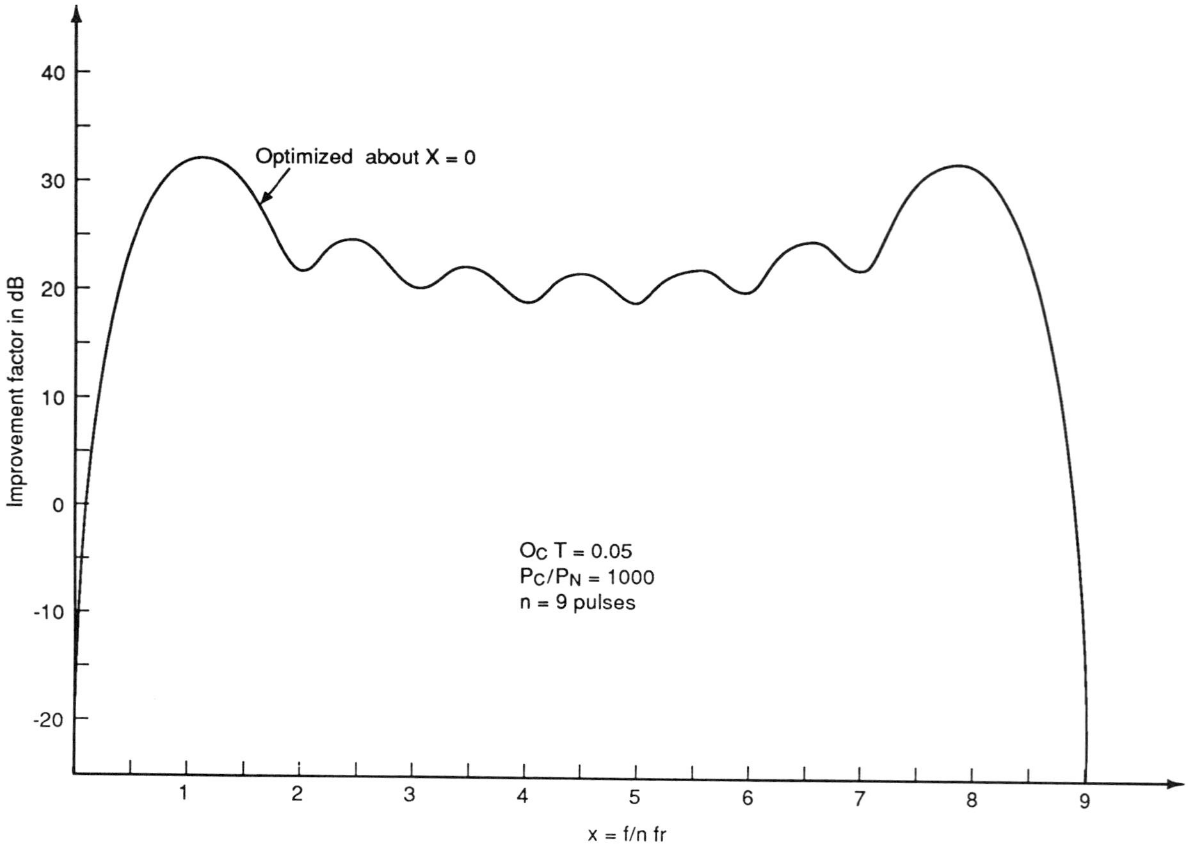

Figure 5.15 Improvement factor for equal interval optimized transversal filter ($x = 0$).

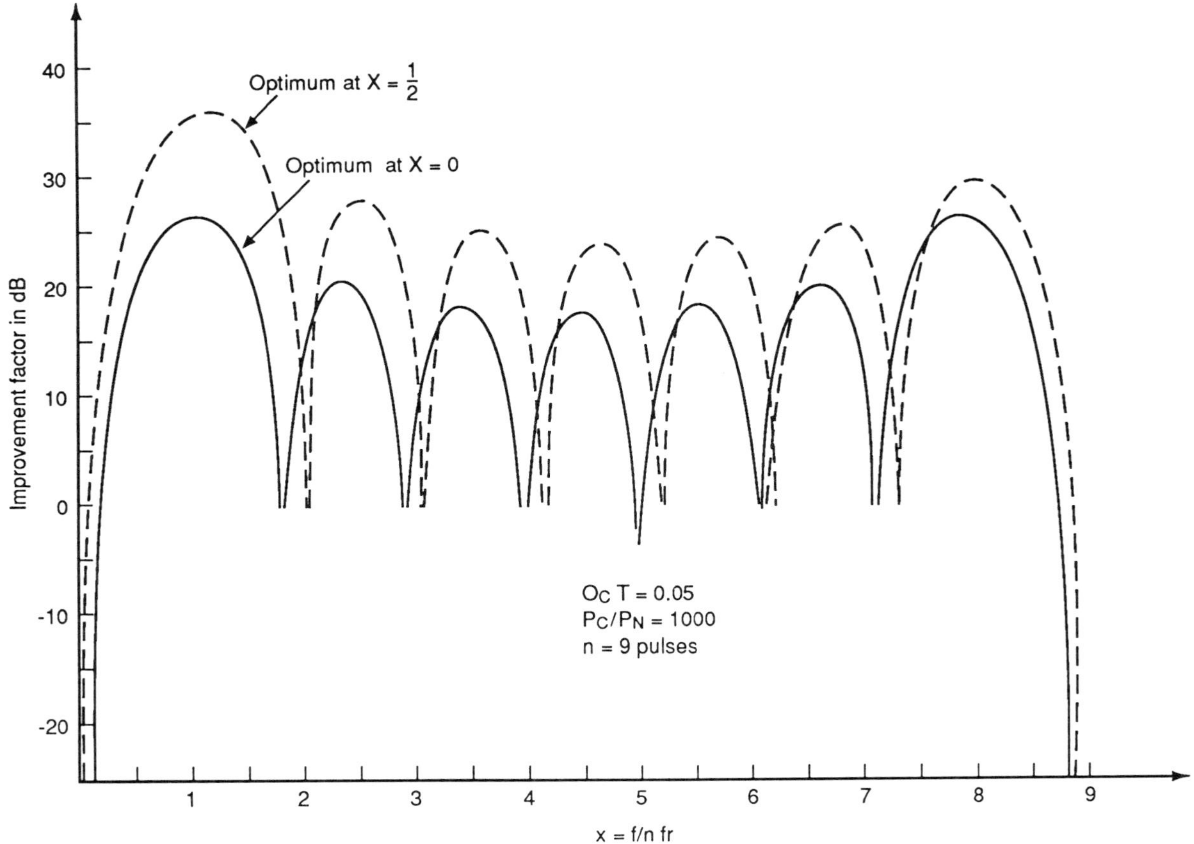

Figure 5.16 Improvement factor for point-optimized transversal filters ($x = 0, {}^1/_2$).

5.2 CASCADED MTI-COHERENT INTEGRATOR

The theory of optimal radar doppler processor design presented in the previous section prescribed the use of transversal filters, the weights of which are determined by the assumed target doppler distribution and interference covariance matrix. An almost optimum alternative doppler processor uses an MTI followed by a bank of coherent integrators [16].

The two-stage cascaded MTI-coherent integrator uses the MTI to enhance the S/C and the coherent integrator primarily to enhance the S/N. The design criterion for each element is relatively straightforward. The MTI is designed to attenuate the worst expected clutter spectrum, and the coherent integrator is matched to the expected dwell time on target.

However, the MTI-coherent integrator provides an inherently less efficient structure than the optimal transversal filter when the interference covariance matrix is known. The inefficiency occurs in the MTI portion of the filter, where only one of the m pulses used in the MIT effectively builds up the S/N. Therefore, if a total of n radar pulses are available for doppler processing, only $n - m + 1$ pulses are available for coherent integration. As a general design principle, the higher is the order of the MTI filter, the fewer pulses available for coherent integration and the less enhancement of S/N available in the processor. Practical designs of this processor usually employ single or double MTI cancellers for this reason.

A popular version of a cascaded MTI-coherent integrator uses the quadrature channel MTI, depicted in Figure 5.9, followed by a coherent integrator implemented with the FFT algorithm. This processor is shown in Figure 5.17. The complex form of the MTI output (quadrature channel) is ideally suited to the FFT processor, which accepts complex inputs. The outputs of the FFT coherent integrator as illustrated by Figure 5.18, are n envelope detected channels that span the doppler region between

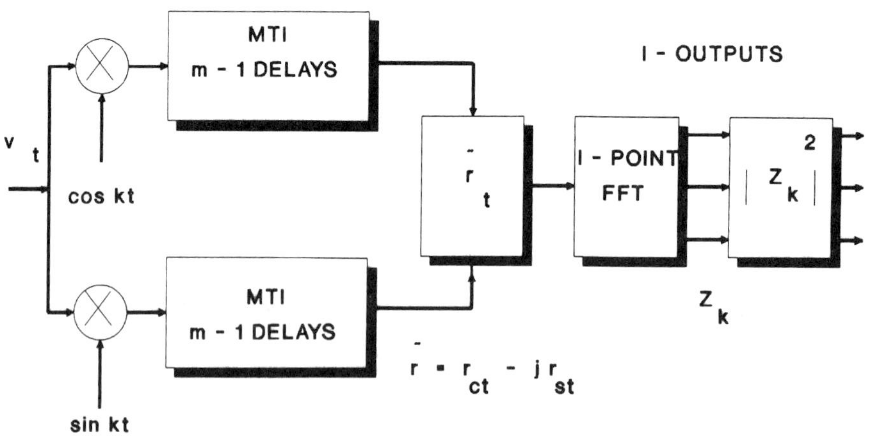

Figure 5.17 Cascaded MTI-FFT coherent integrator.

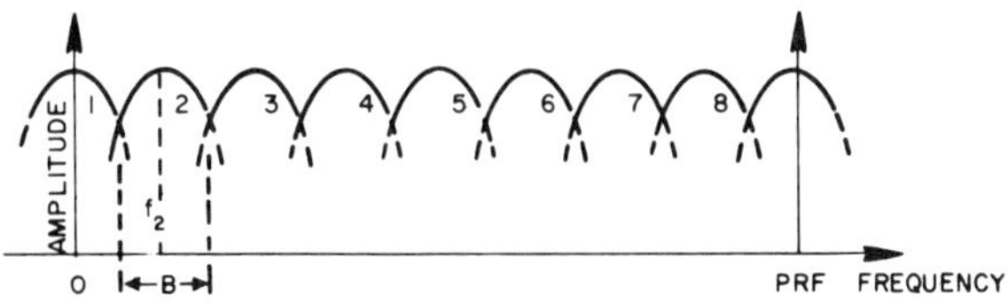

Figure 5.18 FFT coherent integrator filter bank.

PRF lines. Each channel can have a separate gain or CFAR control allowing suppression of a portion of the doppler band that contains excess interference. This capability is important in controlling weather clutter, chaff returns, and electronic interference.

In this section, the essential characteristics of the cascaded MTI-coherent integrator will be derived. Initially, a time-domain approach will be followed because this allows derivation of the weights of an equivalent transversal filter. These weights could then be substituted into (5.11) to determine the improvement factor for the cascaded processor. However, this procedure results in somewhat complex multiple summations [17,18] and as an alternative we provide a relatively straightforward frequency-domain analysis that allows calculation of the improvement factor.

The purpose of pursuing the time-domain analysis is that it illustrates a distinctive feature of the cascade processor, which is important in the actual design of this type of doppler processor. As is apparent by referring to the quadrature channel MTI shown in Figure 5.9, the first MTI output is obtained after m pulses have entered the MTI. Subsequent MTI outputs, which are applied to the l-point FFT coherent integrator, are obtained so long as input pulses to the processor are available. The output of the processor is then a random variable z, obtained after $m + l - 1$ pulses are applied to the cascaded processor, but where only the last l of the MTI output pulses are applied to the integrator. Analytically, all of this is accommodated by a frequency-domain analysis, thereby significantly reducing the possibility of error, for which several examples can be found in the published literature.

The *discrete Fourier transform* (DFT) of a sequence of complex input samples $\tilde{r}_i = (\tilde{r}_0, \tilde{r}_1, \ldots, \tilde{r}_{l-1})$ obtained from the output of the MTI is given by [19, 20]

$$\tilde{s}(k) = \sum_{i=0}^{l-1} \tilde{r}(i)\, \mathrm{e}^{-j(2\pi ik/l)} \tag{5.80}$$

where $k = 0, 1, 2, \ldots, l - 1$ is the output filter number and l is the number of coherently integrated pulses. The similarity to (5.1), which describes the output of a transversal filter, is apparent, where the complex weight vector is identified as

$$a_k^i = \mathrm{e}^{-j(2\pi ik/l)} \tag{5.81}$$

The equivalent transversal filter that represents the DFT function is depicted in Figure 5.19.

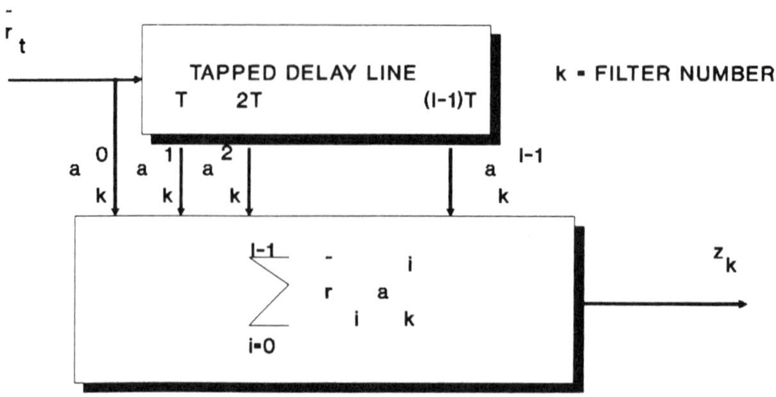

Figure 5.19 Equivalent transversal filter for DFT.

The impulse response of the l-point DFT as represented by the transversal filter of Figure 5.19 is given by

$$h(t) = \sum_{i=0}^{l-1} a_k^i \delta(t + iT) \tag{5.82}$$

The transfer function is obtained as the Fourier transform of the impulse response:

$$H(j\omega) = \sum_{i=0}^{l-1} a_k^i \, \mathrm{e}^{+j\omega_d iT} = \sum_{i=0}^{l-1} [\mathrm{e}^{-j(2\pi k/l - \omega_d T)}]^i \tag{5.83}$$

which can be written as

$$H(j\omega) = \frac{\mathrm{e}^{-j(2\pi k/l - \omega_d T)l} - 1}{\mathrm{e}^{-j(2\pi k/l - \omega_d T)} - 1} \tag{5.84}$$

where the series $\Sigma_{i=0}^{l-1} b^i = b^l - 1/b - 1$ is used. The power transfer function is obtained from (5.84) as

$$|H(j\omega)|^2 = \left|\frac{\sin(l\omega_d T/2 - k\pi)}{\sin(\omega_d T/2 - \pi k/l)}\right|^2 \tag{5.85}$$

The power transfer function for the first filter ($k = 0$) is given by

$$|H(\omega)|^2 = \left|\frac{\sin(l\omega_d T/2)}{\sin(\omega_d T/2)}\right|^2; \quad k = 0 \tag{5.86}$$

which has zeros located at

$$\omega_d T = 2\pi p/l, \quad p = 1, 2, \ldots, l \tag{5.87}$$

and maxima at

$$\omega_d T = \frac{2\pi}{l}(p + 1/2), \quad p = 1, 2, \ldots, l \tag{5.88}$$

The value at $\omega_d = 0$ is given by

$$|H(0)|^2 = \lim_{\omega_d \to 0} \left| \frac{\sin(l\omega_d T/2)}{\sin(\omega_d T/2)} \right|^2 = l^2 \tag{5.89}$$

Therefore, the normalized power transfer function as found by expanding the numerator of (5.85) and dividing by the maximum value is given by

$$|H(j\omega)|_n^2 = \left| \frac{\sin(l\omega_d T/2)}{l \sin(\omega_d T/2 - \pi k/l)} \right|^2 \tag{5.90}$$

which is depicted in Figure 5.20 for a 16-point DFT. The generalized maximum of the power transfer function of the l-point DFT filter occurs at

$$\omega_d T = 2\pi k/l, \quad k = 0, 1, \ldots, l - 1 \tag{5.91}$$

with normalized value equal to 1.

Thus, a bank of filters can be produced as depicted in Figure 5.18 by using the l available output ports of the DFT (see (5.80)). Each output is formed by an equivalent transversal filter of the form shown in Figure 5.19, which has a maximum at a normalized doppler frequency $f_d/f_r = k/l$ and a power transfer function given by equation (5.90).

The DFT is efficiently implemented by using the fast Fourier transform algorithm [19,20]. The FFT digital filter allows forming the full coherent-integration bank to minimize the number of multiplication and addition operations. This will be discussed more fully in the section on general MTI doppler filters, but note that this efficient implementation is the principal reason for the popularity of the cascaded MTI-coherent integrator processor discussed in this section.

The power transfer function for the FFT coherent integrator given by (5.85) can be approximated by replacing the sin(·) function in the denominator with its argument, resulting in

$$|H(j\omega)|^2 \approx \frac{\sin^2(l\omega_d T/2 - k\pi)}{(\omega_d T/2 - \pi k/l)^2} \tag{5.92}$$

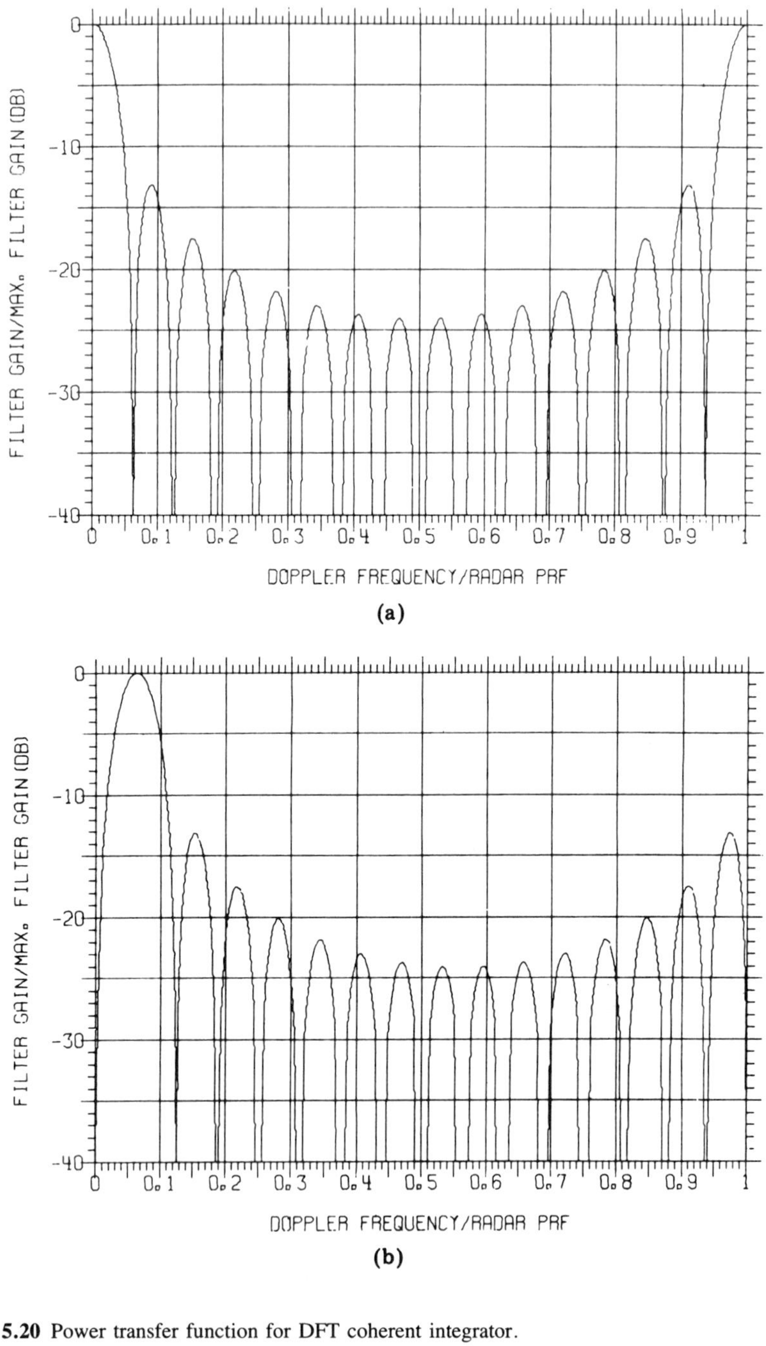

Figure 5.20 Power transfer function for DFT coherent integrator.

The $(\sin^2 x)/x^2$ form of (5.92) illustrates two characteristics of the FFT coherent filter bank. First, the sidelobe structure associated with each coherent integration filter is relatively high (see Figure 5.20). The analogy is to the pattern of a uniformly illuminated antenna aperture. The relative amplitude of the power transfer function sidelobes from (5.85) and (5.92) are listed for reference purposes in Table 5.2. Note that the approximation of (5.92) becomes poorer in the higher order sidelobes as the argument x increases.

Table 5.2
Peak Sidelobe Response of Power Transfer Function for a Coherent Integration Filter

Sidelobe Number	*Peak Amplitude*	$(\sin x)/x$ *Approximation*
1	−13.3 dB	−13.5 dB
2	−17.6 dB	−17.9 dB
3	−20.8 dB	−20.8 dB
4	−21.8 dB	−23.0 dB
5	−23.0 dB	−24.8 dB
6	−23.7 dB	−26.2 dB

Second, the uniformity of the overall response of the FFT coherent-integration doppler filter bank is poor. From (5.92), we can see that adjacent filters cross over when the argument $x = \pi/2$. The magnitude of each individual filter is approximately −3.9 dB below its peak response at this point. The effect of this lack of uniformity is a relatively large straddling loss, the maximum value of which occurs when the target's doppler frequency is located in the region midway between the peak responses ($\omega_d = 2\pi k/lT$) of the coherent integration filter bank.

The sidelobe and cross-over characteristics of the FFT coherent integrator are due to the use of a rectangular window where all input samples within the data window are accepted with equal weight. By analogy to antenna radiation pattern theory, we know that sidelobes can be reduced and the main pattern lobe broadened by using an appropriate amplitude taper applied to the illumination function. This can be accomplished by using one of the standard window functions employed in the design of digital filters [21]. We shall concentrate on two particular types of windows, known as the Hamming and Hann (or hanning) windows, which are depicted in Figure 5.21. Both these windows can be represented by the equation:

$$g(t) = (1 - \alpha) + \alpha \cos(2\pi t/lT) \tag{5.93}$$

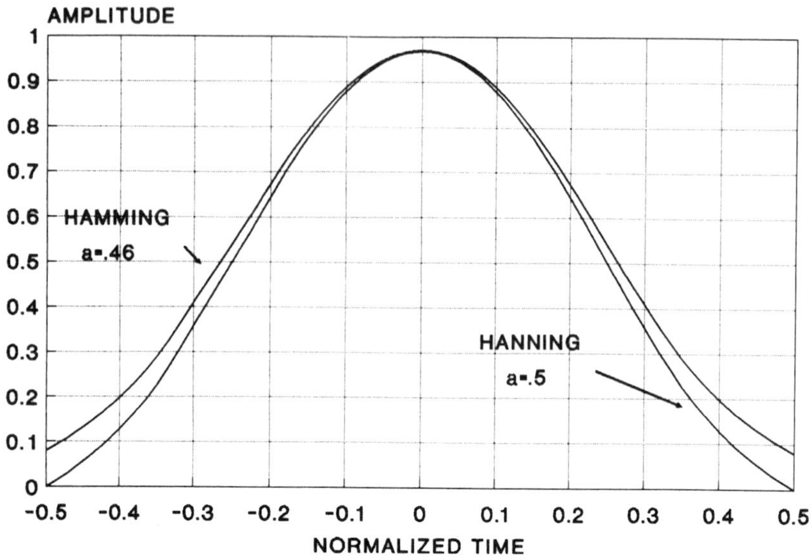

Figure 5.21 Hamming and hanning window functions.

where $-lT/2 \leq t \leq lT/2$ and $\alpha = 0.5$ for the hanning window and $\alpha = 0.46$ for the Hamming window [22].

The application of a window or amplitude weighting function modifies the expression for the DFT of (5.80) to

$$\tilde{s}_w(k) = \sum_{i=0}^{l-1} \tilde{r}(i)g(i)e^{-j[2\pi ik/l]} \tag{5.94}$$

where $g(i)$ is the sampled data equivalent to the continuous window function of (5.93). The impulse response of the DFT given by (5.82) can be written as

$$h_w(t) = g(t)\sum_{i=0}^{l-1} a_k^i \delta(t + iT) = g(t)h(t) \tag{5.95}$$

where $g(t)$ is a continuous periodic version of (5.93) that assumes a value only at the sampling instants where $\delta(t + iT)$ is nonzero. The Fourier transform of $g(t)$ is given by

$$G(\omega) = 2\pi(1 - \alpha)\delta(\omega) + \alpha\pi\left[\delta\left(\omega - \frac{2\pi}{lT}\right) + \delta\left(\omega + \frac{2\pi}{lT}\right)\right] \tag{5.96}$$

The voltage transfer function of the DFT coherent integrator with a Hamming or hanning window can be found by Fourier transforming equation (5.95), which prescribes the convolving of the frequency functions [23] given by equations (5.84) and (5.96) $[H(\omega)*G(\omega)]$, resulting in

$$H_w(j\omega) = 2\pi(1 - \alpha)H_k(j\omega) + \pi\alpha[H_{k+1}(j\omega) + H_{k-1}(j\omega)] \tag{5.97}$$

where the subscripts on the transfer functions indicate that the k, $k + 1$, and $k - 1$ transfer functions are to be combined as indicated. Equation (5.97) can be simplified as

$$H_w(j\omega) = H_k(j\omega) + \alpha[H_{k+1}(j\omega) + H_{k-1}(j\omega)] \tag{5.98}$$

where the weights are $\alpha = 0.5$ for hanning weighting and $\alpha = .426$ for Hamming weighting.

The interpretation of (5.98) is that an effective weighting function window can be applied to the input data of the FFT coherent integrator by using symmetrical postcombining weighting of the adjacent filters combined with the filter under consideration. The sidelobes [21] with hanning weighting (−31 dB) and Hamming weighting (−41 dB) are considerably better than for uniform weighting (−13 dB). For hanning weighting, adjacent filters cross over at −0.4 dB and for Hamming weighting at −0.84 dB, which is a considerable improvement with respect to uniform weighting (−3.9 dB). Note that the broadening of the coherent integrators mainlobe response with weighting increases the amount of receiver noise power in the filter's output, which has the deleterious effect of decreasing the output S/N. This will be considered in more detail at the end of this section when we compare the detection performance of various doppler processors.

Consider now the MTI portion of the cascaded MTI-coherent integrator. From Section 5.1 on optimum radar doppler processors, we know that the MTI is a specialized form of transversal filter that has only real weights. The MTI is completely specified by defining these real weights. The optimal weights for a two-pulse MTI are $\mathbf{w} = (1, -1)$ resulting in an impulse response given by

$$h_{\text{MTI}}(t) = \delta(t) - \delta(t - T) \tag{5.99}$$

and a voltage transfer function:

$$|H_{\text{MTI}}(\omega)| = \left|2 \sin\left(\frac{\omega_d T}{2}\right)\right| \tag{5.100}$$

For the cascaded MTI-coherent integrator, we are interested in the lowest order m-pulse MTI that provides the desired improvement factor. Therefore, to use binomial weights is appropriate because the loss of improvement factor for a low-order MTI is minimal compared to optimal weights (see Figure 5.11) and the design procedure is simplified. Binomial weights (see (5.72)) can be obtained by successive convolution of the optimal weights for a two-pulse MTI (single canceller). Thus, the power transfer function for an MTI utilizing binomial weights can be found by simply multiplying (5.100) with itself $m - 1$ times and squaring, resulting in

$$|H_{\text{MTI}}(j\omega)|^2 = |2\sin(\omega_d T/2)|^{2(m-1)} \tag{5.101}$$

The peak power gain of the binomial MTI is found by evaluating (5.101) at $\omega_d T = \pi$, resulting in $\bar{G}_s = 2^{2(m-1)}$. Normalization of (5.101) to its peak value results in the normalized power transfer function for a binomial MTI:

$$|H_{\text{MTI}}(j\omega)|_N^2 = |\sin(\omega_d T/2)|^{2(m-1)} \tag{5.102}$$

where m is the number of pulses processed by the MTI and T is the radar's interpulse interval.

The next step in the development is to derive an expression for the weights of a transversal filter that represents the kth output port of the combined MTI-FFT coherent integrator. Consider the cascade of an m-pulse MTI processor with an l-point FFT coherent integrator, depicted in Figure 5.22. The l-point FFT coherent integrator is represented by an equivalent transversal filter that provides the kth output port. In accordance with (5.80), the coherent integrator output z_k is given by

$$z_k = \sum_{i=m-1}^{l+m-2} a(i - m + 1)\tilde{r}(i - m + 1) = \mathbf{a}_k^T\tilde{\mathbf{r}} \tag{5.103}$$

where $\mathbf{a}_k^T = (a_0, a_1, a_2, \ldots, a_{l-1})$ are the complex weights of the kth output of the FFT; $\tilde{\mathbf{r}} = (\tilde{r}_0, \tilde{r}_1, \tilde{r}_2, \ldots, \tilde{r}_{l-1})$ are the complex outputs of the MTI that occur in sequence after m input pulses are applied to the MTI; and i is an index that counts the number of input pulses applied to the MTI-coherent integrator cascade. Each output of the MTI in this index system is given by

$$\tilde{r}(j) = \sum_{i=j}^{m+j-1} w(i - j)\tilde{v}(i) = \mathbf{w}^T\tilde{\mathbf{v}} \tag{5.104}$$

so that the total output of the cascaded MTI-FFT coherent integrator is given by

$$z_k = \sum_{i=m-1}^{n-1} \sum_{j=i-m+1}^{i} a(i - m + 1)w(j - i + m - 1)v(j) = \mathbf{b}^T\tilde{\mathbf{v}} \tag{5.105}$$

where $n = l + m - 1$ is the total number of input pulses applied to the processor, $\tilde{\mathbf{v}} = (\tilde{v}_0, \tilde{v}_1, \ldots, \tilde{v}_{n-1})$ is the complex input voltage and $\mathbf{b}^T = (b_0, b_1, \ldots, b_{l+m-1})$ is the weight vector of an equivalent transversal filter for the total processor. From (5.105), the equivalent weight vector can be expressed as [17]

$$\mathbf{b} = \hat{\mathbf{A}}\mathbf{a}_k \tag{5.106}$$

where $a_k^T = (a_0, a_1, \ldots, a_{l-1})$ is the FFT weight vector and $\hat{\mathbf{A}}^T$ is an $m + l - 1$ by l matrix formed from the MTI weight vector as follows:

$$\hat{\mathbf{A}}_1 = \begin{bmatrix} w_o \\ \\ w_{m-1} \\ 0 \\ 0 \\ 0 \end{bmatrix} \left.\begin{matrix} \\ \\ \\ \end{matrix}\right\} \begin{matrix} l-1 \\ \text{zeros} \end{matrix}, \quad \hat{\mathbf{A}}_2 = \begin{bmatrix} 0 \\ w_o \\ \\ w_{m-1} \\ 0 \\ 0 \end{bmatrix}, \quad \hat{\mathbf{A}}_l = \begin{bmatrix} 0 \\ 0 \\ 0 \\ \\ w_o \\ w_{m-1} \end{bmatrix} \left.\begin{matrix} \\ \\ \\ \end{matrix}\right\} \begin{matrix} l-1 \\ \text{zeros} \end{matrix} \tag{5.107}$$

As an example, consider a single-canceller MTI ($m = 2$), followed by a four-point FFT coherent integrator ($l = 4$). The four weight vectors of the FFT are given by $\mathbf{a}_k = (1, 1, 1, 1)$, $(j, 1, -j, -1)$, $(-1, 1, -1, 1)$, and $(-j, 1, j, -1)$, whereas the weight vector of the MTI is given by $\mathbf{w} = (1, -1)$. If we select the second FFT weight vector ($k = 1$), we have

$$\hat{\mathbf{A}} = \begin{bmatrix} 1 & 0 & 0 & 0 \\ -1 & 1 & 0 & 0 \\ 0 & -1 & 1 & 0 \\ 0 & 0 & -1 & 1 \\ 0 & 0 & 0 & -1 \end{bmatrix} \tag{5.108}$$

$$\mathbf{b} = \begin{bmatrix} 1 \\ -1-j \\ j-1 \\ 1+j \\ -j \end{bmatrix} \tag{5.109}$$

which is the weight vector of the equivalent transversal filter, the output which is given by (5.105). This equivalent transversal filter is depicted in Figure 5.22, which also shows the timing involved with the various indices used in (5.103), (5.104), and (5.105).

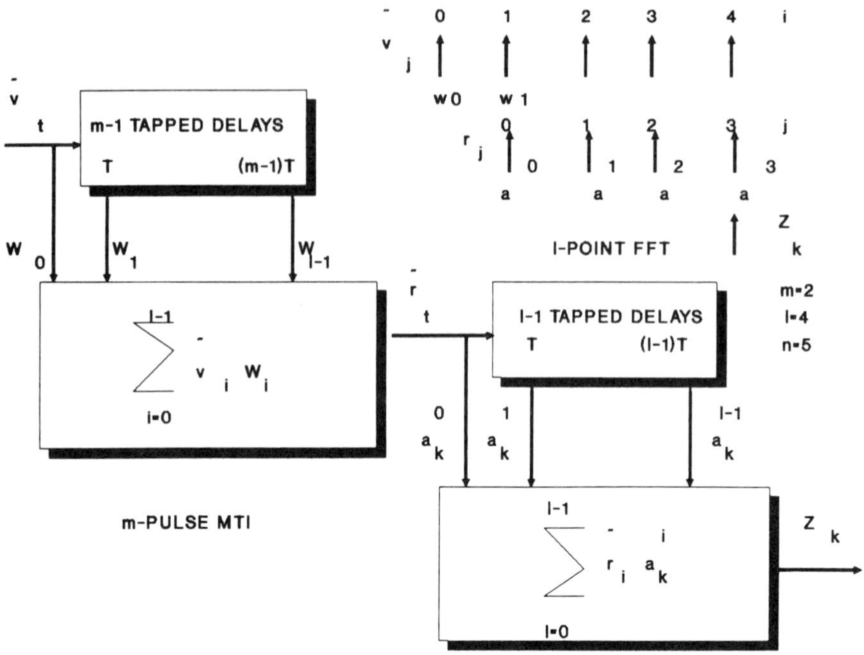

Figure 5.22 Cascaded MTI-equivalent FFT transversal filter.

Alternatively, the normalized power transfer function of the combined MTI-coherent integrator is given by the product of (5.90) and (5.102):

$$|H(j\omega)|_N^2 = \frac{[\sin(\omega_d T/2)]^{2(m-1)} \sin^2(l\omega_d T/2 - k\pi)}{l \sin(\omega_d T/2 - \pi k/l)} \tag{5.110}$$

The improvement factor can be found from

$$I_f = \frac{|H(j\omega)|_N^2 \displaystyle\int_0^{\omega_r} S_I(\omega)\,\mathrm{d}\omega}{\displaystyle\int_0^{\omega_r} S_I(\omega)|H(j\omega)|_N^2\,\mathrm{d}\omega} \tag{5.111}$$

where the interference power spectral density for a Gaussian-shaped clutter spectrum and receiver noise is given by

$$S_I(\omega) = \frac{P_n}{\omega_r} + \frac{P_c}{\sqrt{2\pi}\sigma_c}[\mathrm{e}^{-\omega^2/2\sigma_c^2} + \mathrm{e}^{-(\omega-\omega_r)^2/2\sigma_c^2}];\ 0 \le \omega \le \omega_r \tag{5.112}$$

where P_n is the receiver noise power, P_c is the clutter power, σ_c^2 is the variance of the Gaussian-shaped clutter spectrum's radian frequency response, and ω_r is the pulse repetition radian frequency. A convenient way to normalize (5.112) is to divide by the receiver noise power P_n and convert to f_d as the independent parameter, resulting in

$$I_f = \frac{|H(jf)|_N^2 \int_0^{f_r} S_I(f)\,\mathrm{d}f}{\int_0^{f_r} S_I(f)|H(jf)|_N^2\,\mathrm{d}f} \tag{5.113}$$

$$S_I(f) = \frac{1}{f_r} + \frac{F}{\sqrt{2\pi}\sigma_f}\,[\mathrm{e}^{-f_d^2/2\sigma_f^2} + \mathrm{e}^{-(f_d - f_r)^2/2\sigma_f^2}] \tag{5.114}$$

where F is the clutter-to-noise power ratio, f_r is the radar's pulse repetition frequency, and σ_f^2 is the variance of the clutter's Gaussian-shaped spectrum.

Figure 5.23 depicts the improvement factor for a cascaded single-canceller MTI–eight-point FFT coherent integrator with no integrator weighting. Responses for adjacent filters centered at $k = 2$ and $k = 3$ are shown. This figure illustrates the relatively high cross-over loss and high sidelobes associated with a rectangular window coherent integrator. The improvement factor at the integrator center frequency is 38.5 dB, which compares to 39.1 dB for a transversal filter design (see Figure 5.6) with the same number of nine processed pulses. Note, however, that the peak improvement factor, which is skewed to the right due to the effect of the MTI response, is on the order of 39 dB.

Figure 5.24 depicts the improvement factor for a cascaded single-canceller MTI–eight-point FFT coherent integrator with Hamming weighting ($\alpha = .426$). Again, responses for adjacent filters centered at $k = 2$ and $k = 3$ are shown. Note the broadened response that reduces the cross-over loss and the reduced sidelobe structure. However, the improvement factor is reduced to 37.9 dB at the integrator center frequency, whereas the peak improvement factor is reduced to 38.3 dB. Therefore, the effect of the coherent-integrator weighting function is to provide a more uniform overall doppler frequency response with minimal sidelobes at the expense of a reduced peak improvement factor.

A comparison between the point-optimized transversal filter doppler processor, a cascaded single-canceller MTI–eight-point FFT coherent integrator, and a cascaded single-canceller MTI–eight-point FFT coherent integrator with Hamming weighting is given in Table 5.3. The basis of the comparison is on the averaged improvement factor of a single output port ($k = 2$) where the average is taken between cross-over points of adjacent filters. We can see that, as expected, the point-optimized transversal

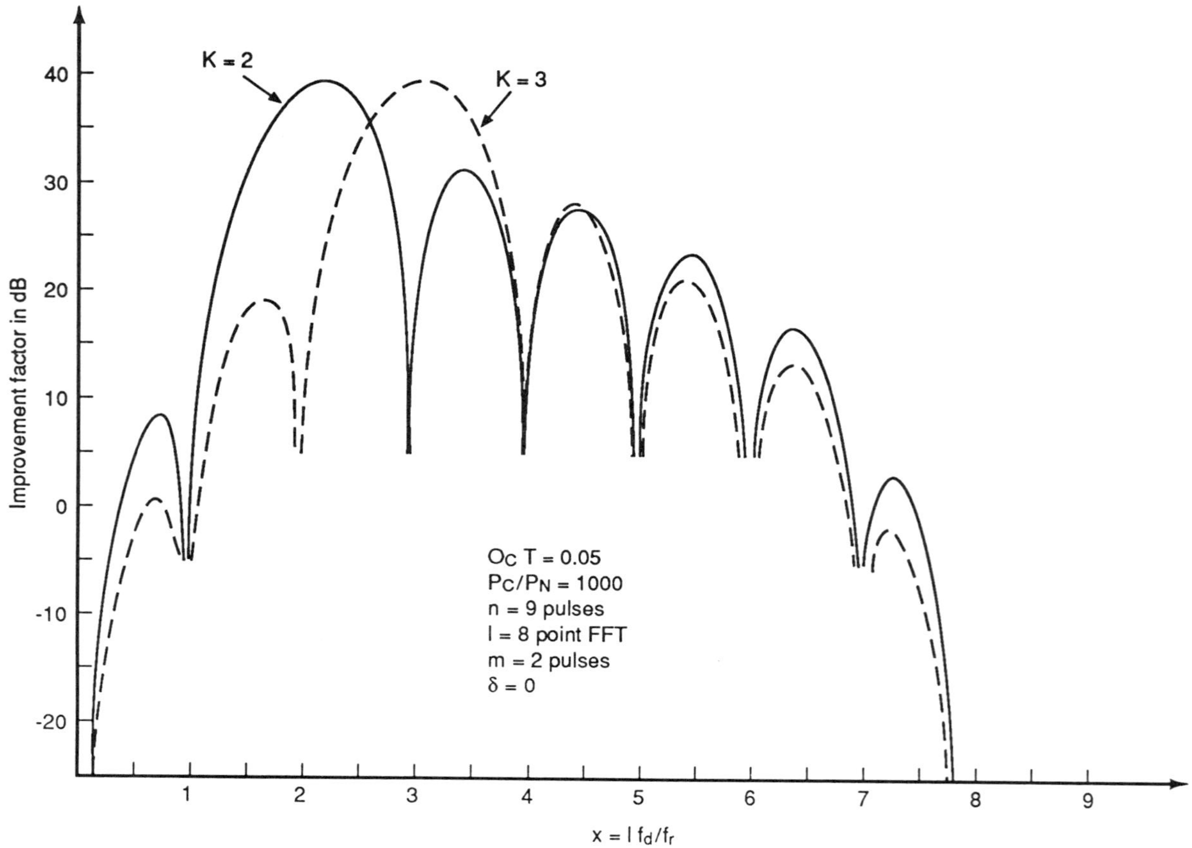

Figure 5.23 Improvement factor for a single-canceller MTI-coherent integrator.

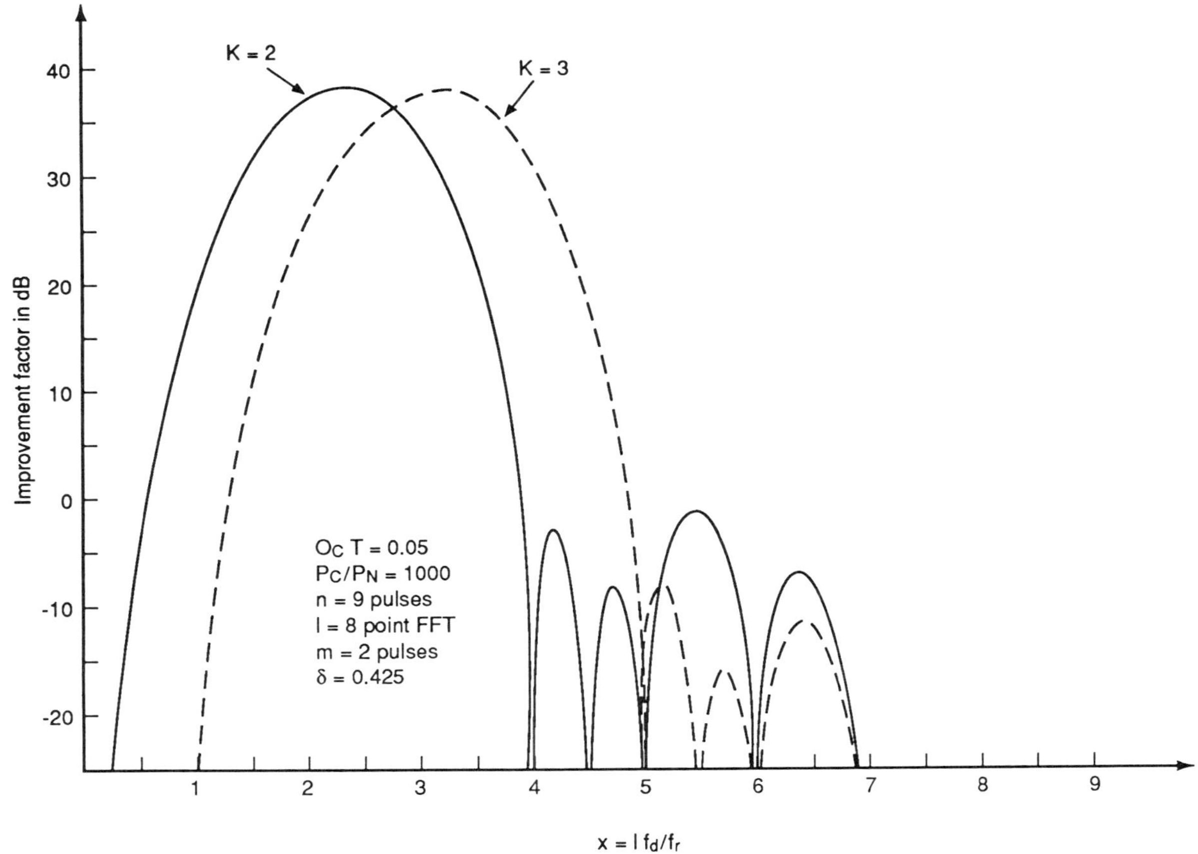

Figure 5.24 Improvement factor for single-canceller MTI-coherent integrator with Hamming weighting.

Table 5.3
Performance Comparison of Doppler Processors

Doppler Processor	*Average Improvement Factor between Cross-Overs*
Point-Optimized Transversal Filter	38.03 dB
MTI-Coherent Integrator FFT	37.63 dB
MTI-Weighted Coherent Integrator	37.61 dB

Note: $n = 9$, $m = 2$, $l = 8$, $\alpha = .426$, $\sigma_c T = .05$, $F = 1000$.

filter provides an advantage of approximately 0.4 dB for the situation considered, whereas the Hamming weighting results in minimal loss of performance.

The significance of the 0.4 dB loss in improvement factor can be determined by using (5.45), which converts the improvement factor into detection performance on the basis of probability of detection. Equation (5.45) is applicable for any transversal filter in a Gaussian-distribution interference. Because we have seen that the cascaded MTI-FFT has a transversal filter representation (see (5.106)) the applicability of (5.45) is established. As an example, if the transversal-filter doppler processor provides a $P_d = 0.9$ with a $P_{fa} = 10^{-6}$, the MTI-coherent integrator provides a $P_d = 0.83$ for the same parameters.

5.3 CASCADED MTI-INCOHERENT INTEGRATOR

A doppler processor used extensively in practice is an MTI followed by an incoherent integrator. The significant feature of this processor, as compared to the cascaded MTI-coherent integrator described in the previous section, is a considerable reduction in hardware. However, the processor can have severe detection losses compared to the doppler processors previously described, the performance of which approached optimal limits.

The detection loss associated with the cascaded MTI-incoherent integrator is due to the number of pulses processed in the MTI and the residual clutter that passes through the MTI. Initially, let us consider the effect of receiver noise on the processor, as this establishes the maximum performance that can be obtained. Also available for this case is an approximate analysis that allows a simple calculation of detection performance [2,24].

The cascaded MTI-incoherent integrator is depicted in Figure 5.25. This implementation uses a quadrature MTI (see Figure 5.9), although a single-channel

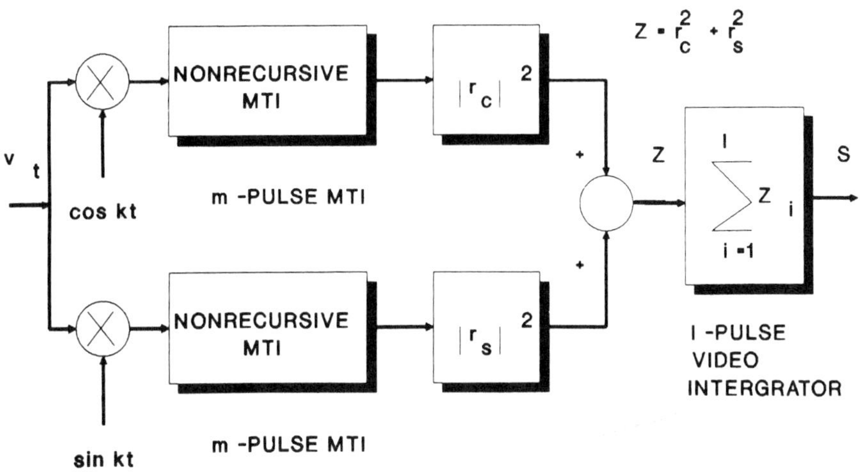

Figure 5.25 Cascaded MTI-incoherent integrator.

MTI can be used with additional detection loss [25]. The MTI correlates the receiver noise, which has the effect of reducing the number of independent noise samples available for integration. This effect reduces the integration gain of the integrator which is proportional to the number of independent samples processed.

An approximate solution can be obtained by assuming that the detected output (z) of the MTI depicted in Figure 5.25 has a Gaussian distribution [2]. The output of the integrator then has a Gaussian distribution due to its linear nature. The detection performance of the processor is then determined by the S/N at the output of the integrator.

To emphasize the linear nature of the integrator, it is depicted in Figure 5.26 as a transversal filter with weights $\mathbf{a}^T = (1, 1, 1, \ldots, 1)$ which is excited by the

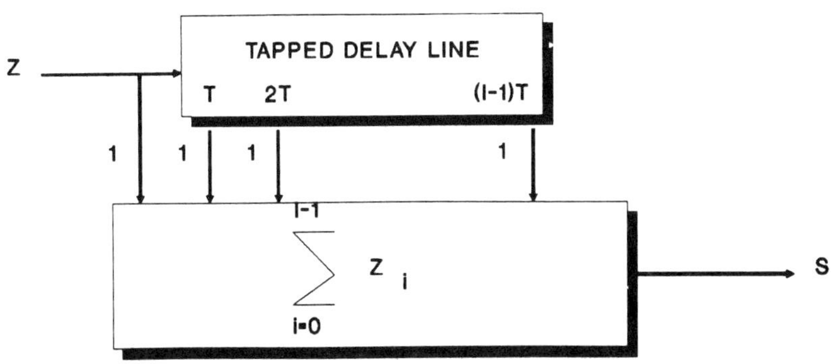

Figure 5.26 Transversal filter integrator.

Gaussian-distributed vector $\mathbf{z}^T = (z_0, z_1, \ldots, z_{l-1})$. The S/N at the integrator's output can be determined by applying the relationships for the transversal filter developed in Section 5.1.

The signal power output of the transversal filter integrator can be found by applying (5.5) with normalized covariance matrix (see (5.48) with $\omega_d = 0$):

$$\mathbf{M}_s = \begin{bmatrix} 1 & \cdots & 1 \\ \vdots & & \vdots \\ 1 & \cdots & 1 \end{bmatrix} \tag{5.115}$$

whereas the noise power output is found by using (5.3) with normalized covariance matrix (see (5.67)):

$$\mathbf{R}_n = \begin{bmatrix} 1 & \rho_1^2 & \cdots & \rho_{l-1}^2 \\ \rho_1^2 & \ddots & \ddots & \vdots \\ \rho_{l-1}^2 & \cdots & \cdots & 1 \end{bmatrix} \tag{5.116}$$

where $\rho^2(j)$, $j = 1, 2, l - 1$, are the correlation coefficients of the squared output of the MTI. The factor by which the integrator improves the output S/N is given by

$$n_e = \frac{l^2}{l + 2\sum_{j=1}^{l-1}(l-j)\rho^2(j)} \tag{5.117}$$

where

$$\rho(j) = \frac{\sum_{i=0}^{m-1}\sum_{k=0}^{m-1} w_i w_k \delta(k - i + j)}{\sum_{i=0}^{m-1} w_i^2}; \quad \delta(0) = 1 \tag{5.118}$$

and $\mathbf{w}^T = (w_0, w_1, \ldots, w_{m-1})$ is the MTI weight vector. The factor n_e is called the *equivalent number of independent samples* because it determines the integration gain in an equivalent manner, just as if n_e independent samples were applied to the integrator. The integration loss for the approximate Gaussian statistics is given by [4]

$$L = 10\log(l/n_e)^{1/2} \tag{5.119}$$

If we examine (5.117) for the extreme condition of uncorrelated samples ($\rho = 0$), we find that $n_e = l$, which is appropriate for independent Gaussian samples. At the other extreme of completely correlated samples ($\rho = 1$), we find, as expected, that $n_e = 1$ and the integrator has no effect. For a single canceller $\mathbf{w} = (1, -1)$, $n_e/l \approx 2/3$($L = 0.88$ dB), whereas for a double canceller $\mathbf{w} = (1, -2, 1)$, $n_e/l \approx 18/35$($L = 1.444$ dB) [2].

The approximate analysis using Gaussian statistics can be expected to provide accurate results only when l is large, which causes the input distribution to the integrator to approach a Gaussian distribution. Table 5.4 lists the integration loss for m = 2-, 3-, 4-, and 5-pulse MTI binomial cancellers obtained by using importance-sampling simulation techniques for a linear envelope detector averaged for l = 4, 8, 16, and 32 pulses [4]. Also shown is the approximate loss computed by an n_e calculation using standard detection curves with identical parameters.

Table 5.4
MTI Noise Integration Loss ($P_d = 0.9$, $P_{fa} = 10^{-6}$)

MTI Pulses	*Average Integration Loss* (dB)	*Loss Calculated Using n_e* (dB)
2	1.0	1.1
3	1.8	1.8
4	2.2	2.4
5	2.5	2.8

The approximate analysis just described indicates that the correlating effect of the MTI on receiver noise causes a detection loss when using video integration: the noise increases with the number of pulses processed by the MTI. However, the function of the MTI is to allow detection of targets embedded in clutter, and a complete analysis must consider the effect of an interfering background that consists of both receiver noise and clutter. An exact analysis of this problem is available [5,26,27] and will follow.

The square-law detected, video-integrated output of the cascaded MTI-incoherent integrator depicted in Figure 5.25 is given by

$$s = \sum_{i=0}^{l-1} z_i = \sum_{i=0}^{l-1} r_{ci}^2 + r_{si}^2 \tag{5.120}$$

This problem is identical in form to the analysis developed in Appendix B for the detection of a steady target in receiver noise, except that in this case the noise is correlated due to both the MTI action and the clutter component of the interfering background. The quadrature output vectors of the MTI, $\mathbf{r}_c = (r_{c0}, r_{c1}, \ldots, r_{cl-1})$ and $\mathbf{r}_s = (r_{s0}, r_{s1}, \ldots, r_{sl-1})$, are Gaussian distributed because they are the result of a linear process on a Gaussian vector. As is well known [28], the correlated Gaussian vectors, $\mathbf{r}_c$ and $\mathbf{r}_s$ can always be represented by an equivalent set of uncorrelated components obtained through a transformation that diagonalizes the covariance matrix. Application of the uncorrelated representation of vectors $\mathbf{r}_c$ and $\mathbf{r}_s$ to (5.120) determines that the random variable s has a noncentral chi-squared distribution when a signal is present and a chi-squared distribution for the case of no signal [29]. Application of the statistics of the integrator output (s) in a standard probability of detection calculation allows determination of the detection performance of the cascaded MTI-incoherent integrator.

The $2l$-dimensional Gaussian probability density function (see equation (A.57)) of the quadrature channel MTI is given by

$$p(\mathbf{r}_c, \mathbf{r}_s) = \frac{\exp\{-1/2[(\mathbf{r}_c - \mathbf{s}_x)^T\mathbf{C}^{-1}(\mathbf{r}_c - \mathbf{s}_x) + (\mathbf{r}_s - \mathbf{s}_y)^T\mathbf{C}^{-1}(\mathbf{r}_s - \mathbf{s}_y)]\}}{(2\pi)^l|\mathbf{C}|} \tag{5.121}$$

where the quadrature mean vectors $\mathbf{s}_x^T = (s_{x0}, s_{x1}, \ldots, s_{xl-1})$ and $\mathbf{s}_y^T = (s_{y0}, s_{y1}, \ldots, s_{yl-1})$ are given by (5.27) and (5.28) (with $P_s = A^2/2$ and $\phi = 0$):

$$s_{xk} = A \sum_{j=0}^{m-1} w_j \cos[\omega_d(t_k - jT)] \tag{5.122}$$

$$s_{yk} = A \sum_{j=0}^{m-1} w_j \sin[\omega_d(t_k - jT)] \tag{5.123}$$

The l by l covariance matrix $\mathbf{C}$ is given by equations (A.59), (A.60), and (A.61) of Appendix A:

$$\mathbf{C} = \begin{bmatrix} C_{11} & \cdots & C_{1l} \\ \vdots & \ddots & \vdots \\ C_{l1} & \cdots & C_{ll} \end{bmatrix} \tag{5.124}$$

with elements:

$$c_{kq} = \sum_{i=0}^{m-1} \sum_{j=0}^{m-1} w_i w_j E[v_{ci+k-1} v_{cj+q-1}] \tag{5.125}$$

where $\mathbf{w}^T = (w_0, w_1, \ldots, w_{m-1})$ is the MTI weight vector and $\mathbf{v}_c = (v_{c0}, v_{c1}, \ldots, v_{cl+m-2})$, or, equivalently, $\mathbf{v}_s = (v_{s0}, v_{s1}, \ldots, v_{sl+m-2})$ is the quadrature MTI input interference vectors.

The key step in the development is to rotate the coordinate system by using a linear transformation (**L**) such that **C** has elements only on the main diagonal. The rotated MTI quadrature outputs then become $\mathbf{r}_c' = \mathbf{L}\mathbf{r}_c$ and $\mathbf{r}_s' = \mathbf{L}\mathbf{r}_c$ where **L** is formed by the eigenvectors $\mathbf{e}_k^T = (e_{k1}, \ldots, e_{kl})$ of **C**:

$$\mathbf{L} = \begin{bmatrix} e_{11} & e_{21} & \cdots & e_{l1} \\ \vdots & \vdots & & \vdots \\ e_{1l} & e_{2l} & \cdots & e_{ll} \end{bmatrix} \tag{5.126}$$

and the diagonal matrix ($\boldsymbol{\lambda}$) has elements corresponding to the eigenvalues (λ_k) of **C** [28]:

$$\boldsymbol{\lambda} = \begin{bmatrix} \lambda_1 & 0 & 0 \\ 0 & \ddots & 0 \\ 0 & 0 & \lambda_l \end{bmatrix} \tag{5.127}$$

The elements of equations (5.126) and (5.127) are found by solving the matrix equation:

$$(\mathbf{C} - \lambda_k \mathbf{I})\mathbf{e}_k = \mathbf{0}; \quad k = 1, 2, \ldots, l \tag{5.128}$$

and can be checked by using

$$\boldsymbol{\lambda} = \mathbf{L}^T \mathbf{C} \mathbf{L} \tag{5.129}$$

The quadrature mean vectors at the MTI ouput are also rotated by the transformation, resulting in $\mathbf{s}_x' = \mathbf{L}\mathbf{s}_x$ and $\mathbf{s}_y' = \mathbf{L}\mathbf{s}_x$.

Equation (5.121) can now be written in terms of the rotated coordinate system as

$$p(\mathbf{r}_c', \mathbf{r}_s') = \frac{\exp\left[-1/2 \sum_{k=1}^{l} \dfrac{(r_{ck-1}' - s_{xk-1}')^2 + (r_{sk-1}' - s_{yk-1}')^2}{\lambda_k}\right]}{(2\pi)^l \prod_{k=1}^{l} \lambda_k} \tag{5.130}$$

where the rotation has decorrelated the quadrature MTI output vectors, which in turn has made the components independent; and hence the overall probability density function is simply the product of the individual component density functions. The characteristic function of the square-law detected, video-integrator output (s) of the cascaded processor can then be found from equation (5.130) by using standard techniques (see [29] for derivation) as

$$M(p) = \prod_{k=1}^{l} \exp\left(-\frac{p g_k^2}{1 + 2p\lambda_k}\right) \bigg/ (1 + 2p\lambda_k) \tag{5.131}$$

where $p = -j\omega$, and g_k^2 are the projections of the signal vector onto the kth eigenvector:

$$g_k^2 = \sum_{i=1}^{l} \sum_{j=1}^{l} (s_{xi-1} s_{xj-1} + s_{yi-1} s_{yj-1}) e_{ki} e_{kj} \tag{5.132}$$

Detection thresholds can be found by transforming (5.132) with $g_k = 0$ to obtain the no-signal probability density function ($s \geq 0$):

$$p(s) = \sum_{k=1}^{l} C_k \exp(-s/2\lambda_k)/2\lambda_k \tag{5.133}$$

where

$$C_k = \prod_{\substack{j=1 \\ j \neq k}}^{l} \left(\frac{\lambda_k}{\lambda_k - \lambda_j} \right) \tag{5.134}$$

Integration of (5.133) to find the false alarm probability in terms of the detection thresholds (s_T) results in [5,26,27]

$$P_{fa} = \sum_{k=1}^{l} \left\{ \prod_{\substack{j=1 \\ j \neq k}}^{l} \left(\frac{\lambda_k}{\lambda_k - \lambda_j} \right) \right\} \exp(-s_T/2\lambda_k) \tag{5.135}$$

Normalized detection thresholds for a single canceller MTI ($m = 2$) and an eight-pulse video integrator ($l = 8$) are plotted in Figure 5.27 for various Gaussian-shaped clutter spectra ($\sigma_c T = 0.01, 0.05, 0.1$) and clutter-to-receiver noise powers ($F = 0$, 5000, 10,000, 20,000).

The procedure for finding the probability of detection is more complex than that for finding the detection thresholds. The difficulty arises because direct inversion of the characteristic function (5.131) to obtain the probability density function is apparently not available in a closed form. Therefore, a numerical solution must be employed [5,27]. The most straightforward numerical solution uses the Gram-Charlier series approach.

The cumulant generating function can be defined as [30]

$$h(p) = \ln M(p) \tag{5.136}$$

The cumulants are then obtained by evaluating

$$\chi_k = (-1)^k \frac{\partial^k}{\partial p^k} [h(p)]\big|_{p=0} \tag{5.137}$$

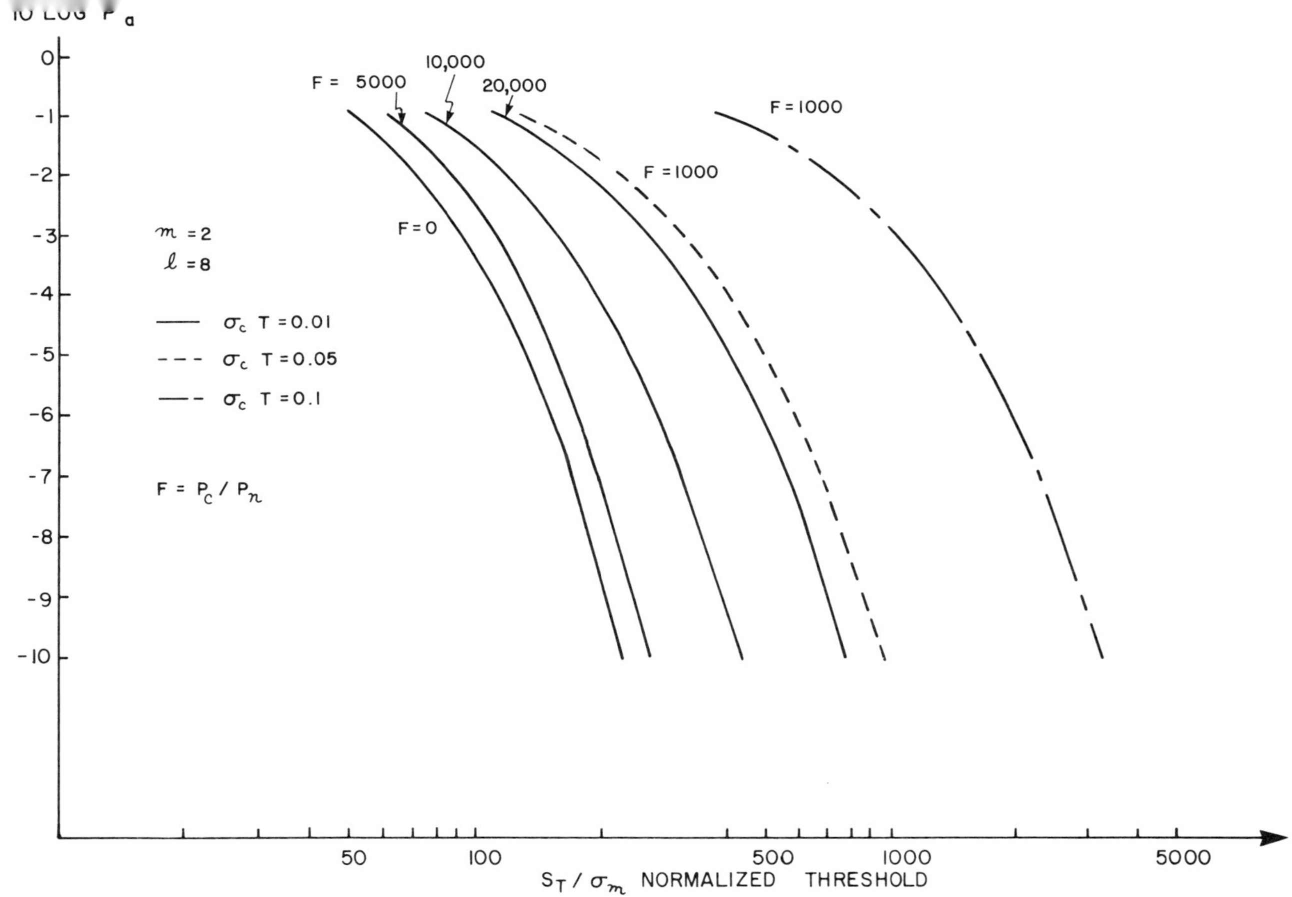

Figure 5.27 Normalized thresholds for cascaded MTI-incoherent integrator.

A convenient way to evaluate the cumulants is to express (5.131) as

$$M(p) = \prod_{k=1}^{l} \mathrm{e}^{-s_k}\, \mathrm{e}^{\alpha_k s_k/p+\alpha_k} \cdot \frac{\alpha_k}{p + \alpha_k}$$

where $s_k = g_k^2/2\lambda_k$ and $\alpha_k = 1/2\ \lambda_k$. The cumulant generating function thus becomes

$$h(p) = \sum_{k=1}^{l} C_k + \frac{\alpha_k s_k}{p + \alpha_k} - \ln(p + \alpha_k) \tag{5.139}$$

The cumulants hence become

$$\chi_i = \sum_{k=1}^{l} (2\lambda_k)^{i-1}(i - 1)! \quad [ig_k^2 + 2\lambda_k] \tag{5.140}$$

The cumulants are now used to obtain a Gram-Charlier series [28] for the detection probabilities of the cascaded MTI-incoherent integrator. A three-term series with Edgeworth grouping was found to provide good accuracy for the case of a single-canceller in cascade with an eight-pulse incoherent integrator [5].

The general procedure for obtaining the detection characteristics of the cascaded MTI-incoherent integrator is approached as follows:

- Form the covariance matrix using equation (5.125), which is a function of the MTI order (m), the number of pulses incoherently integrated (l), and the spectral characteristics of the clutter ($\sigma_c T$);
- Evaluate the eigenvalues (λ_k) and eigenvectors ($\mathbf{e}_k^T$) associated with the covariance matrix ($\mathbf{C}$);
- Find the detection thresholds from (5.135), which involve the eigenvalues (λ_k), the desired false alarm probability (P_{fa}), and the clutter-to-receiver noise power ratio (F);
- Evaluate the transformed signal magnitude g_k^2 from (5.132), which involves the eigenvectors ($\mathbf{e}_k^T$), the signal amplitude (A), the doppler frequency (ω_d), and the MTI weights ($\mathbf{w}^T$). Use (5.122) and (5.123) to evaluate the quadrature MTI signal vectors $\mathbf{s}_x^T$ and $\mathbf{s}_y^T$;
- Use equation (5.140) to evaluate the processor cumulants from g_k^2 and λ_k and then substitute the cumulants into a Gram-Charlier series to evaluate detection probabilities.

The evaluation of the detection probabilities of the cascaded MTI-incoherent integrator is best accomplished by using a computer program generated from this procedure. A computer library program that generates eigenvalues and eigenvectors from a covariance matrix input is a very useful tool in implementing a computer program for the procedure. Before giving results of detection calculations for the MTI-incoherent integrator processor, we will present several illustrations of calculations involving these equations in addition to a special case for which a closed-form solution of eigenvalues and eigenvectors is available.

Consider the case where a single-canceller quadrature channel MTI ($m = 2$) is followed by a four-pulse video integrator ($l = 4$). The covariance matrix is given by

$$\mathbf{C} = \sigma_n^2 \begin{bmatrix} 2 + 2F(1 - \rho_1) & -1 + F(2\rho_1 - \rho_2 - 1) & F(2\rho_2 - \rho_1 - \rho_3) & F(2\rho_3 - \rho_2 - \rho_4) \\ -1 + F(2\rho_1 - \rho_2 - 1) & 2 + 2F(1 - \rho_1) & -1 + F(2\rho_1 - \rho_2 - 1) & F(2\rho_2 - \rho_1 - \rho_3) \\ F(2\rho_2 - \rho_1 - \rho_3) & -1 + F(2\rho_1 - \rho_2 - 1) & 2 + 2F(1 - \rho_1) & -1 + F(2\rho_1 - \rho_2 - 1) \\ F(2\rho_3 - \rho_2 - \rho_4) & F(2\rho_2 - \rho_1 - \rho_3) & -1 + F(2\rho_1 - \rho_2 - 1) & 2 + 2F(1 - \rho_1) \end{bmatrix} \tag{5.141}$$

where σ_n^2 is the receiver noise power, $\rho(i)$ denote the clutter correlation coefficients, and F is the clutter-to-receiver noise power ratio. The covariance matrix is simplified when no clutter is present ($F = 0$):

$$\mathbf{C} = \sigma_n^2 \begin{bmatrix} 2 & -1 & 0 & 0 \\ -1 & 2 & -1 & 0 \\ 0 & -1 & 2 & -1 \\ 0 & 0 & -1 & 2 \end{bmatrix} \tag{5.142}$$

For this case, an analytic solution is available for the eigenvalues and eigenvectors associated with the covariance matrix [26,31]:

$$\lambda_k = 4 \sin^2 \frac{k\pi}{2(l + 1)} \tag{5.143}$$

$$e_{kj} = \sqrt{\frac{2}{l + 1}} \sin \frac{kj\pi}{l + 1}; \quad j = 1, 2, \ldots, l \tag{5.144}$$

The quadrature mean vectors from (5.122) and (5.123) are

$$s_{xk} = -2A \sin\left(\frac{\omega_d T}{2}\right) \sin\left[\omega_d T\left(k - \frac{1}{2}\right)\right] \tag{5.145}$$

$$s_{yk} = 2A \sin\left(\frac{\omega_d T}{2}\right) \cos\left[\omega_d T\left(k - \frac{1}{2}\right)\right] \tag{5.146}$$

The signal amplitude from (5.132) then is

$$g_k^2 = \sum_{i=1}^{l} \sum_{j=1}^{l} 4A^2 \sin^2\left(\frac{\omega_d T}{2}\right) \cos[\omega_d T(i - j)] e_{ki} e_{kj} \tag{5.147}$$

Detection curves for $\omega_d T = 1/4$ are evaluated in [27].

Figures 5.28, 5.29, and 5.30 depict detection curves for a cascaded single-canceller MTI and an eight-pulse video integrator ($\omega_d T = 1/4$) with narrow ($\sigma_c T = 0.01$), medium ($\sigma_c T = 0.05$), and wide ($\sigma_c T = 0.1$) Gaussian-shaped clutter spectra [5]. Also shown for comparison purposes are detection curves for an optimal nine-pulse transversal filter and a quasi-optimum cascaded single-canceller MTI–eight-pulse coherent integrator.

In Figure 5.28, for a narrow clutter spectrum ($\sigma_c T = 0.01$), the MTI improvement factor (see Table 5.1) is 20,000 (43 dB). The detection loss for the optimal transversal filter is on the order of 0.2 dB between no clutter ($F = 0$) and a clutter-to-noise power ratio of $F = 20{,}000$. The detection performance in heavy clutter ($F = 20{,}000$) of the MTI-coherent integrator at maximum response is essentially that of an eight-pulse coherent integrator, indicating that the processor was effective in removing the clutter. The effect of the MTI and clutter on the incoherent integrator results in a loss of 1.1 dB with noise alone and 4.3 dB when the clutter residual power equals the noise power.

In Figure 5.29, the detection performance is shown for a medium clutter spectrum ($\sigma_c T = 0.05$). The single-canceller MTI improvement factor for this case is 29 dB. The detection loss of the optimal transversal filter is on the order of 0.4 dB between no clutter ($F = 0$) and a clutter-to-noise power ratio of $F = 1000$. The peak detection performance of the cascaded MTI-coherent integrator is within the order of 0.15 dB of an eight-pulse coherent integrator in noise, which indicates good performance in clutter. However, at the cross over between adjacent doppler filters the composite performance (considering detection probabilities in both filters) is almost −2 dB below the peak. Use of Hamming weighting of the coherent integrator response considerably smooths the overall doppler response, but the overall performance for $F = 1000$ is uniformly on the order of −1.4 dB below that of an ideal eight-pulse coherent integrator in noise. The detection loss of the MTI-incoherent integrator is

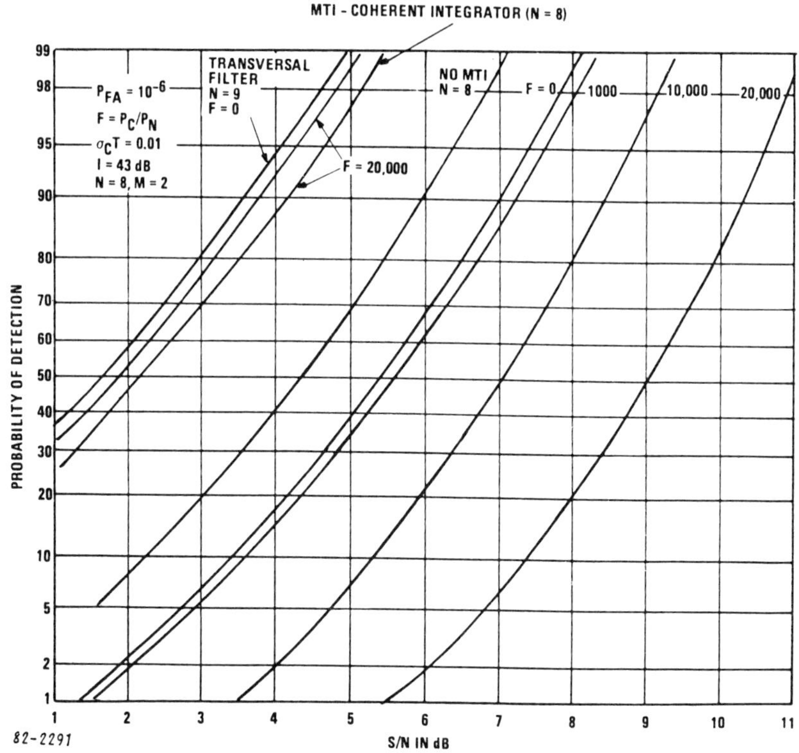

Figure 5.28 Probability of detection for a single-canceller quadrature MTI with video integration ($N = 8$).

again 1.1 dB for noise alone and 3.5 dB when residual clutter power is equal to noise power.

Figure 5.30 depicts the detection performance for a wide clutter spectrum ($\sigma_c T = 0.1$). The single-canceller MTI improvement factor for this case is 23 dB. Performance of the optimum transversal filter again is good over strong clutter ($F = 1000$) with the reduction on the order of 0.8 dB with respect to noise-only performance ($F = 0$). The performance of the single-canceller coherent integrator for $F = 1000$ is poor, being on the order of -2.7 dB below that of an eight-pulse coherent integrator in noise, which indicates that a better MTI filter is required. Use of a double canceller, $\mathbf{w}^T = (1, -2, 1)$, instead of a single canceller, reduces the

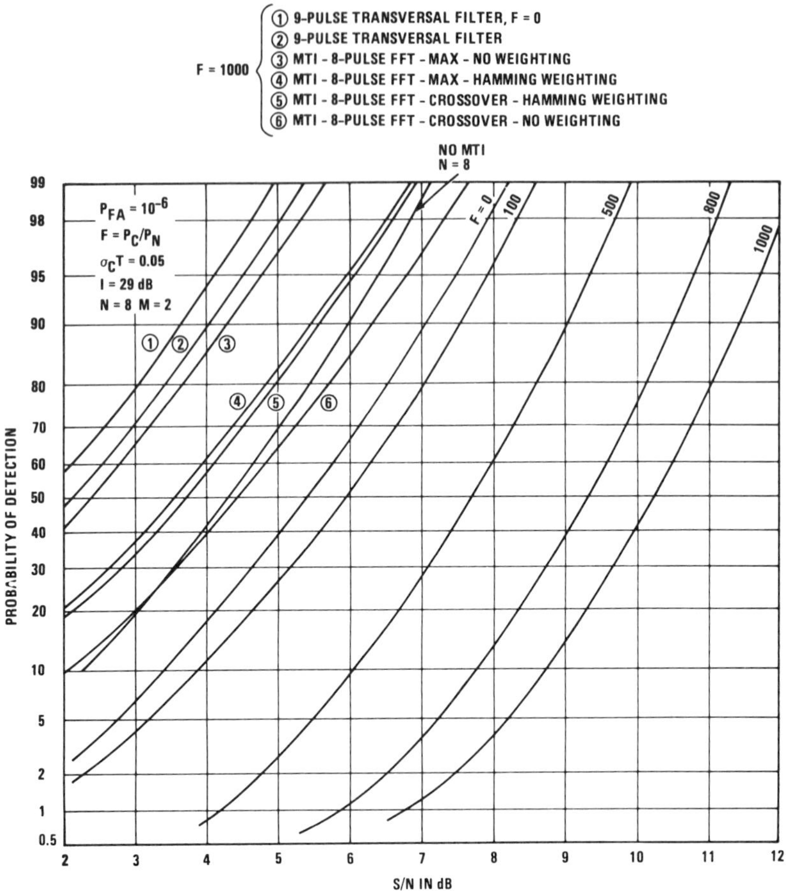

Figure 5.29 Probability of detection for a single-canceller quadrature MTI with video integration (N = 8).

detection loss for F = 1000 to within 0.5 dB of an eight-pulse coherent integrator in noise. Detection loss for the MTI-incoherent integrator ranges from 1.1 dB for noise alone to 7.5 dB for F = 1000. This loss is plotted in Figure 5.31 as a function of residual clutter power (F/I). The curve indicates that residual clutter power must be made significantly less than receiver noise power if good detection performance is to be realized by using the MTI-incoherent integrator doppler processor.

The performance comparison in Figures 5.28 through 5.30 can be summarized as follows:

- The optimal complex weight transversal filters provide the best detection performance, especially in heavy clutter environments;

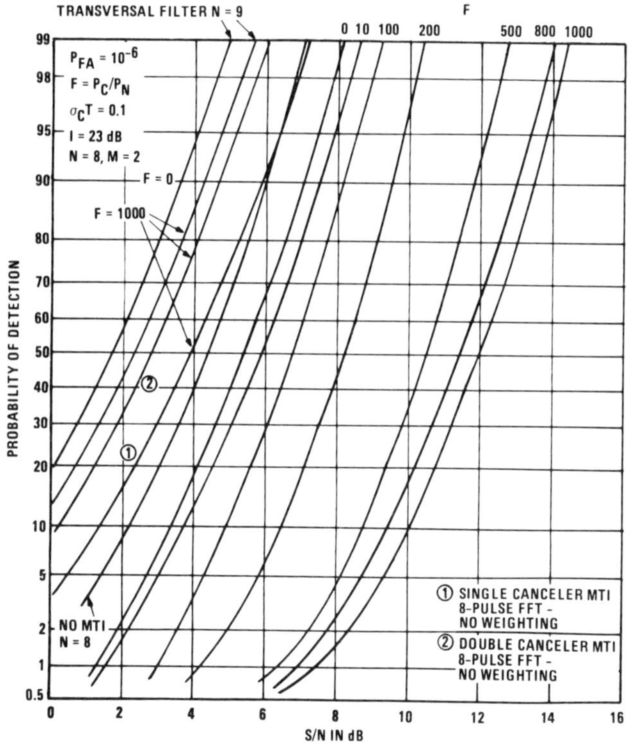

Figure 5.30 Probability of detection for a single-canceller quadrature MTI with video integration ($N = 8$).

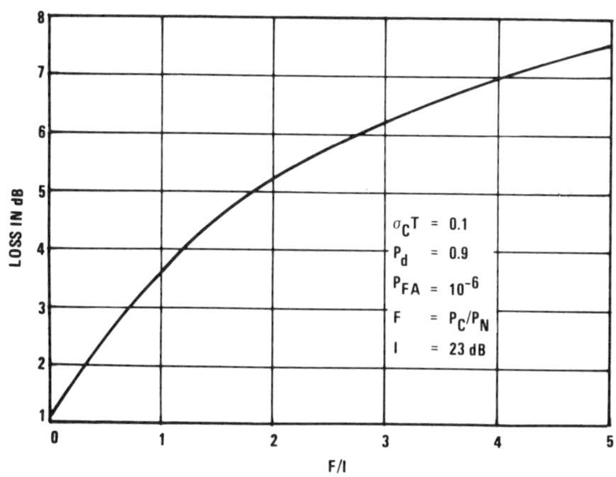

Figure 5.31 Video integration loss with a single-canceller MTI.

- The performance of cascaded MTI-coherent integrators is inherently somewhat less than that of transversal filters for an equal number of processed pulses;
- Coherent integrator weighting is required to provide a uniform overall doppler processing response;
- Higher order MTI filters in combination with coherent integrators are required for wide clutter spectra to provide good detection performance;
- MTI-incoherent integrator doppler processors are inherently lossy. For a single-canceller MTI with eight-pulse incoherent integration, losses range from 1.1 dB for noise-only to 3.5–4 dB when residual clutter equals noise and to over 7.5 dB when residual clutter is 5 times the noise power.

REFERENCES

1. J. DiFranco and W. Rubin, *Radar Detection,* Artech House, Norwood, MA, 1980.
2. W. Hall and H. Ward, "Signal-to-Noise Loss in Moving Target Indicator," *IEEE Proc.*, Vol. 56, No. 2, February 1968.
3. G. Trunk, "MTI Integration Loss," *IEEE Proc.*, Vol. 65, No. 11, November 1977.
4. G. Trunk, "MTI Noise Integration Loss," NRL Rep. 8132, July 1977.
5. D.C. Schleher, "Performance Comparison of MTI and Coherent Doppler Processors," *IEE Int. Radar Conf.*, London, November 1982.
6. D.C. Schleher, ed., *MTI Radar,* Artech House, Norwood, MA, 1978.
7. R. Emerson, "Some Pulsed Doppler MTI and AMTI Techniques," Rand Corporation Rep. R-274, March 1954.
8. J. Capon, "Optimal Weighting Functions for the Detection of Sampled Signals in Noise," *IEEE Trans.*, Vol. IT-10, No. 2, April 1964.
9. G. Andrews, "Optimum Radar Doppler Filtering Techniques," NRL Rep. No. 7727, May 1974.
10. H. Van Trees, *Detection, Estimation, and Modulation Theory, Part III,* John Wiley and Sons, New York, 1971.
11. F. Ayres, *Theory and Problems of Matrices,* Schaum Publishing, New York, 1962.
12. R. Bellman, *Introduction to Matrix Analysis,* McGraw-Hill, New York, 1970.
13. D.C. Schleher and D. Schulkind, "Optimization of Nonrecursive MTI," *IEE Int. Radar Conf.*, London, October 1977.
14. G. Dantzig, *Linear Programming and Extensions,* Princeton University Press, Princeton, NJ, 1963.
15. G. Andrews, "Comparison of Radar Doppler Filtering Techniques," NRL Rep. No. 7811, October 1974.
16. R. McAulay, "A Theory of Optimal Moving Target Indicator (MTI) Digital Signal Processing," Supplement 1, MIT Lincoln Laboratory Rep., Lexington, MA, October 31, 1972.
17. G. Andrews, "Performance of Cascaded MTI and Coherent Integration Filters in a Clutter Environment," NRL Rep. No. 7533, March 1973.
18. G. Dillard, "Signal-to-Noise Ratio Loss in an MTI Cascaded with Coherent Integration Filters," *IEEE Radar Conf.*, Washington, DC, April 1975.
19. E. Brigham and R. Murrow, "The Fast Fourier Transform," *IEEE Spectrum,* Vol. 4, No. 12, December 1967.
20. G. Bergland, "A Guided Tour of the Fast Fourier Transform," *IEEE Spectrum,* Vol. 6, No. 7, July 1969.

21. A. Oppenheim and R. Schafer, *Digital Signal Processing,* Prentice-Hall, Englewood Cliffs, NJ, 1975.
22. L. Rabiner and B. Gold, *Theory and Application of Digital Signal Processing,* Prentice-Hall, Englewood Cliffs, NJ, 1975.
23. M. Schwarz and L. Shaw, *Signal Processing: Discrete Spectral Analysis, Detection and Estimation,* McGraw-Hill, 1975.
24. F. Nathanson, *Radar Design Principles,* 1st Ed., McGraw-Hill, New York, 1969.
25. M. Weiss and I. Gertner, "Loss in Single-Channel MTI with Post-Detection Integration," *IEEE Trans.,* Vol. AES-18, No. 2, March 1982.
26. I. Kanter, "A Generalization of the Detection Theory of Swerling," *IEEE EASCON Conf.,* Washington, DC, September 1974.
27. G. Dillard and J. Richard, "Performance of an MTI Followed by Incoherent Integration for Nonfluctuating Signals," *IEEE Int. Radar Conf.,* Washington, DC, April 1980.
28. H. Cramer, *Mathematical Methods of Statistics,* Princeton University Press, Princeton, NJ, 1946.
29. A. Whalen, *Detection of Signals in Noise,* Academic Press, New York, 1971.
30. M. Kendall and A. Stuart, *The Advanced Theory of Statistics,* Vol. 11, Hafner Publishing Co., New York, 1963.
31. A. Jain, "Fast Karhunen-Loeve Transform for a Class of Random Processes," *IEEE Trans.,* Vol. COM-24, September 1976.

Chapter 6
MTI Systems

Functionally, an MTI radar allows the detection of targets imbedded in clutter. The basic principle involves filtering the undesired clutter return by exploiting the differential doppler frequency shift that exists when there is relative radial motion between the target scatterers and the clutter scatterers. The precise meaning of the term *MTI radar* is somewhat ubiquitous, but is generally defined as follows.

First, an MTI radar generally operates at a sufficiently low-PRF so that unambiguous range measurements are possible. At microwave frequencies, doppler measurements are highly ambiguous, creating so-called blind speeds, which occur in small regions about the radial target velocities corresponding to multiples of the radar's PRF. The objective of the MTI is to provide a comb filter response in the doppler frequency domain that ideally has a stopband in regions of heavy clutter concentration and passes unattenuated all other target velocities. As shown in Section 5.1.3, this comb filter is the optimal linear filter when all target doppler frequencies are assumed equally likely.

In general, the objective of the MTI processor is to remove the clutter, thereby improving the signal-to-clutter ratio. However, it does not improve the single-hit signal-to-noise ratio. As described in Sections 5.2 and 5.3, it is possible to follow the MTI with either a coherent or incoherent (after detection) integrator. The MTI-incoherent integrator can either improve or deteriorate the single-hit S/N ratio, depending upon the situation and must be applied with caution.

6.1 MTI CONFIGURATIONS

A classic form of a ground-based surveillance radar type of coherent MTI is depicted in Figure 6.1. The coherent transmission is formed by mixing a stable local oscillator reference source with a coherent oscillator at an *intermediate frequency* (IF). The transmission is then at the up-converted sum of the STALO and COHO frequencies. The received target signals are down-converted to IF by mixing with the STALO,

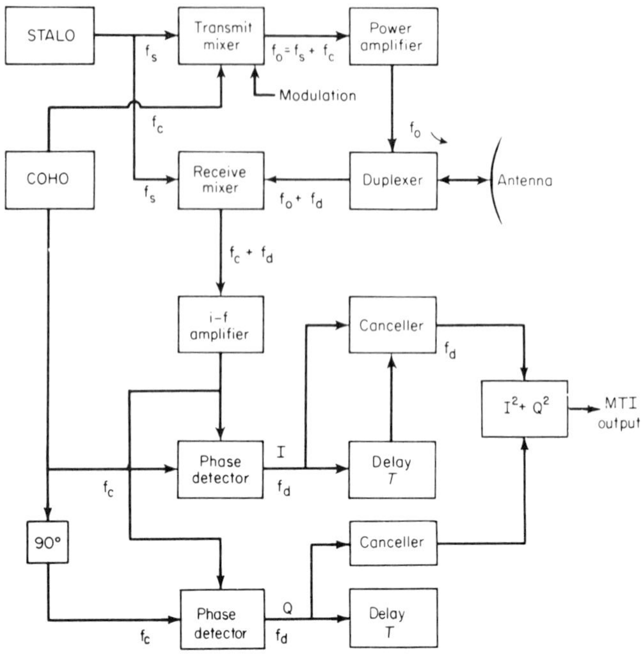

Figure 6.1 Ground-based coherent MTI system.

where they are further converted into baseband signals by comparison with the COHO in a phase detector. The output of the phase detector is a bipolar video signal at the target's doppler frequency (see Section 2.3.1).

The ground and weather clutter spectrum for this coherent ground-based MTI is depicted in Figure 6.2. Note that the ground clutter spectral lobes (which are spread primarily by scanning modulation) appear at multiples of the radar's PRF. This is due to the sampled data aspect of the pulse radar waveform (see Appendix A.3). Weather clutter, which in general has a nonzero radial velocity, is shown between the PRF lines with greater spectral width (due to wind shear) than for ground clutter (see Figure 1.15). The range of target doppler frequencies usually encompasses multiple PRF regions, thereby causing the target doppler to be ambiguous. When the target doppler falls into those regions occupied by the ground clutter lobes, the MTI cannot detect the target. This is the blind speed effect in MTI systems, where the MTI is unable to detect targets with radial velocities that create doppler frequencies corresponding to multiples of the radar's PRF. Also note that the weather clutter spectrum is also spread into multiple PRF regions, thereby potentially causing target responses with higher ambiguous radial speeds to be aliased by the weather clutter.

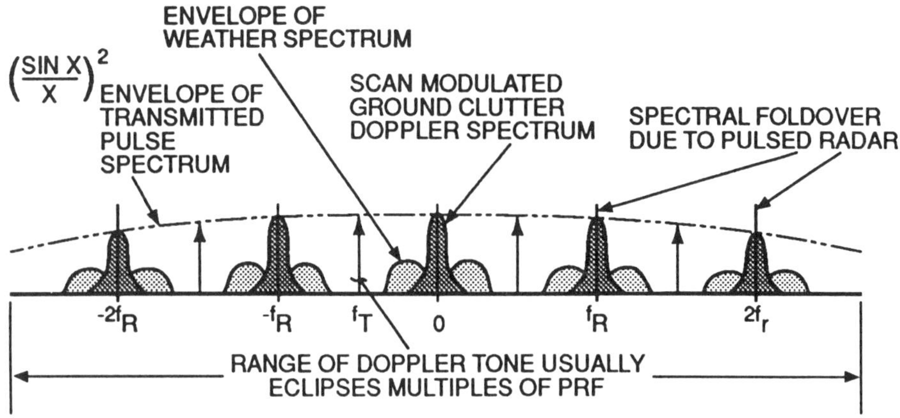

Figure 6.2 Ground and weather clutter spectrum for a coherent ground-based MTI.

A quadrature type of MTI canceller that employs in-phase and quadrature phase detector channels is shown in Figure 6.1. This type of MTI processor preserves the coherent nature of the signals and conforms to the structure of the optimum MTI described in Section 5.1.3. An alternative configuration that uses a single phase detector is sometimes employed, particularly in older designs. With this type of design, both positive and negative frequencies are folded into a band between zero frequency and half the PRF ($f_r/2$). This results in the blind phase phenomenon where signals in quadrature with the COHO reference are rejected.

The use of a single-channel coherent MTI results in a sensitivity loss against receiver noise as compared to fully coherent MTI systems that process both in-phase and quadrature channels. The most severe loss occurs for nonfluctuating targets with high detection probabilities when no postdetection integration is used (see Section 2.3.1). The asymptotic loss using postdetection integration with a large number of pulses approaches 1.5 dB. The number of degrees of freedom for a fluctuating target is reduced by a factor of two, which increases the fluctuation loss when compared to fully coherent processing.

The function of the MTI canceller is to attenuate the clutter returns by applying a comblike filter response, as depicted in Figure 6.3 for single- and double-cancellation units. The type of response shown has deep nulls about the PRF lines and hence is appropriate for use against ground clutter. In general, the greater the number of pulses processed in the MTI canceller, the wider the rejection notch and the greater the clutter attenuation (see Figures 2.25 and 3.5). Also note that in a ground-based surveillance radar, the width of the ground clutter lobe is determined primarily by antenna scanning modulation (see Section 3.2.2). Thus, the slower the antenna scans, the narrower is the clutter lobe and the greater is the clutter attenuation for a given canceller complexity (i.e., number of pulses processed).

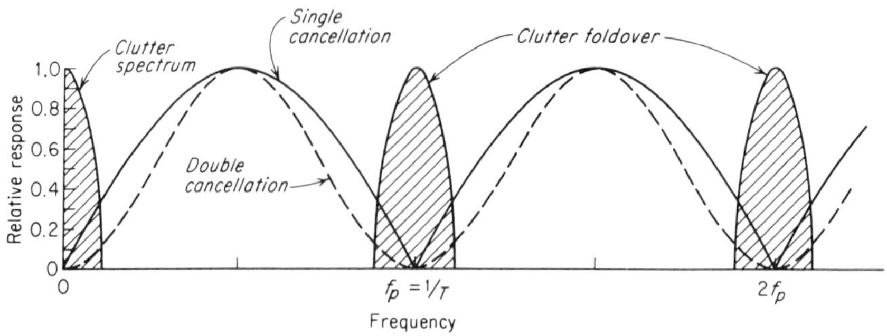

Figure 6.3 MTI filter response for single and double cancellers.

A second type of coherent MTI applicable for radars using power oscillator type transmitters (e.g., magnetrons) functions by phase-locking the COHO to each transmitter pulse (see Figure 2.21). The COHO can be locked by synchronizing its phase to the transmitter phase for each transmitted pulse. Alternatively, a continuously oscillating COHO can be utilized, and the phase of the transmitted pulse may be measured and used to compensate the phase detector output, as depicted in Figure 6.4 [1]. The latter approach is generally preferred in digital MTI implementations (using a digital phase shifter) because a continuously running COHO is generally more stable than one which must be continuously synchronized.

The coherent-on-receive MTI is incapable of cancelling multiple-time-around clutter returns that occur at ranges greater than those corresponding to a basic pulse repetition interval. Thus this type of MTI is not appropriate to situations where multiple-time-around clutter returns are anticipated (i.e., from cities, mountains, *et cetera*). When this occurs, the fully coherent (see Figure 6.1) MTI must be used. Also, if PRF staggering is employed to compensate for blind speeds, block-to-block staggering must be used rather than pulse-to-pulse staggering to cancel ambiguous clutter [2].

Frequently, an MTI must cope with clutter having spectra with a nonzero central doppler frequency. This occurs for fixed-based radars that experience wind-driven airborne clutter such as that associated with precipitation or chaff and also when sea clutter is present. Doppler frequency shift of the clutter also occurs when the MTI is located on a moving platform such as a naval ship or an airborne or space vehicle. In these cases, some form of reference frequency adjustment is required to move the null of the coherent MTI canceller over the clutter spectra. This problem is compounded when clutter spectra with different central frequencies occupy the same range resolution cell (e.g., weather clutter over ground clutter). Solutions to this problem include (1) cascading MTI cancellers with different null frequencies (e.g., see Section 1.1.3) and (2) following the MTI with a bank of doppler filters where the filters containing

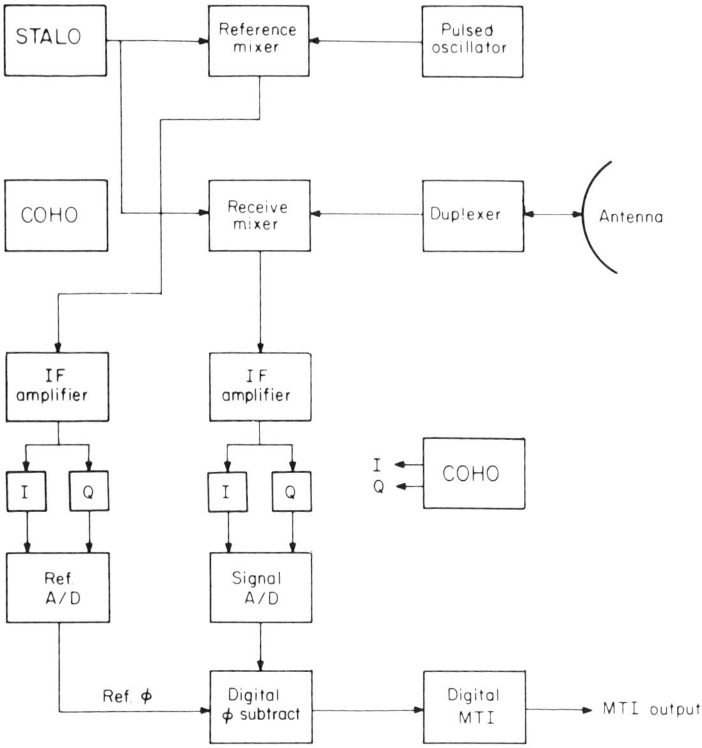

Figure 6.4 Coherent-on-receive MTI with digital phase storage.

moving clutter returns are desensitized, thereby removing the undesired clutter (see Section 1.4; the MTD).

Compensation for platform motion can be accomplished in the coherent MTI by offsetting the COHO's frequency as depicted in Figure 6.5. With this approach, the COHO frequency is mixed with the output of a compensating voltage controlled oscillator (VCO), the frequency of which is adjusted to bring the center frequency of the clutter spectra to zero. The difficulty of this method lies in deriving an accurate control signal for the VCO. In some applications (e.g., naval surface radar), we can derive an open-loop correction from the antenna's azimuth angle, the ship's true motion from its navigation system, and the antenna's pitch and roll motion from gyro sensors. However, in most applications we need an adaptive correction derived by locking the offset oscillator to the clutter's center frequency. In principle, the clutter locking must be accomplished in each range cell because the velocity vector from the radar to the clutter patch generally differs for each cell. However, in practice, we face the problem of finding an uncontaminated sample of the clutter signal to

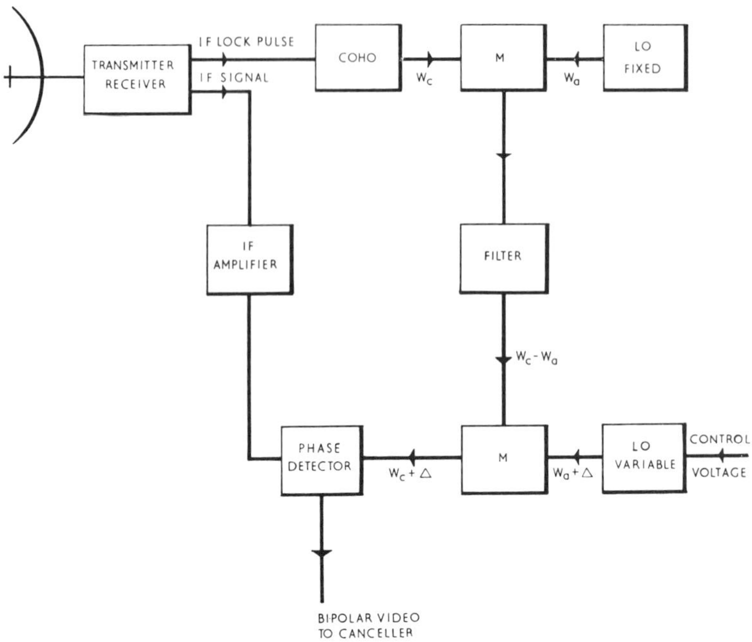

Figure 6.5 Velocity compensated MTI.

lock the frequency. This generally restricts the number of clutter-locking points in the range dimension to three or four, which provides adequate performance (see Section 3.2.7 and Figure 3.33) in most applications.

The time-averaged-clutter coherent airborne radar technique provides an architecture often used in clutter-locking situations. Figure 6.6 depicts the general form of this type of system (also see Figure 3.30) as applied to a coherent-on-receive MTI using an IF canceller. The chief feature of TACCAR is the use of a VCO in a phase-locked loop where the correction for the clutter doppler is obtained from the averaged measurement of the clutter doppler frequency within a sampled range interval.

TACCAR automatically centers the clutter rejection notch at the averaged doppler frequency of the clutter within the sampled range interval. TACCAR can be applied to cancel any fixed or moving clutter source and hence is capable of cancelling ground, sea, weather, and chaff clutter within any gated range interval for radars that are either fixed (e.g., land-based) or on a moving platform (e.g., naval or airborne). The system cannot, however, simultaneously cancel two different clutter sources within a particular range interval. For example, if weather clutter is superimposed on ground clutter return, the TACCAR cannot simultaneously cancel both signals.

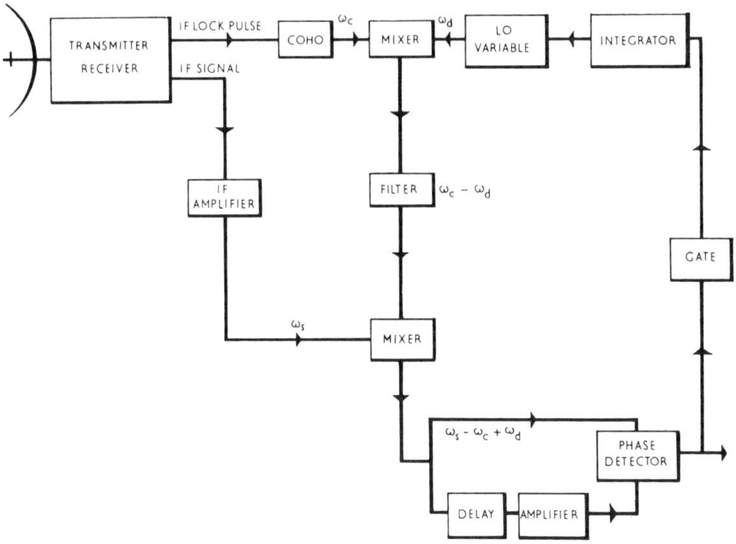

Figure 6.6 Clutter-locking MTI system.

The primary application of TACCAR is in *airborne MTI* (AMTI) or when MTI is used on a naval radar where in both cases the primary clutter component (i.e., land or sea clutter) is relatively narrow in spectral width and the doppler frequencies are to be cancelled one at a time. Another application [3] is used on a land-based MTI system where ground clutter is cancelled close in while weather and chaff clutter that occurs at different dopplers is cancelled at ranges beyond the ground clutter.

An alternative method for controlling the null frequency of a coherent MTI canceller is to insert a phase shifter in the direct channel as depicted in Figure 6.7. The phase shifter for a constant velocity target (v_t) will be adjusted to a value given by

$$\Delta\phi = 2v_t/\lambda f_r \tag{6.1}$$

where λ is the radar's wavelength and f_r is the radar's PRF. The phase shifter, by changing the phase of each radar return pulse by an amount $\Delta\phi$, effectively tunes the canceller null frequency to the average doppler frequency of the clutter.

The coherent MTI uses the phase fluctuations in the return signal to recognize the doppler component produced by a moving target (see Section 2.3.1). The operation of this type of MTI depends on a reference signal at the radar receiver that is coherent with the transmitter signal. Alternatively, a noncoherent MTI (see Section 2.3.1)

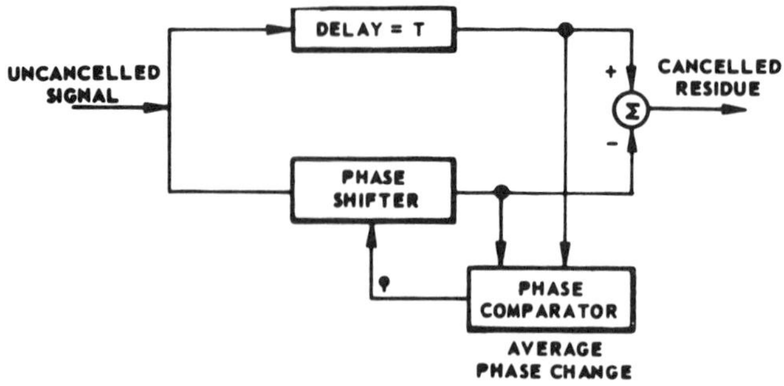

Figure 6.7 Phase-shifter clutter-locking MTI canceller.

that processes the envelope-detected target return signal can be employed. This type of MTI is also called a *clutter-referenced MTI* because the relative rotation of the target phasor with respect to the clutter phasor provides the essential action that allows this type of MTI to function (see Figure 2.16). Hence, for this MTI, the clutter return itself, in the vicinity of the target return, becomes the reference that replaces the function of the COHO in a coherent type MTI. A limitation of this technique is that moving target echoes in a region not containing clutter do not induce the characteristic modulation onto the received envelope at the target's doppler frequency, which allows the target to be detected. Because of this, the noncoherent MTI is sometimes described as being "blind in the clear." One solution is to use a clutter gate that switches the output from normal operation (no MTI) to MTI operation in the presence of clutter. The switching process is difficult to control properly when the clutter appears intermittently.

A block diagram of a typical noncoherent MTI system is depicted in Figure 6.8. The transmitter may be either a power amplifier or a pulsed oscillator, and a STALO is used in the receiver to translate the signal to IF. The stability of the transmitter, for a given MTI performance level, must be comparable to that of a coherent MTI, although the frequency stability of the STALO can be somewhat less. The IF signals are envelope detected in a *linear-logarithmic* (lin-log) receiver that provides a dynamic range compatible with that of the MTI canceller. Alternatively, *sensitivity time control* (STC) can be employed separately or in conjunction with the lin-log characteristic to compress the dynamic range of the system. The transition between the linear and logarithmic characteristics of the receiver is set equal to the expected value of the improvement factor of the MTI. In theory, for Rayleigh distributed clutter, the use of the logarithmic characteristics results in a constant residue, because the fluctuation component of the clutter (i.e., rms value) is proportional to its amplitude (i.e., mean value). However, when the clutter is significantly non-Gaussian or when significant stationary fixed targets (e.g., water towers, large fixed buildings) exist, the residue has significant fluctuations.

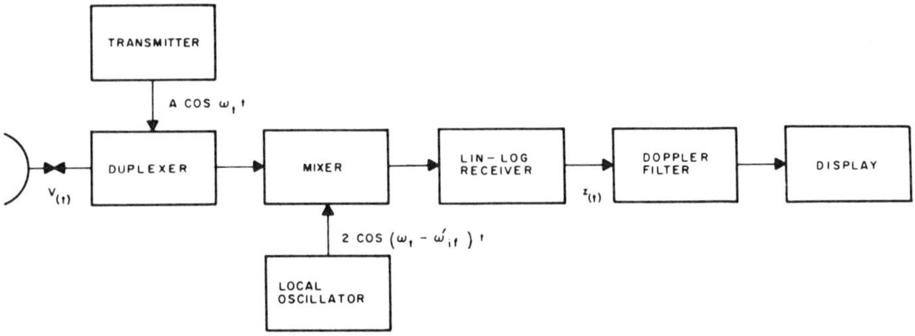

Figure 6.8 Noncoherent MTI.

The envelope-detected signal of the noncoherent MTI is then processed through a conventional single-channel MTI canceller (the quadrature channel of a coherent MTI is not present) that filters the doppler modulation produced by the beating of the target return and the clutter. Because this system depends on the beating of the target signal with the clutter signal, the difference in strength between the two component signals must be less than the subclutter visibility of the MTI for the system to function. For weak clutter signals, clutter gating partially overcomes this limitation.

The advantage of the noncoherent MTI is its simplicity (particularly that no motion compensation is required), which has led to its use in airborne applications where weight and size are at a premium. Its principal disadvantage with respect to a coherent MTI is its reduced performance (see Section 3.1.1), which is due in part to the lack of spectral purity of the clutter when used as a reference signal and also to the spatial patchiness of clutter, which causes gaps in the MTI's coverge. Other limitations on the performance of a noncoherent MTI involve

1. intermodulation products of the clutter and target spectra in the envelope detection, causing spectral spreading;
2. blind speed limitations due to the use of a single MTI channel;
3. the inability of cancelling multiple-time-around clutter if a power oscillator transmitter is used;
4. an increased sensitivity to MTI stability limitations that cause amplitude jitter.

The block diagram of a phase-sensitive noncoherent MTI that removes the requirement for clutter gating is depicted in Figure 6.9 [3]. In this configuration, the reference signal for the MTI is derived by hard limiting the returns (consisting of clutter or receiver noise) from the range cell adjacent to that containing the target. If clutter is present, it provides a zero doppler reference and target phase variation

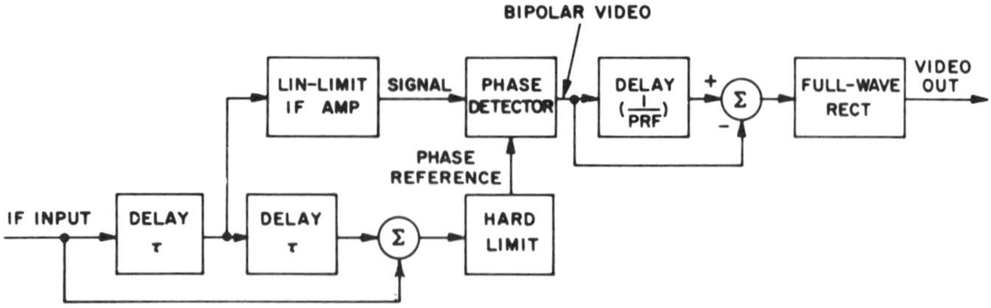

Figure 6.9 Phase-sensitive noncoherent MTI.

at the doppler frequency is detected and passed to the MTI canceller. If only noise is present, the target output appears with a random phase and passes through the MTI canceller.

The phase-sensitive noncoherent MTI provides outputs for moving targets in clutter and point targets in the clear. Otherwise, the same limitations that apply for the conventional noncoherent MTI also apply to the phase-sensitive noncoherent MTI. A further anomaly associated with this type of system is that the three signal paths contained within the MTI can stretch the target when detected over clutter, which can lead to uncertainty in determining the range to the target.

6.2 MTI PERFORMANCE CONSIDERATIONS

MTI performance is generally determined by two distinct factors. The first is related to the ability of the MTI to filter or attenuate the clutter return. This is determined by the number of return pulses (n) processed by the MTI and the width and shape of the clutter spectrum. The second factor involves the overall system stability, which sets an upper limit on MTI system performance. This is a function of the many factors that result in pulse-to-pulse noise jitter in the system with the principal one being transmitter frequency instability (see Section 3.1.2).

The clutter spectral width generally is a function of antenna scanning modulation, wind driven clutter motion, radar platform motion, and radar transmitter instability (see Section 3.2). For a conventional ground-based air surveillance radar, the ground clutter lobe's spectral width is determined primarily by scanning modulation. Hence, the filtering performance of a nonrecursive ground-based MTI filter can be related to the time the antenna dwells on the target and the number of pulses processed by the MTI (see Section 3.2.2). For an airborne radar, the ground clutter lobe's spectral width is determined primarily (particularly as the antenna scans off the aircraft's ground track) by the platform motion effect. This results in the airborne radar MTI filter's performance being determined by the aircraft's velocity, the azimuth scan angle, the antenna aperture in the azimuth dimension, and the number of pulses processed by the MTI (see Section 3.2.3). A further limitation imposed on the performance of an airborne MTI involves the doppler sidelobe returns within the passband of the MTI, which are proportional in magnitude to the two-way antenna sidelobe levels (see Section 3.2.3).

6.2.1 MTI Performance for Ground-Based Surveillance Radar

The detection performance of an MTI is generally related to the MTI improvement factor (I_c), which is defined as the ratio of the output signal-to-clutter ratio of the MTI to the input signal-to-clutter ratio (see Sections 3.1 and 5.1.3). For a conventional ground-based surveillance radar, using a coherent MTI with an n-pulse binomial canceller, where the scanning modulation limitation dominates, the improvement factor (see Figure 3.7) can be approximated in terms of the number of pulses per antenna 3 dB beamwidth (n_B) as

$$I_c = n_B^{2(n-1)}/(1.388)^{n-1}(n-1)! \quad (6.2)$$

For a single canceller ($n = 2$), (6.2) results in an improvement factor $I_c = n_B^2/1.388$, whereas for a double canceller, the result is $I_c = n_B^4/3.853$.

As an example, the ASR-9 radar ($\dot{\theta} = 12.5$ rev/min, $\theta_{3\text{ dB}} = 1.3°$, PRF = 1200 Hz, $n = 3$) results in $n_B = 20.8$ pulses per 3 dB beamwidth and an improvement factor of 46.9 dB.

This improvement factor is further reduced by the instabilities associated with the MTI system. To consider this case, the system instabilities are sometimes measured separately from the limitations caused by clutter spectral spreading resulting in an MTI *cancellation ratio* (CR) limitation. This can be combined (see Section 3.1.2) with the improvement factor limitation (called I_{ideal}) to result in an overall improvement factor (I_t) given by

$$I_t = 1/[(I_{\text{ideal}})^{-1} + (\text{CR})^{-1}] \quad (6.3)$$

As an example, the ASR-9 has a system instability specified as $CR = 60$ dB resulting in an overall improvement factor of $I_t = 46.7$ dB, which compares to a measured value of 45 dB [4].

6.2.2 MTI Performance for Airborne Surveillance Radar

The detection performance of low-PRF airborne MTI radar is generally dominated by platform motion considerations. The two primary effects are the spreading of the mainlobe ground clutter platform motion component as the radar antenna scans away from the clutter's ground track and the translation of antenna sidelobe ground clutter into the MTI's passband (see Figures 2.4 and 2.5).

Generally, the result of increasing the spectral width of the ground clutter's mainlobe component is to diminish the MTI's detection probability. Described in Section 2.2.1, this results from reducing the ratio of the clear region within which the MTI must operate as compared to the blind region that contains mainlobe clutter components spread by platform motion (see Figure 2.6). To examine the magnitude of this effect, consider the spectral spreading (Hz) for the mainlobe ground clutter platform motion component (see Section 3.2.3) given by

$$\sigma_{pm} = 0.528 v_{ac} \sin\theta_0 \cos\phi_0 / D \tag{6.4}$$

where v_{ac} is the aircraft's velocity (m/s), θ_0 is the azimuth angle with respect to the aircraft's ground track, ϕ_0 is the antenna's depression angle to the clutter patch, and D is the antenna's aperture width (m). A high-performance aircraft generally has a small antenna aperture and a large aircraft velocity, which tends to maximize the spectral spreading. The example given in Section 2.2.1 shows that for v_{ac} = 1500 knots and D = 2.5 ft, the PRF must be increased to 4000 Hz and the scan angle restricted to ±30° for equal blind and clear regions, which restricts the nonambiguous range to on the order of 20 nmi. The design of a long-range AMTI radar with 360° scan capability involves using a larger antenna aperture in a lower speed aircraft (see Figure 2.6).

The improvement factor for an n-pulse coherent binomial MTI canceller with a platform motion limitation is given by

$$I_c = 2^{n-1} / (3.318 v_{ac} T \sin\theta_0 \cos\phi_0 / D)^{2(n-1)} (n-1)! \tag{6.5}$$

where $T = 1/f_r$ is the radar's interpulse period. For a single canceller this equals $I_c = 1/5.5\ (v_{ac} T \sin\theta_0 \cos\phi_0 / D)^2$, and for a double canceller it equals $I_{c2} = 1/60.6\ (v_{ac} T \sin\theta_0 \cos\phi_0 / D)^4$. For the high-performance aircraft AMTI example given in Section 2.2.1 (v_{ac} = 1500 knots, f_r = 4000 Hz, θ_0 = 30°, ψ_0 = 0°, D = 2.5 ft), the improvement factor becomes 10.5 dB for a single canceller and 18.1 dB for a double canceller. This calculation illustrates the difficulty of achieving high detection performance with low-PRF AMTIs in a high-performance aircraft.

The severe limitations imposed by platform motion effects has led to the development of motion compensation techniques to provide increased performance in low-PRF AMTIs. The philosophy of these techniques is to remove the effects of platform motion before MTI processing. When this is accomplished, the limitations on MTI performance associated with fixed-based MTIs (i.e., antenna scanning modulation, transmitter instability) would also apply to AMTIs. However, we must recognize that, in the AMTI case, the returns from antenna sidelobes are aliased throughout the entire MTI frequency band, resulting in an improvement factor limitation equal to the ratio of the integrated energy entering the system through the mainlobe to that entering through the sidelobes for the two-way antenna pattern.

Motion compensation techniques are generally applied in the coherent RF or IF portion of the MTI receiver to correct the spectral translation and spreading before MTI processing. It is natural to follow the motion compensation with a coherent MTI for maximum detection performance. TACCAR is used to compensate the coherent MTI for the mean doppler shift induced by platform motion, and a displaced phase-center antenna is used to correct for platform motion spectral spreading (see Section 3.2.7).

For example, consider the case of a low-PRF AMTI radar designed to detect airborne targets over both water and land [2,5]. The pertinent parameters for the AEW radar are a PRF of 300 Hz, an aperture width of 25 ft, a platform velocity of 300 knots, and a transmission frequency at UHF (e.g., 425 MHz). Figure 6.10 illustrates the MTI improvement factor for an N-delay canceller ($N = n - 1$) as a function of the clutter spectral spread (σ_c – Hz) to the PRF (f_r – Hz). The system characteristics that limit the MTI improvement factor can be related to an effective clutter spectral spreading and to a shift in the mean of the spectrum. The total clutter spectral spread can be determined (assuming statistical independence) by taking the square root of the sum of the squares of all the individual spectral widths (see Section 3.2.8).

The achievable MTI improvement factor for an AMTI system is ultimately limited by the internal motion of the clutter itself, the system stability, and the integrated mainlobe-to-sidelobe ratio of the two-way antenna pattern. However, in practice the significant limitations arise from (1) platform motion, (2) antenna sidelobe levels, (3) antenna scanning modulation, and (4) system stability. The expected value for

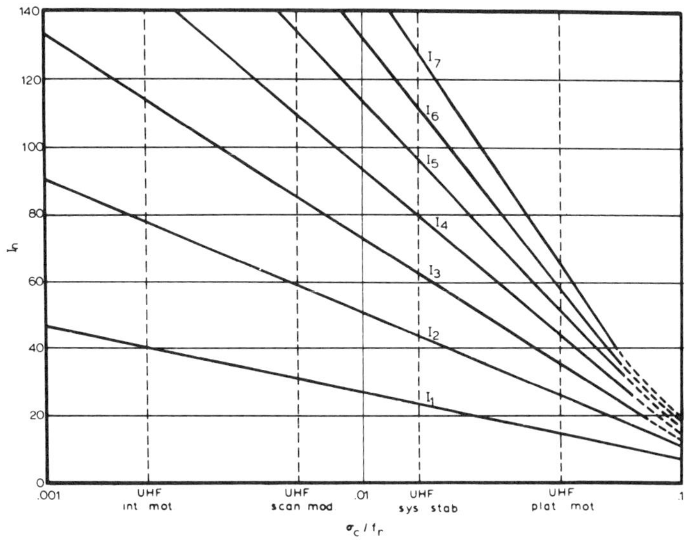

Figure 6.10 MTI improvement factor for an N-delay canceller with system limitations.

each of these components for an AEW radar are superimposed on Figure 6.10 to allow the relative importance of each component to be determined [2,5].

In Figure 6.10, a typical value for the spectral width due to internal clutter motion of $\sigma_v = 0.2$ m/s is used (see Section 3.2.1). At UHF, this results in an rms spectral spread of 0.57 Hz ($\sigma_f = 2\sigma_v/\lambda$) and a normalized value $\sigma_f/f_r = .002$. Figure 6.10 shows that the MTI improvement factor for this condition is 40 dB for a single canceller, 78 dB for a double canceller, and 114 dB for a triple canceller. If the system were limited by internal clutter motion, all practical requirements could be achieved easily by the use of a double canceller.

In general, platform motion can be resolved into two components, one parallel and one perpendicular to the antenna aperture (see Section 3.2.7). These components have the effect of both spreading the spectrum of the clutter returns (see (6.4)) and shifting the mean of the spectrum. Considering only the platform motion spreading effect, the rms value from (6.4) (v_{ac} = 300 knots, $\theta_0 = 90°$, $\phi_0 = 0°$, D = 25 ft) equals σ_{pm} = 10.7 Hz, which results in a normalized value $\sigma_{pm}/f_r = 0.0357$. Figure 6.10 shows that the MTI improvement factor is about 15 dB for a single canceller, 25 dB for a double canceller, and 35 dB for a triple canceller with the antenna aperture parallel to the velocity vector. For this case, many practical requirements cannot be met with single, double, or triple cancellers, thereby determining the requirement for motion compensation processing.

With the antenna aperture perpendicular to the velocity vector, the dominant effect is the shifting of the mean of the clutter spectrum. In general, the mean of the doppler spectrum is determined by the platform velocity, the transmitted frequency, and the angle to the clutter patch under consideration ($f_c = 2v_{ac} \cos\phi_0 \cos\psi_0/\lambda$). Motion compensation of this effect is generally required for successful MTI performance.

Because the antenna sidelobes essentially cover 360°, the doppler shifts received through the sidelobes range from $-v_{ac}$(v_{ac} is the aircraft's velocity) to $+v_{ac}$. This doppler spread can cover the entire MTI ambiguous frequency band. As a result, these sidelobe returns are not cancelled and the MTI improvement factor is limited by the ratio of the integrated energy entering the system through the mainlobe to that entering through the sidelobes for the two-way antenna pattern.

The effect of antenna scanning modulation is related to the antenna's scan rate and beamwidth ($\sigma_f = 0.265\, \dot{\theta}/\theta_B$). This results in an antenna scanning modulation spectral spread for the example ($\dot{\theta}$ = 12 rev/min, $\theta_B = 6.4°$) of σ_f = 1.5 Hz, which has a normalized value $\sigma_f/f_r = 0.005$. The scanning modulation spectral width is affected by the antenna scan rate, the antenna aperture, and the transmitted frequency, but generally most requirements can be met without compensation for this effect.

System stability is greatly simplified through the use of digital processing, which allows a single clock to control the complete system, including the MTI delays. However, the pulse-to-pulse phase stability of the transmitter is not eliminated by digital processing, so that the performance of the system is generally limited by the transmitter stability. For this type of system, the interpulse transmitter stability relative to the transmitter frequency is estimated at 10^{-8}, resulting in a normalized stability of $\Delta f / f_r = .0142$. The corresponding MTI improvement factor (see Figure 6.10) is 22 dB for a single canceller, 42 dB for a double canceller, and 62 dB for a triple canceller. Most requirements can be met with a stability of this magnitude.

The conclusion of the preceding discussion is that the factors limiting the improvement factor for this type of AMTI radar are the antenna sidelobes and platform motion. Motion correction techniques therefore are necessary for adequate performance in conjunction with a low-sidelobe antenna.

A TACCAR type of correction is generally required to correct for the mean shift in the doppler spectrum due to platform motion. This correction is made by phase-locking the MTI's COHO to the clutter's mean frequency. The time constant of this phase-locked loop must be large compared with the interpulse period of the radar to maintain the pulse-to-pulse coherency needed for successful MTI operation. This leads to a single correction for all ranges. However, for an aircraft flying above the earth's surface, the mean doppler shift to each clutter cell is different as it is proportional to the angle-of-arrival, which is a function of range. Hence, the TACCAR correction is generally incomplete, although this could be improved through the use of multiple correction circuits (see Section 3.2.7).

The effect of single-point TACCAR correction results in a smaller MTI improvement factor which is a function of the correction error and the clutter spectral spread (see Figures 3.31 and 3.32). For a typical set of parameters, these losses due to incomplete motion compensation when the antenna is pointed in the direction of the aircraft's velocity vector are depicted in Figure 6.11 [2,5].

Use of a displaced phase-center antenna is a technique that compensates for the component of velocity parallel to the antenna aperture. This is accomplished by either physically or electrically displacing the phase center of the antenna in a direction opposite to the velocity component of the aircraft (see Section 3.2.7). A single canceller can be almost perfectly compensated by DPCA with essentially no loss. For higher-order cancellers (e.g., double, triple, *et cetera*) only the first canceller stage is effectively compensated. This results in an imperfect compensation so that the effects of this platform motion component are not completely eliminated. Figure 6.12 illustrates for typical parameters the MTI improvement factor that includes the effect of DPCA losses for single, double, and triple cancellers. The MTI improvement factor for a single canceller with DPCA coincides with the perfectly compensated curve. The improvement factors for double and triple cancellers are much less than the perfectly compensated curves.

Figure 6.12 indicates that for the example system with DPCA and TACCAR motion compensation, the limiting factor involves the effect of system instabilities. If system stability can be increased to 10^{-9}, then scanning modulation becomes the system limitation. However, the UHF transmitter frequency selected for the radar has the effect of increasing the antenna sidelobes due to excitation of the aircraft's

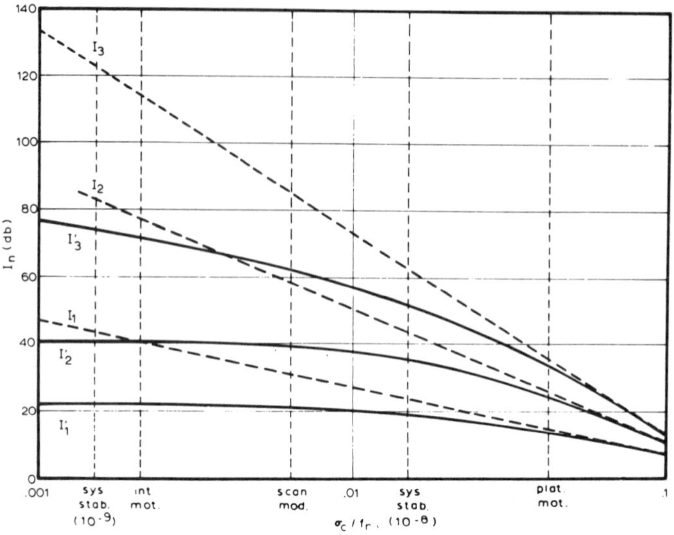

Figure 6.11 MTI improvement factor with TACCAR losses for an N-delay canceller at UHF.

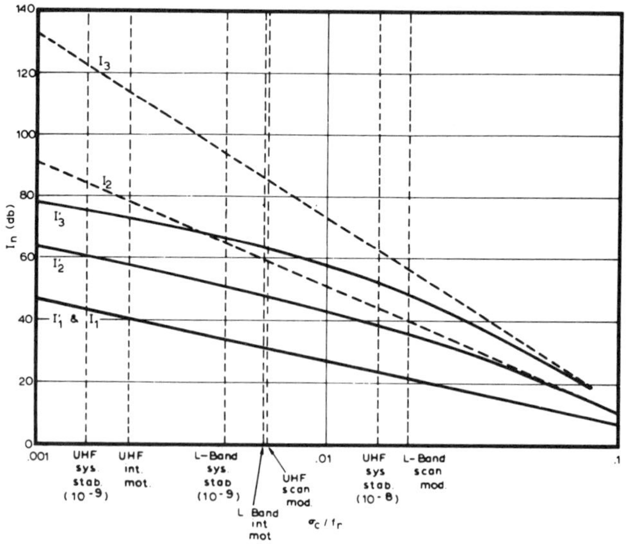

Figure 6.12 MTI improvement factor with DPCA losses for an N-delay canceller.

structure in this frequency range [2,5]. An increase to L-band (i.e., f = 1200 MHz) would substantially reduce the antenna sidelobe level. However, a change to L-band would result in several changes to the system performance limitations depicted in Figures 6.10, 6.11, and 6.12.

First, at an *L*-band frequency, the scanning modulation component spread would increase in direct proportion to the change in frequency from UHF (see Section 3.2.2). As depicted in Figure 6.12, this would then become the dominant limitation, assuming that the system stability could be maintained at 10^{-9}. Notice that the internal clutter motion limitation would also be increased at an L-band frequency. However, a more serious limitation would be the increased TACCAR losses that occur at the L-band frequency, as depicted in Figure 6.13. These occur due to the higher doppler shifts at the increased transmitter frequency. For a double canceller, these limitations are about 25 dB for scanning modulation and TACCAR losses (antenna at 90°). If the TACCAR losses and the DPCA losses could be eliminated, then the limitation to the MTI improvement factor caused by scanning modulation at L-band would be about 55 dB for a triple canceller (see Figure 6.13). Reduction of these losses consists of making TACCAR corrections at additional points in range and applying a DPCA correction to all cancellers [6,7].

To summarize, in order to insure adequate performance of the example long range low-PRF AMTI radar, the following system features are desirable [2,5]:

- Integrated mainlobe-to-sidelobe ratios for the two-way antenna pattern of the order of −60 dB;

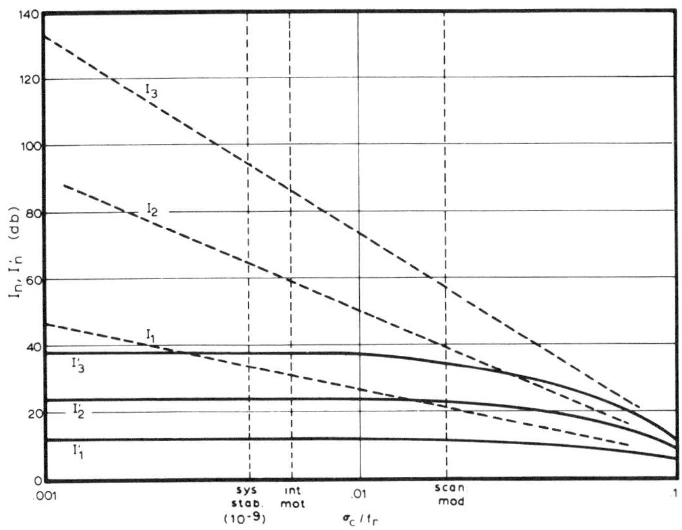

Figure 6.13 MTI improvement factor with TACCAR losses for an *N*-delay canceller at *L*-band.

- Pulse-to-pulse transmitter stability factor equal to 10^{-9} of the transmitted frequency;
- Triple canceller MTI processor;
- TACCAR correction at multiple range points (three minimum);
- DPCA correction of all three cancellers.

6.2.3 MTI Performance for a Nonlinear Receiver

An important requirement for a pulsed doppler radar receiver is to maintain linear operation over a wide dynamic range in order to minimize clutter spectral spreading (see Section 3.2.5). In general, the radar receiver is required to operate on very weak signals and convert them to a voltage sufficient to drive the doppler processor. Typically, about 100 dB of gain is required to convert weak signals at the receiver input to an acceptable voltage at the output. Because the range of input signals (including clutter and target returns) may also exceed 100 dB, if truly linear, the receiver must amplify a 100 dB range of input signals by 100 dB, which, in general, is an impossible task. Therefore, receivers are nonlinear in the sense that they use some technique (e.g., limiter, log amplifier, AGC, or STC) to compress the total range of input signals to a usable range at the output. Because the process of amplification and detection in the receiver always results in some degree of distortion of the input signal, this is described in terms of spurious signal levels, compression, or other appropriate terminology.

In Section 3.2.5, the receiver was modeled as a linear device followed by a limiter, which was described by an error function transfer function. The transfer function for this type of receiver is plotted in Figure 6.14 on a decibel scale and represents a severe limiting action that may be employed, for example, if the receiver is followed by a digital processor employing an A/D converter (i.e., the dynamic range is approximately 6B dB, where B is the number of bits in the A/D converter). However, in general, receivers do not exhibit this severe limiting action, and in this section we will develop MTI improvement factor relationships by traditional receiver distortion theory using intermodulation products characterized by third-order intercept points. An advantage of this approach is that standard methods are available to measure receiver nonlinearity, which then can be used to specify receiver characteristics in terms of achievable MTI improvement factors.

A general expression relating output to input for many electronic devices is a power series of the form:

$$e_0 = k_1 e_i + k_2 e_i^2 + k_3 e_i^3 + \ldots \tag{6.6}$$

This expression describes the transfer characteristics of virtually all electronic devices including mixers and amplifiers, which are the key elements of most receivers. When

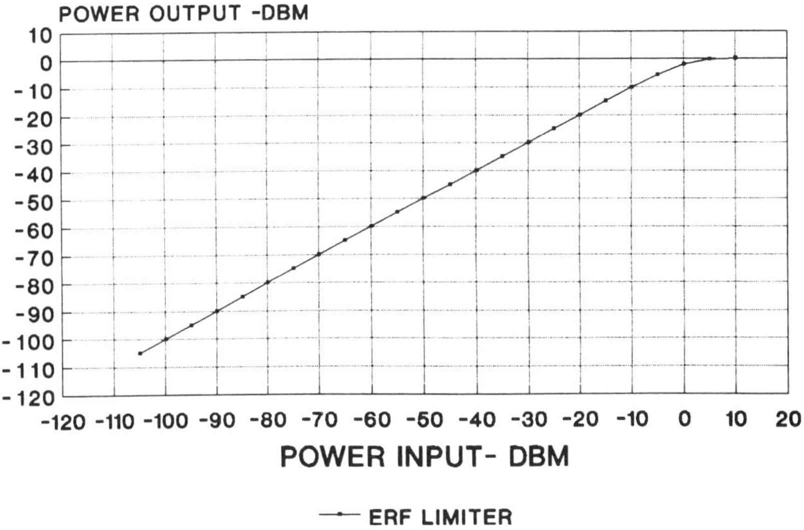

Figure 6.14 Nonlinear receiver with error function transfer characteristic.

the input signal contains multiple frequency components (i.e., clutter), the input can be represented as

$$e_i = e_1 \cos\omega_1 t + e_2 \cos\omega_2 t + e_3 \cos\omega_3 t + \ldots \tag{6.7}$$

When this signal is applied to the receiver represented by (6.6), the resulting output consists of a large number of intermodulation terms (see [8] for all terms through the fifth order). Examination of the intermodulation terms reveals that the amplitudes of fundamental terms (e.g., $e_0 = k_1 e_1 \cos\omega_1 t$) are directly proportional to the level of the input signal; the amplitudes of second-order responses (e.g., $e_0 = k_2 e_1 e_2 \cos(\omega_1 t - \omega_2 t)$) are proportional to the square of the input signal's amplitudes; and the amplitudes of third-order responses (e.g., $e_0 = k_3 e_1 e_2^2 \cos(2\omega_1 t - \omega_2 t)$) are proportional to the cube of the input signal's amplitudes [9].

A plot of these responses against the input signal level on a log-log graph will be straight lines with slopes corresponding to the order of the response, as depicted in Figure 6.15. The fundamental output will increase 1 dB for every dB that the input increases. The second-order response when stimulated with two equal amplitude input tones differing in frequency will increase 2 dB for every dB that the input tones are increased. The third-order response when stimulated with two equal-amplitude input tones differing in frequency will increase by 3 dB for every dB that the input tones are increased.

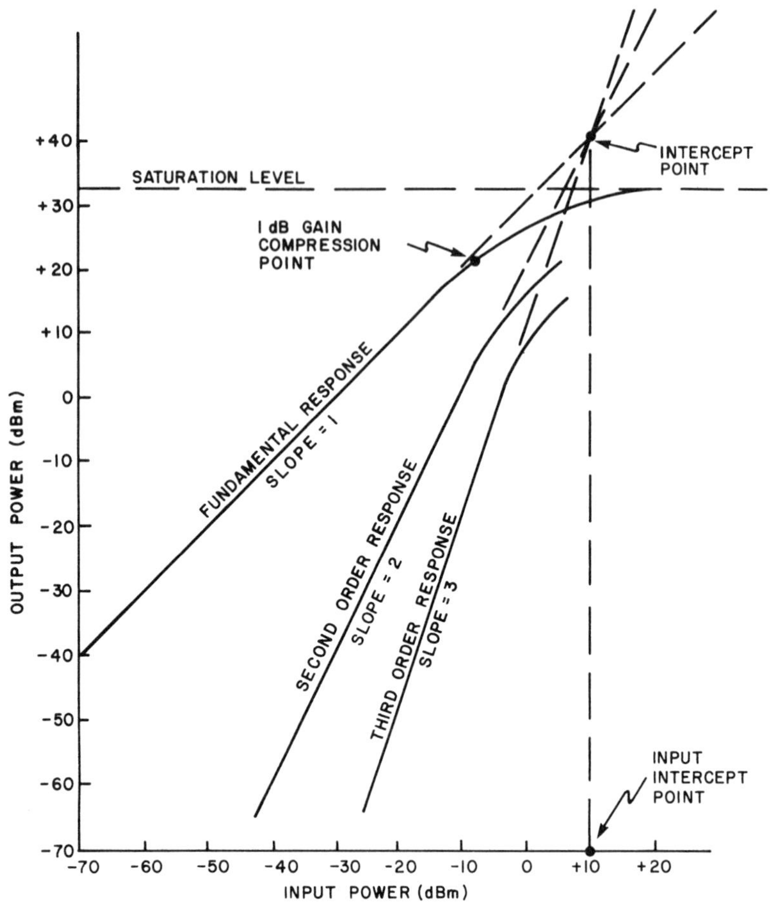

Figure 6.15 Nonlinear receiver second- and third-order intercept points.

The intersection of a projection of the straight-line portion of the second- or third-order response curves with a projection of the linear portion of the fundamental response curve is called an *intercept point*. The second- and third-order plots may or may not intercept the fundamental line at the same point, depending on system design. These characteristics (the known slopes of the response curves and the second- and third-order intercept points) can be used to determine the intermodulation response of the receiver.

In many practical receivers, even-order responses are suppressed using balanced designs (e.g., balanced mixers) resulting in domination by the third-order response. For this situation, the receiver characteristic is given by

$$e_0 = k_1 e_i + k_3 e_i^3 \tag{6.8}$$

where the power gain is given by $G = k_1^2$ and k_3 is the third-order distortion coefficient. The autocorrelation function at the receiver output is given by

$$R_0(\tau) = E\,[e_0(t)e_0(t + \tau)] \tag{6.9}$$

resulting in

$$R_0(\tau) = GR_i(\tau) + 6k_3^2R_i^3(\tau) \tag{6.10}$$

where $R_i(\tau)$ is the input ACF to the receiver. The normalized receiver output ACF can then be found from

$$\rho_0(\tau) = \frac{R_0(\tau)}{R_0(0)} \tag{6.11}$$

and assuming a Gaussian-shaped ACF at the receiver input given by

$$R_i(\tau) = P_c e^{-(\sigma_c\tau)^2/2} \tag{6.12}$$

where P_c is the clutter's power, σ_c (rad/s) is the clutter's spectral spread, and τ is the separation in time between samples. The resulting normalized ACF is then given by

$$\rho_0(\tau) = F_1\,e^{-(\sigma_c\tau)^2/2} + F_2\,e^{-3(\sigma_c\tau)^2/2} \tag{6.13}$$

where

$$F_1 = \frac{1}{1 + \gamma} \tag{6.14}$$

$$F_2 = \frac{\gamma}{1 + \gamma} \tag{6.15}$$

and where

$$\gamma = \frac{6k_3^2P_c^2}{G} \tag{6.16}$$

The third-order intercept point at the output can be found by equating the fundamental and third-order responses, resulting in

$$P_{IPO} = e_0^2 = \frac{4k_1^3}{3k_3} \tag{6.17}$$

If we solve for k_3 and substitute into (6.16), we have

$$\gamma = 10.7 \left(\frac{P_c G}{P_{\mathrm{IPO}}} \right)^2 \tag{6.18}$$

which expresses the nonlinear receiver's normalized ACF in terms of the ratio of the clutter power level to the third-order intercept power level, both at the receiver output.

The improvement factor for a single canceller is given by

$$I_1 = \frac{1}{1 - \rho_0(T)} \tag{6.19}$$

which results in

$$I_{\mathrm{NL1}} = \frac{1}{1 - F_1\, \mathrm{e}^{-(\sigma_c T)^2/2} - F_2\, \mathrm{e}^{-3(\sigma_c T)^2/2}} \tag{6.20}$$

For $\sigma_c T \ll 1$, the resulting MTI improvement factor can be reduced to

$$I_{\mathrm{NL1}} = \frac{I_1}{F_1 + 3F_2} \tag{6.21}$$

where $I_1 = 2/(\sigma T)^2$ is the MTI improvement factor for a single-canceller MTI using a linear receiver. Similarly, the MTI improvement factor for a nonlinear receiver and a double canceller can be found to be

$$I_{\mathrm{NL2}} = \frac{I_2}{F_1 + 9F_2} \tag{6.22}$$

where $I_2 = 2/(\sigma_c T)^4$. Extension to a triple canceller results in

$$I_{\mathrm{NL3}} = \frac{I_3}{F_1 + 27F_2} \tag{6.23}$$

where $I_3 = 4/3(\sigma_c T)^6$.

Examination of (6.18) indicates that, as the third-order intercept point is made large compared to the clutter power, $\gamma \rightarrow 0$ causing the nonlinear receiver MTI improvement factors to approach those achieved with a linear receiver. Conversely, when the clutter power is large with respect to the third-order intercept point (i.e., the receiver is saturated), $F_1 \rightarrow 0$ and $F_2 \rightarrow 1$, resulting in a loss in MTI improvement factor of 4.8 dB for the single canceller, 9.5 dB for the double canceller, and 14.3 dB for the triple canceller.

In summary, the receiver nonlinear characteristic can be determined through measurement (and specification) of the intercept points of various orders. These can then be related to the MTI improvement factor in terms of the relative levels of clutter power to the intercept points for different orders of MTI cancellers and arbitrary clutter spectral shapes. In the preceding section, we have developed the appropriate relationships for a typical case where third-order distortion dominates the receiver characteristic and the clutter spectrum exhibits a Gaussian shape.

6.3 MTI RADAR FILTER DESIGN

MTI filters can be either recursive or nonrecursive. Nonrecursive filters often are preferred because of their superior transient response.

6.3.1 Nonrecursive MTI Filters

The nonrecursive or transversal filter, as shown in Figure 6.16, consists of a tapped delay line, where the output is formed by the addition of weighted signals from each tap (see Sections 2.3.2 and 5.1.1). The impulse response of such a filter is given by

$$h(t) = \sum_{k=0}^{n-1} w_k \delta(t - kT) \tag{6.24}$$

where n is the number of pulses processed, w_k are the filter weights and $\delta(\cdot)$ is an impulse function. The Laplace transform of the impulse response that provides the

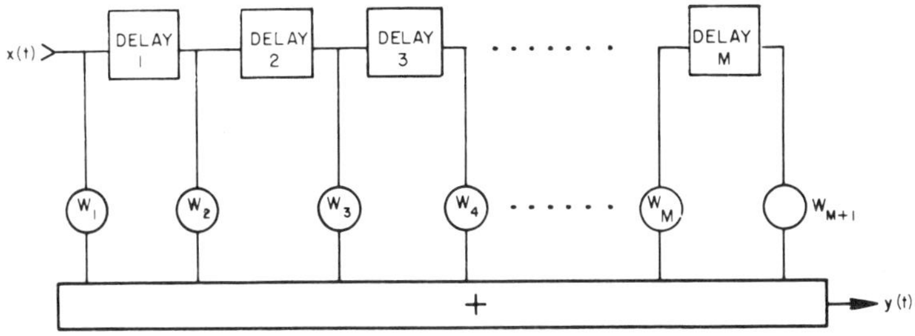

Figure 6.16 MTI transversal or nonrecursive filter.

filter's transfer function is given by

$$H(p) = \sum_{k=0}^{n-1} w_k \, \mathrm{e}^{-kpT} \tag{6.25}$$

where p is a complex variable and T is the radar's interpulse period. The z-transform (see Appendix C) is given by $z = \exp(pT)$, which results in

$$H(z) = \sum_{k=0}^{n-1} w_k z^{-k} = \frac{\sum_{k=0}^{n-1} w_k z^{n-1-k}}{z^{n-1}} \tag{6.26}$$

indicating that the nonrecursive MTI has $n - 1$ poles (equal to the number $N = n - 1$ of delay sections) at the origin in the z-plane.

The simplest form of transversal filter is the single canceller (see Figure 2.18) consisting of a single delay-and-subtract function with weights $w = (1, -1)$ and $n = 2$. The transfer function of the single-canceller MTI (6.26) is

$$H(z) = \frac{z - 1}{z} \tag{6.27}$$

indicating a pole at the origin and a zero located on the unit circle at $z = 1$, as depicted in Figure 6.17. A cascade of $N = n - 1$ single cancellers, as depicted in Figure 6.18, has a transfer function given by

$$H(z) = \left(\frac{z - 1}{z}\right)^{n-1} \tag{6.28}$$

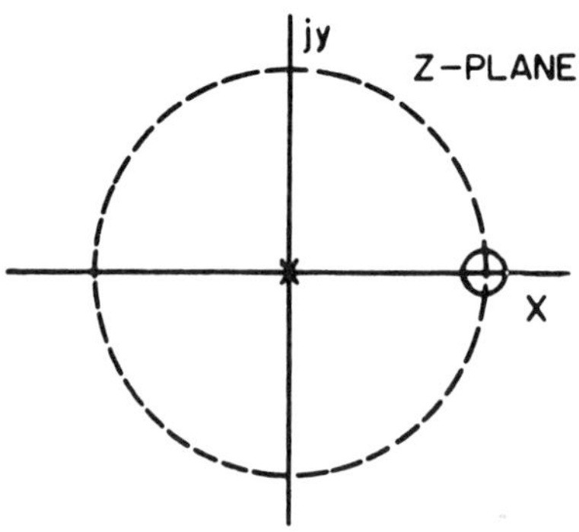

Figure 6.17 Pole zero pattern of a single canceller.

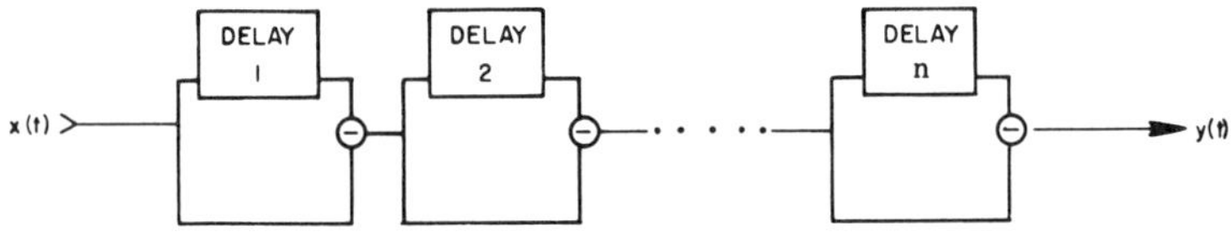

Figure 6.18 An N-stage MTI canceller.

indicating $n - 1$ poles at the origin and $n - 1$ zeros on the unit circle at $z = 1$. This type of MTI canceller is known as a *binomial canceller* because the weights have alternate binomial coefficients given by

$$w_k = (-1)^k \binom{n}{k} = \frac{(-1)^k n!}{k!(n-k)!} \tag{6.29}$$

The frequency response of the single canceller can be found from the diagram shown in Figure 6.19 by using the relationship for $|H(j\omega)|$, which is found by first taking the product of vector lengths from zeros to points on the unit circle and then dividing by the product of vector lengths from poles to points on the unit circle (see equation (C.16)). This results in a transfer function for a single canceller given by

$$|H(j\omega)| = \sqrt{\sin^2(\omega T) + [1 - \cos(\omega T)]^2} = 2\sin(\omega T/2) \tag{6.30}$$

whereas that of the general n-pulse binomial canceller is given by

$$|H(j\omega)| = |2\sin(\omega T/2)|^{n-1} \tag{6.31}$$

where n is the number of pulses processed by the MTI.

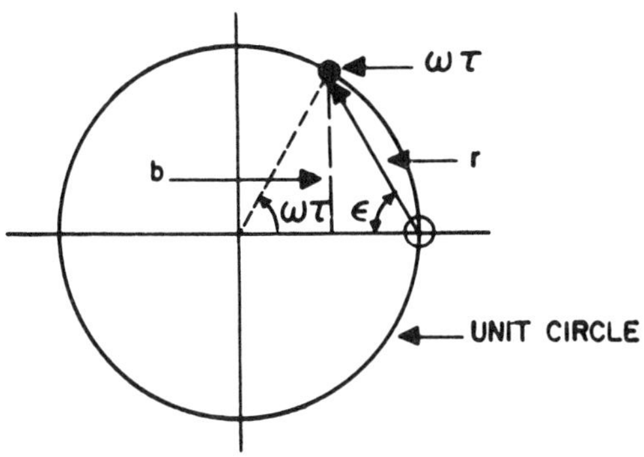

Figure 6.19 MTI frequency response of single canceller from the pole-zero diagram.

Another nonrecursive MTI canceller uses optimal weights rather than binomial weights [2,10,11]. The optimal weights are those which maximize the MTI improvement factor under the constraint that all target velocities have equal probability of occurrence (see Section 5.1.3). The MTI improvement factor (see Section 3.1) is given by

$$I_f = \frac{\sum_{j=0}^{n-1} w_j^2}{\sum_{j=0}^{n-1} \sum_{k=0}^{n-1} w_j w_k \rho_c(j-k)} \tag{6.32}$$

where w_j denotes the MTI weights and $\rho_c(\cdot)$ is the clutter's correlation coefficient. The power gain of the MTI is given by $\bar{G} = \Sigma_{j=0}^{n-1} w_j^2$, which represents an amplification of the target signal component by the MTI and must be normalized when assessing the MTI's clutter attenuation effect.

First, consider the optimal weights for a two-pulse MTI canceller. Equation (6.32) for the improvement factor can be specialized to

$$I_{f_1} = \frac{w_0^2 + w_1^2}{w_0^2 + w_1^2 + 2\rho_1 w_0 w_1} \tag{6.33}$$

where ρ_1 is the single lag correlation coefficient. Next, the improvement factor can be maximized subject to the constraint that the MTI gain $\Sigma_{j=0}^{n-1} w_j = \bar{G}$ is a constant by forming the object function:

$$J = w_0^2 + w_1^2 + 2\rho_1 w_0 w_1 - \lambda(w_0^2 + w_1^2) \tag{6.34}$$

where λ is a Lagrangian multiplier [12]. Then, using the calculus of variations, the stationary points are found, where $\partial J/\partial w_0 = 0$ and $\partial J/\partial w_1 = 0$, resulting in the following linear equations:

$$\begin{aligned} w_0(1 - \lambda) + w_1 \rho_1 &= 0 \\ w_1(1 - \lambda) + w_0 \rho_1 &= 0 \end{aligned} \tag{6.35}$$

Elimination of w_0 and w_1 from these equations results in an expression for the Lagrangian multiplier as

$$\lambda^2 - 2\lambda + 1 - \rho_1^2 = 0 \tag{6.36}$$

The solution of this equation results in

$$\lambda = 1 - \rho_1 \tag{6.37}$$

for the value of λ that maximizes the improvement factor. Substitution of this value into (6.35) results in $w_0 = -w_1$ for the optimal weights, which, when normalized, produce a weight vector $\mathbf{w} = (1, -1)$ and an improvement factor $I_1 = 1/1 - \rho_1$. This demonstrates that the binomial MTI is optimum for the case when only a single-delay MTI is involved.

For higher order MTIs, the procedure used for the single-delay MTI can be followed but the optimum weights are different than those of the binomial MTI and the solution becomes more complex. Rather than using Lagrangian multipliers, the procedure described in Section 5.1.3 is generally used to determine the optimal weights for higher order MTIs. This procedure solves the matrix equation:

$$(\mathbf{R}_n - \lambda_{\min}\mathbf{I})\mathbf{w} = 0 \tag{6.38}$$

by determining the eigenvalues ($\lambda_{\min}$) of

$$|\mathbf{R}_n - \lambda_{\min}\mathbf{I}| = 0 \tag{6.39}$$

where $\mathbf{R}_n$ is the normalized covariance matrix and $\mathbf{I}$ is the identity matrix. Hence, the optimal weights are found as the eigenvector that corresponds to the minimum eigenvalue. The improvement factor is then given by

$$I_f = 1/\lambda_{\min} \tag{6.40}$$

where the eigenvalues $\lambda_{\min}$ are identical to the Lagrangian multiplier of the calculus of variations approach.

To demonstrate the eigenvalue approach, the determinant of equation (6.39) for the single delay is given by

$$\begin{vmatrix} 1 - \lambda & \rho_1 \\ \rho_1 & 1 - \lambda \end{vmatrix} = 0 \tag{6.41}$$

This results in (6.36), which can be solved for $\lambda_{\min} = 1 - \rho_1$. The optimal weights can then be found from (6.35) as the eigenvector $\mathbf{w} = (1, -1)$. Solutions for optimum weights and improvement factors for double- and triple-delay MTI cancellers are given in Table 5.1.

The optimal weight MTI canceller provides a greater improvement factor than the binomial MTI canceller when greater than two pulses (one delay section) are processed. Figure 6.20 depicts the improvement factor provided with both optimal

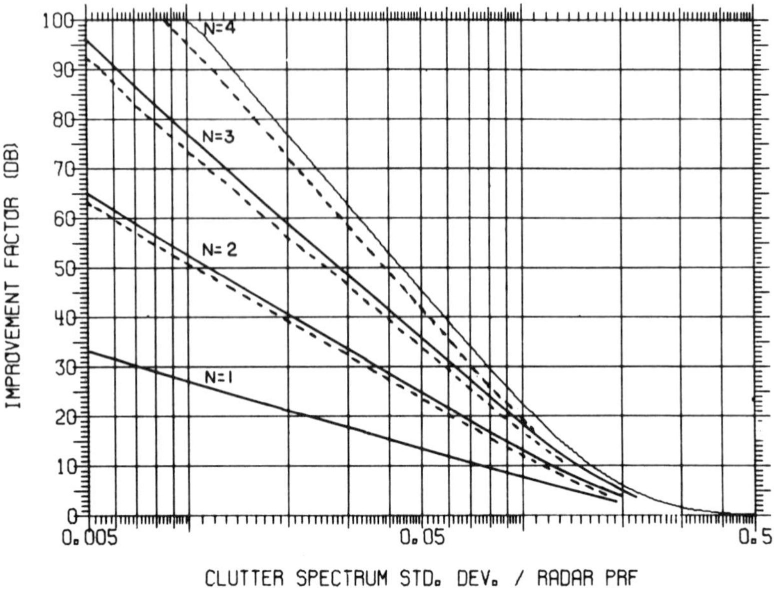

Figure 6.20 MTI improvement factor for optimum and binomial weights.

and binomial weights. If the standard deviation of the clutter spectrum is 0.01 times the PRF, an additional improvement factor can be achieved with optimal weights on the order of 2 dB for the double canceller ($n = 3$), 3.5 dB for the triple canceller ($n = 4$), and 5 dB for quadruple canceller ($n = 5$). To achieve this performance, accurate weights are required and the clutter spectral width should be narrow with respect to the radar's PRF [13].

The z-plane transfer function of the nonrecursive MTI given by (6.26) can also be written in pole-zero form as

$$H(z) = k \frac{(z - z_1)(z - z_1^*) \ldots}{z^{n-1}} \tag{6.42}$$

when the number of delays (cancellers) is even, and

$$H(z) = k \frac{(z - z_0)(z - z_1)(z - z_1^*) \ldots}{z^{n-1}} \tag{6.43}$$

when the number of delays is odd, where z_0, z_1, and z_1^* are the zero locations in the z-plane, and the poles are located at the origin ($z = 0$). Note that for an even number of delays (e.g., a double canceller) the zeros generally occur in conjugate pairs except for the binomial canceller, where they coalesce at $z = 1$. For an odd number of delays, one or more zeros are placed at $z = 1$ (causing a null at zero frequency and multiples of the PRF), and the remaining zeros generally occur in conjugate pairs, except for the binomial canceller. Also, for a nonrecursive MTI canceller, the number of zeros and poles is equal to the number of delay sections.

To illustrate the design of nonrecursive MTI filters in the z-plane, consider the case of the double-delay ($n = 3$) MTI canceller. The normalized ($k = 1$) z-plane transfer function is given by

$$H_2(z) = \frac{z^2 - (z_1 + z_1^*)z + z_1 z_1^*}{z^2} \tag{6.44}$$

where z_1 and z_1^* are the complex conjugate zero locations and $|z_1|^2 = z_1 z_1^*$ is the vector length of the zero from the origin. Comparison of this representation with (6.26) allows the normalized weights to be determined in relation to the zero locations as $w_0 = 1$, $w_1 = -(z_1 + z_1^*) = -2 \operatorname{Re} z_1$, $w_2 = z_1 z_1^* = |z_1|^2$. Solution of the matrix equation $(\mathbf{R}_n - \lambda_{\min}\mathbf{I})\mathbf{w} = \mathbf{0}$ for the optimal weights (i.e., finding the eigenvector) leads to the conclusion that, for the single-delay canceller, $w_0 = -w_1$ (see (6.35)), whereas for the double canceller $w_0 = w_2$, and for the triple canceller $w_0 = -w_3$, $w_1 = -w_2$ (see Table 5.1). This can be generalized to the conclusion that the optimal filter weights are symmetrical for an even number of delay sections (e.g., double canceller) and antisymmetrical for an odd number of delay sections (e.g., single and triple cancellers). Thus, from (6.44), we can further deduce that, when the weights are normalized such that $w_0 = 1$, the largest indexed weight has absolute value $|w_{n-1}| = 1$, resulting in the zeros of the optimal weight MTI canceller being located on the unit circle in the z-plane (i.e., $|w_{n-1}| = |z_{n-1}| = 1$), as depicted in Figure 6.21 for the double-delay canceller.

Another consequence of the weight pattern just described is that the transfer function of the nonrecursive MTI canceller can be determined from the filter weights as

$$|H(j\omega)| = \left| \sum_{i=0}^{(N-1)/2} (-1)^i \, 2|w_i| \sin\left(\frac{N}{2} - i\right)\omega T \right| \tag{6.45}$$

for N odd (e.g., single or triple canceller) and

$$|H(j\omega)| = \left| (-1)^{N/2}|w_{N/2}| + \sum_{i=0}^{(N-2)/2} (-1)^i \, 2|w_i| \cos\left(\frac{N}{2} - i\right)\omega T \right| \tag{6.46}$$

for N even (e.g., double canceller), where N is the number of MTI delay sections.

BLOCK DIAGRAM	T(z)	P-Z DIAGRAM
7) Frequency (or Velocity) Rejection 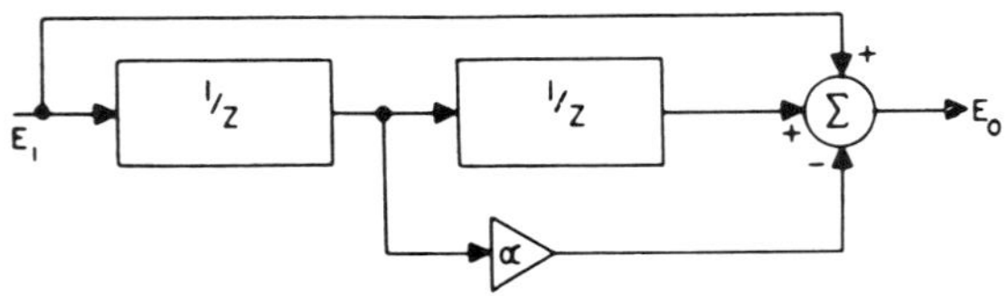	$\frac{(z-z_1)(z-z_2)}{z^2}$ $z_1, z_2 = \frac{\alpha \pm \sqrt{\alpha^2-4}}{2}$	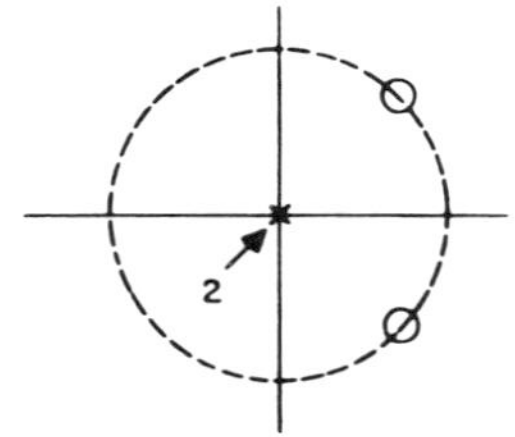
8) General Double Delay Filter 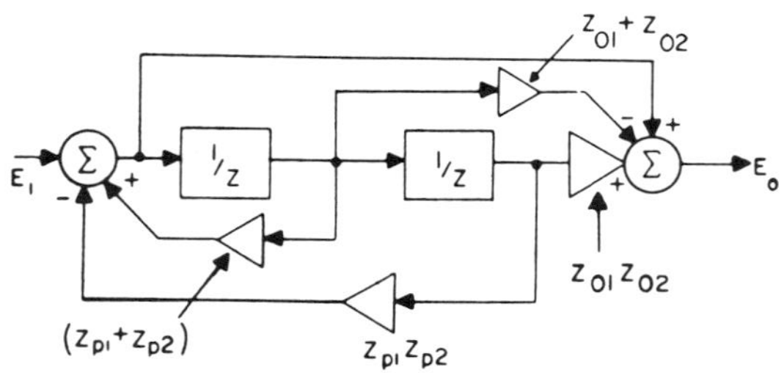	$\frac{(z-z_{01})(z-z_{02})}{(z-z_{p1})(z-z_{p2})}$	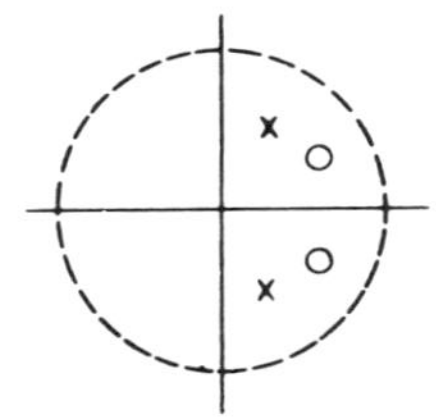 Poles and zeros real or conjugate complex pairs. Poles inside circle.

Figure 6.21 Pole-zero pattern for the double-delay canceller with optimum weights.

For example, the transfer function for a single canceller can be determined for (6.45) as $|H(j\omega)| = 2 \sin\omega T/2$, whereas the transfer function for a double canceller from (6.46) is $|H(j\omega)| = |-|w_1| + 2|w_0| \cos\omega T|$. For a binomial double canceller with weights $\mathbf{w} = (1, -2, 1)$ this becomes $|H(j\omega)| = 2(1 - \cos\omega T) = 4 \sin^2(\omega T)/2$.

From Figure 6.21, the transfer function for the double-delay optimal weight MTI can be written as

$$H_2(z) = \frac{z^2 - 2\cos\phi\, z + 1}{z^2} \tag{6.47}$$

where $\phi = \omega_z T$. The weights of the MTI can be deduced from (6.26) as $\mathbf{w} = (1, -2\cos\phi, 1)$. The improvement factor is given by

$$I_f = \frac{1 + 2\cos^2\phi}{1 + 2\cos^2\phi - 4\rho_1 \cos\phi + \rho_2} \tag{6.48}$$

where ρ_1 and ρ_2 are the correlation coefficients evaluated at $t = T$ and $2T$. If we let $\cos\phi = x$ and find the maximum I_f by setting $dI_f/dx = 0$, we have

$$x = \cos\phi = \frac{\rho_2 + \sqrt{\rho_2^2 + 8\rho_1^2}}{4\rho_1} \tag{6.49}$$

The optimal weight is then given by $\mathbf{w} = (1, w_1, 1)$ where

$$w_1 = -(r/2 + \sqrt{2 + r^2/4}) \tag{6.50}$$

and $r = \rho_2/\rho_1$, which for a Gaussian-shaped correlation coefficient is $r = \exp[-3(\sigma_c T)^2/2]$. The optimal weights for a triple-delay MTI canceller (see Table 5.1) can be found in a similar manner, where the transfer function is given by

$$H_3(z) = \frac{(z - 1)(z^2 - 2\cos\phi\, z + 1)}{z^3} \tag{6.51}$$

resulting in an optimal weight vector equal to $\mathbf{w} = [1, -(1 + 2\cos\phi), 1 + 2\cos\phi, -1)$.

There are several limitations to the optimal weight MTI canceller. First, note (see (6.42)) that in the design of MTI cancellers with an even number of delay sections (e.g., double-delay canceller), the zeros occur at other than zero frequency (i.e., $z = 1$). This means that fixed discrete targets are not being fully cancelled by the MTI filter. Also, as the number of pulses processed by the MTI increases, the design procedure results in significant variations in the MTI passband (see Figure 2.25) and a larger improvement factor than can be supported by MTI stability considerations. A quadratic programming algorithm has been developed that allows trading the excess improvement factor for greater velocity coverage consistent with the cancellation of fixed discrete targets [2,14].

6.3.2 Recursive MTI Filters

The most general form of MTI canceller is the recursive type, the block diagram of which is depicted in Figure 2.26. Recursive MTI filters employ feedback loops as well as the feed-forward configurations associated with nonrecursive filters. The feedback loops allow shaping the MTI velocity response, which, in principle, makes possible the synthesis of almost any frequency response function. However, the recursive filters are susceptible to transients that may circulate through the MTI filter for many radar interpulse intervals, depending on the degree of shaping employed in the MTI. Transients in the MTI may be caused by large clutter returns that act as a step input, the use of step-scan antenna beam programming (e.g., in phased arrays), or by intentional or unintentional nonsynchronous interference. When recursive MTI filters are employed, some form of transient protection generally must be used as well [2,15].

The z-plane transfer function for a recursive MTI is given by

$$H(z) = k\frac{z^{n-1} + a_0 z^{n-2} + \ldots + a_{n-1}}{z^{n-1} + b_0 z^{n-2} + \ldots + b_{n-1}} \tag{6.52}$$

where n is the number of pulses processed by the MTI filter and a_i, $i = 0, 1, 2, \ldots, n - 1$ are the feed-forward coefficients and b_i, $i = 0, 1, 2, \ldots, n - 1$ are the feedback coefficients. The transfer function can be written in pole-zero form as

$$H(z) = k\frac{(z - z_{01})(z - z_{02}) \ldots (z - z_{on-1})}{(z - z_{p1})(z - z_{p2}) \ldots (z - z_{pn-1})} \tag{6.53}$$

where z_{oi}, $i = 1, 2, \ldots, n - 1$ are the filter zeros and z_{pi}, $i = 1, 2, \ldots, n - 1$ are the filter poles.

In practice, most recursive MTI filters are synthesized as cascades with each cascade having no more than two delay sections. Thus, no feedback or feed-forward path needs to span more than two delay sections.

A basic single-delay building block is depicted in Figure 6.22 [2,16]. The transfer function for this filter is given by

$$H(z) = k_f\frac{z + 1/k_f}{z - k_b} \tag{6.54}$$

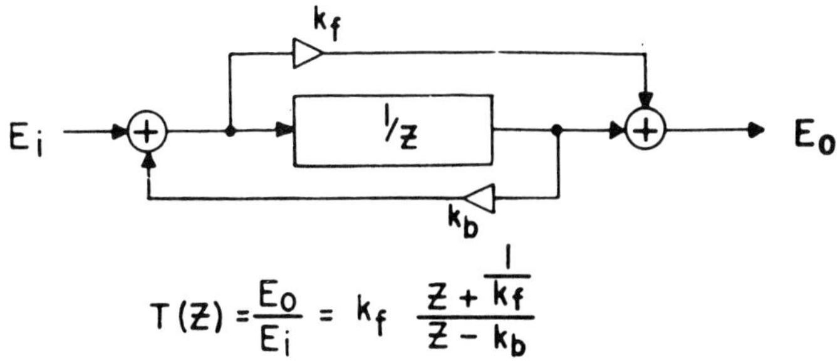

Figure 6.22 One-delay recursive MTI filter with separate pole-zero control.

where the zero is located at

$$z_0 = -1/k_f \tag{6.55}$$

and the pole at

$$z_p = k_b \tag{6.56}$$

For MTI filters used in fixed-based radars operating against ground clutter, we generally want a zero located at zero frequency, which is achieved in the single-delay configuration by setting $k_f = -1$.

The basic double-delay building block is depicted in Figure 6.23 [2,16]. The transfer function for this filter is given by

$$H(z) = k\frac{z^2 + (k_1/k_2)z + 1/k_2}{z^2 - k_3 z - k_4} \tag{6.57}$$

where the zeros are located at

$$z_0 = \frac{1}{2k_2}(-k_1 \pm \sqrt{k_1^2 - 4k_2}) \tag{6.58}$$

and the poles at

$$z_p = \frac{1}{2}(k_3 \pm \sqrt{k_3^2 + 4k_4}) \tag{6.59}$$

This double-delay configuration is the only one where the feedback factors involved

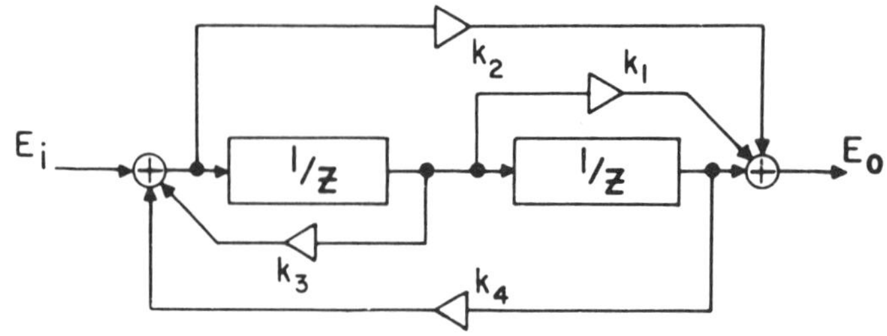

$$T(z) = k_2 \frac{z^2 + \frac{k_1}{k_2} z + \frac{1}{k_2}}{z^2 - k_3 z - k_4}$$

$$\text{ZEROS AT } 1/2k_2 \left[-k_1 \pm \sqrt{k_1^2 - 4k_2}\right]$$

$$\text{POLES AT } 1/2 \left[k_3 \pm \sqrt{k_3^2 + 4k_4}\right]$$

Figure 6.23 Two-delay recursive MTI filter with separate pole-zero control.

in the location of the poles do not affect the zeros and *vice versa*. The pole locations are controlled solely by the feedback loops, whereas the two feed-forward loops govern only the zero locations. If, in the synthesis procedure of the overall MTI filter, the number of poles exceeds the number of zeros, the last poles can be obtained with the basic circuit by dropping either one or both of the zeros. One zero is lost by setting k_1 or $k_2 = 0$ and both zeros are dropped by letting $k_1 = k_2 = 0$.

An example of a triple-canceller MTI designed by using the basic single-delay and double-delay building blocks is depicted in Figure 6.24 [3]. The single-delay section is designed with $k_f = -1$ and $k_b = \alpha_4$, whereas the double-delay section has $k_1 = -2\alpha_1$, $k_2 = 1$, $k_3 = \alpha_2$, and $k_4 = \alpha_3$. The single-delay zero is then at $z_0 = 1$ and the pole is at $z_p = \alpha_4$ (see (6.55) and (6.56)). The double-delay zeros are located at $z_0 = \alpha_1 \pm \sqrt{\alpha_1^2 - 1}$ whereas the poles are located at $z_p = (1/2)(\alpha_2 \pm \sqrt{\alpha_2^2 + 4\alpha_3})$ (see (6.58) and (6.59)). The resulting transfer function is then given by

$$H(z) = \frac{(z - 1)(z^2 - 2\alpha_1 z + 1)}{(z - \alpha_4)(z^2 - \alpha_2 z - \alpha_3)} \tag{6.60}$$

The pole-zero pattern is illustrated in Figure 6.24 and has three zeros located on the unit circle at frequencies $\omega_o T = 0, \pm \cos^{-1}\alpha_1$ with a single pole located on the x-axis of the z-plane at $x = \alpha_4$ and a conjugate pair of poles located at a radius equal to $\sqrt{\alpha_3}$ with angles $\theta = \pm \cos\alpha_2/2\sqrt{\alpha_3}$. The diagram illustrates how the velocity response

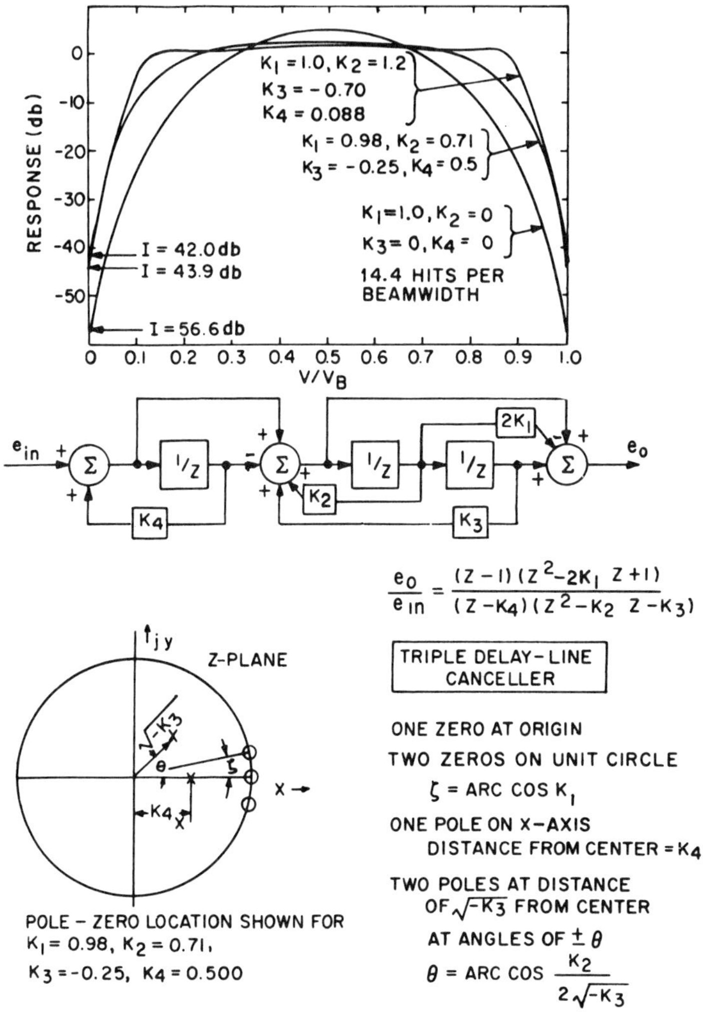

Figure 6.24 Triple MTI canceller design example.

of the filter can be widened at the expense of a reduced improvement factor by the appropriate adjustment of the filter's parameters. Note that the filter's transfer function can be found from the pole-zero diagram by finding the ratio of the product of the zero vectors to the product of the pole vectors for different points (i.e., $\theta = \cos\omega T$) around the unit circle. The improvement factor is then found from the relationship $I_f = \bar{G} \cdot \mathrm{CA}$ as

$$I_f = \frac{\overline{|H(f)|^2} \displaystyle\int_0^{\mathrm{PRF}} S(f)\,\mathrm{d}f}{\displaystyle\int_0^{\mathrm{PRF}} |H(f)|^2 S(f)\,\mathrm{d}f} \tag{6.61}$$

where $H(f)$ is the filter's transfer function and $S(f)$ is the clutter spectrum.

Another recursive filtering configuration commonly found in analog-shaped canceller implementations is depicted in Figure 6.25. This configuration is similar to a binomial double canceller (see (6.28)), except that the velocity response can be widened by using the feedback paths, which move the poles from the origin in the z-plane. The advantage of this configuration in analog implementations is that, if one filter section drifts so that its zero in the z-plane moves away from $z = 1$, the other filter section is still available to cancel fixed, discrete targets at zero frequency.

The normalized transfer function of the shaped double-delay canceller is given by

$$H(z) = \frac{(z-1)^2}{z^2 - (\beta_1 + \beta_2)\,z + \beta_2} \tag{6.62}$$

where a double zero is located at $z = 1$ and the poles are located at

$$z_p = \frac{1}{2}\left[(\beta_1 + \beta_2) \pm \sqrt{(\beta_1 + \beta_2)^2 - 4\beta_2}\right] \tag{6.63}$$

The poles are located in the z-plane at an angle given by $\phi = \cos^{-1}[(\beta_1 + \beta_2)/2\sqrt{\beta_2}]$ at a distance $|z| = \sqrt{\beta_2}$ from the origin.

An example of the shaped double-delay canceller design is depicted in Figure 6.26 [3]. The figure illustrates different canceller shapes that result from selecting various values of β_1 ranging from -0.6 to $+0.6$ and values of β_2 from 0 to 0.5. The values for a double-delay binomial canceller are $\beta_1 = \beta_2 = 0$, which results in an MTI improvement factor of 39.9 dB for the situation depicted. Note that the velocity response of the MTI canceller can be widened or narrowed by selecting the appropriate filter parameters with an accompanying decrease or increase in the MTI improvement factor.

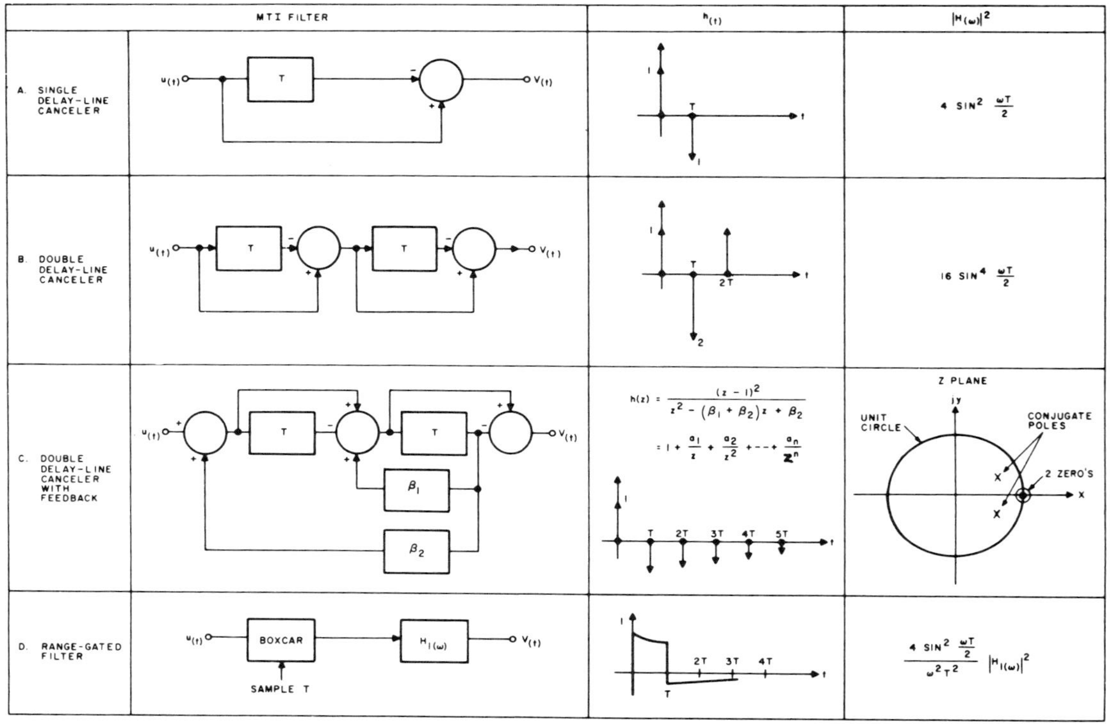

Figure 6.25 Shaped double canceller MTI filter.

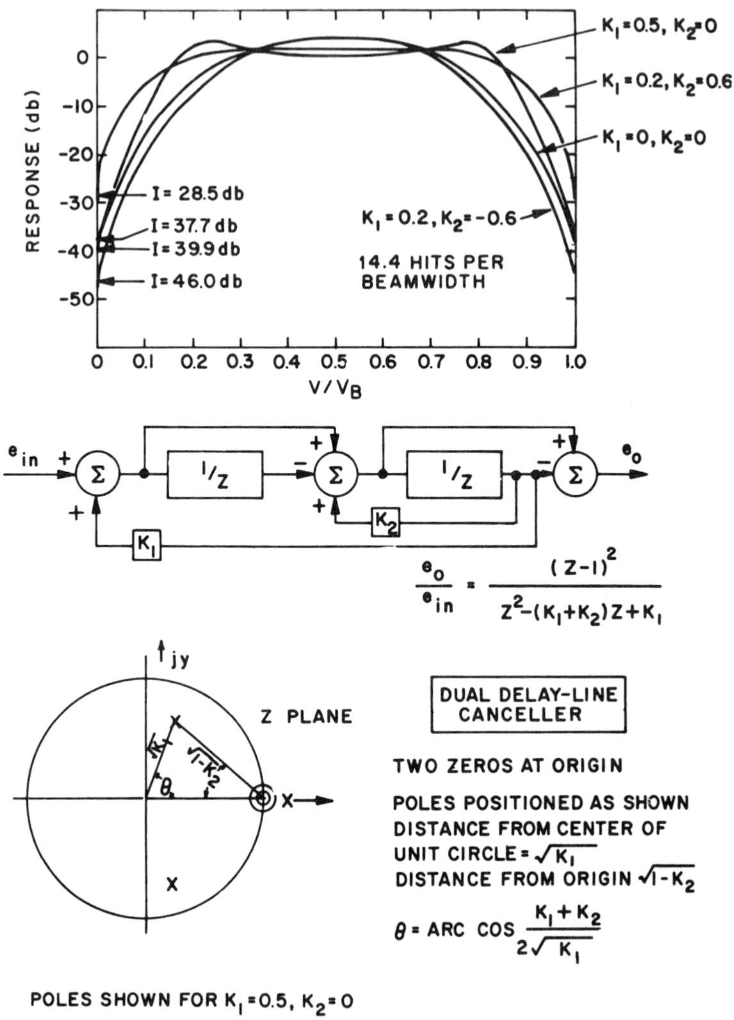

Figure 6.26 Shaped double-delay MTI design example.

In some cases, we might want the filter of Figure 6.25 to provide two separate and distinct zeros on the unit circle. A configuration that allows this is depicted in Figure 6.27. The transfer function for this configuration is given by

$$H(z) = \frac{z^2 - z(2 - a_1) + 1}{z^2 - z(\beta_1 + \beta_2) + \beta_2} \tag{6.64}$$

where the zeros are located at an angle:

$$\phi_0 = \cos^{-1}[(2 - a_1)/2] \tag{6.65}$$

on the unit circle and the poles are given by (6.63).

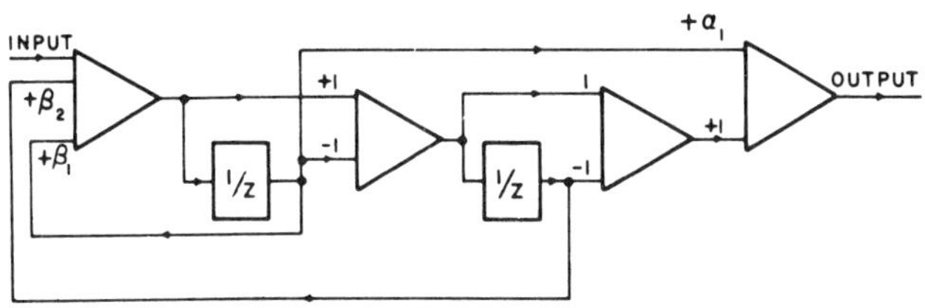

Figure 6.27 Shaped double-delay MTI canceller with adjustable zeros.

6.3.2.1 Synthesis of Recursive MTI Filters

The use of a recursive MTI filter provides the designer the flexibility for adjusting the MTI velocity response according to changing clutter conditions. One example of the need for this capability occurs in fixed-based air search radar, where the clutter's spectral spread is determined primarily by scanning modulation (see Section 6.2.1). In some radars, the antenna's scan rate is varied to provide more hits on target, which allows detection of weaker targets. Figure 6.28 depicts four velocity responses obtained by using the shaped double-delay canceller configuration of Figure 6.25. The velocity shapes are set to provide a 30 dB MTI improvement factor against ground clutter for the antenna scan rates listed (i.e., 3.3 rev/min to 10 rev/min) [2,17]. This capability is also useful in airborne surveillance radar, where the clutter's spectral spread is determined by the aircraft's velocity and the antenna's azimuth scan position (see Section 6.2.2).

The synthesis procedure for designing recursive MTI filters involves selecting a filter shape that provides the maximum velocity coverage consistent with the required MTI improvement factor. Conventional filter synthesis theory can be translated into the sampled data domain by employing a bilinear transformation that translates the

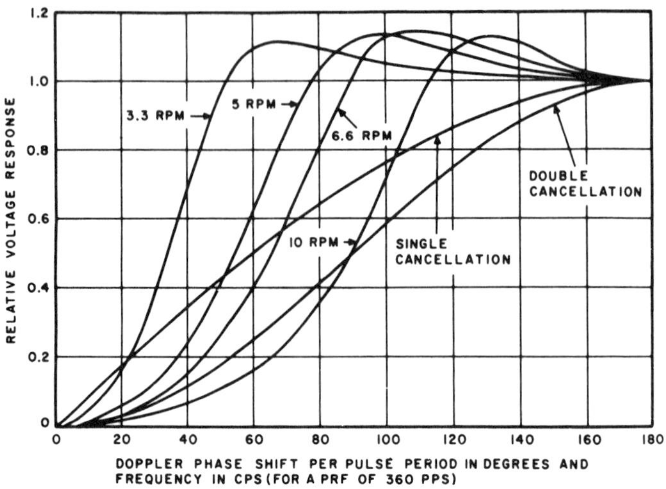

Figure 6.28 Velocity response of shaped double cancellers set for various antenna scan rates.

Laplace transform domain (p-plane) into the z-transform domain (see Appendix C). This allows translating conventional filter designs that use the Butterworth, Chebyschev, or elliptic transfer functions into a sampled-data MTI filter configuration. Once the filter type is selected (e.g., Butterworth), the synthesis procedure involves a selection of the number of filter poles and the cut-off frequency.

To aid in the selection of filter poles and cut-off frequency, examining the performance of an asymptotic filter is useful, which is defined as

$$H_A(\omega)^2 = \begin{cases} \left(\dfrac{\omega}{\omega_{co}}\right)^{2n}; & \omega \leq \omega_{co} \\ 1 \quad ; & \omega > \omega_{co} \end{cases} \tag{6.66}$$

where n is the number of filter poles and ω_{co} is the radian doppler frequency dividing the stopband and passband. This filter is essentially flat in the passband and falls off in the stopband according to the number of filter poles. The clutter attenuation for a Gaussian-shaped spectrum is then given by

$$\text{CA} = \frac{\sqrt{\pi\sigma_c^2/2}}{\displaystyle\int_0^{\omega_{co}} e^{-\omega^2/2\sigma_c^2}\left(\frac{\omega}{\omega_{co}}\right)^{2n} d\omega + \int_{\omega_{co}}^{\pi/T\to\infty} e^{-\omega^2/2\sigma_c^2}\, d\omega} \tag{6.67}$$

where σ_c is the clutter's spectrum standard deviation, n is the number of filter poles, and ω_{co} is the filter's cut-off frequency. This equation is plotted in Figure 6.29, where the filter's cut-off frequency is normalized to the clutter's spectral width (ω_{co}/σ_c). For example, to achieve 30 dB of clutter attenuation with a double-delay design (i.e., two poles) would require the filter's cut-off frequency to be designed at about seven times the clutter's standard deviation width. The best that can be achieved with an infinite number of delay sections $(n \rightarrow \infty)$ is to set the filter's cut-off frequency at about three times the clutter's spectral width. Note that the MTI improvement factor is related to the clutter attenuation by the relationship $I = \bar{G} \cdot \text{CA}$, where $\bar{G}$ is the MTI gain. The MTI gain is given for various binomial MTI cancellers in Table 3.1.

The type of filter shape is determined by the desired velocity response characteristic. The Butterworth design provides a monotonic amplitude *versus* frequency response, where the amplitude is maximally flat about the PRF/2 point in the comb filter response (see Figure 1.6). By allowing a ripple in the passband, the Chebyschev filter, using the same number of poles and zeros, can achieve sharper cut-offs. Elliptic filters yield even sharper cut-offs than the Chebyschev filter for the same network complexity, but allow both passband and stopband ripple. In Figure 6.30, several designs using two-delay Butterworth and Chebyschev responses are depicted for various normalized filter cut-off frequencies ranging from $f_{co}/\text{PRF} = 0.05$ to $f_{co}/\text{PRF} = 0.25$.

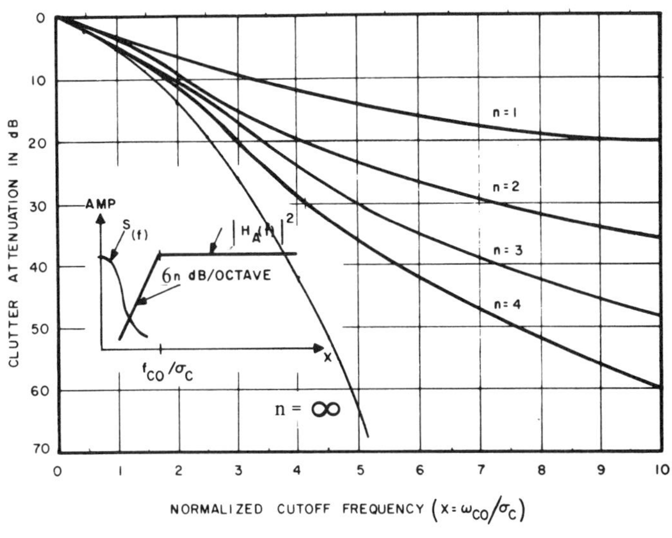

Figure 6.29 Clutter attenuation for various filter cut-off rates.

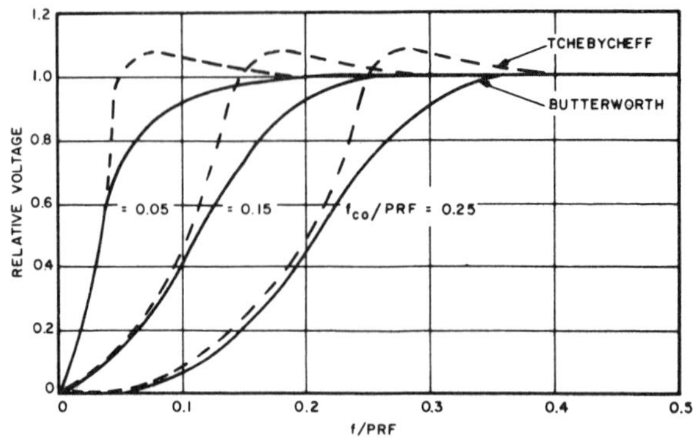

Figure 6.30 Butterworth and Chebyschev shaped double-delay MTI filter amplitude response.

The bilinear transformation most useful for MTI cancellers takes a conventional low-pass filter design and transforms it into an MTI comb filter response. This transformation is given by

$$z = \frac{p + \Omega}{p - \Omega} \tag{6.68}$$

where p is the pole or zero position in the p-plane, Ω is a scale factor related to the filter's cut-off frequency, and z is the transformed pole or zero position in the z-plane. The converse transformation from the z-plane into the p-plane is given by

$$p = \Omega \frac{z + 1}{z - 1} \tag{6.69}$$

Figure 6.31 depicts the transformations between the z- and p-planes where the pole-zero pattern of a two-delay Butterworth design is shown in the p-plane. The scale factor is determined by noting that the imaginary point in the p-plane at $+j$ corresponds to the point in the z-plane labeled (1). By using (6.69), we obtain

$$p = +j = \Omega \frac{\cos\phi_0 - j \sin\phi_0 + 1}{\cos\phi_0 - j \sin\phi_0 - 1} \tag{6.70}$$

where $\phi_0 = \omega_{co}T$. Solution of (6.70) determines the scale factor as

$$\Omega = \tan(\phi_0/2) \tag{6.71}$$

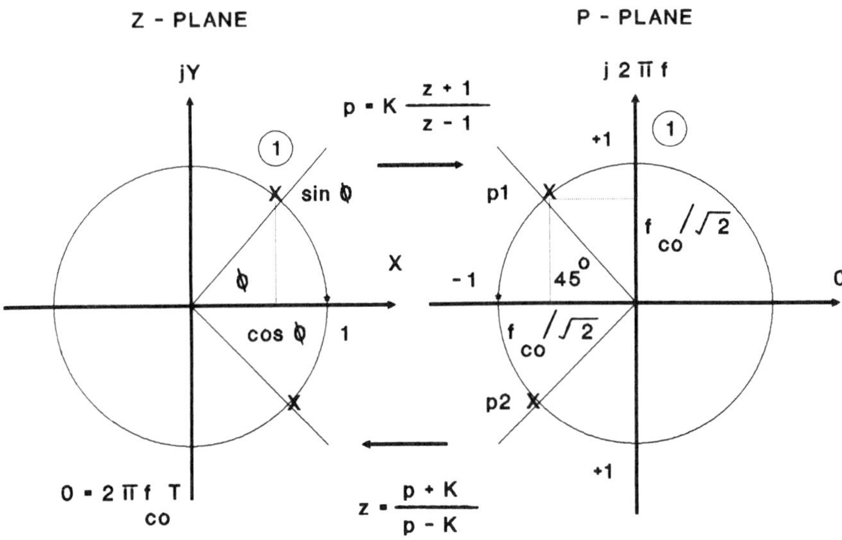

Figure 6.31 Bilinear transformation between the p- and z-planes.

where $\phi_0 = \omega_{co}/\text{PRF}$ represents the fractional angle in the z-plane that corresponds to the cut-off frequency in the p-plane. The scale factor of (6.71) forces the point $j\omega = \pm 1$ in the p-plane to map onto the unit circle in the z-plane at an angle $\mp\phi_0$ that represents a frequency of $f_{co} = \phi_0/2\pi T$.

The poles for a low-pass Butterworth filter are located around a circle in the p-plane at

$$P_k = -\sin\frac{(2k-1)\pi}{2n} + j\cos\frac{(2k-1)\pi}{2n} \tag{6.72}$$

where n is the number of poles (delay sections) and k is from 1 to n. The poles are normalized to the cut-off frequency (ω_{co}) so that the 3 dB point is at $j\omega = 1$. For example, at $n = 2$, the poles are located at $p_{12} = (-1 \pm j)/\sqrt{2}$, whereas for $n = 3$ they are located at $p_{13} = (-1 \pm j\sqrt{3})/2$ and $p_2 = -1$. The Butterworth poles for a shaped double-delay canceller ($n = 2$) would then be located at

$$z_{12} = \frac{1 - \sqrt{2}\tan\phi_0/2 \mp j}{\sqrt{2}\tan\phi_0/2 + 1 \mp j} \tag{6.73}$$

where $\phi_0 = \omega_{co}T$. For example, if the normalized cut-off frequency was $f_{co}/\text{PRF} =$

0.1, then the poles would be located at $z_{12} = 0.5715 \pm 0.2936j$ with magnitude $|z| = 0.6425$ at an angle $\phi = 27.2°$. In the shaped double-delay canceller configuration of Figure 6.25, the parameter $\beta_2 = |z|^2 = 0.4128$ whereas the parameter $\beta_1 = 2 \text{ Re } z_1 - \beta_2 = 0.7302$. The locus of Butterworth poles for the shaped double-delay canceller is depicted in Figure 6.32.

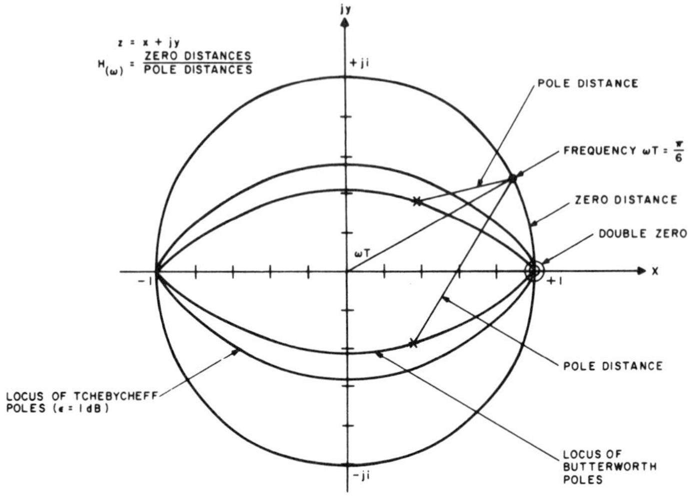

Figure 6.32 Pole-zero pattern of shaped double-delay Butterworth and Chebyschev MTI cancellers.

Alternately, the poles for a low-pass Chebyschev filter are located on a ellipse in the p-plane given by

$$P_k = -\sin\frac{(2k-1)\pi}{2n}\sinh\gamma + j\cos\frac{(2k-1)\pi}{2n}\cosh\gamma \tag{6.74}$$

where n is the number of poles and $k = 1, 2, \ldots, n$. The peak-to-valley power ratio of the ripple in the filter passband is equal to $1 + \epsilon^2$ resulting in the ripple amplitude (δ) being equal to

$$\delta = 1 - \frac{1}{\sqrt{1+\epsilon^2}} \tag{6.75}$$

where ϵ is a filter parameter. The ripple amplitude can be expressed on a dB scale resulting in the values of ϵ given in Table 6.1. The values of

$$\sinh\gamma = \frac{(\alpha+\beta)^{1/n} - (\alpha-\beta)^{1/n}}{2} \tag{6.76}$$

Table 6.1
Chebyschev Filter Ripple Parameters

20 log δ	ϵ	ϵ^2
$\frac{1}{2}$ dB	.3493	.1220
1 dB	.5088	.2589
2 dB	.7648	.5848
3 dB	.9976	.9953

$$\cosh\gamma = \frac{(\alpha + \beta)^{1/n} + (\alpha - \beta)^{1/n}}{2} \tag{6.77}$$

where

$$\alpha = \sqrt{1 + \frac{1}{\epsilon^2}} \tag{6.78}$$

$$\beta = 1/\epsilon \tag{6.79}$$

can then be found, which allows the value of the Chebyschev filter poles to be determined. The point $j\omega = 1$ is now the point where the response has fallen off by the allowed ripple tolerance rather than by 3 dB as with the Butterworth filter. The locus of Chebyschev poles for the double-delay-shaped canceller is depicted in Figure 6.32.

As an example of a Chebyschev filter design, consider the problem of designing a three-pole MTI filter with 1 dB passband ripple and a normalized cut-off frequency at $f_{co}/\text{PRF} = 0.125$ [2,17]. From Table 6.1, the value of $\epsilon = .5088$ is obtained for 1 dB ripple, resulting in $\alpha = 2.2051$ and $\beta = 1.9654$. The values of $\sinh\gamma = .4942$ and $\cosh\gamma = 1.1154$ are found resulting in $p_{13} = -0.2471 \pm j\,0.9659$ and $p_2 = -0.4942$. Applying the bilinear transformation (equation (6.68)) with $\gamma = \tan(45/2)°$ $= 0.4142$ results in z-plane poles equal to $z_{13} = 0.6000 \pm j\,0.5838$ with magnitude $|z| = 0.8375$ at an angle of 44.2° and $z_2 = 0.0881$, as depicted in Figure 6.33, which also shows the amplitude response of the low-pass Chebyschev filter.

The complete canceller can be implemented with a cascade formed by using the single-delay configuration given in Figure 6.22 and the shaped double-delay canceller depicted in Figure 6.25. The parameters for the single-delay canceller are $k_B = 0.0881$ and $k_f = 0$. For the shaped double-delay canceller, $\beta_2 = |z_1|^2 = 0.7012$ and $\beta_1 = 2\,\text{Re}\,z_1 - \beta_2 = 0.4994$. The double-delay canceller can also be implemented

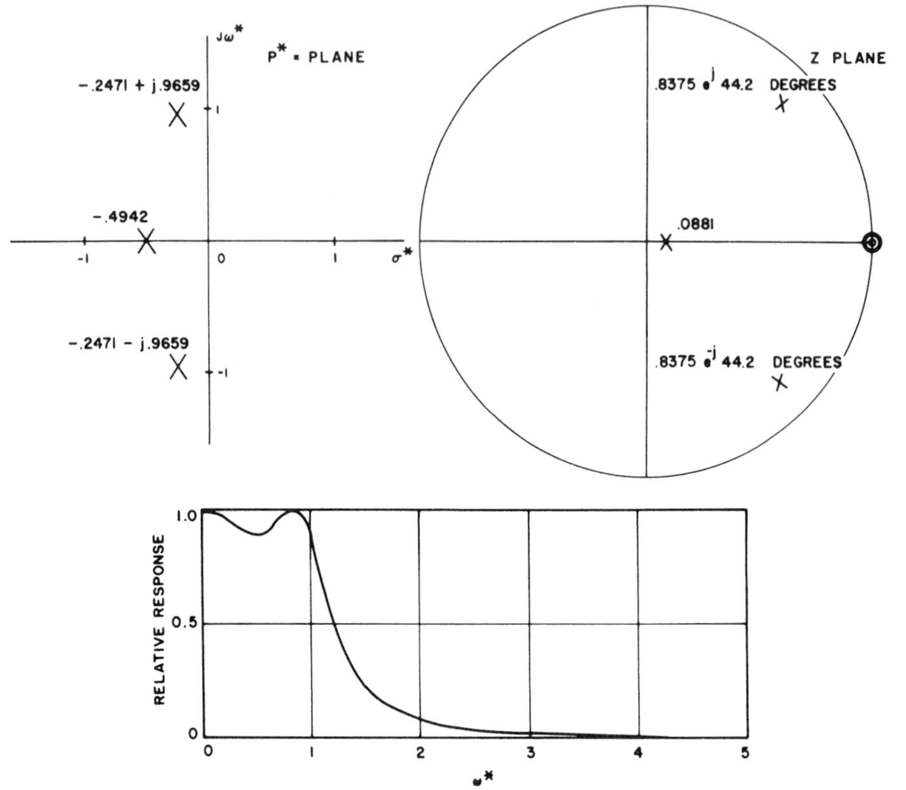

Figure 6.33 Pole-zero plot and frequency response of Chebyschev MTI canceller.

in the canonical form depicted in Figure 6.23. If we set $k_2 = 1$ and solve (6.58), we have $k_1 = -2$. This selection results in a double zero at $z = 1$. Then, from (6.59), we can determine that $k_3 = 2 \text{ Re } z_1 = 1.2001$. Solution of (6.59) then results in $k_4 = -.7012$. The complete triple-delay canceller is depicted in Figure 6.34 with transfer function equal to

$$H(z) = \frac{(z - 1)^2}{(z - 0.0881)(z^2 - 1.2001\, z + .7012)} \tag{6.80}$$

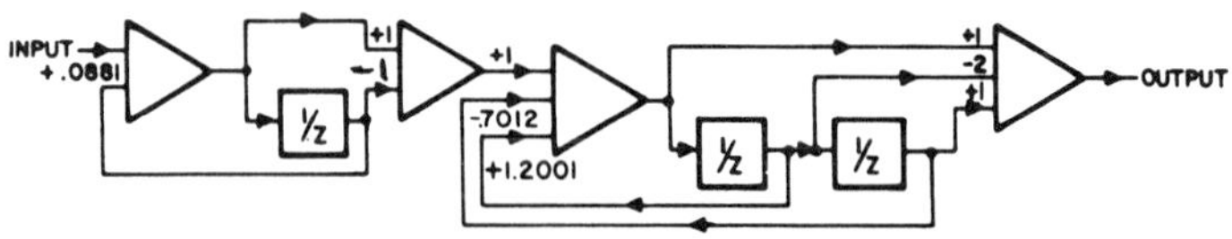

Figure 6.34 Triple delay MTI canceller with Chebyschev filter response.

6.3.3 Doppler Filter Banks as Transversal Filters

Doppler filter banks often are employed, either in conjunction with MTI filters (see Section 5.2) or, separately, to detect targets imbedded in clutter and noise. These filter banks are implemented most efficiently by using digital signal processing techniques that employ the FFT algorithm (see Sections 2.3.3.2 and 2.3.3.3). Most treatments of this subject focus on the computational details of the FFT processing rather than on the basic filtering action of the filter bank. By using a transversal filter representation for the FFT processing, we can gain insight into the properties of such doppler processing [18].

The kth term of the discrete Fourier transform is given by

$$F(k) = \sum_{i=0}^{n-1} f(i)\, e^{-j(2\pi/n)ik} \tag{6.81}$$

where $f(i)$ is the signal sample at time points $i = 0, 1, \ldots, n - 1$; n is the number of sample points (i.e., radar pulses processed); and $F(k)$ is the output transform at point $k = 0, 1, \ldots, n - 1$. If the exponential term of the transform is treated as a complex filter weight given by

$$w_{i,k} = e^{-j(2\pi/n)ik} \tag{6.82}$$

then (6.81) has the form of a transversal filter (see (5.1)). In this representation, depicted in Figure 6.35, the filter outputs are given by

$$F(i, k) = \sum_{i=0}^{n-1} f(i) w_{i,k} \tag{6.83}$$

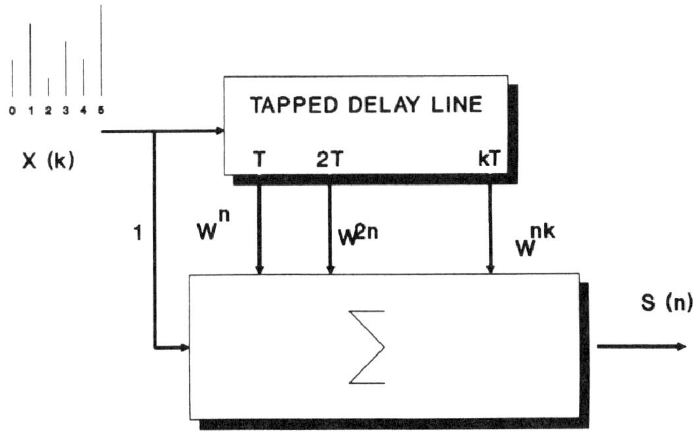

Figure 6.35 Transversal filter representation of FFT processor.

where the index k is the filter number and i is the delay line tap number. Note that n transversal filters are required to complete the representation where $k = 0, 1, \ldots, n - 1$ specifies the particular bandpass filter under consideration.

Expansion of the transversal filter complex weights into

$$w_{i,k} = e^{-j(2\pi/n)ik} = a_n^{ik} \tag{6.84}$$

where $a_n = \exp[-j(2\pi/n)]$, allows the z transform of the transversal filters transfer function to be written as

$$H_k(z) = \left(\frac{z}{a_n^k}\right)^{-(n-1)} \left[\left(\frac{z}{a_n^k}\right)^{n-1} + \left(\frac{z}{a_n^k}\right)^{n-2} + \ldots \frac{z}{a_n^k} + 1\right] \tag{6.85}$$

where the first term has a magnitude of 1 and represents a time delay. Expansion of the second term by using the relationship $1 + r^2 + \ldots + r^{n-1} = (r^n - 1)/(r - 1)$ results in

$$H_k(z) = \frac{\left(\frac{z}{a_n^k}\right)^n - 1}{\left(\frac{z}{a_n^k}\right) - 1} \tag{6.86}$$

However, $a_n^{-kn} = \exp(j2\pi k) = 1$, allowing the amplitude transfer function of the kth filter of the FFT to be expressed as

$$|H_k(z)| = \frac{z^n - 1}{z - a_n^k} \tag{6.87}$$

The pole-zero pattern then consists of n zeros, equally spaced on the unit circle, which correspond to the nth roots of unity and one pole. The pole occurs at $z_p = a_n^k = \exp(-j2\pi k/n)$, which is located on the unit circle at the same position as one of the zeros. The particular pole location at one of the numerator zeros depends upon the filter position (i.e., value of k) selected. For example, if the filter position $k = 0$ is selected, then $z_p = 1$ and the pole zero pattern consists of n zeros equally spaced around the unit circle as depicted in Figure 6.36 with the zero at $z_o = 1$ annihilated by the pole.

The DFT creates each filter passband by selectively deleting the kth zero lying at each of the n complex roots of unity, yielding n passbands. Thus n samples yield n filters, each of order $n - 1$. The stopband in each case is formed by the remaining $n - 1$ zeros evenly spread along the circumference of the unit circle that corresponds to the frequency band between PRF lines.

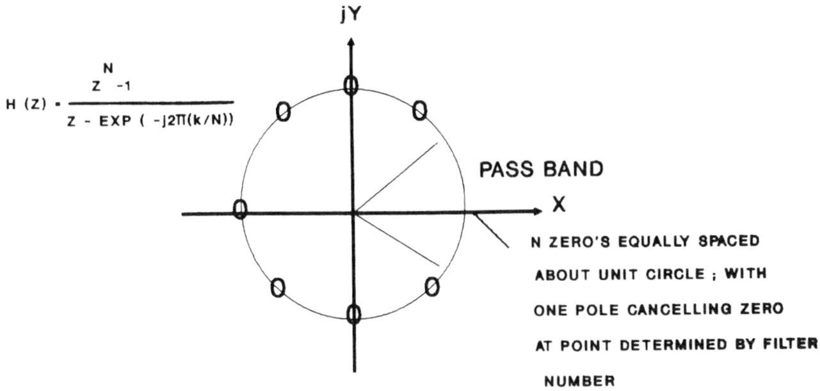

Figure 6.36 Pole-zero pattern of an FFT filter.

The transfer function of the kth FFT filter can be evaluated from (6.87) by letting $z = \exp(j\omega T)$, resulting in

$$|H_k(j)| = \frac{\sin(n\omega T/2)}{\sin(\omega T/2 - k\pi/n)} \tag{6.88}$$

where n is the number of FFT points (delay lines) and T is the radar's interpulse period. The transfer function of the FFT filter that provides for coherent integration against fixed targets can be evaluated by setting $k = 0$, resulting in

$$|H_o(j\omega)| = \frac{\sin(n\omega T/2)}{\sin(\omega T/2)} \tag{6.89}$$

Evaluating this function in the limit as $f \rightarrow 0$ results in $|H_o(0)| = n\omega T/\omega T = n$, which demonstrates that the coherent integration gain equals the number of pulses (n) being coherently integrated. Figure 6.37 depicts the normalized power transfer function ($|H(j\omega)|^2$) for the first ($k = 0$) and second ($k = 1$) filters of a coherently integrating FFT filter. Note that the normalized FFT response (see (6.89)) is often approximated by $|H_o(j\omega)| = |\sin(n\omega T/2)/(n\omega T/2)|$, which results in the peak sidelobes being underestimated for far-out sidelobes (see Table 5.2 and Section 5.2).

The sidelobe levels of the filters shown in Figure 6.37 result from using a rectangular (uniform) weighting function and can be reduced by using a weighted data window, as depicted in Figure 6.38 for Chebyschev weightings. As described in Section 5.2, the weightings can be either applied to the input data or obtained by symmetrical postcombining of adjacent filters on either side of the filter under consideration.

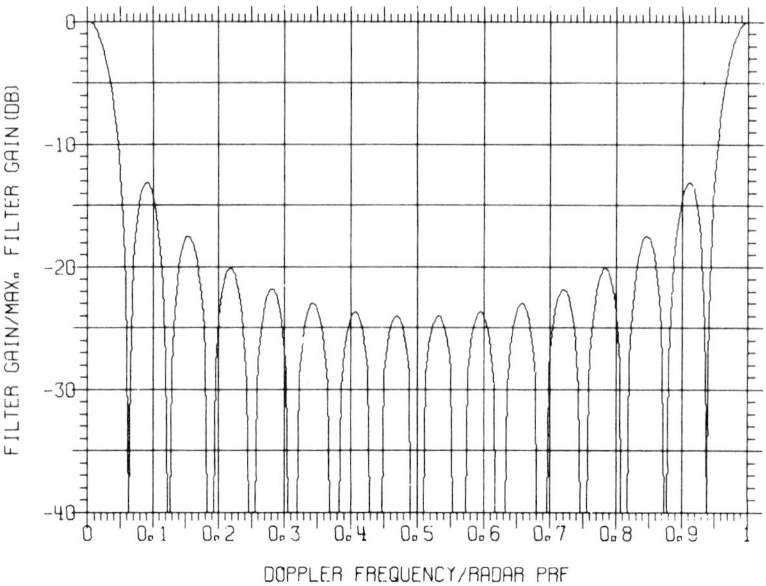

(a)

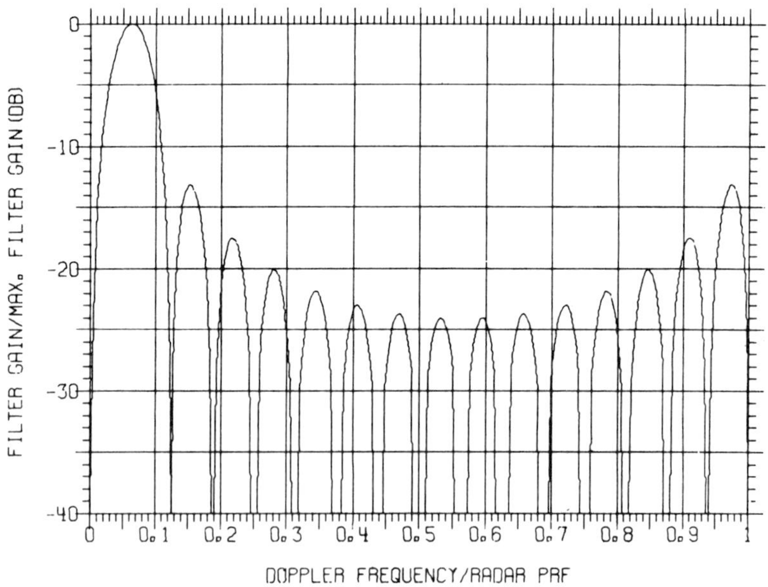

(b)

Figure 6.37 Normalized power transfer function for FFT coherent integrator.

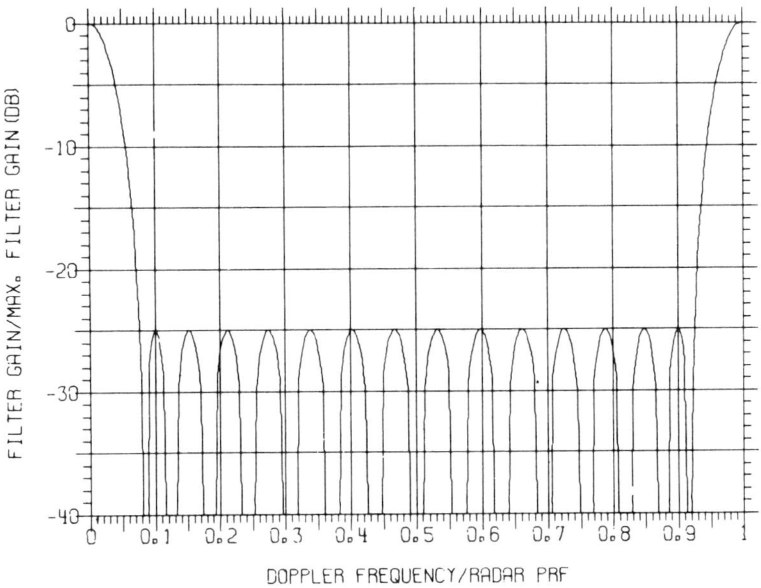

(a)

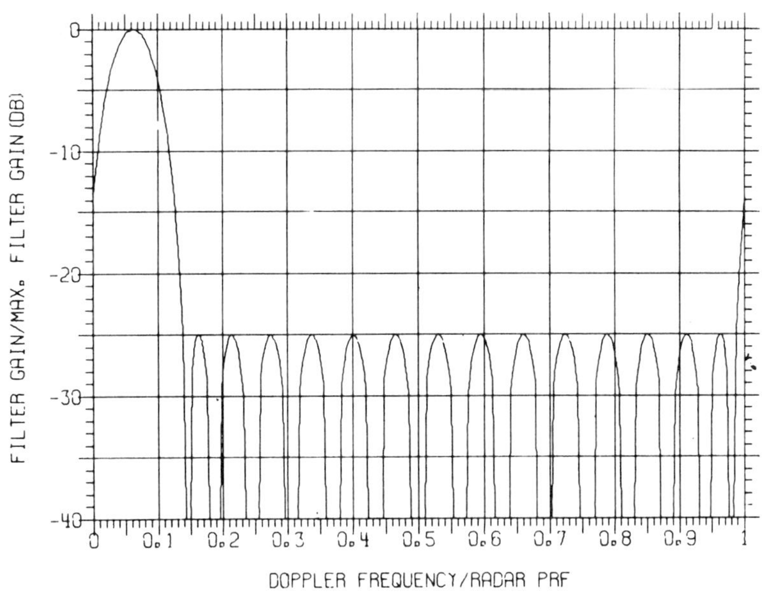

(b)

Figure 6.38 Normalized power transfer function for FFT coherent integrator with Chebyschev filter weightings for a −25 dB sidelobe level.

With symmetrical postcombining the transfer function in the z-plane for the FFT transversal filter (see (5.98)) can be written as

$$H_k(z) = (z^n - 1)\left(\frac{1}{z - a_n^k} + \frac{\alpha}{z - a_n^{k+1}} + \frac{\alpha}{z - a_n^{k-1}}\right) \tag{6.90}$$

where $\alpha = 0.5$ for Hann or hanning weighting and $\alpha = 0.426$ for Hamming weighting. For example, when a four-point FFT filter is employed, the pole-zero pattern from (6.25) consists of zeros deployed about the unit circle at $z_0 = \pm 1$ and $z_0 = \pm j$, where one zero is annihilated by a pole determined by the filter location ($k = 0, 1, 2, 3$). Consider the second filter ($k = 1$), which for uniform weighting has zeros located at $z_0 = \pm 1$ and $z_0 = +j$. Evaluation of (6.90) for symmetrical postcombining weighting results in a transfer function ($k = 1$) given by

$$H_1(z) = (z - j)\cdot\{z^2 + [2\alpha j/(1 + 2\alpha)]z - [1/(1 + 2\alpha)]\} \tag{6.91}$$

where α is the weighting function. For no weighting ($\alpha = 0$), the zeros are located on the unit circle as expected. For hanning weighting ($\alpha = 0.5$), the zeros located on the real axis at $z_0 = \pm 1$ migrate to $z_0 = (\pm\sqrt{3} - j)/4$.

This analysis indicates that the number of zeros of the z-plane transfer function of the weighted FFT filter depends upon the number of samples processed (i.e., number of delay sections). The effect of the weighting is merely to redeploy some of the zeros. In the example for hanning weighting, the zeros on the real axis in the z-plane migrate into the upper half of the z-plane within the unit circle, which has the effect of reducing the filter's sidelobes while broadening the filter's mainlobe.

6.4 STAGGERED PRF

PRF staggering involves the changing of the radar's interpulse period on a pulse-to-pulse basis to provide higher blind speeds or a better low-frequency cut-off of the MTI filter than may be obtained with a constant PRF design by using a binomial canceller. Rather than employing a pulse-to-pulse stagger, we can also use a block-to-block stagger, where the PRF is varied after a number of radar pulses are transmitted at a constant PRF. With a block stagger, the overall MTI velocity response becomes the average of the velocity responses of the individual constant PRF responses.

The advantages of the block stagger (e.g., see the MTD processor, Section 1.4) are that multiple-time-around clutter responses can be cancelled using a coherent MTI system and a system stability (particularly the transmitter) that is easier to achieve than for a system with pulse-to-pulse agility. The composite frequency response of an MTI system using a block stagger with two separate PRFs, in the ratio of 5:4, is depicted in Figure 6.39. Note that the first blind speed of the composite response

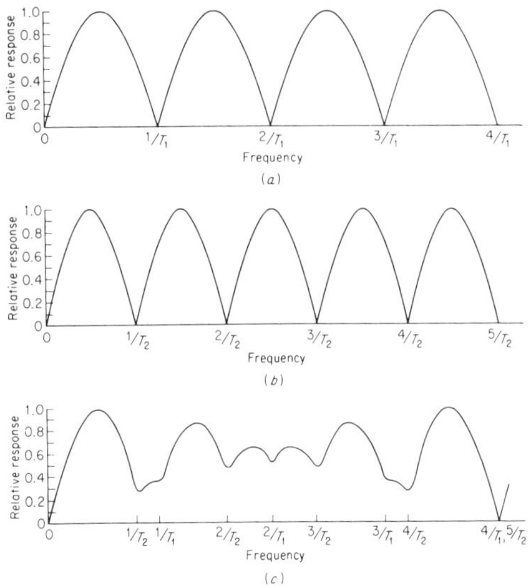

Figure 6.39 Frequency response of a single delay-line canceller for separate PRFs and composite response (PRFs in ratio 5 : 4).

was increased several times over what it would be for a radar operating on only a single PRF. Zero-velocity response occurs only when the blind speeds of each PRF coincide (i.e., at a frequency of $4/T_1 = 5/T_2$ in Figure 6.39). Although the first blind speed may be extended by using more than one PRF, regions of low sensitivity often appear within the composite passband.

The closer the ratio $T_1 : T_2$ approaches unity, the greater is the value of the first blind speed and the first null in the vicinity of $f_d = 1/T_1$ becomes deeper. Therefore, the choice of T_1/T_2 is a compromise between the value of the first blind speed and the depth of the nulls within the filter's passband. The depth of the nulls can be reduced and the first blind speed increased by operating with more than two interpulse periods. Four interpulse periods (see Figure 3.27) are generally considered a good design choice for providing a low ripple passband response (i.e., moderate nulls) and a reasonable first blind speed [3].

For many applications, pulse-to-pulse staggering is advantageous because it provides a favorable velocity response compared to the block-to-block stagger approach [3]. For example, consider a double canceller without feedback that has a 36% wide rejection notch and employs a block stagger approach on a scan-to-scan basis. Of the targets, 36% would be lost on each scan, assuming that all targets were uniformly distributed in doppler over the velocity range of interest, which might well be unacceptable [3]. Nonetheless, with pulse-to-pulse staggering, good response can be obtained on all dopplers of interest on each radar scan. In addition, a better velocity response can be obtained at some dopplers of interest than either of the block stagger intervals provide, which results from the modulation, induced on the target by the nonuniform sampling, at a frequency close to the maximum response of the MTI canceller. Pulse-to-pulse staggering may degrade the MTI improvement factor (see Section 3.2.6) attainable in some systems, but in many systems this restriction is not significant or can be eliminated by using time-varying weights in the MTI canceller [3]. An additional advantage of pulse-to-pulse staggering is that it generally removes the need for shaping the MTI canceller response, which considerably simplifies the canceller design and provides superior transient response.

6.4.1 Blind Speeds with PRF Stagger

When PRF staggering is employed, the radar's interpulse interval is changed on a pulse-to-pulse basis. However, the stagger pattern is periodic in that it repeats after a number of radar pulses are transmitted (see Figure 6.40). The overall period of the n-period stagger is given by

$$T_p = \sum_{i=1}^{n} T_i \tag{6.92}$$

where T_i are the n interpulse periods between the pulses that make up the stagger period. For example, in the two-period ($n = 2$) PRF stagger depicted in Figure 3.24, the two interpulse periods are given by $T_1 = T + T_s$ and $T_2 = T - T_s$, where T_s is the incremental delay forming the stagger sequence, T is the average (nonstagger) period, and $T_p = 2T$ is the overall stagger period. For the three-period stagger ($n = 3$), the interpulse periods are $T_1 = T + T_s$, $T_2 = T$, and $T_3 = T - T_s$ and the overall stagger period is $T_p = 3T$.

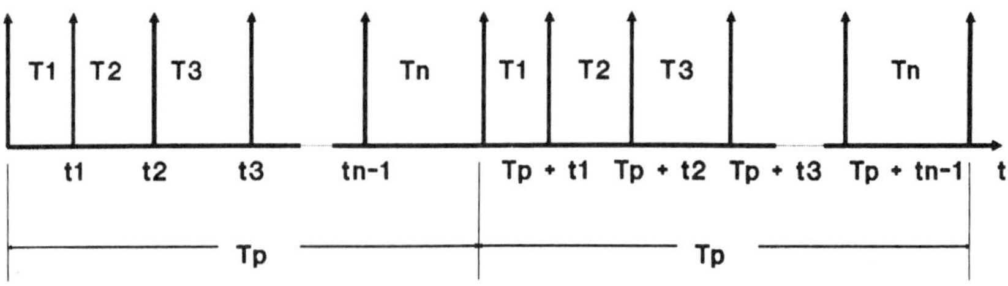

Figure 6.40 General PRF staggering pattern.

The period (T_b) associated with the first blind speed (v_{bs}) of the stagger is given by

$$T_b = \frac{1}{f_{bs}} \tag{6.93}$$

where $f_{bs} = 2\ v_{bs}/\lambda$ is the doppler frequency associated with the first null in the velocity response. The interpulse periods associated with the stagger are then given by

$$T_i = n_i\, T_b \tag{6.94}$$

where n_i are integers defining the stagger ratio. For example, a 5 : 4 stagger would have interpulse periods equal to $T_1 = 5\ T_b$ and $T_2 = 4\ T_b$. The average period of an n-period stagger is then given by

$$T = \frac{n_1 + n_2 + \ldots + n_n}{n} T_b \tag{6.95}$$

If the unstaggered first blind-speed doppler frequency is given by $f_b = 1/T$, then the resulting stagger blind-speed doppler frequency from (6.95) is given by

$$f_{bs} = \frac{n_1 + n_2 + \ldots + n_n}{n} f_b \tag{6.96}$$

For example, with a two-period ($n = 2$) 5:4 stagger ($n_1 = 5$, $n_2 = 4$) ratio, the staggered blind speed would be $5 + 4/2 = 4.5$ times the unstaggered first blind speed. The three-period stagger depicted in Figure 3.24 with periods $T_1 = T + T_s$, $T_2 = T$, and $T_3 = T - T_s$, which are symmetrical about the center period, have a first blind speed equal to the middle stagger ratio times the unstaggered first blind speed. Hence a 5:4:3 ($n_1 = 5$, $n_2 = 4$, $n_3 = 3$) stagger has a first blind speed equal to $f_{bs} = n_2 f_b = 4 f_b$. The four-period stagger (25:30:27:31) velocity response depicted in Figure 3.27 has a first blind speed of 28.25 times the unstaggered first blind speed.

For two- and three-period staggers that use an incremental stagger delay line (see Figure 3.24), it is sometimes convenient to express the stagger using a parameter that expresses the fractional delay represented by the stagger delay. This can be expressed as

$$e = \frac{T_s}{T} \tag{6.97}$$

where T_s is the stagger delay period and T is the average interpulse period. The stagger periods for a two-period stagger ($n = 2$) then become $T_1 = T(1 + e)$ and $T_2 = T(1 - e)$ and the stagger ratio (γ) becomes

$$\gamma_2 = \frac{n_1}{n_2} = \frac{1 + e}{1 - e} \tag{6.98}$$

For a three-period stagger, the interpulse periods become $T_1 = T(1 + e)$, $T_2 = T$, and $T_3 = T(1 - e)$, so that the stagger ratio again can be expressed as

$$\gamma_3 = \frac{n_1}{n_3} = \frac{1 + e}{1 - e} \tag{6.99}$$

The fractional stagger period parameter in terms of the stagger ratio can be obtained from (6.98) or (6.99) as

$$e = \frac{\gamma_{2,3} - 1}{\gamma_{2,3} + 1} \tag{6.100}$$

For example, with a two-period 5:4 stagger, equation (6.100) can be solved for $e = 1/9$ so that the stagger delay equals $T_s = T/9$. For a three-period 5:4:3 stagger, $\gamma_3 = 5/3$, which when substituted into (6.100) results in $e = 1/4$ so that $T_s = T/4$.

6.4.2 PRF Stagger with a Single-Canceller MTI

The PRF staggering and destaggering block diagram depicted in Figure 3.24 shows that the destaggered signal presented to the MTI canceller occurs at the average rate (i.e., interpulse period equals T) associated with the stagger. Therefore, the destaggered signal can be processed by a single, dual, or triple canceller (with or without shaping) with no modification to the canceller.

In this section, we derive the velocity response when an n-period stagger is applied to a single canceller. The general approach utilized is appropriate to any order nonrecursive or recursive MTI canceller [19,20], although we discuss only the nonrecursive case represented by binomial and optimal weight cancellers.

The essence of the method is to break the nonperiodic stagger waveform into an equivalent set of n-periodic waveforms that are hence applied to the MTI canceller. The response of the canceller to each of the n-periodic waveforms is then combined in the canceller output to form the composite response of the canceller. For example, the two-period stagger waveform depicted in Figure 3.24 can be decomposed into two waveforms each with period $T_p = 2T$, but separated in time by $T_1 = T + T_s =$

$T(1 + e)$. Similarly, the three-pulse stagger can be decomposed into three waveforms, each with period $T_p = 3T$ separated in time by $T_1 = T + T_s = T(1 + e)$ and $T_2 = T$. Finally, the general n-period stagger can be decomposed into n-periodic waveforms each with period $T_p = nT$ separated in time by $T_1, T_2, T_3, \ldots, T_n$. The waveforms for the two- and three-period staggers are depicted in Figure 6.41.

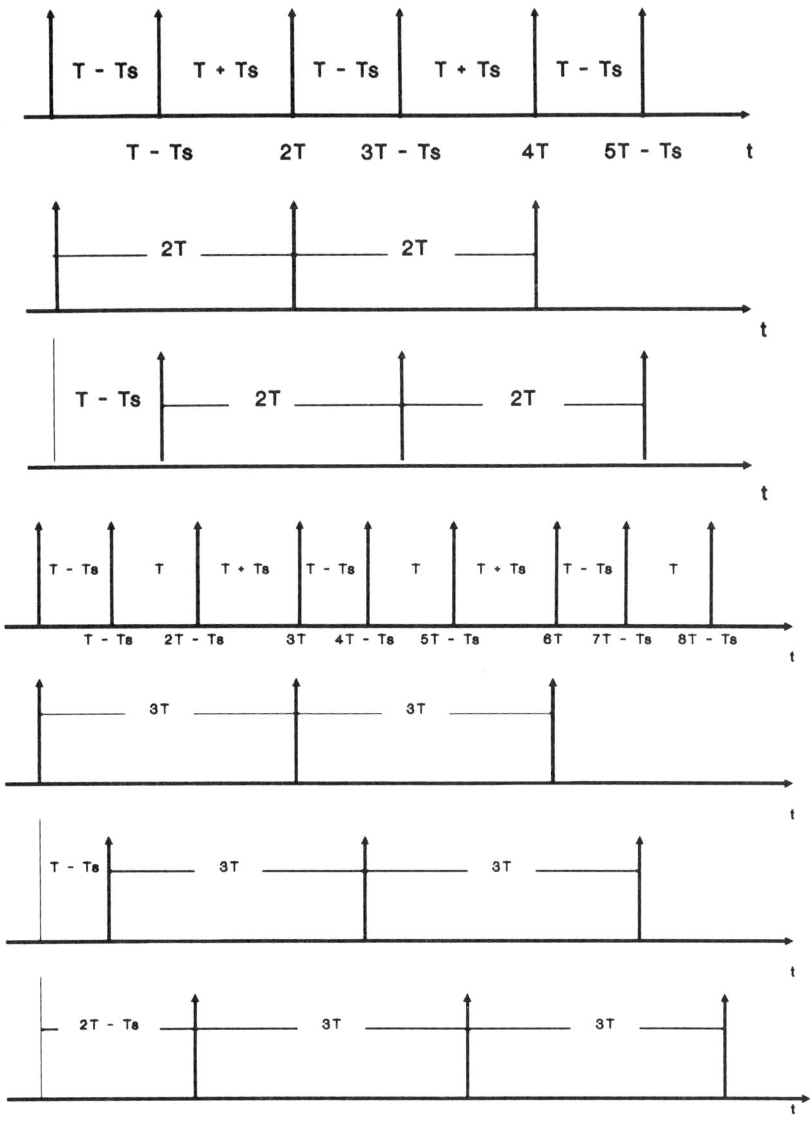

Figure 6.41 Decomposition of two- and three-period stagger waveforms.

The response of a linear filter to an input waveform can be determined using the convolution integral given by

$$g(t) = \int_{-\infty}^{t} r(\tau)\, h(t - \tau)\, \mathrm{d}\tau \tag{6.101}$$

where r_t is the input waveform and $h(t)$ is the impulse response of the filter. The impulse response of an m-pulse nonrecursive MTI is given by (see Section 2.3.2)

$$h(t) = \sum_{k=0}^{m-1} w_k\, \delta(t - kT_k) \tag{6.102}$$

where w_k are the filter weights (e.g., binomial coefficients with alternating signs for a binomial canceller) and T_k are the delay periods associated with the canceller weightings (see Table 3.1 for binomial weights and Table 5.1 for optimal weights). For example, the impulse response of a single canceller ($m = 2$) is given by $h(t) = \delta(t) - \delta(t - T_k)$, where $\mathbf{w} = (w_0, w_1) = (1, -1)$. Note that the delay period associated with the MTI filter weights would generally be given by $T_k = T$ for a conventional (i.e., nonstaggered MTI) canceller, but must coincide with the actual interpulse periods (i.e., $T_i = n_i\, T_b$) for a staggered-PRF MTI in order for it to cancel fixed targets. This synchronization is accomplished automatically in the staggered-PRF MTI by the stagger-destagger action, as depicted in Figure 3.24.

The first step in the derivation of the staggered PRF's velocity response is to compute the response of the single canceller to the first periodic segment of the decomposed staggered waveform. Let us first consider the two-period stagger shown in Figure 6.41. In accordance with the convolution integral given by (6.101), the MTI filter's output response to a sinusoidal input $r(t) = \exp(j\omega t)$ results from the weighted sum of those past input samples generated by the MTI filter. In the case of a single canceller, two output samples are generated by each input sample, where, due to the time reversal in the convolution integral (i.e., $h(t - \tau)$), the second sample occurs in the past and hence contributes to the output. Furthermore, due to the destaggering action of the PRF stagger procedure, the second sample must coincide with the primary sample associated with the second waveform of the decomposed stagger set. The complex frequency response of the first waveform of the decomposed stagger set with a single MTI canceller can then be written as

$$H_1(j\omega) = \mathrm{e}^{j\omega k T_P}(1 - \mathrm{e}^{-j\omega T_2}) \tag{6.103}$$

where T_p is the stagger period equal to $2T$ for the two-period stagger, T_2 is the period separating the two decomposed waveforms of the stagger set equal to $T - T_s$, and $k = 0, 1, 2, \ldots$. In a similar manner, the complex frequency response to the second waveform of the decomposed stagger set equals

$$H_2(j\omega) = \mathrm{e}^{j\omega(kT + T_1)}(1 - \mathrm{e}^{-j\omega T_1}) \tag{6.104}$$

where T_1 is the period that allows the second waveform to be aligned with the first and equals $T + T_s$ for the two-period stagger.

The frequency responses generated by the two decomposed waveforms represented by (6.102) and (6.103) consist of sinusoidal samples with magnitudes given by $|H_i(j\omega)|$, where $i = 1, 2$. The power transfer function can be found by adding the powers from each component as

$$|H(j\omega)|^2 = \frac{1}{2}\sum_{i=1}^{n} |H_i(j\omega)|^2 \tag{6.105}$$

where n is the number of periods in the stagger. Therefore, the power transfer function for the frequency response of a two-period stagger with a single canceller is given by

$$|H(j\omega)|^2 = \frac{1}{2}|1 - e^{-j\omega T_1}|^2 + \frac{1}{2}|1 - e^{-j\omega T_2}|^2 \tag{6.106}$$

This can be written as

$$|H(j\omega)|^2 = 2 - \cos(\omega T_1) - \cos(\omega T_2) \tag{6.107}$$

which further simplifies to

$$|H(j\omega)|^2 = 2[1 - \cos(\omega T)\cos(\omega Te)] \tag{6.108}$$

where T is the average stagger period (i.e., unstaggered period) and $e = T_s/T$ is the fractional stagger delay determined by the stagger ratio (see (6.99). Note that, from (6.108), the nonstaggered transfer function ($e = 0$) becomes $|H(j\omega)|^2 = 4\sin^2(\omega T/2)$ as expected for a single canceller.

Equation (6.106) can be extended to find the normalized power transfer function for an n-period PRF stagger followed by a single canceller:

$$|H(j\omega)|^2 = 2\left[1 - \frac{1}{n}\sum_{i=1}^{n}\cos(\omega T_i)\right] \tag{6.109}$$

Specializing to a three-period PRF stagger where $T_1 = T(1 + e)$, $T_2 = T$, and $T_3 = T(1 - e)$ results in

$$|H(j\omega)|^2 = 2\left[1 - \frac{2\cos(\omega T_e) + 1}{3}\cos\omega T\right] \tag{6.110}$$

Note that this reduces to the single canceller transfer function for the unstaggered case ($e = 0$). The normalized frequency transfer function for a 5:4:3 PRF stagger and a single canceller is depicted in Figure 6.42, where the response is compared against the unstaggered response using the average period. The variation in the passband is on the order of 3 dB for this configuration where the maximum variation occurs at the band edges.

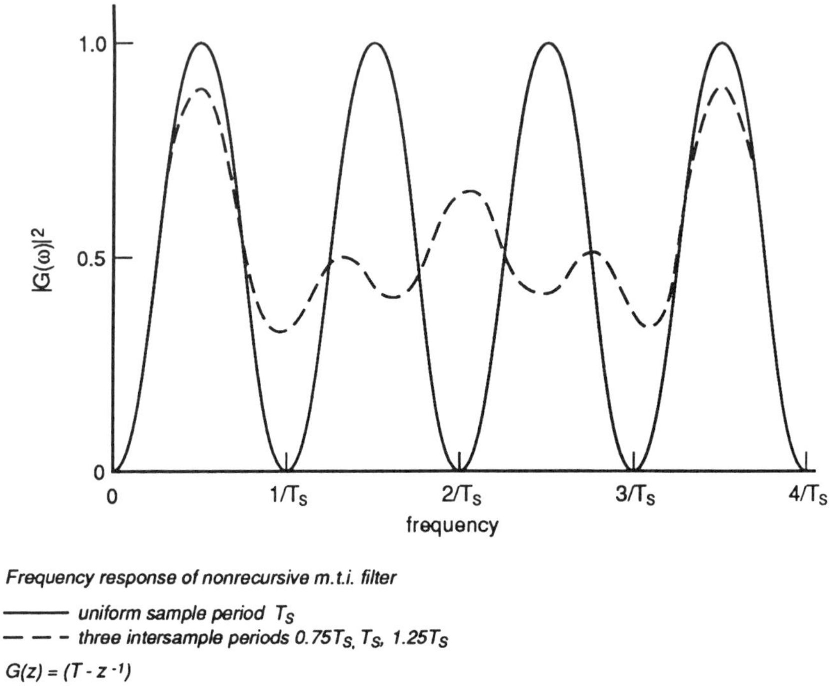

Figure 6.42 Power transfer function of a single-canceller MTI with a 5 : 4 : 3 stagger ratio.

6.4.3 PRF Stagger with a Dual-Delay-Canceller MTI

When a dual-delay MTI canceller is employed with a PRF stagger, three weighted pulses are involved in determining the output response of each segment of the decomposed stagger waveform. The impulse response of a dual-delay line canceller in the stagger configuration is given by $h(t) = w_0\,\delta(t) + w_1\,\delta(t - T_k) + w_2\,\delta(t - T_j)$, where the weight set $\mathbf{w} = (w_0, w_1, w_2)$ is equal to $\mathbf{w} = (1, -2, 1)$ for a double canceller and optimal weights are given in Table 5.1.

The power transfer function response for a double-stagger PRF applied to a dual-delay MTI canceller is then given by

$$|H(j\omega)|^2 = \frac{1}{2}\left|w_0 + w_1\,\mathrm{e}^{-j\omega T_2} + w_2\,\mathrm{e}^{-j\omega(T_1+T_2)}\right|^2 + \frac{1}{2}\left|w_0 + w_1\,\mathrm{e}^{-j\omega T_1} + w_2\,\mathrm{e}^{-j\omega(T_1+T_2)}\right|^2 \tag{6.111}$$

where $T_1 = T + T_s$ and $T_2 = T - T_s$. Specialization to a double canceller ($w_0 = 1$, $w_1 = -2$, $w_2 = 1$) and simplification results in

$$|H(j\omega)|^2 = 6 - 4\cos\omega(T + T_s) - 4\cos\omega(T - T_s) + 2\cos 2\omega T \tag{6.112}$$

which further reduces to

$$|H(j\omega)|^2 = 6 - 8\cos\omega T\cos\omega Te + 2\cos 2\omega T \tag{6.113}$$

where $e = T/T_s$ is the fractional stagger delay determined by the stagger ratio (see (6.100)). Note that for the unstaggered configuration ($e = 0$), (6.113) reduces to $|H(j\omega)|^2 = 16\sin^4(\omega T/2)$, which is the conventional double-canceller transfer function. The velocity response for a 63:65 dual-period stagger into a double canceller is depicted in Figure 6.43 [3]. Note the deep nulls at the passband edges, which are characteristic of stagger systems using a small number of stagger periods.

The power transfer function response for a triple-stagger PRF applied to a double-canceller MTI is given by

$$|H(j\omega)|^2 = \left|1 - 2\,\mathrm{e}^{-j\omega T_3} + \mathrm{e}^{-j\omega(T_2+T_3)}\right|^2 + \left|1 - 2\,\mathrm{e}^{-j\omega T_2} + \mathrm{e}^{-j\omega(T_1+T_2)}\right|^2 + \left|1 - 2\,\mathrm{e}^{-j\omega T_1} + \mathrm{e}^{-j\omega(T_1+T_3)}\right|^2 \tag{6.114}$$

which for a symmetrical stagger $T_1 = T + T_s$, $T_2 = T$, and $T_3 = T - T_s$ results in

$$|H(j\omega)|^2 = \frac{1}{3}[18 - 8\cos\omega(T - T_s) - 8\cos\omega(T + T_s) - 8\cos\omega T + 2\cos\omega(2T - T_s) + 2\cos\omega(2T + T_s) + 2\cos 2\,\omega T] \tag{6.115}$$

This can be further simplified by expressing the periods in terms of the fractional

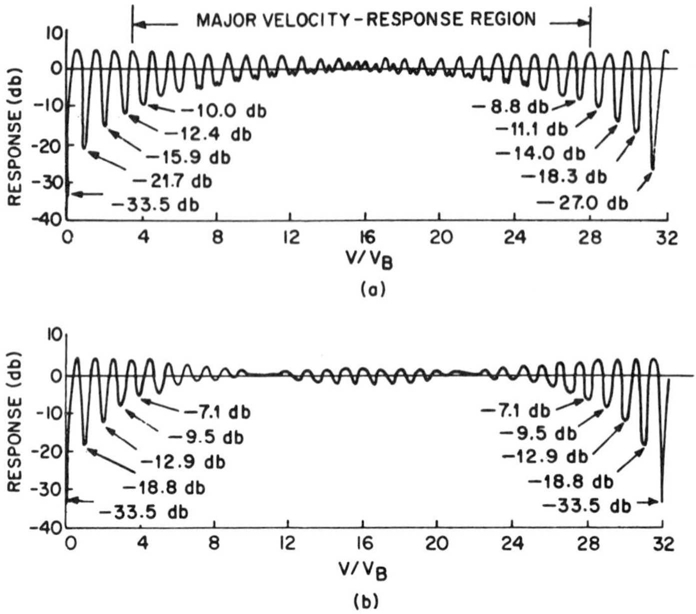

Figure 6.43 Power transfer function of a double-canceller MTI with 63 : 65 and 31 : 32 : 33 stagger ratios.

delay stagger $e = T_s/T$ as $T_1 = T(1 + e)$, $T_2 = T$, and $T_3 = T(1 - e)$. The resulting power transfer function for a triple-stagger PRF and a double canceller is given by

$$|H(j\omega)|^2 = \frac{1}{3}[18 - 8(1 + 2\cos\omega Te)\cos\omega T + 2(1 + 2\cos\omega Te)\cos 2\,\omega T] \tag{6.116}$$

where e can be found in terms of the stagger ratio using equation (6.100). Note that (6.116) reduces to the power transfer function for a double canceller for the unstaggered case ($e = 0$). The velocity response for a 31:32:33 stagger into a double canceller is depicted in Figure 6.43 [3]. Comparison with the double-period stagger configuration shown in the figure indicates that the passband ripple is reduced by the additional stagger period.

For the general n-period PRF stagger with a double MTI canceller, the power transfer response can be expressed as

$$|H(j\omega)|^2 = \frac{1}{2}|1 - 2\,e^{-j\omega T_1} + e^{-j\omega(T_1+T_n)}|^2 + \sum_{i=2}^{n}\frac{1}{2}|1 - 2\,e^{-j\omega T_i} + e^{-j\omega(T_i+T_{i-1})}|^2 \tag{6.117}$$

This can be simplified to

$$|H(j\omega)|^2 = 3 - 2\cos\omega T_1 + \cos\omega(T_1 + T_n) - 2\cos\omega T_n + \sum_{i=2}^{n} [3 - 4\cos\omega T_i + \cos\omega(T_i + T_{i-1}) - 2\cos\omega T_{i-1}] \quad (6.118)$$

where T_i, $i = 1, 2, \ldots, n$ are the periods that make up the overall stagger period (i.e., $T_p = \Sigma_{i=1}^{n} T_i$).

The optimum choice of the stagger ratio depends on the velocity range over which there must be no blind speeds and the permissible depth of the first null in the velocity response curve. A four-period stagger ratio is considered a good choice for some applications [3]. It is advantageous to alternate the long and short stagger periods in a sequence to keep the transmitter's duty cycle as constant as possible. The velocity response for a 51 : 62 : 53 : 61 : 58 PRF stagger ($n = 5$) into a double canceller is depicted in Figure 6.44. Note that the passband ripples are reduced from the two- and three-period staggers depicted in Figure 6.44 and the four-period stagger shown in Figure 3.27.

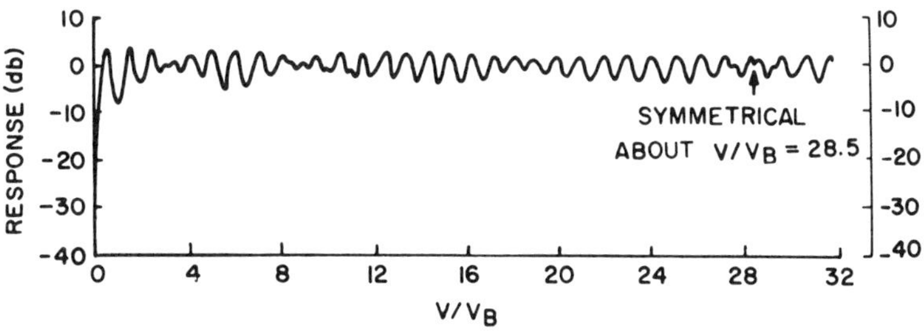

Figure 6.44 Power transfer function of a double-canceller MTI with a 51 : 62 : 53 : 61 : 68 stagger ratio.

6.4.4 Synthesis of Staggered MTI Cancellers

The previous sections described the response of simple nonrecursive MTI filters with PRF sampling. Several velocity responses (see Figures 6.41, 6.44, and 3.27) were shown that indicated the effects of varying the stagger order and ratios. It is apparent that different bandpass shapes can be achieved by varying the stagger parameters as well as by eliminating blind speeds within the expected velocity range of the target.

Several procedures have been proposed that allow achieving a specified transfer function by using a staggered-PRF MTI canceller [20–22]. However, the procedures generally are complicated and a computer must be used to synthesize a specified staggered MTI design. This is in contrast to the shaped canceller design procedure described in Section 6.3.2.1, where conventional filter theory can be translated into the sampled-data domain by using a simple bilinear transformation. The advantage of the staggered-PRF configuration over the shaped canceller approach is that it provides a superior transient response due to the use of a simple filter (i.e., single or double canceller). The disadvantage of the staggered-PRF design approach is that multiple-time-around clutter is not cancelled.

Some insight into the synthesis of staggered-PRF MTI cancellers can be gained by examining their pole-zero response in the z-plane [20]. First, for the nonuniform sampling that occurs with staggered-PRF designs, the z-transform must be redefined to coincide with the periodic component of the stagger that has period $T_b = 1/f_{bs}$, where f_{bs} is the doppler frequency of the first blind speed. The z-transform is then defined as

$$z_b = e^{j\omega T_b} \tag{6.119}$$

The first blind-speed period is related to the average period (unstaggered period) by

$$T = \frac{n_2 + n_2 + \ldots + n_n}{n} T_b \tag{6.120}$$

where n_i are the stagger numbers and n is the number of stagger periods. Specialization to a symmetrical triple stagger the average period is

$$T = n_2 T_b \tag{6.121}$$

so that the z_b-transform of a single canceller $[H(z) = 1 - z^{-1}]$ is given by

$$H(z_b) = (1 - e^{-j\omega n_2 T_b}) = 1 - z_b^{-n_2} \tag{6.122}$$

For example, for a 3 : 4 : 5 stagger ratio, $H(z) = 1 - z_b^{-4}$ and four poles and zeros would be distributed around the unit circle producing null responses at $f_b = 1/T$.

Next, consider the pole-zero pattern of a three-period staggered PRF design with a single canceller. This can be determined from the extension to (6.106) as

$$|H(z_b)|^2 = |1 - z_b^{-n_1}|^2 + |1 - z_b^{-n_2}|^2 + |1 - z_b^{-n_3}|^2 \tag{6.123}$$

Specialization to a 3 : 4 : 5 stagger results in

$$|H(z_b)|^2 = |1 - z_b^{-3}|^2 + |1 - z_b^{-4}|^2 + |1 - z_b^{-5}|^2 \tag{6.124}$$

The zeros of this function can then be found by solving the resulting polynomial:

$$|H(z_b)|^2 = H(z_b)\, H^*(z_b) = 0 \tag{6.125}$$

When this is done, we find that the zeros move off the unit circle, which provides for extension of the blind speeds, but also causes a deterioration of the clutter attenuation in the vicinity of $z_b = 1$ (see Section 3.2.6) [20].

This discussion indicates that the effect of the PRF stagger is to move the filter zeros away from the unit circle, which is desirable except in the case of the zeros located at $z_b = 1$. This situation can be corrected using time-variable weights for each stagger interpulse period [3,20]. The effect of this on the staggered-PRF transfer function is given by

$$|H(z_b)|^2 = |w_0 + w_1 z_b^{-n_1}|^2 + |w_0' + w_1' z_b^{-n_2}|^2 + |w_0'' + w_1'' z_b^{-n_3}|^2 \tag{6.126}$$

The synthesis procedure is then to adjust the weights (time variable) for each set of the decomposed stagger (e.g., w_0, w_1; w_0', w_1'; w_0'', w_1''), so that a zero is produced at $z_b = 1$, which provides a deep null for fixed ground clutter [20]. Using the multiple zeros produced by the PRF stagger we also can adjust the weighting coefficients to produce the optimum weights (see Table 5.1) that maximize the MTI improvement factor under the condition that all target velocities are equally likely [20].

REFERENCES

1. D.K. Barton, *Modern Radar System Analysis*, Artech House, Norwood, MA, 1988.
2. D.C. Schleher, ed., *MTI Radar*, Artech House, Norwood, MA, 1978.
3. W. Shrader, "MTI Radar," in M.I. Skolnik, ed., *Radar Handbook*, 1st Ed., McGraw-Hill, New York, 1970.
4. D.C. Schleher, *Introduction to Electronic Warfare*, Artech House, Norwood, MA, 1986.
5. T. Ap Rhys and G. Andrews, "AEW Radar Antennas," *Antennas for Avionics*, AGARD Conf. Preprint No. 139, 1974.
6. G. Andrews, "Airborne Motion Compensation Techniques, Evaluation of TACCAR," NRL Rep. No. 7407, Washington, DC, April 12, 1972.
7. G. Andrews, "Airborne Motion Compensation Techniques, Evaluation of DPCA," NRL Rep. No. 7426, Washington, DC, July 20, 1972.
8. H. Sachs and J. Krstansky, "Controlling RFI Susceptibility in Receivers," *Electronic Industries*, September 1960.
9. L. Lemley, W. Gleason, and G. Gross, "RF Receivers for Electronic Warfare," NRL Rep. 8737, Washington, DC, May 1986, ADB105737.
10. R. Emerson, "Some Pulsed Doppler MTI and AMTI Techniques," Rand Corp. Rep. R-274, March 1954.
11. J. Capon, "Optimum Weighting Functions for the Detection of Sampled Signals in Noise," *IEEE Trans.*, Vol. IT-10, April 1964.
12. R. Bellman, *Introduction to Matrix Analysis*, McGraw-Hill, New York, 1960.
13. G. Andrews, "Optimal Radar Doppler Processors," NRL Rep. 7727, Washington, DC, May 1974.
14. D.C. Schleher and D. Schulkind, "Optimization of Nonrecursive MTI," *IEE Int. Radar Conf.*, London, October 1977.

15. C. Fowler, A. Uzzo, and A. Ruvin, "Signal Processing Techniques for Surveillance Radar Sets," *IEEE Trans.*, Vol. MIL-5, No. 2, April 1961.
16. D. Linden and B. Steinberg, "Synthesis of Delay Line Networks," *IEEE Trans.*, Vol. ANE-4, No. 1, March 1957.
17. W. White and A. Ruvin, "Recent Advances in the Synthesis of Comb Filters," *IRE Nat. Conv. Rec.*, 1957.
18. G. Bagley, "A Survey of Cancellation Versus Integration for Radar Clutter Rejection," NRL Rep. No. 2880, Washington, DC, August 1974.
19. H. Thomas and N. Lutte, "Z-Transform Analysis of Nonuniformly Sampled Digital Filters," *IEE Proc.*, Vol. 119, No. 11, November 1972.
20. H. Thomas, N. Lutte, and M. Jelffs, "Design of MTI Filters with Staggered PRF, a Pole-Zero Approach," *IEE Proc.*, Vol. 121, No. 12, December 1974.
21. R. Roy and O. Lowenschuss, "Design of MTI Detection Filters with Nonuniform Interpulse Periods," *IEEE Trans.*, Vol. CT-17, No. 4, November 1970.
22. P. Prinsen, "Elimination of Blind Velocities of MTI Radar by Modulating the Interpulse Period," *IEEE Trans.*, Vol. AES-9, No. 5, September 1973.

Chapter 7
Pulsed Doppler Systems

By definition, a doppler radar is one that uses the Doppler effect, associated with the radial component of relative radar-target velocity, to select targets having particular radial velocities. A pulsed transmitter utilized in a doppler radar is referred to as a *pulsed doppler radar*. The signals returned from targets illuminated by a pulsed doppler radar are generally processed in a set of narrowband filters that coherently integrate the signals to which they are tuned. The filters both resolve and enhance signals within a particular velocity band. In search radar applications, the bank of contiguous doppler filters is used to cover the velocity band of interest to extract specific targets and reject clutter and targets outside that velocity band. Once the target is acquired, only one filter may be needed for tracking.

The pulsed doppler approach differs from MTI (see Chapter 6) in that the doppler filter performs coherent integration of the selected signal within a narrow spectral band, whereas the MTI passes targets through a broad comb filter response band. From a theoretical standpoint, the PD radar is the optimum processor for detecting a doppler-shifted return signal in Gaussian clutter and receiver noise when the doppler frequency of the target is known *a priori* (see Chapter 5). This is contrasted to the MTI approach where all target velocities are assumed equally likely.

Pulsed doppler radars can be categorized as low, medium, or high PRF, according to whether doppler (low-PRF), range (high-PRF), or range and doppler (medium-PRF) are ambiguous (see Chapter 2). High- and medium-PRF radars have been used primarily in airborne applications, where platform motion clutter spectral spreading causes mainlobe ground clutter to extend over tens of knots and where short-range clutter effects are minimized by the aircraft altitude. Use of these range-ambiguous waveforms in surface-based radar is made difficult by the presence of short-range clutter, which competes with long-range targets and has a large power advantage over the targets due to the nth-power increase of clutter power range (i.e., $C/S = k\,(R_t/R_c)^n$, where $n = 3$ for distributed surface clutter and $n = 4$ for discrete clutter), with respect to the target power. This has caused low-PRF PD radars to assume an increasingly important role in surface-based surveillance because of their superior performance in clutter rejection and the decreasing cost of modern digital processors [1]. An example of a low-PRF PD radar is the moving detector described in Section 1.4.

The most difficult choice among the various PD waveforms involves the airborne case, where high, low, and medium PRFs might be advantageous depending on the specific application. For example, in AEW applications (e.g., the E-2C and AWACS radars) that involve large antenna installations (i.e., on the order of 25 to 30 ft), both low (the E-2C) and high PRF (the AWACS) have been successfully implemented to detect small RCS airborne targets over both land and water [2]. The high-PRF design has the advantages of rejecting low-speed targets due to ground vehicular traffic and higher angular resolution (through the use of a higher transmitter frequency), which provides an ECCM advantage and better rejection of chaff. A medium-PRF design [2–4] has been attempted (i.e., the Nimrod radar) for this application but was apparently unsuccessful due in part to the use of a relatively small azimuth antenna aperture (i.e., on the order of 8 ft) that increased both the clutter power and spectral spreading (see (6.4)) and the computer complexities of handling the large clutter intake while resolving the various range and doppler ambiguities.

The unambiguous range (R_u in nmi) as a function of PRF (in kHz) is given by

$$R_u = \frac{80.91}{\mathrm{PRF_{kHz}}} \tag{7.1}$$

This relationship, plotted in Figure 7.1, illustrates how rapidly the range decreases with an increase in PRF. For example, from an unambiguous range of 400 nmi at a PRF of 200 Hz, the unambiguous range decreases to 10 nmi at a PRF of 8 kHz. The unambiguous velocity (in knots) is also related to the PRF (in kHz) as

$$v_u = 971\lambda \ \mathrm{PRF} \tag{7.2}$$

where λ is the wavelength in meters. By combining (7.1) and (7.2), the result is

$$R_u \, v_u = \frac{2.3565 \times 10^4}{f_{\mathrm{GHz}}} \tag{7.3}$$

where R_u is in nmi, v_u is in knots, and the radar's frequency is in GHz. Equation (7.3) can be expressed as an inequality given by $R_u \, v_u \leq 2.3565 \times 10^4 / f_{\mathrm{GHz}}$, which establishes the unambiguous range and velocity boundary for the low-PRF waveform.

As an example, consider a low-PRF AEW design that operates at 425 MHz with an azimuth aperture of 25 ft. Consider that we want an unambiguous range of 300 nmi, which from (7.3) results in a first blind speed of 185 knots. The spectral spreading in velocity units at the 3 dB point can be expressed as

$$\Delta v = v_{ac} \theta_B \cos\phi_0 \sin\theta_0 \tag{7.4}$$

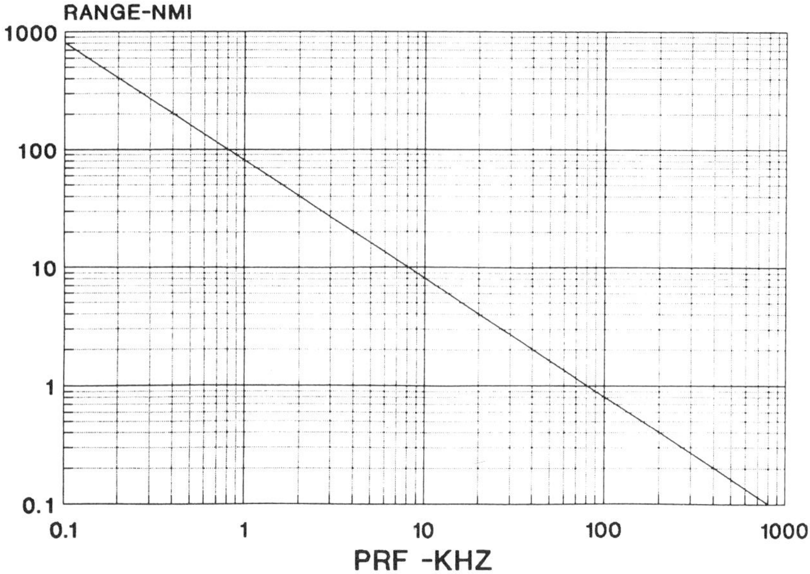

Figure 7.1 Unambiguous range-velocity contour.

where v_{ac} is the aircraft's velocity, θ_B is the azimuth beamwidth in rad, ϕ_0 is the depression angle, and θ_0 is the azimuth pointing angle with respect to the ground track. For a 25-ft aperture at UHF, the resulting beamwidth is approximately $\theta_B = \lambda/D = 1/10.8$ rad, resulting in a worst-case ($\phi_0 = 0$, $\theta_B = 90°$) velocity spread for a 300-knot aircraft speed of 28 knots. This thus represents a velocity blockage of $2 \times 28/185$ or 0.3 of the velocity region covered by the radar. The radar on the average would then be capable of only detecting 70% of the targets if they were spread equally in velocity over the entire range of possible target velocities (i.e., 600 knots for aircraft targets). This could be improved by the use of a displaced phase center antenna, which would compensate for the motion-induced spectral spreading (see Sections 3.2.7 and 6.2.2).

As a further example, consider an X-band AI radar with an azimuth aperture of 2.5 ft. Consider that an unambiguous range of 30 nmi is desired in a low-PRF mode. From (7.2), this results in a first blind speed of about 80 knots. The ground clutter spectral spreading ($\psi_0 = 0$, $\theta_B = 30°$) in velocity units for an interceptor speed of 1500 knots is about 30 knots. This would block 75% of the velocity coverage, which is not acceptable from a design standpoint. To accommodate this situation, we customarily increase the radar's PRF, resulting in the use of medium- or high-PRF waveforms for AI applications.

This example illustrates the problem of employing a low-PRF doppler radar design for a small aperture high-speed AI radar due to the excessive spreading of mainlobe ground clutter about the PRF lines, which blocks a significant portion of the velocity region associated with the target. The other significant effect associated with a low-PRF design concerns the doppler shift induced on the ground clutter that enters the radar through the antenna sidelobes. This sidelobe clutter has velocity components that extend up to the aircraft's ground speed and, in a low-PRF design, generally are aliased throughout the velocity region occupied by the target. The maximum improvement factor of the low-PRF doppler radar against clutter is thus limited to the integrated two-way antenna mainlobe-to-sidelobe ratio (see Section 6.2.2). This is significant in many airborne installations due to the difficulty of achieving a low sidelobe antenna design in this restrictive environment.

The limitation of the airborne low-PRF PD radar led to the development of airborne high-PRF designs whose performance in the case of closing targets does not depend upon the level of antenna sidelobes [5]. The ground clutter spectrum returned to a high-PRF design is depicted in Figure 7.2 and extends to doppler frequencies corresponding to the aircraft's ground speed given by $f_d = \pm 2v_r/\lambda$. The major responses in the spectrum corresponds to the altitude return (from beneath the aircraft) and the mainlobe clutter response. The clutter spectrum is ambiguous about the PRF lines in accordance with the sampling theorem for pulsed waveforms. From the illustration, we can see that the high-PRF design provides a clutter-free region that extends from the most positive part of the central line of the clutter spectrum (due to the aircraft's forward motion) to the most negative part of the next ambiguous line of the clutter spectrum (due to the antenna's backlobes). In this clutter-free region, target detection is limited only by receiver noise. The minimum PRF is then determined (see Figure 7.2) as the sum of the doppler frequency of the target most rapidly closing ($f_{dt} = 2\dot{R}_{max}/\lambda$) plus the maximum sidelobe clutter frequency ($f_{da} = 2v_r/\lambda$), which is determined by the radar's velocity (v_r). Note, however, that, if both closing and receding targets must be resolved, then the minimum PRF must be increased to $PRF_{min} > 2f_{dt} + f_{da}$. Alternatively, this can be stated as $PRF_{min} > 2f_{da} + f_{dt}$ as the sidelobes extend both ahead of and behind the aircraft. This ensures that, for example, a closing target response associated with the central line can be distinguished from a receding target response associated with the next ambiguous line of the spectrum. The shorter the wavelength, the higher are the doppler frequencies of the clutter and

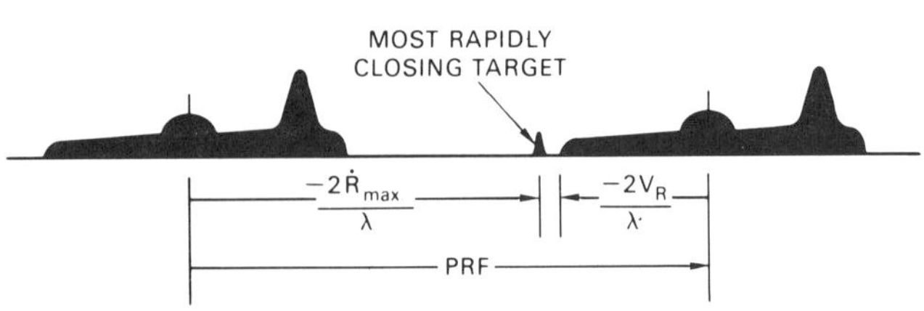

Figure 7.2 Ground clutter spectrum for an airborne high-PRF PD radar.

the target and, hence, the higher the required PRF. Typically, in airborne interceptor (AI) applications at X-band frequencies, the PRF is on the order of 100 to 300 kHz.

A clutter and target vector diagram that shows the effects of the various types of clutter (i.e., mainlobe, altitude line, sidelobe) for different target headings is depicted in Figure 7.3. An aircraft with a scanning antenna is flying in the direction shown. Two targets are indicated, one directly ahead of the aircraft and one at an azimuth angle of 45°. Each target is located at the center of the circles. The three circles, labeled 1, 2, and 3, represent particular ratios of target ground velocity to interceptor ground velocity. A target vector showing velocity ratio and direction is drawn from the center of each target position. The tip of the vector will show whether the target is in the clear or is obscured by clutter. In Figure 7.3, the two target vectors show the targets to be outside the clutter. If both targets were traveling in the opposite direction, the one straight ahead would be obscured by opening sidelobe clutter, and the target on the 45° bearing, because of its high velocity, would be clear of opening clutter.

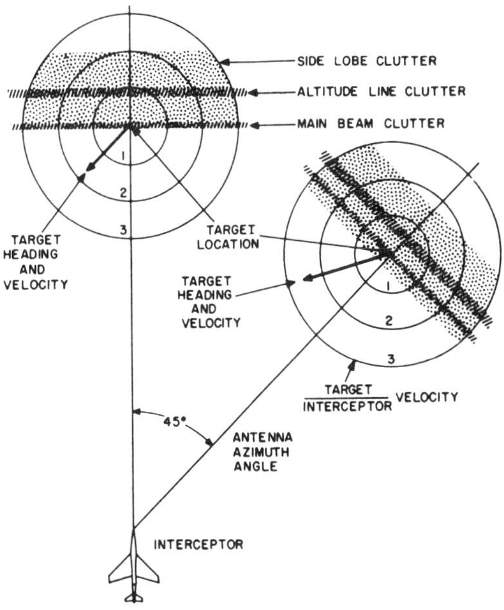

Figure 7.3 Clutter and target vector diagram for a high-PRF doppler radar.

Figure 7.3 shows that closing targets and high-speed receding targets (those moving at twice the interceptor speed) can be detected in the clutter-free region. All other velocity targets must compete with clutter to be detected in a high-PRF design. The clutter in the high-PRF case is generally severe due to the large number of ambiguous ranges that results from the waveform design which causes short-range clutter to fall in the same range cell as a long-range target. The result of this effect is that the high-PRF design is ineffective under tail-chase conditions, which may be a major difficulty in some applications.

The medium-PRF airborne PD radar was developed to get around some of the limitations of low- and high-PRF designs in fighter applications [5, 6]. In particular, the medium-PRF airborne PD radar is designed to work for tail chases as well as closing engagements, thereby providing good all-aspect coverage. For operation, at X-band, medium-PRFs range from about 8 to 16 kHz. If the maximum operating range is not exceptionally long, the PRFs can be set high enough to provide adequate separation between the periodic repetitions of the mainlobe clutter spectrum without incurring particularly severe range ambiguities. Mainlobe clutter can then be isolated from the bulk of the target return on the basis of doppler frequency, and individual targets can be isolated from the bulk of the sidelobe clutter through a combination of range and doppler resolution. Also, ground-based moving targets, being close to the frequency of the mainlobe clutter, can be rejected along with it, without rejecting an unacceptable amount of possible target return (see Figure 7.4).

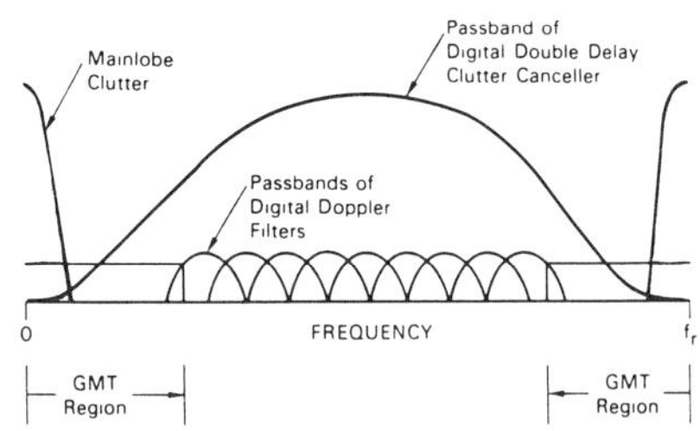

Figure 7.4 Rejection of ground moving targets in a medium-PRF PD radar.

Due to the ambiguous range and velocity responses associated with a medium-PRF design, the radar must contend with blind zones both in the doppler spectrum and the range interval being searched. Doppler blind zones are less severe with medium PRFs than with low PRFs because mainlobe clutter covers a much smaller portion of the doppler frequency spectrum. Range blind zones result when target echos are suppressed by sidelobe clutter simultaneously received from shorter ranges or eclipsed by transmitter pulses. Figure 7.5 depicts how blind zones occur when target echos at long ranges compete with range ambiguous sidelobe clutter responses from shorter ranges due to the ambiguous range response. Blind zones can be suppressed by switching among a number of different PRFs, judiciously selected to provide clear zones in both the doppler and range domains. Typically, the radar may be cycled through a fixed number of fairly widely spaced PRFs. For example, if 8 PRFs are used, the target is in the clear on any three of these, and the echos exceed the detection

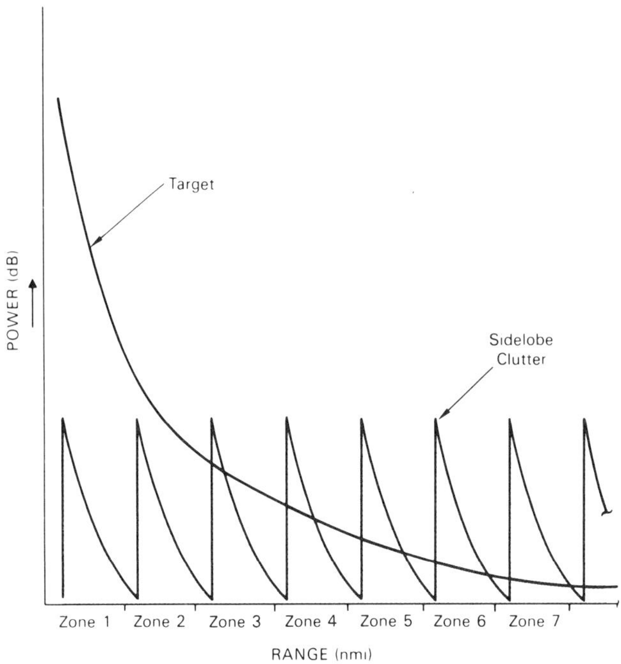

Figure 7.5 Variation of target and sidelobe clutter response in a medium-PRF design.

threshold on all three, the target will be deemed to have been detected. The judicious selection of the various PRFs associated with a medium-PRF radar is one of the major system decisions associated with the design of this type of radar.

An important class of sidelobe returns that are particularly troublesome to medium-PRF airborne PD radars comes from structures on the ground (i.e., water towers, large buildings) that have an exceptionally large radar cross section (typically from 20 to 60 dB/m^2). The return from these discrete clutter targets, although reduced by the two-way antenna sidelobes, can be as large as those of desired targets. The strength of these targets is enhanced in medium-PRF designs by ambiguous range, which allows a short-range discrete target to fall in the same range cell as a long-range airborne target. Furthermore, radar platform motion induces a doppler shift onto these discrete targets, which translates them into the passband of the target doppler filters where they are detected as if they were aircraft in the radar's mainlobe.

One way of dealing with these unwanted targets is to provide the radar with a guard channel. This consists of a separate receiver-processor with its input supplied by a small horn antenna mounted on the radar antenna. The width of the horn's mainlobe is sufficient to encompass the entire region illuminated by the radar antenna's principal sidelobes, and the gain of the horn's mainlobe is greater than any of the sidelobes. Any detectable target in the radar antenna's sidelobes will therefore produce a stronger output from the guard receiver than from the main radar receiver. When this occurs, the output of the main radar receiver is blanked, thereby eliminating the unwanted target.

7.1 PULSED DOPPLER CONFIGURATIONS

There are many different implementations of PD radars depending on the application and whether low-, medium-, or high-PRF waveforms are utilized. Implementations for low-PRF (MTI) configurations were discussed in Section 6.1, and block diagrams of MOPA and coherent-on-receive types were given in Figures 2.20 and 2.21. In this section, we shall describe high-PRF configurations that apply to airborne and surface applications; medium-PRF configurations will also be discussed.

A block diagram of an airborne high-PRF configuration is depicted in Figure 7.6 [7]. A master frequency control is utilized to control the carrier and PRFs and to provide coherent references for the receiver mixing processes (see Figure 1.10). This process is identical to the MOPA configuration described in Figure 2.20 and

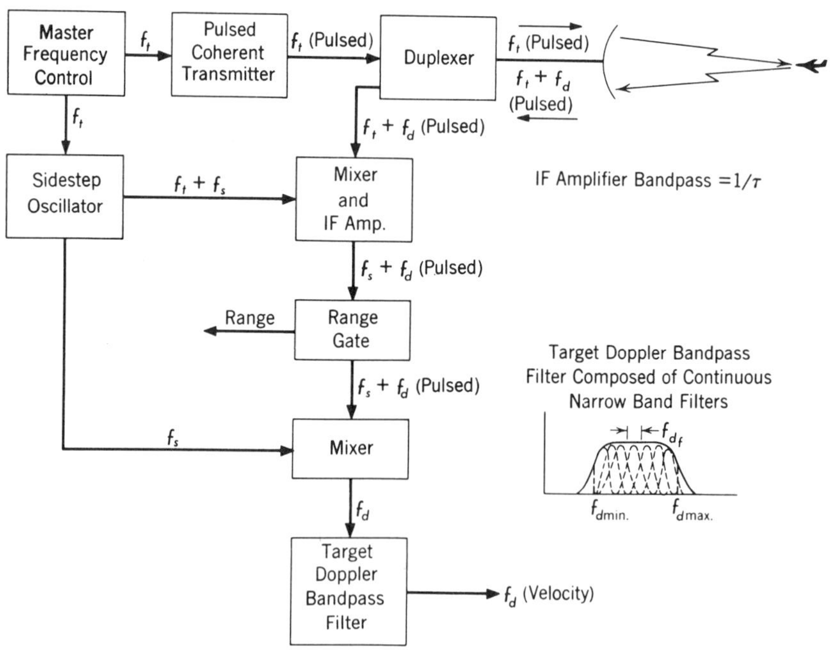

Figure 7.6 Block diagram of a high-PRF PD radar.

Section 6.1 for low-PRF radars. However, the stability requirements are far more severe when used in a high-PRF PD due to the translation of short-range clutter components into target doppler filters, resulting from the modulation effects of transmitter noise sidebands, which are referred to as *transmitted clutter* (see Section 3.3).

The master frequency control (STALO) is generally derived from a highly stable crystal oscillator that generates a low fundamental frequency multiplied up to form the STALO operating frequency. The crystal oscillator is also typically employed to generate timing signals (e.g., digital clock) and the sidestep oscillator (COHO) frequency so that all signals in the radar are locked onto a single stable reference. The STALO is then applied to a pulsed coherent transmitter that amplifies this low-powered signal to the power level required for transmission. The power amplifier generally consists of a solenoid-focused gridded traveling wave tube (TWT) or a high-powered klystron amplifier.

The coherent pulsed transmitter is connected to the antenna through a duplexer, which is a high-powered switch that serves the dual function of (1) passing the high-power transmitted pulses to the antenna while blocking their path to the receiver on transmission and (2) passing the low-power returns to the receiver while blocking their path to the transmitter on receive. The duplexer is one of the critical components in high-PRF radar because the recovery time of this element determines the maximum duty factor of the high-PRF waveform.

In high-PRF airborne designs, PRFs might range from 100 to 300 kHz, whereas, in medium-PRF airborne radars, the PRF might range from 10 to 30 kHz. The duty factor of the transmitter is given by the ratio of the radar's pulsewidth (τ) to its interpulse period (T) as

$$d_u = \frac{\tau}{T} = \tau \cdot \text{PRF} \tag{7.5}$$

In high-PRF airborne radars, the duty factor might approach 0.5, which is the upper limit set by the necessity that the receiver must be blanked during transmission and the duplexer that has a finite recovery time must gate the receiver on during reception. Comparable duty factors for low-PRF designs might range from 0.001 to a maximum of 0.2 if a large pulse compression factor is employed [6]. For medium-PRF radars, the duty factor might range from 0.01 to 0.03, which reflects the moderate pulse compression ratios that are sometimes employed to provide higher range resolution [8].

The average power transmitted by the radar is given by the peak power multiplied by the duty factor:

$$P_A = P_p \, d_u \tag{7.6}$$

In high-PRF designs, we typically use only the power returned to the radar in the central spectral line of the transmitted spectrum for detection in the receiver. This is equivalent to a CW radar that transmits an average power (P_{CW}) of

$$P_{\text{CW}} = P_A\, d_u = P_p\, d_u^2 \tag{7.7}$$

Note that there is no dilution of signal-to-noise ratio in this process because the receiver noise power is diminished in exactly the same ratio (d_u) as the signal power when only the central line of the spectrum is processed.

The receiver portion of the airborne high-PRF PD radar shown in Figure 7.6 extracts the doppler-shifted target return from a background that consists of the received clutter and the inevitable receiver noise. The clutter (i.e., primarily from the ground) can be resolved into components resulting from (1) mainlobe clutter, (2) altitude-line clutter, and (3) sidelobe clutter (see Figure 1.14). Mainlobe clutter results from clutter returns (i.e., surface, weather, chaff), which enter the receiver through the mainlobe antenna pattern and are large in magnitude (e.g., 60 to 90 dB above receiver noise), but are concentrated in a relatively narrow spectral band, corresponding to the closing velocity of the radar to those scatterers located within the main antenna lobes. Altitude-line clutter results from surface reflections directly beneath the aircraft, has large magnitude, and its spectrum is centered about zero frequency. Sidelobe clutter is attenuated from mainlobe clutter by the ratio of the relative sidelobe-to-mainlobe antenna response, but exhibits a wide spectrum extending over a frequency range that corresponds to the aircraft's closing and receding (from backlobes) velocity with respect to the surface. The receiver and associated processor must be linear over the full dynamic range of the mainlobe clutter, which may be as large as on the order of 80 to 90 dB above receiver noise to avoid spectral spreading of mainlobe clutter (see Sections 3.2.5 and 6.2.3).

The first step in the receiving process is to translate the return signal to IF by mixing with the sidestep oscillator, which is cohered to the master reference oscillator. The signal is then range-gated into range cells which generally correspond to the range resolution of the radar (i.e., the effective pulsewidth of the radar). Range-gating serves to reduce the clutter intake into the radar by excluding all returns that are not folded into the range cell occupied by the expected target return (see Figure 1.13). Note that in a high-PRF radar design, the range is highly ambiguous causing many clutter returns at different ranges than the target to compete with the target return.

Multiple range gates can be distributed throughout the interpulse interval depending upon the radar's duty factor. When duty factors approaching 0.5 are utilized, only one range gate is employed, and the return from different ranges cannot be isolated with range gates. In this case, receiver blanking serves the function of a single range gate. However, if the duty factor is much less than 0.5, a significant advantage can be gained by employing multiple range gates in a high-PRF design.

Note, however, that range gating must precede doppler filtering so that each range gate added entails adding a significantly larger number of doppler filters. This is particularly significant in a high-PRF design (compared to a medium- or low-PRF design) as a great many more doppler filters are required to provide the same doppler resolution for high-PRF operation as for a medium-PRF radar.

After range-gating, the next step in the receiving chain for a high-PRF design is to remove mainlobe and altitude clutter returns. One method employs notch filters centered at the mainlobe and altitude-line clutter center frequencies [9]. However, we may want to process only the doppler-clear region, that is, the doppler region extending from $f_d = 2v_r/\lambda$, particularly if the radar utilizes interlaced medium- and high-PRF waveforms (the medium-PRF design is more effective in the sidelobe clutter region). Advantageous in this case is to employ a bandpass filter that passes the signals having doppler frequencies which occur in the clear region. A design that uses an analog-filter configuration to accomplish this function is depicted in Figure 7.7 [6].

In the analog processor shown in Figure 7.7, the range-gated IF signal is shifted in frequency to place the highest central-line sidelobe clutter frequency just below the band of frequencies that the filters are designed to pass. This frequency is equal to $f_d = 2v_r/\lambda$ and can be supplied by using information from the aircraft's inertial navigation system. The first filter (the analog-clear-region filter) eliminates all but the central-line frequencies of the doppler spectrum. As a result of eliminating all

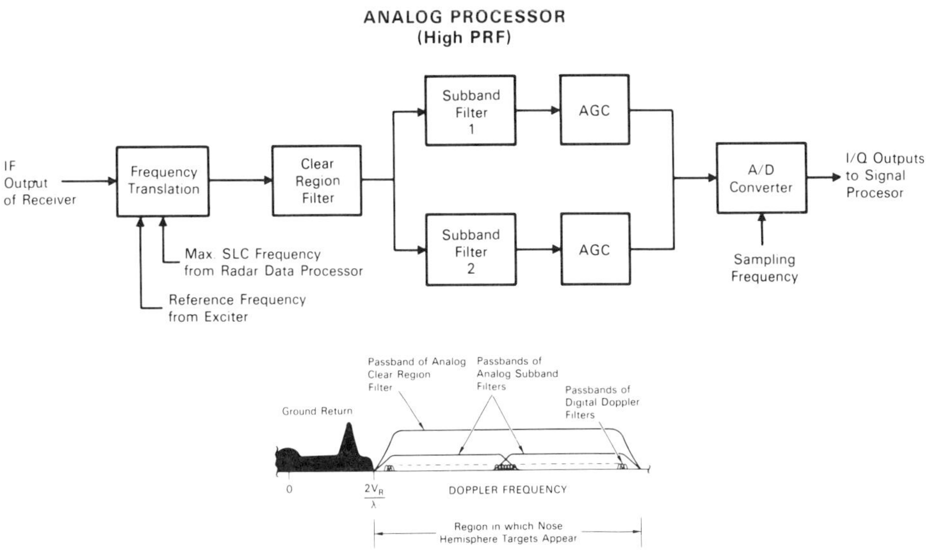

Figure 7.7 High-PRF clear-region doppler bandpass filter.

but the central-line return, the output of the analog-clear-region filter is a continuous signal. Because the filter's passband spans only the clear region of the central-line return, this signal consists primarily of target echos immersed in receiver noise and mainlobe clutter residue.

Two analog subband filters further attenuate the clutter and divide the clear region into two contiguous bands. Automatic gain control is applied separately to each subband so that a strong return in one (e.g., clutter residuals) will not desensitize the other. The filter output after AGC is then digitized in an A/D converter for subsequent digital processing. The filters must provide over 80 dB of attenuation to reject the mainlobe clutter. This function is generally performed by using analog processing to reduce the dynamic range required of the A/D converter. To avoid creating doppler ambiguities, the sampling rate of the A/D converter must be at least equal to the width of the doppler-clear region in which targets are expected to appear. This width is equal to the doppler frequency that corresponds to the fastest anticipated target.

Once digitized, the I and Q components of the return are supplied to the digital signal processor depicted in Figure 7.8. These returns received during the period the main antenna lobe dwells on the target (T_d) are amplitude weighted and supplied to a bank of doppler filters. The optimal bandwidth of each doppler filter is given by $B_D = 1/T_d$ so that the resulting number of doppler filters in the clear region is equal to

$$n_f = \frac{f_{dt}}{B_D} = \frac{2v_t T_d}{\lambda} \tag{7.8}$$

where v_t is the velocity of the fastest expected target. Note that the coherent dwell often is divided into several coherent processing intervals. The number of doppler filters in the clear region is then reduced to $n_f = 2v_t\,\text{CPI}/\lambda$ and postdetection integration of $n' = T_d/\text{CPI}$ samples is generally employed. Weighting reduces the return's spectral sidelobes, thereby providing additional noise and clutter rejection in each doppler filter and preventing strong targets at one frequency being detected in adjacent filters.

The output of the doppler filters usually formed using the FFT algorithm are then envelope detected, normalized using CFAR processing, and threshold detected to form a target detection. At this point, target detections are highly ambiguous in range, a problem that must be resolved to complete the detection process. In Figure 7.8, an FM ranging procedure is indicated, which resolves the range ambiguities by changing the frequency of the radar transmitter (see Section 7.3.2). Alternatively, the range ambiguities could be resolved by transmitting multiple PRFs during the period the mainlobe of the antenna dwells on the target.

An alternative high-PRF receiver provides detection of doppler-shifted targets using all-digital processing; it is depicted in Figure 7.9. In this configuration, mainbeam clutter is eliminated by using an MTI canceller (comb) filter, rather than the analog

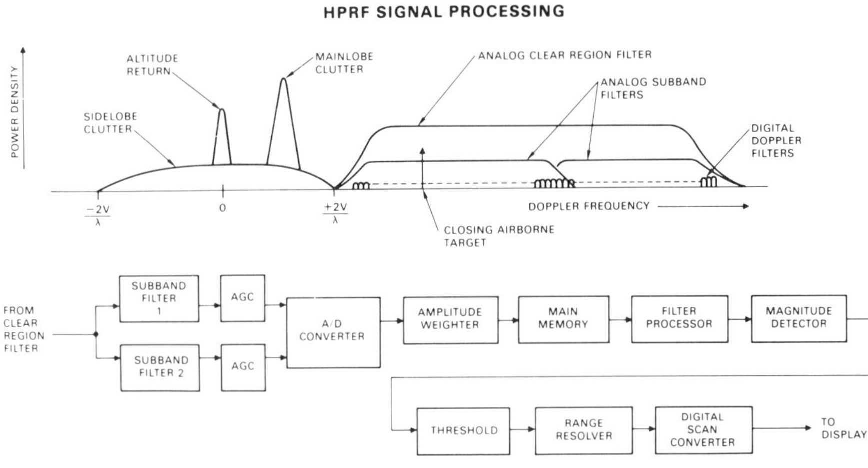

Figure 7.8 High-PRF doppler radar digital-signal processor.

design depicted in Figure 7.7. The A/D converter in this design must handle the full dynamic range of the clutter signals so that on the order of 14 bits plus sign are required. The mainbeam clutter canceller must also carry the full dynamic range, whereas the FFT processor need have only enough dynamic range to accommodate sidelobe clutter and the target's dynamic range.

A block diagram of a surface-based PD radar receiver is depicted in Figure 7.10 [10]. This receiver uses range-gated bandpass filters to form the doppler filters. The filters provide both rejection of clutter returns and detection of doppler-shifted target returns. In this design, both predetection coherent integration (bandpass filtering) and postdetection integration are performed. The equivalent predetection doppler bandwidth for this configuration is given by

$$B_e = B_{RF}^{1-\alpha} (2B_v)^\alpha \tag{7.9}$$

where B_{RF} is the bandpass filter's bandwidth, B_v is the postdetection bandwidth, and α is a parameter dependent on the ratio $B_{RF}/2B_v$, which generally ranges from 0.6 to 0.7 for most practical situations [11]. For large values of $B_{RF}/2B_v$, the parameter α approaches 0.5, which results in an equivalent doppler bandwidth of

$$B_e = \sqrt{2\, B_{RF}\, B_v} \tag{7.10}$$

The bandpass filters utilized in the receiver depicted in Figure 7.10 must provide high attenuation in the stopband while being precisely tuned to a precise doppler frequency. Generally, crystal filters are needed to satisfy this type of requirement,

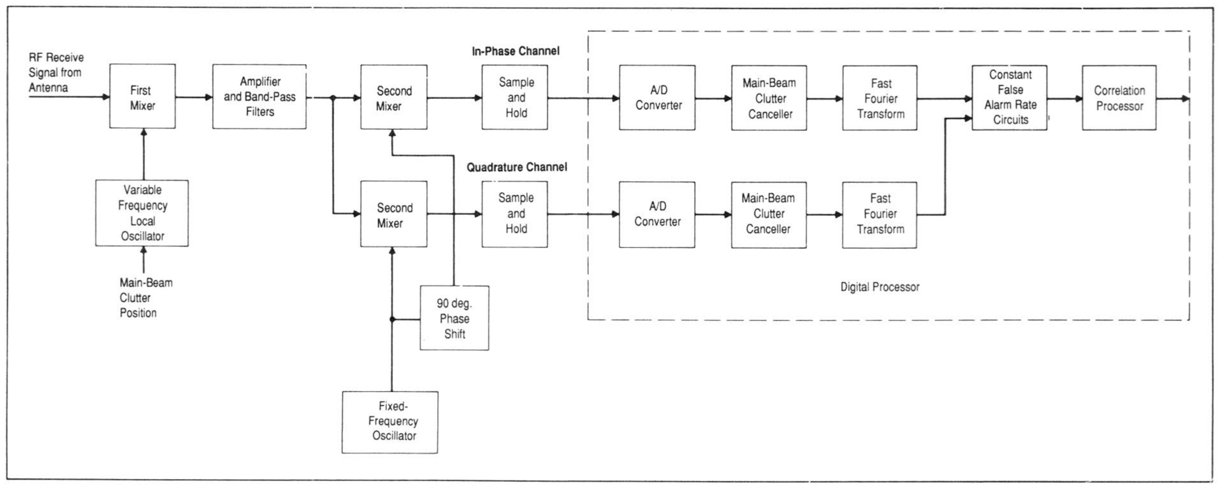

Figure 7.9 High-PRF PD radar receiver block diagram.

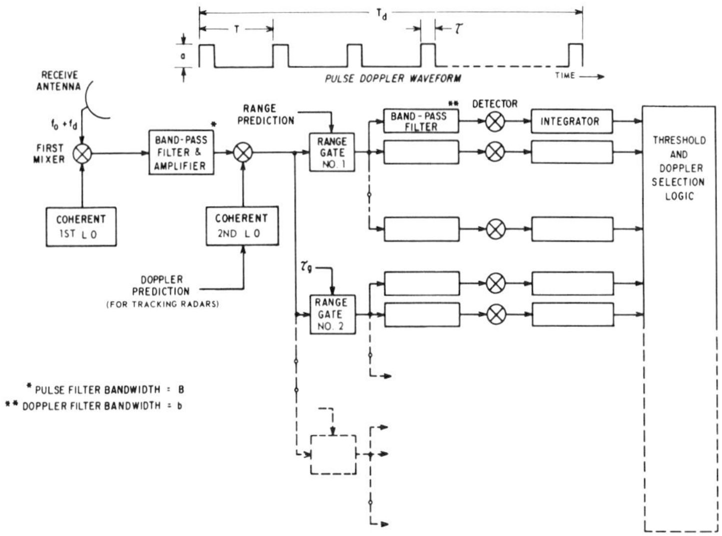

Figure 7.10 PD receiver using range-gated bandpass filters.

but they have practical limitations when used in such a design [10]. These limitations have led to the use of range-gated homodyne (zero IF) PD receivers.

A homodyne PD receiver is depicted in Figure 7.11, which is configured for a tracking radar application. A range-gated single-sideband (SSB) mixer translates the microwave RF signal to baseband (zero frequency) in a single step and provides the in-phase and quadrature channels necessary to retain the doppler information contained in the RF signal. A contiguous doppler filter bank is synthesized by offsetting the RF reference oscillators into the SSB mixer by the frequency separation desired between the filters. The output of the SSB mixer will be a unipolar video signal when the reference oscillators match the doppler-shifted RF signal. The low-pass filters (video integrators) that follow the SSB mixer then coherently integrate the doppler-shifted target echos. The final output is formed by squaring and adding the outputs of the quadrature I and Q channels. The equivalent doppler bandwidth of the homodyne receiver is thus twice the bandwidth of the low-pass filter ($B_D = 2B_v$).

The homodyne PD receiver provides a flexible method of controlling both the filter separation and bandwidth of the doppler filter bank. Note that, in this implementation, the video integrator attenuates the clutter returns. Hence, the low-pass filter must have a sufficient number of poles to provide the desired attenuation of the clutter return. In general, this type of filter is implemented using an active low-pass filter (i.e., a filter using an integrated circuit amplifier).

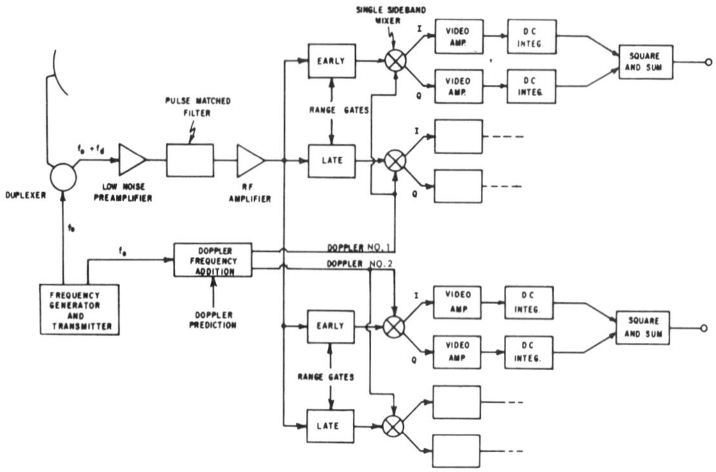

Figure 7.11 PD homodyne receiver.

The homodyne receiver can be implemented digitally, which allows synthesizing a large number of doppler filters. In a digital implementation, the range-gated I and Q returns for each transmitting pulse are applied to an A/D converter, where the digital output is stored in a digital memory. Digital filters can be formed by sequentially rotating the stored return in accordance with the phase pattern expected from a particular doppler (i.e., $f_d = d\phi/2\pi\, dt$). The summed vector output of this phase-rotated output is then compared to a detection threshold (see Section 2.3.3).

The processing employed in medium-PRF PD radar is similar to that utilized in low-PRF PD radars (i.e., the MTD processor) rather than that used in high-PRF processors. A block diagram of a medium-PRF PD radar signal processor is illustrated in Figure 2.13. Medium-PRF processors generally use a clutter canceller followed by a coherent integrator implemented through the FFT algorithm. Additional gain control is required to prevent saturation of the A/D converter, and narrower doppler filters are required in the medium-PRF processor to attenuate the higher clutter experienced as compared to a low-PRF processor. Both range and doppler ambiguities must be resolved in the medium-PRF design, which is generally accomplished by using multiple PRFs.

7.2 WAVEFORM CONSIDERATIONS [6,8,12–14]

A basic decision that must be made in the design of a PD radar is the selection of the PRF. Low, medium, and high PRFs have been used in various applications (see Chapter 2). Generally, low-PRF (unambiguous range) designs are used in surface-based applications and for airborne radar in look-up or beyond-the-clutter-horizon applications. One exception is in airborne applications using antenna motion compensation techniques (see Sections 3.2.7 and 6.2.2) and relatively large antenna apertures (e.g., the E-2C radar). High-PRF (unambiguous doppler) configurations are used primarily in airborne applications that look down into heavy clutter and are designed to detect closing targets. Medium-PRF (ambiguous range and doppler) designs are used primarily in airborne applications that require detection of tail chase targets looking down into heavy clutter with moderate performance against closing targets. Interlaced medium- and high-PRF modes have also been used to provide good all-aspect performance against both closing and chasing targets.

Figure 7.12 depicts the three PRF types in range-doppler space for an AI application [8]. In general, the interpulse period is divided into range cells (range

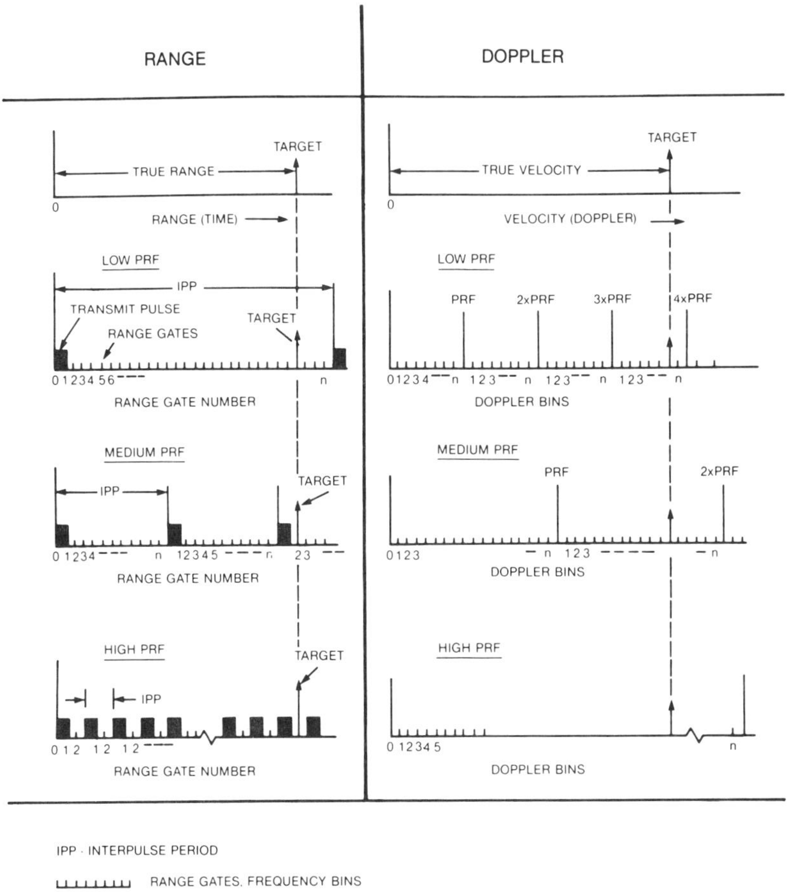

Figure 7.12 PD PRF radars in range-doppler space.

gates) that correspond to the radar's transmitted bandwidth ($\tau_e = 1/BW$), and the doppler space is divided into doppler cells (doppler filters), the bandwidth of which is inversely proportional to the dwell time of the radar antenna's mainbeam on the target. The number of range gates is greatest for a low-PRF radar, whereas only a small number is used for a high-PRF radar. Conversely, only a small number of doppler filters are used in a low-PRF design, whereas a large number are used in a high-PRF design. Although the interpulse period is generally filled with range gates for all types, the low-PRF type is the only design that fills the frequency region between PRF lines with doppler filters. In medium-PRF designs, those doppler frequencies occupied by ground vehicular traffic are generally excluded from the doppler frequency coverage region (see Figure 7.4). Generally, the high-PRF design fills only the doppler-clear region ($2v_r/\lambda < f_d < \text{PRF} - 2v_r/\lambda$) with doppler filters, whereas the medium-PRF design excludes the region occupied by moving vehicular targets from the doppler coverage (see Figure 7.4).

The principal advantage of a low-PRF design is its ability to restrict clutter intake to just that which is in the same range cell as the target. For example, a rain or chaff clutter cell several miles in extent would not interfere with targets not directly in the cell, whereas, in a medium- or high-PRF design, this same rain cell could clutter the entire unambiguous range interval. Also, sidelobe and mainlobe clutter can be restricted by increasing the range resolution of the radar and are not magnified by the folding of ambiguous clutter returns from short range into the same range cell as the target.

Other features of the low-PRF design include the ability to use a sensitivity-time-control function that allows normalization of the clutter return over the full range of the radar and the fact that range can be directly measured without generating any spurious targets. STC cannot be used in medium- or high-PRF designs as it would attenuate long-range targets of interest as well as near-range clutter. This results in a low-PRF system processer requiring less dynamic range, which allows better control of false alarms. In medium- and high-PRF designs, a range correlation system is required to resolve the resulting ambiguous ranges. When multiple targets are present, they can confuse the range correlator and generate spurious targets called *ghosts*.

On the negative side, in a low-PRF design, unless mainlobe clutter is separated in range from the targets the radar encounters, it can be rejected only on the basis of differences in doppler frequency. The highly ambiguous doppler response causes a lack of doppler visibility as depicted in Figure 7.13. Here, the frequency of the mainbeam ground clutter is translated to zero frequency. The spread of the mainlobe clutter spectrum is determined primarily by the aircraft's velocity, azimuth antenna beamwidth, and azimuth look angle. At azimuth angles significantly off the aircraft's ground track (i.e., 30° to 90°), the mainlobe spectrum generally occupies a large fraction of the frequency region between PRF lines. The wide doppler notch blinds the low-PRF PD radar to many targets of interest.

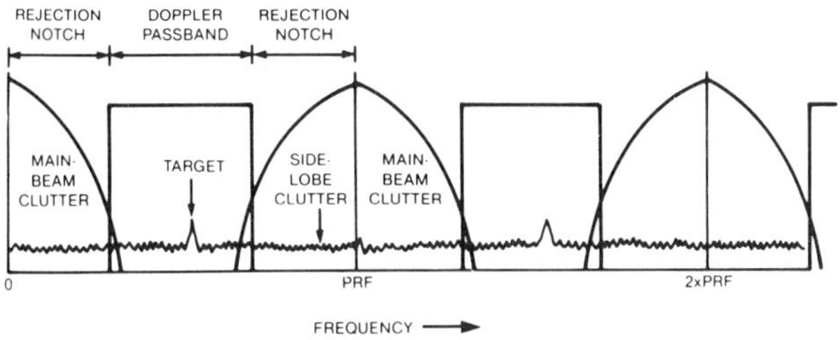

Figure 7.13 Low-PRF PD radar spectrum.

In an airborne search radar, the criterion that determines the doppler rejection notch is primarily that required to reject slow-moving vehicular ground targets rather than mainbeam clutter rejection because slow-moving target rejection requires a wider notch. In a low-PRF radar design operating over land, ground targets can be confused with low-flying airborne targets due to the aliasing effect of the ambiguous doppler response. A typical rejection notch width is ±55 knots with respect to the ground, which translates to a doppler notch width of ±1870 Hz at X-band, ±625 Hz at S-band, ±225 Hz at L-band, and ±80 Hz at UHF. This can be compared to mainlobe clutter widths by using (7.4). For example, a UHF AEW radar (aperture width equal to 25 ft) moving at 300 knots has a clutter spread of 28 knots, whereas an X-band interceptor radar (aperture width equal to 2.5 ft) moving at 600 knots has a clutter width of 32 knots (see the introduction to Chapter 7). A possible exception where the clutter spread exceeds the ground target notch is the X-band interceptor cited earlier, which moves at supersonic speeds (e.g., 1500 knots), inducing a clutter spread of 80 knots. Note that the clutter spread due to platform motion can be reduced by using antenna processing techniques (see Sections 3.2.7 and 6.2.2), but the doppler notch requirement to reject ground targets is difficult to overcome. However, in some applications, ground targets can be resolved by using PRF switching techniques because they generally occur in the first ambiguity zone of the doppler response (see Figure 7.14).

For the airborne interceptor radar at X-band, using a doppler notch width of ±1870 Hz (i.e., ±55 knots), requires that 3740 Hz of the doppler spectrum between PRF lines be blind to targets of interest. Because typical low PRFs for airborne radars range from 1000 to 4000 Hz, depending on the maximum range of interest, essentially no usable doppler spectrum remains. In summary, whereas a low-PRF system provides accurate and ghost-free range information, the visible Doppler spectrum available after clutter and rejection of ground-based moving targets is generally too limited for airborne look-down applications.

In an airborne high-PRF design, target dopplers that appear clear of the sidelobe clutter can be detected in a noise-limited environment, as depicted in Figure 7.15, which shows a simplified down-converted version of a high-PRF airborne radar

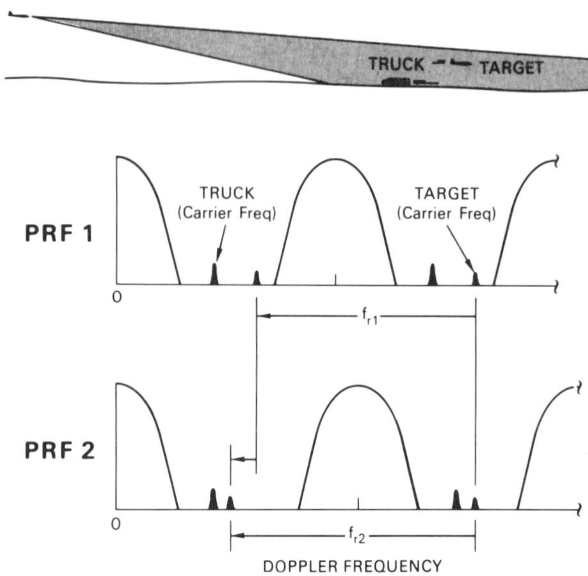

Figure 7.14 Resolving ground targets in a low-PRF doppler radar using PRF switching.

Figure 7.15 High-PRF PD airborne radar spectrum.

spectrum. The sidelobe clutter extends to $\pm 2v_r/\lambda$ around the PRF lines, where v_r is the aircraft's ground speed. The mainlobe clutter center frequency changes according to the radar's look angle with respect to the aircraft's velocity vector. A closing target is shown having a total doppler shift that is the target's line-of-sight doppler plus the mainbeam clutter doppler along the line of sight. Thus, whenever the target's doppler along the line of sight is greater than the maximum sidelobe clutter doppler, the target is clear of the sidelobe clutter and can be detected against a receiver noise background.

This condition occurs as a function of the ratio of the target's velocity to the radar's velocity and the antenna look angle, as depicted in Figure 7.3. For look angles directly along the radar's velocity vector, any target with a line-of-sight velocity component that is toward the radar becomes clear of sidelobe clutter. As the radar scans to a 90° look angle, the target's velocity component along the line of sight must be greater than the radar's ground speed to clear the sidelobe clutter.

A feature of the high-PRF waveform in an airborne application is that slow-moving ground-based targets can be eliminated because such a wide doppler spectrum is available for target processing. For example, in the X-band high-speed interceptor example (i.e., ground speed equal to 1500 knots), the sidelobe clutter spectrum will extend out to 50 kHz. If the expected targets close on the interceptor at a maximum of 600 knots, then an additional 20 kHz of clear doppler must be provided. If a PRF of 150 kHz, for example, is selected, the lower zone of 80–100 knots can be discarded to eliminate ground targets. The result would have only minimal effect in the detection of high-speed closing aircraft targets.

High-PRF airborne radar designs generally operate in a velocity-only search detection mode for initial detection. In this mode range, ambiguities are not resolved and a maximum detection range equivalent to that of a CW radar (i.e., with an average power of $P_A = P_p d_u$) is achieved. When range information is desired, the high-PRF design must employ frequency modulation of the RF carrier (e.g., linear or sinusoidal frequency modulation) or multiple PRFs during the radar's dwell time on target to resolve the range. This requirement may cause an estimated 20–25% degradation in detection range compared to that achieved when range information is not required [8]. A significant limitation of the high-PRF waveform for airborne applications is its limited ability to detect targets moving at velocities that cause their doppler frequencies to appear in the sidelobe clutter regions. As shown in Figure 7.3, the clutter-free detection regions occur in the forward or closing aspects of the target, whereas, for rear or chasing aspects, the target must compete with large sidelobe clutter returns, particularly at low altitudes.

With the high-PRF waveform, range folding causes short-range clutter to compete with long-range targets. In particular, all the sidelobe clutter folds into a small ambiguous range interval due to the many range ambiguities. For example, with a PRF of 150 kHz, the unambiguous range is only 0.5 nmi and, because sidelobe clutter can be well above receiver noise out to beyond 10 nmi range, more than 19 ambiguities of clutter will be folded into the PRF interval. Hence, at low altitudes, sidelobe clutter generally will obscure targets located within its velocity domain (i.e., $f_d = \pm 2v_r/\lambda$).

In summary, a high-PRF waveform provides ghost-free target doppler, the capability to reject slow-moving targets, and a clutter-free target detection region that generally facilitates detection of closing targets. However, the limited all-target-aspect detection, especially at low altitudes in tail chases, generally restricts the use of this waveform to closing targets. In addition, high-PRF designs require complicated methods for resolving range ambiguities and may also suffer from eclipsing problems, where target returns are lost, either totally or partially, because the receiver is blanked when the target is received due to the high duty factor of the waveform (see Section 7.2.1).

The medium-PRF waveform combines some of the attributes of both high- and low-PRF waveforms, which for airborne applications can provide slow-moving target rejection, all-target-aspect coverage, sidelobe clutter-free regions, and accurate range information. The range and doppler ambiguities of a medium-PRF design cause the returns from short-range sidelobe clutter to fold into the same doppler regions occupied by target returns. However, the essence of medium-PRF waveform design is to establish clear regions within the range-doppler space by the judicious placement of multiple medium PRFs.

The first step in medium-PRF design is to set the doppler visibility, defined as the percentage of the interval between PRF lines not occupied by mainlobe clutter (see Figure 2.6). Figure 7.16 shows that the improvement in the velocity-clear region increases from 13 to 70% at an antenna look angle of 30° and an aircraft speed of 400 m/s as the PRF is increased from 2 to 6 kHz. The improvement would be even greater for a 10 kHz PRF, which would provide a clear region of about 83%.

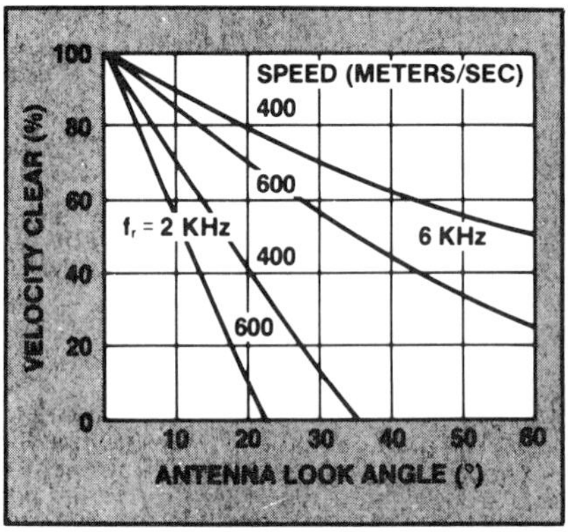

Figure 7.16 Velocity-clear performance with PRF variation.

When using a PRF of 6 kHz, the maximum unambiguous range is about 25 km. Search for targets at long range results in multiple-time-around returns, where, for example, a target at 150 km would arrive after 6 pulses have been transmitted. The most important result of the ambiguity is the presence of sidelobe clutter that arises from short ranges, which competes with targets at larger ranges (see Figure 2.12). Because the sidelobe clutter region is most intense at the minimum time delay point, the dominant return is generally caused by the first doppler surface within the clutter region, as depicted in Figure 7.17 for an airborne application.

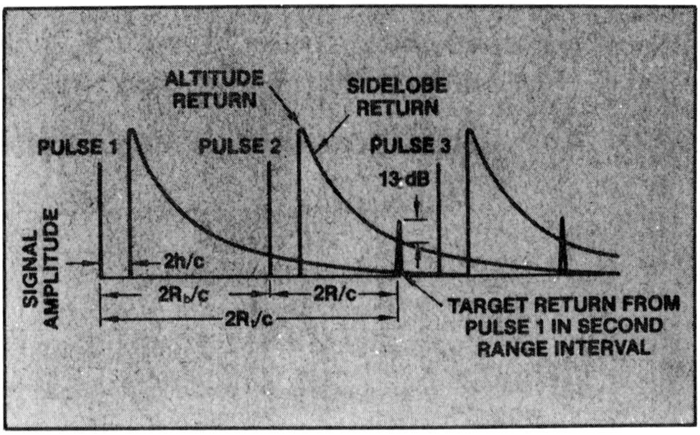

Figure 7.17 Sidelobe return for a medium-PRF PD radar.

In a medium-PRF design, the sidelobe clutter that arises from the transmitted pulse immediately preceding the target echo arrival competes with the target signal. This clutter comes from a ring on the ground at a constant range (R_1) from the aircraft (see Figure 2.11). In general,

$$R_1 = R_t - (n - 1)R_b \tag{7.11}$$

where n is the ambiguous range interval and R_b is the unambiguous range interval determined by the PRF ($R_b = c/2\text{PRF}$). Figure 7.17 illustrates an idealized clutter profile for the case where $n = 2$, and only the clutter from the first ambiguous interval (at R_1) is of concern because all other intervals by comparison return negligible power. The clutter from the first interval decays rapidly as range increases, beginning as an altitude signal (specular return) and changing to a diffused sidelobe return. When the target appears to be out by several ambiguous range intervals in range, a blind zone will occur in the range dimension where the target signal is not sufficiently larger than the sidelobe clutter to provide detection. This effect is depicted in Figure 7.18, where the target return power from the initial transmitted pulse is superimposed on the ambiguous sidelobe clutter response. As the target moves farther out in range (i.e., becomes highly ambiguous), the range blind zones spread until the target is completely submerged in ground clutter.

In addition to the range blind zones due to strong sidelobe clutter are the blind zones due to eclipsing. While the radar is transmitting (and for the duplexer recovery time thereafter), the receiver is blanked. Consequently, if a target's echos are received during these periods, the target return is blanked. If the radar transmits narrow pulses, then the receiver blocking may be inconsequential; however, if the pulses are long, as they are in some medium-PRF radars (e.g., with pulse compression), the additional

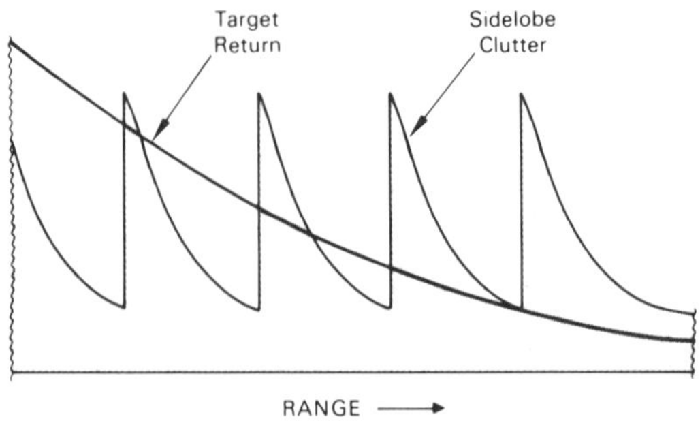

Figure 7.18 Target and sidelobe clutter returns in a medium-PRF airborne radar.

range blanking due to eclipsing can be significant. Figure 7.19 depicts the relative range blanking for a representative medium-PRF radar due to sidelobe clutter and eclipsing.

To resolve both range and doppler blind zones occurring in medium-PRF airborne radars, the radar's PRF is cycled through a fixed number of fairly widely spaced PRFs while the radar's mainlobe dwells on the target. For example, the radar might be cycled through eight different PRFs. The blind doppler zones for these PRFs due to sidelobe clutter are depicted in Figure 7.20. If a target is in the clear on any three PRFs and its echos exceed the detection threshold on all three, the target will be deemed detected. The range ambiguities and all spurious target ranges will then be resolved. The optimum PRF code is a function of the operational situation such as radar altitude, clutter levels, and speeds that must be determined for each specific radar.

For a target to be detected during a particular transmitted PRF of a medium-PRF code, it must be simultaneously in both a doppler-clear region (see Figure 7.20) and in a range-clear zone (see Figure 7.19). If the target's doppler frequency falls in a doppler blind zone, its return will not pass through a doppler filter and be detected,

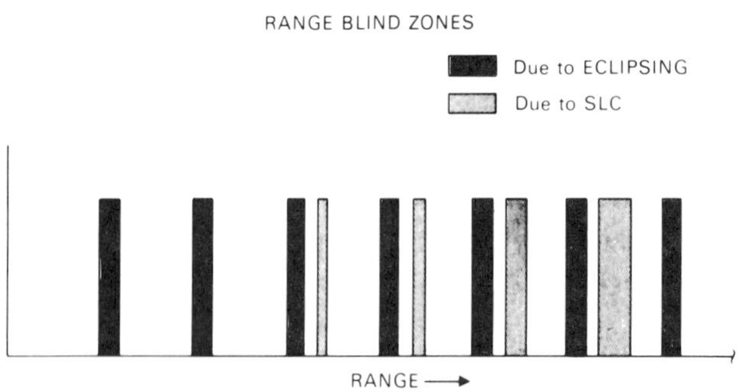

Figure 7.19 Combined range zones due to eclipsing and sidelobe clutter in a medium-PRF radar.

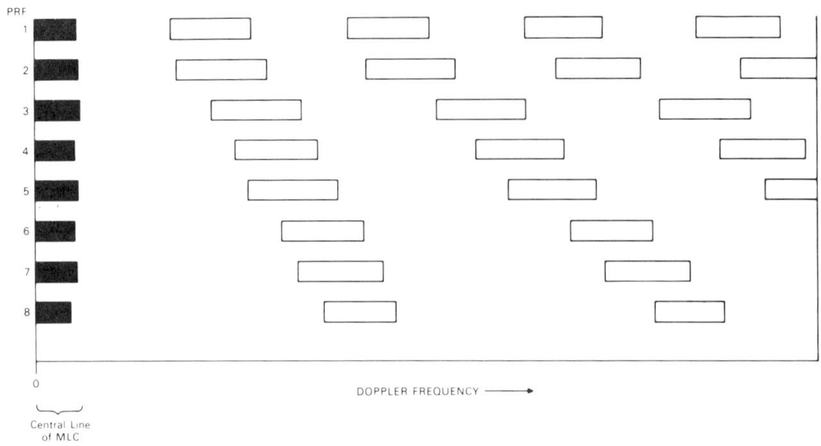

Figure 7.20 Doppler blind zones with multiple PRFs for medium-PRF PD radars.

despite the fact that it is in a range-clear zone. If the target is in a range blind zone, although its return may pass through a doppler filter, it still will not be detected because the accompanying sidelobe clutter through the filter will cause the detection threshold to be greater than the target response. Figure 7.21 depicts a range-doppler matrix of a representative medium-PRF radar that shows the zones for which the radar is both range-clear and doppler-clear for at least three of eight widely spaced PRFs. Note that for a tracking radar, the PRF can be adaptively controlled to place the target in clear range and velocity zones, so that only one burst at this PRF need be transmitted, rather than the long, diverse bursts required by a search radar.

In a medium-PRF design, the number of PRFs processed during an antenna dwell time (the time the antenna takes to scan over the target) is constrained on the high side by the available processing time. If the time becomes too long, the antenna scan rate drops below an acceptable value. On the low side, the number of PRFs per dwell must be sufficient not only to resolve range, but also to ensure good doppler visibility.

Figure 7.22 illustrates the problem of generating spurious reports in a multiple-PRF processing sequence [8]. A three-PRF code is used with a two-out-of-three detection criterion. The PRIs for the three PRFs shown are 7, 8, and 9 range gates. For PRF 1, a target in range gate 4 appears in range gates 4, 11, 18, 25, and so on. A target in range gate 26 appears in range gate 26, modulo-PRI so that for PRF 1, the target appears in range gates 5, 12, 19, 26, and so on. Similarly, for PRF 2, the target in gate 4 appears in gates 4, 12, 20, 28, and so on. The correlation process for the three PRFs, as shown in Figure 7.22, identifies the correct targets in range gates 4 and 26, but generates spurious targets (ghosts) in range gates 12 and 18. In a practical situation, more range gates are used (i.e., higher range resolution), thus

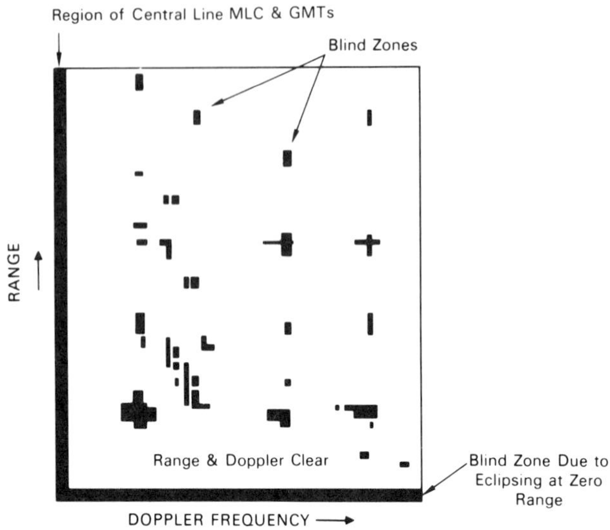

Figure 7.21 Range-doppler regions in a medium-PRF radar that are clear on at least three of eight PRFs.

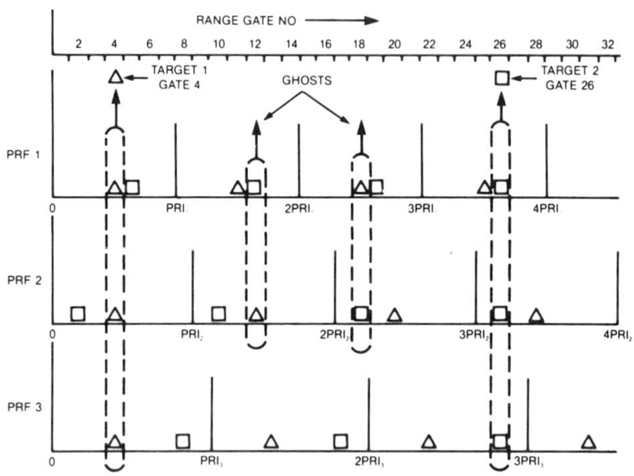

Figure 7.22 Spurious target generation in a medium-PRF radar.

tending to minimize ghosts. The PRFs then must be carefully chosen, not only to resolve the range unambiguously, but also to minimize the number of possible cross-correlations when dealing with multiple targets.

To analyze the probability of detecting targets in the clear regions of a medium-PRF design, a visibility factor (F_a) in range-doppler space can be defined as [1]

$$F_a = (1 - \Delta r/R_u)(1 - \Delta_v/v_b) \tag{7.12}$$

where Δr is the blind range zone, and Δv is the blind velocity zone within a PRF interval, and R_u and v_b are the unambiguous range and velocity responses of the radar (i.e., $R_u = c/2$ PRF, $v_u = \lambda$ PRF$/2$). During the radar's dwell time on target (T_d), m bursts of different PRFs are transmitted with average duration $T_f = T_d/m$. The detection probability (P_1) per burst will then be $F_a P_d$, where P_d is the detection probability per burst during period T_f, assuming full visibility. In an observation interval $T_d = mT_f$, assuming independent trials, the cumulative probability of at least n successes in m attempts is given by

$$P_c = 1 - \sum_{i=0}^{n-1} \binom{m}{i} P_1 i(1 - P_1)^{m-i} \tag{7.13}$$

For example, if the P_d on each look were 80% and the visibility factor was 50%, the cumulative probability for at least three detections in a burst of eight PRFs would be $P_c = 1 - 0.3 = 0.7$. Equation 7.13 illustrates that, as F_a decreases, the number of PRF bursts (m) must be increased to obtain a high cumulative probability. However, shortening each burst reduces the P_d of each trial and may further reduce the visibility factors in range and doppler. Optimization of the waveform and the resulting performance ultimately depends on the intensity of the clutter and the stability of the radar system.

The duration of the coherent burst depends on the number of doppler cells that must be formed between PRF lines and the PRF. Also, coherent processing cannot start until the first echo from mainbeam maximum range clutter is received. This restricts the minimum burst period (T_f), particularly for long-range surface radars that might experience chaff or weather clutter at long range and airborne radars when their beam is pointed toward the horizon [1]. If the maximum clutter time delay is $t_{dc} = 2R_c/c$, then the portion of the burst that contributes to the processed output has a duration of

$$T_f' = T_f - t_{dc} \tag{7.14}$$

and the doppler filter bandwidths and coherent integration gains will depend on T_f' rather than T_f. As a rule of thumb, to use a T_f greater than $4t_{dc}$ is generally desirable [1].

7.2.1 Waveform Duty Factor Considerations [6,8]

In general, a trade-off must be made in each of the waveform choices (i.e., low, medium, or high PRF) between a high peak power and a low peak power transmitter. Because the average transmitter power is generally established by the detection requirements of the radar, this choice determines its duty factor (see 7.5). In high- and medium-PRF radars, the duty factor affects the range visibility factor (Δr) due to target eclipsing which ultimately affects the radar's detection probability (see 7.13). Where mixed modes are required (e.g., high and medium PRF), the choice often is mandated by the requirement for a low-peak power design in the high-PRF mode because building a coherent transmitter that can simultaneously support both high- and low-power designs is difficult. When this is the case, pulse compression is usually dictated, which allows the average power to be attained in the medium-PRF mode while maintaining the desired range resolution.

An advantage of a high-PRF design is that high average transmitted power can be achieved by simply increasing the PRF. This means that pulse compression is generally not required to attain the desired average power. The resulting duty factor in an airborne design might be as large as between 0.3 and 0.5. When operating at these high duty factors, a considerable amount of target return is lost due to eclipsing (echos being received in part or wholly when the radar is transmitting and the receiver is blanked).

Eclipsing sufficiently reduces the effective S/N to leave periodic holes of appreciable size in the radar range coverage. The reduction is determined by the amount of blocking due to the relative timing of the target return and receiver blanking. A target is totally eclipsed only when the period during which the echos are received coincides exactly with the period during which the receiver is blanked.

The reduction in S/N due to eclipsing for an approaching target is depicted in Figure 7.23. A rapidly approaching target will not remain at an eclipsed range for a long period of time. Also, the decrease in range, causing the signal strength to increase, fills the gaps in range coverage. In a scanning high-PRF radar, a rapidly approaching target may move from one eclipsed region to another between scans, thereby creating a "blind spccd" effect [15].

In applications where closing rates may be comparatively low and more continuous detection is required, the length of time that any one range remains eclipsed may be reduced by switching the PRF among different values in a manner analogous to the elimination of blind range zones in a medium-PRF design. In single-target tracking, periodically changing the PRF at appropriate times can generally keep a target in the clear, although the holes in range coverage are not easily eliminated, particularly at short ranges.

Medium-PRF waveforms operate at duty factors that range from 0.01 to 0.03 so that either a high peak power or low peak power transmitter with pulse compression can be used. In either case, a short processed pulsewidth is required in a medium-

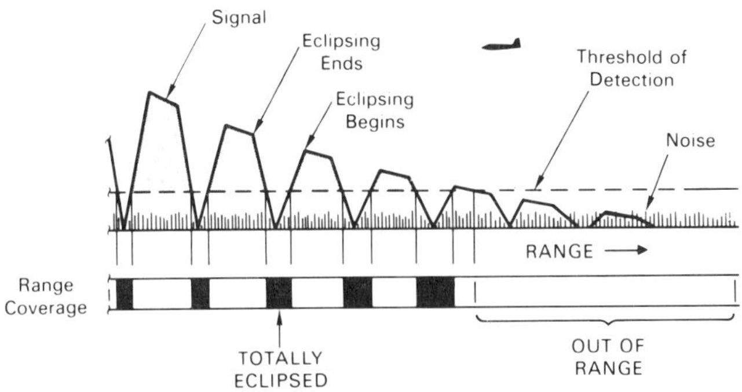

Figure 7.23 Reduction of S/N ratio in a high-PRF radar due to eclipsing.

PRF waveform to resolve range unambiguously, and also for target range resolution requirements, so that pulse compression is required when using long transmitted pulsewidths. Both pulse compression and long transmitted pulses have deleterious effects on radar performance in a medium-PRF waveform.

Most current medium-PRF radars use binary phase coding to implement pulse compression, although these waveforms suffer from being doppler sensitive, causing a difficulty in achieving low-range sidelobes. The doppler sensitivity is caused by the shift in phase of individual bits of the phase code by the target doppler so that, in the extreme, if the last bit is shifted by 180°, the code is no longer matched. A target doppler that causes this shift is known as the *critical velocity*. At this point, large time sidelobes result and the compressed pulse amplitude is reduced. For example, with a 13-bit Barker code using 1 μs bits at X-band, the critical velocity is 1215 knots.

The time sidelobe requirement on the compressed pulse are typically on the order of 30 dB with a processing loss of 1 dB. However, the sidelobe levels are increased during eclipsing, which generates a source of false targets, thereby causing difficulties in the range correlation portion of the medium-PRF processor. Polyphase coding mitigates both the doppler sensitivity losses and the higher sidelobe levels, but is difficult to implement.

Additional problems arise in a medium-PRF design using a long transmitted pulse: the eclipsing of short-range targets and the reduction of the range visibility factor (Δr). The short-range target problem can be alleviated by interlacing a short-pulse mode into the waveform. The visibility problem can be solved by increasing the number of bursts in the PRF diversity code, but this lengthens the overall waveform.

In summary, pulse compression can be used to achieve the required processed short received pulsewidths in a medium-PRF waveform, but the problems with doppler sensitivity, processing losses, time sidelobes, blind ranges, and complexity make this less than an ideal choice. Pulse compression is most compatible with low-PRF designs, where the only fundamental problem is concerned with the detection of short-range targets.

7.2.2 Antenna Sidelobe Requirements [8,16,17]

An important factor limiting the performance of airborne PD radars is the clutter that enters the radar through the antenna's sidelobes. Each possible waveform option responds differently to sidelobe clutter, and hence the performance requirement on antenna sidelobes are different for low-, medium-, and high-PRF waveforms. The essence of the effect is that sidelobe surface clutter returns are spread throughout the frequency region $f_d = \pm 2v_r/\lambda$ by the doppler shifts induced onto the clutter scatterer returns, which are illuminated by the omnidirectional antenna sidelobes. In low-PRF designs, sidelobe clutter is aliased into the target doppler region by the highly ambiguous doppler associated with this waveform. In high-PRF designs, short-range (R_c) sidelobe clutter is magnified by the factor $(R_t/R_c)^4$ as it competes with long-range target returns (R_t) due to the highly ambiguous range associated with this waveform. In a medium-PRF design, sidelobe clutter is aliased into target doppler regions and short-range clutter is magnified, but to a lesser degree than in low- or high-PRF designs. The medium-PRF waveform design problem then becomes to find PRF codes that provide range and doppler regions in which the sidelobe clutter is sufficiently benign to allow target detections.

For a low-PRF design, sidelobe clutter appears in the same range cell as the target. The highly ambiguous doppler response causes the sidelobe clutter power spectral density to be approximately uniformly distributed between PRF lines (see Figures 2.4 and 2.5). The MTI improvement factor (I_f) is the ratio of the signal-to-clutter ratios at the input and output of the MTI canceller ($I_f = (S/C)_i/(S/C)_o$), which can be expressed as $I_f = \bar{G}\cdot\text{CA}$, where $\bar{G}$ is the MTI's power gain (see Table 3.1) and CA is the clutter attenuation. When sidelobe surface clutter is present, the input clutter is the sum of the mainlobe clutter (P_{MLC}) and the sidelobe clutter (P_{SLC}). At the output of an MTI processor, the mainlobe clutter is attenuated (P_{MLC}/CA), whereas the sidelobe clutter is primarily in the passband of the MTI and passes unattenuated through it. The target power (P_t) is enhanced by the power gain of the MTI, resulting in an output target power equal to $\bar{G}P_t$. The resulting MTI improvement factor is thus given by

$$I = \left(\frac{\bar{G}P_t}{P_{\text{MLC}}/\text{CA} + P_{\text{SLC}}}\right) \Big/ \left(\frac{P_t}{P_{\text{MLC}} + P_{\text{SLC}}}\right) \tag{7.15}$$

which for $\text{CA} \to \infty$ and $P_{\text{MLC}}/P_{\text{SLC}} \gg 1$ can be simplified to

$$I = \frac{\bar{G}P_{\text{MLC}}}{P_{\text{SLC}}} \tag{7.16}$$

Further, the ratio $g^2 = P_{\mathrm{MLC}}/P_{\mathrm{SLC}}$ is proportional to the two-way integrated antenna sidelobe power gain of the antenna, which expresses the ratio of the power radiated and received through the antenna's mainlobe to that radiated and received through the antenna's sidelobes. Therefore, the maximum MTI improvement factor is given by

$$I_m = \bar{G}g^2 \tag{7.17}$$

where g is the one-way integrated sidelobe (rms) power gain of the antenna with respect to an isotropic antenna (e.g., ordinary, 0 to -5 dBi; low, -5 to -20 dBi; ultralow, -20 dBi and lower; $g = 1/\bar{G}_{\mathrm{SL}}$).

When a doppler filter bank is employed in cascade with the MTI canceller in a low-PRF PD design, the improvement factor of the processor is further enhanced because the doppler filter containing the target excludes a portion of the sidelobe clutter. If the coherent processing interval is T_c seconds long ($T_c \leq T_d$, the target dwell period), $n_f = 1/\mathrm{PRF} \cdot T_c$ doppler filters can be formed between PRF lines. The resulting maximum improvement factor is thus given by

$$I_m = \frac{\bar{G}g^2}{\mathrm{PRF} \cdot T_c} \tag{7.18}$$

where $\bar{G}$ is the MTI power gain, g is the integrated one-way power antenna sidelobe gain, T_c is the CPI of the radar, and PRF is the radar's pulse repetition frequency.

Note that the improvement factor defines the relative improvement in the S/C ratio due to the processor. The radar's probability of detection can be enhanced, not only by greater processing gain, but also by increasing the signal-to-clutter power ratio into the processor. This is accomplished in a low-PRF PD radar by increasing the transmitter bandwidth that provides finer range resolution, thereby allowing the unambiguous range to be divided into more range cells (see (4.8)).

When using a high-PRF waveform, often only the doppler spectrum clear of sidelobe clutter (see Figure 1.14) is processed, in which case the antenna sidelobe clutter level is of no consequence. In that event, some high-PRF systems use a uniformly weighted antenna aperture to maximize antenna gain and operate with the resulting high sidelobes. If the high-PRF radar operates at relatively high altitudes (e.g., the AEW mission), the sidelobe clutter may be reduced sufficiently to allow detection of targets in this velocity region.

The accurate computation of sidelobe clutter in a high-PRF airborne PD radar involves complex geometrical relationships that generally must be handled through a computer program [18–21]. The computational procedure generally involves the integration of sidelobe returns from the ground (range gated by the radar's pulsed response) that occur between isodops (contours of equal doppler response) corresponding to the frequency extremes of the passband of a doppler filter (see Figure 1.13 and Section 1.1.4.1).

An idea of the magnitude of the sidelobe clutter for a high-PRF design can be obtained through a rough approximation of the isodop area that is developed as follows [15]. Figure 7.24 depicts the geometry for a high-PRF airborne radar by using a flat earth approximation. The area between isodops is approximated as the difference between circular areas (antenna beams projected onto the ground), the diameters of which extend from the intersection of the antenna's boresight with the surface (formed by antenna depression angle ϕ) and two elevation angles ϕ_1 and ϕ_2 that represent the angular difference between two isodops. The clutter area between isodops for the approximation can be shown to equal

$$A_c = \frac{\pi h^2}{4 \sin\phi} \left[\left(\cos\phi - \frac{\sin\phi}{\tan\phi_2} \right)^2 - \left(\cos\phi - \frac{\sin\phi}{\tan\phi_1} \right)^2 \right] \tag{7.19}$$

where h is the radar's altitude, ϕ is the depression angle of the radar to the target, ϕ_1 is the elevation angle to one isodop contour, and ϕ_2 is the elevation angle to a second isodop contour. The average doppler frequency of the doppler filter containing the sidelobe clutter represented by A_c is given by

$$f_{\text{SLC}} = \frac{2v_a \cos\phi_a}{\lambda} \tag{7.20}$$

The doppler frequency range containing the clutter area A_c is given by

$$\Delta f = f_1 - f_2 = \frac{2v_a}{\lambda} (\cos\phi_1 - \cos\phi_2) \tag{7.21}$$

which can be solved for the angular difference $\Delta\phi = \phi_2 - \phi_1$ represented by the doppler filter width $\Delta f = 1/T_c$ as

$$\Delta\phi = \frac{\lambda}{2v_a T_c \sin\phi_a} \tag{7.22}$$

where v_a is the aircraft's velocity, λ is the radar's wavelength, T_c is the coherent processing interval, and ϕ_a is the average elevation angle to the clutter ($\phi_a = \phi_1 + \phi_2/2$).

As an example, consider an AEW radar ($f = 1.3$ GHz) at an altitude of 25,000 ft with a velocity of 250 knots that is to detect a low-flying 1 m^2 target at 150 nmi in the sidelobe clutter region. Referring to Figure 7.24, the depression angle can be calculated as $\phi = \sin^{-1} (h/R_t + R_t/2r_e) = 2.5°$ ($r_e = 4{,}587$ nmi). The doppler shift of the mainbeam clutter is given by $f_{\text{MBC}} = 2v_a \cos\phi/\lambda = 1113$ Hz. Assume that the target doppler is 500 Hz, which would occur for an aircraft moving away from

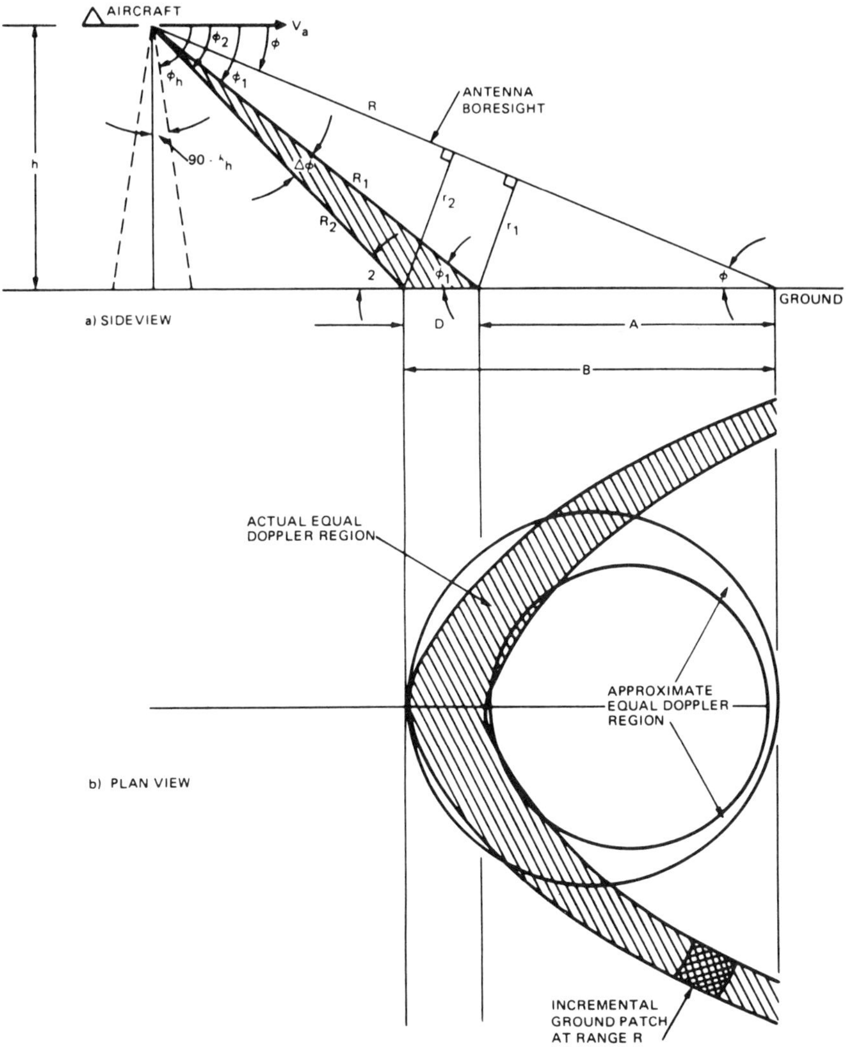

Figure 7.24 Approximation of sidelobe clutter area in a high-PRF PD radar.

the AEW radar at approximately 360 knots. The problem is to compute the S/C ratio as a function of antenna sidelobe levels for a doppler filter set at 500 Hz center frequency in the sidelobe region.

The first step in the approximate calculation is to compute the elevation angle (ϕ_a) to the isodop from (7.20) as $\phi_a = 63.34°$. Because $\phi_a = \phi_1 + \phi_2/2$, (7.22) can be solved to determine ϕ_1 and ϕ_2 using the further assumption that the coherent processing interval is 30 ms. This results in $\Delta\phi = \phi_2 - \phi_1 = 1.91°$, $\phi_1 = 62.4°$, and $\phi_2 = 64.3°$. Equation (7.19) can then be solved for the clutter area that appears within the 33 Hz ($B_D = 1/T_c$)-wide doppler filter at 500-Hz center frequency as $A_c = 3.7 \times 10^6$ m^2. For farmland terrain ($\sigma° = -20$ dB at $\phi = 60°$; see Figure 4.9), the equivalent clutter cross-sectional area is then $\sigma_c = 3.7 \times 10^4$ m^2 modified by the duty factor (d_u) of the PD radar, which is assumed as $d_u = 0.1$, resulting in $\sigma_c d_u = 3.7 \times 10^3$ m^2. The average range to the sidelobe clutter isodop can be found by averaging $R_1 = h/\sin\phi_1$ and $R_2 = h/\sin\phi_2$ as $\bar{R}_c = R_1 + R_2/2 = 4.6$ nmi. This results in an additional clutter magnification because the returns from short-range clutter ($\bar{R}_c = 4.6$ nmi) and the long-range target ($R_t = 125$ nmi) are both folded into the same doppler filter. The magnification is given by $(R_t/\bar{R}_c)^4 = 5.5 \times 10^5$ so that the clutter return in the doppler filter exceeds the target return by $C/S = \sigma_c/\sigma_t = 2 \times 10^9 = 93$ dB, where both clutter and target are illuminated by an omnidirectional antenna. To define the ratio of the one-way power mainlobe antenna gain (G_{ML}) to the one-way power sidelobe antenna gain (G_{SL}) as $\hat{g} = G_{\text{ML}}/G_{\text{SL}}$ would require $\hat{g} =$ 46.5 dB sidelobes be deployed. Sidelobes in this region require the use of ultralow sidelobe antennas (i.e., $G_{\text{SL}}/G_{\text{ML}} > -40$ dB).

In a medium-PRF airborne radar design, sidelobe levels are of great importance and have a direct bearing on system performance. Some insight can be gained by considering the *range extent* of sidelobe clutter for constant antenna sidelobe levels. Figure 7.25 depicts the sidelobe clutter extent for a typical medium-PRF system *versus* radar altitude, where the range extent is measured in terms of 1 μs range gates [8]. Range extent is defined as the difference in range from the altitude line to the range at which sidelobe clutter equals thermal noise. As is evident if we consider that the average number of 1 μs gates in an interpulse period for an AI radar (PRF = 10 kHz) is about 90, when the sidelobe levels go above 45 dB, a significant portion of the unambiguous range interval is covered by sidelobe clutter. For example, with a 35 dB sidelobe level at 10,000 ft altitude, 64 gates or 71% of the unambiguous range interval has sidelobe clutter greater than thermal noise. This would have a severe effect on the detection performance at those target dopplers with minimum PRF visibility.

A further consideration in medium-PRF radars involves the detection of large discrete targets in the radar sidelobes. These targets have doppler frequencies induced by aircraft motion that are aliased into target doppler regions by the ambiguous doppler response of the medium-PRF design. In addition, the ambiguous range of the medium-PRF waveform causes short-range sidelobe targets to fold into the same range gates as long-range targets. This situation is further exacerbated in airborne radars by

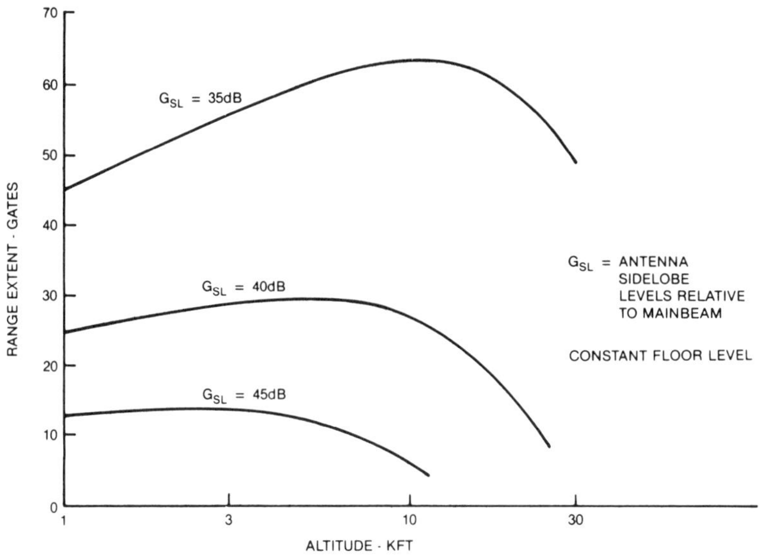

Figure 7.25 Extent of sidelobe clutter in a medium-PRF PD radar.

reflection lobes caused by the main radar beam bouncing off the aerodynamically shaped radome surface into the sidelobe region.

For example, if a medium-PRF airborne radar can detect a 1 m^2 target at 30 nmi, it also can detect a sidelobe target at 2 nmi of about 200 m^2, assuming a relative sidelobe one-way power gain of -35 dB. One method of dealing with these unwanted targets is to provide the radar with a guard channel, as depicted in Figure 7.26. The guard channel receiver blanks the sidelobe target if its response is greater in the guard band channel (the guard band antenna response exceeds the mainlobe antenna's sidelobe response) than in the main channel [6]. Another method of suppressing unwanted sidelobe targets is through an amplitude comparison scheme with a single receiver channel [8]. In this method, the unambiguous range to the sidelobe target is first

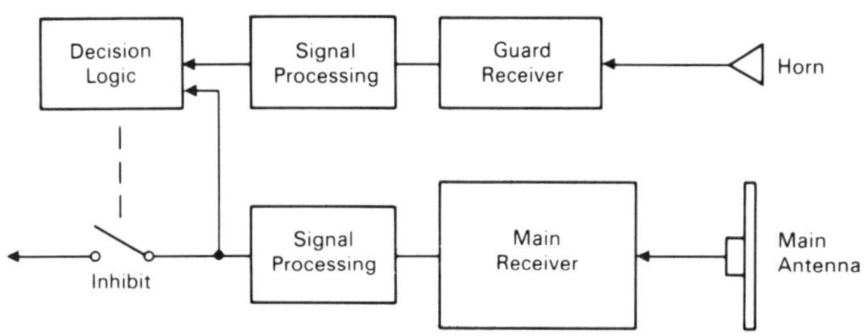

Figure 7.26 Guard channel receiver in a medium-PRF PD radar.

resolved. A discrete target in the sidelobes can generally be detected only at short ranges due to the two-way attenuation property of the antenna sidelobes. However, *bona fide* targets in the mainlobe generally provide a large signal at short ranges. Thus, if the detected signal's resolved range is small and its amplitude is smaller than that expected from a short-range minimum size target (e.g., 1 m^2), then the target is assumed to be a spurious sidelobe discrete target.

7.3 RESOLVING RANGE AMBIGUITIES

A characteristic of medium- and high-PRF PD waveforms is the generation of highly ambiguous range information. When accurate range information is required of these systems, the ambiguous ranges must be resolved. Resolving range ambiguities is not a trivial matter and a degradation in detection range of up to 20–25% can result from this process, which generally involves a modulation of the transmitted signal's frequency or PRF [8].

Range ambiguities are generally resolved by using one of two methods. The multiple-PRF method uses a PRF code composed of a number of fixed PRFs. This type of ranging involves sequential measurement of the ambiguous range in each PRF followed by comparison of the measurements to eliminate ambiguities. The second method involves frequency modulating the transmitter by using either a linear or sinusoidal waveform. The time lag between transmission and reception is converted to a frequency difference. By measuring this frequency difference, the time lag, and hence the range to the target, can be determined. Both methods are susceptible to the generation of spurious targets (i.e., ghosts) when multiple targets are present within the same doppler filter.

The multiple-PRF ranging method generally provides the more accurate range measurement. However, when a high duty factor waveform is employed (e.g., airborne high-PRF design), use of multiple-PRF ranging may not be feasible due to the small number of range gates available to resolve ambiguities. In this case, FM ranging is employed, which is generally less accurate than direct pulse-delay ranging.

Resolution of range ambiguities generally should reduce detection sensitivity from what it would be without ranging. The reduction is primarily due to the need for ranging to divide the total observation time (i.e., the period during which the radar dwells on the target) into a number of segments. For detection, the target generally must be detected independently during each segment of the modulation cycle that resolves the range ambiguities. The time-on-target is thus significantly reduced from the total available observation time. In addition, the probability of detection for each segment must be increased for adequate probability of detecting the target on at least two of the different segments. Both effects cause a significant reduction in the overall probability of detection. A further reduction of observation time may also result from discarding target returns during transient periods associated with changing of the modulation cycle during ranging (e.g., changing of the FM slope during linear FM ranging).

7.3.1 Multiple High-PRF Ranging [15,22]

In high-PRF airborne PD radars, the use of high PRFs (e.g., 100 to 300 kHz), necessary to prevent clutter from obscuring target doppler returns, results in highly ambiguous range information. This situation is depicted in Figure 7.27, which shows radar returns for two targets (both ambiguous and unambiguous) that are widely separated in range. Due to the ambiguous range, there is no way of discerning which transmitted pulse caused which target return; hence, true target range is unknown. Another situation that can occur when using a high PRF is the eclipsing of target returns by subsequent transmitted pulses (see Section 7.2).

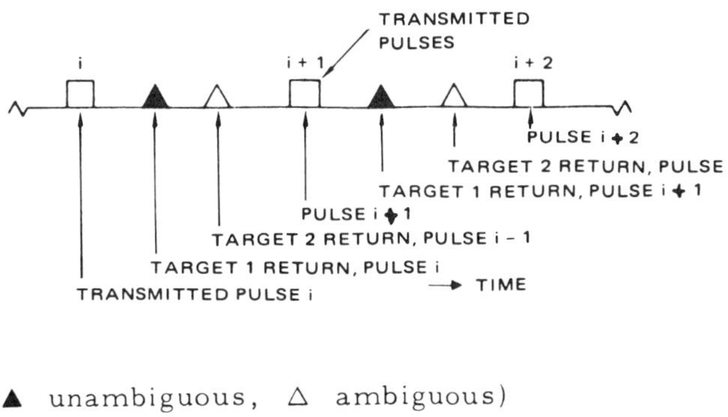

Figure 7.27 Ambiguous target returns in a high- or medium-PRF PD radar.

To overcome these difficulties, multiple PRFs are used sequentially. By observing the target interpulse position in each of the PRFs, the true target range can be deduced. Furthermore, by selecting an appropriate PRF value, the start of an eclipse can be detected and the PRF may be switched to a value that causes the target return to fall within a clear range zone, thereby preventing the eclipse.

The principles involved in a two-PRF ranging system are depicted in Figure 7.28. The PRFs are chosen to have a common submultiple frequency (f_B) with an unambiguous period $T_u = 1/f_B$. The ambiguous range to the target is determined first in PRF_1 and then redetermined for PRF_2. By taking a time comparison of the two ambiguous ranges, made possible by synchronization of the PRFs, we obtain a coincidence pulse that is a measure of the true range.

The common submultiple frequency (f_B) is chosen so that its interpulse period is equal to, or greater than, the expected target range. The minimum value of PRF

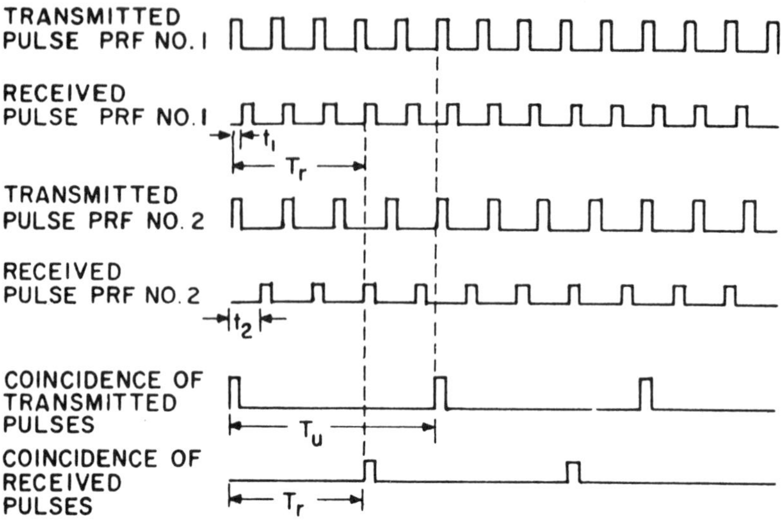

Figure 7.28 Two-PRF ranging system.

is chosen to allow a clear-doppler region free of clutter in which to make a target detection (PRF $\geq 2f_d + f_t$). The ratio of the PRF to the ranging frequency is set equal to an integer, N, by adjustment of either frequency. The second PRF is then set equal to $(N + 1)$ times the ranging frequency. Other values for the second PRF are possible, but this value gives the widest allowed pulsewidth and closest PRF values.

The transmitted pulsewidth has a maximum allowable value, depending on the PRF values and the required resolvable range, depicted in Figure 7.29. From the figure, we can see that

$$(\mathrm{PRF}_1)^{-1} = (\mathrm{PRF}_2)^{-1} + \tau_2 \tag{7.23}$$

where τ_2 is the radar's pulsewidth. The submultiple frequency is given by

$$f_B = \frac{\mathrm{PRF}_1}{N} = \frac{\mathrm{PRF}_2}{N + 1} \tag{7.24}$$

where N is an integer. By solving (7.23) and (7.24), the maximum allowable pulsewidth can be determined as

$$\tau_{2\ \max} \leq \frac{1}{N\,(n + 1)\,f_B} = \frac{1}{(N + 1)\ \mathrm{PRF}_1} \tag{7.25}$$

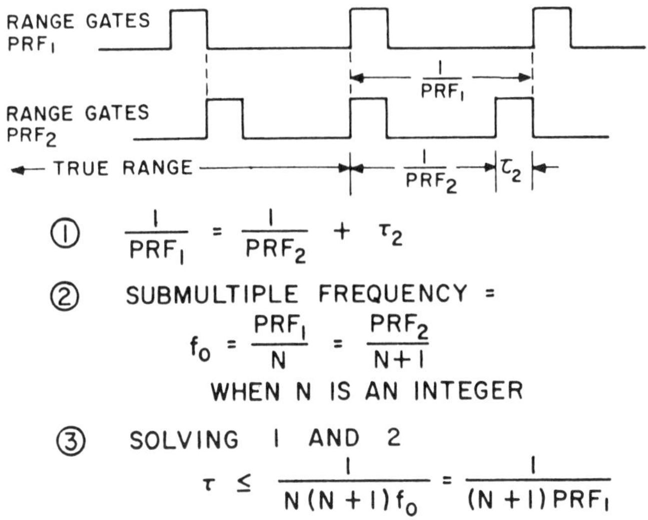

Figure 7.29 Maximum allowable pulsewidth in a high-PRF PD radar.

This value pulsewidth prevents the pulses adjacent to the true coincidence from overlapping and causing a false coincidence. The resulting maximum pulsewidth is dependent on the range to be resolved and the ratios of the PRFs to the ranging frequency.

As an example of a two-PRF ranging system, consider a medium-PRF radar with a maximum operating range (R_u) of 100 nmi [15]. The common submultiple frequency (f_B) can then be found from

$$f_B = \frac{c}{2R_u} \tag{7.26}$$

where $c = 3 \times 10^5$ km/s and R_u (km) is the maximum desired unambiguous range. For the example, the basic PRF can be determined as approximately $f_B = 810$ Hz. If the ranging parameter is selected as $N = 17$, $\text{PRF}_1 = Nf_B = 13.77$ kHz and $\text{PRF}_2 = (N + 1) f_B = 14.58$ kHz (see (7.24)). Further, the unambiguous range (R_{u1}) for the first PRF equals

$$R_{u_1} = \frac{c}{2\text{PRF}_1} \tag{7.27}$$

where $c = 3 \times 10^5$ km/s resulting in $R_{u_1} = 10.893$ km and $R_{u_2} = 10.288$ km.

For a target at 60 km, the ambiguous range measured by the first PRF equals $R_{a_1} = 5.535$ km $(60 - 5 \cdot 10.893)$ whereas that measured by the second PRF equals $R_{a_2} = 8.560$ km. If we define the ratio:

$$x = \frac{R_{a_1}}{R_{u_1}} \tag{7.28}$$

where x is the fractional range represented by the first ambiguous measurement, the result is $x = 0.508$. Similarly, $y = R_{a_2}/R_{u_2} = 0.832$. Then, from these measurements, the unambiguous target range can be determined by using

$$R_t = (n_1 + x)R_{u_1} \tag{7.29}$$

$$R_t = (n_2 + y)R_{u_2} \tag{7.30}$$

where n_1 and n_2 are integers selected by trial and error so that (7.29) and (7.30) provide the same range (R_t). In the example, $n_1 = n_2 = 5$, resulting in a measured range equal to $R_t = 5.508 \cdot 10.893 = 5.832 \cdot 10.288 = 60$ km.

Alternatively, the Chinese remainder theorem can be used to determine the true range from the two ambiguous measurements [9]. From (7.24), the unambiguous ranging period (T_u) is given by

$$T_u = NT_1 = (N + 1)\, T_2 \tag{7.31}$$

where T_1 is the PRI associated with PRF_1 and T_2 is the PRI associated with PRF_2. Dividing by the common multiple $N \cdot (N + 1)$ results in

$$\frac{T_u}{N(N + 1)} = \frac{T_1}{N + 1} = \frac{T_2}{N} \tag{7.32}$$

The choice of $m_1 = N + 1$ and $m_2 = N$ implies that the total unambiguous range is divided into 306 range cells so that

$$m_1 = \frac{T_u}{T_2} = N + 1 \tag{7.33}$$

$$m_2 = \frac{T_u}{T_1} = N \tag{7.34}$$

for the two-PRF ranging code. In accordance with the Chinese remainder theorem, the unambiguous target range measured in range units ($1 \le \bar{R}_c \le m_1 m_2$) is given by

$$\bar{R}_c = C_1 a_1 + C_2 a_2 \text{ modulo}(m_1 m_2) \tag{7.35}$$

where

$$a_1 = \mathrm{INT}(m_1 x) \tag{7.36}$$

$$a_2 = \mathrm{INT}(m_2 y) \tag{7.37}$$

The constants C_1 and C_2 are determined from

$$C_1 = b_1 m_2 = (1)\ \mathrm{modulo}(m_1) \tag{7.38}$$

$$C_2 = b_2 m_1 = (1)\ \mathrm{modulo}(m_2) \tag{7.39}$$

where b_1 is the smallest positive integer that, when multiplied by m_2 and divided by m_1, gives unity as a remainder. The unambiguous range is then given by

$$R_t = \frac{\bar{R}_c R_u}{m_1 m_2} \tag{7.40}$$

For the example $m_1 = N + 1 = 18$ and $m_2 = N = 17$. From (7.38), $b_1 = 17$, and from (7.39), $b_2 = 1$, resulting in $C_1 = 289$ and $C_2 = 18$. The unambiguous target range in range units then becomes

$$\bar{R}_c = 289 \cdot a_1 + 18 \cdot a_2\ \mathrm{modulo}(m_1 m_2) \tag{7.41}$$

which in the example bccomes $\bar{R}_c = 99$. The target range is hence given by $R_t = (99/306) \cdot 185.4 = 60$ km.

The maximum range performance of the radar is determined by the average power transmitted. This implies a pulsewidth as wide as possible to minimize the requirement for excessive peak power. As an example, consider a high-PRF airborne PD system that requires PRFs in excess of 172 kHz (at X-band for a 1400-knot target from an 1800-knot interceptor) and assume that an unambiguous range of 50 nmi is required [22]. This corresponds to an f_B of 1.62 kHz and an $N = 172/1.62 = 106$. the maximum allowable pulsewidth is then $\tau = 1/(106)\ (107)\ (1620) = 0.05\ \mu$s. This will require a relatively high peak power, which leads to the motivation for using three or more PRFs.

A three-PRF ranging system functions by forming two pairs of PRFs with each pair measuring the range to a different subranging frequency. These two subranging frequencies form, in turn, a two-PRF ranging system that measures range to the final ranging frequency. A particular choice of the three-PRF ranging parameters [22] that leads to a relatively simple mechanization results in a submultiple ranging frequency given by

$$f_B = \frac{\mathrm{PRF}_1}{N(N+1)} = \frac{\mathrm{PRF}_2}{N(N+2)} = \frac{\mathrm{PRF}_3}{(N+1)(N+2)} \tag{7.42}$$

The unambiguous range is on the order of N^2 times the submultiple frequency as compared to N for the two-PRF ranging system. This results in a value for N with the three-PRF ranging system that is approximately the square root of its comparable value for a two-PRF ranging system.

The Chinese remainder theorem affords a convenient method for decoding the true range from the several ambiguous measurements in the multiple-channel search system [9]. In this method, the period between PRF transmissions is expressed in units of pulsewidth (range cells) by dividing the period into a number of contiguous range cells. Each cell is numbered sequentially as 1, 2, 3, ..., m_i; $i = 1, 2, 3$. In the three-PRF ranging system, this results in each PRI being divided into a number of range cells, designated by m_1, m_2, and m_3 ($m_1 > m_2 > m_3$). The three ambiguous range cell numbers, designated by a_1, a_2, a_3, which correspond to the target's measured range for each PRF, are then determined. The target's unambiguous range cell number ($1 \leq \bar{R}_c \leq m_1,m_2,m_3$) is thus given by

$$\bar{R}_c = (C_1 a_1 + C_2 a_2 + C_3 a_3) \text{ modulo}(m_1 m_2 m_3) \tag{7.43}$$

where

$$\begin{aligned} C_1 &= b_1 m_2 m_3 = (1) \text{ modulo}(m_1) \\ C_2 &= b_2 m_1 m_3 = (1) \text{ modulo}(m_2) \\ C_3 &= b_3 m_1 m_2 = (1) \text{ modulo}(m_3) \end{aligned} \tag{7.44}$$

and b_1 is the smallest positive integer that, when multiplied by $m_2 m_3$ and divided by m_1, gives unity as the remainder (b_2 and b_3 are determined in the same manner).

The procedure using the Chinese remainder theorem can be summarized as follows. Once m_1, m_2, and m_3 have been chosen to provide the desired unambiguous range, the constants C_1, C_2, and C_3 are determined (see (7.44)) for this ranging code. The ambiguous range measurements a_1, a_2, and a_3 are determined by measuring the target range in range units for each PRF. Then, the unambiguous target range is determined by multiplying each constant by its ambiguous range (see (7.43)) and dividing by m_1, m_2, m_3 as many times as possible to determine the remainder ($\bar{R}_c$) that equals the unambiguous range in range units. In the three-PRF multiple-ranging system, the maximum allowable pulsewidth is given by [22]

$$\tau_{3\,\max} \leq \frac{1}{N(N+1)(N+2)f_B} = \frac{1}{(N+2)\,\text{PRF}_i} \tag{7.45}$$

where $f_r = N(N+1)(N+2)f_B$ represents the lowest common multiple of all three PRFs. For the previously discussed example of the two-PRF system ($\text{PRF}_1 = Nf_B =$

171.72 kHz, $PRF_2 = (N + 1)f_B = 173.34$ kHz) the value of N now becomes 10 so that $PRF_1 = (N)(N + 1)f_B = 178.2$ kHz, $PRF_2 = (N)(N + 2)f_B = 194.4$ kHz, and $PRF_3 = (N + 1)(N + 2)f_B = 213.84$ kHz. However, the principal advantage of the three-PRF ranging system is that the maximum allowable pulsewidth $\tau_{3\ max} = 0.5$ μs, as compared to the value $\tau_{2\ max} = 0.05$ μs for the two-PRF ranging system. Thus, ten times the average power may be obtained by using the same peak power with three PRFs instead of two. This advantage is partially offset by the need for three detections of the target, which has a lower probability of success than the two detections required of the two-PRF system. Also, the probability on each trial is reduced because three PRFs must be generated during the time that the radar dwells on the target rather than two PRFs for the two-PRF ranging method.

The block diagram of a three-PRF range measuring system is depicted in Figure 7.30. The common multiple frequency $f_r = (N)(N + 1)(N + 2)f_B$ is generated by using a crystal controlled oscillator. The three PRFs are then generated by dividing

THREE PRF RANGING EQUATIONS

① SUBMULTIPLE FREQUENCY =

$$f_0 = \frac{PRF_1}{N(N+1)} = \frac{PRF_2}{N(N+2)} = \frac{PRF_3}{(N+1)(N+2)}$$

② MAXIMUM PULSE WIDTH =

$$\tau_3 = \frac{1}{N(N+1)(N+2)f_0} = \frac{1}{(N+2)PRF_1}$$

③ LOWEST COMMON MULTIPLE FREQUENCY =

$$f_r = N(N+1)(N+2)f_0$$

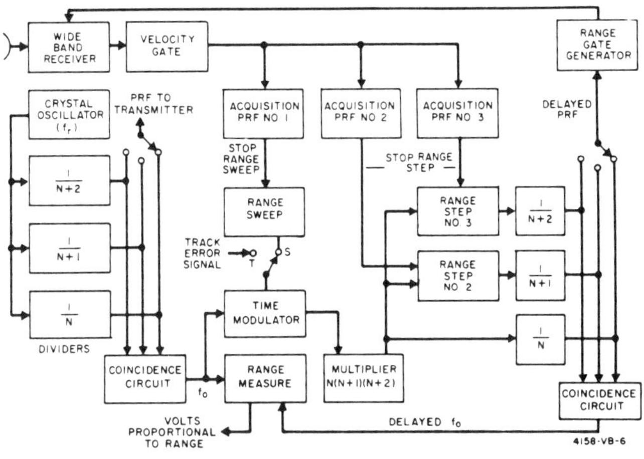

Figure 7.30 Block diagram of a three-PRF range-measuring system.

the common multiple frequency (f_r) by N, $N + 1$, and $N + 2$. This guarantees that all PRFs are phase locked (i.e., synchronized). Any one PRF can be used to key the transmitter by using the PRF selector. All three PRFs are compared in a coincidence circuit, the output of which provides pulses at the ranging frequency (f_B). This is used as a time base for measuring range in a manner analogous to the way that the transmitter pulse is used in a low-PRF radar to measure range.

The actual measurement of range is accomplished as follows. The first PRF (PRF_1) is transmitted and the range gates are searched by means of the range sweep and time modulator. On detection of the target, the range sweep is stopped and the time modulator maintains its delay. The next PRF (PRF_2) is transmitted and the range gates are searched incrementally in time steps corresponding to the reciprocal of f_r. On target detection, the search stops, leaving the counter at the proper phase to position PRF_2 range gates over the target. The same procedure is then used in PRF_3 to obtain the third ambiguous range measurement. At this time, as all range gates line up with the target, the true range can be measured.

7.3.2 FM Ranging [6,23]

As the PRF is increased in a high-PRF PD radar, a point eventually is reached where the range is so ambiguous that it cannot be resolved by using the multiple-PRF ranging system. If a range is required, it must be measured indirectly, as in a CW radar, by frequency modulating the radar transmitter. Both sinusoidal and linear FM have been employed for this purpose [23]. However, we will primarily describe linear FM ranging, which is the more common of the two types [6].

Linear FM ranging in its simplest form consists of transmitting with a single high-PRF waveform that is periodically ramp modulated in frequency, as depicted in Figure 7.31. The received signal is delayed in time and translated in frequency by the target's doppler shift as shown in the figure. The linear modulation continues for a period at least several times as long as the round-trip transit time for the most distant target of significance. Over the course of this period, the instantaneous difference between the frequency of the received echos and the frequency of the transmitter is measured. During the second-half of the ranging cycle, the transmitter frequency remains fixed and, after a short receiver "dead" time, the difference between the transmitted and received signals is again measured.

During the non-FM half of the ranging cycle, the signal frequency is a function of the target velocity; whereas during the FM half, the signal frequency is proportional to the algebraic sum of the target velocity and range. Demodulated returns from targets with high velocities are shifted higher in frequency, and returns from targets with large ranges are shifted lower in frequency during the FM mode. The target range is then obtained as

$$R_{\mathrm{FM}} = \frac{c(f_{d_1} - f_{d_2})}{2\, k_{\mathrm{FM}}} \tag{7.46}$$

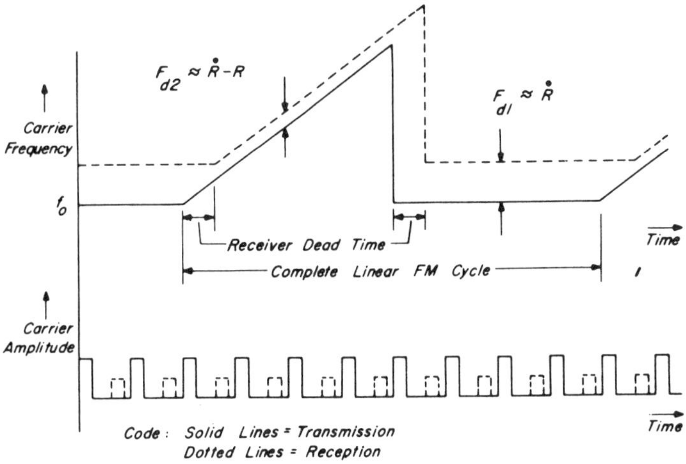

Figure 7.31 Transmitter and receiver frequency in a linear-FM ranging system.

where f_{d_1} (Hz) is the target's doppler shift measured during the non-FM part of the ranging cycle, f_{d_2} (Hz) is the measured difference in frequency during the FM part of the ranging cycle, and k_{FM} is the rate of linear FM sweep in Hz/s. A block diagram of a high-PRF PD radar using linear-FM ranging is depicted in Figure 7.32.

An alternate two-slope waveform used in linear FM ranging is depicted in Figure 7.33. This waveform uses equal up-slope and down-slope linear ramps, where the magnitude of the slope is equal to that of the rising frequency slope of the previously described waveform (see Figure 7.31). If the target is closing (i.e., it has a positive doppler), the difference between the frequency of the transmitter and that of the received echos will be decreased by f_d during the rising frequency segment ($\Delta f_1 = k_{FM}t_r - f_d$) and increased by f_d during the falling frequency segment ($\Delta f_2 = k_{FM}t_r + f_d$). Consequently, if the frequency differences for the two segments are added ($\Delta f = \Delta f_1 + \Delta f_2 = 2k_{FM}t_r$), the doppler frequency shift is cancelled and only the frequency difference ($2k_{FM}t_r$) due to the round-trip transit time to the target remains. Thus, the target range can be determined from

$$t_r = \frac{\Delta f_1 + \Delta f_2}{2k_{FM}} \tag{7.47}$$

where t_r is the round-trip transit time, Δf_1 is the frequency difference between the transmitter and echo frequencies during the rising slope, Δf_2 is the similar frequency difference during the falling slope, and k_{FM} is the rate of change of the transmitter frequency.

For example, assuming no doppler shift, the measured difference is 10 kHz on each segment, and the transmitter has a slope equal to 10 Hz/μs. From (7.47), the

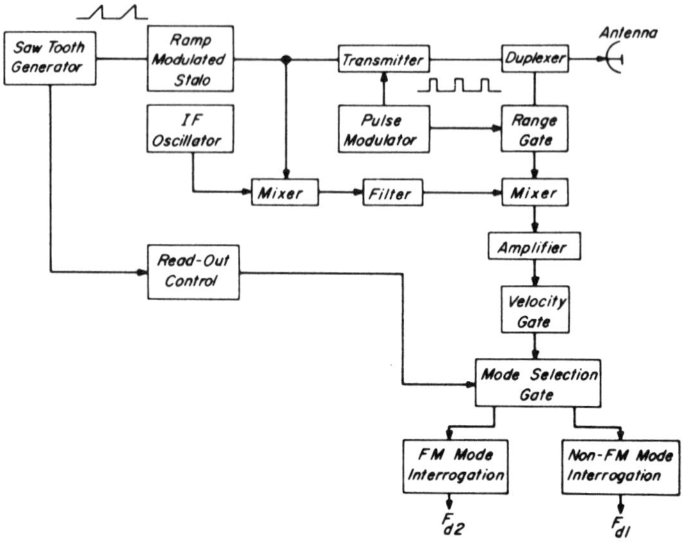

Figure 7.32 Block diagram of a linear-FM ranging high-PRF radar.

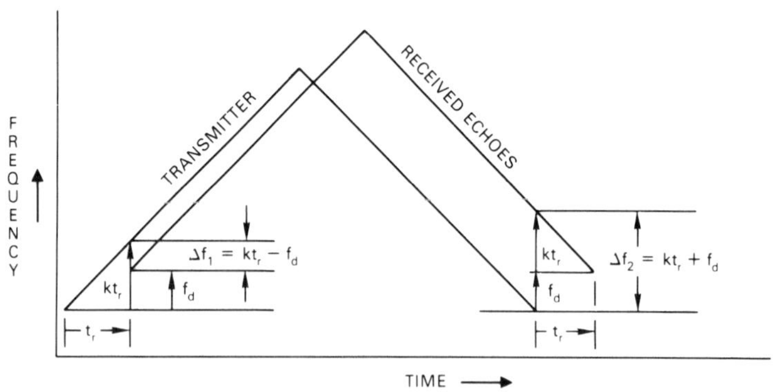

Figure 7.33 A two-slope FM ranging waveform.

transit time is 1000 μs, resulting in a range equal to 81 nmi. Now assume that the target has a doppler shift of 3 kHz. On the up-slope, the measured frequency difference will be 7 kHz, whereas on the down-slope, it will be 13 kHz. Again applying (7.47), the transit time still will be 1000 μs, illustrating the independence of the waveform on the doppler shift. Note, although the example considers a positive doppler shift, the waveform works equally well on negative doppler shifts.

If the antenna beam illuminates two targets at the same time, there will be two frequency differences during the first segment of the modulation cycle and two during the second, as depicted in Figure 7.34. Also note that the same situation occurs for

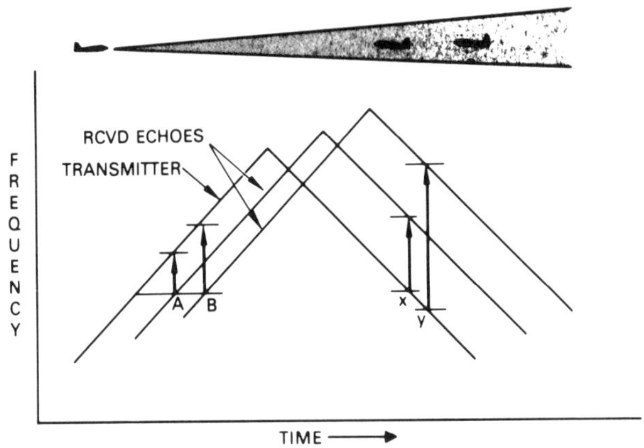

If two targets are detected simultaneously, two frequency differences will be measured during each segment of the modulation cycle.

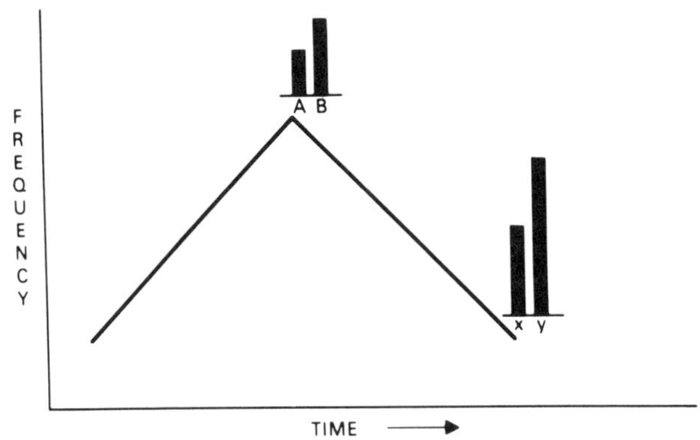

All the radar sees are two frequency differences at the end of each segment. Radar has no way of telling whether A should be paired with x or y.

Figure 7.34 FM ranging with two targets.

the saw-toothed waveform depicted in Figure 7.31. Without further information, to determine which pair of difference frequency measurements (i.e., A and x or y) pertains to the same target would be impossible. This ambiguity can be resolved by modifying the ranging waveform to transmit a three-slope ranging waveform, as depicted in Figure 7.35.

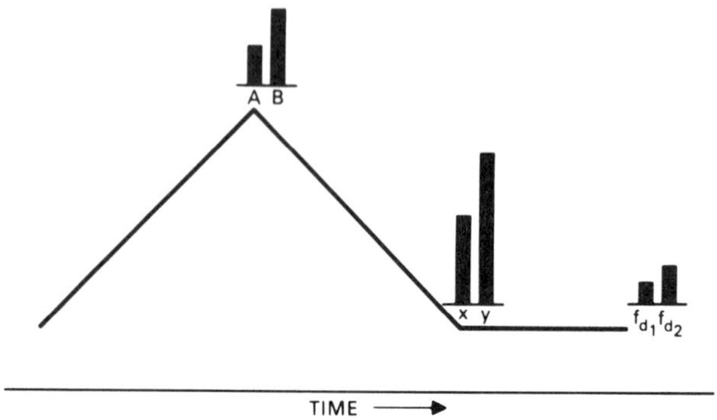

Figure 7.35 Three-slope FM ranging waveform.

The basic three-slope ranging waveform illustrated in Figure 7.35 consists of the two-slope waveform (see Figure 7.33) complemented with a constant frequency segment that allows determining the target's doppler frequencies. The target's doppler frequencies also can be determined by subtracting the two frequency differences measured during the up-slope and down-slope portions of the ranging waveform (i.e., $\Delta f_2 - \Delta f_1 = (k_{FM}t_r + f_d) - (k_{FM}t_r - f_d) = 2f_d$). Therefore, by subtracting A (or B) from x (or y) and comparing the results with the measured doppler frequencies, association of the correct pair of doppler frequency upslope-downslope measurements with each other is possible. For example, if $(x - A)$ were twice one of the measured doppler frequencies, the pairing would be x with A and y with B. Otherwise, y would be paired with A and x with B.

In air-to-air applications, the doppler frequency shift can be greater than the frequency difference due to the ranging timc ($k_{FM}t_r$). This occurs because the rate of change of frequency (k_{FM} in Hz/s) is generally made low enough that the maximum value of $k_{FM}t_r$ will be only a small fraction of the highest doppler frequency encountered. This situation is depicted in Figure 7.36, where return echos are higher than the transmitted frequency for positive doppler rather than being less than the transmitted frequency, as depicted in Figure 7.33.

The relationships between the measured frequency differences for two or more simultaneously detected targets is interpreted more easily when presented on a separate line for each slope, as depicted in Figure 7.37. The difference frequencies on the rising segment of the FM ranging waveform ($\Delta f_1 = k_{FM}t_r - f_d$; $f_d > k_{FM}t_r$) appear as negative frequencies on the diagram (i.e., A and B), whereas difference frequencies

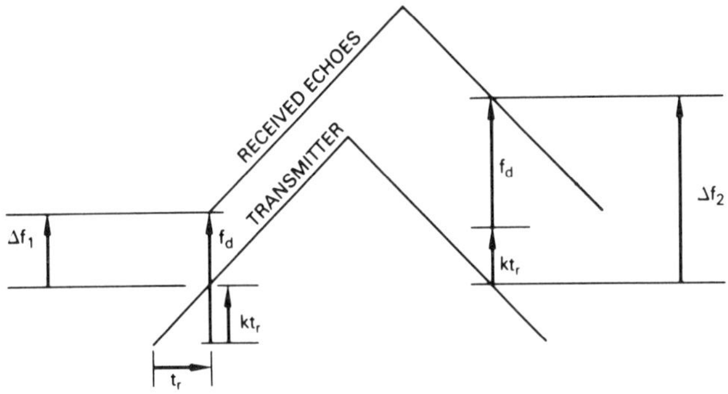

Figure 7.36 Two-slope FM ranging waveforms with doppler shift greater than ranging frequency difference.

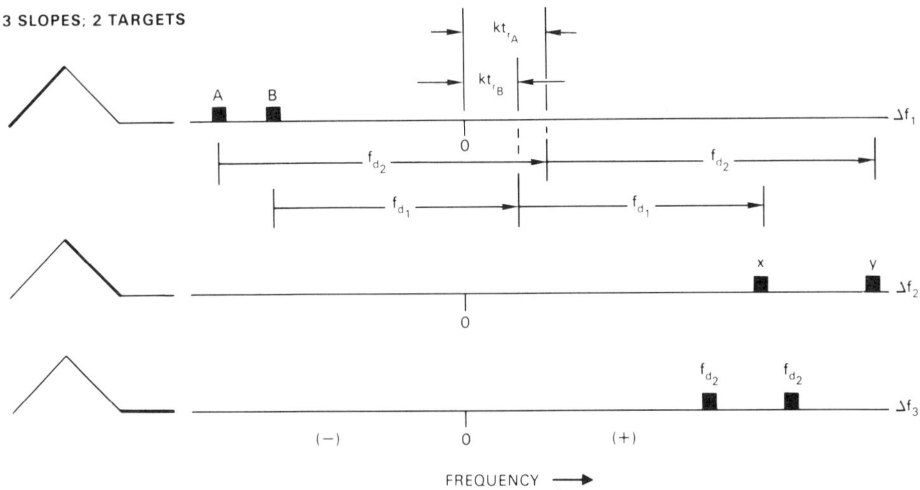

Figure 7.37 Frequency difference plot for FM ranging with two targets.

on the falling portion of the waveform (i.e., x and y) and the assumed doppler frequencies (i.e., f_{d_1} and f_{d_2}) appear as positive frequencies on the diagram. The differences can be paired by using the doppler frequencies measured on the third segment and connecting those pairs separated by twice the doppler frequencies f_{d_1} and f_{d_2}(A and y, B and x on the diagram). The transit times can then be found from the diagram as $t_{ra} = (f_{d_2} - A)/k_{\text{FM}}$ and $t_{rb} = (f_{d_1} - B/k_{\text{FM}})$.

When three or more targets are present, we still can pair corresponding targets on the up- and down-slopes using the procedure described in Figure 7.37; however, certain combinations of range and doppler frequencies may allow multiple pairings (i.e., ghosts). Ghosts can be eliminated by adding more slopes to the waveform code.

As with multiple PRFs, if N is the number of slopes, all possible combinations of $N - 1$ simultaneously detected target can be "deghosted." The problem of ghosts is generally much less severe with FM ranging than with multiple PRF ranging because neither range nor doppler frequency is ambiguous and, in practice, a three-slope waveform code is generally used.

The accuracy of FM ranging is proportional to the ratio of the frequency resolution of the doppler filter bank to the rate of change of the transmitter frequency ($k_{FM} = \dot{f}$), as depicted in Figure 7.38. The range accuracy for FM ranging is then given by

$$R_{acc} = \frac{cB_D}{2k_{FM}} \tag{7.48}$$

where B_D is the doppler filter's 3 dB bandwidth, and k_{FM} is the rate of change of the transmitter frequency. For example, if the doppler filter's bandwidth is 100 Hz and k_{FM} = 3 MHz/s, then the range accuracy from (7.48) is 5 km.

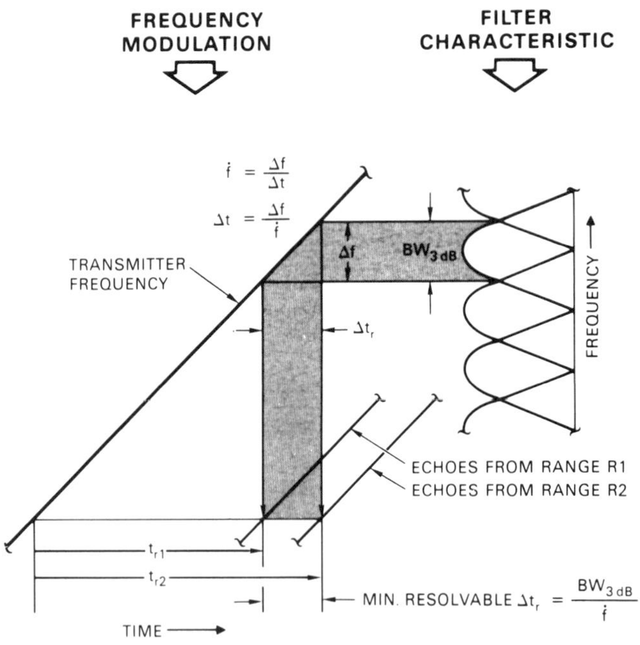

Figure 7.38 Minimum resolvable difference in ranging time.

From (7.48), we can see that the FM ranging accuracy depends on the doppler filter bandwidth (B_D) and the rate at which the transmitter frequency is changed (k_{FM}). The greater k_{FM} is, the greater the frequency difference a given transit time will produce and the greater range accuracy that can be achieved. Similarly, the smaller is the doppler filter bandwidth (B_D), the greater the range accuracy. However, the transmitter frequency FM slope (k_{FM}) cannot be made arbitrarily steep nor the doppler bandwidth arbitrarily small to increase range accuracy.

The minimum doppler filter bandwidth is determined by the coherent processing interval of the radar, which in a search radar is related to the time the radar antenna illuminates the target (i.e., t_o is the radar observation time). With a three-slope FM ranging code, this limits the minimum possible observation time to $t_o/3$, which, in turn, limits the minimum doppler filter bandwidth to $B_D = 3/t_o$ (Hz).

In air-to-air applications, the transmitter's FM rate (k_{FM}) is limited by the smearing of ground clutter returns (which may be received out to ranges greater than 100 nmi) into the doppler region occupied by target returns. This is the same effect that limits the performance of high-PRF PD radars due to transmitter instability (see Section 3.3). The detection of targets within the doppler region occupied by the spread clutter spectrum is generally inhibited as target returns are obscured by the clutter. The rate of transmitter frequency change (k_{FM}) must then be held to a small fraction of the maximum target doppler shift to provide a doppler-clear region sufficiently wide to allow target detection.

The clutter spectral situation for the detection of long-range targets with various degrees of transmitter FM is depicted in Figure 7.39. Two long-range targets are shown, one with a high doppler frequency (*B*) and the other (*A*) with a doppler frequency only slightly higher than the highest ground clutter frequency. The first profile shows the normal ground clutter spectrum for a high-PRF airborne PD radar with the two targets in the clear-doppler region. The second profile shows what happens when the frequency shift corresponding to the target's range is comparable to the doppler shift of the higher closing rate target (*B*). In this case, mainlobe clutter spreads into the normally clutter-clear region and obscures both targets. In the third profile, the frequency shift corresponding to the target's range is a small fraction of the target's doppler shift. This relationship results in a limited spread of the clutter's spectral width, which has little effect on target detection.

In general, the constraints on transmitter frequency change (k_{FM}) and doppler filter bandwidth (B_D) are such that, in a typical airborne interceptor application, the range accuracy is on the order of a few miles [6]. This is substantially poorer than can be obtained with pulse-delay ranging. If maximum detection range is desired in a high-PRF PD radar, a sequential detection mode can be employed. Initially, a velocity search mode is employed in which range is not measured. Then, the FM-ranging mode is activated to establish the target's range.

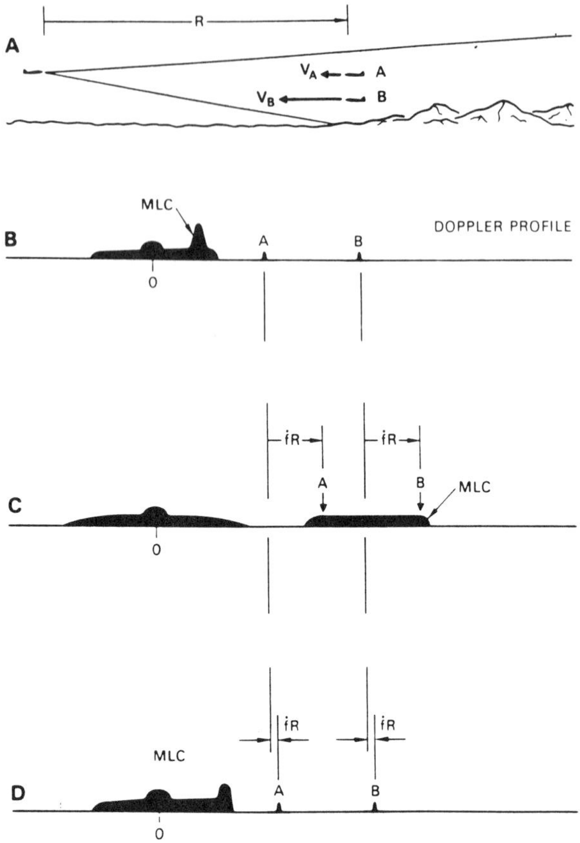

Figure 7.39 Ground clutter spectral spreading due to transmitter FM in a high-PRF radar.

7.4 MAINLOBE CLUTTER

A major factor determining the performance of PD radars is the ability of the radar to attenuate mainlobe surface clutter (i.e., ground and sea clutter). Highly range-ambiguous radar waveforms (i.e., high- and medium-PRFs) accentuate mainbeam clutter, requiring higher clutter attenuation for a given level of performance by these types of waveform design. The basic limits on performance of surface-based high- and medium-PRF designs is generally set by the phase and frequency instabilities of the radar's reference oscillator (STALO) and transmitter, which spreads the mainlobe clutter into the frequency region occupied by the target's doppler filters (see Section

3.3). In addition to clutter spectral spreading due to oscillator instability effects, mainlobe clutter is also spread by radar platform motion, antenna scanning, and random variations in signal amplitude and phase, as discussed in Section 3.2. These latter effects generally determine the doppler-clear region available for detection of moving targets in low- and medium-PRF PD designs. In airborne designs, sidelobe clutter can be the limiting factor for low- and medium-PRF designs (see Section 7.2.2), whereas oscillator instability generally determines performance in clutter for a high-PRF design.

Another effect that can limit the performance of PD radars is the spreading of the mainlobe clutter spectrum into the doppler region occupied by targets due to receiver nonlinearity. This problem was covered in Sections 3.2.5 and 6.2.3 for MTI systems. However, in high-PRF PD designs, mainlobe clutter can exceed the receiver noise level by 80 to 100 dB due to the folding of range-ambiguous clutter into the target's range cell. Figure 7.40 depicts the spectral spreading of Gaussian-shaped mainbeam clutter down to the 140 dB level for a soft-limiting type of receiver (see Figure 6.14). This allows the detection performance of high-PRF PD radars to be determined against weak targets in the presence of receiver nonlinearities.

To compute the performance of those designs limited by mailobe clutter, we need to compute the clutter power contained within that portion of the clutter spectrum.

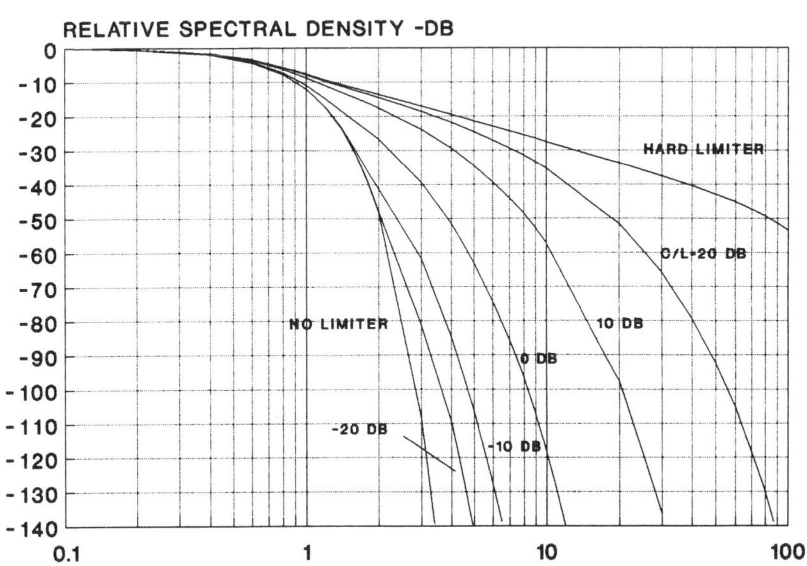

Figure 7.40 Spectral spreading for a nonlinear receiver.

The incremental clutter power received from a small patch on the ground (see Figure 7.41) from a flat earth can be expressed as

$$dP_c = \frac{P_t G^2(\theta_{az}, \phi_{el}) \lambda^2 \sigma^\circ \, ds \, d\theta}{(4\pi)^3 R_c^4 L_c} \tag{7.49}$$

where $G^2(\theta, \phi)$ is the antenna gain in the direction of the clutter cell, θ_{az} is the azimuth angle, ϕ_{el} is the depression angle, σ° is the normalized clutter reflectivity per unit area, R_c is the range to the clutter patch, and L_c is the radar system clutter losses. For a low-PRF radar, the mainlobe clutter RCS(σ_c) = $\sigma^\circ \int_s ds\, d\theta$) can be expressed as

$$\sigma_c = \sigma^\circ R_c \theta_{az} (c\tau_e/2) \sec\phi_{el} \tag{7.50}$$

where θ_A is the radar's two-way power antenna beamwidth and τ_e is the radar's effective pulsewidth. Substituting this into (7.49) and defining $G^2(\theta_{az}, \phi_{el}) = G_t^2 F_c^4$, where F_c^2 is the clutter pattern propagation factor and G_t is the peak antenna gain, $R_\tau = c\tau_e/2$ and assuming ϕ_{el} is small results in

$$P_{c\,\mathrm{ML}} = \frac{P_t G_t^2 \lambda^2 F_c^4 \sigma^\circ \theta_{az} R_\tau}{(4\pi)^3 R_c^3 L_c} \tag{7.51}$$

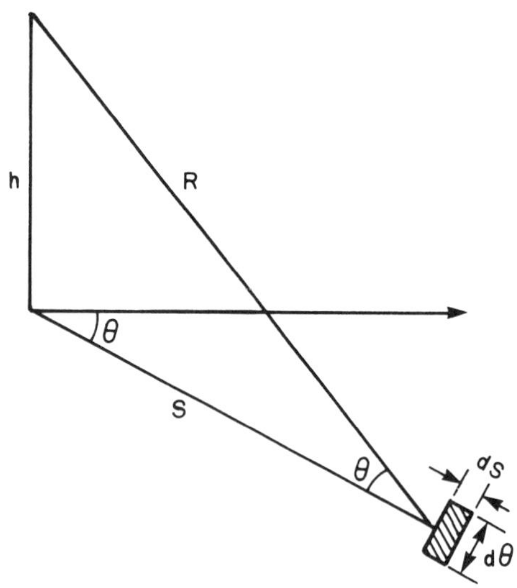

Figure 7.41 Geometry of clutter cell.

For range ambiguous clutter, this can be expressed as

$$P_{c\,\mathrm{ML}} = \frac{P_t G_t^2 \lambda^2 \theta_{\mathrm{az}} R_T}{(4\pi)^3 L_c} \sum_{i=1}^{\infty} \frac{\sigma_i^\circ F_{ci}^4}{R_{ci}^3} \tag{7.52}$$

where σ_i°, F_{ci}^4, and R_{ci}^4 are the respective values associated with each of the ambiguous clutter elements. Note that in a range-ambiguous radar, the clutter elements closest to the radar contribute the most incremental clutter power. In cases where reflectivity, propagation factor, and attenuation remain relatively constant, an effective clutter range (R_{ce}):

$$R_{ce}^3 = \frac{1}{\sum_{i=1}^{n} (1/R_{ci}^3)} \tag{7.53}$$

can be substituted into (7.52) [24].

In the summation of (7.52), both the clutter reflectivity (σ_i°) and the pattern propagation factor generally vary with the range (R_{ci}) to the clutter patch. By using the constant-gamma reflectivity model (i.e., $\sigma^\circ = \gamma \sin\phi_{\mathrm{el}}$) allows the clutter reflectivity to be removed from the summation. For small depression angles, the relationship $\phi_{\mathrm{el}} = h/R$ applies, allowing the clutter power to be evaluated as

$$P_{c\,\mathrm{ML}} = \frac{P_t G_t^2 \lambda^2 \theta_{\mathrm{az}} R_T \gamma h_a}{(4\pi)^3 L_c} \sum_{i=1}^{\infty} \frac{F_{ci}^4}{R_{ci}^4} \tag{7.54}$$

where h_a is the antenna's height above the surface and γ is the constant reflectivity coefficient. For high- and medium-PRF PD airborne radars, the antenna has sufficient height (h_a) that clutter in the first several ambiguous range intervals is not illuminated by the antenna's mainlobe. The summation in (7.52), and also in (7.54), if appropriate, will then have a small value of F_{ci} for $i = 1$ and possibly for several successive ambiguities with ranges $R_{ci} = R_{c1} + (i - 1)R_a (i > 1)$, where $R_a = c \cdot \mathrm{PRI}/2$ for a constant PRF.

The clutter magnification properties of range-ambiguous radars (i.e., medium- and high-PRF) can be evaluated by computing the signal-to-clutter power ratio obtained through division of the standard radar equation by (7.52) resulting in

$$S/C = \frac{\sigma_t}{\theta_{\mathrm{az}} R_t R_T L \sum_{i=1}^{\infty} \sigma_i^\circ \left(\frac{R_t}{R_{ci}}\right)^3 F_{ci}^4} \tag{7.55}$$

where σ_t is the target's RCS, R_t is the target range, and $L = L_r/L_c$. When the clutter and target range are equal in the peak of the antenna beam ($F_c = 1$), $S/C = \sigma_t/\sigma_c$, where σ_c is the unambiguous clutter RCS given by $\sigma_c = \sigma^\circ \theta_{az} R_t (c\tau_e/2)$. For ambiguous clutter ranges (R_{ci}), the clutter reflectivity (σ_i°) at each ambiguity is then magnified by the factor $(R_t/R_{ci})^3 F_{ci}^4$, which must be summed over all possible clutter ambiguities to obtain the effective clutter RCS.

For high-PRF radars, mainlobe clutter power can be estimated by evaluating the clutter power returned to a CW radar (with transmitter power P_t) and multiplying this value by the radar's duty factor ($d_u = \tau/T$). The mainlobe clutter power for a CW radar can be obtained by replacing the summation in (7.52) with an integral as [24]

$$P_{c\ \mathrm{ML}} = \frac{P_t G_t^2 \lambda^2 \theta_{az}}{(4\pi)^3 L_c} \int_0^\infty \frac{\sigma^\circ F_c^4}{R^3}\, \mathrm{d}R \tag{7.56}$$

where F_c^4 is the two-way (transmitting-receiving) pattern propagation factor that includes the elevation voltage gain pattern of the antenna. For the case of a Gaussian-shaped elevation beam pattern point at the horizon, the pattern propagation factor can be expressed as

$$F_c^4 = \exp\left(-\frac{5.55\phi_{el}^2}{\phi_B^2}\right) \tag{7.57}$$

where ϕ_B is the antenna's one-way power elevation beamwidth. For small depression angles, $\phi_{el} = h_a/R$ and

$$\int_0^\infty \frac{\exp(-5.55\, h_a^2/\phi_{el}^2 R^2)}{R^3} = \frac{\phi_{el}^2}{(3.33\, h_a)^2} \tag{7.58}$$

The mainlobe clutter power for a PD radar (with duty factor d_u) can then be found from (7.56) with a constant σ° as

$$P_{c\ \mathrm{ML}} = \frac{P_t d_u G_t^2 \lambda^2 \theta_{az} \phi_{el}^2 \sigma^\circ}{(4\pi)^3 L_c (3.33 h_a)^2} \tag{7.59}$$

where P_t is the radar's peak power, d_u is the duty factor, θ_{az} is the azimuth beamwidth, ϕ_{el} is the elevation beamwidth, h_a is the height of the antenna's phase center above the surface, G_t is the antenna's peak gain, and L_c is the loss associated with clutter detection. By using the antenna's gain-beamwidth relationship $G_t = 4\pi/(\theta_{az}\phi_{el}L_n)$, where $L_n = 1.2$ is the azimuth antenna pattern loss factor, $\theta_{az} = 1.5\ \lambda/D_a$, and D_a

is the azimuth antenna's aperture (both D_a and L_n adjusted for low sidelobes), the result is

$$P_{c\,\mathrm{ML}} = \frac{P_t d_u \sigma^\circ \lambda D_a}{6\pi L_c L_n^2 (3.33 h_a)^2} = \frac{P_t \sigma^\circ \lambda D_a d_u}{400 h_a^2} \tag{7.60}$$

where the second term is evaluated for an azimuth antenna's beam-shape loss of $L_c = 1.33$ [24]. For example, consider a surface-based X-band PD radar ($\lambda = 3$ cm) with a 0.1 duty factor, which uses a 1.2 m azimuth aperture with its phase center 2.4 m above a surface with a reflectivity coefficient of $\sigma^\circ = 0.004$, resulting in a $P_{c\,\mathrm{ML}}/P_t = 6.25 \times 10^{-9} = -82$ dB. For a 10 kW transmitter power, the mainlobe clutter power is equal to -12 dBm.

Equation (7.56) can also be evaluated by using a constant-gamma reflectivity model ($\sigma^\circ = \gamma \sin\phi_{\mathrm{el}}$), which, for small depression angles ($\phi_{\mathrm{el}} = h/R$), results in the integral:

$$\int_0^\infty \frac{\sigma^\circ F_c^4}{R^3}\, \mathrm{d}R = \int_0^\infty \{\gamma h_a \exp[-(5.55 h_a^2)/(\phi_{\mathrm{el}}^2 R^2)]/R^4\}\, \mathrm{d}R$$

$$= \frac{\gamma \phi_{\mathrm{el}}^3}{29.4\, h_a^2} \tag{7.61}$$

Substitution of this into (7.56) for a PD radar with duty factor d_u results in

$$P_{c\,\mathrm{ML}} = \frac{P_t d_u G_t^2 \lambda^2 \theta_{\mathrm{az}} \phi_{\mathrm{el}}^3 \gamma}{29.4 (4\pi)^3 L h_a^2} \tag{7.62}$$

This can be further simplified by using $\phi_{\mathrm{el}} = 1.5\lambda/D_e$, where D_e is the elevation aperture and the other simplifications applied in (7.60) resulting in

$$P_{c\,\mathrm{ML}} = \frac{P_t d_u \lambda^2 D_a \gamma}{533 D_e h_a^2} \tag{7.63}$$

Note that the mainlobe clutter power is inversely proportional to the altitude squared, which results in a less stringent requirement for clutter attenuation in an airborne radar application than to a surface radar. For example, consider an X-band surface-based radar with a transmitter power of 10 kW and a duty factor of 0.1 that uses a circular aperture ($D_a = D_e$) operating at 3 m above a terrain with reflectivity 0.1 [1]. The mainlobe clutter power from (7.63) is equal to $P_{c\,\mathrm{ML}} = 1.9 \times 10^{-5}$ W. If we assume a 1 kHz doppler filter bandwidth with a system noise temperature of $T_s = 3000$ K results in a noise power in the filter equal to 4×10^{-17}W. The mainlobe clutter must then be attenuated by about 120 dB in the target's Doppler filter so that detection is noise-limited.

In the airborne case for a circular beam, the clutter RCS can be approximated as

$$\sigma_c = \frac{\sigma^\circ \pi(\theta R)^2}{4} \tag{7.64}$$

where θ is the two-way power antenna beamwidth ($\theta = \theta_{3\text{dB}}/\sqrt{2}$). Substitution of this result into the radar equation yields

$$P_{c\ \text{ML}} = \frac{P_t d_u G_t^2 \theta_{3\text{dB}}^2 \sigma^\circ \lambda}{(\sqrt{2}\ 16\pi)^2 R^2 L} \tag{7.65}$$

where d_u is the radar duty factor, $\theta_{3\text{dB}}$ is the antenna's one-way power beamwidth, R is the range to the mainlobe clutter patch, and L is a system loss term. Substitution of $G = \pi^2/\theta_{3\text{dB}}^2 = (\pi D/\lambda)^2$, where D is the circular antenna aperture diameter results in

$$P_{c\ \text{ML}} = \frac{P_t d_u \sigma^\circ D^2}{50 R^2 L} \tag{7.66}$$

If we assume the same parameters as for the X-band surface PD radar (i.e., P_t = 10 kW, d_u = 0.1, σ° = 0.1) with D = 0.7 m, L = 1, and R = 10 km, the result is a mainlobe clutter power of $P_{cML} = 10^{-8}$ W. The noise power (T_s = 3000 K) in a 1 kHz doppler filter is 4×10^{-17} W resulting in a C/N = 2.5×10^8, requiring 84 dB clutter attenuation of mainlobe clutter for detection to be noise-limited in the doppler filter.

REFERENCES

1. D.K. Barton, *Modern Radar System Analysis*, Artech House, Norwood, MA, 1988.
2. J. Clarke, "Airborne Early Warning Radar," *Proc. IEEE*, Vol. 73, No. 2, February 1985.
3. J. Clarke and C. Stewart, "Airborne Radar, Part 1: Air-to-Surface," *Microwave Journal*, Vol. 29, No. 1, January 1986.
4. J. Clarke and C. Stewart, "Airborne Radar, Part 2: Air-to-Surface," *Microwave Journal*, Vol. 29, No. 2, February 1986.
5. L. Perkins, H. Smith, and D. Mooney, "The Development of Airborne Pulse Doppler Radar," *IEEE Trans. on AES*, Vol. AES-20, No. 3, May 1984.
6. G. Stimson, *Introduction to Airborne Radar*, Hughes Aircraft Co., El Segundo, CA, 1983.
7. D. Povejsil, R. Raven, and P. Waterman, *Airborne Radar*, Van Nostrand, Princeton, NJ, 1961.

8. W. Long and K. Harriger, "Medium PRF for the AN/APG-66 Radar," *Proc. IEEE*, Vol. 73, No. 2, February 1985.
9. D. Mooney and W. Skillman, "Pulse-Doppler Radar," in M.I. Skolnik, *Radar Handbook*, 1st Ed., McGraw-Hill, New York, 1970.
10. F. Nathanson, *Radar Design Principles*, McGraw-Hill, New York, 1969.
11. D.C. Schleher, *Introduction to Electronic Warfare*, Artech House, Norwood, MA, 1986.
12. E. Aronoff and N. Greenblatt, "Medium PRF Radar Design and Performance," 20th Tri-Service Radar Symp., 1974.
13. F. Williams and M. Radant, "Airborne Radar and the Three PRFs," *Microwave J.*, July 1983.
14. S. Hovanessian, "Medium PRF Performance Analysis," *IEEE Trans.*, Vol. AES-18, No. 3, May 1982.
15. S.I. Hovanessian, *Radar System Design and Analysis*, Artech House, Norwood, MA, 1984.
16. W. Morchin, "Antenna Sidelobe Requirements for the Three PRFs of an AEW Radar," *Microwave Journal*, Vol. 31, No. 9, September 1988.
17. G. Morris, "Antenna Sidelobe Requirements for the Medium PRF Mode of an Airborne Radar," *Microwave Journal*, Vol. 32, No. 9, September 1989.
18. J. Farrell and R. Taylor, "Doppler Radar Clutter," *IEEE Trans.*, Vol. ANE-11, September 1964.
19. A. Friedlander and L. Greenstein, "A Generalized Clutter Computation Procedure for Airborne Pulse Doppler Radars," *IEEE Trans.*, Vol. AES-6, No. 1, January 1970.
20. M. Ringel, "An Advanced Computer Calculation of Ground Clutter in an Airborne Pulse Doppler Radar," *NAECON Rec.*, 1977.
21. J. Jao and W. Goggins, "Efficient, Closed-Form Computation of Airborne Pulse-Doppler Radar Clutter," *IEEE Int. Radar Conf.*, Washington, DC, May 1985.
22. W. Skillman and D. Mooney, "Multiple High-PRF Ranging," *IRE 5th Nat. Conv. Military Electronics*, 1961.
23. G. Hetrich, "Frequency Modulation Techniques as Applied to Pulse Doppler Radar," *IRE Conv. Rec.*, Pt. 5, 1962.
24. D.K. Barton, ed., *CW and Doppler Radar*, *Radars*, Vol. 7, Artech House, Norwood, MA, 1978.

Chapter 8

Special Topics in Doppler Radar Systems

In this chapter, we discuss a number of special topics related to the use of the Doppler effect in radar systems. These topics generally involve the application of special doppler signal processing techniques that provide a unique feature within the radar system not achievable otherwise.

Doppler beam sharpening (DBS) is used in airborne radars to provide the capability for real-time surface mapping [1]. The resolution of the surface map is generally sufficient to allow the detection of small fixed targets (e.g., cars, tanks, boats) in the presence of significant background clutter.

Bistatic and multistatic radar operate with their receivers and transmitter physically separated by an appreciable distance. Bistatic operation provides reduced vulnerability to *electronic countermeasures* (ECM) and *antiradiation missiles* (ARMs), the ability to operate with a high-duty cycle transmitter waveform, and increased detection of targets having a bistatic RCS that exceeds their monostatic RCS [1]. Doppler processing is required in bistatic radars, as it is in monostatic radars, to extract targets immersed in clutter.

An important distinction between bistatic and monostatic radar is the need for system synchronization between the receiver and transmitter. When coherent doppler processing is employed, coherence between the transmitter and receiver must be maintained. This can be accomplished by either a direct communication link between the transmitter and receiver, by phase-locking the receiver to transmitter leakage, or by use of a master clock (e.g., a cesium clock) in both the transmitter and the receiver [2]. Alternatively, the receiver can use the bistatically reflected clutter as a reference in the same manner as used in a noncoherent MTI (see Section 2.3.1). The synchronization is most complex for airborne radar, where both receiver and transmitter are in motion with respect to both targets and the surface.

In the monostatic radar case, the doppler shift is maximum when the target is moving in a radial direction with respect to the radar and zero when moving tangential to the radar. For the bistatic case, the doppler shift is given by $f_d = (\dot{R}_t + \dot{R}_R)/\lambda$ where R_t is the target range to the transmitter and R_R is the range from the target to the receiver. Doppler minima occur when $R_R + R_t$ is a constant, indicating that targets moving in elliptical paths with the transmitter and receiver positions as foci exhibit zero doppler frequency shift. Conversely, targets following hyperbolic trajectories exhibit doppler maxima. By employing clutter tuning (i.e., controlling the relative motion of two bistatic aircraft with respect to the target), we can detect moving targets located perpendicular to the aircraft's velocity vector instead of parallel to the vector, as in monostatic doppler airborne radars [3].

The concept of semiactive missile guidance is depicted in Figure 8.1 [4]. The semiactive missile system generally consists of (1) a doppler radar that both tracks and illuminates the target and (2) a missile seeker that bistatically receives and tracks the doppler shifted target-reflected illumination by comparing it to the directly received illumination (i.e., rear signal) via a data link that provides a coherent reference to the missile. The primary advantage of this missile guidance technique is that it becomes increasingly accurate as the target-to-missile range decreases. This leads to a more effective missile system whose size and warhead can be minimized. The ECM vulnerability of a semiactive missile is low due to the passive nature of the missile seeker and the use of very narrow receiver bandwidths in the doppler tracking circuitry [1].

Inverse synthetic aperture radar (ISAR) is a technique used operationally to image targets with rotational motion, such as ships, aircraft, and space objects [5]. The technique also has application to instrumentation radar for evaluating radar cross section of targets and target models.

The concept of reducing RCS to escape radar detection (i.e., low observable technology) ultimately depends on the radar's ability to detect the target in a terrain or clutter backscatter environment rather than on its free-space detection range capability [1]. The capability of the radar to detect low-flying, low-observable targets thus depends on the radar's subclutter visibility, which, in turn, depends on the stability of the radar's transmitter, receiver, and signal processing elements. This generally stresses the level of sophistication in modern doppler radar system design.

When a low-PRF MTI radar system is used to detect low-flying, low-observable targets, consideration must be given to the aliasing of these targets into the doppler frequency region occupied by birds, insects, and other slow-moving clutter. Because the RCS of this angel clutter is of the same magnitude as the low-observable targets, we cannot discriminate between the two unless the first blind speed of the MTI is sufficiently high to allow adjusting the MTI filter's rejection band to reject the angel clutter (this will be examined later, in Figure 8.43). This tends to favor low-frequency MTI radars (e.g., L-band or UHF) for this type of application.

The performance of *two-dimensional* (2-D) surface-based air surveillance radars is often degraded at short ranges due to the presence of land clutter for which the C/N ratio exceeds the improvement factor of the MTI [6]. This generally occurs because the peak of the radar's antenna beam is directed at the horizon (i.e., at low elevation angles) to detect long-range targets whereas short-range targets are detected

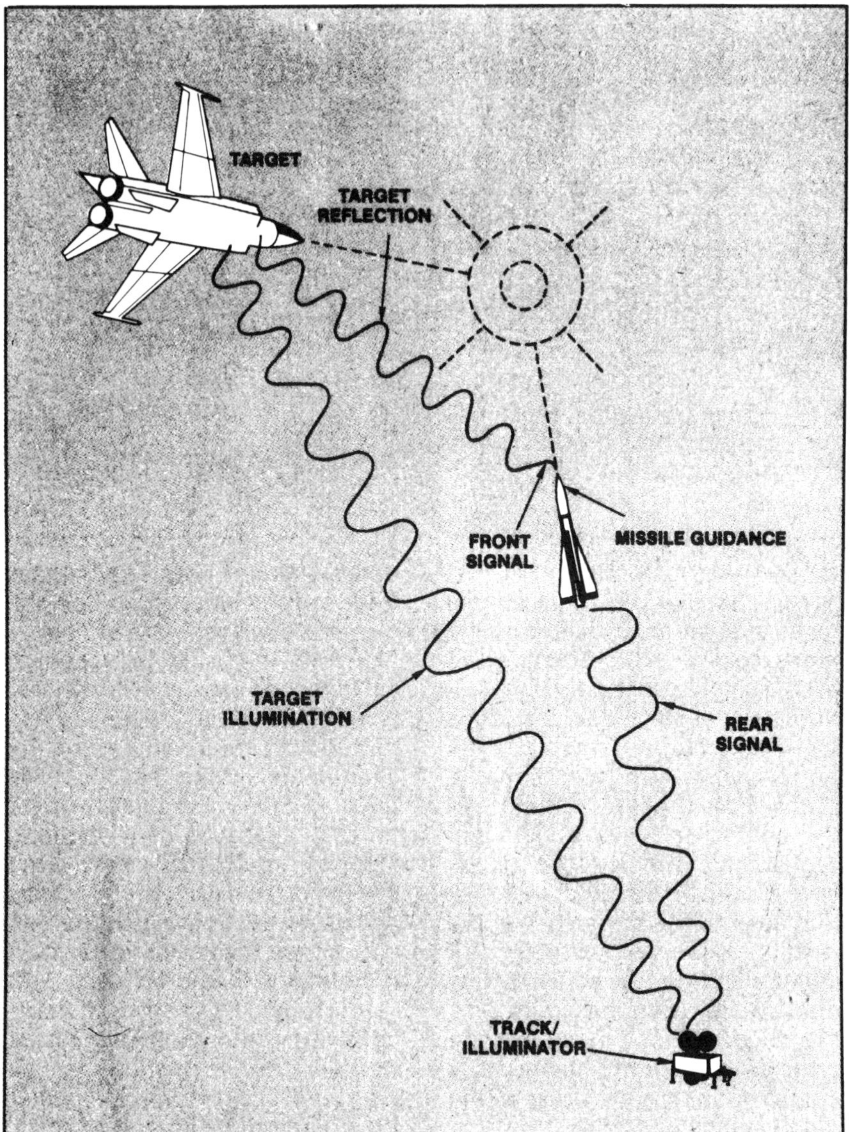

Figure 8.1 Semiactive missile guidance.

at relatively high elevation angles with reduced antenna gain. The net effect is that ground clutter is illuminated with full antenna gain and short-range targets (in the same range cell as ground clutter but at higher elevation angles) are illuminated with reduced antenna gain, thereby diminishing the overall S/C for this type of target.

For this situation, the clutter intake to the 2-D radar can be reduced through the use of a dual-beam antenna [6,7]. The dual-beam antenna provides both low and high elevation beams. The radar transmits on the lower beam, but receives on both the low and high beams. The low beam is activated for long-range targets (i.e., from 50 nmi to a maximum range in a typical ARSR) while the higher beam is activated for short-range targets (i.e., from 0 to 50 nmi). Ground, sea, and angel clutter (e.g., cars, birds, insects, *et cetera*) are decoupled in the high-beam receiving antenna, thereby reducing the improvement factor requirement of the MTI for short-range targets.

The overall detection performance of an MTI system depends both on the clutter intake and the improvement factor available from the MTI processor. Hence, a natural combination is an MTI system that uses a pulse compression processor to reduce the size of the clutter patch (see Section 1.3.2), thereby reducing the overall clutter intake. From a theoretical viewpoint, both MTI and pulse compression processors are linear systems and should be fully compatible. However, pulse compression networks produce range sidelobes that, due to system instabilities, may not be completely cancelled in the MTI filter [8]. This situation is exacerbated by the sensitivity of some pulse compression waveforms to doppler shifted signals (the pulse compression network is generally matched to the radar's carrier frequency), which cause the sidelobe levels to increase from their nominal value (see Section 7.2.1).

The PD weather radar is the one sensor capable of probing inside a storm's shield of clouds to reveal the storm's internal structure [9]. This capability provides the capability for detecting weather phenomena such as severe thunderstorms, tornadoes, wind shear conditions and other meteorological conditions.

The PD weather radar functions by measuring the first three moments ($m_n = \int_v v^n S(v)\, dv$) of the velocity spectrum associated with reflections backscattered from hydrometeor particles contained within each volumetric resolution cell examined by the radar. The zeroeth moment ($n = 0$) of the doppler spectrum is proportional to the storm's intensity within each cell and derived by finding the area beneath the power spectrum. The first moment ($n = 1$) represents the mean velocity of the particles and hence provides information of the storm's movement in a radial direction with respect to the radar. The second moment ($n = 2$) measures the variance of the velocity spectrum for the particles contained within the radar's volumetric resolution cell, which is proportional to the relative velocity between particles within each cell, and provides an indication of turbulence and wind shear effects.

8.1 DOPPLER BEAM SHARPENING

Doppler beam sharpening is a method for reducing surface clutter returns in an airborne radar by using doppler filtering to narrow the antenna's azimuth beamwidth synthetically. Aircraft motion causes a unique doppler frequency to be associated with the surface clutter returned from each angular direction within the antenna's mainbeam pattern (see Section 3.2.3). This causes a spectral spreading of the doppler clutter return that is a function of the aircraft's velocity, azimuth antenna beamwidth, and the antenna's pointing angle relative to the aircraft's velocity vector. Conversely, the return from a target has a unique doppler frequency associated with its angular direction. If a narrowband doppler filter (bandwidth inversely proportional to dwell time) can be placed over the target, only the clutter returned from the angular directions associated with the filter's bandwidth compete with the target. Under favorable conditions, a significant increase in S/C ratio can be achieved for either stationary or slowly moving targets. The effect is equivalent to reducing the antenna's beamwidth; therefore, the descriptive term, *doppler beam sharpening* [1,7].

In a forward-looking airborne radar that illuminates the earth's surface, all returns from scatterers within a basic resolution cell (see figure 3.16 and Section 3.2.3) combine to form the radar return. The cell dimensions are proportional to the effective transmitter pulsewidth and the antenna azimuth beamwidth, which is relatively large in airborne applications due to the limited space available for antenna aperture.

The clutter RCS backscattered from the surface for a small beamwidth can be expressed as

$$\sigma_c = \sigma^\circ R_c \theta_{az} (c\tau_e/2) \sec\psi \tag{8.1}$$

where σ° is the backscatter coefficient, R_c is the range to the clutter patch, θ_{az} is the azimuth beamwidth, τ_e is the radar's effective pulsewidth, and ψ is the grazing angle from the radar to the clutter cell. The single-pulse S/C ratio for a surface target located in the clutter cell is then given by

$$S/C = \frac{\sigma_t}{\sigma^\circ R_c \theta_{az} (c\tau_e/2) \sec\psi \, L_c} \tag{8.2}$$

where σ_t is the target's RCS and L_c is a loss term primarily due to beam shape and range-straddling effects.

For example, assume that a 5 m^2 fixed target (i.e., small car) is to be detected at 10 nmi from a K_u-band airborne radar (λ = 1.85 cm) with a 3 ft horizontal aperture (i.e., θ_{az} = 1.5°) and a 0.2 μs effective pulsewidth. The clutter area for this situation is $A_c = 1.46 \times 10^4$ m^2. For low grazing angles, (i.e., ψ = 2°) over farmland terrain, the backscattering coefficient is $\sigma^\circ = -30$ dB, resulting in $S/C = -4.6$ dB. MTI will not be effective in this situation as there is no relative motion between the fixed target and the ground clutter. Hence, the target cannot be generally detected with this type of system without the addition of a doppler beam-sharpening mode.

The general geometry for doppler beam sharpening is depicted in Figure 8.2. The doppler shift to the *n*th point within the range slice is given by

$$f_{dn} = \frac{2v_{ac}}{\lambda} \cos\alpha_n \tag{8.3}$$

where v_{ac} is the aircraft velocity and α_n is the angle between the line from the *n*th point to the radar and the aircraft's velocity vector. As is apparent from (8.3), except for isodops ($\alpha_n = k$), each point on the surface produces a unique doppler shift. Now assume that the azimuth dimension of the clutter cell is subdivided into a number of smaller cells, each with equally sized angles so that the frequency difference within each cell is given by

$$\Delta f_n = \frac{2v_{ac}}{\lambda} (\cos\alpha_{n1} - \cos\alpha_{n2}) \tag{8.4}$$

Letting $\Delta\theta_n$ be the differential angle represented by each subcell such that $\alpha_{n1} = \alpha_m - \Delta\theta/2$ and $\alpha_{n2} = \alpha_m + \Delta\theta/2$, where α_m is the median angle to the center of the clutter cells results in $\cos\alpha_{n1} - \cos\alpha_{n2} = \Delta\theta \sin\alpha_m$. Setting the doppler filter bandwidth equal to $B_D = \Delta f_n$ and solving for the doppler beam-sharpened azimuth angle results in

$$\Delta\theta = \frac{\lambda B_D}{2v_{ac} \sin\alpha_m} \tag{8.5}$$

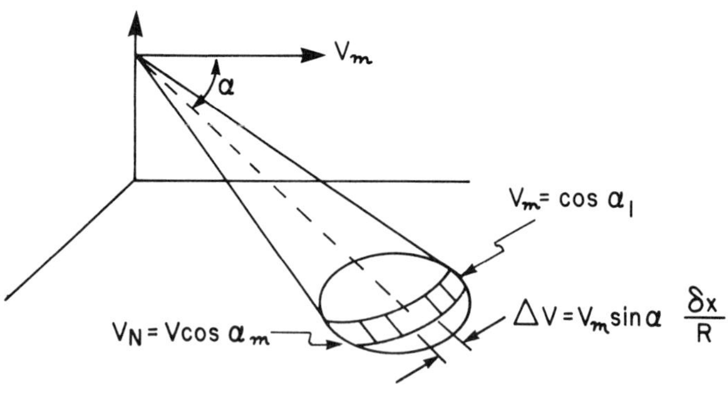

Figure 8.2 Doppler beam-sharpening geometry.

which can also be expressed in terms of cross-track range resolution as

$$\delta_{az} = \frac{R_c \lambda B_D}{2v_{ac} \sin\alpha_m} \tag{8.6}$$

As is apparent from (8.5) and (8.6), the highest resolutions are achieved with the largest aircraft velocities and when the beam is positioned perpendicular to the aircraft's ground track (i.e., $\alpha_m = 90°$). For the beam positioned along the ground track (i.e., $\alpha_m = 0°$), there is no improvement in resolution due to doppler beam sharpening.

For a scanning surveillance radar, the minimum doppler resolution is a function of the time that the radar beam dwells on the target ($B_D = 1/T_d$). Hence, the angular resolution (rad) can be expressed as

$$\Delta\theta = \frac{\lambda}{2v_{ac}T_d \sin\alpha_m} \tag{8.7}$$

where T_d is the target dwell time. The denominator of (8.7) is equivalent to an effective synthetic aperture equal to

$$D_e = 2v_{ac}T_d \sin\alpha_m \tag{8.8}$$

so that the improvement in azimuth resolution and equivalent S/C ratio is given by

$$I = \frac{D_e}{D_{az}} = \frac{2v_{ac}T_d \sin\alpha_m}{D_{az}} \tag{8.9}$$

where D_{az} is the horizontal width of the antenna.

Returning to the previous example, assume that $v_{ac} = 400$ knots and the radar scans at 20°/s [10]. Then, the effective aperture $D_e = 72$ ft at a scan angle of 45° results in an improvement factor of $I = 14.6$ dB and a $S/C = +10$ dB. Alternatively, (8.2) can be modified by using the synthetic aperture beamwidth, resulting in

$$S/C = \frac{2\sigma_t v_{ac}T_d \sin\alpha_m}{\sigma^\circ R_c \lambda (c\tau_e/2) \sec\psi \, L_c} \tag{8.10}$$

The cross-track resolution cell width, from (8.6), is equal to $\delta_{az} = 53$ ft and, hence, additional beam sharpening and higher range resolution can improve the S/C on the fixed target. For example, if the pulsewidth is reduced to 0.1 μs (50 ft), the S/C would be increased to 13 dB.

The spectral spreading of the ground clutter due to platform motion for the situation depicted in Figure 8.2 is given by

$$\Delta f_{3\text{dB}} = \frac{2v_{ac}\Delta\theta_{3\text{dB}}\sin\alpha_m}{\lambda} \tag{8.11}$$

For the example given earlier (v_{ac} = 400 knots, α = 45°, D_{az} = 3 ft), the spectral spreading can be evaluated as $\Delta f_{3\text{dB}}$ = 400 Hz and the doppler filter bandwidth is $B_D = 1/T_d$ = 13.33 Hz. To achieve doppler beam sharpening, the range of possible doppler returns from the surface clutter must be filled by a bank of contiguous doppler filters. This filter bank is depicted in Figure 8.3 for the example problem and consists of 30 doppler filters uniformly distributed ±200 Hz about a center doppler frequency of 15 kHz. The center doppler frequency is given by

$$f_{dm} = \frac{2v_{ac}\sin\alpha_m}{\lambda} \tag{8.12}$$

Note that the bandwidths of the doppler filters depend on the target dwell time and hence are constant for a given antenna azimuth scan rate. However, the filters must cover the total spectral spread given by (8.11), and hence a sufficient number of doppler filters are needed to cover the spectral spread due to the maximum expected aircraft velocity and scan angle.

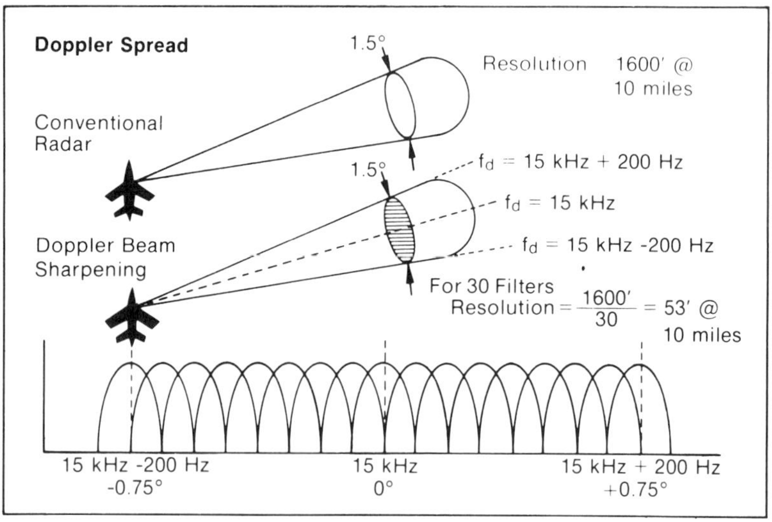

Figure 8.3 Doppler beam-sharpening filter bank.

Note that the analysis generally should include other clutter spectral spreading effects such as antenna scanning modulation, internal clutter motion, and transmitter frequency instability (see Section 3.2). These effects tend to widen further the clutter spectrum over that caused by platform motion, but they typically are much smaller than platform motion spectral spreading and thus can be neglected. If the target has motion, it will generally translate its doppler frequency and may cause the target doppler to fall in a different doppler filter than its surrounding clutter return.

The doppler beam sharpening resolution provided for the example problem as a function of range and various scan angles is depicted in Figure 8.4. Note that the resolution is better for shorter ranges and larger scan angles. Insight into the doppler filtering requirements for doppler beam sharpening can be deduced from these curves. For example, at 10 nmi and a scan angle of 45°, 30 doppler filters (B_D = 13.33 Hz) are required to span the clutter spectrum induced by platform motion. Each filter provides an equivalent azimuth cross-track resolution of 1600/30 = 53 ft. At 10 nmi and a scan angle of 20.7°, the spectral width is halved, requiring only 15 doppler filters to cover the surface clutter spectrum. The cross-track azimuth resolution is 1600/15 = 106 ft for this situation. Therefore, a changing number of doppler filters must be produced for a constant data time window (T_d), where the number of filters is proportional to the aircraft velocity and the sine of the antenna's azimuth scan angle.

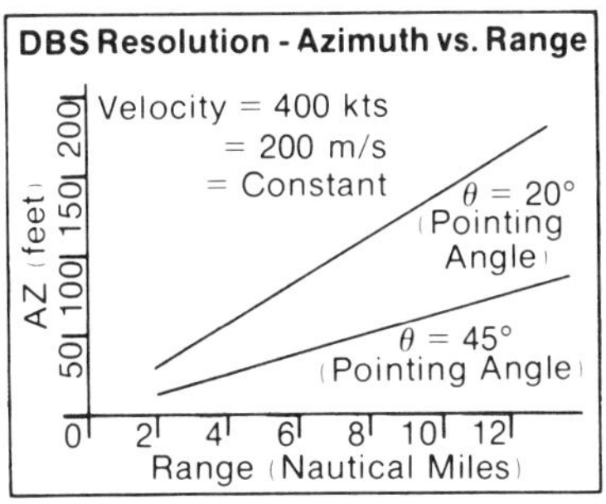

Figure 8.4 Doppler beam-sharpening cross-track resolution.

8.2 MTI FOR BISTATIC RADAR

The basic elements of both monostatic and bistatic radars are depicted in Figure 8.5. In the bistatic configuration, the receiver is located away from the transmitter, and detection is based on target reflections received at the receiver site.

Consider first the measurement of target range. In a monostatic radar, the measurement of time delay (ΔT) between the transmitted and received pulses is directly converted into target range (i.e., $R = c\Delta T/2$). The situation is somewhat different

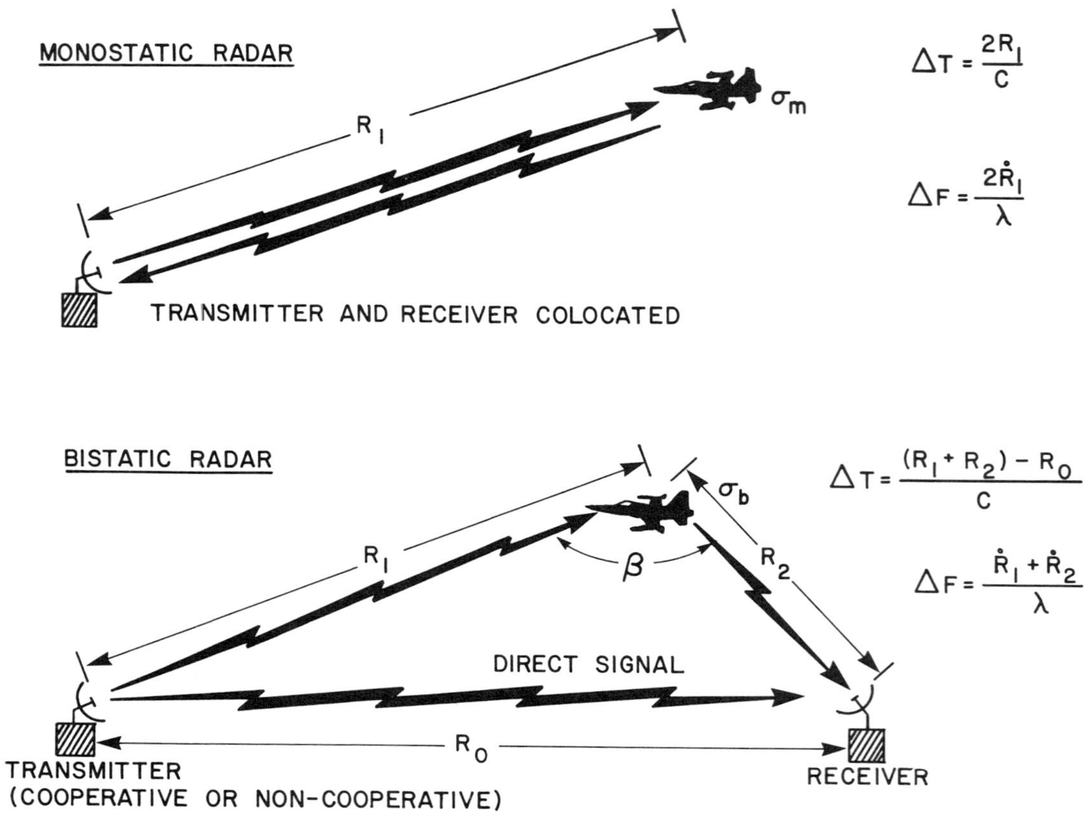

Figure 8.5 Basic elements of monostatic and bistatic radar systems.

for a bistatic radar, where the measurement of time delay initially entails the reception of a synchronization signal from the transmitter and then knowledge of any of the three quantities within the bistatic triangle depicted in Figure 8.6 to determine target range. For example, we measure (1) the differential time delay (ΔT) between the transmitted and received signals, given by $\Delta R = c\Delta T = (R_1 + R_2) - R_0$; (2) the distance ($R_0$) between the transmitter and receiver sites; and (3) the angle (θ) of the intercepted target reflection at the receiver, which then allows the target range (R_2) to be calculated from the law of cosines as [11]

$$R_2 = \frac{\Delta R(\Delta R + 2R_0)}{2[\Delta R + R_0(1 - \cos\theta)]} \tag{8.13}$$

If the foci of an ellipse are located at the transmitter and receiver sites (see Figure 8.6), the loci of all target points for which $R_1 + R_2$ is a constant fall on the elliptical contour. Hence, the location of the bistatic range from the receiver site to the target can be interpreted as the intersection of a line from the receiver in the direction of the target with the ellipse defined by ΔR and R_0 as depicted in Figure 8.7. From the definition of an ellipse, the contours of constant bistatic range ($R_1 + R_2 - R_0 = \text{constant}$) follow the equation:

$$\frac{4x^2}{(R_1 + R_2)^2} + \frac{4y^2}{(R_1 + R_2)^2 - R_0^2} = 1 \tag{8.14}$$

where R_1 is the range from the transmitter to the target and R_2 is the range from the receiver to the target (see Figure 8.6) [12]. Note that, as the target moves on the elliptical contour defined by (8.14), the bistatic range is a constant and hence the doppler frequency shift measured at the receiver is zero. The doppler shift as the target moves along the baseline between the transmitter and receiver is also zero. This results because the magnitude of the doppler shifts with respect to the transmitter and the receiver are equal along the baseline, but opposite in sign, thereby counterbalancing each other.

The bistatic radar equation is given by [11]

$$P_r = \frac{P_t G_t G_r \lambda^2 \sigma_b}{(4\pi)^3 R_1^2 R_2^2 L_s} \tag{8.15}$$

where σ_b is the bistatic RCS, G_t and G_r are the transmitter and receiver antenna gain, R_1 and R_2 are the transmitter and receiver range, and L_s is the system loss that includes both transmitter and receiver propagation losses. The equation is identical to the standard monostatic radar equation if $G^2 = G_t G_r$, $\sigma_m = \sigma_b$, and $R^4 = R_1^2 R_2^2$. The range

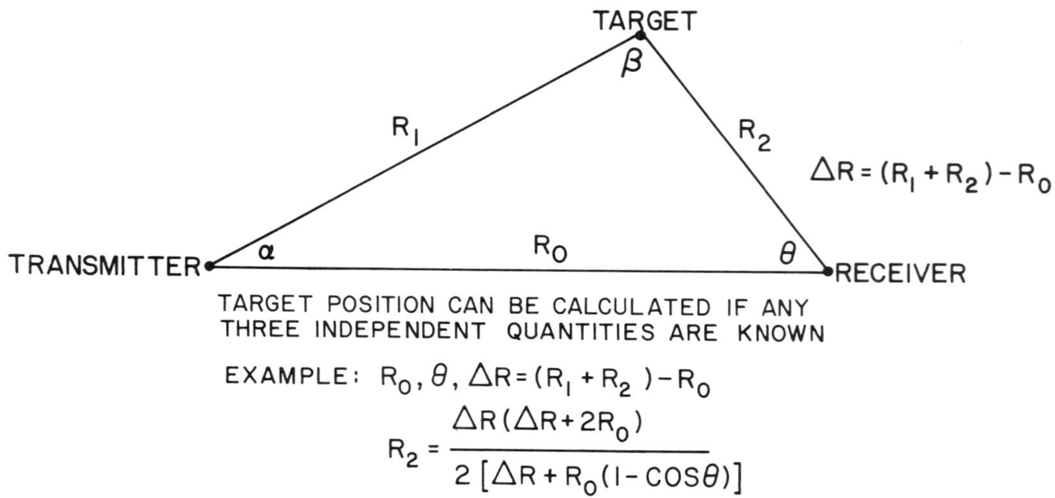

Figure 8.6 The bistatic range measurement triangle.

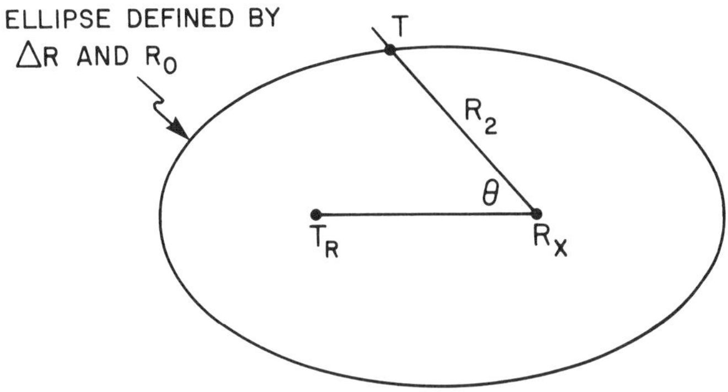

Figure 8.7 Physical interpretation of the bistatic range as intersection of an ellipse and a straight line.

dependence of the bistatic radar can be expressed as

$$P_R = \frac{k}{R_1^2 R_2^2} \tag{8.16}$$

which can be solved for $R_1 R_2 = \sqrt{k/P_r} = c_0^2$. From analytic geometery, we identify the locus of the vertex of the bistatic triangle (see Figure 8.6), having a constant base and the product of its other two sides equal to a given constant, called the *ovals of Cassini*. This curve, depicted in Figure 8.8, provides contours of equal power returned to the bistatic receiver. Note that the region between the transmitter and receiver is one of high sensitivity.

The bistatic doppler shift is given by

$$f_d = \frac{\dot{R}_1 + \dot{R}_2}{\lambda} \tag{8.17}$$

where $\dot{R}_1$ is the range rate in the direction from the transmitter to the target, and $\dot{R}_2$ is the range rate in the direction from the receiver to the target (see Figure 8.5). The doppler shift resulting from an arbitrary velocity target is given by

$$f_d = \frac{v}{\lambda}\,[\cos(\psi - \phi_R) + \cos(\psi - \phi_T)] \tag{8.18}$$

where v is the target's velocity, ψ is the angle of the target's velocity vector with respect to the baseline connecting the receiver and transmitter, ϕ_T is the transmitter angle, and ϕ_R is the receiver angle of the bistatic triangle shown in Figure 8.9. If

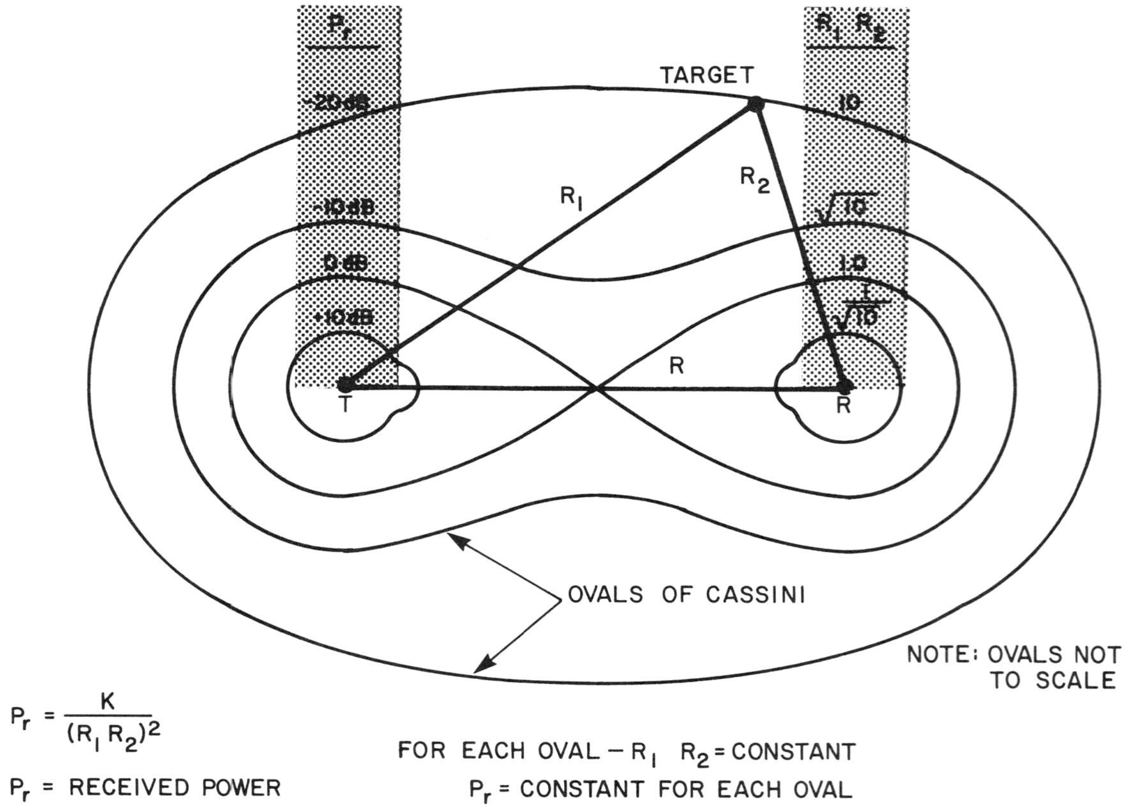

Figure 8.8 Constant received power contour in a bistatic radar.

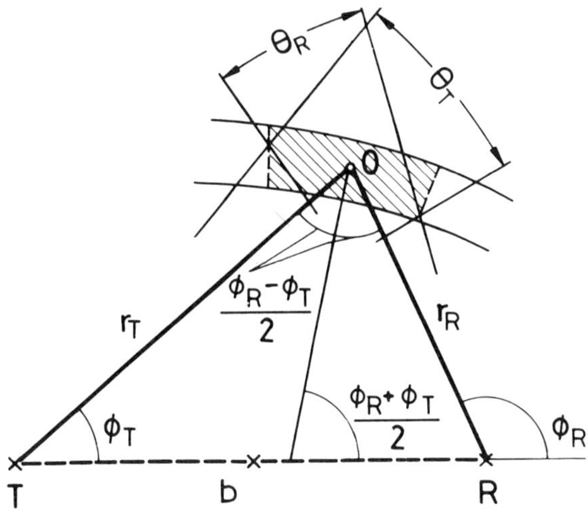

Figure 8.9 Bistatic triangle geometry.

we differentiate this relationship to determine the angle for the maximum doppler frequency, the result is the equation $\sin(\psi - \phi_R) + \sin(\psi - \phi_T) = 0$, which can be solved to determine

$$\psi_{\max} = \frac{\phi_R + \phi_T}{2} \tag{8.19}$$

The line formed by this angle is shown in Figure 8.9 and bisects the bistatic angle $\beta = \phi_R - \phi_T$. From analytic geometry, we know that this line which bisects the bistatic angle is the normal to the constant range ellipse that contains the target point. Therefore, the maximum doppler frequency occurs when the target velocity vector is normal to the constant range ellipse. The analogy to the monostatic radar is evident if its circular geometry (maximum doppler on normal radial) is replaced with elliptical geometry (maximum doppler on normal to ellipse).

A useful approach is to decompose (8.18) for the bistaic doppler response into three terms [13]:

$$f_d = \left(\frac{2v}{\lambda}\right) \cdot \cos\left(\frac{\phi_R - \phi_T}{2}\right) \cdot \cos\left(\psi - \frac{\phi_R + \phi_T}{2}\right) \tag{8.20}$$

The first term $2v/\lambda$ corresponds to the maximum doppler frequency associated with a target moving at velocity v observed by a radar with wavelength λ. The second term $(\cos(\phi_R - \phi_T)/2)$ is a doppler reduction factor related to the bistatic radar triangulation effect that causes a position-sensitive reduction in doppler frequency shift, as depicted in Figure 8.10. The third term ($\cos\delta$; $\delta = \psi - (\phi_R + \phi_T)/2$) relates to the angle the target velocity vector makes with respect to the normal of the constant range ellipse (i.e., contour of zero doppler frequency). The overall interpretation of (8.21) is that maximum doppler frequency is achieved when the target moves along a trajectory that is normal to the constant range ellipses (i.e., a hyperbolic trajectory), but that the overall doppler frequency varies from a maximum corresponding to its monostatic radar value at long range to zero when it crosses the bistatic radar baseline (see Figure 8.10).

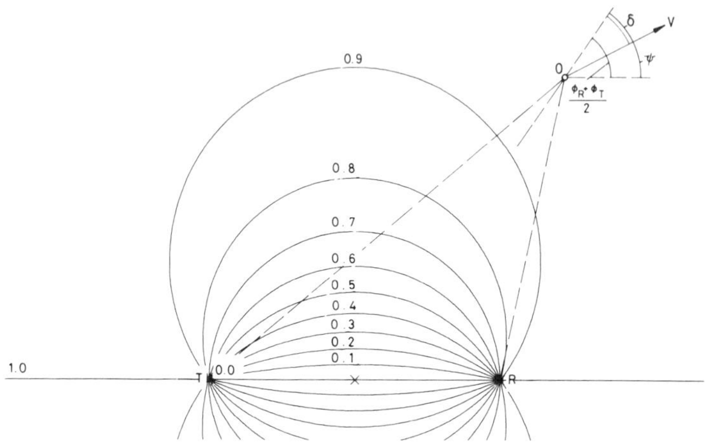

Figure 8.10 Contours of constant doppler frequency reduction in a bistatic radar.

The doppler sensitivity factor affects the design of MTI filters in a bistatic radar. The unambiguous-target doppler area and resolution is generally fixed because the pulse repetition frequency and the number of pulses per beamwidth cannot be changed for different range elements in the same transmitting direction. This requires position-dependent doppler filter characteristics to cope with the variable-target doppler and clutter situation, as depicted in Figure 8.11. The ability to adjust the bistatic radar baseline (e.g., in an airborne situation) with respect to the target's position to minimize the clutter doppler spread is referred to as *clutter tuning* [3,14]. This effect can be exploited to detect moving targets (with radial velocity that is less than the spread of the clutter across a monostatic radar beam) located perpendicular to the aircraft velocity vector instead of parallel to the vector, as is done with monostatic airborne radars. In this case, the trajectories of the transmitter and receiver platforms are coordinated so that the doppler frequency spread of ground clutter located perpendicular to the velocity vector of the aircraft (rather than along the velocity vector) is zero.

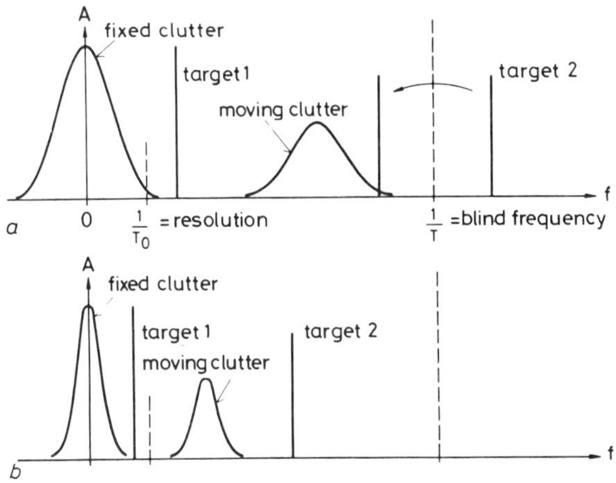

Figure 8.11 Doppler spectrum for target and clutter in a bistatic radar.

Bistatic doppler shift results from the target velocity component normal to the constant range ellipsoid containing the target point. From analytic geometry, we know that the normal to an ellipse is equal to the tangent of a hyperbola with the same foci as the ellipse. Hence, the bistatic doppler shift arises from hyperbolic contours normal to the constant range elliptical contours. The hyperbolic surfaces are related to the bistatic triangle as the locus of target points that have constant range differences $R_1 - R_2$. The equation for the hyperbolic constant doppler frequency contours is given by

$$\frac{4x^2}{(R_1 - R_2)^2} - \frac{4y^2}{R_0^2 - (R_1 - R_2)^2} = 1 \tag{8.21}$$

which are plotted parametrically in Figure 8.12 along with the elliptical constant bistatic range contours. The parametric bistatic radar constant range and doppler curves depicted in Figure 8.12 form a range-doppler bistatic radar-coordinate system.

Another factor of interest in bistatic doppler radars is the size of the clutter patch, which, when multiplied by the bistatic reflection coefficient, determines the clutter power returned to the radar. The clutter patch is a function of both the transmitter and receiver beamwidths and the bistatic angle. The area of the clutter patch is given by [15]

$$A_{CB} = \frac{c\tau(R_1\theta_T)(R_2\theta_R)\sec^2(\beta/2)}{\sqrt{2[(R_1\theta_T)^2 + (R_2\theta_R)^2]}} \tag{8.22}$$

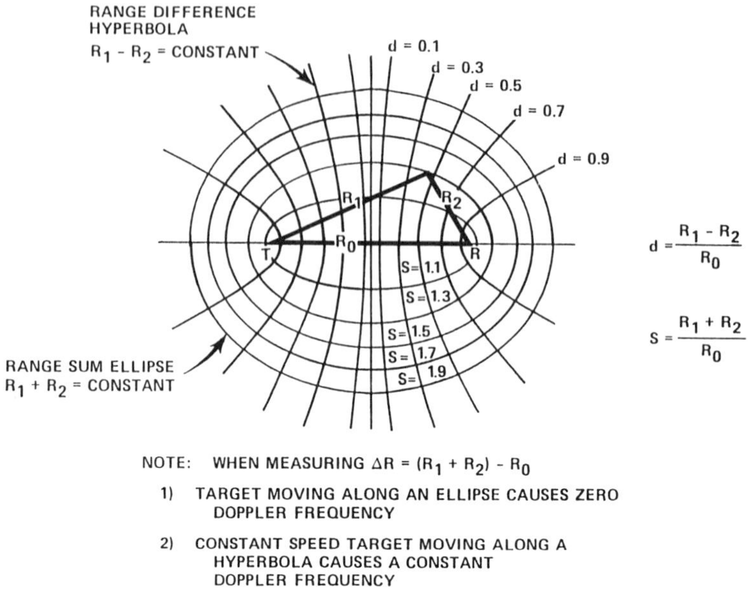

Figure 8.12 Bistatic radar constant range and doppler curves.

where τ is the radar pulsewidth, R_1 and R_2 are the transmitter and receiver target range, θ_T and θ_R are the transmitter and receiver beamwidths, and β is the bistatic angle (see Figure 8.9). Note that for the monostatic case (i.e., $R_1 = R_2$, $\theta_T = \theta_R$, $\beta = 0$), the clutter patch area reduces to $A_c = (c\tau/2)\theta_T R$ as expected. At long ranges, there is little difference between the bistatic and monostatic radar patches. The bistatic area is smaller only when at least one of the bistatic ranges is significantly less than the monostatic range.

A fundamental consideration with bistatic radars is the method employed to ensure that the receiver is synchronized to look at the same instantaneous volume of space as illuminated by the transmitter. One solution, employed in semiactive missile guidance, is to have both the transmitter and receiver site track the target. However, this is not applicable to surveillance radar, where the target location is uncertain.

Basically, four different bistatic surveillance radar configurations can be employed. The first two configurations involve floodlight antennas for one element of the system, which constantly look at the field of interest. The other element of this type of system employs a narrow scanning-beam antenna. The narrow beam provides the bistatic radar with its angular resolution capability necessary to separate multiple targets within the field of view. In addition, the angle measurement in a bistatic radar of this type is an essential parameter used in determining target range.

The preferred type of floodlight system employs a wide, fixed transmitting beam antenna and a narrow scanning-beam receiver. This configuration provides the greatest advantage against ECM because of its narrow receiving antenna beam. However, the use of a wide floodlight beam (possibly omnidirectional) for one element of this system generally compromises its sensitivity compared with an equivalent monostatic radar. The reduction of sensitivity is proportional to the ratio of the antenna gain of the floodlight beam to that of the antenna gain of the monostatic radar (see (8.15)). The sensitivity reduction factor can be significant for wide-area coverage systems. In addition, bistatic radars tend to be less sensitive than their monostatic counterparts due to the geometry, which tends to extend the round-trip path from the transmitter to target to receiver. This is particularly true when the bistatic transmitter is in a sanctuary location to reduce vulnerability to an ARM attack.

The next, more complicated configuration employs a multiple-beam receiver and a narrow scanning-beam transmitting antenna. The multiple-beam antenna simultaneously looks at all possible ranges within the narrow scanning beam's field of illumination that might provide a return. The disadvantage of this configuration is that a separate receiver must be provided for each multiple antenna beam (e.g., possibly hundreds of receivers).

The alternate to the multiple-beam receiver uses a rapid electronically scanning antenna beam that follows the pulse as it propagates within a particular narrow transmitter-antenna beam position. The advantage of this configuration is that only a single receiver is required, but the antenna is quite complex, as is the nonlinear scanning pattern necessary to implement this system. This type of bistatic system, depicted in Figure 8.13, is called a *pulse chasing* system. Pulse chasing requires selection of the receive angle where possible target echoes can arrive at the receive antenna, depending on the delay time (see Figure 8.13).

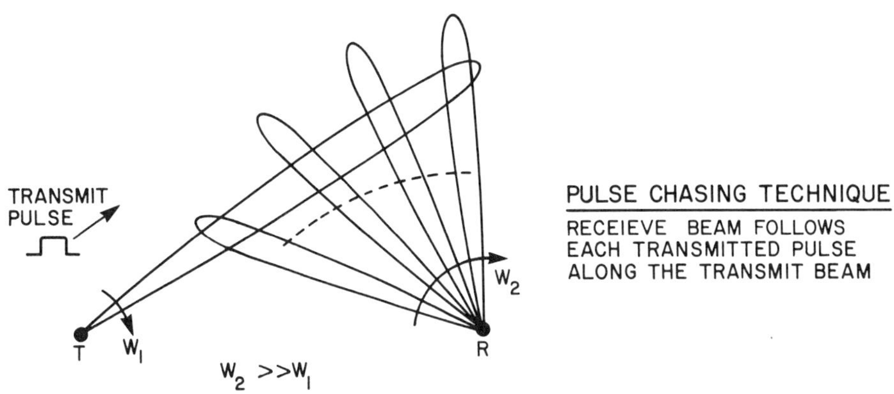

Figure 8.13 Geometry of pulse-chasing bistatic radar.

Pulse chasing requires a nonlinear scanning of the receiver beam to instantaneously point in the direction where possible target echos can arrive at the receive antenna. The receiving beam position depends on the overall time delay from the transmitter to the scatterer and then to the receiver. The motion of the center of the area that the receiving antenna beam must point to receive the reflected return is given by

$$-\frac{d\phi_R}{dt} = -\frac{\partial\phi_R}{\partial R_1}\frac{\partial R_1}{\partial t} \tag{8.23}$$

where ϕ_R is the complementary receiving beam angle, ϕ_T is the transmitting beam angle, and R_1 is the range from the transmitter to the scatterer (see Figure 8.13). From the law of sines and the bistatic triangle (see Figure 8.9):

$$R_1 = \frac{R_0 \sin\phi_R}{\sin(\phi_R - \phi_T)} \tag{8.24}$$

Differentiating (8.24) and recognizing that $\partial R_1/\mathrm{d}t = c$ because the wave propagates at the speed of light results in

$$-\frac{\partial\phi_R}{\partial t} = \frac{\sin^2(\phi_R - \phi_T)}{\sin\phi_T}\frac{c}{R_0} \tag{8.25}$$

which defines the receiving antenna scanning pattern as a function of time. For example, if $\phi_T = 30°$ with a baseline of 100 km, the scanning rate would be 86°/ms at $\phi_R = 180°$ ($\theta = 0°$) whereas, at $\phi_R = 120°$ ($\theta = 60°$), the scanning rate would be a maximum of 344°/ms. This example points out the dependence of the scanning rate on the parameters of the bistatic radar and its nonlinear nature. Note that the receiving antenna beam has to look in the direction of possible targets, but also has to remain in that position for the echo duration determined by the transmitting pulse length.

Rapid scanning by the receiving beam would initially appear to cause large beam-scanning losses, producing both detection losses and clutter spectral spreading. However, the receiving antenna beam acts primarily as a sampling function that collects the target and clutter energy reflected through illumination by the transmitter beam. Hence, beam-scanning losses and clutter spectral spreading will be determined primarily by the transmitting beam scanning characteristics.

Doppler processing techniques are applicable to bistatic radar, where the principles of operation are basically the same as for monostatic radar with the provision that the observed motion is orthogonal to the range sum contours rather than radial in the transmitter-to-target direction. The doppler shift is useful for discrimination of clutter and resolution of targets on a velocity basis. Passive systems can also provide doppler discrimination. When the complete bistatic geometry is known and the target's location is determined by bistatic techniques, the projection of the target velocity vector on the plane formed by the two sites and the target can be determined. Use of two-dimensional modeling allows us to estimate the horizontal velocity vector. If only the doppler difference between the two received signals is available, then the motion of the targets may be detected, but the velocity cannot be determined.

The areas of doppler sensitivity are different in bistatic and monostatic radars. The bistatic system has an area of low doppler sensitivity surrounding the baseline, whereas the monostatic system has no such area anywhere in its coverage. Both systems have, in addition, low doppler sensitivity for targets, anywhere in their coverage areas, moving along contours of constant range difference (which are circles for monostatic and ellipses for bistatic radar). There is a distinction between having an area of the coverage where targets cannot be detected regardless of direction and the general situation, where the direction of the target velocity vector is the determining factor.

The different doppler sensitivities between bistatic and monostatic radars open the possibility of complementary radar operation. For example, a bistatic doppler receiver could be used with a monostatic radar to provide MTI coverage for tangential targets. Also, clutter tuning can be used in airborne bistatic radars to maximize sensitivity in selected target areas of high interest.

The primary problem with bistatic radar doppler processing concerns the method for achieving coherence between the transmitted and received signals. Where a coherent data link is available between the transmitter and receiver, or when the transmitter and receiver are cohered to a precision master clock, the situation is similar to that in a monostatic radar. However, when the transmitter and receivers are independent, we may have great difficulty obtaining a coherent reference suitable for MTI operation.

Two situations occur with respect to coherent references for an independent bistatic radar MTI. In the most favorable one, the noncooperative transmitter uses a coherent transmitter. The problem for the independent bistatic MTI radar receiver is then to lock an oscillator to the coherent transmitter with sufficient stability to allow MTI operation. For example, with the Marconi S264 surveillance radar, an estimated oscillator stability of one part in 10^9 is required at 600 MHz to provide acceptable MTI operation (see Section 3.3) [2].

Measurement of a radar signal to the required frequency accuracy requires an integrated S/N approximately equal $1/(\tau \cdot \Delta f)^2$, where τ is the radar's pulsewidth (4 μs). and Δf is the rms frequency error (0.6 Hz). The required integrated S/N is on the order of 110 dB, which must be achieved by receiving the transmitter pulse either directly or through the radar's sidelobe radiations. To achieve the required S/N by interception of the mainbeam radar signal transmissions appears feasible. The problem is then to maintain the required frequency stability between mainbeam interceptions that occur for most long-range surveillance radars on the order of every 5–10 s. This order of frequency stability appears feasible using a rubidium oscillator as a timing source (one part in 10^{10} error) [16]. The storage could be achieved digitally in a system similar to that of a digital RF memory. However, any drift in the radar's STALO during the 5–10 s interscan period would result in an error in the measured doppler shift of the bistatic radar, which, in turn, could result in the inability to cancel fixed targets and ground clutter.

If the radar uses a power oscillator transmitter, then the independent bistatic MTI must adjust its reference oscillator on a pulse-to-pulse basis. This requires using the direct RF transmissions of the radar to synchronize the bistatic MTI's COHO to the radar's transmissions. From the previous analysis, we see that the sidelobe leakage from the radar does not provide sufficient S/N at the bistatic receiver to allow MTI operation. An alternate scheme is to use reflections from close-in clutter scatterers as the reference signal that locks the bistatic MTI's COHO [2]. A block diagram of such a bistatic MTI system is depicted in Figure 8.14. A clutter attenuation on the order of 20 dB was achieved with a practical system of this type [2]. Performance was limited mainly by pulse-to-pulse fluctuations in the phase of the clutter signals used to prime the MTI COHO.

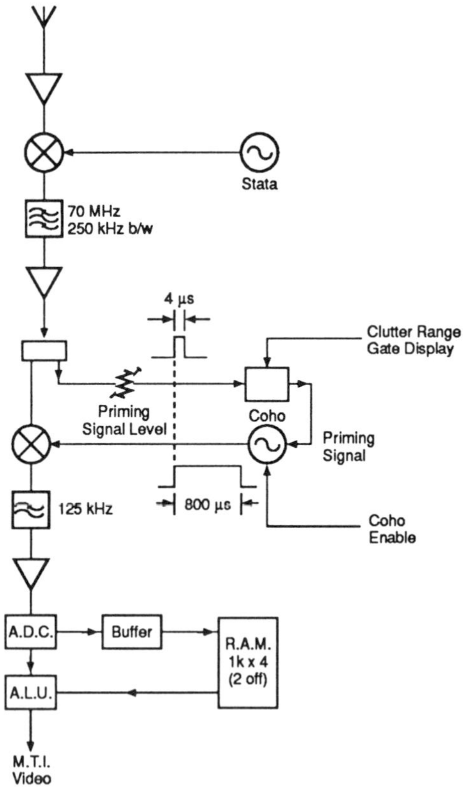

Figure 8.14 Bistatic radar MTI system.

Another option for MTI processing in an independent bistatic radar is the use of a noncoherent MTI, covered in Section 2.3.1. The principles of operation for a bistatic noncoherent MTI are identical to those for a monostatic MTI. Although in a monostatic radar, the performance expected of a noncoherent MTI is less than that available from a coherent MTI, it may be the best available technique for achieving MTI operation in a bistatic radar, considering the many problems with implementing a coherent independent bistatic MTI radar.

The radar cross section (σ_b) of a target illuminated by a bistatic radar is a measure of the energy scattered in the direction of the receiver. In accordance with the monostatic-bistatic hypothesis [6,11,12] for targets that are sufficiently smooth, the bistatic and monostatic radar cross sections will be similar when averaged over all target aspects (except for forward scattering at a bistatic angle of 180°). Because in general all aspects will not be equally likely, the cross sections presented to bistatic and monostatic radars are not the same but can be expected to range over similar values. This hypothesis does not cover the case where the bistatic angle (β) equals 180° (forward scatter). The forward-scatter cross section can be many times the monostatic cross section.

The forward-scatter bistatic radar cross-section enhancement is due to an interference phenomenon that creates a shadow or hole in the wavefront, thereby creating a beam pattern which is the same as the radiation pattern of a uniformly illuminated aperture antenna with a shape the same as that of the shadow cast by the object [3].

The forward-scatter RCS of a target lying directly on a line between the transmitter and the receiver ($\beta = 180°$) is given by

$$\sigma_f = \frac{4\pi A^2}{\lambda^2} \tag{8.26}$$

where A is the geometric area of the target intercepting the transmitter beam and λ is the radar wavelength. This RCS can be contrasted with the backscattered RCS of a sphere in the optical region equal to the projected area of the sphere $A = \pi a^2$, where a is the radius of the sphere. For example, if a sphere has a backscattered RCS of 0.25 m^2 at $\lambda = 0.1$ m (S-band), the forward-scatter RCS is 78.5 m^2, which corresponds to an RCS enhancement of about 25 dB [6].

The forward-scatter RCS enhancement falls off as the bistatic angle diminishes from 180°, ultimately reaching the backscatter RCS as the bistatic angle decreases to zero. However, the cross section enhancement can be significant for bistatic angles on the order of 165° or greater [6].

Because the magnitude of an object's forward scatter does not depend on its material composition, bistatic radar operating in the forward scatter region may be able to detect targets whose monostatic RCS has been specially reduced by absorbing materials. For low-observable vehicles, the forward-scatter RCS will remain high, although the monostatic RCS has been significantly reduced.

8.3 SEMIACTIVE DOPPLER RADAR GUIDANCE

Semiactive missile doppler radar (see Figure 8.1) is one of the most effective methods for providing missile guidance and, hence, is extensively employed operationally. From a radar viewpoint, it combines the features of both a pulsed doppler (or CW) and a bistatic radar. In effect, having a semiactive seeker is the same as having a complete target-tracking radar aboard the missile, that generates its own navigation commands as the missile "homes" onto the target without the weight and cost penalty of carrying a heavy transmitter. A primary advantage of the semiactive homing guidance system is that it becomes increasingly accurate as the target-to-missile range decreases. This leads to an effective missile system for which the size and warhead can be minimized. The ECM vulnerability of semiactive missiles is low due to the passive nature of the missile seeker and the use of very narrow receiver bandwidths in the doppler tracking circuitry. The unique advantages of semiactive guidance over active homing guidance systems are that the illumination is generally much stronger than that available in active seekers, the seeker is simpler than the active seeker, and it presents no signal emissions indicating its location.

The air-to-air semiactive missile homes onto the scatter from a target that is illuminated by a radar carried in the launch aircraft. The missile receives the doppler-shifted target-reflected illumination and compares it to a reference developed in the seeker. This reference is developed either through a signal intercepted from the radar through a rear antenna (see Figure 8.1) or by synchronizing a stable oscillator to the radar transmitter before launch and in effect using the oscillator as a coherent memory of the radar signal.

A block diagram of an early version of a missile seeker using a rear reference antenna configuration is depicted in Figure 8.15. In this version, the rear signal, after conversion to IF, closes the AFC-loop around the microwave local oscillator

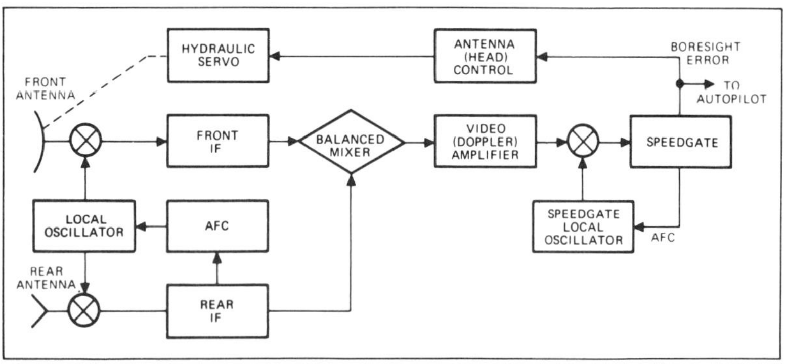

Figure 8.15 Semiactive missile seeker using rear reference antenna.

and acts as the reference for the IF coherent detector. The target signal, received in the front antenna, is heterodyned to IF and amplified in a relatively wideband amplifier. The signal is then converted to the target's doppler band in the coherent detector. The doppler signal is amplified in the video amplifier, which has a bandwidth equal to the total range of possible dopplers, and is heterodyned with the *speed-gate* LO, which tracks the desired signal in the narrow speed gate.

Target acquisition is accomplished by sweeping the frequency of the speed-gate LO over the doppler bandwidth. In essence, this examines the spectrum by moving it past the frequency window of the speed gate. When a signal exceeds the detection threshold, the search is stopped and the signal is examined to verify that it is a coherent target rather than a false alarm due to noise. A valid target is then tracked in frequency and guidance commands are extracted from this signal.

A modern variation of this type of seeker is called the *inverse receiver,* which was made possible by the availability of narrowband, highly selective crystal filters at IF and the development of low-noise tunable solid-state microwave oscillators [17,18]. The inverse receiver, depicted in Figure 8.16, gets its name from the inverse placement of the narrow speedgate in the IF relative to earlier seekers. This configuration allows the transmitter feedthrough, clutter, and jamming signals, which are on the orders of magnitude greater than the desired signal, to be rejected early in the signal path, before they distort the desired signal due to amplifier dynamic range saturation effects.

The semiactive radar seekers depicted in Figures 8.15 and 8.16 are ideally configured for operation with CW waveforms. They are adaptable to pulsed doppler waveforms, which are generally used in modern design, through use of the center line of the doppler spectrum for tracking the target. However, for sampled data operation, the primary difference is in the doppler acquisition scheme. Because targets are illuminated only in short bursts, the use of a sweeping gate for acquisition would

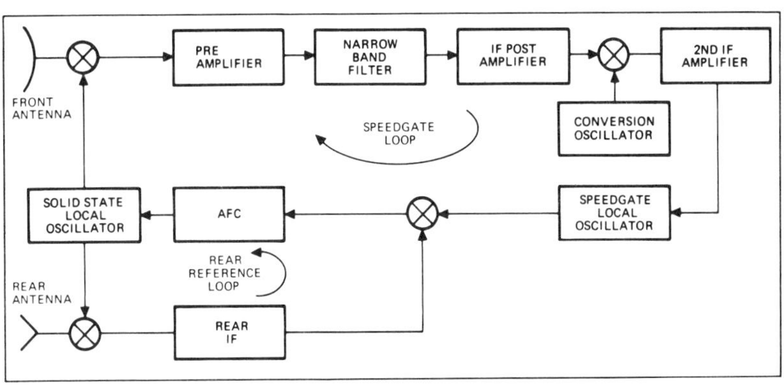

Figure 8.16 The semiactive radar seeker inverse receiver.

result in excessively long acquisition times. The doppler uncertainty region must be examined in parallel by a bank of doppler filters. A high duty cycle PD waveform can be used (typically 30 to 50% duty factor), the PRF of which is chosen to be unambiguous in doppler for the highest closing velocity target, thereby providing a clutter-free region of target doppler frequencies for acquisition and tracking. Figure 8.17 depicts a general configuration for a semiactive missile seeker using a PD illumination radar waveform that also employs an internal coherent reference.

The semiactive missile homes onto the scatter from a target illuminated by a radar carried in the launch vehicle. Therefore, the doppler frequency of the target as seen by the missile may be quite different from that seen by the illuminating radar.

This is illustrated in Figure 8.18 for a simple colinear case, which results in a Doppler frequency as seen by the missile with respect to the illuminating radar's transmitted frequency as

$$f_{dm} = \frac{v_r + 2v_t + v_m}{\lambda} \tag{8.27}$$

where v_r, v_t, and v_m are the velocities of the radar, target, and missile, respectively. This is the doppler frequency that would be experienced in the semiactive seeker of

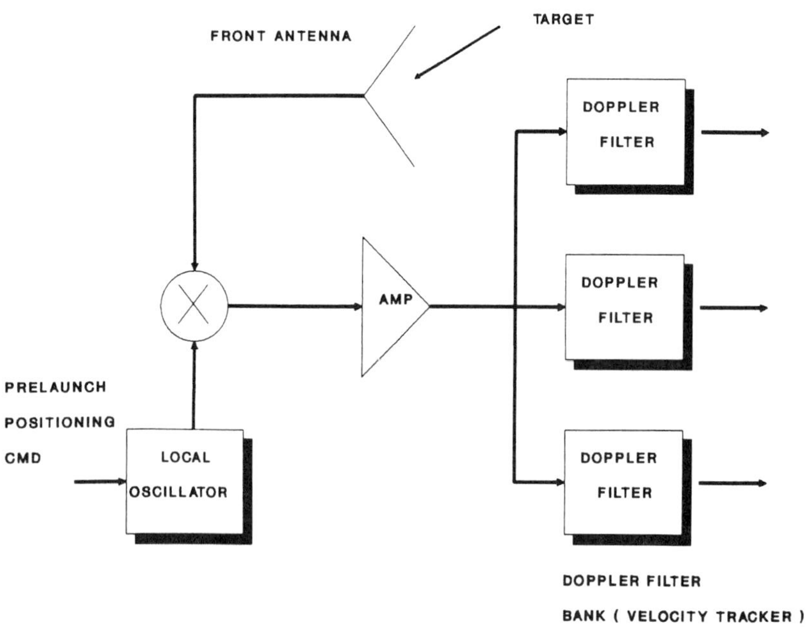

Figure 8.17 PD semiactive seeker with internal reference.

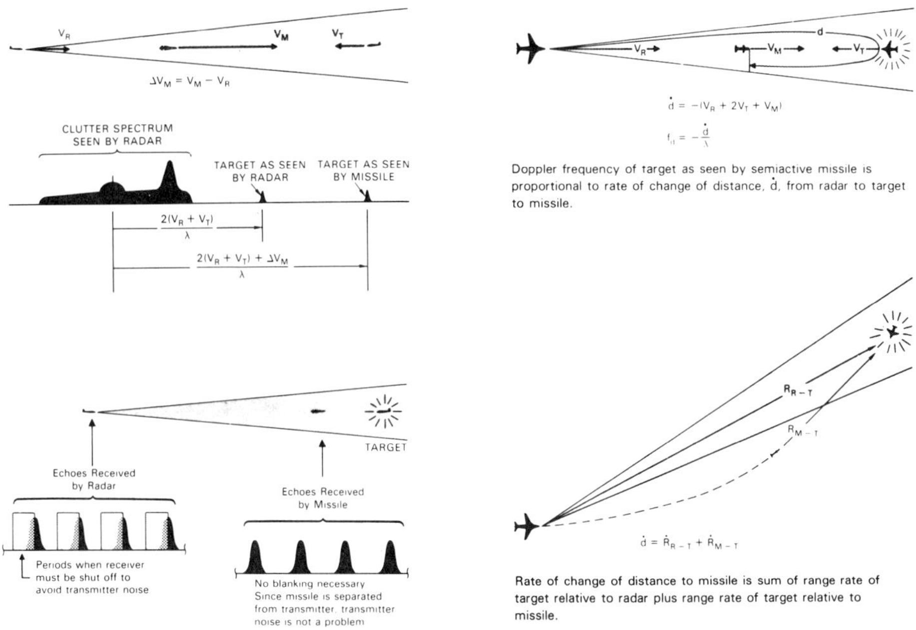

Figure 8.18 Doppler frequency of the target as seen by a semiactive missile.

Figure 8.17, which employs an internal reference. The missile velocity v_m equals the radar velocity plus the incremental velocity (Δv_m) of the missile relative to the radar ($v_m = v_r + \Delta v_m$). The relative velocity is then equal to $f_{dm} = 2(v_r + v_t)/\lambda + \Delta v_m/\lambda$, where the first term is the target doppler seen by the illuminating radar (i.e., without the missile) and the second term is the doppler of the missile with respect to the radar ($f_m = (v_m - v_r)/\lambda$). These equations apply only if all the velocities are colinear and the missile is on the line of sight from the radar to the target.

When a rear reference antenna is used, the doppler processed in the semiactive missile seeker is the difference between the doppler at the missile received from the target (see (8.27)) and that received at the missile directly from the radar $f_{mr} = (v_r - v_m)/\lambda$. Hence, with respect to the reference antenna, the semiactive missile doppler is given by $f_m = f_{mr}$, which equals

$$f_{dt-r} = \frac{2v_t + 2v_m}{\lambda} \tag{8.28}$$

In the more general case depicted in Figure 8.19, the doppler frequency of the target as seen at the missile equals the doppler of the target relative to the radar plus

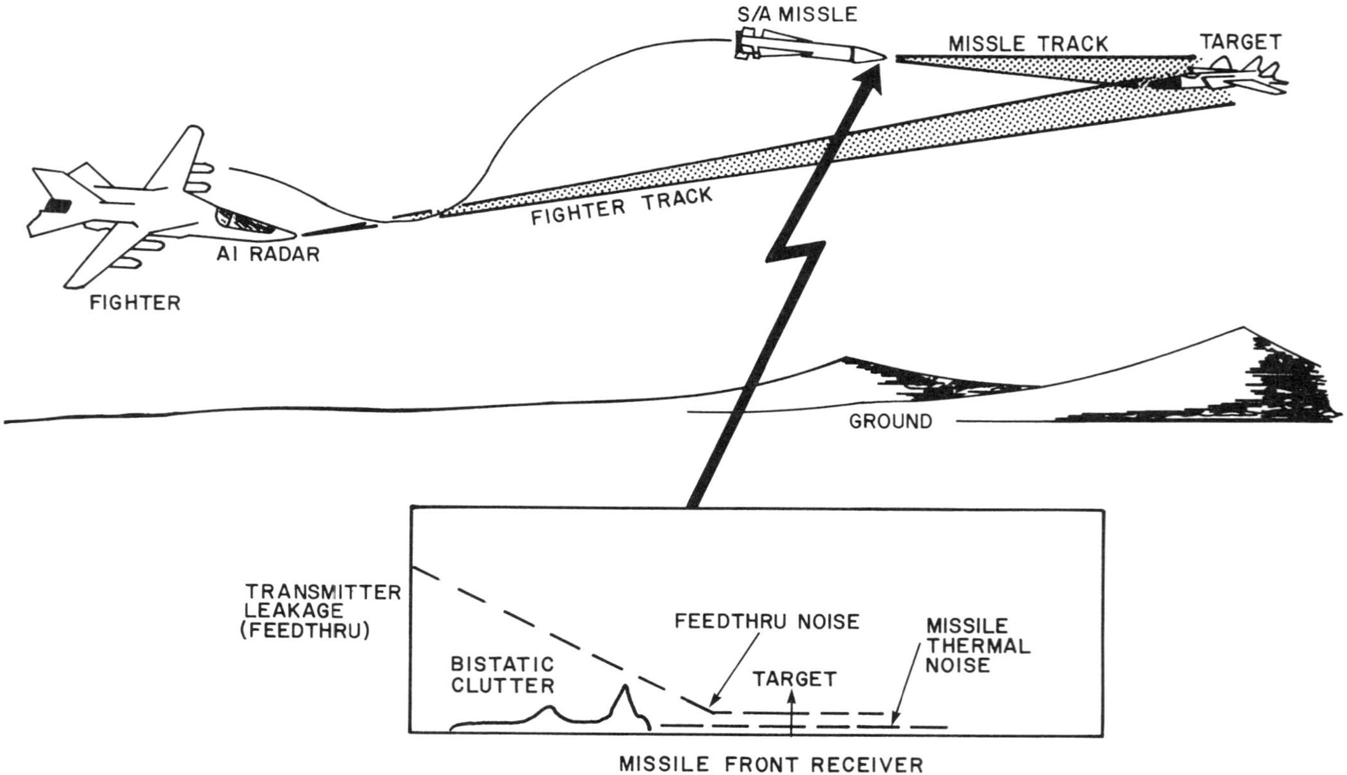

Figure 8.19 Semiactive missile illumination geometry.

the doppler of the target relative to the missile. If the geometry is resolved with respect to the lines of sight between the radar and target (LOS_{rt}) and the missile to the target (LOS_{tm}), then the doppler can be expressed as

$$f_{dm} = \frac{v_r \cos\alpha + v_t \cos\beta + v_t \cos\phi + v_m \cos\theta}{\lambda} \quad (8.29)$$

where α is the angle between the radar and LOS_{rt}, β is the angle between the target and LOS_{rt}, ϕ is the angle between the target and LOS_{tm}, and θ is the angle between the missile and LOS_{tm}.

The central ground clutter spectrum seen by the illuminating PD radar and the relative target dopplers as seen in the radar and missile are depicted in Figure 8.18 for an air-to-air semiactive missile encounter. The corresponding ground clutter spectrum as seen in the missile seeker (referenced to the signal received in the rear antenna) for both an air-to-air and a surface-to-air encounter is depicted in Figure 8.20. Note that the maximum differential ground clutter doppler velocity as seen in the missile for the surface-to-air encounter is proportional to v_m (i.e., $f_{d\max} = 2v_m/\lambda$). Feed-through leakage from the illuminating radar's transmitter received in

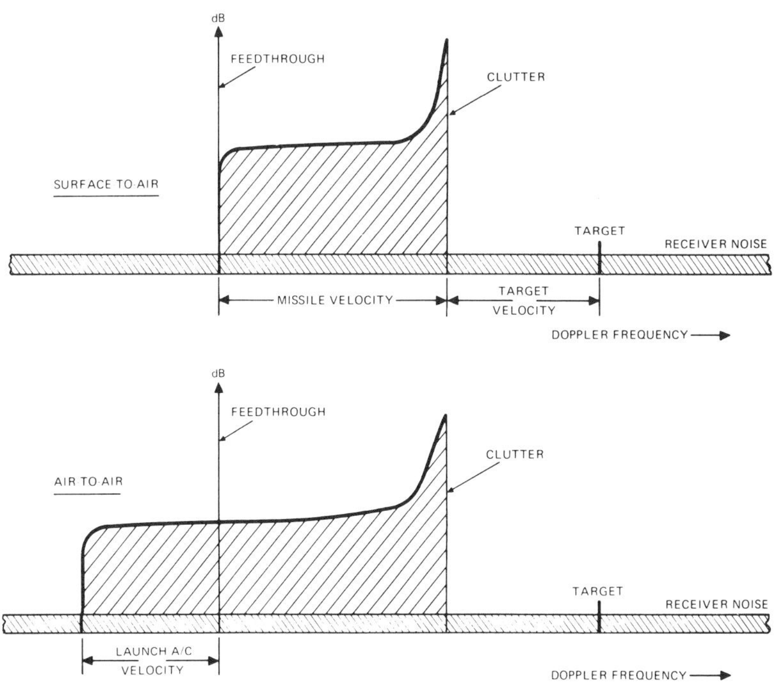

Figure 8.20 Semiactive radar missile ground-clutter spectrum.

the missile antenna's back- and sidelobes is at the same frequency as the reference signal received through the rear antenna. The air-to-air situation is more complicated. The maximum differential doppler shift with respect to the reference signal (i.e., frequency proportional to $v_m - v_r$) is still proportional to the missile velocity v_m (i.e., $f_d = 2v_m/\lambda$), but in addition, there is a negative frequency component proportional to the illuminating radar's aircraft velocity v_r (i.e., $f_d = -2v_r/\lambda$) that is reflected from ground clutter illuminated by the radar antenna's backlobes and sidelobes.

In the design of a high-PRF PD radar used as an illuminator in a semiactive missile system, the waveform must be selected to accommodate the target doppler frequencies (see Figure 8.18) and the ground clutter spectrum (see Figure 8.20) as seen by the missile. For example, to ensure that the velocity data obtained by the missile is unambiguous, in computing the minimum acceptable PRF, we must add to the maximum target closing rate half the velocity of the missile relative to the radar. Also, in the design of the missile's mainlobe clutter-rejection filters, we must be cognizant of the higher missile velocity relative to that of the launch aircraft. This is particularly important if the missile is drawn into a tail chase that decreases the apparent target doppler frequency as seen by the missile.

The presence of transmitter leakage in the missile seeker is a major consideration in the system design. The spectrum shown in Figure 8.20 is generally broadened by FM noise on either the illuminating radar transmitter or the seeker's receiver local oscillator (see Figure 8.19). Therefore, unless very low-noise sources are used, the feed-through and clutter FM noise components at the target doppler frequency could mask the target signal. In addition, the dynamic range of the seeker receiver must be sufficiently high to prevent suppression of the target signal by the feed-through and clutter, as well as prevent generation of crossmodulation and intermodulation products that compete with the target signal.

An example of the semiactive missile illumination radar transmitter noise requirements for an internal reference system in an air-to-air application are tabulated in Table 8.1. A typical transmitter noise specification that results from this requirement with 1 kHz speed-gate bandwidth is depicted in Figure 8.21.

Table 8.1
Radar Transmitter Noise Requirement for Internal Reference Semiactive Missile

Requirement Factor	*Value*
Typical Leakage Power at Acquisition 10 s after Launch	−66 dBm
Typical Missile Receiver Noise Power (BW = 1 kHz)	−137 dBm
Leakage Power-to-Noise Power Ratio	−71 dB
Leakage Suppression Factor	−10 dB
Radar Transmitter Noise in 1 kHz Speed-Gate Bandwidth	−81 dB

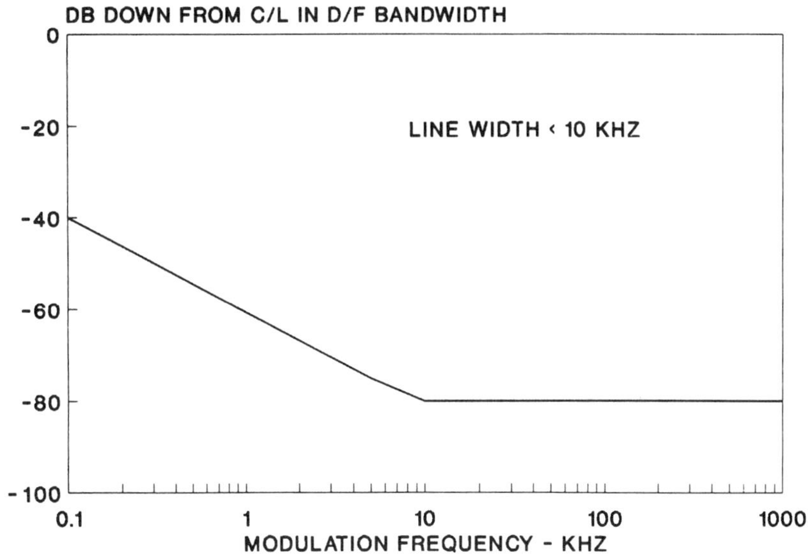

Figure 8.21 Typical illuminating radar transmitter noise specification for an internal reference missile seeker.

As a further example, consider the external reference semiactive air-to-air missile system receiver depicted in Figure 8.22. The FM noise component of the illuminating radar's transmitter leakage will be partially cancelled due to the correlation effect between the reference (i.e., the transmitter sample received through the rear antenna) and the transmitter noise induced onto the ground clutter (see Section 3.3). Typical radar transmitter noise requirements (in a 1 kHz doppler bandwidth) for external reference semiactive missile seekers are tabulated in Table 8.2. A typical transmitter noise specification for an external reference semiactive missile seeker is depicted in Figure 8.23. Note that, due to cancellation of FM noise, the requirement for AM noise becomes increasingly important in meeting the overall noise requirement. AM noise is generally much less than FM noise because the transmitter is almost invariably operated in saturation, whereas balanced mixers and other design techniques are used in the receiver to suppress AM noise.

Another critical element of the missile seeker receiver is the voltage-controlled local oscillator, which translates the signals received through the missile seeker antenna into the doppler filter (i.e., speed gate), where missile guidance information is derived. Any noise sideband components on this oscillator will translate the received transmitter leakage and ground clutter in frequency, thereby requiring a noise specification for this component in a manner similar to that for the illuminating radar's transmitter. A specification for a typical local oscillator FM noise requirement for both CW and

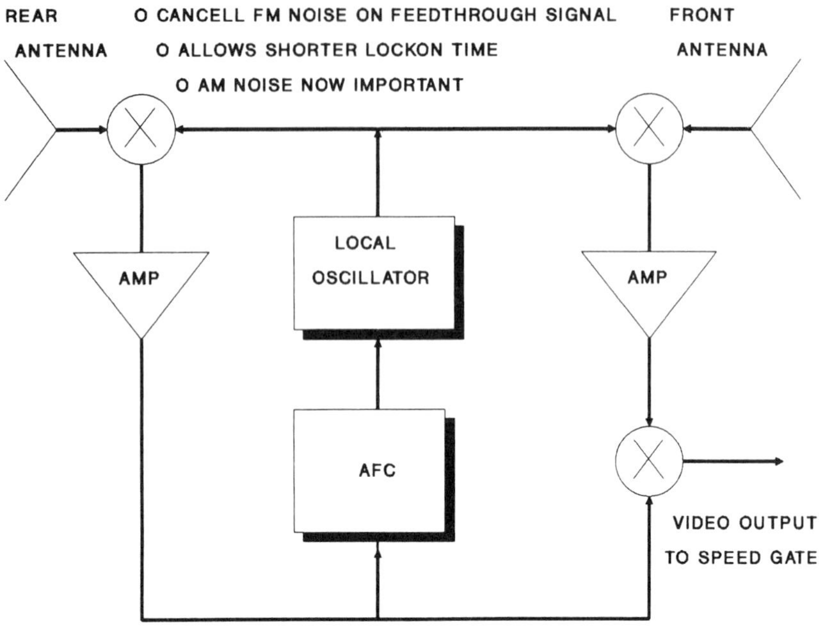

Figure 8.22 Block diagram of an external reference semiactive missile seeker.

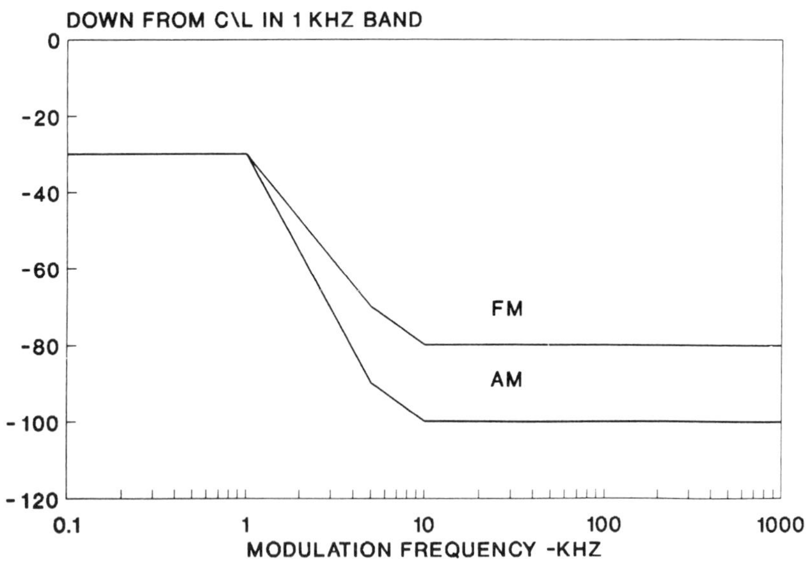

Figure 8.23 Typical illuminating radar transmitter noise specification for external missile seeker.

Table 8.2
Radar Transmitter Noise Requirement for External Reference Semiactive Missile

Requirement Factor	*Value*
Typical Leakage Power at Acquisition 1 s after Launch	−50 dBm
Typical Missile Receiver Noise Level (*BW* = 1 kHz)	−137 dBm
Leakage Power to Noise Power Ratio	−87 dB
Leakage Suppression Factor	−10 dB
Radar Transmitter AM Noise Specification in 1-kHz Speedgate Band	−97 dB
Typical Missile FM Cancellation Ratio	+15 dB
Typical Transmitter FM Noise Specification in 1 kHz Speed-Gate Band	−82 dB

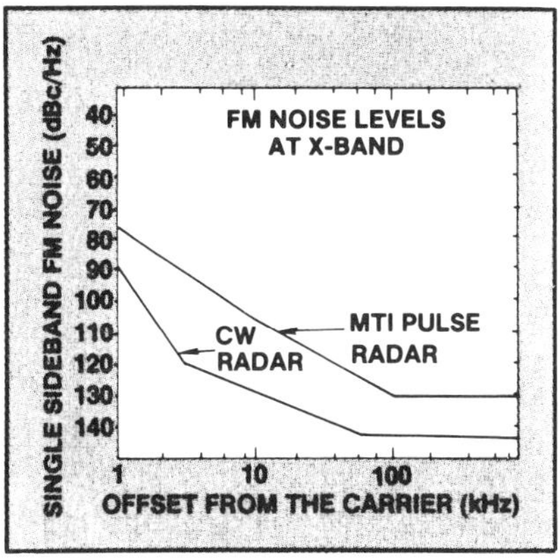

Figure 8.24 Local oscillator FM noise requirement.

PD radars is depicted in Figure 8.24 [18]. Note that this specification is defined in dBc/Hz, whereas the curves given in Figures 8.21 and 8.23 utilize a 1 kHz noise bandwidth and, hence, are defined in dBc/(kHz).

8.4 INVERSE SYNTHETIC APERTURE RADAR [5]

A *synthetic aperture radar* (SAR) achieves high resolution in the cross-range dimension by taking advantage of the motion of the vehicle carrying the radar to synthesize the effect of a large antenna aperture. The basic action of the SAR is to produce a azimuth angle resolution higher than can be achieved with a conventional radar. As the name "synthetic aperture" implies, this can be viewed as the result of synthetically creating the equivalent of a large linear array antenna. Alternatively, this can be viewed as dividing the conventional antenna's azimuth cell into a number of doppler cells, each associated with a particular azimuth segment (i.e., doppler beam sharpening; see Section 8.1). Both models are equivalent, and either can be used to determine the characteristics of a synthetic aperture radar.

The classic SAR is appropriate for providing high-resolution ground maps but ill suited for imaging targets such as ships and aircraft that have rotational motion. For unless the differential doppler shifts, which such motions produce, are accurately predicted and compensated, they tend to defocus the array and blur the image. With slightly different algorithms, these shifts, rather than those due to the radar's relative velocity, can be used to provide the angular resolution needed for imaging. This technique is called *inverse synthetic aperture radar*.

Consider the conventional strip map SAR geometry depicted in Figure 8.25. The azimuth beamwidth of a uniformly illuminated antenna with aperture width D_{az} at wavelength λ in rad is given by $\theta_{\mathrm{az}} = \lambda/D_{\mathrm{az}}$. By direct analogy, the beamwidth of a synthetic array of length L_e is given by

$$\theta_s = \frac{\lambda}{2L_e} \tag{8.30}$$

where the factor 2 in the denominator relates to the two-way propagation path from the antenna element to the target, compared with the one-way path of a conventional antenna. The antenna elements in the synthetic array (see Figure 8.25) are formed due to vehicle motion between consecutive radar transmitted pulses, and hence are separated by the distance v_pT, where v_p is the vehicle velocity and T is the radar's interpulse period. The cross-range resolution of the synthetic array is then given by

$$\delta_x = \theta_s R = \frac{\lambda R}{2L_e} \tag{8.31}$$

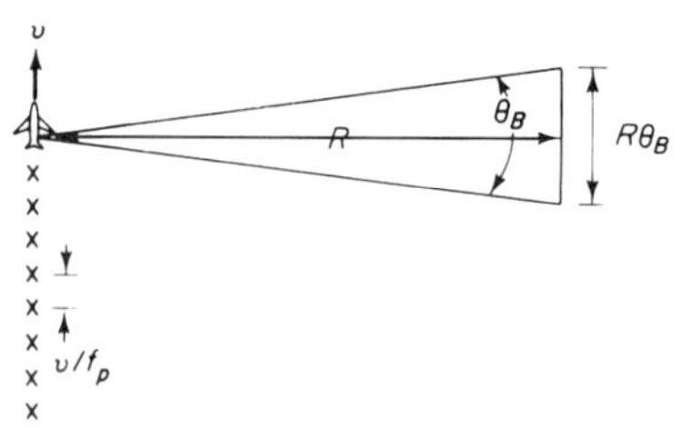

Figure 8.25 Synthetic aperture radar geometry.

The synthetic array contains n elements, each separated by a distance $v_p T$, resulting in a total length $L_e = nv_pT$. The total integration time required to form the synthetic array is given by $t_i = nT$, which can be defined in terms of the doppler bandwidth as $B_D = 1/nT$, resulting in a cross-range resolution of

$$\delta_x = \frac{\lambda R B_D}{2v_p} \tag{8.32}$$

Note that (8.32) is identical to (8.6) evaluated at a squint angle $\alpha_m = \pi/2$, demonstrating that the SAR relationships derived from synthetic array and doppler beam-sharpening considerations result in identical formulations.

The cross-range resolution of an unfocused array (i.e., quadratic phase variation at each array element are not corrected) can be obtained by defining the synthetic array beamwidth as the beamwidth that results when the difference between the two-way minimum and maximum paths as depicted in Figure 8.26 result in a quarter-wavelength path-length difference [11]. From this geometry, $R^2 + (L_e/2)^2 = (R + \lambda/8)^2$, which can be solved for the unfocused synthetic aperture length as

$$L_{eu} = \sqrt{R\lambda} \tag{8.33}$$

Substitution of this into (8.31) results in a further expression for the cross-range resolution of an unfocused synthetic aperture array as

$$\delta_{xu} = \frac{\sqrt{\lambda R}}{2} \tag{8.34}$$

For example, if $R = 100$ km with an X-band radar ($\lambda = 3$ cm), the cross-range resolution would be 27 m. Note that the cross-range resolution (unfocused) is a function of both the range and the radar's frequency, but is independent of the actual antenna's aperture.

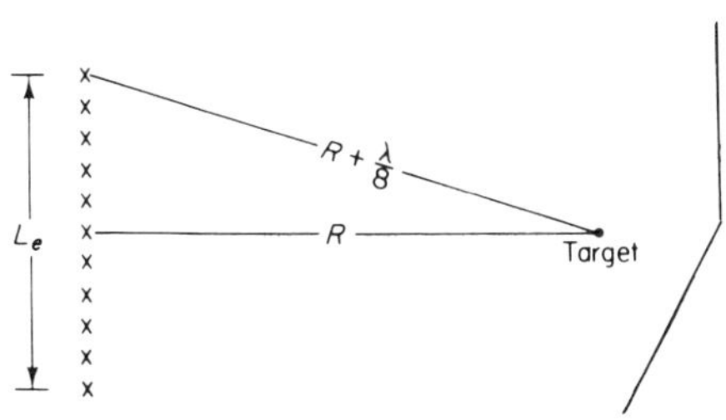

Figure 8.26 Geometry of the unfocussed synthetic aperture radar.

The synthetic array can be focused at range R by applying a quadratic phase correction at each position (i.e., for all elements in L_e) of the array equal to $\Delta\phi = 2\pi x^2/\lambda R$, where x is the distance from the center of the synthetic array to the corrected element. A different correction must be applied for each range R and, when focused, all the return echo signals from a scatterer at that range are in phase. The resulting synthetic array is focused over a strip equal to L_e (see Figure 8.26) that must be illuminated by each element of the array, which has a real beamwidth equal to $\theta = \lambda/D$. Hence, the real beamwidth must be equal to $\theta = L_e/R = \lambda/D$, resulting in $L_e = R\lambda/D$. Substitution of this into (8.30), which applies for the synthetic array, results in $\theta_s = D/2R$ and a focused cross-range resolution of

$$\delta_{xf} = R\theta_s = \frac{D}{2} \tag{8.35}$$

Note that the focused cross-range resolution is independent of both the radar's transmitted frequency and the range at which it is focused, but depends on the real aperture dimensions.

One fundamental limitation of a synthetic array is that the element positions must be spaced less than a half-wavelength apart to prevent grating lobes. Because the element positions depend on the vehicle velocity (v_p) and the radar's interpulse period ($T = 1/\text{PRF}$), so that $v_pT < \lambda/2$, the resulting PRF must satisfy the inequality

$$\text{PRF} > \frac{2v_p}{\lambda} \tag{8.36}$$

where $f_d = 2v_p/\lambda$ is the doppler shift of the radar-carrying vehicle.

The side-looking antenna position in both the unfocused and focused synthetic aperture arrays may limit the cross-resolution capability of these systems. For example, consider an S-band ($\lambda = 0.1$ m) synthetic aperture radar that operates at a 100-km range with an azimuth aperture width of 3.33 m from an aircraft with 300 m/s velocity. The array length for the unfocused case is $L_{eu} = \sqrt{\lambda R} = 100$ m, whereas the cross-range resolution $\delta_x = \sqrt{\lambda R}/2 = 50$ m. Because the vehicle has a velocity of 300 m/s, it takes $t_i = 0.33$ s to form the array. In the focused case, the array length is $L_e = R\lambda/D = 3$ km, so that the integration time $t_i = 10$ s. The resolution in the focused case is $\delta_x = D/2 = 1.67$ m. The cross-range resolution is thus inversely proportional to the integration or processing time, as can be deduced by applying (8.6) or (8.32), and can be rewritten as

$$\delta_x = \frac{\lambda R}{2v_p t_i \sin\alpha} \tag{8.37}$$

where t_i is the total integration time and α is the angle between the vehicle's velocity vector and the antenna's boresight ($\alpha = 90°$ in the side-looking case).

This example illustrates that the cross-range resolution capability of a synthetic aperture radar depends on the integration or processing time required to form the synthetic array, which, in turn, is determined by both the parameters and the geometry of the array and the type of processing utilized (i.e., focused or unfocused). One method for increasing the processing time is to use a spotlight mode SAR in which the real antenna boresight is continually realigned to track a particular target area of interest [19]. The geometry for spotlight SAR is depicted in Figure 8.27. Theoretically, resolutions better than $D/2$ can be obtained with this mode and the only requirements are that the real antenna beamwidth must be large enough to adequately illuminate the desired area to be imaged, and the duration of the illumination must be long enough to obtain sufficient effective rotation of the scene for the desired cross-range resolution (see (8.37)). Another benefit of this mode is that the scene to be imaged is observed over a range of incidence angles that averages the speckle (specular radar returns associated with a particular viewing aspect) and makes a smoother image.

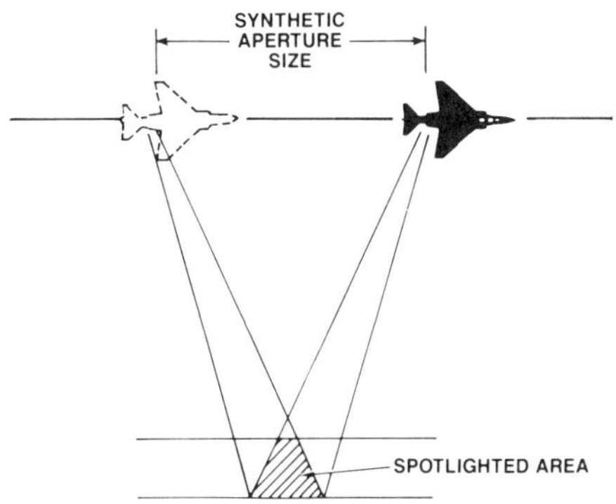

Figure 8.27 Spotlight SAR.

One interpretation of spotlight SAR is that it is equivalent to the synthetic array radar flying a circular path around the target area, as depicted in Figure 8.28. This follows from the fact that, once the synthetic array is corrected for the quadratic phase caused by its straight-line path, then the phase of the signal returned from a fixed scatterer at the focused range has constant value. A constant phase return would also be returned if the SAR moved on a locus of constant range with respect to the target and hence on a circular path. The equivalent circular arc angle ($\Delta\phi$), which the SAR moves to provide a specified cross-range resolution, can be found by calculating the integration time (t_i) from (8.37) and using this in the relationship

$$\Delta\phi = \frac{v_p t_i}{R} \tag{8.38}$$

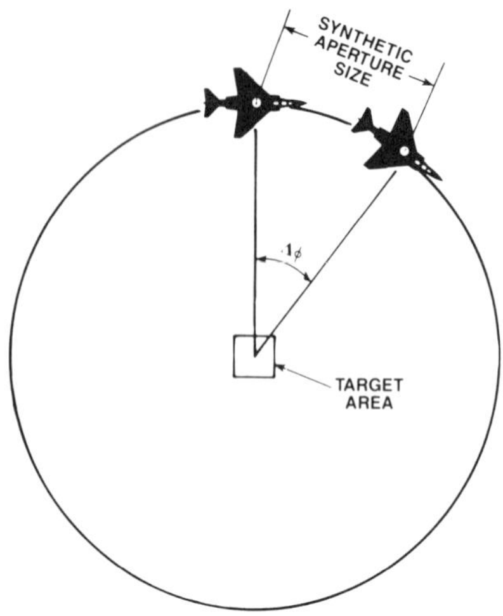

Figure 8.28 Spotlight SAR with a circular flight path.

For this example (v_p = 300 m/s, t_i = 10 s, R = 100 km), the resulting angle would be $\Delta\phi = 3/100$ rad = 1.8°.

From Figure 8.28, it is apparent that the same data would be collected if the radar were stationary and the target area rotated. When this occurs, the system is called an *inverse synthetic aperture radar*. ISAR can also function from a moving platform, as depicted in Figure 8.29. As illustrated, with SAR the differences in doppler frequency that enable us to obtain fine angular resolution are due to the motion of the radar-bearing aircraft. By contrast with ISAR, the differences are due to the angular rotation of the target (e.g., yaw, pitch, and roll) as seen by the radar. In Figure 8.29, the radar's speed is slightly greater than the target's, resulting in point P_1 on the target's tail having a slightly negative rate. Because the target is turning away, point P_2 on the nose has a positive range rate. Thus the doppler frequencies of the returns from P_2 and all points between it and P_1 differ in proportion to their distances from P_1.

To image the target, phase corrections are first made to provide motion compensation that removes the variations due to the radar's and target's displacement. The returns from each resolvable range increment are then applied to a separate bank of doppler filters, and an image is produced from their outputs just as in conventional SAR. The image's cross-range dimension is not necessarily horizontal but rather is perpendicular to the axis about which the target happens to be rotating at the moment. No image can be formed if that axis is colinear with the radar's line of sight or if the radar has no rotational motion as viewed from the radar.

A basic two-dimensional ISAR image can be generated as depicted in Figure 8.30. A series of range profiles is generated by a high-range-resolution radar as it

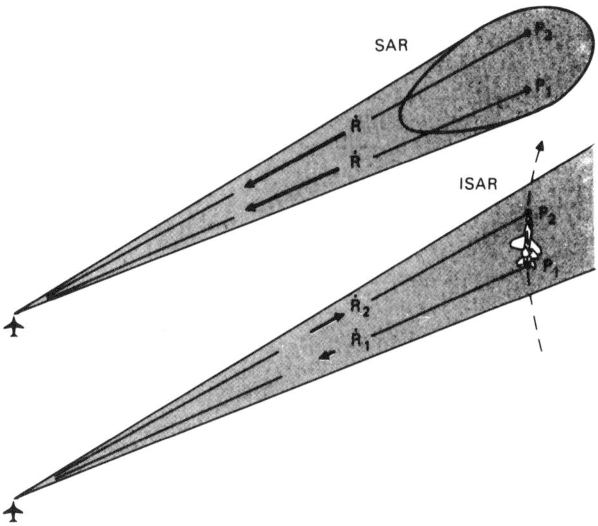

Figure 8.29 Comparison of SAR and ISAR from an airborne platform.

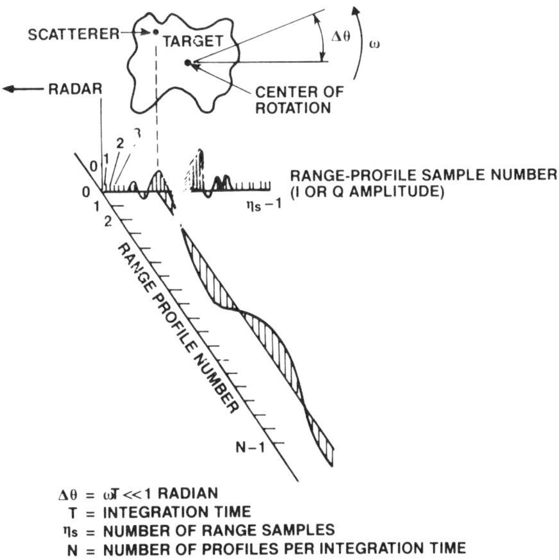

Figure 8.30 ISAR range-doppler sampling of a target with rotational motion.

observes a rotating target. The range profiles are sampled at a rate that corresponds to the radar's PRF. There are then n range cells, where each range cell has a resolution $\Delta r_s = c\tau_e/2$ ($\tau_e = 1/B_R$, where B_R is the radar's transmitted bandwidth). The n range cells each contain $N = 1/\text{PRF} \cdot t_i$ range profiles, where t_i is the integration time required to produce the desired cross-range resolution as the target rotates through $\Delta\theta$ rad.

The N range profile samples in each range cell are doppler filtered to extract cross-range image data. Each target scatterer on the rotating target produces a doppler signature with a frequency that is proportional to the cross-range distance between the scatterer and the center of target rotation. The magnitude of each scatterer-related doppler signature is proportional to the reflectivity of that scatterer. The rotating target reflectivity is thus mapped in orthogonal coordinates consisting of slant range and cross range, where the cross-range scale factor is dependent on target-aspect angular rotation rate. Orientation of the rotation axis relative to the radar's line of sight establishes the orientation of the range-doppler image plane.

The doppler frequency gradient required to obtain high cross-range resolution is generated by the motion of the object relative to the radar. This motion can be generated in a variety of ways, which generally can be related to the simplified case depicted in Figure 8.31 of a stationary monostatic radar illuminating a rotating object.

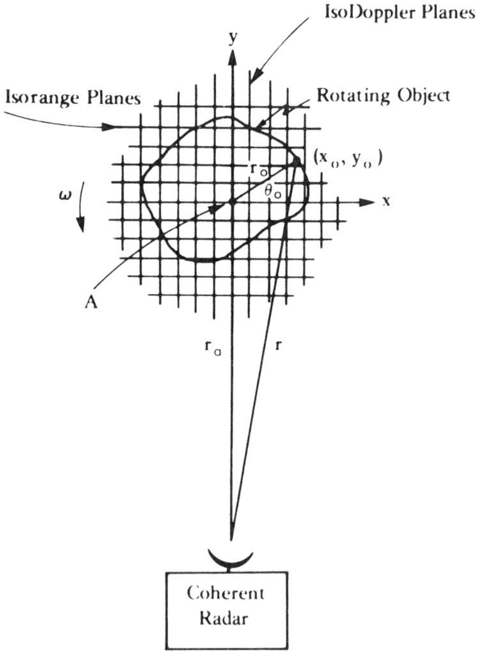

Figure 8.31 ISAR target geometry for range-doppler imaging.

The object rotates with uniform angular motion in the *x*-*y* plane about the *z*-axis. The object is contained within a coordinate system in range-doppler space, shown as *isodoppler* and *isorange* lines, where the isorange lines are parallel to the radar-target LOS and the isodoppler lines are perpendicular to the isorange lines. The *x*-axis represents the cross-range dimension, whereas the *y*-axis is the slant-range dimension.

Consider a uniformly rotating scatterer located at the point (x_0, y_0) with angular rate ω and initially positioned ($t = 0$) at angle θ_0 in the coordinate system of Figure 8.31 [20]. The range to this scatterer is given by

$$r = R + v_r t + r_0 \cos(\omega t - \theta_0) \tag{8.39}$$

where R is the range to the center of the object, v_r is its radial velocity, r_0 is the distance of the scatterer to the center of the object, ω is the scatterer's angular rotation rate, and θ_0 is the initial angle of the scatterer with respect to the cross-range axis. The motion compensated range (i.e., the object's translational velocity doppler component is removed at the radar) of the scatterer is then given by

$$r = R + x_0 \sin\omega t + y_0 \cos\omega t \tag{8.40}$$

where $x_0 = r_0 \sin\theta$ and $y_0 = r_0 \cos\theta$. The doppler frequency (f_d) of the returned signal is hence given by

$$f_d = \frac{2}{\lambda}\frac{\mathrm{d}r}{\mathrm{d}t} = \frac{2x_0\omega}{\lambda}\cos\omega t - \frac{2y_0\omega}{\lambda}\sin\omega t \tag{8.41}$$

If the radar data are processed over a very small time interval about $t = 0$, the target range and doppler can be approximated as

$$r \approx R + y_0 \tag{8.42}$$

$$f_d \approx \frac{2x_0\omega}{\lambda} \tag{8.43}$$

Therefore, by measuring the returned radar signal in terms of slant range ($r_s = R + y_0$) and doppler frequency (f_d) that allows determining the cross range ($r_c = x_0 = \lambda f_d/2\omega$), we can estimate the position (x_0, y_0) of the point scatterer.

The result of ISAR processing is then to map or image all scatterers contained within the rotating target in the range-doppler coordinate system depicted in Figure 8.31. The presence of the object rotation rate (ω) in determining the cross-range component of the scatterer implies that to obtain a properly scaled image of the object the magnitude of ω must be known. Most techniques for estimating the rotation rate depend on *a priori* knowledge or analysis of periodicities in the radar signal level.

Another implicit assumption is that the distance (R) from the radar antenna to the center of the object is constant and known, which implies that any time-varying components are removed by using appropriate motion compensation techniques.

High resolution in slant range can be achieved by a variety of methods involving either very short pulses or coded pulses of long duration that are compressed into short equivalent pulses. The key parameter in this regard is the transmitted radar bandwidth (B_R), which allows a slant range resolution (Δr_s) given by

$$\Delta r_s = \frac{c}{2B_R} \tag{8.44}$$

To achieve high cross-range resolution (Δr_c) implies a resolution of doppler frequency (Δf_d) given by

$$\Delta f_d = \frac{2\omega \Delta r_c}{\lambda} \tag{8.45}$$

Because a frequency resolution $\Delta f_d(\Delta f_d = B_D)$ requires a coherent processing time interval, approximately equal to $t_i = 1/\Delta f_d$, this results in a cross-range resolution given by

$$\Delta r_c = \frac{\lambda}{2\omega t_i} = \frac{\lambda}{2\Delta\theta} \tag{8.46}$$

where $\Delta\theta = \omega t_i$ is the angle through which the object rotates during the coherent processing time. This relationship is equivalent to that obtained for spotlight SAR, where the target object remains stationary, but the radar moves with velocity (v_p) through an angle $\Delta\phi = v_p t_i / R$ (see (8.38)), which, when substituted into the relationship $\Delta r_c = \delta_x = \lambda R / 2 v_p t_i$ (see (8.37)), results identically in $\Delta r_c = \lambda / 2\Delta\phi$. This illustrates that the relative rotational motion between the target and radar determines the cross-range resolution in both SAR and ISAR systems.

Any target aspect rotation relative to the radar produces a doppler gradient across scatterers that are spread out in the cross-range dimension. Aspect rotation is produced by both target translation in any direction other than along the LOS between the radar and target and that due to target angular motion. The rotation produced by target angular motion is generally resolved into pitch, roll, and yaw components. These rotational components for an aircraft target are depicted in Figure 8.32. For aircraft targets seen from a ground-based radar, the major source of rotational motion up to X-band frequencies is generally the translational tangential component, whereas for ship targets observed at ranges beyond about 10 nmi, target pitch, roll, and yaw are generally the most significant components [5].

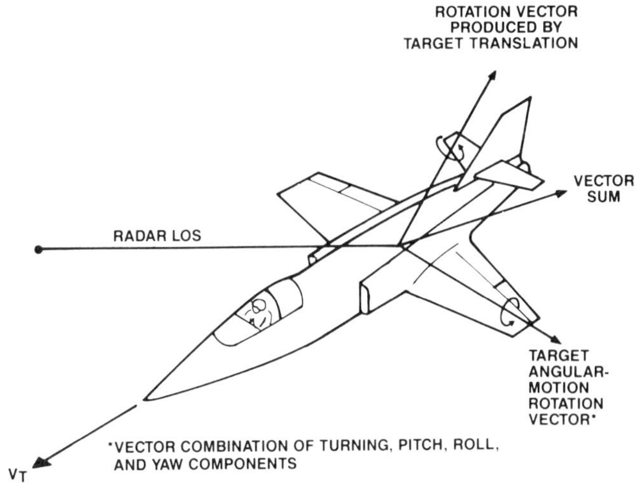

Figure 8.32 Rotational components for an aircraft target.

The target geometry for aircraft, where the target has translational velocity v_t in a direction other than the LOS between the radar and the target, is depicted in Figure 8.33. The ISAR image plane (cross range by slant range) is directed in a plane containing the radar LOS vector and an orthogonal vector that contains the tangential component of target velocity. For ship targets, where pitch, roll, or yaw causes rotational motion (see Figure 8.34), the ISAR image plane is directed in a plane containing the radar LOS vector and the rotational motion component. For example, the ISAR image plane for ship roll is depicted in Figure 8.35.

Although all portions of the ship pitch, roll, and yaw at the same angular velocity, their apparent linear velocities sensed by a radar fixed in space depend on their distance from the ship's center of gravity. These different linear velocities create proportionately different doppler shifts in the echos returned to a radar fixed in space. As a result of the vessel's pitching, for example, the higher an object is on a ship, such as a superstructure for a radar antenna, the greater will be the doppler shift of its echo, as depicted in Figure 8.36.

An ISAR image can be generated from a time history of slant-range profiles by performing a FFT on the data points in each slant range cell. This extracts the target doppler components associated with the cross-range image. The FFT process can be accomplished only if alignment exists along the time history of range profiles. However, the rotational motion of the target generally causes each scatterer to move its position with respect to the radar during image formation (see Figure 8.31). This is also apparent from (8.41) and (8.42), which respectively indicate a time variation in slant range and doppler frequency (proportional to cross range) during the integration

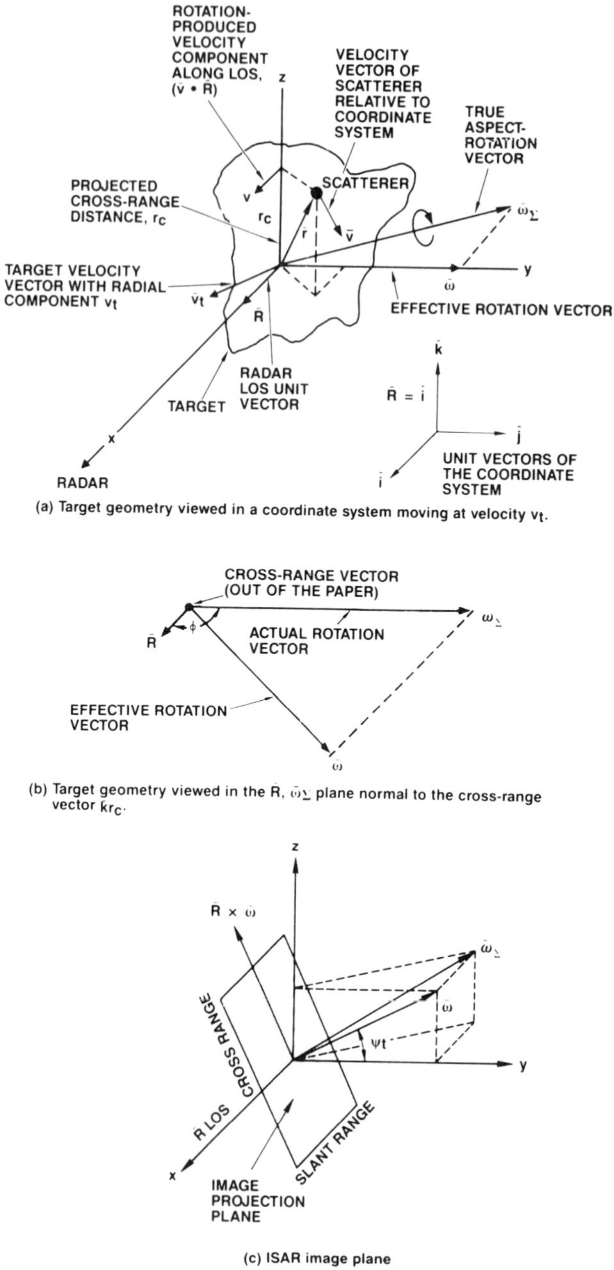

Figure 8.33 Image plane geometry for moving coordinate system.

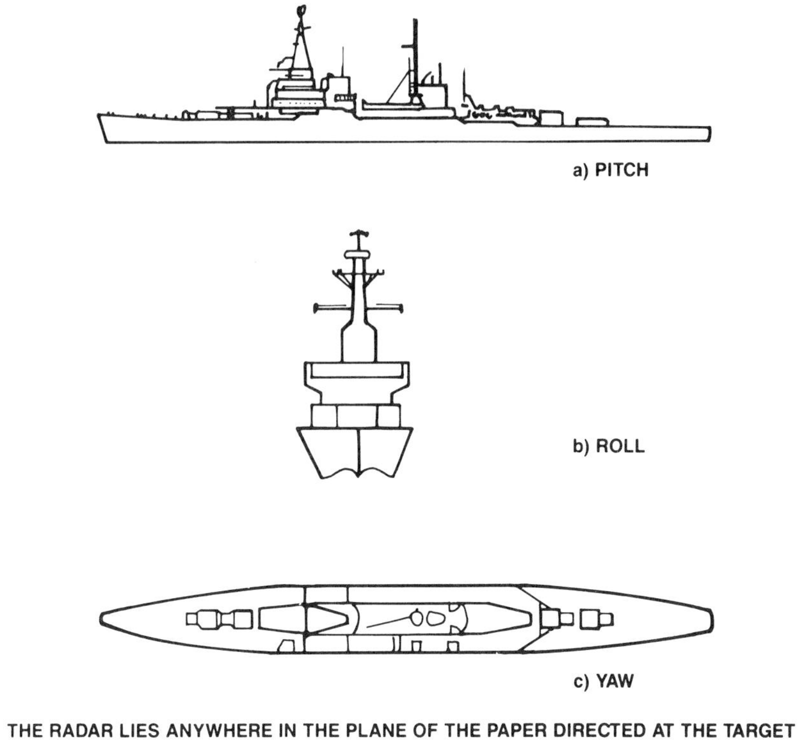

Figure 8.34 ISAR image views produced by target pitch, roll, and yaw.

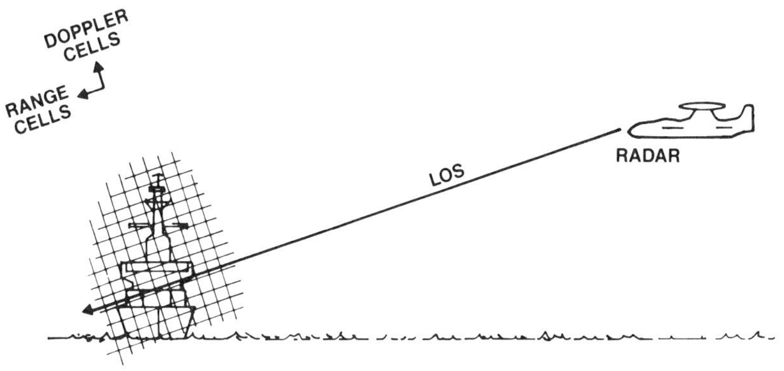

Figure 8.35 ISAR image plane with range and doppler cells for ship target with roll.

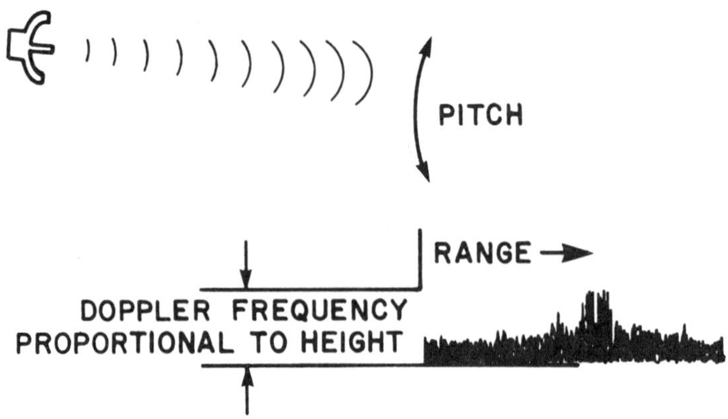

Figure 8.36 ISAR imaging of ship due to pitch rotational motion.

time t_i necessary to form an image. The image aberration caused by motion through resolution cells results in a cross-range-dependent astigmatism called *range walk* and a slant-range-dependent cross-range focus error called *variable range rate* [21].

Fine cross-range resolution implies coherent processing over a large rotational angle ($\Delta\theta$), reflected in relatively large variations in slant range (see (8.41)) and doppler (see (8.42)) for a particular scatterer. Generally, during a processing time interval sufficiently long to give a desired cross-range resolution, points on the rotating object thus may move through several resolution cells. Therefore, the conventional range-delay measurement and doppler frequency analysis implied by (8.42) and (8.43) will result in degraded imagery for the large processing interval case.

To avoid image degradation caused by target migration through multiple resolution cells when using simple range-doppler processing requires limiting the coherent processing time (t_i). Consider the imaging of a body having its cross-range diameter given by D_x and the slant-range diameter of D_y. Hence, points on the body will not move through a slant-range resolution cell (Δr_s) in the processing time (t_i) with angular rotation ($\Delta\theta$) if

$$\Delta r_s > \frac{D_x \Delta\theta}{2} \tag{8.47}$$

Use of the relationship for cross-range resolution (see (8.46)) given by $\Delta r_c = \lambda/2\Delta\theta$ results in

$$\Delta r_c \Delta r_s > \frac{\lambda D_x}{4} \tag{8.48}$$

which imposes a limitation on the product of cross-range and slant-range resolution in terms of the cross-range diameter of the imaged target. Similarly, migration through cross-range resolution cells (Δr_c) is avoided if

$$\Delta r_c > \frac{D_y \Delta\theta}{2} \tag{8.49}$$

resulting in

$$(\Delta r_c)^2 > \frac{\lambda D_y}{4} \tag{8.50}$$

which imposes a limit on cross-range resolution in terms of the slant-range diameter of the body to be imaged. Note that (8.48) and (8.50) require one to choose sufficiently small coherent processing time and small object size to avoid migration of points through resolution cells, which causes smearing of ISAR images.

As an example, assume that an X-band radar ($\lambda = 0.03$ m) is used to image a ship target using pitch rotational motion so that the ISAR image plane consists of the ship length for slant range and the ship's height for cross range (see Figure 8.36). Assume that the ship's length is 200 m (D_y) and its height is 30 m (D_x). Hence, the maximum cross-range resolution (Δr_c) for simple processing is given by $\Delta r_c > \sqrt{(.03)200/4} = 1.5$ m, and the cross range by slant range product is $\Delta r_c \cdot \Delta r_s > (.03)(30)/4 = 0.225$ m. Further, assuming a pitch angular rate of 1°/s (ω) results in an angular excursion of $\Delta\theta = \lambda/2\Delta r_c = 0.01$ rad or 0.6° to achieve the limit cross-range resolution with a processing time of $t_i = \Delta\theta/\omega = 0.6$ s. Table 8.3 lists some predicted cross-range resolutions for ship targets by using an S-band ($\lambda = 0.1$ m) radar [5].

Table 8.3

Predicted Cross-Range Resolutions for Ship Targets Using an S-Band Radar in Sea State 5 ($t_i = 1$ s)

Ship Type	*Motion Type*	*Average Angular Velocity (ω)*	*Average Angular Excursion in 1 s ($\Delta\theta$)*	*Average Cross-Range Resolution (Δr_c)*
Destroyer	pitch	1.01°/s	.0177 rad	2.82 m
	yaw	0.54°/s	.0093 rad	5.35 m
	roll	6.30°/s	.1100 rad	0.46 m
Carrier	pitch	0.16°/s	.0028 rad	17.83 m
	yaw	0.08°/s	.0014 rad	35.54 m
	roll	0.38°/s	.0066 rad	7.56 m

Often, finer resolution is desired than can be achieved using the limits imposed by the cell migration constraints given in (8.48) and (8.50). In this case, objects move through range and doppler resolution cells, and simple range-doppler imagining will yield blurred images; therefore, the effect of motion through resolution cells must be compensated. Several techniques of compensating for this motion through cells have been developed. These range from linear piecewise approximations to account for the motion to sophisticated polar reformatting techniques [21].

Polar reformatting is a process applicable to both SAR and ISAR data. It is most practically implemented by using a "stretch" technique for the radar pulse compression processing [20]. With this pulse compression technique, the receiver local oscillator is chirped at a time corresponding to the object's range at the same rate as the transmitted waveform. The returned signal is thus "deramped" so that the return from each scatterer of the rotating object appears as a unique doppler frequency with respect to the carrier. In the following, we will sketch the fundamental elements of polar reformatting [20].

From Figure 8.31, the complex signal returned from the rotating object can be expressed as

$$f_r(t) = \int_s \sigma(\mathbf{s}_0) u\left(t - \frac{\mathbf{r}}{2c}\right) \mathrm{d}\mathbf{s}_0 \tag{8.51}$$

where $\mathbf{r}$ is the range to the differential scattering surface, $\sigma(\mathbf{s}_0)$ is the target reflectivity observed at the receiver over the surface ($\mathbf{s}_0$), and $u(t)$ is the transmitted pulse compression waveform. The Fourier transform of the received waveform is then given by

$$F_r(\omega) = \int_{-\infty}^{\infty} f_r(t)\, \mathrm{e}^{-j\omega t}\, \mathrm{d}\omega \tag{8.52}$$

If we substitute for $f_r(t)$ in (8.52) the result is

$$F_r(\omega) = \int_s \sigma(\mathbf{s}_0) \int_{-\infty}^{\infty} u\left(t - \frac{\mathbf{r}}{c}\right) \mathrm{e}^{-j\omega t}\, \mathrm{d}\omega\, \mathrm{d}\mathbf{s}_0 \tag{8.53}$$

which can be simplified to

$$F_r(\omega) = \int_s \sigma(\mathbf{s}_0) U(\omega)\, \mathrm{e}^{-j(\omega \mathbf{r}/2c)}\, \mathrm{d}\mathbf{s}_0 \tag{8.54}$$

where $U(\omega)$ is the Fourier transform of $u(t)$. The term $\mathrm{e}^{-j(\omega \mathbf{r}/2c)}$ causes a distortion of the compressed waveform $u(t)$, as $\mathbf{r}$ varies with time over the processing interval required to form the ISAR image. The time varying effects of the range variation **r** can be removed by multiplying the received signal by a reference signal given by $m(t) = e^{+j(\omega \mathbf{r}/2c)}$, which forms the fundamental step in the polar reformatting motion compensation technique.

A block diagram of a simple ISAR system using polar-format processing is depicted in Figure 8.37 [20]. Stretch processing is embodied in the design whereby the pulse compression *waveform generator* (WFG) is used for both the transmitter and receiver local oscillator. After deramping, the signal bandwidth is limited to those doppler frequencies represented by the scatterers within the imaged scene (see (8.41) and (8.43)). The use of the linear FM waveform in a deramping scheme on a pulse-to-pulse basis results in signals with frequency and phase determined by the relative range of a scatterer to scene center, thus establishing the conditions required to perform frequency-domain range-doppler imaging.

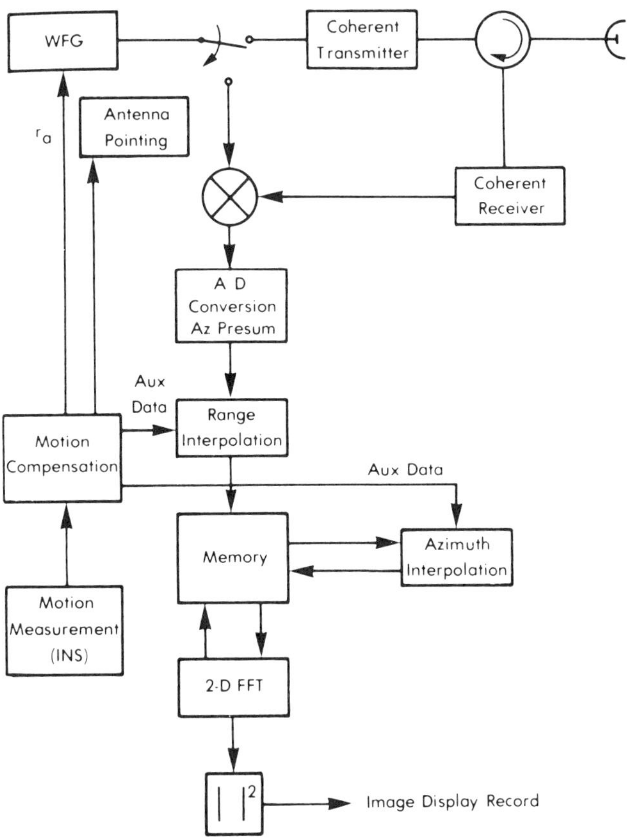

Figure 8.37 Simple SAR system block diagram with polar-format processing.

Figure 8.38 shows the relative positions of the data samples in polar format collected in the ISAR image plane. The samples, represented as black dots, are shown arrayed along radial lines at angle θ_i corresponding to the LOS angles to scene center for each radar transmission, as projected onto a reference plane. The spacing between samples T_r are determined by the original deramped signal projected into the reference plane. The PRF of the radar must be high enough to unambiguously sample the doppler spectrum, which depends on the range extent of the illuminated scene.

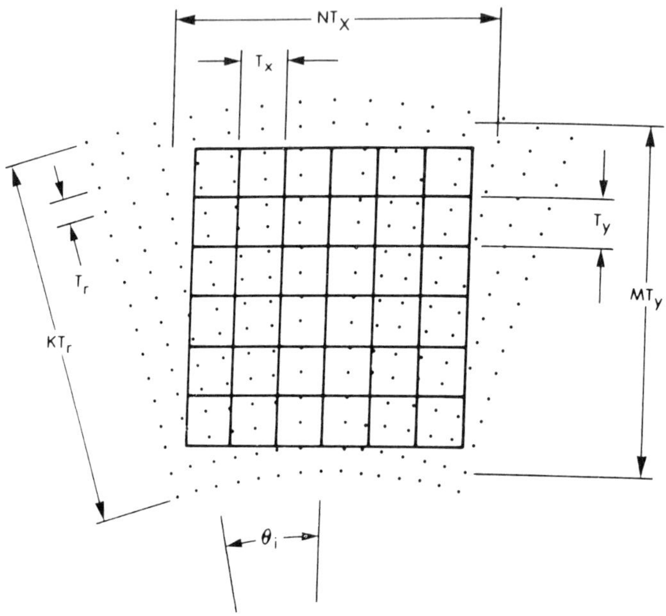

Figure 8.38 Data collection digital polar-format geometry.

The number of samples along each radial line K is determined by the effective duration of the pulse compression ramp. This duration is directly proportional to the effective RF bandwidth used to obtain slant-range resolution.

To form an image, the geometric array of samples in Figure 8.38 must be Fourier transformed in two dimensions. To accomplish this, the data must be resampled into a rectilinear grid with spacings T_x and T_y as depicted in the figure. The number of interpolated grid samples in the range and azimuth dimensions determines the output sample spacings in both dimensions of the image.

The rectilinear sample grid can be formed in two separate one-dimensional interpolation steps. The first step depicted in Figure 8.39 simply resamples each radar pulse so that the new samples fall on positions along the horizontal lines making up the interpolation grid. This may be accomplished digitally where the input function is a discrete set of uniformly spaced samples, whereas the output samples are computed at a lower sampling rate and with some specified delay with respect to the first sample within each radar pulse.

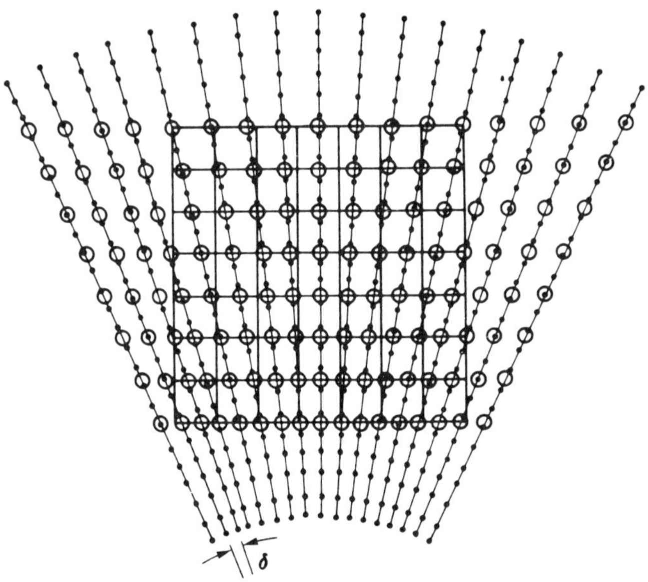

Figure 8.39 Polar-format range interpolation.

The second stage of the polar-format interpolation is depicted in Figure 8.40. The azimuth interpolation operates on the samples produced from the range interpolation, only in an orthogonal direction. This may be accomplished digitally with output samples computed at appropriate times for each row of data. However, the input samples to this process cannot be considered equally spaced and digital processing techniques that take this into account must be used.

After the two-dimensional interpolation to form the rectilinear signal grid, a complex-valued image is formed by using the two-dimensional FFT algorithm. Detection of this array produces the ISAR image.

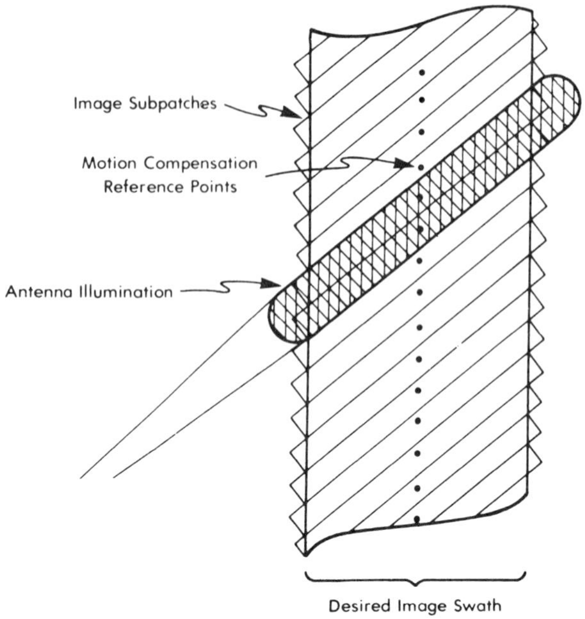

Figure 8.40 Polar-format azimuth interpolation.

8.5 MTI CONSIDERATIONS FOR LOW-OBSERVABLE TARGETS

The reduction of the radar cross section of a vehicle to escape detection falls into a general class of techniques called *low observables*. An aircraft is considered to be a low observable if its characteristics detectable by a radar can be reduced tenfold. For example, reducing the RCS of a fighter-bomber from 10 m^2 to 1 m^2 decreases its free-space detection range by a factor of about 2, which correspondingly reduces the range within which an opposing air defense system has only half as much time to engage the low-flying, fast-moving aircraft [1].

Low-observable technology, carried to its ultimate limit, is called *very low observable* or *stealth technology*. Stealth technology is defined as making an aircraft virtually impossible to detect and intercept. Stealth technology has apparently been applied to a number of aircraft and other vehicles and must be considered by designers of modern radars for military application.

The question as to what constitutes a stealth target is difficult to answer and will not be pursued here. Rather, we will consider several fundamental principles that apply to detection of radar targets with low-RCS signatures. Also, we will discuss RCS reduction techniques and radar techniques that may provide some capability against stealthy targets.

In the free-space radar detection equation that determines the clutter-free range at which a target can be detected, the RCS(σ_t) of the target has the same effect on the detection range as the transmitter power pulse length and antenna gains; that is, $R = k\sigma_t^{1/4}$. Generally, the RCS of the target is a complex quantity that depends on the target size, shape, and radar properties such as (1) the radar frequency, (2) the aspect angle at which the radar views the target, and (3) the radar polarization. For bistatic radar, the RCS is a function of the viewing angles and polarizations of both the transmitter and the receiver.

The RCS of simple geometric shapes can be predicted with reasonable accuracy and extensive lists of formulas are available for these cases [22]. Of this class of target, the perfectly conducting spherical target stands out because its RCS is independent of aspect angle and when its radius (a) is much greater than the incident radiation's wavelength ($a/\lambda \gg 1$), the RCS is only a function of the sphere's surface area given by $\sigma = \pi a^2$. However, the cross sections of most complex targets do not follow simple rules and must be measured to determine their RCS.

Figure 8.41 depicts the RCS of a scaled Boeing 737 aircraft [22]. The RCS patterns of Figure 8.41 were measured at X-band with a 1/15 scale model of the 737, and hence the RCS values are much lower than for the actual aircraft (i.e., they have not been adjusted to predict values observable on the actual aircraft at $\lambda = 0.45$ m). The patterns in the nose-on region are dominated by the re-entrant returns from the jet engine intake ducts. Specular flashes occur within the region 15° on either side of the nose due to the leading edge of the wings. A narrow specular flash occurs broadside to the aircraft whose magnitude provides the maximum aircraft RCS. A pair of broad intense lobes occur in the region 95° to 105° from the nose,

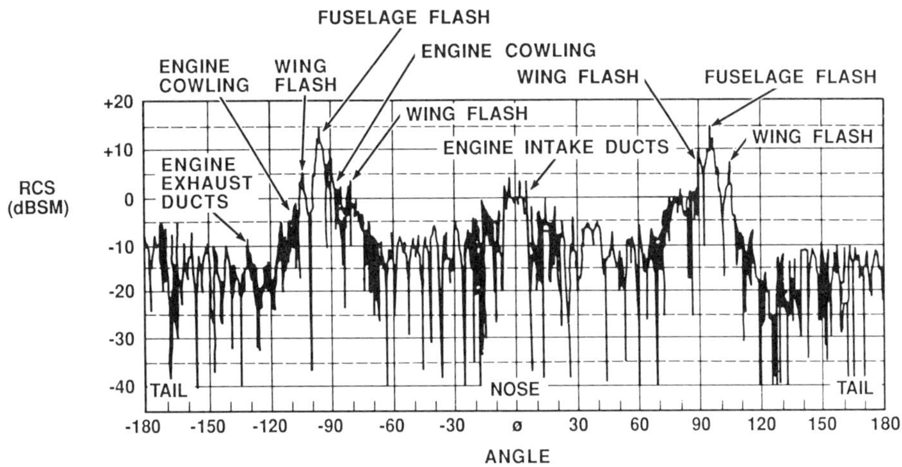

Figure 8.41 RCS of a scaled Boeing 737 aircraft measured at 10 GHz.

which are probably due to the engine cowlings. The lowest RCS values are in the tail region 120° to 150° from the nose. To determine the detection characteristics of such a complex target, we customarily assign a mean (also the median is sometimes used) RCS value and a probability distribution (e.g., Rayleigh) to describe the target. The radar's detection range is then determined for the mean RCS and averaged over the probability distribution to determine the overall detection probability of the radar. Extensive curves of Swerling type targets are available for this purpose.

Given the complexity of practical radar targets, it is interesting to explore what can be done to reduce their RCS. In general, there are only four ways to reduce the radar echo of an object:

1. shaping;
2. radar absorbing materials;
3. passive cancellation;
4. active cancellation.

Each involves a different methodology and exploits different aspects of the overall problem. In applying these techniques, remember that the radar energy captured by the target must be either scattered, absorbed, or cancelled. If it is scattered, the strategy is to direct it away from the radar receivers, whether monostatic or bistatic. Cancellation is practical only if there are a few primary scatterers with characteristics that are well known over the frequency range of interest.

Of the four methods for RCS reduction, shaping is the most useful. Shaping presumes that the radar threat will appear within a definable and limited cone of directions. Radar absorbing materials are typically used to complement shaping techniques or used when shaping is not a useful or implementable option. Deep reduction in target RCS is possible by the use of passive cancellation, but only over narrow aspect angle sectors and for narrow bandwidths. Active cancellation depends on the addition of active signal sources on the target and requires careful control of the reradiated energy. Of the four, only shaping and *radar absorbing materials* (RAM) have apparently been of any tactical use.

The objective of shaping is to orient target surfaces so that they reflect an incident wave in directions other than back at the radar receivers. This principle is difficult to apply to an existing vehicle, such as the scaled 737 aircraft with its RCS depicted in Figure 8.41. However, the result in reducing the average RCS would be apparent in the figure if the large specular returns could be eliminated from the RCS pattern. In new aircraft design, the shaping method has apparently been applied to both the F-117A and B-2 aircraft [23]. The F-117A planform consists of 18 straight-line sections arranged at four main orientations: one angle is used for the leading edges of the wing and tail; another for their trailing edges; wing and tail tips are aligned roughly fore and aft, as are the fuselage sides immediately behind the wing; and the rear section of the wingtip and the aft fuselage are at a common angle [23]. Edges tend to be strong radar reflectors. A radar signal hitting parts of the aircraft such as the wing's leading edge behaves like a light beam hitting a reflective surface. If the radar beam arrives at an angle, it departs at a similar angle in a direction away from the radiating source. A planform with only four angles minimizes the number of directions in which the radar signals are reflected. The planform for the B-2 is apparently even simpler, consisting of 12 straight-line sections, each of which is angled in one of four directions [23].

RCS reduction through shaping is effective only if we are willing to accept higher echo return signals over some aspect angles to gain lower echo signals over critical threat sectors. This is because reorientation of the surfaces of the radar target for favorable reductions over the specified sector is inevitably accompanied by an increase in RCS over other sectors.

As the name implies, *radar absorbers* depend on the absorption of incident energy, thereby reducing the amount reflected back to the radar. Although commercial RAMs of several different varieties have been available for decades (e.g., for anechoic chambers), they are not necessarily suitable for tactical applications. RAM that is useful for tactical applications makes use of the energy exchange properties of carbon and certain magnetic iron compounds. When illuminated by an RF radar signal, the molecular structure of these materials is excited and the excitation converts the RF energy into heat. In this way, the material absorbs a portion of the incident signal, thereby reducing the signal available for reflection back to the radar [24].

Although the principle involved with RAM is relatively simple, its implementation is more complex, especially as applied to high-performance aircraft, where weight and the ability to withstand the stresses of the flight envelope are at least as important as the RAM's electromagnetic capabilities. In general, four types of RAM are used in practice: broadband absorbers; narrowband or resonant absorbers; hybrids, which combine features of the two; and surface paints [24].

Broadband absorbers offer a consistent performance over a range of frequencies and are most useful for aircraft where the need is for a RAM that is integral with airframe components. This might take the form of a honeycombed sandwich with absorber impregnated foam applied to the individual cell walls in such a way as to present logarithmically graded resistance to the incoming signal. Another option for the sandwich filling might be absorbent foam with a pyramidal configuration. For example, a honeycomb arrangement using fiberglass for the skins and comb and carbon black-silver powder as the absorber is estimated to achieve 95% absorption against signals in the 2.5–13 GHz frequency range with a material thickness of 2.5 cm and a weight of about 310 grams per 930 cm^2 of surface area [24]. This material would be suitable for subsonic vehicles, but Kevlar-based materials are generally used in modern RAM applications.

The second major type of RAM is the narrowband, resonant absorber in which an elastomer such as rubber is loaded with absorber, backed by a metal reflector, and applied parasitically to the surface needing protection in sheet or molded form.

Designed for specific threat frequencies, such material can be tuned for peak performance over a narrow range of frequencies by varying the thickness of the sheet, which would ideally be equal to one-quarter of the incoming signal's wavelength. The resonant absorber works by using both absorption and quasioptical techniques that generate equal-amplitude reflections to cancel out the natural reflectivity of the material. Absorbers of this type may be targeted against a broader range of threats by using several layers of material, with each layer tuned to be effective against a particular frequency [24].

Although narrowband absorbers offer the advantage of a relatively small weight penalty, their performance is degraded if the threat signal does not strike them at exactly 90°. The ability to cope with other angles of incidence appears to be the major impetus behind the development of hybrid absorbers that combine features of both narrowband and broadband materials to produce a performance that, although varying with the angle of incidence, betters their individual performances over a spectrum of situations.

Another RAM alternative is radar-absorbent paint that forms a surface finish loaded with an absorberlike iron ferrite. Iron Ball paint of this type is believed to be particularly effective in the suppression of re-emissions generated by surface-current effects. When a structure is illuminated by a radar, electromagnetic currents are generated on any of its surfaces that are conductive. This, in turn, leads to a retransmission effect where the skin acts as an antenna. Even rivit heads if not properly suppressed, materially contribute to a vehicle's RCS. More serious is the surface current effect on the radar's visibility of such areas as wing and tail leading and trailing edges together with any external antennas that it may carry.

RAM can be selectively applied to an aircraft to substantially reduce its RCS, particularly in areas that produce high specular returns. Locations where RAM is effective in reducing the RCS of an aircraft are depicted in Figure 8.42. Note that many of the suggested RAM locations in Figure 8.42 correspond to those elements identified in Figure 8.41 as producing large RCS returns at particular aspect angles.

Antennas can contribute significantly to a vehicle's RCS. In general, the RCS of an antenna consists of two components. One, due to reflections from the antenna and antenna support structure, is termed the *structural mode response*. The second component, due to reradiation of energy captured by the antenna, is termed the *radiation mode response*. Because the illuminating radar transmitter and receiver are normally at the same point, the RCS in the direction of the illuminating radar is usually of most concern. Also, the out-of-band response often can be more troublesome than the in-band response.

The RCS of the radiation mode can be estimated as $\text{RCS} = \rho G\lambda^2/4\pi$, where ρ is a term that depends on the reflection coefficient and efficiency of an antenna with an effective capture area of $A = G\lambda^2/4\pi$. The RCS is highly dependent on polarization. Some antennas such as the dipole basically present no backscatter response for an incident orthogonal polarization. Others like the horn or reflector antenna can

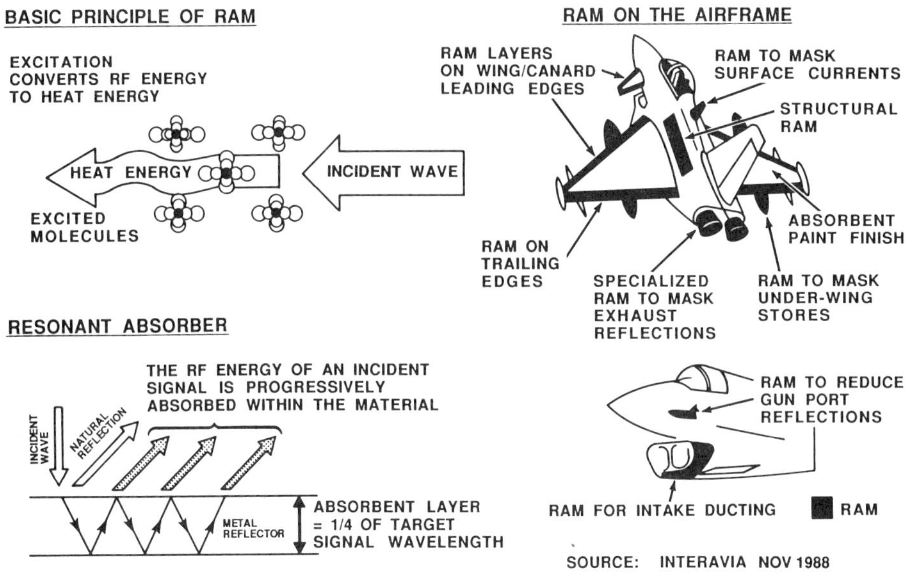

Figure 8.42 RAM application to airframe.

look totally reflective for an orthogonal incident polarization. In general, the RCS of an antenna can be quite large when the gain and reflection coefficient of the antenna are large, as would be the case for a high-gain antenna with poor impedance match.

An antenna with interesting RCS characteristics is the phased array. The ability of the antenna to scan electronically allows orientation of the antenna aperture plane at a significant angle (e.g., on the order of 20° to 30°) with respect to incoming radiation. Thus, the radiation mode component of the antenna's RCS is minimized, except when the antenna is electronically scanned in the direction of the target. A similar concept has been described, whereby the phased array antenna is generally aligned in the direction of the incoming radiation, but nonreciprocal phase shifters are used. This causes the transmitting and receiving antenna patterns to point in different directions, except at boresight (perpendicular to the antenna's aperture). Thus, in general, the radiation mode RCS is that associated with the antenna sidelobe pattern rather than the mainbeam pattern. In a 10–20 GHz phased array with a 12-in^2 aperture, the RCS was reduced from 0.15–0.60 m^2 to 0.004–0.015 m^2 by using this technique.

A range of mean RCS for typical conventional radar targets are depicted in Figure 8.43 [22]. Also shown is the volume of a perfectly conducting reference sphere that exhibits the indicated RCS. The sphere is intended to show that the RCS of an object is roughly proportional to the 2/3 power of the object's volume.

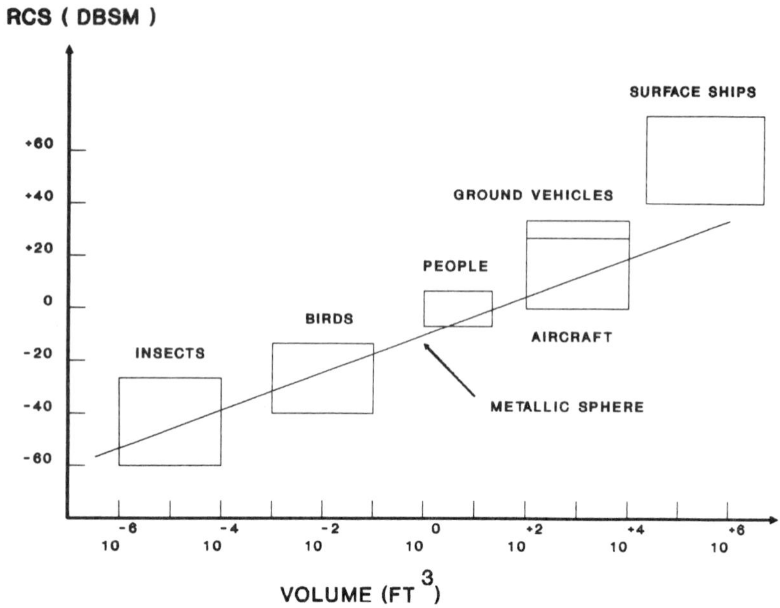

Figure 8.43 Radar cross section of conventional targets.

The basic thrust of current low-observable technology is to cause a mismatch between the radar, which was designed to detect conventional targets, and the target, which is now much smaller in RCS than the radar was designed to handle. The variation of free-space detection range $R = k\sigma^{1/4}$ with RCS favors the radar as a sixteenfold reduction in RCS reduces the detection range by only a factor of 2. This indicates that the RCS of a high-altitude stealth vehicle that attempts to escape detection must be very small if it is to be effective. In addition, the energy radiated by the radar can be increased (e.g., using pulse compression to maintain the radar's resolution) to compensate for the reduced RCS and bring the radar–target detection equation into balance.

One problem with increasing radar energy to detect small RCS targets is that clutter returns are increased in the same proportion as the radar's power (assuming that the radar's resolution is maintained). This may not only present a problem in rejecting the increased clutter return (i.e., increased subclutter visibility), but small targets (e.g., insects, birds, *et cetera*) are now detected because they exhibit RCSs in the same range as the stealth targets (e.g., -20 dB/m^2 to -30 dB/m^2; see Figure 8.43).

This problem is illustrated by an experiment run to determine the effect of clutter on the detection of low-RCS targets [25]. A sensitive S-band surveillance radar (P_p = 500 kW, τ = 0.8 μs, PRF 1040 pulses/s) using a double-canceller MTI with 35 dB improvement factor was set up in a heavy ground clutter environment that also included bird and insect targets. Figure 8.44(a) depicts the normal video ground clutter return that extends out to about 30 nmi. Figure 8.44(b) depicts the same return and MTI processing with an estimated sensitivity to detect targets with 0.1 m^2 cross section in the MTI passband. In Figure 8.44(c), the estimated sensitivity was extended further (by reducing STC and increasing integration gain) so that 0.004 m^2 targets could be detected in the MTI passband. The extensive clutter residuals and angel targets appearing on the display indicates the difficulty in detecting low-RCS targets on the order of -20 dB/m^2 to -30 dB/m^2 in heavy clutter.

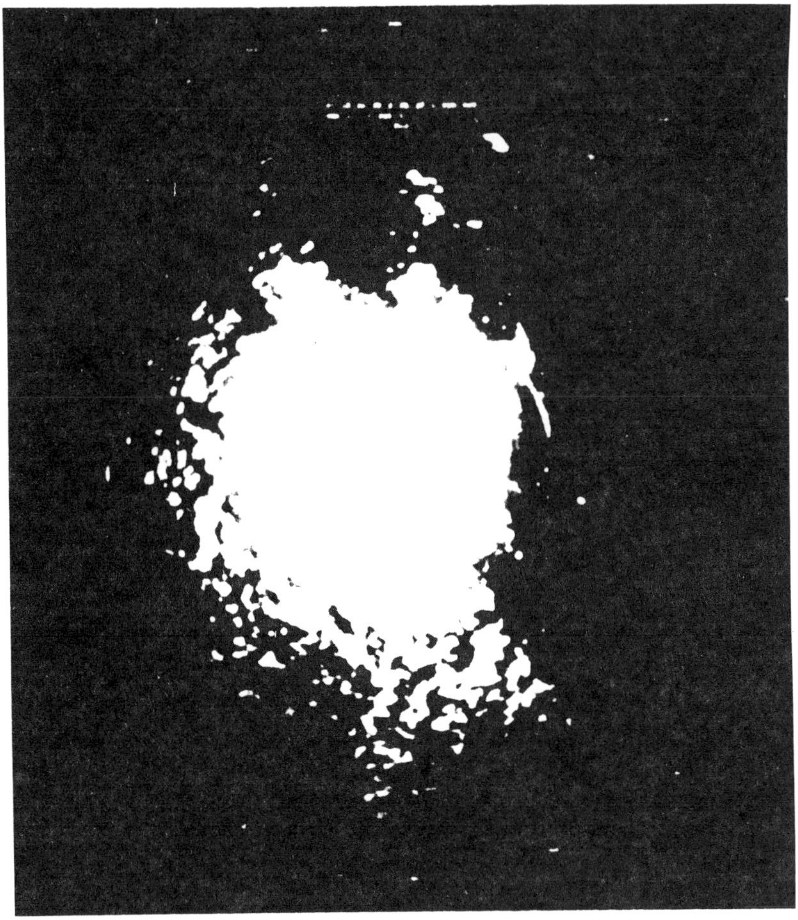

Figure 8.44 Clutter returns for MTI surveillance radar with low-RCS targets.

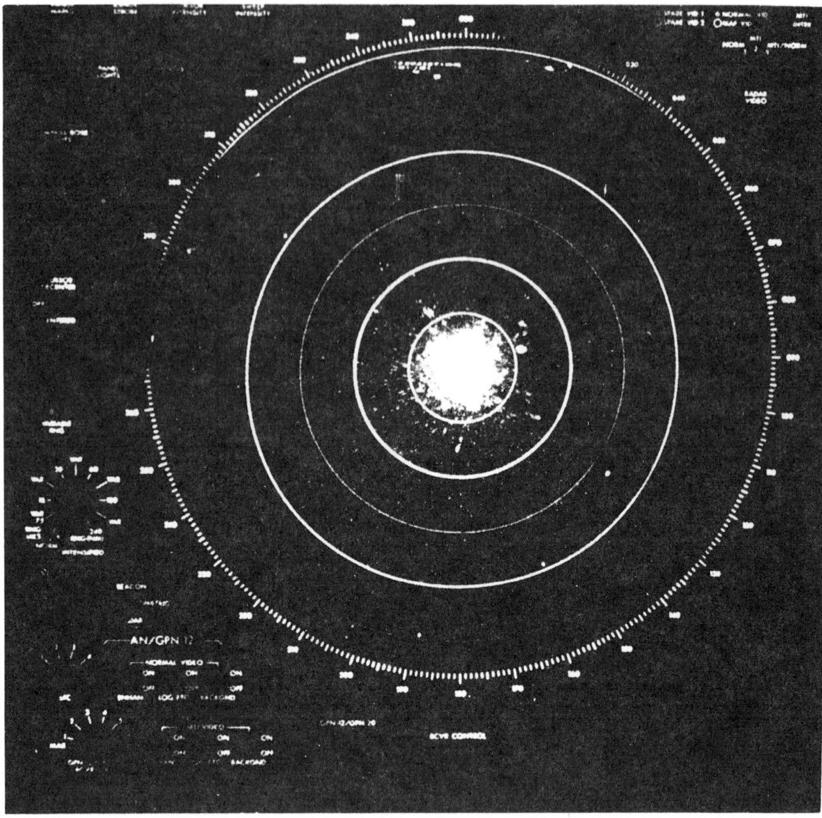

Figure 8.44 continued.

The performance of the MTI system depicted in Figure 8.44 illustrates a critical issue encountered in the detection of low-flying, low-observable targets. Rejection of birds, insects, and other slow-moving clutter sources requires that the MTI's blind speed be high enough to create a clutter notch in the velocity region occupied by such angel clutter. For example, the S-band MTI used in the GPN-21 has a first blind speed on the order of 100 knots. If a 30-knot rejection notch were used, then the doppler visibility (i.e., ratio of doppler region not occupied by clutter to total interval) would be 0.4 (see Section 7.2.1 and Figure 2.6). However, use of a double-canceller MTI filter in the GPN-21 rather than a sharp cut-off filter further compounds the problem. The double-canceller response is equal to $|H(\omega)|^2 = 16$ $\sin^4(\omega T/2)$ and the MTI gain (response to uniformly distributed velocity target) $\bar{G}$ $= 6$. Thus, at 30 knots (the 3 dB point), the response for a uniformly distributed velocity target (possibly aliased from a higher velocity region) is 6/8 or -1.25 dB

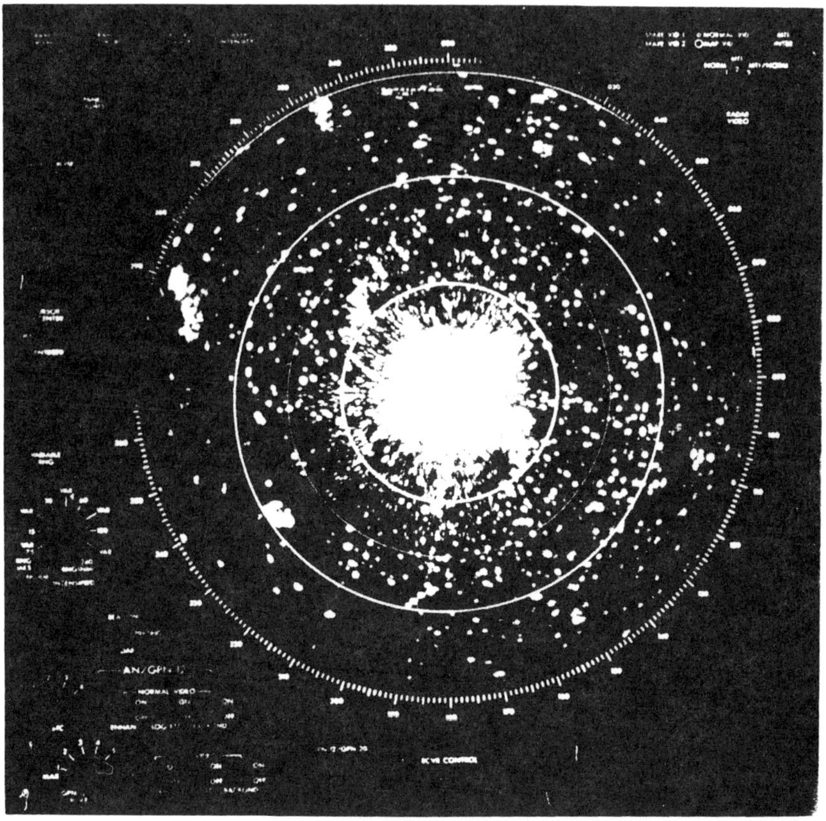

Figure 8.44 continued.

reduced from that of a low-RCS clutter target moving at this radial velocity. This points to the desirability of using a sharp cut-off MTI filter to reject birds, insects, and other slow-moving clutter sources and explains the large amount of clutter targets visible in Figure 8.44(c). In addition, the use of a lower transmitter frequency (i.e., L-band or UHF) further reduces the response to birds and insects.

Next, consider the same MTI design operating at L-band (f = 1.3 GHz). The first blind speed is then 230 knots and the doppler visibility is increased to 0.74. The 3 dB response is increased to 69 knots, and the response at 30 knots for angel clutter is reduced to −8.2 dB with respect to the response for a uniformly distributed velocity target. A further decrease in the transmitter frequency to UHF would increase the performance against birds, insects, and other slowly moving clutter sources. When moving clutter sources are removed, system stability becomes the critical factor in determining the performance of an MTI to detect low-flying, low-observable targets.

The quality factor that determines the ability of an MTI radar to detect a low-RCS target in a clutter background is its subclutter visibility or equivalent MTI improvement factor (see Sections 3.1 and 3.3). The subclutter visibility of a radar is ultimately limited by the frequency and phase stabilities of its reference master oscillator and transmitter (see Figure 3.41). Hence, the detection of low-RCS targets in a cluttered background generally depends on how stable the radar can be made.

As a further example, the free-space detection range and subclutter visibility requirements for an ASR-9 radar are tabulated as a function of the target's RCS in Table 8.4. From Table 8.4, we can see that useful free-space detection ranges are provided on targets as small as on the order of -30 dB/m^2. Note that the ASR-9 is only modestly powered compared to many military type radars of the same class.

The MTI in the ASR-9 provides on the order of 45–50 dB subclutter visibility. This would restrict the capability of the ASR-9 to the detection of low-observable targets immersed in ground clutter (low-level penetrators) on the order of -20 dB/m^2 RCS. The ultimate capability of an MTI radar to detect targets immersed in clutter depends on the stability of the transmitter, receiver, and signal processing elements. In the ASR-9, this limitation is on the order of 60 dB, which is considered representative of the current technology.

The detection of low-observability cross-section targets must also consider that the RCS of birds is on the order of -20 to -40 dB/m^2 at S-band and insects on the order of -40 to -70 dB/m^2. Also, weather clutter at this band is on the order of 0 to -20 dB/m^2. Elimination of these spurious targets becomes more difficult as the RCS of the desired targets becomes comparable, requiring judicious design of the MTI's velocity response and selection of an appropriate transmitter frequency.

Table 8.4
Detection Range and Subclutter Visibility Requirements against Low-Observability Targets

$\sigma(m^2)$	*Free-Space Range (nm)*	*SCV (dB)*
1	62.6	32.7
0.1	35.2	42.7
0.01	19.8	52.7
0.001†	11.1	62.7
0.0001†*	6.2	72.7

Note: ASR-9 parameters
†RCS of birds
*RCS of insects: $\sigma^0 = -30$ dB

8.6 DUAL-BEAM ANTENNA [7]

The detection performance of two-dimensional surface-based air surveillance radars frequently is degraded by the presence of ground clutter and extraneous targets such as birds and automobiles, which is generally due to the use of a fan-shaped antenna beam with its maximum gain directed slightly above the horizon (see Figure 8.45) to detect targets at the radar's maximum range. As a result, ground clutter and extraneous targets at low elevation angles are illuminated with the full antenna gain, whereas short-range targets at high elevation angles are illuminated with reduced antenna gain, thereby diminishing the target-to-clutter ratio.

The solution to this problem is to use a dual-beam antenna system, which provides both a low and high antenna beam. The radar transmits on the lower beam, but receives on both beams. The lower beam is activated when receiving long-range

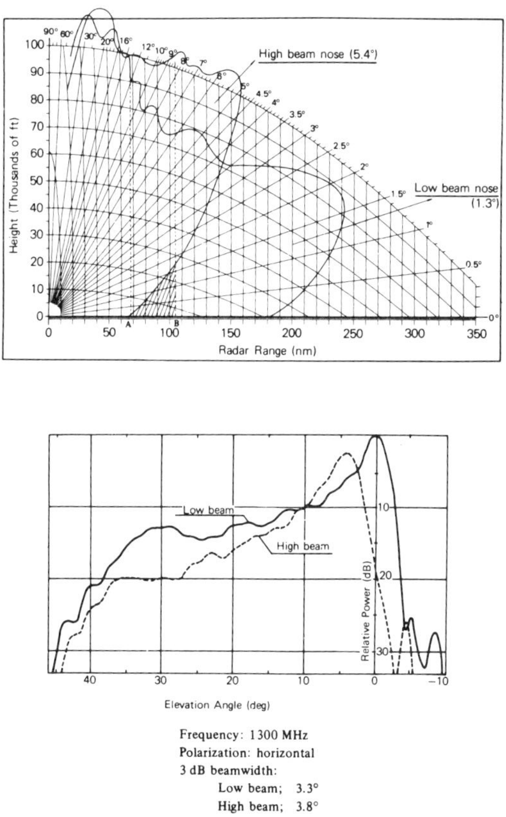

Figure 8.45 Elevation patterns of a dual-beam air-search radar.

targets, and the higher beam is used to detect short-range targets. The net effect is that ground clutter and surface targets are decoupled in the higher receiving beam, thereby enhancing the target-to-clutter ratio. In addition, low-altitude targets such as birds and insects, which are problems at short ranges, are also attenuated in the higher beam.

A dual-beam antenna system pattern for an L-band air-search surveillance radar is depicted in Figure 8.45. The nose of the lower beam is directed close to the horizon (at 1.3° elevation angle) to maximize the probability of detecting small targets at long ranges, which occur at relatively small elevation angles (e.g., a 40,000-ft high aircraft at 210 nmi has a 0.6° elevation angle). The low-beam antenna pattern is shaped so that a target at a 45° elevation angle (e.g., a 6.6-nmi range at a 40,000-ft altitude) has a response attenuated on the order of 30 dB from the peak antenna response. However, the surface clutter (or ground target) return, which occurs in the same range gate as the high-altitude target, is still approximately within the peak response of the low-beam antenna, thereby obscuring the attenuated target by the large surface clutter return. The high-beam antenna response is adjusted to respond to higher elevation targets (which can occur only at short ranges) while attenuating the low-elevation surface-clutter returns, as shown in Figure 8.45. In operation, the radar always transmits on the low beam, switches to the high beam when looking for short range targets that occur at higher elevation angles, and receives on the low beam when looking for long-range targets (i.e., greater than 100 nmi in Figure 8.45). In the example shown in Figure 8.45, small-target visibility over the strongest expected surface clutter is limited in the cross-hatched region (A to B), thereby indicating that a further upward tilting of the high beam in elevation to achieve greater coverage will severely restrict the overall radar detection performance in clutter.

The improvement in received target-to-clutter ratio for typical clutter conditions (i.e., ground, sea, and angel clutter) with a dual-beam antenna as compared to a single-beam system is depicted in Figure 8.46 [1]. For ground clutter (open farmland), an improvement of 15–20 dB is obtained. Target detection is then possible by using an MTI processor (subclutter visibility, SCV $>$ 30 dB) in almost all coverage areas except for short-range ($R < 5$ nmi) high-altitude targets. The improvement for sea clutter (sea state 4) is similar to that for land clutter (i.e., 15–20 dB), except that the target-to-clutter ratios are higher due to the reduced intensity of sea clutter as compared to ground clutter. As with ground clutter, target detection is possible by using a motion compensated MTI processor for all but short-range high-altitude targets. For angel clutter (i.e., primarily birds), an improvement of about 15 dB is obtained. This generally results in over 15 dB of target-to-clutter ratio (except for short-range high-altitude targets), which allows a sensitivity-time-control function to be applied, thereby providing target detectability. A fourth-power attenuation law as a function of range is recommended for the STC to eliminate birds ($\sigma_t = 0.01\ \text{m}^2$) while retaining aircraft ($\sigma_t = 1 - 10\ \text{m}^2$) targets [8].

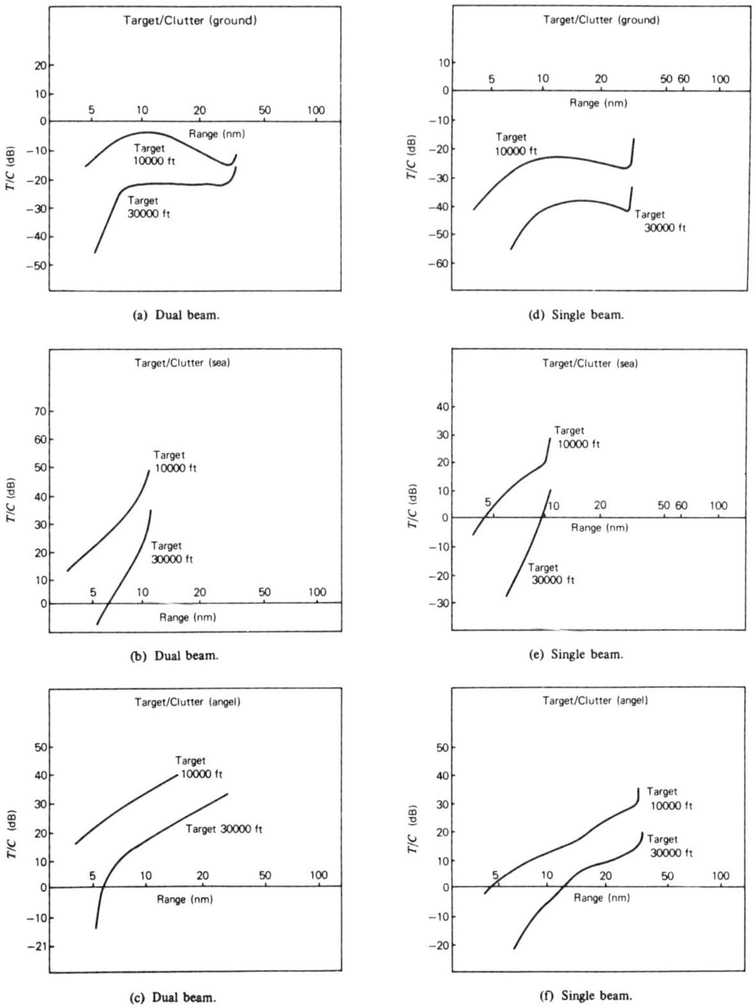

Figure 8.46 Typical received target-to-clutter ratio for a dual-beam antenna system.

8.7 PULSE COMPRESSION WITH MTI

Pulse compression is a natural adjunct to low-PRF and some medium-PRF PD systems. In these systems, pulse compression reduces the effective clutter input to the doppler processor (e.g., MTI) by reducing the range dimension of the clutter resolution cell. With pulse compression, the clutter radar cross section for pulsewidth-limited surface clutter is given by

$$\sigma_c = \sigma^0 R \theta_{\text{az}} \frac{c\tau_e}{2} \sec\psi \tag{8.55}$$

where σ° (m^2/m^2) is the clutter backscatterer coefficient, R is the range to the clutter patch, θ_{az} is the azimuth beamwidth, ψ is the grazing angle, $\tau_e = 1/B$, and B is the pulse compression bandwidth. Alternatively, the clutter RCS can be expressed as

$$\sigma_c = \sigma_{cu}/\beta \tag{8.56}$$

where σ_{cu} (see (4.2) and (4.6)) is the clutter RCS for the uncompressed pulse (τ) and β is the pulse compression factor given by $\beta = B \cdot \tau$. This version of the equation illustrates that the target-to-clutter power ratio into the doppler processor is reduced by the pulse compression factor (β). The same reduction also applies for distributed volume clutter with pulse compression, where the clutter RCS is given by

$$\sigma_c = \eta \frac{\pi}{8} \theta_{\text{az}} \phi_{\text{el}} R^2 \frac{c\tau_e}{2} \tag{8.57}$$

and the parameters are as defined in (8.55), except for η, which is the volume backscattering coefficient (m^2/m^3), and ϕ_{el} = the elevation beamwidth.

The use of pulse compression to reduce the distributed clutter intake into the doppler processor decreases both the subclutter visibility and the improvement factor necessary to provide a specified level of detection performance (see Section 3.1). For example, a radar that required subclutter visibility of SCV = 40 dB against distributed surface clutter and used a 10 μs pulsewidth would require subclutter visibility of only SCV = 20 dB if pulse compression were to provide an effective pulsewidth of 0.1 μs (i.e., β = 100 if the uncompressed pulsewidth were maintained at 10 μs). The penalty for the increased performance would be the increased complexity of the doppler signal processor, which should process 100 times the range samples in the pulse compression case. Also, the requirement for pulse timing and pulsewidth jitter is increased by the pulse compression factor (β) to provide equal cancellation ratios (see Section 3.1.2).

A question that arises in the use of pulse compression MTI systems is whether the pulse compression network ought to preceed or follow the MTI. Because both are linear systems, in theory it should make no difference in which order they were applied. However, practical hardware considerations would often dictate a preferred order. Most modern radars implement the MTI function in a digital signal processor, whereas the pulse compression function can be implemented in either digital or analog form. A major concern in the design of a digital MTI is the requirement for a wide dynamic range. If the A/D converter in the digital MTI can handle the largest signal present without saturating, linear operation can be achieved, thereby providing maximum performance (see Sections 3.2.5 and 6.2.3).

When an analog pulse compression network is used, a natural place is in front of the digital MTI as the signal to be compressed is in analog form. The dynamic range of distributed clutter is maintained through the pulse compression network because the distributed clutter process and receiver noise have similar statistical properties. However, the dynamic range of discrete clutter and point targets is increased by the pulse compression ratio ($\beta = B \cdot \tau$). Therefore, for the dynamic range of the signal into the digital MTI to be increased by 20 dB or more (for pulse compression ratios of 100 or more) is not unusual when the pulse compression network precedes the MTI. However, if the MTI precedes the pulse compression network, echo signals from targets and discrete clutter are dispersed in time (and reduced in amplitude); hence, it may not be needed to handle signals in excess of the average clutter level for time-bandwidth products of 100 or more. The dynamic range can be reduced further with STC in low-PRF MTI systems [26].

As another example, consider a system wherein both pulse compression and MTI functions are implemented digitally (see Figure 2.14). A method that accomplishes digital pulse compression by using the FFT is depicted in Figure 8.47. A digitized version of the input signal at baseband (I and Q channels) is convolved with a digitized reference function that corresponds to the impulse response of the matched filter to the pulse compression signal. This is accomplished in the frequency domain by first

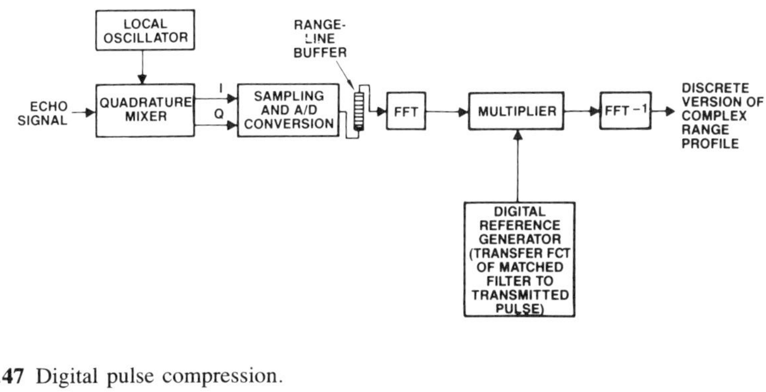

Figure 8.47 Digital pulse compression.

performing an FFT on the input signal, $S(\omega)$, which is then multiplied by the matched filter transfer function, $H(\omega)$, thereby providing the pulse compression output frequency spectrum $G(\omega) = S(\omega)H(\omega)$. This is hence converted into the time domain by performing an inverse FFT on the output spectrum, resulting in $g_t \leftrightarrow G(\omega)$. In this fully digital version, it is advantageous to place the MTI function ahead of the pulse compression function as the removal of clutter components by the MTI reduces the number of bits that must be carried in the complex multiplication of the convolution process (i.e., FFT, multiplication, inverse FFT).

Note that pulse compression generally improves the target-to-clutter ratio only when distributed clutter is involved. For discrete clutter (see Chapter 4), which is prevalent over developed land, pulse compression provides no improvement in target-to-clutter ratio within the range cell in which the discrete clutter is contained. However, pulse compression isolates the discrete clutter into small cells between which targets can be detected over the generally weaker distributed clutter. This property of a radar, termed its *interclutter visibility,* provides a measure of its capability to detect targets between points of strong clutter by virtue of its ability to resolve strong and weak clutter areas (see Figure 4.1) [8].

When the clutter's dynamic range exceeds the clutter attenuation capability of the MTI, we sometimes must employ limiting to reduce clutter residuals to the receiver noise level (see Section 3.2.5). Figure 8.48 depicts a configuration consisting of a pulse compression network followed by an MTI with a linear-limiting amplifier that compresses the clutter's dynamic range [8]. When this is ideally accomplished (see Figure 8.48(a)), the compressed pulse appears as a short pulse and range sidelobes are stable (i.e., they are precisely repeated on a pulse-to-pulse basis). The MTI system then performs as if the pulse compression network were not present and a short pulse

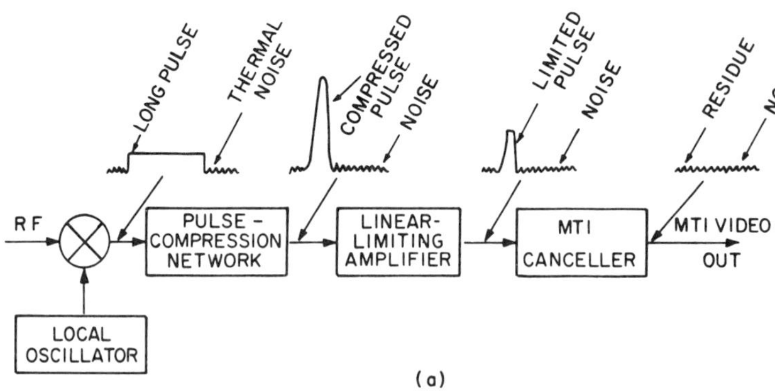

Figure 8.48 Pulse compression with a limiting MTI.

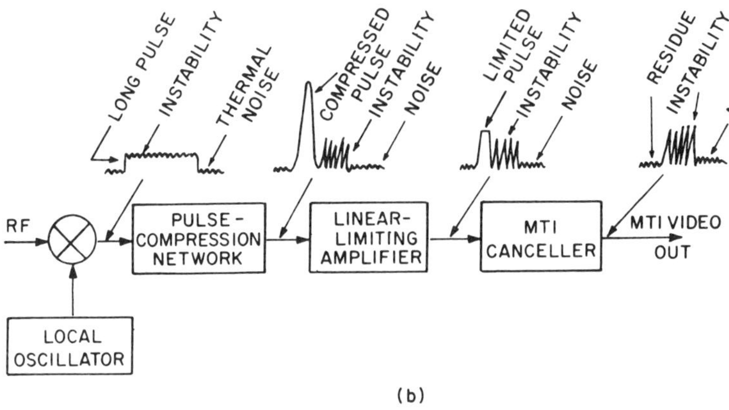

Figure 8.48 continued.

was transmitted. However, oscillator and transmitter instabilities coupled with timing jitter generate random noiselike range sidelobes in the pulse compression network (see Figure 8.46(b)). These sidelobes are not cancelled in the MTI system and are accentuated in the output because they fall beneath the limit level, whereas targets are generally limited, thereby reducing their peak amplitude. For example, assume that the noiselike component of the sidelobes is −40 dB below the peak transmitted signals. This noiselike component will not cancel in the MTI system and, hence, any clutter component that exceeds system noise by 40 dB will produce a residue on the display. If the clutter exceeds the threshold by 60 dB, the residue from the MTI system will exceed the display threshold by 20 dB, eliminating the effectiveness of the MTI.

One cause of noiselike range sidelobes in a pulse compression system is phase modulation of the transmitter or the various reference oscillators located in the receiver. Phase modulation is considered more probable than amplitude modulation in the transmitter because most high-powered transmitters used in MTI or PD radars are operated in a saturated mode, which tends to limit AM effects. To consider this effect, the phase distorted linear FM pulse compression signal is represented as

$$v_t = S_t \, e^{j(\omega_o t + \alpha t^2/2 + \Delta\phi \, \sin\omega_m t)} \tag{8.58}$$

where S_t is the envelope of the pulse compression signal with support $|t| < \tau/2$, ω_o is the radian carrier frequency, α is the chirp slope, $\Delta\phi$ is the peak phase error in rad, τ is the uncompressed pulsewidth, and ω_m is the random frequency of the phase error. The response to this signal by the pulse compression network using the "paired echo" technique is given by

$$g_t = J_0(\Delta\phi)\bar{S}_t \cos\omega_o t$$
$$+ \sum_{n=1}^{\infty} k_n J_n(\Delta\phi)\bar{S}\left(t + \frac{n\omega_m \tau}{\Delta\omega}\right) \cos\left(\omega_o + \frac{n\omega_m}{2}\right)t$$
$$+ (-1)^n \bar{S}\left(t - \frac{n\omega_m \tau}{\Delta\omega}\right) \cos\left(\omega_o - \frac{n\omega_m}{2}\right)t \qquad (8.59)$$

where $\bar{S}_t$ is the compressed pulse envelope function, $\Delta\omega$ is the chirp radian bandwidth, k_n is the weighting factor of the passband frequency response, and $J_n(\cdot)$, $n = 1, 2, \ldots$ are variously ordered Bessel functions [27]. The first term in (8.59) is the desired undistorted pulse compression signal and the other terms represent the distortion terms, consisting of various range sidelobes dispersed in time and frequency. Also apparent is that, if randomly distributed phase-distortion-modulation components are applied, randomly spaced noiselike range sidelobes will result, as depicted in Figure 8.48.

The preceding discussion points to a significant difference between an MTI system that uses pulse compression and one that uses an equivalent short pulse. The pulse compression system has the potential for generating noiselike range sidelobes proportional to the peak clutter returns that may restrict the detection of small moving targets. The approximate magnitude of the noiselike range sidelobes can be deduced from (8.59) by using the small argument approximating $(\Delta\phi < 0.4) J_0(\Delta\phi) = 1$, $J_1(\Delta\phi) = \Delta\phi/2$, and $J_n(\Delta\phi) = 0$ for $n \geq 2$, and the approximation that for small phase deviations the components will add, which results in [27]

$$g_t = \bar{S}_t \cos\omega_o t$$
$$+ \sum_{i=1}^{N} \frac{\Delta\phi}{2} \bar{S}\left(t + \frac{\omega_i \tau}{\Delta\omega}\right) \cos\left(\omega_o + \frac{\omega_i}{2}\right)t$$
$$- \sum_{i=1}^{N} \frac{\Delta\phi}{2} \bar{S}\left(t - \frac{\omega_i \tau}{\Delta\omega}\right) \cos\left(\omega_o - \frac{\omega_i}{2}\right)t \qquad (8.60)$$

The level of each distortion term below the main signal level is depicted in Figure 8.49 as a function of the peak phase distortion $\Delta\phi$.

One solution employed when clutter limiting is required would limit the clutter's dynamic range into the pulse compression network as well as into the MTI [8]. This is depicted in Figure 8.50 for a system with a 30 dB pulse compression ratio, an instability noise level −28 dB below the carrier level, and an MTI system with 30

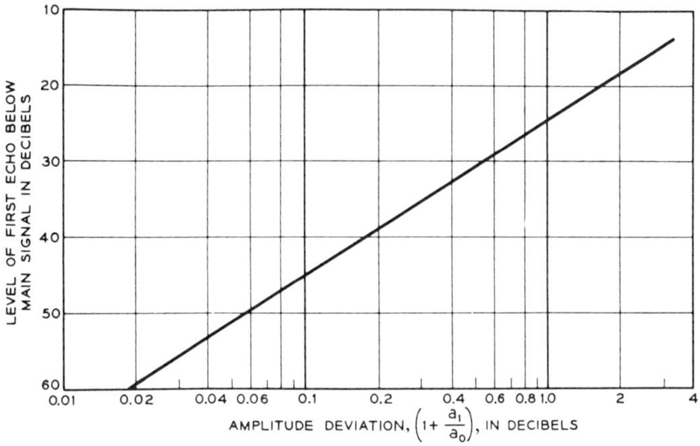

Figure 8.49 Relative level of first-order range sidelobes due to phase distortion.

dB improvement factor due to internal clutter motion. With these adjustments, the range sidelobes due to system instability will be at or below the noise level for either distributed or point (discrete) clutter. The peak signals will vary from 28 dB above the noise level for distributed clutter to 58 dB (i.e., due to the 30 dB increase in dynamic range from pulse compression) above the noise level for strong point clutter. The linear-limiting amplifier preceding the MTI is set to limit at 30 dB above the noise level (i.e., to the cancellation capability of the MTI) so that strong point clutter, which may be as high as 58 dB above the noise level, will not have a residue that exceeds the receiver noise level. If the instability noise were, for example, 43 dB below the carrier level, the first limiter would be set 43 dB above the noise level resulting in 15 dB less target suppression. Thus, target detectability would improve in and near strong clutter areas, although the MTI improvement was still limited to 30 dB by internal clutter motion.

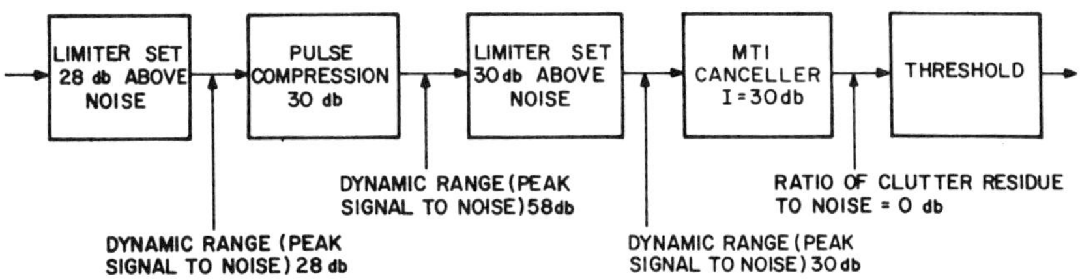

Figure 8.50 Example of limited pulse compression and MTI.

The pulse compression network is generally designed as a matched filter to the transmitted waveform. However, the use of MTI with pulse compression implies the presence of a moving target that will doppler shift the frequency of the received signal into the pulse compression network, thereby causing a mismatch. For example, pulse compression systems that use binary phase coding are generally sensitive to a doppler-shifted signal, causing distortion of it and an increase in range sidelobes. The doppler sensitivity is caused by the shifting in phase of the individual bits of the phase code by the target doppler so that, in the extreme, if the last bit is shifted by 180°, the code is no longer matched. A target doppler that causes this is termed the *critical velocity*. For example, with a 13-bit Barker code using 1 μs bits at X-band, the critical velocity is 1215 knots.

The basic cause of the pulse compression network's sensitivity to doppler-shifted signals is not being tuned to the correct frequency. To circumvent this problem, a bank of contiguous matched filters could be used to cover the range of expected doppler frequency shifts. However, this means excessive hardware, and a better solution would be to employ a transmitted waveform with a pulse compression network that exhibits a tolerance to doppler-shifted echoes with minimum degradation.

A pulse compression waveform that allows a single pulse compression filter to be matched for all doppler frequency shifts is given by

$$s(t) = P(t)\cos\left[\frac{2\pi f_o^2 T}{B}\ln\left(1 - \frac{B_t}{f_o T}\right)\right] \tag{8.61}$$

where $P(t)$ is a rectangular pulse of width T, f_o is the center frequency, T is the pulse duration, and B is the bandwidth [11,28]. This can be demonstrated by finding the spectrum of (8.61) and showing that no dispersive term results for $f = f_o \pm f_d$, where f_d is the doppler shift [28]. Expansion of the natural logarithmic factor in a series results in

$$s(t) = P(t)\cos\left(2\pi f_o t + \frac{\pi B t^2}{T}\frac{2\pi B^2 t^3}{3 f_o T^2} + \ldots\right) \tag{8.62}$$

By neglecting the third- and higher-order terms (i.e., assuming $2\pi B^2 t^3 \ll 3 f_o T^2$) (8.62) is reduced to the linear chirp pulse compression waveform given by $s(t) = P(t)\cos(2\pi f_o t + \pi B t^2/T)$. Thus, the linear FM chirp pulse compression waveform is a practical approximation to the theoretical pulse compression waveform that is invarient to doppler frequency shifts. Differentiation of the argument of (8.61) to determine the instantaneous radian frequency results in $\omega_t = 2\pi f_o^2 T/(f_o T - Bt)$. The period is then given by $T_p = (f_o T - Bt)/f_o^2 T$, which indicates that the doppler-tolerant waveform has linear period modulation.

Figure 8.51 depicts the relative peak amplitude of various pulse compression waveforms as a function of doppler shift [26]. As expected, the linear FM waveform

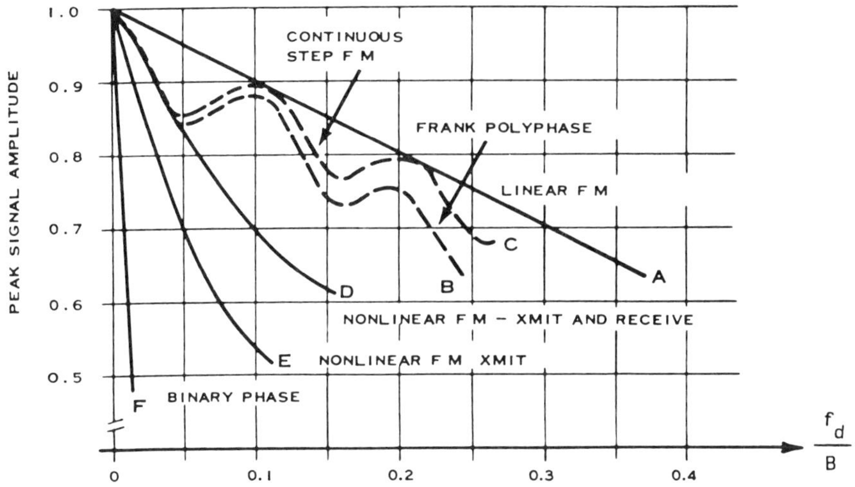

Figure 8.51 Peak response *versus* doppler shift for different pulse-compression waveforms.

exhibits the least degradation with doppler shift. For example, the peak response to a linear FM signal has less than 1 dB loss if the doppler shifts are less than $|f_d| < B/10$ from the tuned frequency. At the other extreme, the binary phase has the greatest sensitivity to doppler shift with nonlinear FM, continuously stepped FM, and Frank polyphase coding with intermediate tolerance to doppler shift.

8.8 DOPPLER WEATHER RADAR [9]

The doppler weather radar detects, maps, and determines the properties of weather disturbances from remote locations. The targets in this case are the various hydrometeors that form the structure of a storm or, in the extreme, exist in clear air, thereby allowing a tracing of wind motion by the radar. The advantage of doppler weather radar is that it penetrates the cloud cover that usually shields the storm from other sensors (e.g., photographic) and provides a measurement of the internal storm structure, which is critical in determining its meteorological significance.

The target echo returned to a doppler weather radar is formed from the backscatter of individual hydrometeors contained within the volumetric resolution cell of the radar (see (4.47)). The large number of reflecting particles contained within the volume resolution cell form a distributed target, the reflectivity of which is characterized by its average backscattering coefficient $\bar{\eta}$ (m^2/m^3). This is one of the primary parameters measured by the doppler weather radar. Representative values of $\bar{\eta}$ as a function of rainfall rate are given in Figure 4.37. Note that the reflectivity generally increases as the fourth power of frequency.

The doppler power spectrum of the received echo generally allows determining the wind fields associated with the weather disturbance. The mean doppler frequency ($\bar{f}_d$) associated with the power spectrum is a measure of the radial velocity of the weather within the volumetric cell under measurement. The spectral width of the doppler power spectrum is a measure of the difference in radial velocity between various hydrometeors within the radar's volumetric cell. The spectral width is generally determined by four effects: wind shear, wind turbulence, beam broadening, and distribution in falling velocity (see Section 4.2.1.2). For broad elevation beams, the primary effect is from wind shear (the change of wind velocity with altitude), whereas, for radars with narrow elevation beams, wind turbulence generally predominates.

The block diagram of a simplified PD weather radar is depited in Figure 8.52. The simplification generally involves the use of a homodyne receiver (see Figure 7.11) rather than the more usual superheterodyne receiver employed in most radars of this type (see Figure 7.6). The functioning of the doppler weather radar is similar to that of the coherent master oscillator power amplifier PD radars described in Chapter 7. However, the nature of the target is different (i.e., distributed rather than point), which requires (1) a modification to the radar range equation; (2) a change in the concept of a matched filter receiver; and (3) a doppler processor, which extracts the moments of the doppler power spectrum rather than one that uses pure doppler filtering techniques. In the following, we will concentrate on these differences in our discussion of doppler weather radar.

The in-phase (I) and quadrature (Q) signals returned from a point target in the homodyne PD radar (see Figure 8.52) are illustrated in Figure 8.53 [9]. The normal range of meterological target velocities is on the order of tens of meters per second, so that the radar PRF must be selected to be at least twice the largest expected doppler frequency ($\bar{f}_d = 2v/\lambda$) to avoid ambiguous velocity measurements. Samples

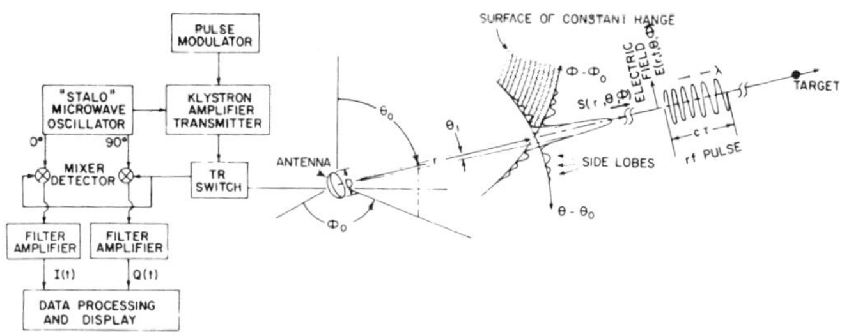

Figure 8.52 Block diagram of PD weather radar.

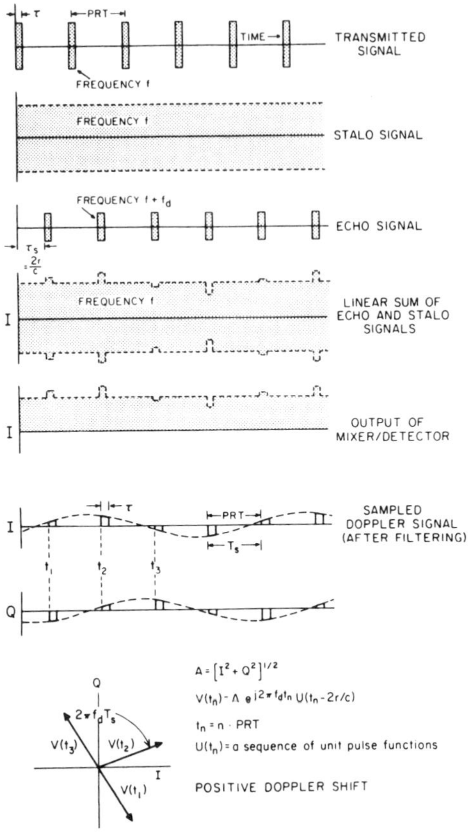

Figure 8.53 Return signals in homodyne PD radar.

of in-phase and quadrature video (bipolar video) are depicted in Figure 8.54 for both fixed and moving targets [9]. The variation in pulse amplitude for the moving target return from one radar transmission to another (three are shown in Figure 8.54) carry the doppler spectral information.

The returns from distributed weather radar targets for a number of radar transmissions is depicted in Figure 8.55 [9]. Returns contained within a time interval corresponding to the radar's pulsewidth (i.e., range cell) are sampled for consecutive pulse transmissions to construct doppler spectra for scatterers located within the range $c\tau_s/2$. The doppler weather echos in sixteen different range cells are depicted in

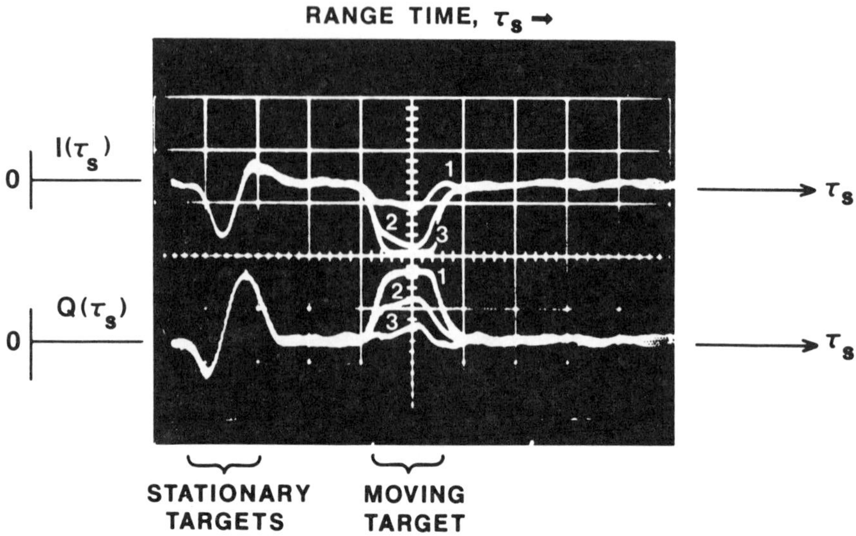

Figure 8.54 Samples of in-phase and quadrature echos for stationary and moving targets.

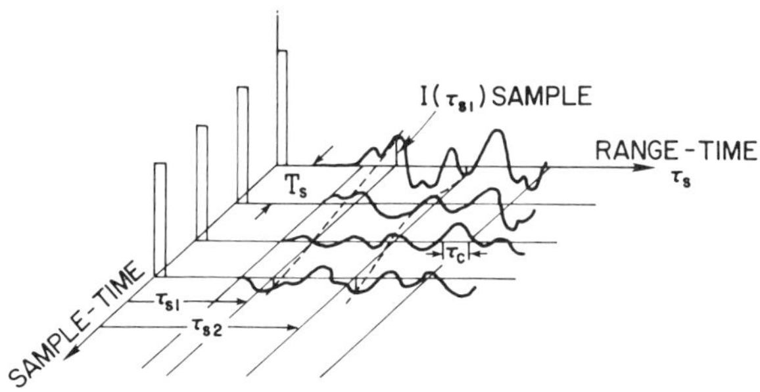

Figure 8.55 Idealized target return for the in-phase component of a PD weather radar.

Figure 8.56 [9]. These data were taken with the radar beam directed vertically and the first gate location at a 600 m range. The first targets displayed are ground targets with relatively constant doppler frequency. Each successive trace encompasses scatterers of about a 170 m depth spaced 600 m apart. The first four or five range-gate locations contain fast-moving targets with relatively high doppler shifts. The transients seen in range gate 12 are due to lightning, which moves at the velocity of the air, and hence produces different dopplers from precipitation that falls through the air. The random nature of the target dopplers are apparent from the figure.

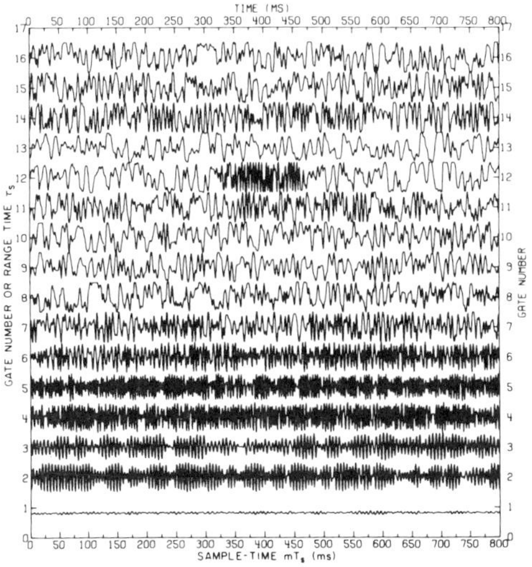

Figure 8.56 PD weather returns for a vertically pointed radar.

The amplitudes of the in-phase and quadrature components of the weather echo will follow a Gaussian probability distribution due to the large number of randomly spaced scatterers that combine to form each return. The distribution of the *I* and *Q* returns are depicted in Figure 8.57 for (a) uncorrelated sample components, (b) correlated sample components, and (c) a weather echo mixed with a ground target echo. In Figure 8.57(a), note that each *I* and *Q* sample (I_m and Q_m) are uncorrelated, but that the *I* and *Q* random processes (I_t and Q_t) are generally correlated (for relatively short sampling times) because the mechanism controlling their phase change with time are related. This is illustrated in the figure by tracing consecutive *I* and *Q* samples that follow the ordered paths projected. When I_m and Q_m samples are correlated, they are still Gaussian distributed but will scatter about some line as depicted in Figure 8.57(b). When weather echos and ground target returns are received in the same range cell, the *I* and *Q* probability distributions are still Gaussian, but the paths of consecutive samples will generally encircle the ground target return as depicted in Figure 8.57(c). When I_m and Q_m samples are combined to form an envelope (i.e., $v = \sqrt{I_m^2 + Q_m^2}$), then the random variable v follows a Rayleigh probability density function, $p(v) = v \exp(-v^2/2\sigma^2)/\sigma$.

One factor affecting the selection of a transmitter frequency for a doppler weather radar is the attenuation of the radar signal due to water absorption. Thus, although high microwave frequencies (e.g., X-band and Ku-band) provide large backscatter from precipitation, they also experience large transmission loss and hence do not penetrate severe weather. For example, an X-band radar ($\lambda = 3$ cm) might experience

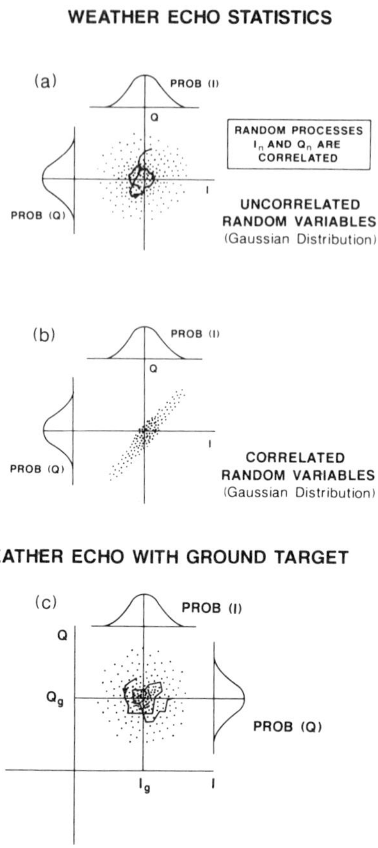

Figure 8.57 Weather echo statistics for uncorrelated and correlated returns.

a two-way transmission loss through heavy precipitation ($r = 16$ mm/hr) of 1 dB/km, whereas at S-band ($\lambda = 10$ cm) the loss would be 0.01 dB/km. If both radars were designed with the same power aperture and pulsewidth, from (4.48), the power returned from the precipitation would be $P_c = k\bar{\eta}/L$, where $\bar{\eta}$ is the backscattering coefficient and L is the attenuation due to water absorption. The backscatter coefficient is larger at X-band than at S-band by the fourth power of frequency. Thus, at a given range, the design at X-band provides greater power than the S-band design by the relationship $P_c = kf^4/L$. The two designs would then be in balance at a range through the precipitation of approximately 20 km with the S-band radar providing greater power at greater ranges. Alternatively, consider that the two radars were designed with the same beamwidths, transmitter power, and pulsewidths; then, the power returned from the precipitation would equal $P_c = kf^2/L$, and the designs would be in balance at a range of approximately 10 km.

The information necessary to determine the power spectral density of the weather return is contained within the I and Q baseband video signal depicted in the block diagram of Figure 8.52. Before doppler processing, the I and Q signals are range gated into time segments that correspond to the radar's effective transmitted pulsewidth as depicted in Figure 8.56. The doppler power spectrum is then determined for each range-gated segment, which is a measure of the turbulence and radial velocity of that particular segment of the storm's structure.

The power spectral density of the time series depicted in Figure 8.56 could be determined by applying the FFT algorithm to the sampled data (see Section 2.3.3.3). However, a simpler procedure, which reduces the data processing, is generally used in doppler weather radars. This procedure, called *pulse-pair processing,* estimates the moments of the power spectral density. Then, assuming that the power spectral density is Gaussian in shape, the complete spectral density is determined by estimating the first moment ($\bar{f}_d$), which establishes the center frequency of the spectrum, and the standard deviation σ_f, which determines the spectral spread or variance. The equivalence of the two methods are displayed in the scattergram shown in Figure 8.58 [29]. The scattergram was taken for a radar with an unambiguous velocity capability of ± 34 m/s, observing a storm at a 100 km range with echos the S/N ratio of which was larger than 10 dB. In the following, we shall give some of the theoretical background associated with pulse-pair processing used to determine the moments of the spectral distribution.

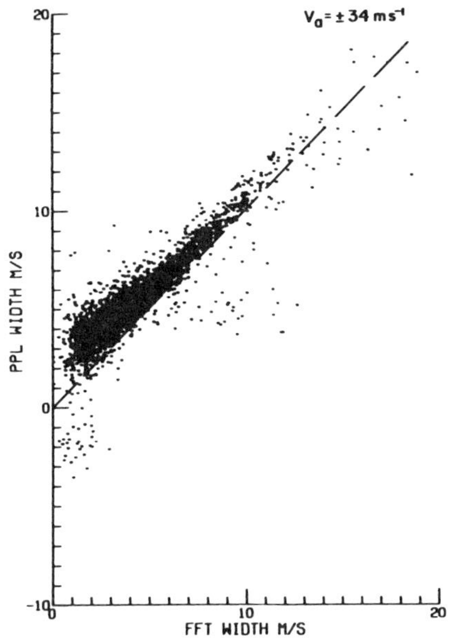

Figure 8.58 Scattergram of weather echo spectral width estimate of FFT *versus* pulse-pair processing ($S/N > 10$ dB).

The autocorrelation function for a zero-mean stationary random process can be expressed as the Fourier transform of the power spectral density:

$$R(\tau) = \int_{-\infty}^{\infty} S(f)\, e^{j2\pi f\tau}\, df \tag{8.63}$$

Differentiating the ACF n times with respect to τ results in

$$R^{(n)}(\tau) = \int_{-\infty}^{\infty} (2\pi jf)^n S(f)\, e^{j2\pi f\tau}\, df \tag{8.64}$$

The normalized spectral moments can be expressed as

$$m_n = \frac{\int_{-\infty}^{\infty} f^n S(f)\, df}{\int_{-\infty}^{\infty} S(f)\, df} = \frac{R^{(n)}(0)}{(2j\pi)^n R(0)} \tag{8.65}$$

where the denominator represents the total power of the weather echo return. The first moment is then given by

$$m_1 = \overline{f}_d = \frac{\dot{R}(0)}{(2j\pi)R(0)} \tag{8.66}$$

and the second moment is

$$m_2 = -\frac{\ddot{R}(0)}{4\pi^2 R(0)} \tag{8.67}$$

The variance of the power spectral density or the spectral spread is given by

$$\sigma^2_{(f)} = m_2 - m_1^2 \tag{8.68}$$

resulting in

$$\sigma^2_{(f)} = -\frac{1}{4\pi^2}\left[\frac{\ddot{R}(0)}{R(0)} - \left(\frac{\dot{R}(0)}{R(0)}\right)^2\right] \tag{8.69}$$

Equations (8.66) and (8.69) imply that the mean frequency of the power spectral density and its rms spectral width can be determined simply through knowledge of the autocorrelation function of the process.

The weather echo power spectral density cannot be observed directly and, therefore, must be estimated in a receiver noise background. Due to the independence of the signal and receiver noise processes, the ACF can be written as

$$R(\tau) = R_s(\tau) + R_n(\tau) \tag{8.70}$$

where $R_s(\tau)$ is the weather echo ACF and $R_n(\tau)$ is the receiver noise ACF. Similarly, the observed power spectral density equals

$$S(f) = S_s(f) + S_n(f) \tag{8.71}$$

The ACFs can be expressed in polar form as

$$R(\tau) = A(\tau)\, e^{2\pi j\phi(\tau)} \tag{8.72}$$

where $A(\tau)$ is a real even amplitude function of the ACF and $\phi(\tau)$ is an odd phase function [30]. For the weather echo, the mean power spectral density is then given by

$$\bar{f}_d = \frac{\dot{R}_s(0)}{(2\pi \mathrm{j})R_s(0)} = \dot{\phi}_s(0) \tag{8.73}$$

and the spectral spread is given by

$$\sigma^2_{(f)} = -\frac{1}{4\pi^2}\left[\frac{\ddot{R}_s(0)}{R_s(0)} - \left(\frac{\dot{R}_s(0)}{R_s(0)}\right)^2\right] = -\frac{1}{4\pi^2}\frac{\ddot{A}_s(0)}{A_s(0)} \tag{8.74}$$

Using the definition of a derivative:

$$\bar{f}_d = \dot{\phi}(0) \approx \frac{\phi_s(\tau) - \phi_s(0)}{\tau} = \frac{\phi_s(\tau)}{\tau} \tag{8.75}$$

which can be expressed as

$$\bar{f}_d = \frac{1}{2\pi\tau}\tan^{-1}\frac{\mathrm{Im}[R_s(\tau)]}{\mathrm{Re}[R_s(\tau)]} \tag{8.76}$$

Thus, an estimate of the mean doppler frequency can be obtained through formation of the ACFs of the I and Q channels of a single lag (τ) that corresponds to the processing of a pulse pair (i.e., two consecutive samples). However, the weather echo cannot be separated from its noise background. Therefore, the estimate becomes

$$\bar{f}_d = \frac{1}{2\pi\tau} \tan^{-1} \frac{\mathrm{Im}[R)(\tau) - R_n(\tau)]}{\mathrm{Re}[R(\tau) - R_n(\tau)]} \tag{8.77}$$

where $R(\tau)$ is the observed ACF of the target echo in noise and $R_n(\tau)$ is the ACF of noise alone. This implies that the noise must be measured in addition to the ACFs to estimate the mean doppler frequency of the spectrum. Equation (8.74) can be converted to a radial velocity estimate by the relation $\bar{v}_r = \lambda \bar{f}_d/2$.

In a manner similar to the preceding, the spectral spread becomes

$$\sigma^2_{(f)} = -\frac{1}{4\pi^2} \frac{\ddot{A}_s(0)}{A_s(0)} \approx \frac{1}{2\pi^2\tau^2} \left[1 - \frac{A_s(\tau)}{A_s(0)} \right] \tag{8.78}$$

Then, the rms value of the spectral spread for pulse-pair processing expressed in terms of the measurable ACFs becomes

$$\sigma_{(f)} = \frac{1}{\sqrt{2}\pi\tau} \left[1 - \frac{|R(\tau) - R_n(\tau)|}{R(0) - R_n(0)} \right)^{1/2} \tag{8.79}$$

Figure 8.59 shows the various combinations of pulse pairs that can be applied to pulse-pair processing [31]. In general, M pulse pairs will be processed. The simplest situation for which this theory applies is when the pulse pairs are spaced and therefore independent. This is wasteful of data because, for example, four pulses are required to form two pulse pairs, whereas, if the pulse pairs are contiguous and therefore correlated, only three pulses are necessary to form two pulse pairs. In addition, to estimate the ACFs at two points instead of one may be advantageous. This requires the formation of multiple lags, which can be obtained by staggering the PRF.

The ACF of the sampled I and Q channels (see Figure 8.52) can be estimated by processing the pulse pairs formed by

$$\hat{R}(\tau) = \frac{1}{M} \sum_{i=0}^{M-1} v_{2i} v^*_{2i+1} \tag{8.80}$$

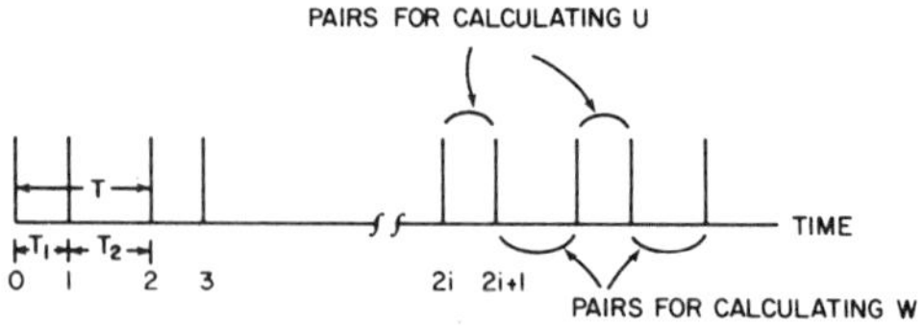

Figure 8.59 Pulse-pair processing by using contiguous, independent, and staggered pulses.

where $v_{2i} = v(iT)$ and $v^*_{2i+1} = v^*(iT + \tau)$ are the complex (I and Q channels) and complex conjugate (*) voltage returns received by the radar, T is the separation between pulse pairs, and τ is the pulse-pair lag. The total power estimate of the pulse train is given by

$$\hat{P}_{s+n} = \frac{1}{L}\sum_{k=0}^{L-1} |v_k|^2 \tag{8.81}$$

Then, the estimate of the mean frequency ($\bar{f}_d$) is formed as

$$\hat{f}_d = \frac{1}{2\pi\tau} \arg[\hat{R}(\tau)] \tag{8.82}$$

where $\arg[\hat{R}(\tau)]$ is defined as in (8.77) and the spread is estimated as

$$\hat{\sigma}^2_{(f)} = \frac{1}{2\pi^2\tau^2}\left[1 - \frac{|\hat{R}(\tau)|}{\hat{P}_{s+n} - N}\right] \tag{8.83}$$

where N is the noise power. Specializing to a Gaussian-shaped signal ACF in a white noise background results in an ACF given by

$$R(\tau) = S\rho(\tau)\,\mathrm{e}^{j\omega_0\tau} + N\delta(\tau) \tag{8.84}$$

where S is the signal power, N is the noise power, ω_0 is the spectrum's mean radian frequency, and

$$\rho(\tau) = \mathrm{e}^{-2\pi^2 w^2 \tau^2} \tag{8.85}$$

where w is the spectral spread. With this ACF, the general expression for spectral spread can be shown to be [31]

$$\hat{\sigma}^2_f = \frac{1}{8\pi^2\tau^2\rho^2(\tau)}\left\{\frac{1}{M^2}[1 - \rho^2(\tau)]\sum_{-(M-1)}^{M-1}\rho^2(mT)(M - |m|)\right.$$
$$\left. + \frac{N^2}{MS} + \frac{2N}{MS}\left[1 - \left(1 - \frac{1}{M}\right)\delta(T - \tau)\right]\right\} \tag{8.86}$$

For contiguous correlated pulse pairs ($T = \tau$) and $\delta(T - \tau) = 1$, and for spaced independent pulse pairs ($T \rightarrow \infty$), resulting in $\delta(T - \tau) = 0$. The independent pulse-pair spectral spread then becomes

$$\hat{\sigma}^2_f = \frac{1}{8\pi^2\tau^2 M}\frac{(1 + N/S)^2 - \rho^2(\tau)}{\rho^2(\tau)} \tag{8.87}$$

Equation (8.86) can be further approximated for large M as

$$\hat{\sigma}_f^2 = \frac{1}{8M\pi^2\tau^2\rho^2(\tau)}\left\{[1 - \rho^2(\tau)] \sum_{-(M-1)}^{M-1} \delta(T - \tau)\rho^2(mT)\right.$$

$$\left. + (N/S)^2 + 2(N/S)[1 - \rho(2T)\delta(T - \tau)]\right\} \tag{8.88}$$

which yet again simplifies [31] for $wT < 1/2\pi$ as

$$\hat{\sigma}_f^2 = \frac{1}{8\pi^2 M\rho^2(\tau)\tau^2}\left\{\frac{2\pi^{3/2}w\tau^2}{T} + \frac{N^2}{S^2}\right.$$

$$\left. + 2(N/S)[1 - \rho(2T)\delta(T - \tau)]\right\} \tag{8.89}$$

For large signal-to-noise ratios ($S/N \gg 1$), (8.89) becomes

$$\hat{\sigma}_f^2 = \frac{w}{4\sqrt{\pi}MT\rho^2(\tau)} \tag{8.90}$$

Figure 8.60 plots (8.86) for the standard deviation of the estimate of normalized mean frequency ($\bar{f}_d\tau$) as a function of the normalized spectral width ($w\tau$), S/N, and the number (M) of contiguous pulse pairs ($T = \tau$). These results apply to single-lag processing with the lag time (τ) equal to the pulse repetition interval (T). As is apparent from Figure 8.60, the estimate of mean frequency becomes more precise for high S/N values and smaller normalized spectral widths.

Figure 8.60 allows the effect on the standard deviation of using higher-order lags to be assessed. For example, if the S/N is 0 dB and the normalized width is 0.05, the normalized frequency error is 0.16. If a second lag is used, the normalized width is 0.1 and the normalized frequency error increases to 0.24. To scale these normalized frequency errors to the actual velocity error, we must multiply the normalized error by ($\lambda/2\pi\tau\sqrt{M}$). Therefore, using the second lag instead of the first halves the scaling factor. However, the normalized error for the second lag is not doubled (see the dashed line), and hence the actual velocity error is less for the second lag. In fact, from the figure (see the dashed line), we observe that the standard error of estimate for the mean velocity becomes smaller for longer lags until the normalized width ($nw\tau$) becomes greater than about 0.13 for an S/N of 0 dB and somewhat larger for larger S/N values.

The predicted performance of the contiguous pulse-pair estimator for various lags is given in Figure 8.61(a) for $S/N = -6$ dB and in Figure 8.61(b) for $S/N =$ 0 dB. The X-band radar ($\lambda = 3.2$ cm) uses a pulse repetition interval of 350 μs and

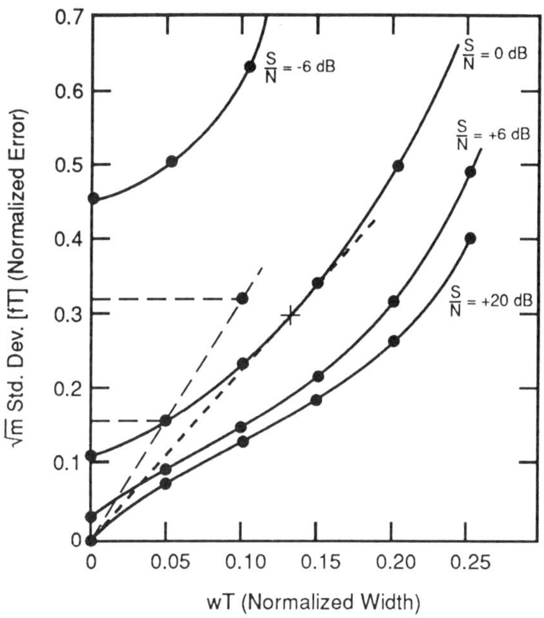

Figure 8.60 Standard deviation of the estimate of the normalized mean doppler frequency by single pulse-pair processing.

a dwell time of 45 ms (128 radar samples). The standard deviation of the mean velocity estimate is improved for all lags and S/N values when the width of the spectrum is less than 1 m/s. However, when the width of the spectrum is 4 m/s, the standard error of mean velocity estimate increases rapidly for $n > 2$. For spectral widths of 2 m/s or less, the second and third lags always give better estimates than the first [32].

Matched filters (or suitable approximations) are generally used in most radars to maximize the S/N for each returned pulse (see Appendix B, Section B.5). These filters are usually implemented by adjusting the receiver bandwidth (B_R) so that $B_R \tau_e \approx 1$, where τ_e is the transmitter's effective pulsewidth. The intuitive reasoning that justifies the matched filter concept against a point target is as follows.

If a rectangular pulse is transmitted and reflected from a point scatterer, the return will have a $(\sin x)/x$ voltage spectrum in which the first null occurs at $f = 1/\tau_e$. For small bandwidths ($B_R \ll 1/\tau_e$), the voltage output will be proportional to the receiver bandwidth (B_R), resulting in the signal power increasing as the receiver bandwidth squared (B_R^2). The noise power increases in direct proportion to the receiver bandwidth (B_R) so that, in this region, the S/N increases as the receiver bandwidth is increased. As the bandwidth approaches the first null, $B_R = 1/\tau_e$, the signal power

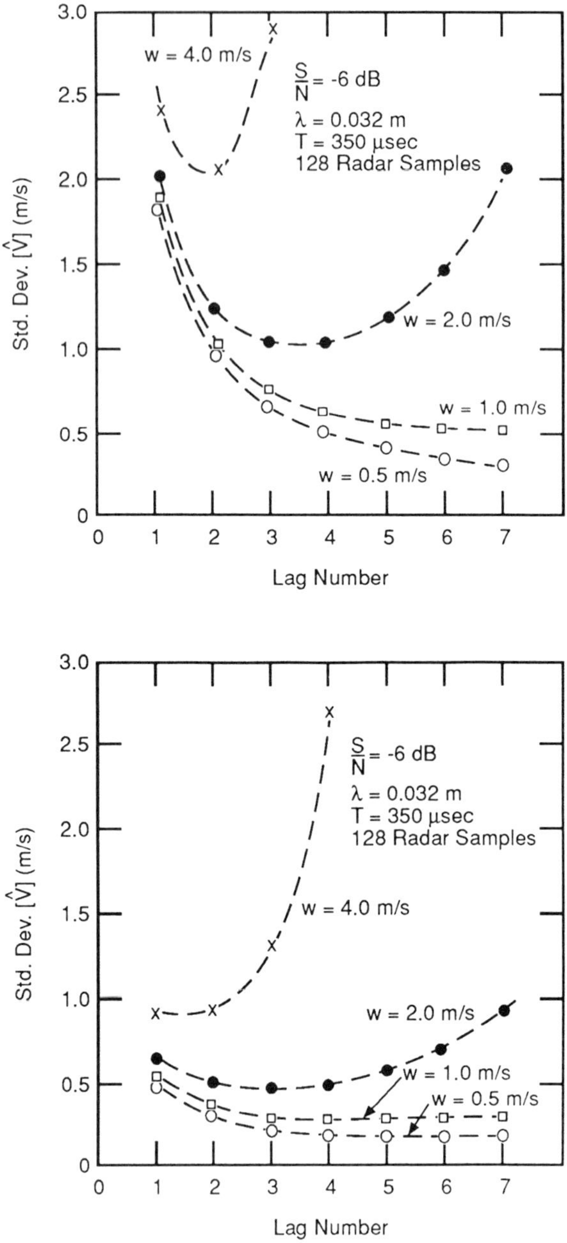

Figure 8.61 Standard deviation of the mean doppler velocity estimate by a single pulse pair as a function of lag number: (a) $S/N = -6$ dB; (b) $S/N = 0$ dB.

begins to increase less rapidly and, as the bandwidth increases beyond this point ($B_R > 1/\tau_e$), the signal power remains approximately constant (the pulse is reproduced with good fidelity). However, the noise power increases continually in direct proportion to the receiver bandwidth. The result is an optimum ratio that maximizes the S/N at about $B_R = 1/\tau_e$.

With a distributed target such as a weather echo for small bandwidths (i.e., $B_R \ll 1/\tau_e$), the target power from each scatterer still tends to increase as B_R^2, but the resolution also increases in direct proportion to B_R, resulting in fewer scatterers (i.e., a diminished volume-resolution cell) contributing to the total target power. As a result, in this region, both signal and noise increase in the same proportion (i.e., with B_R) as the bandwidth is increased, and no maximum can be achieved as with point targets. However, as will be discussed later, most PD weather radars use $B_R = 1/\tau_e$ (the point target matched filter condition) to provide improved velocity estimates. The resulting loss in S/N for a weather echo when a point-target matched filter condition is used (i.e., $B_R = 1/\tau_e$), has been estimated to be between 1.8 to 2.3 dB, depending on the particular transmitted waveform and receiver shapes employed [33].

The power returned to the radar from a point scatterer is given by

$$P_{ri} = \frac{P_t G_t^2 \lambda^2 \sigma_{bi} F^4(\theta, \phi)}{(4\pi)^3 R^4 L} \tag{8.91}$$

where σ_{bi} is the scatterer's RCS and $F^4(\theta, \phi)$ is the two-way pattern propagation factor. The total power return is obtained by summing (8.91) over all possible scatterers, resulting in

$$\bar{P}_r = \frac{P_t G_t^2 \lambda^2}{(4\pi)^3 L} \sum_i \frac{\sigma_{bi} F^4(\theta_i, \phi_i) w_i^2}{R_i^4} \tag{8.92}$$

where w_i^2 is a weighting factor that represents the combined effect of the finite duration transmitter pulse waveform convolved with the receiver impulse function (i.e., the effect of each point scatterer is to apply a delayed replica of the transmitter pulse to the receiver with its response to this signal being summed to form the output waveform). The sum of the individual scatterer RCS can be replaced by the reflectivity factor for volume clutter $\bar{\eta} = \Sigma_v \sigma_{bi}/\Delta V$, where ΔV is the differential volume. To convert the summation to an integral then results in the following form for the weather radar equation:

$$\bar{P}_r = \frac{P_t G_t^2 \lambda^2 \bar{\eta}}{(4\pi)^3 L} \int_V \frac{F^4(\theta, \phi) w^2(R)}{R^4} \, \mathrm{d}V \tag{8.93}$$

where $\bar{\eta}$(m^2/m^3) is assumed constant over the differential volume, and the explicit dependence of the weighting function $w(R)$ on range is shown. The differential volume resolution cell is given by $dv = R^2\, d\theta\, d\phi\, dR$, so that equation (8.93) can be written as

$$\bar{P}_r = \frac{P_t G_t^2 \lambda^2 \bar{\eta}}{(4\pi)^3 R^2 L} \int_0^\infty w^2(R)\, dR^2 \int_0^\pi \int_0^\pi F^4(\theta, \psi)\, d\theta\, d\phi \qquad (8.94)$$

The second double integral has been evaluated for a Gaussian-shaped beam pattern and circular aperture antenna by Probert-Jones as equal to $I = \pi\theta_1\phi_1/8\ \ln 2$, where θ_1 and ϕ_1 are the 3 dB, one-way azimuth and elevation beamwidths [26]. The resulting weather radar equation is then equal to

$$\bar{P}_r = \frac{P_t G_t^2 \lambda^2 \bar{\eta} \theta_1 \phi_1}{512\pi^2 \cdot \ln 2\ L R^2} \int_0^\infty w^2(R)\, dR \qquad (8.95)$$

where $w(R)$ is the range weighting factor. When the receiver bandwidth is much larger than the reciprocal transmitter pulsewidth ($B_R \gg 1/\tau_e$) and the transmitter pulses are rectangular in shape, the reflected returns for the individual scatterers pass through the receiver with high fidelity and $\int_0^\infty w^2(R)\, dR$ becomes equal to $c\tau_e/2$, the range resolution of the radar. The weather equation hence becomes [11]

$$P_r = \frac{P_t G_t^2 \lambda^2 \bar{\eta} \theta_1 \phi_1 c \tau_e}{1024\pi^2 \ln 2\ L R^2} \qquad (8.96)$$

However, this generally produces an optimistic estimate and, if additional precision is required, the range-weighting factor must be evaluated for the transmitter waveform and receiver shape used in the radar. In the following, the range-weighting factor is evaluated for a rectangular-shaped transmitter pulse and a Gaussian-shaped receiver bandpass [33].

The receiver output (v_o) for a rectangular pulse into a Gaussian-shaped receiver can be found from the convolution integral:

$$v_o(t) = \int_{-\infty}^{t} v_i(\tau) h(t - \tau)\, d\tau \qquad (8.97)$$

where $v_i(t)$ is the transmitter pulse and $h(t)$ is the impulse response of the Gaussian-shaped receiver. The Gaussian-shaped filter's normalized frequency response is given by

$$G(f) = e^{-(4\ln 2) f^2 / B_6^2} \qquad (8.98)$$

where B_6 is the bandwidth at which the voltage response is down to one-half its peak value and the power response is -6 dB. The impulse response of this filter is found by Fourier transforming $G(f)$, resulting in

$$h(t) = \sqrt{\frac{\pi}{4\ln 2}} B_6 \, e^{-(\pi B_6 t)^2/4\ln 2} \tag{8.99}$$

If we substitute into the convolution integral, the receiver output results as

$$v_o = \frac{A_o}{2}\left\{\mathrm{erf}\left[aB_6\left(t + \frac{\tau_e}{2}\right)\right] - \mathrm{erf}\left[aB_6\left(t - \frac{\tau_e}{2}\right)\right]\right\} \tag{8.100}$$

where τ_e is the transmitter pulsewidth, B_6 is the receiver 6 dB bandwidth, and $a = \pi/2\sqrt{\ln 2}$. The receiver response is depicted in Figure 8.62, resulting in a weighting factor $w(R)$ equal to

$$w(R) = \frac{1}{2}\left[\mathrm{erf}(x + b) - \mathrm{erf}(x - b)\right] \tag{8.101}$$

where $x = \pi B_6 \Delta\tau_x/2\sqrt{\ln 2}$, $b = \pi B_6(\tau_e/2)/2\sqrt{\ln 2}$, and $\Delta\tau_x = \tau_s - \tau_r - 2R/C$ is the time interval from the peak response to the sample amplitude. The effect of the weighting factor must be determined numerically because the integral $\int w^2(R)\, dR$ is not available in closed form. However, the effect can be closely approximated through the empirical substitution $\mathrm{erf}(\cdot) = \tanh(\cdot)$, which allows a closed-form solution [33]. The weighting function is then obtained as [9]

$$\int_0^\infty w^2(R)\, dR = \frac{c\tau_e}{2}\left[\coth(\chi) - 1/\chi\right] \tag{8.102}$$

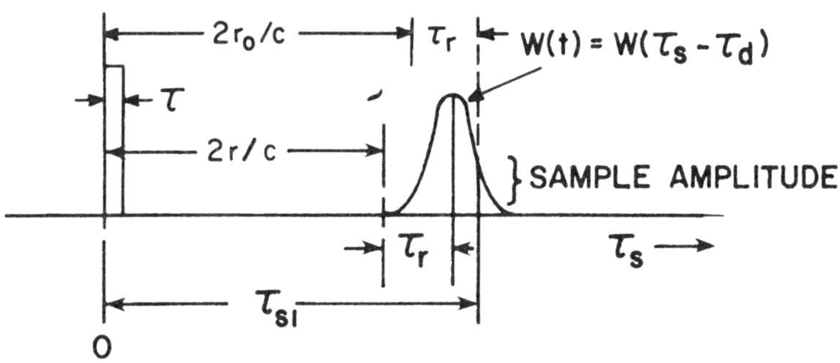

Figure 8.62 Basic receiver response to point target echo of a rectangular transmitter pulse.

where $\chi = \pi B_6\tau_e/2\sqrt{\ln 2}$. Note that as $B_6\tau_e$ become large ($B_6\tau_e \gg 1$), the integral approaches $c\tau_e/2$ as expected. The weather radar equation then becomes

$$P_r = \frac{P_t G_t^2 \lambda^2 \bar{\eta} \theta_1 \phi_1 c\tau_e[\coth(\chi) - 1/\chi]}{1024\pi^2 \ln 2 L R^2} \tag{8.103}$$

From (8.101), a loss term can be defined that accounts for the deviation of the weighting from the range resolution $c\tau_e/2$, given by

$$l_r = \coth(\chi) - 1/\chi \tag{8.104}$$

This is plotted in Figure 8.63 as a function of $B_6\tau_e$ and shows an increasing loss as the product is made less than 1. The loss for the matched filter condition ($B_6\tau_e = 1$) is $l_r = -2.3$ dB, which is obtained by numerical integration (the approximate loss from the curve is $l_r = -2.87$ dB) [33].

Also useful is to define the range resolution (R_6) for the Gaussian receiver case [33]. We define

$$f_w = \frac{\text{erf}(x + b) - \text{erf}(x - b)}{\text{erf}\, b} \tag{8.105}$$

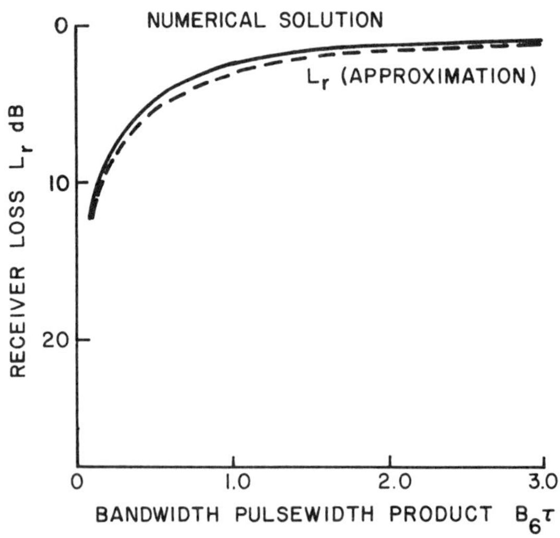

Figure 8.63 Receiver signal power loss due to finite bandwidth.

as the normalized weighting function and then solve for the range R_6 for which $f_w^2 = 1/4$ results in

$$R_6 = \frac{c\tau_e/2}{aB_6\tau_e} \cosh^{-1}[2 + \cosh(aB_6\tau_e)] \tag{8.106}$$

where $a = \pi/2\sqrt{\ln 2}$. This is plotted in Figure 8.64 normalized to the transmitter pulsewidth range equivalent $(2R_6/C\tau_e)$ as a function of the receiver bandwidth-pulsewidth product $(B_6\tau_e)$.

The S/N for weather echos can be deduced from (8.103) as

$$S/N = \frac{C_o\bar{\eta}}{R^2} [\coth(\chi) - 1/\chi] \frac{\tau_e^2}{B_6\tau_e} \tag{8.107}$$

where $C_o = P_t G_t^2 \lambda^2 \theta_1 \phi_1 c/1024\pi^2 \ln 2LkT_s$ relates to parameters associated with the radar. If $B_6\tau_e$ is made a constant, which optimizes radar performance, S/N is proportional to the square of the transmitter pulsewidth (τ_e). Also, as B_6 approaches zero, the maximum S/N is obtained. Thus, as previously discussed, we cannot construct

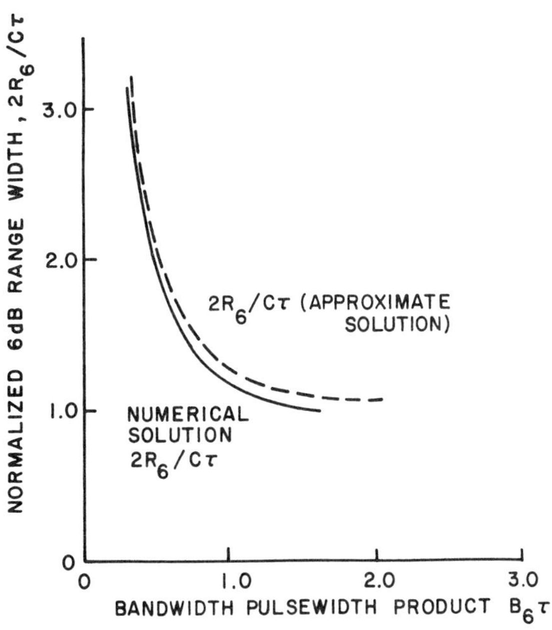

Figure 8.64 Normalized 6 dB range width for a Gaussian receiver.

a matched filter for distributed weather echos in the same manner as for a point target. However, Figure 8.60 indicates that the standard deviation of the normalized mean doppler frequency estimate using pulse-pair processing is sensitive to the pulse repetition interval (T). Thus, if $B_6\tau_e$ is held constant, τ_e must increase if B_6 is reduced. If the average power of the radar is constrained, T must increase in direct proportion to τ_e to maintain a constant duty cycle. Inspection of Figure 8.60 (for high S/N) indicates that this results in higher error variance in the mean velocity estimate. Therefore, for this situation, a balance must be achieved between range resolution and S/N. The optimization condition in a weather radar for determining the receiver-bandwidth–transmitter-pulsewidth product ($B_6\tau_e$) considering both the S/N and resolution requirements should then be the product of $S/N \times R_6^{-1}$ [33]. Figure 8.65 depicts a plot of $S/N \times R_6^{-1}$ *versus* the receiver-bandwidth–transmitter-pulsewidth product ($B_6\tau_e$), which shows a maximum on the order $B_6\tau_e = 1$. This result is consistent with design criteria used in current operational weather radars.

The pulse-pair processing method is presently the most widely used technique for producing large area velocity maps of weather radar data. Because the technique does not use frequency selective techniques (e.g., FFT spectral analysis) to obtain mean frequency and spectral width estimates, pulse-pair measurements are susceptible to ground clutter, which could be returned either locally or due to anomalous propagation [34]. If ground clutter is mixed with weather echos, the desired velocity estimates are affected because the pulse-pair measurement provides averages of the overall spectrum presented at its input.

Because ground clutter returns are generally narrower in spectral width than weather returns and centered at zero frequency, they can usually be removed by a suitable MTI. Ideally, the MTI must be designed to provide a notch at zero frequency, wide enough to eliminate the ground clutter, and with a uniform passband shape that passes the weather echoes unattenuated (see Figure 8.66). Attenuation of the weather

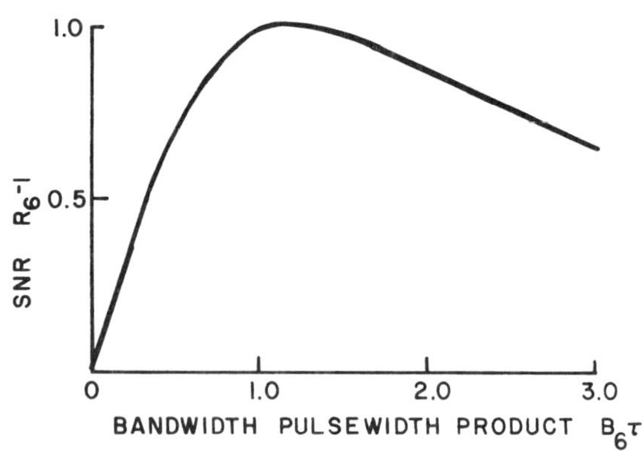

Figure 8.65 S/N resolution product as a function of the product of receiver bandwidth-pulsewidth.

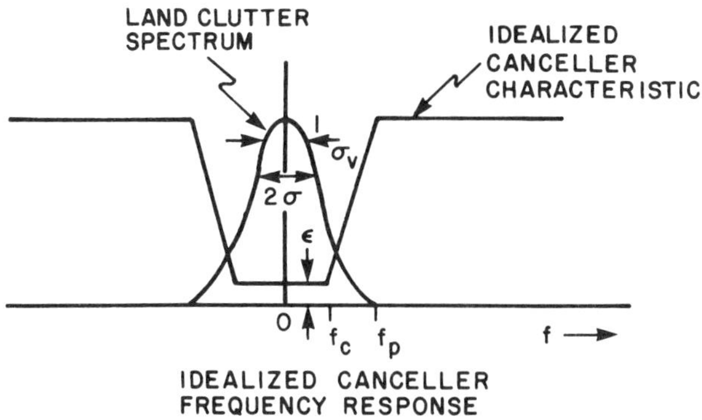

Figure 8.66 Idealized MTI canceller shape for weather radar with ground clutter.

echo signal degrades the S/N and ripples in the MTI passband distort the measurement of the weather echos ACF used in pulse-pair processing.

Accuracy is a crucial issue in the design of an MTI canceller for weather radar design. For meteorological purposes, the least measurable weather velocity needed is about 1 m/s [34]. Because velocities in excess of 50 m/s are rare, the spread of doppler velocities to be measured is relatively small. A nonuniform amplitude response of the clutter filter has the effect of reducing the S/N, which affects the amplitude accuracy with pulse-pair processing. In addition, this response can also cause significant shifts in mean frequency as well as degradation of the width measurement. Both effects are functions of the signal spectral shape so that the effect cannot simply be corrected by adjusting the output results. Also clear is that an imbalance of the MTI responses in the I and Q channels can also distort frequency estimates and has the same effect as errors of the quadrature channels themselves.

In general, cancellation ratios in excess of 50 dB and clutter attenuation of 40 dB are required for adequate performance of a weather radar MTI against ground clutter [34]. A basic limitation in achieving the desired clutter attenuation is the internal-motion (i.e., wind-induced) clutter spectrum, which for a 40-knot wind has a velocity standard deviation on the order of 0.3 m/s (see Chapter 3). An MTI design employing a three-pole elliptical MTI filter achieved 50 dB of clutter attenuation within the stopband for velocities less than 0.32 m/s and less than 1 dB ripple within the passband for velocities greater than 1.4 m/s [34].

The requirements on passband ripple in the MTI to meet a velocity measurement accuracy by using pulse-pair processing can be derived as follows [34]. The measurement of mean velocity (i.e., doppler frequency) by contiguous pulse-pair processing solves the equation:

$$\int_{-\infty}^{\infty} S(f)\, e^{-j2\pi T(f-\hat{f}_d)}\, df = 0 \tag{8.108}$$

where $S(f)$ is the doppler spectrum, T is the radar's interpulse period, and $\hat{f}$ is the mean frequency estimate. Equation (8.108) can be written as (see (8.63))

$$R(T)\,\mathrm{e}^{j2\pi T\hat{f}_d} = 0 \tag{8.109}$$

Differentiation with respect to T and solution for $\hat{f}_d$ results in

$$\hat{f}_d = \frac{\dot{R}(T)}{j2\pi R(T)} \tag{8.110}$$

which is identical to (8.66). In a similar manner, when an MTI canceller is used, the pulse-pair processing solves the equation:

$$\int_{-\infty}^{\infty} S(f)|H(f)|^2\,\mathrm{e}^{-j2\pi T(f-\hat{f})}\,\mathrm{d}f = 0 \tag{8.111}$$

where $|H(f)|^2$ is the power transfer function of the MTI filter. Because we are interested in small perturbations in the passband, the MTI response is represented as

$$|H(f)|^2 = 1 + k(f - f_o) \tag{8.112}$$

where f_o is the true center frequency of the spectrum, $|H(f)|^2$ has unity gain at f_o (i.e., $|H(f_o)|^2 = 1$), and k represents a small constant slope over the significant parts of the signal spectrum. Substitution of (8.112) into (8.111) results in

$$\mathrm{e}^{j2\pi T(\hat{f}-f_o)}\int_{-\infty}^{\infty} S(f)[1 + k(f - f_o)]\,\mathrm{e}^{-j2\pi T(f-f_o)}\,\mathrm{d}f = 0 \tag{8.113}$$

To define the ACF about the true center frequency as

$$R(T) = \int_{-\infty}^{\infty} S(f)\,\mathrm{e}^{-j2\pi(f-f_o)T}\,\mathrm{d}f \tag{8.114}$$

results in

$$\Delta f = \hat{f} - f_o = \frac{k\dot{R}(T)}{j2\pi R(T)} \tag{8.115}$$

which can be expressed as (see (8.76))

$$\Delta f = \frac{1}{2\pi T} \tan^{-1}\left(\frac{k\dot{R}(T)}{2\pi R(T)}\right) \tag{8.116}$$

Equation (8.116) allows determining the frequency error, which is incurred because of the slope error in the MTI passband. The ACF for a Gaussian-shaped frequency spectrum can be written as (see (8.85))

$$R(\tau) = \mathrm{e}^{-(2\pi w\tau)^2/2} \tag{8.117}$$

where w is the spectrum's spectral width. Then, the frequency error for small errors can be approximated as

$$\Delta f = \frac{1}{2\pi T} \tan^{-1}\left[\frac{k(2\pi wT)^2}{2\pi T}\right] \approx kw^2 \tag{8.118}$$

To define the amplitude tolerance as 2Δ results in a slope:

$$k = \frac{2\Delta}{1/T} = 2\Delta T \tag{8.119}$$

so that the normalized frequency error can be expressed as

$$\Delta fT = 2\Delta(wT)^2 \tag{8.120}$$

where Δf is the frequency error, T is the radar's interpulse period, Δ is the amplitude ripple in the MTI's passband, and w is the doppler spectrum's width.

For example, assume that a $\Delta = \pm 0.5$ dB amplitude ripple exists in the MTI's passband over the unambiguous bandwidth $1/T$ and the maximum width suitable for analysis is $wT = 0.3$ (see Figure 8.60). Then, from equation (8.120), the normalized error is $\Delta fT = 0.024$. This corresponds to the resolution of 40 pulses if the spectrum is analyzed by using an FFT spectral analysis and indicates that less than 1 dB ripple is allowable for pulse-pair measurements involving 40 or more pulses.

Alternatively, this can be viewed as a bias error in the spectral estimation of the mean frequency. For example, Figure 8.60 indicates that, for high S/N and $wT = 0.25$, the normalized standard deviation of the mean frequency error for contiguous pulse-pair processing results in $\sqrt{M}\hat{\sigma}_f T = 0.4$. From (8.120), the bias error for 1 dB ripple for this case is $\Delta fT = 0.017$. The bias error then represents the frequency error for pulse-pair processing using $M = 553$ pulse-pair samples. This indicates that an MTI canceller with a 1 dB ripple in the passband limits the pulse-pair estimator of mean frequency to fewer than 553 pulses for weather spectra having a 1σ width of 25% of the unambiguous doppler bandwidth. At that point, the bias error due to the ripple exceeds the noise error in the estimation process.

REFERENCES

1. D.C. Schleher, *Introduction to Electronic Warfare,* Artech House, Norwood, MA, 1986.
2. H. Griffiths and S. Carter, "Provision of Moving Target Indication in an Independent Bistatic Radar Receiver," *Radio and Electronic Engineer,* Vol. 54, No. 7–8, July–August 1984.
3. J. Glasser, "Fifty Years of Bistatic and Multistatic Radar," *IEE Proc.,* Vol. 133, Pt. F, No. 7, December 1986.
4. A. Ivanov, "Semi-Active Radar Guidance," *Microwave Journal,* Vol. 26, No. 9, September 1983.
5. D. Wehner, *High Resolution Radar,* Artech House, Norwood, MA, 1987.
6. D.K. Barton, *Modern Radar System Analysis,* Artech House, Norwood, MA, 1988.
7. D.C. Schleher, ed., *MTI Radar,* Artech House, Norwood, MA, 1978.
8. W. Shrader, "MTI Radar," Chapter 17 in M. Skolnik, ed., *Radar Handbook,* 1st Ed. McGraw-Hill, New York, 1970.
9. R. Doviak and D. Zrnić, "The Doppler Weather Radar," in E. Brookner, ed. *Aspects of Modern Radar,* Artech House, Norwood, MA, 1988.
10. M. Shohat, "Doppler Beam Sharpened Attack Radar Uses Magnetron," *Military Electronics/ Countermeasures,* October 1979.
11. M.I. Skolnik, *Introduction to Radar Systems,* 2nd Ed., McGraw-Hill, New York, 1980.
12. J. Caspers, "Bistatic and Multistatic Radar," in M.I. Skolnik, ed., *Radar Handbook,* 1st Ed. McGraw-Hill, New York, 1970.
13. E. Hanle, "Survey of Bistatic and Multistatic Radar," *IEE Proc.,* Vol. 133, Pt. F, No. 7, December 1986.
14. J. Glaser, "Bistatic Radars Hold Promise for Future Systems," *Microwave System News,* October 1984.
15. M. Jackson, "The Geometry of Bistatic Radar Systems," *IEE Proc.,* Vol. 133, Pt. F, No. 7, December 1986.
16. G. Smith, "A Layman's Guide to Precision Timing," *EW Design Engineers' Handbook,* Horizon House, Norwood, MA, 1989–1990.
17. J. Long and A. Ivanov, "Radar Guidance of Missiles," *Electronic Progress,* Vol. 16, No. 3, Fall 1974.
18. A. Ivanov, "Radar Guidance of Missiles," *IEEE Int. Radar Conf.,* Washington, DC, 1975.
19. E. Brookner, "Synthetic Aperture Radar Spotlight Mapper," in *Radar Technology,* Artech House, Norwood, MA, 1977.
20. D. Ausherman, A. Kozma, J. Walker, H. Jones, and E. Poggio, "Developments in Radar Imaging," *IEEE Trans.,* Vol. AES-20, No. 4, July 1984.
21. J. Walker, "Range-Doppler Imaging of Rotating Objects," *IEEE Trans.,* Vol. AES-16, No. 1, January 1980.
22. E.F. Knott, "Radar Cross Section," in E. Brookner, ed., *Aspects of Modern Radar,* Artech House, Norwood, MA, 1988.
23. D. Richardson, "Is Stealth Misleading?" *Interavia Aerospace Rev.,* October 1989.
24. M. Streetly, "Hiding from Radar," *Interavia Aerospace Rev.,* November 1988.
25. R. Brown, E. Starczewski, and M. Wicks, "Clutter Considerations in Modern Radar Systems," RADC-TR-88-252, Rome Air Development Center, Rome, NY, October 1988.
26. F. Nathanson, *Radar Design Principles,* McGraw-Hill, New York, 1st Ed., 1969.

27. C. Cook, "Transmitter Phase Modulation Errors and Pulse Compression Waveform Distortion," *Microwave Journal,* Vol. 6, No. 5, May 1963.
28. R. Thor, "A Large Time-Bandwidth Product Pulse-Compression Technique," *IEEE Trans.,* Vol. MIL-6, April 1962.
29. D. Zrnić, "Spectrum Width Estimates for Weather Echoes," *IEEE Trans.,* Vol. AES-15, No. 5, September 1979.
30. K. Miller and M. Rochwarger, "A Covariance Approach to Spectral Moment Estimation," *IEEE Trans.,* Vol. IT-18, No. 5, September 1972.
31. D. Zrnić, "Spectral Moment Estimates for Correlated Pulse Pairs," *IEEE Trans.,* Vol. AES-13, No. 4, July 1977.
32. R. Strauch, R. Kropfli, W. Sweezy, W. Moninger, and R. Lee, "Improved Doppler Velocity Estimates by the Poly-Pulse-Pair Method," *Proc. 18th Radar Meteorology Conf.,* Boston, 1978.
33. R. Doviak and D. Zrnić, "The Weather Radar Equation—Receiver Bandwidth Considerations," *Proc. 18th Radar Meteorology Conf.,* Boston, 1978.
34. H. Groginsky and K. Glover, "Weather Radar Canceller Design," *Proc. 18th Radar Meteorology Conf.,* Boston, 1978.

Appendix A
Representation and Processing of Radar Signals

Performance measures for radar systems are generally expressed in probabilistic terms. Such a statistical detection theory approach recognizes that radar signals resulting from targets, clutter, receiver noise, and other interferring noise are random in nature and cannot be defined precisely. This has led to the use of statistical detection theory as the primary method for designing the signal processing function of radar systems.

Statistical detection theory is an abstract process derived from the field of statistical hypotheses testing [1]. In a practical sense, the theory defines how radar signals can be optimally processed from a detection viewpoint.

First, let us consider how to put the radar detection problem into a statistical hypothesis testing framework [2]. Assume an abstract source that generates either of two hypotheses: one is that a target is present (H_1) and the other is that a target is absent (H_0). Then, consider that through a probabilistic transition mechanism this source properly induces the information onto the radar signals observed in each radar-resolution cell.

The observer as represented by the radar detector cannot view the source directly, but is aware of the probabilistic transition mechanism represented by the conditional probability density functions of the observed samples under either hypothesis.

Statistical hypothesis testing theory then prescribes that the optimal procedure is to form the likelihood ratio by using the observed radar signal samples and to compare this quantity against a decision threshold. If the decision threshold is exceeded, a target is declared; otherwise, we assume that no target is present. The decision threshold is formed by using the Neyman-Pearson criterion, which fixes the false alarm probability while maximizing the probability of detection.

The statistical detection process, depicted in Figure A.1, is further described in several later sections of this appendix. Let us now focus our attention on the signal processing function performed in the likelihood ratio computer and the decision comparator.

The function of the radar signal processor is to process the observed sample data set ($\mathbf{x} = x_1, x_2, \ldots, x_n$) in accordance with the optimum operations defined by

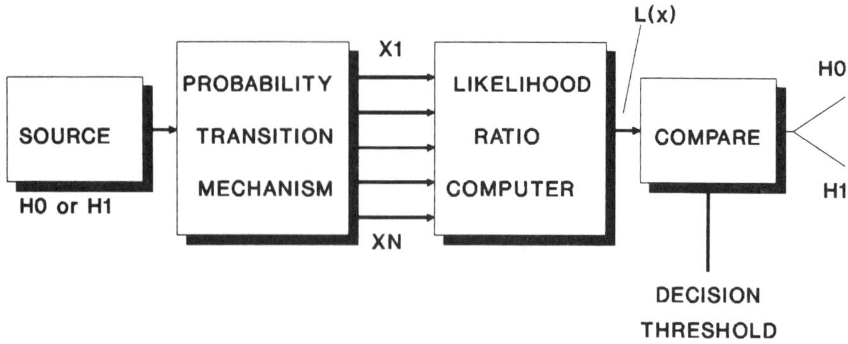

Figure A.1 The statistical detection process.

statistical detection theory. The performance of the signal processor must be determined by considering the detailed operations performed on the radar signals. To accomplish these performance analyses, we must represent mathematically the various signals, clutter, and noise as they propagate through the signal processor.

First, we will describe the narrowband or bandpass representation of radar signals. As the name implies, this representation applies for signals with their frequency content confined to a relatively narrow region about the carrier (or center) frequency. This topic will be useful when considering the coherent processing that occurs in the RF and IF sections of the radar signal processor.

Then, complex notation will be introduced and applied to narrowband signals. This topic is necessary to the understanding of much of the literature on signal processing and also provides the most convenient way to consider the envelope of radar signals. Envelope detection is utilized to extract the amplitude information carried in radar signals and in the noncoherent processing function of the radar signal processor.

Next, we shall consider the radar as a sampled data system. This becomes important in digital signal processing because we must work with samples of radar signals rather than the actual signals. The primary topic covered here will be the sampling theorem that states the conditions under which signal samples can be used instead of the actual signals without any loss of information.

The final topic concerns the statistical description of radar signals in the form of probability distributions. In general, the multidimensional probability density function at all sample points is necessary for completely describing a random process. This is conveniently accomplished for Gaussian or normal processes, as will be discussed. Further simplification is possible for purely random and Markoff processes, which we shall also discuss.

A.1 NARROWBAND REPRESENTATION [3–7]

To introduce the topic of narrowband signals, let us first consider a deterministic signal of finite energy commonly employed in radar applications. This is the sine-wave modulated rectangular pulse depicted in Figure A.2.

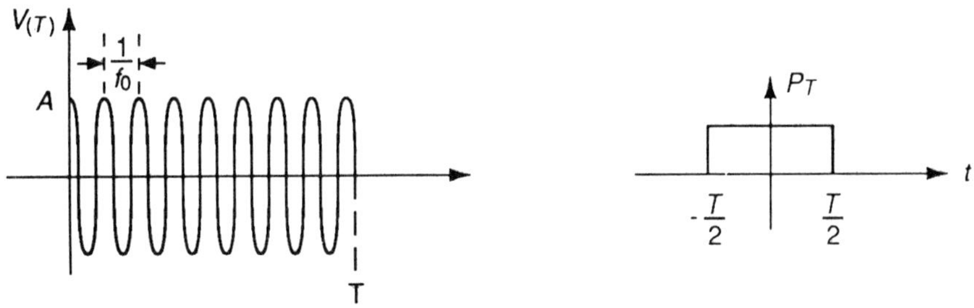

Figure A.2 Modulated rectangular pulse.

The signal is represented mathematically as

$$v_t = A\, P_T\,(t - T/2)\, \cos\omega_0\, t \tag{A.1}$$

where $P_T(t)$ is the rectangular pulse function [5]

$$P_T(t) = \begin{cases} 1, & |t| \le T/2 \\ 0, & \text{otherwise} \end{cases} \tag{A.2}$$

The Fourier transform of $AP_T\,(t - T/2)$ is given by

$$P_t(j\omega) = AT\,\frac{\sin(\omega T/2)}{\omega T/2}\,\mathrm{e}^{-j\omega T/2} \tag{A.3}$$

and the magnitude squared is given by

$$|P_t(j\omega)|^2 = (AT)^2\,\frac{\sin^2(\omega T/2)}{(\omega T/2)^2} \tag{A.4}$$

The effect of the cosine modulation on the video spectrum of (A.3) is to shift the spectrum into a narrowband bandpass signal:

$$G(\omega) = \frac{1}{2}\,[P_l(\omega - \omega_0) + P_l(\omega + \omega_0)] \tag{A.5}$$

The resulting narrowband signal spectrum is given by

$$G(\omega) = \frac{AT}{2}\,e^{-j(\omega+\omega_0)T/2}\,\frac{\sin(\omega + \omega_0)T/2}{(\omega + \omega_0)T/2} + \frac{AT}{2}\,e^{-j(\omega-\omega_0)T/2}\,\frac{\sin(\omega - \omega_0)T/2}{(\omega - \omega_0)T/2} \tag{A.6}$$

The magnitude squared can be found from

$$|G(j\omega)|^2 = G(j\omega)\,G(-j\omega) \tag{A.7}$$

and is given by

$$|G(j\omega)|^2 \approx \left(\frac{AT}{2}\right)\left\{\frac{\sin^2(\omega + \omega_0)T/2}{[(\omega + \omega_0)T/2]^2} + \frac{\sin^2(\omega - \omega_0)T/2}{[(\omega - \omega_0)T/2]^2}\right\} \tag{A.8}$$

where the carrier frequency (ω_0) has been assumed sufficiently high that the cross-product term can be neglected. The resulting symmetrical spectrum is depicted in Figure A.3.

This example illustrates that modulation by a sinusoidal carrier translates the lowpass video spectrum to a carrier frequency. If the amplitude of the spectrum of a signal is zero except in a region about the relatively high carrier frequency, the signal is said to be a narrowband signal. The concept of a narrowband signal can be extended to random processes, where a band-limited low-pass process is modulated by a sinusoidal carrier. Such random narrowband signals can be used to represent clutter and noise processes in the analysis of radar signal processing.

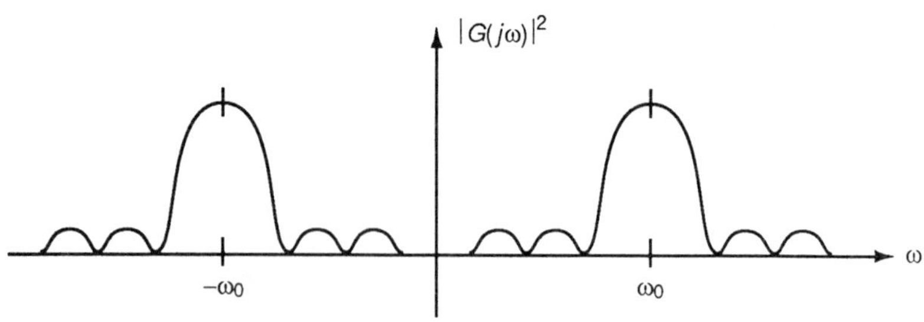

Figure A.3 Spectrum of rectangular pulse.

A random narrowband process is depicted in Figure A.4. A *random process* is defined as narrowband if its power spectral density $S(\omega)$ is zero outside the band $\pm\omega_c$ about the carrier frequency ω_0, as depicted in Figure A.4 [8].

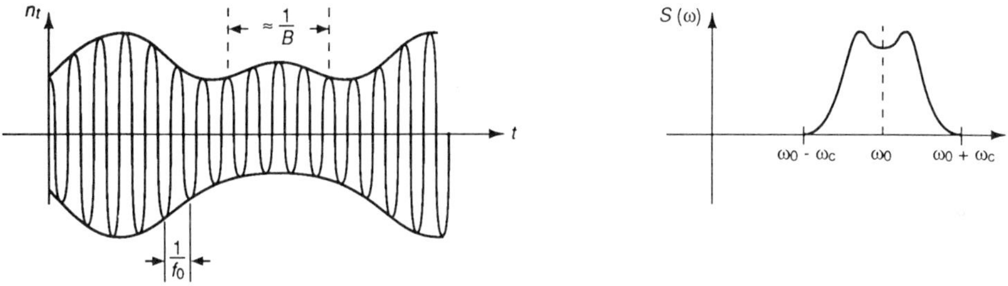

Figure A.4 Random narrowband process.

A narrowband random process $n(t)$ can be written by using Rice's representation [A3] as

$$n_t = x_t \cos\omega_0 t - y_t \sin\omega_0 t \tag{A.9}$$

or equivalently as

$$n_t = r_t \cos(\omega_0 t + \theta_t) \tag{A.10}$$

where

$$E(x_t) = E(y_t) = 0 \tag{A.11}$$

$$E(n_t^2) = E(x_t^2) = E(y_t^2) = \sigma^2 \tag{A.12}$$

The slowly varying envelope (r_t) is given by

$$r_t = \sqrt{x_t^2 + y_t^2} \tag{A.13}$$

and the phase is given by

$$\theta_t = \tan^{-1}\frac{x_t}{y_t} \tag{A.14}$$

The low-pass ($|\omega| < \omega_c$) processes x_t and y_t are called *quadrature* (or in-phase and quadrature, I and Q) components, given by

$$x_t = r_t \cos\theta_t \tag{A.15}$$

$$y_t = r_t \sin\theta_t \tag{A.16}$$

The process n_t is wide-sense stationary (that is, the mean is a constant and autocorrelation depends only on time difference, $\tau = t_1 - t_2$), provided (1) the autocorrelation functions and power spectral densities of low-pass processes x_t and y_t are equal; (2) the means of x_t and y_t equal zero; and (3) the cross-correlation functions $R_{xy}(\tau) = -R_{yx}(\tau)$. We can show that n_t is strict-sense stationary (that is, the statistics of n_t are not affected by a time shift) if x_t and y_t are low-pass Gaussian processes in addition to the earlier conditions [9].

By using the criteria for n_t to be wide-sense stationary, we can show the following [6]:

1. If n_t is wide-sense stationary, x_t and y_t are also wide-sense stationary;
2. Because $R_{xx}(\tau) = R_{yy}(\tau)$, $R_{xx}(0) = R_{yy}(0)$, and the variance of the x_t and y_t processes must be equal;
3. Because $R_{xy}(\tau) = -R_{yx}(\tau)$, $R_{xy}(0) = R_{yx}(0) = 0$ and the x_t and y_t processes are not correlated at the same instant of time. If x_t and y_t are Gaussian distributed, they are statistically independent at the same instant of time;
4. The ACF of n_t equals

$$R_{nn}(\tau) = R_{xx}(\tau) \cos\omega\tau + R_{xy}(\tau) \sin\omega\tau \tag{A.17}$$

so that $R_{nn}(0) = R_{xx}(0) = R_{yy}(0)$, and the variances of the x_t, y_t, and n_t processes are equal;

5. If the power spectral density of n_t is symmetrical about the carrier frequency for positive frequencies resulting in the low-pass processes, x_t and y_t, being symmetrical about $\omega = 0$, then $R_{xy}(\tau) = 0$ and the processes x_t and y_t are orthonormal (uncorrelated for all time). If x_t and y_t are Gaussian processes, they are independent random processes.

A.2 COMPLEX REPRESENTATION [4]

The characteristics of narrowband signals developed in the previous section can be developed in a compact manner by using complex notation. The advantage of this notation is that in many cases the properties of narrowband signals can be expressed by a single complex quantity, rather than a pair of real variables.

Consider first a deterministic narrowband signal analogous to that of (A.1), which is given by

$$v_t = f_t \cos(\omega_c t + \phi_t) \tag{A.18}$$

This signal can be decomposed into quadrature components (I and Q) as depicted in Figure A.5. The quadrature components are given by

$$v_{ct} = f_t \cos\phi_t \tag{A.19}$$

$$v_{st} = f_t \sin\phi_t \tag{A.20}$$

All the information contained in the signal is apparently represented by the quadrature components because the signal can be reconstructed by multiplying the in-phase component by $\cos\omega_c t$ and the quadrature component by $\sin\omega_c t$ and then adding.

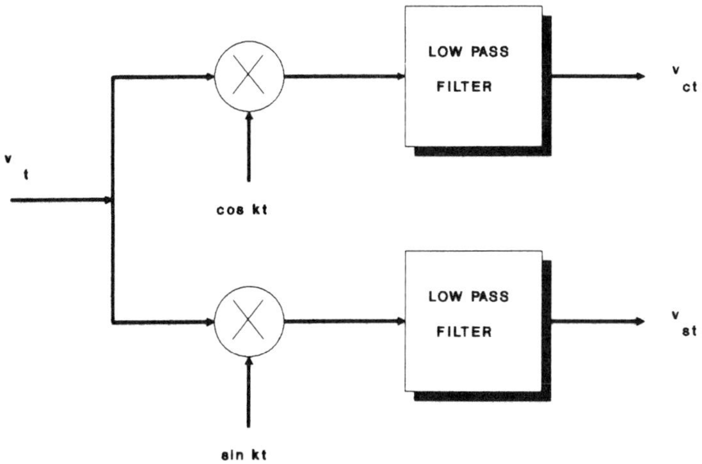

Figure A.5 Decomposition into quadrature components.

The narrowband signal can be represented more compactly as

$$\tilde{v}_t = v_{ct} - jv_{st} \tag{A.21}$$

where $\tilde{v}_t$ is called the *complex envelope*. The complex envelope can also be represented as

$$\tilde{v}_t = |\tilde{v}t| \, e^{j\phi t} \tag{A.22}$$

where

$$|\tilde{v}_t| = \sqrt{v_{ct}^2 + v_{st}^2} \tag{A.23}$$

is the envelope of the real signal, and the phase is given by

$$\phi_t = \tan^{-1}\frac{v_{st}}{v_{ct}} \tag{A.24}$$

The actual narrowband signal is given by

$$v_t = \text{Re}[\tilde{v}_t\, e^{j\omega_c t}] = |\tilde{v}_t|\, e^{j(\omega_c t + \phi_t)} \tag{A.25}$$

where $\omega_c t + \phi_t$ is the instantaneous phase.

Random narrowband processes can also be decomposed into quadrature components, as depicted in Figure A.5. Band-limited ($\pm\omega_c$ about ω_o) random noise processes can be represented as

$$\tilde{n}_t = n_{ct} - jn_{st} \tag{A.26}$$

where the actual signal is given by

$$n_t = \text{Re}[\tilde{n}_t\, e^{j\omega_0 t}] \tag{A.27}$$

The ACF of the real signal can be found in terms of the autocorrelation function of the complex envelope as

$$R_{nn}(\tau) = \frac{1}{2}\,\text{Re}[\tilde{R}_{\tilde{n}\tilde{n}}(\tau)\, e^{j\omega_0 \tau}] \tag{A.28}$$

where

$$\tilde{R}_{\tilde{n}\tilde{n}}(\tau) = E[\tilde{n}_t\, \tilde{n}^*_{t-\tau}] \tag{A.29}$$

The power spectral density can be found from the Fourier transform of the ACF as

$$S_n(\omega) = \frac{1}{4}\,[\tilde{S}_{\tilde{n}}(\omega - \omega_0) + \tilde{S}_{\tilde{n}}(-\omega - \omega_0)] \tag{A.30}$$

where $\tilde{S}_{\tilde{n}}(\omega)$ is the Fourier transform of $\tilde{R}_{\tilde{n}\tilde{n}}(\tau)$.

These relations obtain the statistics of the bandpass process from those of the complex process and *vice versa*. Note that the spectrum of the complex process is even only if the bandpass process is symmetric about ω_0. The spectrum of the actual process ($S_n(\omega)$ and the complex process ($\tilde{S}_{\tilde{n}}(\omega)$) are depicted in Figure A.6 for an asymmetrical spectrum.

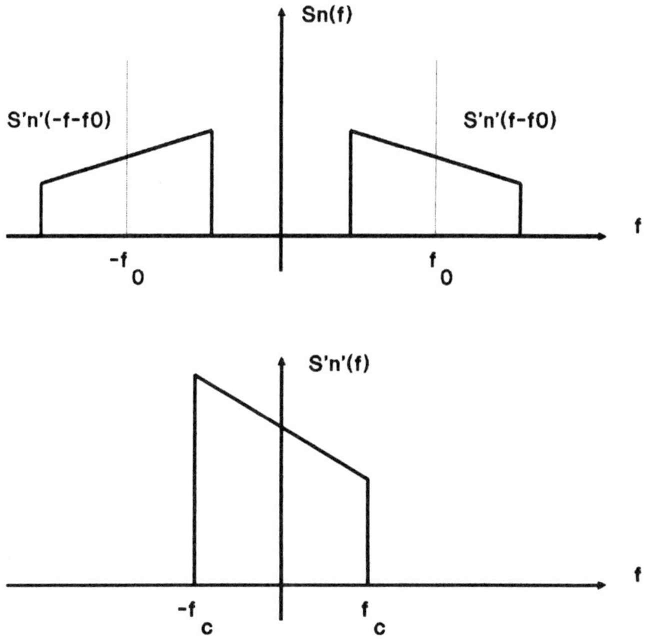

Figure A.6 Spectrum of actual and complex process for an asymmetrical spectrum.

To determine the autocorrelation function of $\tilde{R}_{\tilde{n}\tilde{n}}(\tau)$ in terms of the autocorrelation and cross correlation functions of the quadrature components is instructive. Thus,

$$\tilde{R}_{\tilde{n}\tilde{n}}(\tau) = E\,[(n_{ct} - jn_{st})(n_{ct-\tau} + jn_{st-\tau})] \tag{A.31}$$

This can be expanded into

$$\tilde{R}_{\tilde{n}\tilde{n}}(\tau) = R_c(\tau) + R_s(\tau) + j\,[R_{cs}(\tau) - R_{sc}(\tau)] \tag{A.32}$$

Because $R_c(\tau) = R_s(\tau)$, then

$$R_c(\tau) = R_s(\tau) = \frac{1}{2}\,\mathrm{Re}[\tilde{R}_{\tilde{n}\tilde{n}}(\tau)] \tag{A.33}$$

$$R_{cs}(\tau) = -R_{sc}(\tau) = \frac{1}{2}\,\mathrm{Im}[\tilde{R}_{\tilde{n}\tilde{n}}(\tau)] \tag{A.34}$$

The power spectral densities can be determined from (A.33) and (A.34) as

$$S_c(\omega) = S_s(\omega) = \frac{1}{4}\,[\tilde{S}_{\tilde{n}}(\omega) + \tilde{S}_{\tilde{n}}(-\omega)] \tag{A.35}$$

$$S_{cs}(\omega) = \frac{1}{4j}\,[\tilde{S}_{\tilde{n}}(\omega) - \tilde{S}_{\tilde{n}}(-\omega)] \tag{A.36}$$

If the narrowband process is symmetrical about the carrier frequency, $\tilde{S}_{\tilde{n}}(\omega) = \tilde{S}_{\tilde{n}}(-\omega)$ and

$$S_c(\omega) = S_s(\omega) = \frac{1}{2}\,\tilde{S}_{\tilde{n}}(\omega) \tag{A.37}$$

$$S_{cs}(\omega) = S_{sc}(\omega) = 0 \tag{A.38}$$

indicating that the quadrature processes are orthogonal and uncorrelated.

If the bandwidth of the narrowband process is much greater than other bandwidths of interest, this is called a *white process*. The complex envelope can be written as in (A.26) as

$$\tilde{w}_t = w_{ct} - jw_{st} \tag{A.39}$$

where the autocorrelation function is given by

$$\tilde{R}_{\tilde{w}\tilde{w}}(\tau) = N_0\,\delta(\tau) \tag{A.40}$$

and $N_0/2$ is the power spectral density of the actual white process.

In many cases, n_{ct} and n_{st} of (A.26) are zero-mean, stationary, jointly Gaussian random processes with identical covariance functions. Then, if

$$E\,[\tilde{n}_t\,\tilde{n}_{t-\tau}] = 0 \tag{A.41}$$

$\tilde{n}_t$ is a zero-mean stationary complex Gaussian process. If the power spectral density of the actual process (n_t) is symmetrical about the carrier, the quadrature components n_{ct} and n_{st} are statistically independent.

A.3 SAMPLING THEOREM [5,10–12]

The sampling theorem is a concept extensively utilized in radar system analysis. The theorem defines the minimum rate (called the *Nyquist rate*) at which a band-limited signal must be sampled to preserve the information in the signal. The samples then completely characterize the signal and can be conveniently formed into a signal vector. The signal vector can be utilized to reconstruct the continuous waveform by utilizing a suitable set of interpolation functions.

A typical pulsed radar system can be visualized as a sampled data system whereby the radar operates on a set of samples obtained from continuous representations of the target, clutter, and noise processes. The problem is thus in an ideal form for digital signal processing.

In the following, we shall consider the application of the sampling theorem to deterministic, random video, and bandpass signals. A simple proof will be given for deterministic video signals and results will be stated for random and bandpass signals.

Consider first the sampling theorem as applied to the video band-limited signals that may occur after matched filtering and envelope detection. The general form of the sampler and digital signal processor is depicted in Figure A.7. The band-limited analog waveform and its associated sampled representation are depicted in Figure A.8. Here, we assume that the sampling waveform can be represented by a series of impulses whose magnitude equals the value of the analog waveform at the sampling

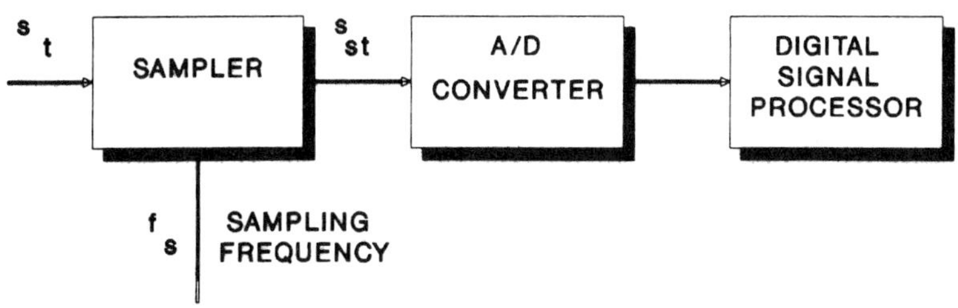

Figure A.7 Sampler and digital signal processor.

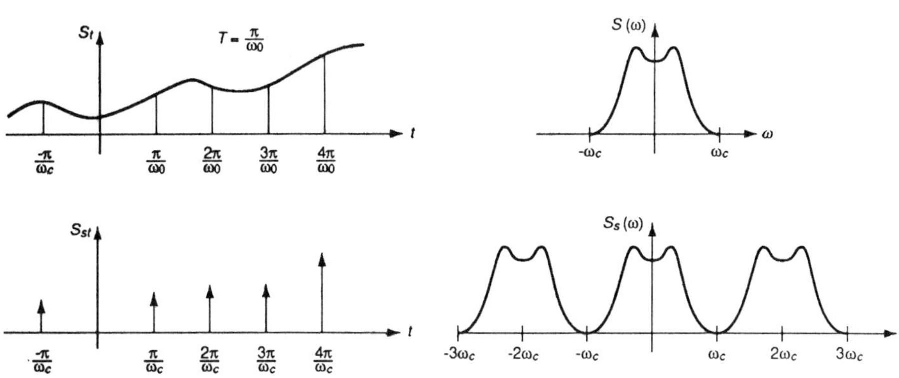

Figure A.8 Bandlimited sampled signal and spectrum.

point. Note that the sampling spreads the spectrum into a number of periodic replicas of the original displaced in frequency by twice the sampling rate.

The sampled waveform is given by

$$s_{st} = T \sum_{n=-\infty}^{\infty} s(nT)\delta(t - nT) \tag{A.42}$$

where T is the sampling period, $\delta(\cdot)$ is an impulse function, and $s(\cdot)$ is a continuous band-limited waveform ($S(\omega) = 0$; $|\omega| \geq \omega_c$). To find the spectrum of (A.42) by using the Fourier transform shifting theorem, the result is

$$S_s(\omega) = T \sum_{n=-\infty}^{\infty} s(nT)\, \mathrm{e}^{-jnT\omega} \tag{A.43}$$

This can be further simplified using Poisson's sum formula [5] to

$$S_s(\omega) = \sum_{n=-\infty}^{\infty} S\left(\omega + \frac{2\pi n}{\pi}\right) \tag{A.44}$$

which is depicted in Figure A.8. The sampling period is related to the band-limited frequency by

$$\omega_c = \pi/T \tag{A.45}$$

From Figure A.8, we can see that the original spectrum can be recovered by using an ideal low-pass filter of bandwidth $\pm\omega_c$ on the sampled waveform. The ideal low-pass filter is represented mathematically as

$$P_{\omega_c}(\omega) = \begin{cases} 1, & |\omega| \leq \omega_c \\ 0, & \text{otherwise} \end{cases} \tag{A.46}$$

and the spectrum of the continuous waveform is given by

$$S(\omega) = S_s(\omega)P_{\omega_c}(\omega) \tag{A.47}$$

Finding the inverse transform of (A.47) provides the sampling theorem given by

$$s_t = \sum_{n=-\infty}^{\infty} s(nT) \frac{\sin\omega_c(t - nT)}{\omega_c(t - nT)} \tag{A.48}$$

Equation (A.48) illustrates that the original continuous waveform can be reconstructed from its samples, $s(nT)$, using the $(\sin\cdot)/(\cdot)$ interpolation functions. Note that at the sampling instant only one interpolation function contributes to the output, whereas between sampling points an infinite number of interpolating functions is required to exactly reproduce the original waveform. Therefore, in a practical situation where only a finite number of samples are available, the theorem indicates that the sampling representation will be exact only at the sampling instants and will be in error between them.

A general interpretation of the sampling theorem is that $2f_c t_o$ independent pieces of information are needed for completely specifying a video band-limited signal that is t_o seconds in length. In signal space, a waveform t_o seconds long can be defined by $2f_c t_o$ samples. The components of a signal vector, $\mathbf{s}$, are given by

$$\mathbf{s} = (s_1, s_2, \ldots, s_{2 f_c t_o}) \tag{A.49}$$

where

$$s_k = s\left(\frac{k}{2f_c}\right) \tag{A.50}$$

The next question of interest is to find how many independent samples are required to specify bandpass signals, such as those represented by (A.18). We can show that the high-frequency sampling theorem is given by [10]

$$s_t = \sum_{n=-\infty}^{\infty} [s_I(nT)\cos\omega_0 t - s_Q(nT)\sin\omega_0 t]\cdot\frac{\sin\omega_c(t - nT)}{\omega_c(t - nT)} \tag{A.51}$$

where $s_I(\cdot)$ and $s_Q(\cdot)$ are low-pass in-phase and quadrature signals and ω_0 is the carrier frequency. Examination of (A.51) shows that $4f_c t_o$ independent samples are required for completely characterizing a narrowband, band-limited signal. This is a considerable saving from the apparent $2(f_c + f_o)t_o$ samples that are obtained by a casual examination of the video sampling theorem.

If x_t is a low-pass random process, the spectrum of which is band-limited to $\pm\omega_c$, x_t can be represented by [6, 9, 12]

$$x_t = \sum_{n=-\infty}^{\infty} x(nT)\frac{\sin\omega_c(t - nT)}{\omega_c(t - nT)} \tag{A.52}$$

Equation (A.52) is identical to the sampling theorem for deterministic signals, except that the right-hand side is understood to be a limit in the quadratic mean sense as the number of samples approaches infinity.

A.4 STATISTICAL SIGNAL DESCRIPTION [13–16]

The signals received in a radar system can be characterized as random processes. Roughly speaking, a random process is a process in which the dependent variable (signal amplitude or phase) does not depend in a completely definitive way on the independent variable (time).

To quantify this discussion, consider an ensemble of records of the same random process. If the records are sampled at the same instant of time, each will indicate a different value. This range of values satisfies the definition of a random variable and therefore is described by a probability density function [8].

Next, consider sampling the ensemble at a different instant of time. If the two instants are close together, we should expect that the values from each record would be related or correlated. Therefore, we must describe this situation by a two-dimensional probability density function. To extend this reasoning to the n-sample case, we obtain the n-dimensional probability density function, given by

$$p(\mathbf{x}, t) = p(x_1, x_2, \ldots, x_n; t_1, \ldots, t_n) \tag{A.53}$$

which in a statistical sense completely defines the random process, provided that n is consistent with the minimum required by the sampling theorem.

The situation can be simplified if we assume that the statistics of the random process are stationary (independent of the time origin). Then, (A.53) can be written as

$$p(\mathbf{x}) = p(x_1, x_2, \ldots, x_n) \tag{A.54}$$

This also means that we can determine the statistics of the process from a single time record because ensemble statistics and time statistics are identical.

In general, to work with the n-dimensional probability density function is difficult and much of the radar literature is concerned with specialized situations of (A.54).

The most elemental random process is one in which the various samples are statistically independent. This type of process is called a *purely random process* [13], and its probability density function is given by

$$p(\mathbf{x}) = p(x_1)\, p(x_2) \ldots p(x_n) \tag{A.55}$$

This model is used extensively in the Marcum-Swerling approach to radar statistical detection theory.

The next more complicated random process is called a *Markoff random process* [13]. Here, the value of the process at present is dependent only on the value of the process at the previous sample time and on a transition probability. This can be expressed as [8]

$$p(\mathbf{x}) = p(x_1, x_2, \ldots, x_n) = p(x_n|x_{n-1}) \ldots p(x_2|x_1)\, p(x_1)$$
$$= p(x_1) \prod_{i=2}^{n} p(x_i|x_{i-1}) \tag{A.56}$$

The transition probabilities, $p(x_n|x_{n-1})$ are called *conditional probabilities* and reflect the basic mechanism of the Markoff process.

From (A.56) and the definition, it is clear that the properties of the Markoff process depend on only the two-dimensional probability density function. The Markoff concept can be extended to the third- and higher-dimensional probability density functions by considering projections of Markoff processes where each projection is a Markoff process, but the overall process is not Markovian. This type of process is sometimes called a *vector Markoff process*, but it is beyond the scope of the present discussion [14].

Rather than decomposing the process, a considerable simplification can be made if the process is a Gaussian random process. When this is the case, the n-dimensional probability density function is easily determined, and all the properties of the process depend on its power spectral density or equivalent autocorrelation function. This has led to the extensive use of the Gaussian random process in radar analysis.

A.4.1 Gaussian Random Processes

A random process x_t is Gaussian if the variable $x_1, x_2, \ldots, x_n$ have a jointly (n-dimensional) Gaussian probability density function for any set of values $t_1, t_2, \ldots, t_n$ and any value of n. The joint density function is given by

$$p(x_1, \ldots, x_n) = \frac{1}{(2\pi)^{n/2}\sqrt{|\mathbf{C}|}} \exp\left[-\frac{1}{2}(\mathbf{x}-\mathbf{m})^T\mathbf{C}^{-1}(\mathbf{x}-\mathbf{m})\right] \tag{A.57}$$

where the variable ($\mathbf{x}$) and mean ($\mathbf{m}$) vectors are given by

$$\mathbf{x} = \begin{bmatrix} x_1 \\ \vdots \\ x_n \end{bmatrix}, \quad \mathbf{m} = \begin{bmatrix} m_1 \\ \vdots \\ m_n \end{bmatrix} \tag{A.58}$$

and the covariance matrix ($\mathbf{C}$) is

$$\mathbf{C} = \begin{bmatrix} c_{11} & \cdots & c_{1n} \\ \vdots & & \\ c_{n1} & \cdots & c_{nn} \end{bmatrix} \tag{A.59}$$

where the elements are

$$c_{ij} = E\{(x_i - m_i)(x_j - m_j)\} = \sigma_i\sigma_j\rho_{ij}$$
$$= R_x(t_i, t_j) - m_i m_j \tag{A.60}$$

The correlation coefficient is given by

$$\rho_{ij} = \frac{c_{ij}}{\sigma_i\sigma_j} \tag{A.61}$$

As is apparent from (A.57), all finite dimensional distributions of a Gaussian process are completely determined once its mean and covariance function are specified [13]. If the Gaussian process is stationary, we have

$$m_i = m_j = m$$
$$\sigma_i = \sigma_j = \sigma \tag{A.62}$$
$$R_x(t_i, t_j) = R_x(t_i - t_j) = R_x(\tau) \tag{A.63}$$

If the set of values $t_1, \ldots, t_n$ are chosen so that $x_1, x_2, \ldots, x_n$ are uncorrelated:

$$c_{ij} = \begin{cases} \sigma_i^2, & i = j \\ 0 & i \neq j \end{cases} \tag{A.64}$$

then $\mathbf{C}$ is a diagonal matrix

$$\mathbf{C} = \begin{bmatrix} \sigma_1^2 & \cdots & 0 \\ \vdots & \sigma_2^2 & \vdots \\ 0 & \cdots \ddots & \sigma_n^2 \end{bmatrix} \tag{A.65}$$

The inverse matrix can be found as

$$\mathbf{C}^{-1} = \begin{bmatrix} 1/\sigma_1^2 & \cdots & 0 \\ \vdots & 1/\sigma_2^2 \ddots & \vdots \\ 0 & \cdots & 1/\sigma_n^2 \end{bmatrix} \tag{A.66}$$

resulting in

$$p(x_1, \ldots, x_n) = \frac{1}{\left(2\pi \prod_{i=1}^{n} \sigma_i^2\right)^{n/2}} \exp\left[-\frac{1}{2}\sum_{i=1}^{n} \frac{1}{\sigma_i^2}(x_i - m_i)^2\right] \tag{A.67}$$

Equation (A.67) is the product of n first-order density functions, implying that the samples at $x_1, \ldots, x_n$ are independent. Hence, for a Gaussian random process where the samples are selected to be uncorrelated, they are also independent. For stationary white noise, the random variables $x_1, \ldots, x_n$ are uncorrelated and hence independent for any choice of $t_1, \ldots, t_n$.

The joint characteristic function of samples $x_1, \ldots, x_n$ is

$$M_n(\omega_1, \ldots, \omega_n; t_1, \ldots, t_n) = \exp\left(j\mathbf{m}^T\boldsymbol{\omega} - \frac{1}{2}\boldsymbol{\omega}^T\mathbf{C}\boldsymbol{\omega}\right) \tag{A.68}$$

where

$$\boldsymbol{\omega} = \begin{bmatrix} \omega_1 \\ \vdots \\ \omega_n \end{bmatrix} \tag{A.69}$$

For a zero-mean Gaussian process, this can be written as

$$M_n(\boldsymbol{\omega}) = \exp\left(-\frac{1}{2}\sum_{i=1}^{n}\sum_{k=1}^{n} \omega_k\, c_{ki}\, \omega_i\right) \tag{A.70}$$

The moments can be found from the characteristic function as [6]

$$E\,(x_1^{b_1} \ldots x_n^{b_n}) = (-j)^\beta \frac{\partial^\beta}{\partial\omega_1^{b_1} \ldots \partial\omega_n^{b_n}} M\,[\omega_1 \ldots \omega_n], \quad \text{all } \omega_i = 0 \tag{A.71}$$

where

$$\beta = \sum_{i=1}^{n} b_i$$

An important case occurs for $n = 4$:

$$E[x_1x_2x_3x_4] = E(x_1x_2)E(x_3x_4) + E(x_1x_3)E(x_2x_4) + E(x_1x_4)E(x_2x_3) \tag{A.72}$$

In many cases, the analysis of problems involving correlated Gaussian random variables can be simplified by using a linear transformation that transforms the variables into statistically independent random variables [6]. To consider this case, define

$$\mathbf{Q} = \mathbf{x}^T\mathbf{C}\mathbf{x} \tag{A.73}$$

where $\mathbf{Q}$ is the quadratic form associated with the covariance matrix $\mathbf{C}$. Then form the determinant $|\mathbf{C} - \lambda\mathbf{I}|$, which is called the characteristic polynomial of the matrix $\mathbf{C}$. Setting

$$|\mathbf{C} - \lambda\mathbf{I}| = \mathbf{0} \tag{A.74}$$

results in a polynomial equation in λ of degree n. The roots of this equation, designated as λ_i, are called *eigenvalues* of $\mathbf{C}$, and the vectors $\mathbf{x}^{(i)}$ associated with the eigenvalues are called *eigenvectors*. A matrix $\mathbf{A}$ of the eigenvectors:

$$\mathbf{A} = \begin{bmatrix} x_1^{(1)} & \cdots & x_1^{(n)} \\ \vdots & & \vdots \\ x_n^{(1)} & \cdots & x_n^{(n)} \end{bmatrix} \tag{A.75}$$

can be formed where the eigenvectors are normalized so that the eigenvectors are orthonormal. Then,

$$\mathbf{CA} = \mathbf{A}\boldsymbol{\lambda} \tag{A.76}$$

$$\mathbf{A}^T\mathbf{CA} = \boldsymbol{\lambda} \tag{A.77}$$

where

$$\boldsymbol{\lambda} = \begin{bmatrix} \lambda_1 & \cdots & & 0 \\ \vdots & \lambda_2 & & \vdots \\ & & \ddots & \\ 0 & \cdots & & \lambda_n \end{bmatrix} \tag{A.78}$$

is a diagonal matrix and $\mathbf{A}^T$ is a linear transformation that transforms the correlated set of Gaussian random variables with covariance matrix $\mathbf{C}$ into a statistically independent set of Gaussian random variables with covariance matrix $\boldsymbol{\lambda}$.

The narrowband representation of equation (A.9) can be specialized to the case where n_t is a sample function of a stationary Gaussian process with zero mean:

$$n_t = x_t \cos\omega_0 t - y_t \sin\omega_0 t \tag{A.79}$$

Hence, x_t and y_t are also sample functions of stationary Gaussian processes with zero mean. We have previously shown that for symmetrical power spectral densities:

$$E(x_t) = E(y_t) = E(n_t) = 0$$

$$E(x_t^2) = E(y_t^2) = E(n_t^2) = \sigma^2$$

$$R_{xy}(\tau) = 0 = R_{xy}(0) \tag{A.80}$$

By using the preceding, we can write the multidimensional probability density function of the variables:

$$x_1 = x_t, \quad x_2 = x_{t+\tau}$$
$$x_3 = y_t, \quad x_4 = y_{t+\tau} \tag{A.81}$$

in terms of the covariance matrix:

$$\mathbf{C} = \sigma^2 \begin{bmatrix} 1 & \rho(\tau) & 0 & 0 \\ \rho(\tau) & 1 & 0 & 0 \\ 0 & 0 & 1 & \rho(\tau) \\ 0 & 0 & \rho(\tau) & 1 \end{bmatrix} \tag{A.82}$$

The determinant of **C** is given by

$$|\mathbf{C}| = \sigma^8(1 - \rho^2)^2 \tag{A.83}$$

The four-dimensional joint characteristic function is given by

$$M(\boldsymbol{\omega}) = \exp\left\{ -\frac{1}{2} \sigma^2 \left[\omega_1^2 + \omega_2^2 + \omega_3^2 + \omega_4^2 + 2\rho(\omega_1\omega_2 + \omega_3\omega_4)\right] \right\} \tag{A.84}$$

The bandpass process of (A.79) is sometimes demodulated by passing it through a square-law device and low-pass filter as depicted in Figure A.9. The output is given by

$$z_t = x_t^2 + y_t^2 \tag{A.85}$$

The correlation function of z_t is given by

$$R_z(\tau) = E(z_t\, z_{t+\tau}) = E(x_1^2x_2^2) + E(x_3^2x_4^2)$$
$$+ E(x_1^2x_4^2) + E(x_2^2x_3^2) \tag{A.86}$$

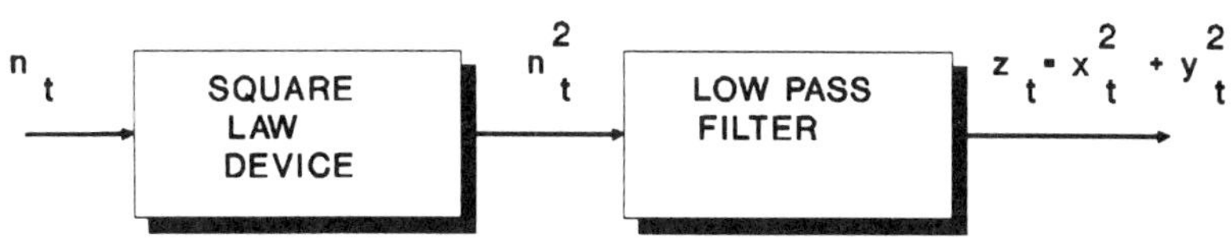

Figure A.9 Square-law demodulator.

By using (A.84) in (A.71), we can find that

$$E(x_1^2x_2^2) = E(x_3^2x_4^2) = \sigma^4\,(1 + 2\rho^2)$$
$$E(x_1^2x_4^2) = E(x_2^2x_3^2) = \sigma^4 \tag{A.87}$$

resulting in

$$R_z(\tau) = 4\sigma^4\,[1 + \rho^2(\tau)] \tag{A.88}$$

Let us return to our discussion of Markoff processes, where we can show [13] that a stationary Gaussian process can be a Markoff process only if its autocorrelation function is given by

$$R(\tau) = \exp(-\alpha|\tau|) \tag{A.89}$$

The power spectral density is given by

$$S(\omega) = \frac{2\alpha}{\omega^2 + \alpha^2} \tag{A.90}$$

The elements of the covariance matrix are given by

$$c_{ij} = \exp(-\alpha|i - j|T) = r^{|i-j|} \tag{A.91}$$

where T is the intersample period, and

$$r = \exp(-\alpha T) \tag{A.92}$$

An advantage of using the Markoff process in analytical work is the relatively simple inverse covariance matrix, given by [17]

$$\mathbf{C}^{-1} = \frac{1}{1 - r^2}\begin{bmatrix} 1 & -r & 0 & 0 & 0 \\ -r & 1 + r^2 & -r & 0 & 0 \\ 0 & -r & 1 + r^2 & 0 & 0 \\ 0 & 0 & 0 & 1 + r^2 & -r \\ 0 & 0 & 0 & -r & 1 \end{bmatrix} \tag{A.93}$$

The properties of Gaussian random processes can be summarized as follows [6]:

1. A linear transformation of a multivariate Gaussian vector produces another Gaussian vector. This includes as special cases the integral of a Gaussian process and the sum of dependent jointly Gaussian vectors;
2. If Gaussian variables are not correlated, they are also statistically independent;
3. For variables that have a multivariate Gaussian distribution and a positive definite covariance matrix, a linear transformation can be found that transforms the dependent variable into a new set of statistically independent Gaussian variables;
4. For zero-mean Gaussian variables x_1, x_2, x_3, x_4 the following mixed moment relationship holds

$$E(x_1x_2x_3x_4) = E(x_1x_2)E(x_3x_4) + E(x_1x_3)\,E(x_2x_4) + E(x_1x_4)E(x_2x_3)$$

5. If a Gaussian process is stationary in the wide sense, it is also stationary in the strict sense;
6. The first- and second-order probability density functions are given by

$$p(x) = \frac{1}{\sqrt{2\pi}\sigma}\exp\left[-\frac{1}{2\sigma^2}(x-m)^2\right]$$

$$p(x_1x_2) = \frac{1}{2\pi\sigma_1\sigma_2\sqrt{1-\rho^2}}\exp\left\{-\frac{1}{1-\rho^2}\left[\frac{(x_1-m_1)^2}{2\sigma_1^2}\right.\right. \tag{A.94}$$

$$\left.\left. -\rho\frac{(x_1-m_1)(x_2-m_2)}{\sigma_1\sigma_2} + \frac{(x_2-m_2)^2}{2\sigma_2^2}\right]\right\} \tag{A.95}$$

REFERENCES

1. E. Lehmann, *Testing of Statistical Hypothesis*, John Wiley and Sons, New York, 1959.
2. A. Sage and J. Melsa, *Estimation Theory with Applications to Communications and Control*, McGraw-Hill, New York, 1971.
3. S. Rice, *Mathematical Analysis of Random Noise*, in N. Wax, "Selected Papers on Noise and Stochastic Processes," Dover, Mineola, NY, 1954.
4. H. Van Trees, *Detection, Estimation and Modulation Theory, Part III*, John Wiley and Sons, New York, 1971.
5. A. Papoulis, *The Fourier Integral and Its Applications*, McGraw-Hill, New York, 1962.
6. A. Whalen, *Detection of Signals in Noise*, Academic Press, New York, 1971.
7. M. Schwartz, *Information Transmission, Modulation and Noise*, McGraw-Hill, New York, 1980.
8. A. Papoulis, *Probability, Random Variables and Stochastic Processes*, McGraw-Hill, New York, 1984.

9. W. Lanning and R. Battin, *Random Processes in Automatic Control*, McGraw-Hill, New York, 1956.
10. J. DiFranco and W. Rubin, *Radar Detection*, Artech House, Norwood, MA, 1980.
11. M. Schwartz, W. Bennett, and S. Stein, *Communication Systems and Techniques*, McGraw-Hill, New York, 1966.
12. I. Selin, *Detection Theory*, Princeton University Press, NJ, 1965.
13. M. Chen and G. Uhlenback, *On the Theory of Brownian Motion II*, in N. Wax, "Selected Papers on Noise and Stochastic Processes," Dover, Mineola, NY, 1954.
14. E. Wong, *Stochastic Processes in Information and Dynamical Systems*, McGraw-Hill, New York, 1971.
15. W. Davenport and W. Root, *An Introduction to the Theory of Random Signals and Noise*, McGraw-Hill, New York, 1958.
16. R. Schwartz and B. Friedland, *Linear Systems*, McGraw-Hill, New York, 1965.
17. L. Wainstein and V. Zubakov, *Extraction of Signals from Noise*, Prentice-Hall, Englewood Cliffs, NJ, 1962.

Appendix B
Optimum Radar Signal Processing

The structure of an optimal signal processor generally depends on the optimization criteria employed. Many statistical optimality criteria used in radar and communication lead to likelihood ratio processors. The maximum signal-to-noise ratio criterion is extensively used in matched filter design. A minimum variance signal estimator criterion is used to design Wiener and Kalman filters.

We shall concentrate exclusively on likelihood ratio processors, which result from the application of the Neyman-Pearson criterion to the radar detection problem. The Neyman-Pearson decision rule maximizes the radar's probability of detection while maintaining a fixed probability of false alarm. Hence, the signal processor is optimized relative to the performance measures commonly used in radar design. Additionally, we can show [1] that the likelihood ratio processor, the maximum S/N processor, and the minimum variance signal estimator processor are all identical for the detection of nonfluctuating targets in a background of white Gaussian noise.

The detection problem for a conventional two-dimensional (range and azimuth angle) search radar is illustrated in Figure B.1. Returns are examined in each range-resolution cell for a period that corresponds to the scanning antenna's dwell time on the target. The waveform of the target return consists of a train of doppler-frequency-shifted RF pulses modulated by both the antenna's directivity pattern and the radar's transmitted waveform. The target return waveform must be detected in the presence of an interferring background that consists of receiver noise and clutter. This situation represents a complex detection problem, even for the relatively simple models generally employed in radar analysis.

This section analyzes a classical version of the radar detection problem. The classical version involves a number of simplifying assumptions, relative to the practical radar detection problem, which allow a complete solution to be obtained. Extensive performance curves [2] for the classical radar problem are available that plot S/N *versus* probability of detection (P_d) with probability of false alarm (P_{fa}) as a parameter. Results for practical situations are developed using the detection loss concept. *Detection*

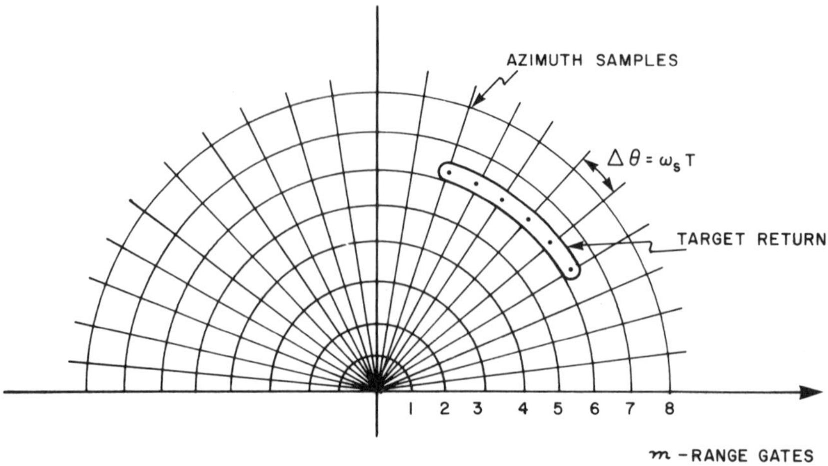

Figure B.1 The search radar detection problem.

loss refers to the additional S/N required to achieve a specified performance level under conditions that deviate from the ideal contemplated by the classical theory.

The simplifying assumptions associated with the classic radar detection problem are as follows:

1. The directivity pattern of the antenna is constant within the main beam (3 dB beamwidth), and no transmission or reception occurs outside this beam;
2. The background interference consists only of receiver noise, which can be modeled as a Gaussian white noise process (see Section A.4.1);
3. The parameters of the received signal and noise are completely known with the possible exception of a uniformly distributed random phase associated with the target return, and is due to the uncertain transit time between the target and the radar;
4. The target doppler frequency shift is known.

Under these assumptions, the classic radar detection problem consists of detecting a train of m equal-amplitude pulses of known energy (E_r) imbedded in a background of white Gaussian noise of known power spectral density ($N_0/2$). The phase of the RF carrier can be coherent within the pulse train with random starting phase. Alternatively the RF phase can be noncoherent or random from pulse to pulse within the pulse train. Even if the transmitter is noncoherent, the receiver pulse train can be made coherent by phase locking a reference oscillator (COHO) to the transmitter on a pulse-by-pulse basis. Processing that makes use of both the phase and amplitude information across the whole pulse train is called *coherent signal processing*. Noncoherent signal processing makes use of only the phase and amplitude information for each pulse within the pulse train.

Two operations essential to signal processing are matched filtering and envelope detection. Matched filtering reduces the noise power at the output of the signal processor consistent with maximizing the S/N at a particular point in time. Envelope detection extracts amplitude information from the radar signal, but destroys all phase information. Noncoherent signal processing employs matched filtering on an individual return pulse, followed by envelope detection and then weighted combination with other pulses in the pulse train. Coherent signal processing employs a weighted combination of both amplitude and phase information across the entire pulse train in combination with matched filtering, which is then followed by envelope detection. The utilization of the complete phase structure of the entire pulse train allows higher detection performance with coherent processing compared with noncoherent processing. However, coherent signal processing is more complex, requires highly stable circuitry, and targets must be correlated over the full processing time for coherent processing to be totally effective.

In this appendix, we will derive the essential results of classical radar detection theory from the viewpoint of digital signal processing. By allowing the samples to become dense, we shall show that the digital results correspond to the well known analog theory. Operations involving matched filters are currently performed by using analog circuitry, but the rapid advancement in *very-high-speed (digital) integrated circuits* (VHSIC) and GaAs digital circuits ought to allow these operations to be performed digitally. The reader is referred to the many excellent texts on this subject for the derivation of the well established analog theory [2–8].

B.1 LIKELIHOOD RATIO PROCESSORS

Consider the radar detection problem just described, specialized to the situation where the background consists of only receiver noise. The statistical detection problem is to examine the radar waveform (r_t) in a radar-resolution cell to determine which of two hypotheses is true. Under the first hypothesis (H_1), a target is present and the radar waveform consists of signal plus noise. Under the second hypothesis (H_0), the target is absent and only receiver noise is present. The problem can be compactly written as

$$\begin{aligned} r_t &= s_t + n_t \quad &: H_1 \\ r_t &= n_t &: H_0 \end{aligned} \tag{B.1}$$

The conditional probability density function completely describes the received waveform in a statistical sense under either alternative, where

$p(\mathbf{r}|H_1)$ = Conditional probability density function with target present.

$p(\mathbf{r}|H_0)$ = Conditional probability density function with target absent.

and $\mathbf{r} = (r_1, \ldots, r_n)$ is a vector in signal space (see Section A.4) that represents a sampled version of the continuous waveform.

Two performance measures are commonly used in radar. The probability of detection is the probability that the target is present and declared:

$$P_d = \int_{R_0}^{\infty} p(\mathbf{r}|\mathrm{H}_1)\, d\mathbf{r} \tag{B.2}$$

where R_0 is the decision boundary. The probability of false alarm is the probability that no target is present, but the decision is made in error and a target is declared:

$$P_{fa} = \int_{R_0}^{\infty} p(\mathbf{r}|\mathrm{H}_0)\, d\mathbf{r} \tag{B.3}$$

Obviously, a decision rule should be selected to maximize P_d while restricting the P_{fa}. The simplest rule of this class, which is extensively used in radar, is the Neyman-Pearson criterion. This criterion specifies a decision boundary that maximizes the probability of detection (P_d) while maintaining a fixed probability of false alarm (P_{fa}). The detection problem under the Neyman-Pearson criterion can be written as follows.

$$\text{Maximum: } P_d = \int_{R_0}^{\infty} p(\mathbf{r}|\mathrm{H}_1)\, d\mathbf{r}$$

$$\text{Under Constraint: } P_{fa} = \int_{R_0}^{\infty} p(\mathbf{r}|\mathrm{H}_0)\, d\mathbf{r} = \gamma \tag{B.4}$$

The optimum decision region can be found by using the calculus of variation employing Lagrange multipliers [9,10]. Form the auxiliary function:

$$G = P_d + \lambda(\gamma - P_{fa}) \tag{B.5}$$

where λ is a Lagrange multiplier. This can be written as

$$G = \int_{R_0}^{\infty} p(\mathbf{r}|\mathrm{H}_1)\, dr + \lambda\left(\gamma - \int_{R_0}^{\infty} p(\mathbf{r}|\mathrm{H}_0)\, d\mathbf{r}\right) \tag{B.6}$$

Clearly, maximizing G will maximize P_d:

$$\frac{\partial G}{\partial \bar{r}} = -p(R_0|\mathrm{H}_1) + \lambda p(R_0|\mathrm{H}_0) = 0 \tag{B.7}$$

Therefore, to maximize P_d the constant Lagrange multiplier equals

$$\lambda = \frac{p(R_0|H_1)}{p(R_0|H_0)} \tag{B.8}$$

Define the likelihood ratio as the ratio of the conditional probability densities under alternative hypotheses:

$$LR(\mathbf{r}) = \frac{p(\mathbf{r}|H_1)}{p(\mathbf{r}|H_0)} \tag{B.9}$$

The function G can be written as

$$G = \lambda\gamma + \int_{R_0}^{\infty} [L(r) - \lambda]\, p(\mathbf{r}|H_0)\, d\mathbf{r} \tag{B.10}$$

where maximum G, and hence maximum P_d, are clearly achieved by selecting the decision boundary so that hypothesis H_1 is chosen when $L(\mathbf{r})$ exceeds λ (this increases the value of the integral) and hypothesis H_0 is chosen otherwise.* The optimum test then becomes

$$L(\mathbf{r}) = \frac{p(\mathbf{r}|H_1)}{p(\mathbf{r}|H_0)} \mathop{\gtrless}_{H_0}^{H_1} \lambda \tag{B.11}$$

In many cases, use of the log-likelihood ratio instead of the actual likelihood ratio is convenient. The log-likelihood ratio is given by

$$l(\mathbf{r}) = \ln L(\mathbf{r}) = \ln \frac{p(\mathbf{r}|H_1)}{p(\mathbf{r}|H_0)} \tag{B.12}$$

Equation (B.10) becomes

$$G = \lambda\gamma + \int_{R_o}^{\infty} (e^{l(\mathbf{r})} - e^{\ln\gamma})\, p(\mathbf{r}|H_0)\, d\mathbf{r} \tag{B.13}$$

*This selection of the decision boundary includes values of $L(\mathbf{r})$ that will increase the value of the integral while excluding values of $L(\mathbf{r})$ that will decrease the integral.

and the optimum test becomes

$$l(\mathbf{r}) = \ln \frac{p(\mathbf{r}|\mathrm{H}_1)}{p(\mathbf{r}|\mathrm{H}_0)} \underset{\mathrm{H}_0}{\overset{\mathrm{H}_1}{\gtrless}} \ln\lambda = \Gamma \tag{B.14}$$

because this selection maximizes the integral in (B.13). Equation (B.13) illustrates that the choice between hypotheses H_0 and H_1 can be based not only on the likelihood ratio $L(\mathbf{r})$, but also on any monotonic function $f(\mathbf{r}) = f[L(\mathbf{r})]$ of the likelihood ratio.

Consider the case where multiple independent samples of the signal-plus-noise and noise waveforms are available under the alternative hypotheses. The conditional probability density functions under the alternate hypotheses are then of the form represented by (A.55), resulting in a log-likelihood function given by

$$l(\mathbf{r}) = \ln \frac{p(r_1|\mathrm{H}_1) \ldots p(r_n|\mathrm{H}_1)}{p(r_1|\mathrm{H}_0) \ldots p(r_n|\mathrm{H}_0)} = \sum_{i=1}^{n} \ln[L(r_i)] \tag{B.15}$$

The log-likelihood ratio test then becomes

$$l(\mathbf{r}) = \sum_{i=1}^{n} l(r_i) \underset{\mathrm{H}_0}{\overset{\mathrm{H}_1}{\gtrless}} \Gamma \tag{B.16}$$

where $l(r_i) = \ln L(r_i)$. Equation (B.16) indicates that the optimal signal processor for multiple independent samples forms the log-likelihood ratio for each individual sample and then combines this result in a linear integrator. For Gaussian random variables, this result applies for uncorrelated samples as well as independent samples.

In general, the likelihood ratio for a set of coherent radar returns is a function of the unknown amplitude, phase, and doppler-shifted frequency of the radar return. When the distribution of these parameters is known *a priori*, the best test variable is the likelihood ratio averaged over the unknown parameters [11]. The average or generalized likelihood ratio test [11, 12] hence becomes

$$\bar{L}(\mathbf{r}) = \int L(\mathbf{r}|\boldsymbol{\Omega})\, p(\boldsymbol{\Omega})\, d\boldsymbol{\Omega} \tag{B.17}$$

where $L(\mathbf{r}|\boldsymbol{\Omega})$ is the likelihood ratio conditioned on the unknown vector $\boldsymbol{\Omega}$ (e.g., phase, amplitude, doppler frequency) and $p(\boldsymbol{\Omega})$ is the probability density function of the unknown parameters.

In nearly all radar situations of practical interest, the phase of the return signal is uniformly distributed over $(0, 2\pi)$ with respect to an arbitrary reference signal.

Target amplitude distributions are usually modeled by Rayleigh, Rician, chi-squared, or log-normal probability distributions.

Optimal signal processors based on the averaged likelihood ratio for phase and amplitude distributions have been developed [11] and lead to practical results. The situation for the averaged likelihood ratio processor when the target's doppler frequency is uniformly distributed over the band of interest is far more complex and, in this situation, a maximum likelihood processor (a bank of contiguous doppler filters) is usually employed [11,12].

B.2 LIKELIHOOD RATIOS FOR GAUSSIAN PROCESSES

In principle, likelihood-ratio processors can be developed for any situation where the necessary probability distributions are known. However, in practice, most likelihood ratio processors have been developed for additive Gaussian noise. A general formulation for Gaussian noise follows, and it will be specialized in later sections to matched filters, correlation detectors, synchronous detectors, and quadrature channel detectors.

Consider the following radar detection problem, where the received target signal within a range resolution cell under alternative hypotheses is given by

$$\begin{aligned} r_t &= \sqrt{2E_r} f_t \cos(\omega_o t + \theta) + n_t \quad &: \mathrm{H}_1 \\ r_t &= n_t &: \mathrm{H}_0 \end{aligned} \tag{B.18}$$

$$0 \le t \le T_p$$

and

$$\int_0^{T_p} f_t^2 \, \mathrm{d}t = 1 \tag{B.19}$$

where E_r is the average energy in the received signal, f_t is the normalized transmitter envelope, n_t is a Gaussian narrowband noise process, θ is the phase of the signal, ω_c is the carrier radian frequency, and T_p is the receiver processing time.

The detection problem can be compactly written in complex notation (see Section A.2) as

$$\begin{aligned} r_t &= \mathrm{Re}\{[\sqrt{2E_r} \tilde{f}_t \, \mathrm{e}^{-j\theta} + \tilde{n}_t] \, \mathrm{e}^{-j\omega_c t}\} \quad &: \mathrm{H}_1 \\ r_t &= \mathrm{Re}[\tilde{n}_t \, \mathrm{e}^{-j\omega_c t}] &: \mathrm{H}_0 \end{aligned} \tag{B.20}$$

$$0 \le t \le T_p$$

where $\tilde{n}_t$ and $\tilde{f}_t$ are complex envelope processes. The complex signal process is given by

$$\tilde{s}_t = \sqrt{2E_r}\,\tilde{f}_t \tag{B.21}$$

where

$$\int_0^{T_p} |\tilde{f}_t|^2\, \mathrm{d}t = 1 \quad \int_0^{T_p} |\tilde{s}_t|^2\, \mathrm{d}t = 2E_r \tag{B.22}$$

The detection problem described by (B.20) consists of examining the received waveform r_t and determining whether it consists of signal plus noise or noise alone. The optimal signal processor, as previously described, forms the likelihood ratio and compares this against a threshold. To determine the likelihood ratio, we need to find the conditional probability density functions under the alternative hypotheses. Because the conditional probability density functions completely define the received signal in a statistical sense, not surprisingly, they contain the information to form an optimal processor.

Clearly, the received waveform is a Gaussian random process under either hypothesis. Under hypothesis H_0, the received waveform is just the Gaussian noise process, whereas, under hypothesis H_1, the received waveform is a shifted version of the Gaussian noise process. Also recall from the sampling theorem discussion (see Section A.3), that it is possible to represent a random process (limit in quadratic mean sense) by a set of samples that takes the form of a vector in signal space. The number of samples (k) is a function of the bandwidth (B) occupied by the signal and the noise, given by

$$k = 2B\,T_p \tag{B.23}$$

Given these conditions, the conditional probability distributions under either hypothesis are given by the kth-dimensional Gaussian probability density function (see (A.57) for real variables). The kth-dimensional probability density function for complex envelope samples is given by [10]

$$p(\tilde{\mathbf{n}}) = \frac{1}{(2\pi)^k|\mathbf{R}_{\tilde{n}}|}\exp\left(-\frac{1}{2}\,\tilde{\mathbf{n}}^{\dagger}\mathbf{R}_{\tilde{n}}^{-1}\tilde{\mathbf{n}}\right) \tag{B.24}$$

where the symbol † denotes the complex conjugate transpose, and the elements of the covariance function are

$$R_{\tilde{n}}(kl) = \frac{1}{2}\,E(\tilde{n}_k\tilde{n}_l) \tag{B.25}$$

The detection problem can be stated in sampled-data form by letting $\tilde{\mathbf{r}}^T = (\tilde{r}_1, \tilde{r}_2, \ldots, \tilde{r}_k)$ be a vector of k complex receiver samples, $\tilde{\mathbf{s}}^T = (\tilde{s}_1, \tilde{s}_2, \ldots, \tilde{s}_k)$ be a vector of k complex signal samples, and $\tilde{\mathbf{n}}^T = (\tilde{n}_1, \tilde{n}_2, \ldots, \tilde{n}_k)$ be a vector of k complex Gaussian noise samples, where the symbol T represents the vector transpose. The problem can be written in vector form as follows:

$$\begin{aligned} \tilde{\mathbf{r}} &= \tilde{\mathbf{s}}\, e^{-j\theta} + \tilde{\mathbf{n}} \quad &: H_1 \quad 0 \leq t \leq T_p \\ \tilde{\mathbf{r}} &= \tilde{\mathbf{n}} \quad &: H_0 \end{aligned} \tag{B.26}$$

where θ is uniformly distributed $(0, 2\pi)$. By using (B.24), the likelihood ratio can be written as

$$\mathrm{LR}(\tilde{\mathbf{r}}|\tilde{s}, \theta, \omega) = \frac{\dfrac{1}{(2\pi)^k|\mathbf{R}_{\tilde{n}}|} \exp\{-1/2[(\tilde{\mathbf{r}} - \tilde{\mathbf{s}}\, e^{-j\theta})^{\dagger}\mathbf{R}_{\tilde{n}}^{-1}(\tilde{\mathbf{r}} - \tilde{\mathbf{s}}\, e^{-j\theta})]\}}{\dfrac{1}{(2\pi)^k|\mathbf{R}_{\tilde{n}}|} \exp[-1/2(\tilde{\mathbf{r}}^{\dagger}\mathbf{R}_n^{-1}\tilde{\mathbf{r}})]} \tag{B.27}$$

which can be simplified by combining terms, finding the log-likelihood ratio, and simplifying the notation by suppressing the known signal and doppler frequency dependence. The resulting log-likelihood ratio is

$$l(\tilde{\mathbf{r}}|\theta) = -\frac{1}{2}\tilde{\mathbf{s}}^{\dagger}\mathbf{R}_n^{-1}\tilde{\mathbf{s}} + \frac{\tilde{\mathbf{r}}^{\dagger}\mathbf{R}_n^{-1}\tilde{\mathbf{s}}\, e^{-j\theta} + \tilde{\mathbf{s}}^{\dagger}\, e^{j\theta}\mathbf{R}_n^{-1}\tilde{\mathbf{r}}}{2} \tag{B.28}$$

which can be further simplified to

$$l(\tilde{\mathbf{r}}|\theta) = -\frac{1}{2}\tilde{\mathbf{s}}^{\dagger}\mathbf{R}_n^{-1}\tilde{\mathbf{s}} + R_e[e^{+j\theta}\,\tilde{\mathbf{s}}^{\dagger}\mathbf{R}_n^{-1}\tilde{\mathbf{r}}] \tag{B.29}$$

The log-likelihood ratio of (B.29) can be adapted for coherent processing of the received radar samples by incorporating the assumed known phase dependence into the signal term. This is equivalent to choosing $\theta = 0$ and results in an optimum signal processor that computes

$$\mathrm{Re}(\tilde{\mathbf{s}}^{\dagger}\mathbf{R}_{\tilde{n}}^{-1}\tilde{\mathbf{r}}) \underset{H_0}{\overset{H_1}{\gtrless}} \frac{1}{2}\tilde{\mathbf{s}}^{\dagger}\mathbf{R}_{\tilde{n}}^{-1}\tilde{\mathbf{s}} + \gamma' = \gamma \tag{B.30}$$

Because the signal ($\tilde{\mathbf{s}}$) and inverse noise covariance matrix ($R_{\tilde{n}}^{-1}$) are presumed known, the optimal processor can be viewed as a digital filter with complex weights given by

$$\mathbf{w}_o = \mathbf{R}_{\tilde{n}}^{-1}\tilde{\mathbf{s}}$$

and the optimal processor tests for

$$\mathrm{Re}(\mathbf{w}_o^\dagger \tilde{\mathbf{r}}) \underset{H_0}{\overset{H_1}{\underset{\leq}{>}}} \gamma \tag{B.31}$$

An equivalent form of the optimal processor in terms of the individual samples is given by

$$\mathrm{Re}\left(\sum_{i=1}^{k} w_i^\dagger \tilde{r}_i\right) \underset{H_0}{\overset{H_1}{\underset{\leq}{>}}} \gamma \tag{B.32}$$

which illustrates that the optimal coherent processor forms a weighted sum of the complex received signal, followed by extraction of the real part, which is tested against a fixed threshold.

When the phase between samples or among a group of samples is unknown, the optimal processor forms the averaged likelihood ratio in accordance with (B.17). The averaged likelihood ratio is given by

$$\overline{\mathrm{LR}}(\tilde{\mathbf{r}}) = \frac{1}{2\pi}\int_0^{2\pi} \exp\left[-\frac{1}{2}\,\tilde{\mathbf{s}}^\dagger \mathbf{R}_n^{-1}\tilde{\mathbf{s}} + |\tilde{\mathbf{s}}^\dagger \mathbf{R}_n^{-1}\tilde{\mathbf{r}}| \cos(\theta - \phi)\right] d\theta \tag{B.33}$$

resulting in a log-likelihood ratio of

$$\bar{l}(\tilde{\mathbf{r}}) = -\frac{1}{2}\,\tilde{\mathbf{s}}^\dagger \mathbf{R}_{\tilde{n}}^{-1}\tilde{\mathbf{s}} + \ln[I_0(|\tilde{\mathbf{s}}^\dagger \mathbf{R}_n^{-1}\tilde{\mathbf{r}}|)] \tag{B.34}$$

where $I_0(\cdot)$ is the zero-order Bessel function of the first kind with imaginary argument. The optimal signal processor for radar signals with an unknown phase becomes

$$|\tilde{\mathbf{s}}^\dagger \mathbf{R}_{\tilde{n}}^{-1}\tilde{\mathbf{r}}| \underset{H_0}{\overset{H_1}{\underset{\leq}{>}}} \gamma \tag{B.35}$$

This processor can be interpreted as a digital filter with complex weights given by

$$\mathbf{w}_o = \mathbf{R}_{\tilde{n}}^{-1}\tilde{\mathbf{s}} \tag{B.36}$$

followed by an envelope detector, and the optimal test becomes

$$|\mathbf{w}_0^\dagger \tilde{\mathbf{r}}| \underset{H_0}{\overset{H_1}{\gtrless}} \gamma \tag{B.37}$$

The digital filter complex weights for the known and unknown phase cases are identical. However, the introduction of the unknown phase condition results in the need for an envelope detector. The performance of the signal processor with an unknown phase is somewhat less than that of the signal processor when the phase is known, as we shall determine later in this section. This is intuitively satisfying because we expect that the loss of phase information would result in lower performance.

In this appendix, we consider only the Gaussian white noise case, leaving the more complex correlated Gaussian noise case to Chapter 5, on optimum radar doppler processing. The covariance function for the complex white noise process in matrix form is $R_{\tilde{n}} = N_0\mathbf{I}$ (A.40), where the power spectral density is given by $N_0/2$ (W/Hz) and $\mathbf{I}$ is the identity matrix. Because the inverse of the identity matrix is another identity matrix, the inverse covariance function is given by

$$\mathbf{R}_{\tilde{n}}^{-1} = \frac{1}{N_0}\mathbf{I} \tag{B.38}$$

The term in (B.30) and (B.34) involving the signal terms is then easily evaluated as

$$\frac{1}{2}\tilde{\mathbf{s}}^\dagger \mathbf{R}_{\tilde{n}}^{-1}\tilde{\mathbf{s}} = \frac{1}{2N_0}\tilde{\mathbf{s}}^\dagger \mathbf{I}\tilde{\mathbf{s}} = \frac{1}{2N_0}\sum_{i=1}^{k}|\tilde{s}_i|^2 = \frac{E_r}{N_0} \tag{B.39}$$

where the final result follows from an application of Parseval's theorem [7]. The transposed complex conjugate of the weights of (B.36) is given by

$$\mathbf{w}_o^\dagger = \left[\frac{1}{N_0}\mathbf{I}\tilde{\mathbf{s}}\right]^\dagger = \frac{1}{N_0}\tilde{\mathbf{s}}^\dagger\mathbf{I} = \frac{1}{N_0}\tilde{\mathbf{s}}^* \tag{B.40}$$

where the asterisk indicates a complex conjugate operation. The completely known signal test becomes

$$\text{Re}\left(\frac{1}{N_0}\sum_{i=1}^{k}\tilde{s}_i^*\tilde{r}_i\right) \underset{H_0}{\overset{H_1}{\gtrless}} \frac{E_r}{N_0} + \gamma' = \gamma \tag{B.41}$$

and the unknown phase test is given by

$$\left| \frac{1}{N_0} \sum_{i=1}^{k} \tilde{s}_i^* \tilde{r}_i \right| \underset{H_0}{\overset{H_1}{\gtrless}} \gamma \tag{B.42}$$

B.3 EXACTLY KNOWN SIGNAL—THE SYNCHRONOUS DETECTOR

The optimal digital signal processor for an exactly known signal (i.e., phase, amplitude, and doppler frequency are known) in a background of white Gaussian noise is given by (B.41). The analog version of this test can be derived by using Parseval's theorem [7]:

$$\sum_{i=1}^{k} \tilde{s}_i^* \tilde{r}_i = \int_0^{T_p} \tilde{s}_t^* \tilde{r}_t \, dt \tag{B.43}$$

resulting in

$$\mathrm{Re}\left(\frac{1}{N_0} \int_0^{T_p} \tilde{s}_t^* \tilde{r}_t \, dt \right) \underset{H_0}{\overset{H_1}{\gtrless}} \gamma \tag{B.44}$$

The signal envelope can be selected as

$$\tilde{s}_t^* = s_{ct} = s_t = \sqrt{2E_r} f_t \tag{B.45}$$

and the phase is known. Then, the test becomes

$$\frac{1}{N_0} \int_0^{T_p} r_{ct} s_t \, dt \underset{H_0}{\overset{H_1}{\gtrless}} \gamma \tag{B.46}$$

The received signal can be written in terms of its quadrature components:

$$r_t = r_{ct} \cos\omega_c t - r_{st} \sin\omega_c t \tag{B.47}$$

from which its real part can be derived by using

$$r_{ct} = (2 r_t \cos\omega_c t)_{\mathrm{LP}} \tag{B.48}$$

where the subscript LP indicates that only the low-pass component is utilized. The final signal processor for the case of a completely known signal becomes

$$\frac{2}{N_0}\int_0^{T_p} r_t s_t \cos\omega_t \, dt \underset{H_0}{\overset{H_1}{\gtrless}} \gamma \tag{B.49}$$

A block diagram for this processor is depicted in Figure B.2. This processor is sometimes called a *synchronous detector* because it requires a coherent reference in synchrony with the carrier of the input signal.

The requirement for a synchronous reference makes this detector extremely difficult to realize in practice. Rather, the detector is used to set a standard of performance, as it represents the optimal processor when all signal parameters are exactly known.

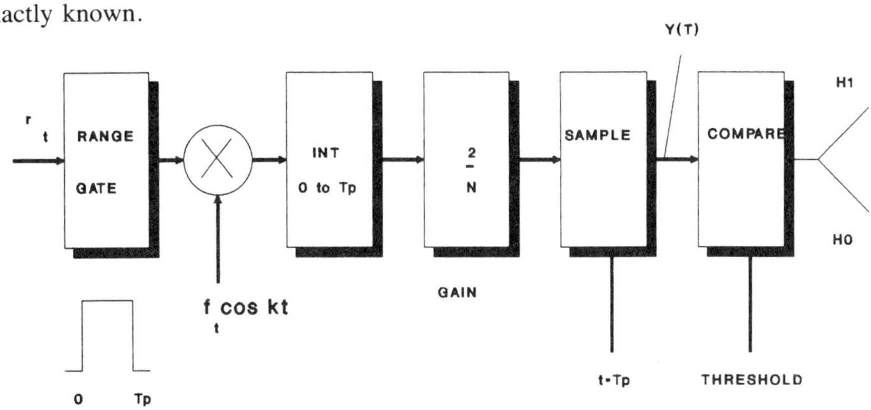

Figure B.2 Block diagram of synchronous detector.

To determine the performance of the exactly known signal, a useful approach is to consider the quadrature components of (B.20):

$$\begin{aligned} r_t &= (\sqrt{2E_r} f_t + x_t)\cos\omega_c t - y_t \sin\omega_c t \quad &: H_1 \\ r_t &= x_t \cos\omega_c t - y_t \sin\omega_c t &: H_0 \end{aligned} \tag{B.50}$$

Note that the quadrature noise term y_t will be suppressed by the synchronous detector action. The test statistic (y_t), as shown in Figure B.2, is given by

$$y_t = \int_0^T \sqrt{2} f_t r_t \cos\omega_c t \, dt \tag{B.51}$$

Because y_t is the output of a linear operation on a Gaussian distributed process, it is a Gaussian random variable. Therefore, the detector performance is determined by the mean and variance of y_t under the alternative hypotheses.

Under hypothesis H_0, the mean is given by

$$E(y_t|H_0) = \int_0^T \frac{1}{\sqrt{2}} f_t E(x_t)\, dt = 0 \tag{B.52}$$

The variance is given by

$$E(y_t)^2 = \int_0^T \int_0^T \frac{1}{2} f_\alpha f_\beta E(x_\alpha x_\beta)\, d\alpha\, d\beta \tag{B.53}$$

The autocorrelation of the white noise is given by

$$E(x_\alpha x_\beta) = \frac{N_0}{2}\, \delta(\alpha - \beta) \tag{B.54}$$

resulting in

$$\text{var}(y_T) = \frac{N_0}{4} \tag{B.55}$$

Under hypothesis H_1, we need to evaluate only the mean because the variance will be the same as under hypothesis H_0. The mean is given by

$$\begin{aligned} E(y_t|H_1) &= \int_0^T \sqrt{2} f_t \cos\omega_c t \cdot [\sqrt{2E_r} f_t + E(x_t)] \cos\omega_c t\, dt \\ E(y_t|H_1) &= \sqrt{E_r} \end{aligned} \tag{B.56}$$

The resulting Gaussian probability distributions under the alternative hypotheses are shown in Figure B.3.

By using the probability distribution under hypothesis H_0, the probability of false alarm can be computed as

$$P_{fa} = \frac{1}{2}[1 - \text{erf}(\gamma_1)] \tag{B.57}$$

where γ_1 is the threshold given by

$$\gamma_1 = \text{erf}^{-1}(1 - 2P_{fa}) \tag{B.58}$$

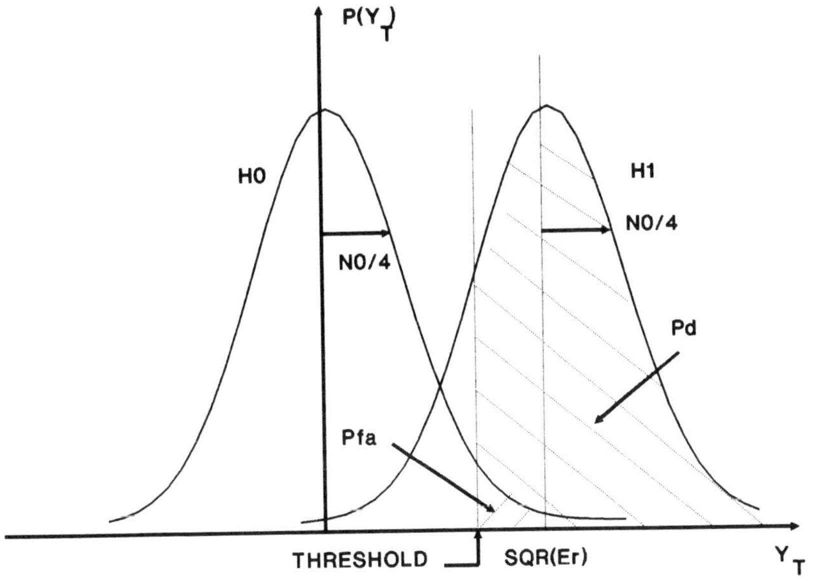

Figure B.3 Exactly known signal probability distributions.

and the error function is defined as

$$\text{erf}z = \frac{2}{\sqrt{\pi}} \int_0^z e^{-t^2}\, dt \tag{B.59}$$

The probability of detection can be computed using the probability distribution under hypothesis H_1 resulting in

$$P_d = \frac{1}{2}\left[1 + \text{erf}\left(\sqrt{\frac{2E_r}{N_0}} - \gamma'\right)\right] \tag{B.60}$$

where γ is the normalized threshold. The average S/N ratio can be defined as

$$R_A = \frac{2E_r}{N_0} \tag{B.61}$$

so that (B.60) can be written as

$$P_d = \frac{1}{2}[1 + \text{erf}(\sqrt{R_A} - \gamma')] \tag{B.62}$$

However, some authors [2] use the peak S/N rather than the average ratio in their performance curves. The rationale for this can be understood by reference to Figure B.4, which depicts the relationship between the peak and average amplitudes of the energy waveform for a rectangular pulse. The relationship between the two definitions is

$$R_p = 2R_A = \frac{4E_r}{N_0} \tag{B.63}$$

Note that the mean-square of y_t divided by its variance under hypothesis H_1 results in R_p rather than R_A. The performance curve given in Figure B.5 for an exactly

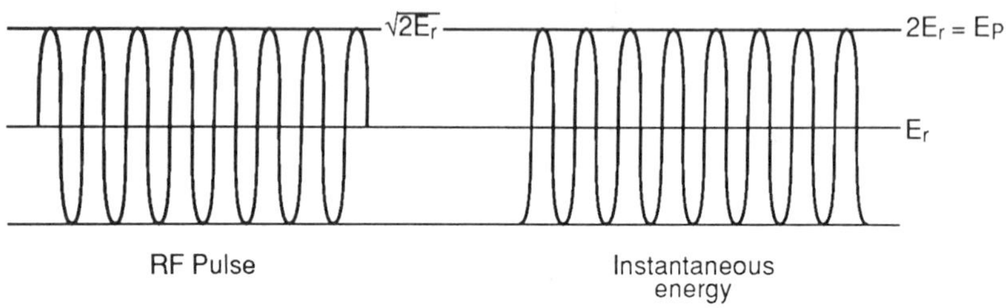

Figure B.4 Peak and average amplitudes for the rectangular pulse.

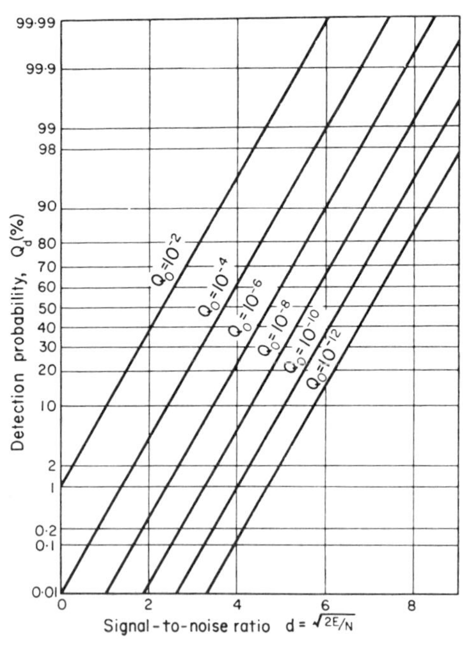

Figure B.5 Exactly known signal detection characteristic.

known signal is plotted so that either the peak or the average S/N can be used as a parameter.

B.4 EXACTLY KNOWN SIGNAL—CORRELATORS AND MATCHED FILTERS

The structure of the optimal processor for an exactly known signal can be viewed in a number of different ways. Of course, the results are all equivalent and the performance is given by Figure B.5.

In the synchronous detector, the coherent reference was identified explicitly as a phase reference associated with the receiver; the reference also can be viewed as associated with a knowledge of the signal structure. In fact, for more complex signals (i.e., pulse compression or coded waveforms), this is a more fundamental viewpoint, utilized extensively in the design of radar systems. Such processors are called *correlators* as they involve a multiplication of the received waveform by a replica of the transmitted waveform. It is intuitively obvious that the processing mechanism is based on the complete correlation of the signal part of the received waveform with the replica, whereas the noise portion is totally uncorrelated, resulting in a build-up of the signal with respect to the noise in the integrator portion of the correlator.

The exactly known signal detection problem can be put into a correlator configuration by simply expressing the signal of (B.45) with the carrier included:

$$v_t = \sqrt{2E_r}\, f_t \cos\omega_c t \tag{B.64}$$

The optimal test of (B.49) becomes

$$\frac{2}{N_0}\int_0^{T_p} r_t v_t \, dt \underset{H_0}{\overset{H_1}{\gtrless}} \gamma \tag{B.65}$$

which is the correlator version of the optimal processor. This is depicted in block diagram form in Figure B.6.

A practical problem with both the synchronous detector and the correlator in radar applications involves the need to generate an active replica of the signal in exact phase synchrony with the received signal. The matched filter, however, is a passive device that has imbedded in its design information relative to both the amplitude and phase structure of the signal. In addition, the matched filter also includes information relative to the statistical parameters of the noise. The advantage of the matched filter is that it processes the target signal as it occurs, and therefore does not require precise time synchronization of a signal reference.

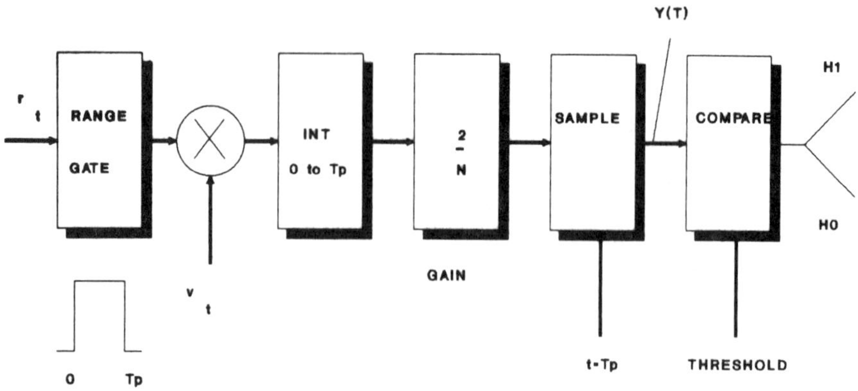

Figure B.6 Correlation detector.

The digital version of the matched filter has already been derived and is given by the complex weights (phase and amplitude), which multiply the complex received signal samples (see (B.40)). However, an instructive exercise is to derive the analog version of a matched filter.

The output at time T of a filter is obtained by convolving its impulse response, $h(t)$, with the input signal, $r(t)$. The filter output at time T is given by

$$y_t = \int_0^T h(T - \tau)r(\tau)\,d\tau \tag{B.66}$$

Comparing this equation with the correlator expression indicates that

$$h(T - \tau) = v(\tau) \tag{B.67}$$

Substitution of $t = T - \tau$ results in the impulse response for the matched filter:

$$h(t) = v(T - t) = \sqrt{2E_r} f(T - t) \cos\omega_c(T - t) \tag{B.68}$$

which consists of the signal waveform reversed in time and delayed by the duration of the signal. In the frequency domain, the transfer function is given by

$$H(j\omega) = \int_0^T v(T - t)\, e^{-j\omega t}\, dt \tag{B.69}$$

Substitution of $z = T - t$ results in

$$H(j\omega) = e^{-j\omega T} \int_0^T v(z)\, e^{j\omega z}\, dz = V^*(j\omega)\, e^{-j\omega T} \tag{B.70}$$

when $v(z)$ is real. Thus, the transfer function of the matched filter is the conjugate of the Fourier transform of the signal modified by a time delay factor.

The output waveform of a matched filter excited by a rectangular RF input pulse is depicted in Figure B.7. The same figure also shows a correlator output. Note that the matched filter and correlator outputs coincide only at the positive peaks of the RF carrier. This is consistent with the condition imposed to derive the matched filter's impulse response; namely, that the matched filter output and the correlator output be equal at the time T [4].

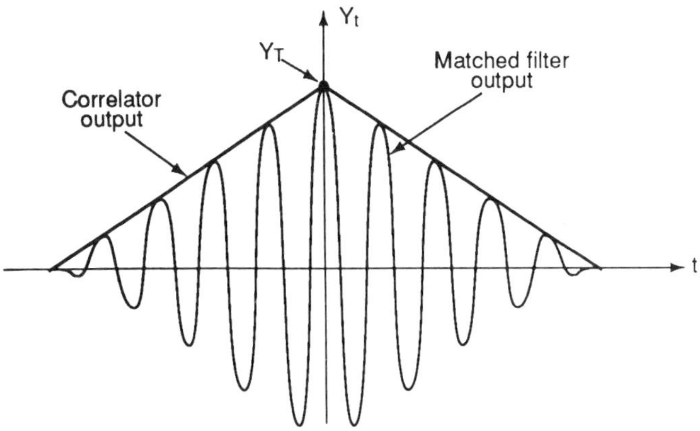

Figure B.7 Matched filter output waveform.

The necessity of sampling the matched filter at time T to obtain its maximum output is equivalent to the synchronization problem of the correlator. This problem is obviated in practical systems by employing an envelope detector, which, in effect, causes the matched filter and correlator waveforms to be identical. Thus, all requirements for RF carrier synchronization in the matched filter are removed, and it can process the signal as it occurs without any requirement for a precision phase reference.

In some problems, use of a complex form of the correlator or matched filter is convenient [10]. This can be obtained from (B.44), which defines the optimal processor in complex form, and (A.25), which converts the complex form into a real form. The resulting optimal test is

$$\left| \frac{1}{N_0} \int_0^{T_p} \tilde{s}_t^* \tilde{r}_t \, dt \right| \underset{H_0}{\overset{H_1}{\gtrless}} \gamma \tag{B.71}$$

and is depicted in Figure B.8. The impulse response of the complex matched filter is similarly given by [10]

$$\tilde{h}(t) = \tilde{s}^*(T - t) \tag{B.72}$$

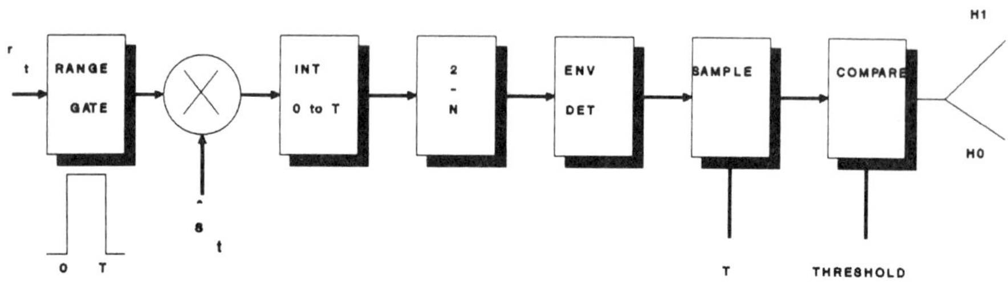

Figure B.8 Optimal processor for complex signals.

B.5 MATCHED FILTER—MAXIMUM S/N RATIO

The matched filter was introduced in the previous section as a natural implementation of the likelihood ratio processor for the detection of a steady signal in white Gaussian noise. This is different from the historical approach which derives the matched filter as the linear filter that maximizes S/N at the filter's output for the preceding problem.

The fact that the two approaches are equivalent for detection of signals in white Gaussian noise can be seen by examination of Figure B.3, which shows the probability distributions for the test statistic of the likelihood ratio processor. The scale of the Gaussian signal-plus-noise distribution is determined by the signal's energy. The width of the Gaussian distributions under either hypothesis is determined by the noise variance. If the scale were increased or the variance reduced while holding the other quantity constant, the probability of detection could be increased for a fixed false alarm probability. However, the probability of detection is maximized, leading to the conclusion that the scale-to-noise variance ratio is maximized or, equivalently, the S/N is maximized. Hence, the likelihood ratio processor that is linear for the Gaussian case and the maximum S/N linear filter are equivalent in this case.

The importance of the matched filter in radar design makes productive a more formal derivation than the heuristic argument. In addition, the derivation that follows is also useful in the design of transversal filters for MTI applications [13] and adaptive antenna arrays [14] utilized to reject interference while maintaining a desired antenna radiation pattern.

Consider the linear transversal filter depicted in Figure B.9, which is driven by either complex signals ($\tilde{s}_k$) or complex noise ($\tilde{n}_k$) samples. The complex filter signal voltage output is given by

$$\tilde{\mathbf{v}}_s = \mathbf{w}^T \tilde{\mathbf{s}} \tag{B.73}$$

and the noise output is given by

$$\tilde{\mathbf{v}}_n = \mathbf{w}^T \tilde{\mathbf{n}} \tag{B.74}$$

where **w** is the complex filter weight vector. The signal power is given by

$$P_s = \frac{1}{2} E|\tilde{\mathbf{v}}_s|^2 = \frac{1}{2} \mathbf{w}^\dagger \tilde{\mathbf{M}}_s \mathbf{w} \tag{B.75}$$

where

$$\tilde{\mathbf{M}}_s = E(\tilde{\mathbf{s}}\tilde{\mathbf{s}}^\dagger) \tag{B.76}$$

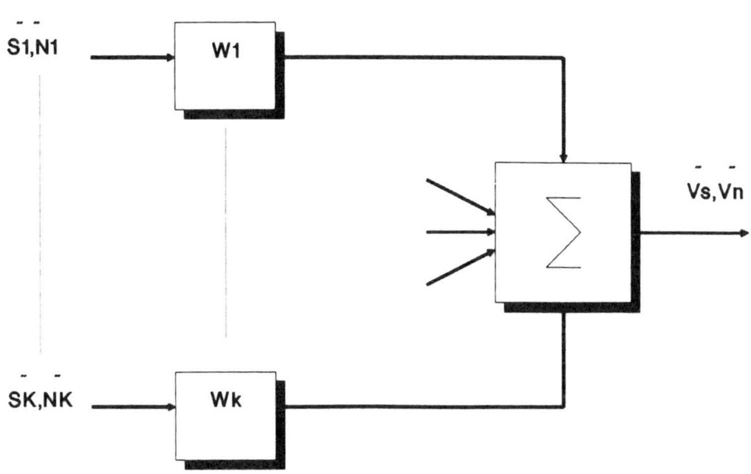

Figure B.9 Transversal filter.

is the signal's covariance matrix and the symbol † represents the complex conjugate transpose. The noise power is similarly given by

$$P_N = \frac{1}{2}\,\mathbf{w}^{\dagger}\tilde{\mathbf{R}}_{\tilde{N}}\mathbf{w} \tag{B.77}$$

where $\mathbf{R}_N$ is the noise covariance matrix. The signal-to-noise power ratio is then given by

$$S/N = \frac{\mathbf{w}^{\dagger}\tilde{M}_s\mathbf{w}}{\mathbf{w}^{\dagger}\tilde{\mathbf{R}}_{\tilde{N}}\mathbf{w}} \tag{B.78}$$

For white Gaussian noise, the covariance matrix is given by

$$\tilde{\mathbf{R}}_{\tilde{N}} = N_0\mathbf{I} \tag{B.79}$$

resulting in a noise power of

$$P_N = \frac{N_0}{2}\,|\mathbf{w}|^2 \tag{B.80}$$

By the Cauchy-Schwarz inequality [15]

$$|v_s|^2 \le |\,\mathbf{w}|^2 \cdot |\tilde{\mathbf{s}}|^2 \tag{B.81}$$

resulting in

$$S/N \le \frac{|\mathbf{w}|^2|\tilde{\mathbf{s}}|^2}{N_0|\mathbf{w}|^2} \le \frac{1}{N_0}\,|\tilde{s}|^2 = \frac{2E}{N_0} \tag{B.82}$$

where the last step follows from (B.22). Equality is achieved for complex weights given by

$$\mathbf{w}_o = \tilde{\mathbf{s}}^* \tag{B.83}$$

which corresponds to (B.40) for the likelihood ratio processor, and the S/N equals

$$S/N = \frac{2E}{N_0} \tag{B.84}$$

Note that the matched filter's S/N depends only on the signal energy and noise power spectral density, and is independent of the signal waveform.

Now consider the case where the noise is correlated. The approach is to apply a linear transformation (see (A.75)) to transform the variables into statistically independent random variables and then to apply the optimal matched filter weights of (B.83) to maximize S/N. The linear transformation (**A**) is called a *prewhitening filter* and is depicted in Figure B.10.

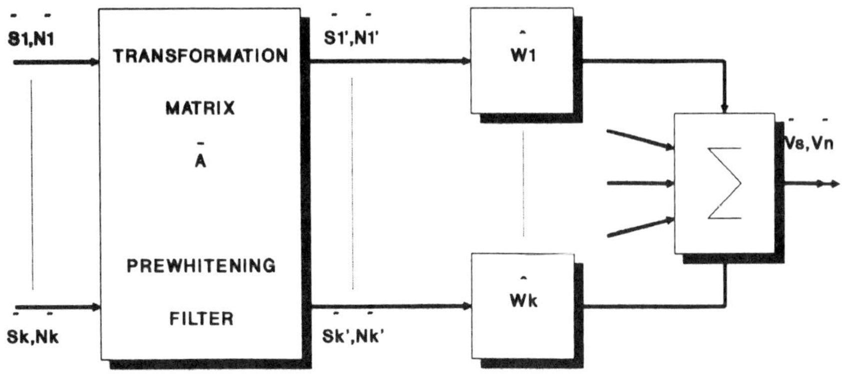

Figure B.10 Prewhitening filter.

The complex signal voltage vector is given by

$$\tilde{\mathbf{v}}_s = \hat{\mathbf{w}}^T\tilde{\mathbf{s}}' = \hat{\mathbf{w}}^T\mathbf{A}\tilde{\mathbf{s}} \tag{B.85}$$

and the complex noise vector is given by

$$\tilde{\mathbf{v}}_n = \hat{\mathbf{w}}^T\tilde{\mathbf{n}}' = \hat{\mathbf{w}}^T\mathbf{A}\tilde{\mathbf{n}} \tag{B.86}$$

where $\hat{\mathbf{w}}$ is the matched filter weights for white noise. The output noise power is given by

$$P_n = \frac{1}{2}\,\hat{\mathbf{w}}^\dagger\mathbf{A}^\dagger\tilde{\mathbf{R}}_{\tilde{N}}\mathbf{A}\hat{w} = \frac{1}{2}\,\hat{\mathbf{w}}^\dagger E(\tilde{\mathbf{n}}'\tilde{n}'^\dagger)\,\hat{w} \tag{B.87}$$

By hypothesis, $E(\tilde{\mathbf{n}}'\tilde{n}'^\dagger) = \mathbf{I}$ and

$$\mathbf{A}^\dagger\tilde{\mathbf{R}}_{\tilde{N}}\mathbf{A} = \mathbf{I};\quad \tilde{\mathbf{R}}_{\tilde{n}}^{-1} = [\mathbf{A}\mathbf{A}^\dagger] \tag{B.88}$$

The optimal weights are given by

$$\mathbf{w}_o = \mathbf{A}^T\hat{\mathbf{w}}_o = \mathbf{A}^T\tilde{\mathbf{s}}'^* = \mathbf{A}^T[\mathbf{A}\tilde{\mathbf{s}}]^* = \tilde{\mathbf{R}}_{\tilde{N}}^{-1}\tilde{\mathbf{s}}^* \tag{B.89}$$

which is identical to (B.36), derived by using the likelihood ratio approach.

B.6 QUADRATURE CHANNEL DETECTOR—UNKNOWN PHASE

In a typical radar application, the range between the target and the radar represents a very large number of transmitted signal wavelengths. This makes specifying the phase of the return signal extremely difficult, and we usually assume that the signal phase of the return signal is a random variable uniformly distributed over an angle of 2π rad.

The optimal test for the unknown phase problem has been derived in the section on likelihood ratios for Gaussian processes and is given by equation (B.42). In this section, we describe several signal processors that implement this optimal test and give detection curves that allow determining the performance of such optimal processors. Signal processors that implement the optimal unknown phase test are called *quadrature-channel* or *I and Q channel processors* because the first operation of the processor is to decompose the signal into its quadrature components.

The optimal test of (B.42) can be expanded into quadrature components by using (B.45) for the known transmitted signal, (A.21) for the unknown phase received signal, and (B.48) to convert into real signals. The resulting digital test is

$$\left(\frac{2}{N_0}\sum_{i=1}^{k} s_{ti} r_{ti} \cos\omega t_i\right)^2 + \left(\frac{2}{N_0}\sum_{i=1}^{k} s_{ti} r_{ti} \sin\omega t_i\right)^2 \underset{H_0}{\overset{H_1}{\gtrless}} \gamma_T \qquad \text{(B.90)}$$

which can be converted into an analog test by using Parseval's theorem (see B.43):

$$\left(\frac{2}{N_0}\int_0^{T_p} r_t s_t \cos\omega t \, dt\right)^2 + \left(\frac{2}{N_0}\int_0^{T_p} r_t s_t \sin\omega t \, dt\right)^2 \underset{H_0}{\overset{H_1}{\gtrless}} \gamma_T \qquad \text{(B.91)}$$

The unknown phase can be interpreted as a quadrature-channel implementation of the known signal processor, where the outputs of the quadrature channels are squared, combined, and tested against a threshold. Figure B.11 depicts the correlator version of this processor and Figure B.12 shows the matched filter implementation. Note that these implementations apply only to situations where the initial phase of the signal is unknown and only approximately to situations where multiple independent return signals, each with an unknown phase, are combined in a postdetection integrator. This results because (B.34) must be used for the multiple independent signal case, rather than (B.35), as we shall discuss more fully in the next section.

A convenient way to determine the performance of the signal processors depicted in Figures B.11 and B.12, is to write the detection problem by using Rice's representation:

$$\begin{aligned} r_t &= s_x \cos\omega_c t - s_y \sin\omega_c t \quad &: H_1 \\ r_t &= x_t \cos\omega_c t - y_t \sin\omega t \quad &: H_0 \end{aligned} \qquad \text{(B.92)}$$

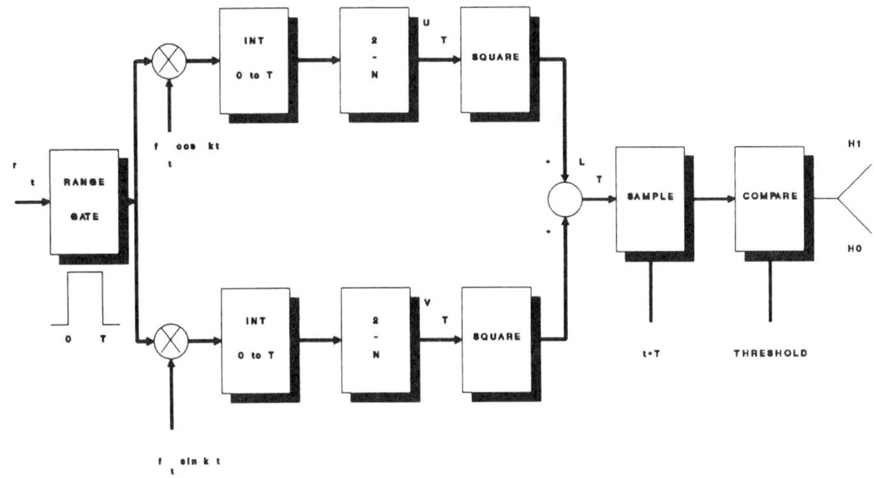

Figure B.11 Quadrature-channel detector using a correlator.

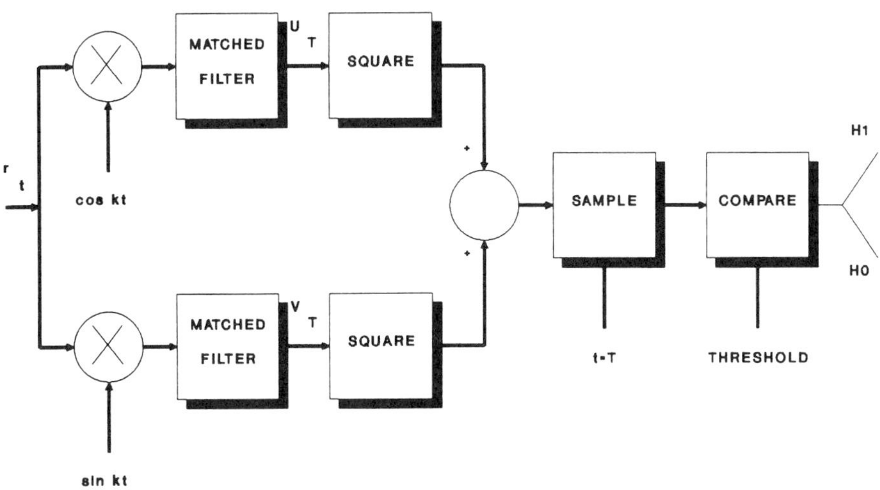

Figure B.12 Quadrature-channel detector using a matched filter.

where

$$s_x = s_t \cos\theta + x_t$$
$$s_y = s_t \sin\theta + y_t \tag{B.93}$$
$$s_t = \sqrt{2E_r} f_t$$

The detection statistic is given by

$$l_T = u_T^2 + v_T^2 \tag{B.94}$$

where u_T and v_T are Gaussian random variables because they result from linear operations on Gaussian processes:

$$u_T = \int_0^T r_t f_t \cos\omega_c t \, dt$$

$$v_T = \int_0^T r_t f_t \sin\omega_c t \, dt \tag{B.95}$$

The mean of the Gaussian variables:

$$E(u) = \sqrt{2E_r} \cos\theta = m_x$$
$$E(v) = \sqrt{2E_r} \sin\theta = m_y \tag{B.96}$$

The variance is given by

$$\mathrm{var}(u) = \mathrm{var}(v) = N_0/2 = \sigma^2 \tag{B.97}$$

The joint Gaussian density function (see (A.95)) is given by

$$p(uv|\mathrm{H}_1) = \frac{1}{2\pi\sigma^2} \exp\left[-\frac{(x - m_x)^2}{2\sigma^2} - \frac{(y - m_y)^2}{2\sigma^2} \right] \tag{B.98}$$

This can be transformed to find the probability distribution of the test statistic l_T:

$$p(l|\mathrm{H}_1) = \frac{1}{2\sigma^2} \exp\left(-\frac{l + m^2}{2\sigma^2} \right) I_o\left(\frac{\sqrt{l}\, m}{\sigma^2} \right) \tag{B.99}$$

$$p(l|\mathrm{H}_0) = \frac{1}{2\sigma^2} \exp\left(-\frac{l}{2\sigma^2} \right) \tag{B.100}$$

The distribution under the null hypothesis (H_0) is easily integrated to find

$$P_{fa} = \exp\left(-\gamma_T/2\sigma^2\right) \tag{B.101}$$

where γ_T is the decision threshold. The probability of detection can be found as

$$P_d = Q\left[\sqrt{4E_r/N_o},\quad \sqrt{2\ln(1/P_{fa})}\right] \tag{B.102}$$

where $Q(\cdot,\ \cdot)$ is Marcum's "Q function" given by [16]

$$Q(\alpha, \beta) = \int_{\beta}^{\infty} x \exp\left[-\frac{1}{2}(x^2 + \alpha^2)\right] I_o(\alpha x)\, dx \tag{B.103}$$

By using (B.63), the probability of detection can be written in terms of the average, S/N (R_A) as

$$P_d = Q\left[\sqrt{2R_A},\quad \sqrt{2\ln(1/P_{fa})}\right] \tag{B.104}$$

This is plotted in Figure B.13.

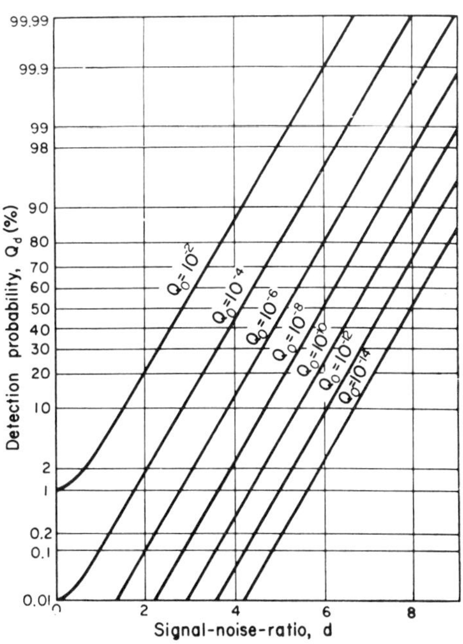

Figure B.13 Unknown phase detection characteristic.

B.7 MATCHED FILTER WITH ENVELOPE DETECTOR—UNKNOWN PHASE

We showed in the previous section that the quadrature channel detector implemented the optimum test for the unknown phase case given by (B.42). Note that this test applies for only a single observation, and a more complex processor results from multiple observations, as can be seen from (B.35) and (B.15). This will be discussed more fully in the next section.

An alternative implementation that consists of a matched filter followed by an envelope detector provides a realization with considerable practical value [7]. This realization is a natural extension of the optimal test for the exactly known signal, as can be seen by visualizing the matched filter output shown in Figure B.7 with an unknown phase, but followed by an envelope detector to obtain the peak output. The equivalence of the matched-filter–envelope-detector realization and the quadrature-channel detector will now be demonstrated formally.

The matched-filter–envelope-detector realization is depicted in Figure B.14.

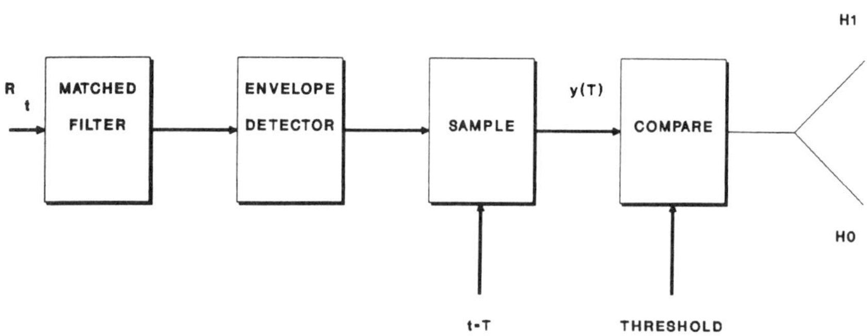

Figure B.14 Matched filter with envelope detection.

The output of the matched filter (y_t) at time T is given by

$$Y(T) = \int_0^T h(T - \tau)r(\tau)\,d\tau \tag{B.105}$$

The bandpass matched-filter impulse response (h_t) can be written in terms of its low-pass equivalent ($h_L(t)$) as

$$h_t = h_L(t)\cos\omega_c t \tag{B.106}$$

Substitution of this into (B.105) and expansion results in

$$Y(T) = \cos\omega_c T \int_0^T r_t h_L(T - t) \cos\omega_c t \, dt$$

$$+ \sin\omega_c T \int_0^T r_t h_L(T - t) \sin\omega_c t \, dt \qquad \text{(B.107)}$$

or, equivalently,

$$Y(T) = Y_c(T) \cos\omega_c T + Y_s(T) \sin\omega_c T \qquad \text{(B.108)}$$

where

$$Y_c(T) = \int_0^T r_t h_L(T - t) \cos\omega_c t \, dt \qquad \text{(B.109)}$$

$$Y_s(T) = \int_0^T r_t h_L(T - t) \sin\omega_c t \, dt \qquad \text{(B.110)}$$

The squared envelope-detected matched-filter output at time T becomes

$$|Y(T)|^2 = Y_c^2(T) + Y_s^2(T) \qquad \text{(B.111)}$$

This replicates the optimal test given by (B.94) if the filter's impulse response corresponds to that of a matched filter:

$$h_L(t) = f(T - t) \qquad \text{(B.112)}$$

where $f(\cdot)$ is the transmitted waveform. The performance of this realization is the same as that of the quadrature-channel detector, given in Figure B.13.

B.8 MULTIPLE OBSERVATIONS

The generalized radar detection problem depicted in Figure B.1 illustrates that multiple target returns are usually available and must be processed to form a detection decision. Optimal processing of the received pulse train depends on the detailed phase structure of the signal return. In general, three possible situations can arise:

- The pulses within the train are all coherently related to each other and the initial phase is known;

- The pulses within the train are all coherently related to each other and the initial phase is unknown;
- The pulses within the train are not coherently related (incoherent or noncoherent) and the initial phase of each pulse is unknown.

The first situation is simply the case of the exactly known signal, and the optimal signal processor is described by (B.41) and (B.44). The process that occurs as samples are accumulated from pulse to pulse within the signal train is called *coherent integration*. The performance for an n-pulse signal train can be determined from Figure B.5 by entering the curve with

$$R_A = nR_{A_1} \tag{B.113}$$

where R_{A_1} is the single pulse average S/N.

The second situation is the unknown phase case where the optimum processor is defined by (B.42) and (B.91). In this case, coherent integration is performed within the pulse train, but the initial phase is uncertain, requiring an envelope detection process. The performance for an n-pulse train can be determined by using Figure B.13 with the average S/N increased n times in accordance with (B.113).

The third situation is the one most common in radar work and results in noncoherent signal processing. The random phase among the samples within the pulse train can be caused by transmitter instability, target decorrelation, receiver instability, or any combination of these effects.

The optimal signal processor for the case of unknown phase and multiple observations can be determined by using (B.34) and (B.15). The resulting optimal test is given by

$$\sum_{i=1}^{n} \ln[I_o(|\tilde{\mathbf{s}}_i^{\dagger}\mathbf{R}_N^{-1}\tilde{\mathbf{r}}_i|)] \underset{H_0}{\overset{H_1}{\gtrless}} \gamma \tag{B.114}$$

which for white Gaussian noise results in

$$\sum_{j=1}^{n} \ln\left(I_0\left|\frac{1}{N_0}\sum_{i=1}^{k} \tilde{s}_{ij}{}^* \, \tilde{r}_{ij}\right|\right) \underset{H_0}{\overset{H_1}{\gtrless}} \gamma \tag{B.115}$$

The analog form of this processor is depicted in Figure B.15 for the quadrature-channel realization and Figure B.16 for the matched-filter–envelope-detector realization.

All operations, with the exception of the $\ln I_o(\cdot)$, specified in the block diagrams of Figure B.15 and B.16 are easily performed. Expansion of the $\ln I_o(\cdot)$ function in

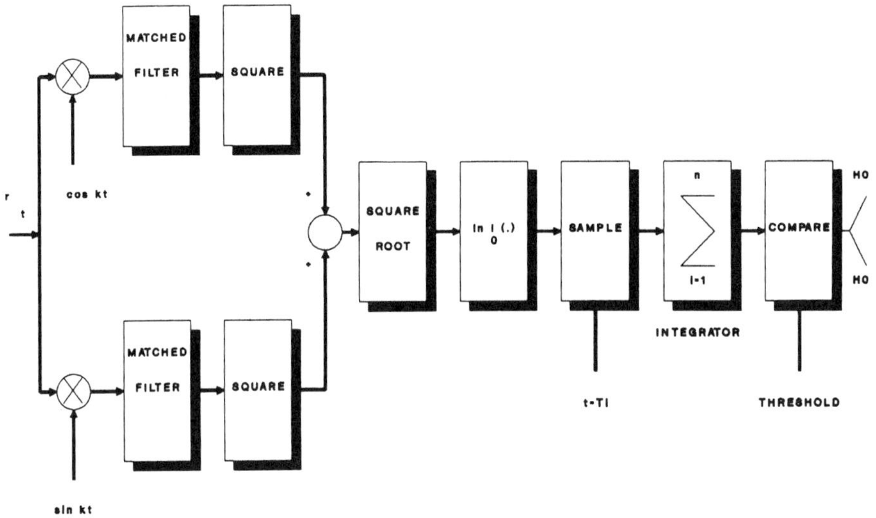

Figure B.15 Unknown phase multiple-signal quadrature-channel processor.

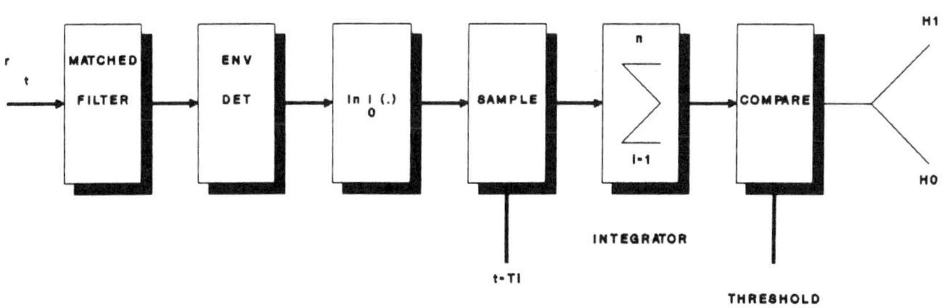

Figure B.16 Unknown phase multiple-signal matched-filter processor.

a power series [2] results in a quadratic approximation for small values of the argument and a linear approximation for large values of the argument. This leads to a square-law detector (square of envelope-detected output followed by a linear integrator) approximation for small S/N and a linear detector (envelope-detected output followed by a linear integrator) for large S/N.

The square-law detector case has been extensively analyzed by Marcum and Swerling [17]. The linear detector case is more difficult to analyze, but extensive curves are nonetheless available for this case [18,19]. In practice, the linear detector has been favored because it is less susceptible to saturation effects than is the square-law detector. Fortunately, there is little difference in detection performance between the square-law and linear detectors [2,17] with the maximum reported deviation being on the order of 0.2 dB.

REFERENCES

Appendix B

1. L. Brooks and I. Reed, "Equivalence of the Likelihood Ratio Processor, the Maximum Signal-to-Noise Ratio Filter, and the Wiener Filter, *IEEE Trans.*, Vol. AES-8, No. 5, September 1972.
2. J. DiFranco and W. Rubin, *Radar Detection*, Artech House, Norwood, MA, 1980.
3. R. Berkowitz, *Modern Radar*, John Wiley and Sons, New York, 1965.
4. A. Whalen, *Detection of Signals in Noise*, Academic Press, New York, 1971.
5. I. Selin, *Detection Theory*, Princeton University Press, NJ, 1965.
6. W. Davenport and W. Root, *An Introduction to the Theory of Random Signals and Noise*, McGraw-Hill, New York, 1958.
7. H. Van Trees, *Detection, Estimation and Modulation Theory, Part I and II*, John Wiley and Sons, New York, 1968.
8. L. Wainstein and V. Zubakov, *Extraction of Signals from Noise*, Prentice-Hall, Englewood Cliffs, NJ, 1962.
9. M. Spiegel, *Advanced Calculus*, Schaum's Outline Series, McGraw-Hill, New York, 1963.
10. H. Van Trees, *Detection, Estimation and Modulation Theory, Part III*, John Wiley and Sons, New York, 1971.
11. L. Brennen, I. Reed, and W. Sollfrey, "A Comparison of Average Likelihood and Maximum Likelihood Tests for Detecting Radar Targets of Unknown Doppler Frequency," *IEEE Trans.*, Vol. IT-14, No. 1, January 1968.
12. I. Selin, "Detection of Coherent Radar Returns of Unknown Doppler Shift," *IEEE Trans.*, Vol. IT-11, No. 3, July 1965.
13. G. Andrews, "Optimum Radar Doppler Filtering Techniques," NRL Rep. No. 7727, Washington, DC, May 1974.
14. S. Applebaum, "Adaptive Arrays," *IEEE Trans.*, Vol. AP-24, No. 5, September 1976.
15. A. Householder, *The Theory of Matrices in Numerical Analysis*, Blaisdell, New York, 1964.
16. C. Helstrom, *Statistical Theory of Signal Detection*, Pergamon, New York, 1968.
17. J. Marcum and P. Swerling, "Studies of Target Detection by Pulsed Radar," *IEEE Trans.*, Vol. IT-6, No. 2, April 1960.
18. L.V. Blake, M.I. Skolnik, ed., *Radar Handbook*, Chapter 2 in 1st Ed., McGraw-Hill, New York, 1970.
19. G. Robertson, "Operating Characteristics for a Linear Detector of CW Signals in Narrow Band Gaussian Noise," *Bell System Techn. J.*, Vol. 46, April 1967.

Appendix C
The *z*-Transform

The z-transform carries information associated with the sampling instants in a sampled-data system. This transform is used extensively in the synthesis and analysis of MTI filters, where it plays the same role as do the Laplace and Fourier transforms for continuous signals. Also, the z-transform is useful in the solution of difference equations used in digital computers and digital signal processors to approximate differential equations.

To examine the MTI filtering problem, consider a video signal $f(t)$ that is uniformly sampled by an impulse function to form a sampled-data signal given by

$$f^*(t) = f(t) \sum_{n=0}^{\infty} \delta(t - nT_s) \tag{C.1}$$

where T_s is the time period between sampling instants. The sampled signal can also be written as

$$f^*(t) = \sum_{n=0}^{\infty} f(nT_s)\delta(t - nT_s) \tag{C.2}$$

The Laplace transform of the sample data signal is given by

$$F^*(p) = \sum_{n=0}^{\infty} f(nT_s)\, \mathrm{e}^{-\mathrm{j}pT_s} \tag{C.3}$$

which can be expressed as

$$F^*(p) = \frac{1}{T_s} \sum_{-\infty}^{\infty} F(p + jn\omega_s) \tag{C.4}$$

where ω_s is the sampling radian frequency [1]. This relationship demonstrates that the effect of the sampling is to cause the spectrum of the continuous signal $f(t)$ to be replicated in a periodic manner throughout the entire frequency spectrum. In accordance with the sampling theorem (see Appendix A, Section A.3), if the continuous signal $f(t)$ is band-limited to the frequency range $|f| \leq 1/2T_s$, the signal can be recovered from the sampled signal $f^*(t)$. This is depicted in Figure C.1, which shows the amplitude spectrum of the sampler output with respect to the spectrum at the sampler input.

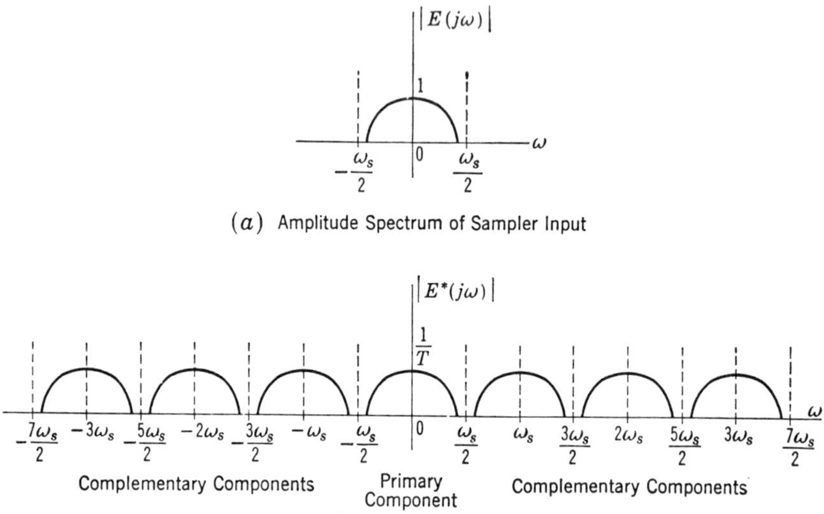

Figure C.1 Sampler transfer characteristics in the frequency domain.

Next, consider an MTI filter with a transfer function $H(p)$ applied to the sampled-data signal. The output spectrum $G(p)$ is found by taking the product of the transforms of the signal and the MTI filter as

$$G(p) = F^*(p)H(p) \tag{C.5}$$

The sampled-data version of the output spectrum $G^*(p)$ is found from (C.4) as

$$G^*(p) = \frac{1}{T}\sum_{-\infty}^{\infty} F^*(p + jn\omega_s)H(p + jn\omega_s) \tag{C.6}$$

However, $F^*(p + jn\omega_s)$ is a periodic function with period ω_s because $f^*(t)$ is the output of a sampler, resulting in

$$F^*(p) = F^*(p + jn\omega_s) \tag{C.7}$$

Thus (C.6) can be written as

$$G^*(p) = F^*(p)\frac{1}{T_s}\sum_{-\infty}^{\infty} H(p + jn\omega_s) = F^*(p)H^*(p) \tag{C.8}$$

This relation indicates that the MTI filter is periodic in ω_s, resulting in the MTI filter assuming the comb filter characteristic, depicted in Figure C.2. Also, the periodic nature of the sampled signals and filter allows the effect of the MTI filter on the signal throughout the entire frequency domain to be determined by analyzing just the effect in the fundamental period (i.e., $|f| \leq 1/2T_s$). The z-transform allows this analysis to be accomplished by essentially stripping the periodic character of the signal or filter function through transformation into a domain where only the pole-zero configuration in the fundamental period ($|f| \leq 1/2T_s$) is retained.

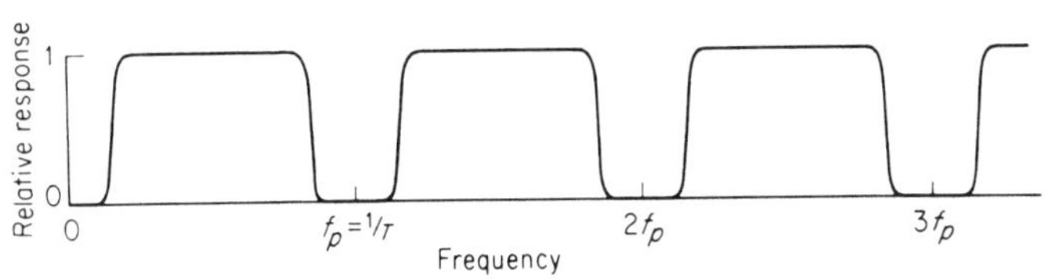

Figure C.2 Comb filter characteristic of MTI filter.

The z-transform of a signal $f(t)$ describes the behavior of the signal at the sampling instants [1]. The signal's z-transform is derived in three steps:

1. The values of the signal are determined at the sampling instants. These values form the starred time function $f^*(t)$ (see (C.2));
2. The Laplace transform of this succession of samples is taken, with the result $F^*(p)$ (see (C.3));
3. The z-transform of $F^*(p)$ is found by using the transformation:

$$z = e^{pT} \tag{C.9}$$

where p is the Laplace transform variable and T is the period of the sampled signal. The reason for using the z-transform is found by considering the sampling function:

$$i(t) = \sum_{n=0}^{\infty} \delta(t - nT_s) \tag{C.10}$$

the Laplace transform of which is given by

$$I(p) = \frac{1}{1 - e^{-pT}} \tag{C.11}$$

The poles of this function (i.e., the values of p that make the denominator equal to zero) occur throughout the imaginary axis ($p = j\omega$) of the p-plane at values $p = j2\pi n/T$, where $n = 0, \pm 1, \pm 2, \ldots$ The z-transform of the sampling function then becomes the rational function:

$$I(z) = \frac{z}{z - 1} \tag{C.12}$$

with a single pole at $z = 1$ that corresponds to the pole at $p = 0$ in the p-plane. Hence, for sampled waveforms, the z-transform converts the periodic multiplicity of poles in the p-plane (which is a result of the periodicity of the sampled signal) into a single pole in the z-plane.

From the preceding, we see that the z-transform depends on the choice of the sampling frequency, although not necessarily on the actual signal, being in sampled form. Any continuous signal possessing a Laplace transform also has a z-transform. If the sampling frequency is at least twice the highest signal frequency, the z-transform carries all the information contained in the Laplace transform.

The z-transform can be visualized as a mapping of the pole-zero pattern of the time function $f^*(t)$ or $h^*(t)$ in the p-plane (Laplace transform domain) into the z-plane (z-transform domain). This mapping transformation is depicted in Figure C.3 and maps the zero to $2\pi/T$ interval along the $j\omega$-axis of the p-plane onto the unit circle in the z-plane. The left half-plane strip, defined by $\sigma < 0$ and $0 \leq \omega \leq 2\pi/T$ is then mapped into the interior of the unit circle and the corresponding right half-plane strip into its exterior. Because the poles of any stable network are restricted to the left half-plane, including its boundary, the poles of the transfer function of an MTI filter network therefore must lie within or on the unit circle of the z-plane.

The transfer function of an MTI filter in the p-plane can be expressed as a rational function of p with real coefficients as

$$H(p) = C\frac{p^n + a_{n-1}p^{n-1} + \ldots + a_0}{p^m + b_{m-1}p^{m-1} + \ldots + b_0} \tag{C.13}$$

This function can be factored into a function with n zeros and m poles. The z-transform then maps this function into the z-plane with the same number of poles and zeros as

$$H(z) = K\frac{(z - z_{01})(z - z_{02}) \ldots (z - z_{0n})}{(z - z_{p_1})(z - z_{p_2}) \ldots (z - z_{pm})} \tag{C.14}$$

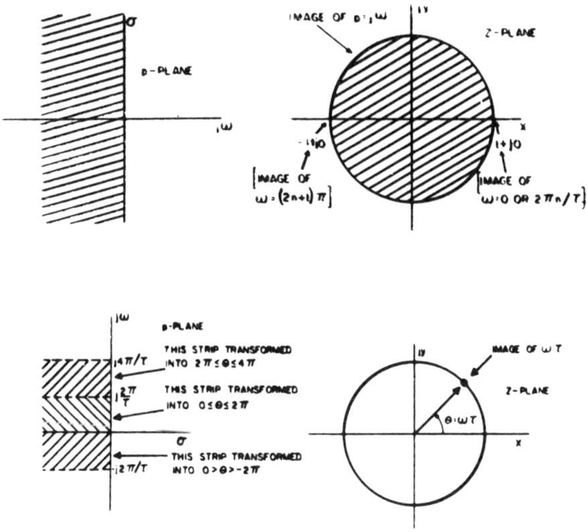

Figure C.3 Mapping of p-plane into z-plane.

or, equivalently,

$$H(z) = K \frac{z^n + \alpha_{n-1} z^{n-1} + \ldots + \alpha_0}{z^m + \beta_{m-1} z^{m-1} + \ldots + \beta_0} \tag{C.15}$$

Hence, (as the coefficients of the transfer function are real) the z-plane poles and zeros of the transfer function must occur either on the real axis or in complex conjugate pairs. The locations of zeros are not otherwise restricted. Also, there is no restriction on the multiplicity of poles or zeros [2–4].

Because the transfer function $H(z)$ (see (C.14)) is a rational function of z, we may visualize it as the ratio of two complex vector products. The numerator is given by the products emanating from the zeros and terminating at the observation point, and the denominator consists of a similar product of pole vectors. The magnitude of the transfer function $|H(j\omega)| = |H(z)|$ at $z = \exp(j\omega T)$ is found as illustrated in Figure C.4 by traversing the unit circle in the z-plane and taking the ratio of the zero vector products divided by the pole vector products. This can be expressed as

$$H(j\omega) = \frac{\text{Product of vector lengths from zeros to unit circle}}{\text{Product of vector lengths from poles to unit circle}} \tag{C.16}$$

The point on the unit circle is defined by the angle $\phi = \omega T$. Because the period T is related to the sampling frequency as $f_s = 1/T$, the angle in radians equals $\phi =$

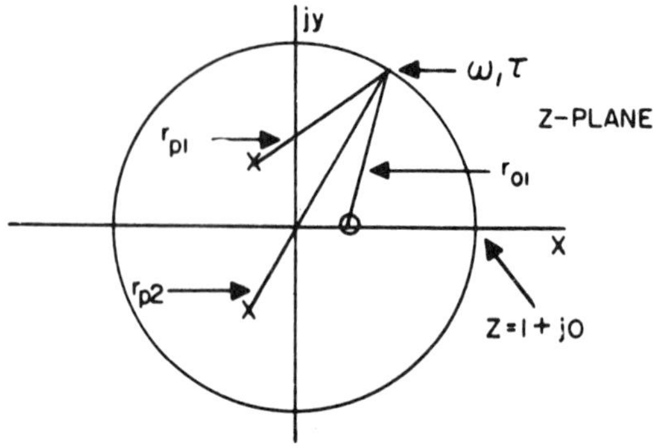

Figure C.4 Finding the magnitude of the transfer function from a pole-zero diagram.

$2\pi f/f_s$. For example, one rotation about the unit circle ($\phi = 2\pi$) corresponds to a linear frequency change from $f = 0$ to $f = f_s$, whereas a half rotation ($\phi = \pi$) corresponds to a frequency change from $f = 0$ to $f = f_s/2$. Note that a pole or zero at the origin has no effect on $|H(j\omega)|$, as the corresponding vector length is unity.

From (C.8), we can see that the sampled transfer function of the MTI filter is given by

$$H^*(p) = \frac{1}{T_s} \sum_{-\infty}^{\infty} H(p + j_n\omega_s) \tag{C.17}$$

The impulse response of the MTI filter can be found by taking the inverse Laplace transform of (C.17) (see (C.2)) as

$$h^*(t) = \sum_{n=0}^{\infty} h(nT_s)\delta(t - nT_s) \tag{C.18}$$

where $h(nT_s)$ denotes values of the impulse response of the sampling instants, which are separated by nT_s, $n = 0, 1, 2, \ldots$ In general, we see from (C.18) that MTI filters are formed by cascades of delay line networks with a fundamental period of the sampling period T_s. The Laplace transform of the filter's impulse response is given by

$$H^*(p) = \sum_{-\infty}^{\infty} h(nT_s)\, e^{-nT_s p} \tag{C.19}$$

resulting in a z-transform of

$$H(z) = \sum_{-\infty}^{\infty} h(nT_s)z^{-n} \tag{C.20}$$

From (C.20), we can deduce that the z-transform of a pure delay is given by z^{-1}.

The output of an MTI filter can be written in terms of its z-transform as

$$G(z) = H(z)F(z) \tag{C.21}$$

which results from (C.8). In most cases, (C.21) can be used to determine the inverse transform of the output using a power series expansion in powers of $1/z$, which allows determining the output timc function.

As a simple example, consider a train of uniform impulses (e.g., a fixed target) applied to an MTI canceller as depicted in Figure C.5. The z-transform of the impulse train is given by (C.12), whereas the z-transform of the single canceller is given by $H(z) = 1 - z^{-1}$. The resulting output can then be determined by using (C.21), resulting in

$$G(z) = \frac{z-1}{z} \cdot \frac{z}{z-1} = 1 \tag{C.22}$$

The inverse z-transform is then a single impulse $\delta(t)$ at $t = 0$ (see Table C.1).

In the preceding discussion, a number of z-transforms have been derived and discussed. Table C.1 gives a short table of z-transforms for some of the more elemental operations.

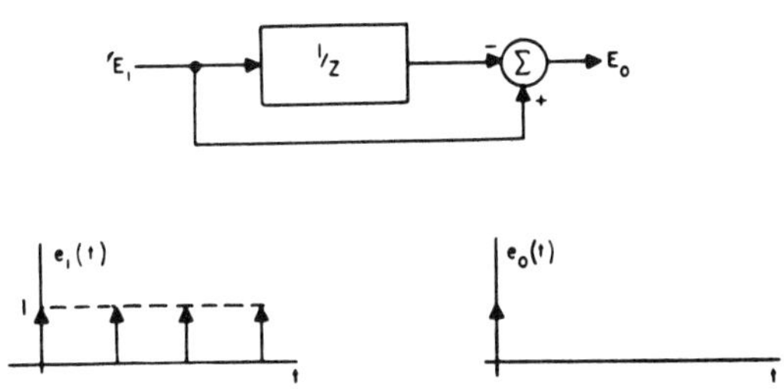

Figure C.5 MTI canceller with a train of uniform pulses at input.

Table C.1
Short Table of z-Transforms

Time Function	*z-Transform*	*Notes*
$\delta(t)$	1	Unit Impulse
$\delta(t - nT)$	z^{-n}	Delayed Impulse
$\sum_{n=0}^{\infty} \delta(t - nT)$	$\frac{z}{z - 1}$	Uniform Pulse Train
$\sum_{n=0}^{\infty} k^n \delta(t - nT)$	$\frac{z}{z - k}$	Exponential Decaying Pulse Train ($k < 0$)
$\sum_{n=0}^{\infty} n\delta(t - nT)$	$\frac{z}{(z - 1)^2}$	Ramp
$\sin\alpha t$	$\frac{z \sin\alpha T}{z^2 - 2z \cos\alpha T + 1}$	Sinusoid
$\exp(-\alpha t)$	$\frac{z}{z - e^{-\alpha T}}$	Exponential

C.1 MTI FILTER TRANSFER FUNCTIONS

The transfer function of an MTI filter given by (C.20) indicates that the filter can be formed by cascading a number of delay line networks. The most general (cannonical) form of a delay line network is depicted in Figure C.6. This form provides both feed-forward paths (α_0, α_1, . . ., α_n) and feedback paths (β_1, β_2, . . ., β_n), which result in an infinite impulse response or recursive MTI filter. When the feedback paths are removed, the filter becomes a finite impulse response or nonrecursive MTI filter (i.e., a transversal filter).

The transfer function of the cannonical form becomes

$$H(z) = \frac{\alpha_0 z^n + \alpha_1 z^{n-1} + \ldots + \alpha_n}{z^n - \beta_1 z^{n-1} - \ldots - \beta_n} \tag{C.23}$$

The numerator coefficients control the filter zeros, and the denominator coefficients control the filter poles. When the denominator coefficients are made equal to zero, the configuration reverts to a tapped delay line type of transversal filter, resulting in a transfer function given by

$$H(z) = \frac{\alpha_0 z^n + \alpha_1 z^{n-1} + \ldots + \alpha_n}{z^n} \tag{C.24}$$

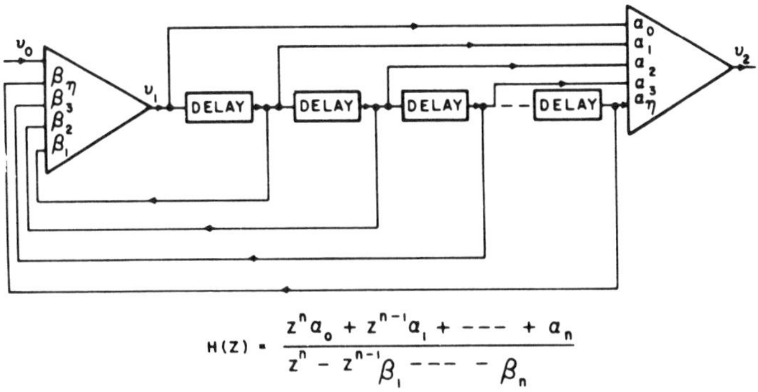

Figure C.6 Cannonical form of an MTI comb filter.

The transfer function of the cannonical configuration can be factored into the pole-zero form given by (C.14). Each delay line provides the capability for generating one pole and one zero. The poles and zeros may occur in either conjugate pairs about the real axis in the z-plane or in multiple form on the real axis. In addition, the poles must lie either within or on the unit circle in the z-plane for the filter to be stable. Due to the conjugate nature of the pole-zero locations, the MTI filter can generally be synthesized by cascading sections using no more than two delay lines [5]. A general form for a two-element network used to place a null at zero frequency (e.g., to attenuate ground clutter) is depicted in Figure C.7 and has a transfer function given by

$$H(z) = \frac{(z - 1)^2}{z^2 - (\beta_1 + \beta_2)z + \beta_2} \tag{C.25}$$

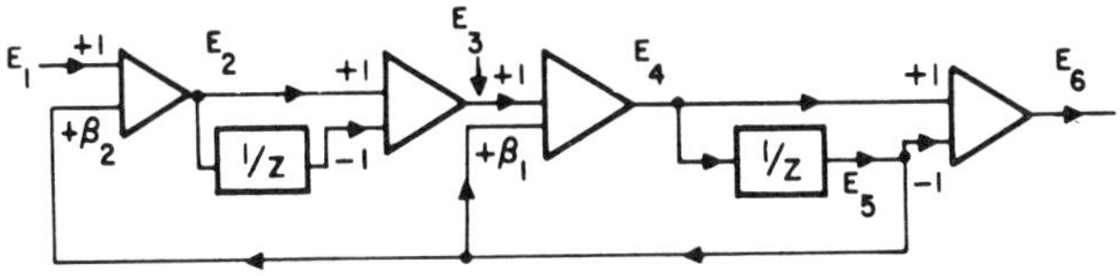

Figure C.7 Double shaped MTI filter with two poles.

The nonrecursive or transversal filter, in which the transfer function is given by (C.24), has an impulse response duration given by NT, where N is the number of delay sections employed in the filter. This short transient response is advantageous in an interference environment. An example of this type of filter is the single MTI canceller depicted in Table C.2 [3]. The signal canceller has a z-transform equal to

$$H_1(z) = \frac{z - 1}{z} \tag{C.26}$$

which has a simple zero at $z = 1$ (at frequencies equal to multiples of the PRF and zero; $f = \pm n/T$, where $n = 0, 1, 2, \ldots$) and a pole at $z = 0$. The transfer function can be found geometrically from Figure C.8 as $|H_1(j\omega)|^2 = \sin^2(\omega T) + [1 - \cos(\omega T)]^2$, resulting in

$$|H_1(j\omega)|^2 = 2[1 - \cos(\omega T)] = 4 \sin^2 \frac{\omega T}{2} \tag{C.27}$$

Table C.2
Periodic Filters

PERIODIC FILTERS

Note: 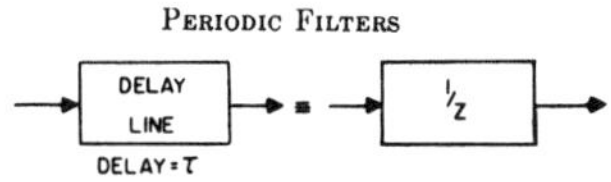

BLOCK DIAGRAM	T(z)	P-Z DIAGRAM
1) Single Canceller 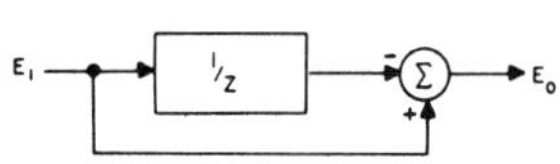	$\frac{z-1}{z}$	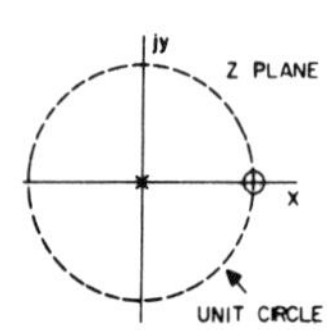
2) Double Canceller 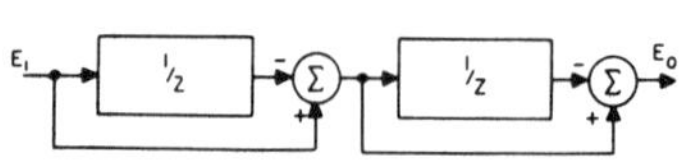	$\left(\frac{z-1}{z}\right)$	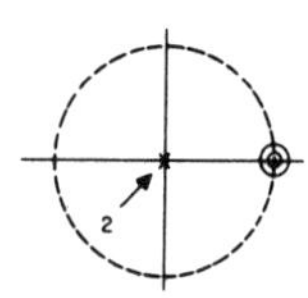

Table C.2
(continued)

BLOCK DIAGRAM	T(z)	P-Z DIAGRAM
3) Video Sweep Integrator 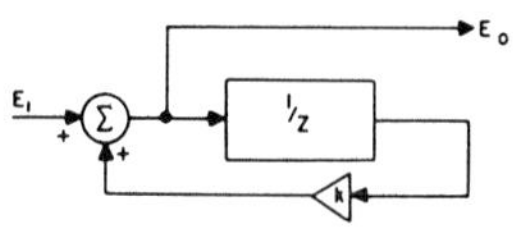	$\frac{z}{z-k}$	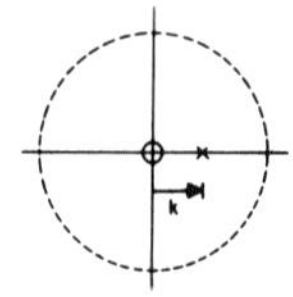
4) Double Loop Sweep Integrator 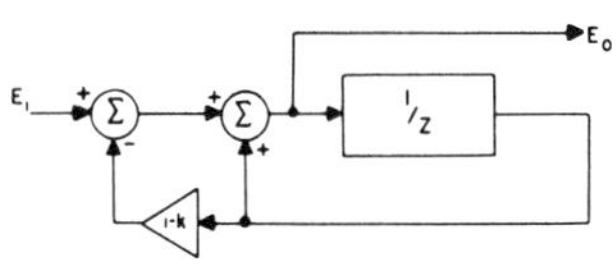	$\frac{z}{z-k}$	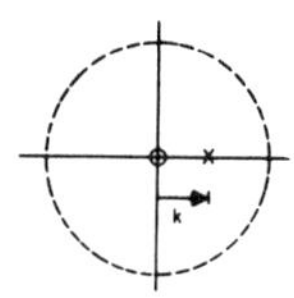
5) Response Shaped Single Canceller 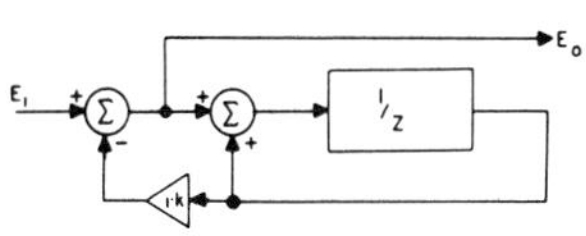	$\frac{z-1}{z-k}$	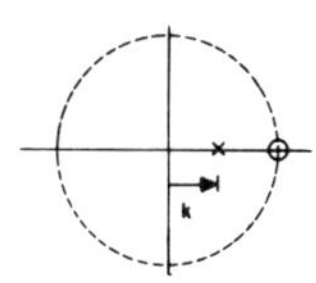
6) Response Shaped Double Canceller 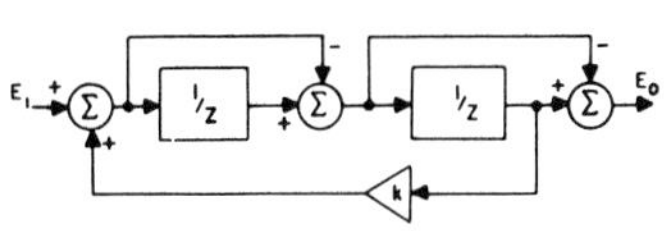	$\frac{(z-1)^2}{(z-z_1)(z-z_2)}$ $z_1, z_2 =$ $\frac{k \pm \sqrt{k^2-4k}}{2}$	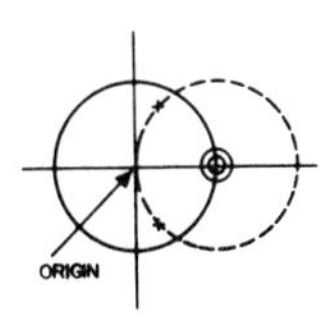

Table C.2
(continued)

BLOCK DIAGRAM	T(z)	P-Z DIAGRAM
7) Frequency (or Velocity) Rejection 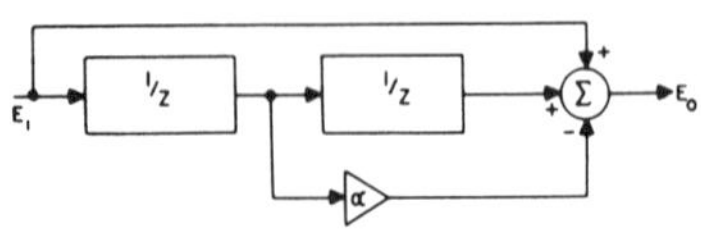	$\frac{(z-z_1)(z-z_2)}{z^2}$ $z_1, z_2 =$ $\frac{\alpha \pm \sqrt{\alpha^2-4}}{2}$	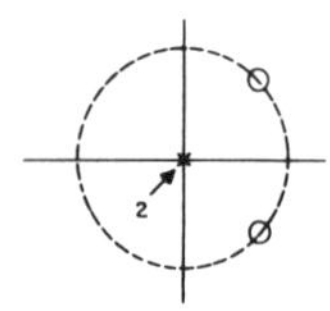
8) General Double Delay Filter 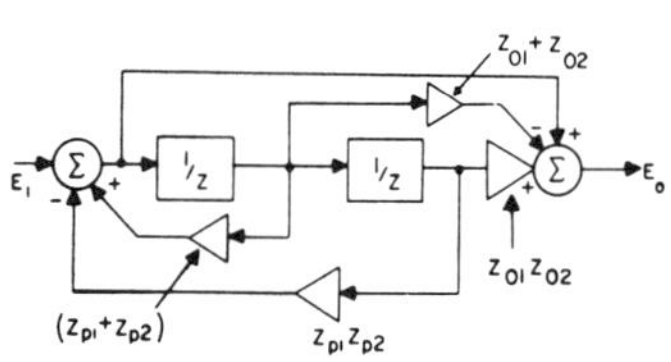	$\frac{(z-z_{01})(z-z_{02})}{(z-z_{p1})(z-z_{p2})}$	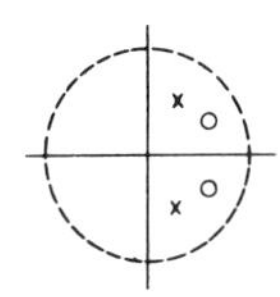 Poles and zeros real or conjugate complex pairs. Poles inside circle.

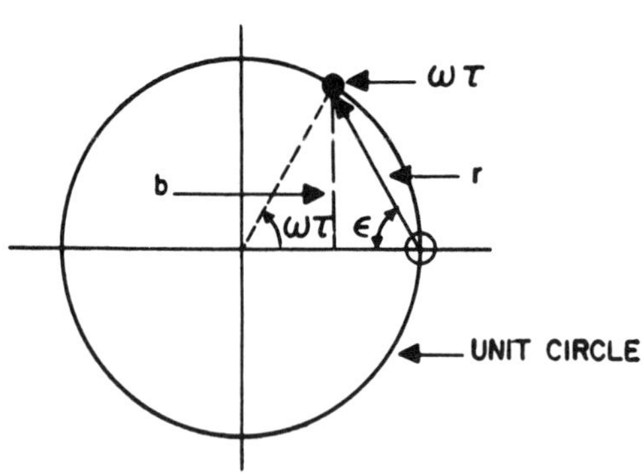

Figure C.8 Pole-zero pattern of a single canceller.

The pole-zero pattern of a double MTI canceller is also cataloged in Table C.2 and results from (C.21), $H_2(z) = H_1^2(z)$, as

$$H_2(z) = \frac{(z-1)^2}{z^2} \tag{C.28}$$

The double canceller has two poles at $z = 0$ and two zeros at $z = 1$. The z-transform of an N-delay MTI canceller can be deduced from the preceding as

$$H_N(z) = \frac{(z-1)^N}{z^N} \tag{C.29}$$

which has N zeros at $z = 1$ and N poles at the origin. The transfer function of an N-delay canceller is then given by

$$|H(j\omega)|^2 = 2^{2N} \sin^{2N}\left(\frac{\omega T}{2}\right) \tag{C.30}$$

Table C.2 also contains the pole-zero patterns of a number of MTI filters and video integrators that employ feedback paths in the delay networks. The effect of the feedback is generally to allow control of the pole placement of the filter. As an example, consider the simple video sweep integrator whose block diagram and pole zero pattern are depicted in Figure C.9. The transfer function of the video sweep integrator is given by

$$H(z) = \frac{z}{z-k} = \sum_{k=0}^{\infty} k^n z^{-n} \tag{C.31}$$

so that the filter's impulse response is given by

$$h(t) = \sum_{k=0}^{\infty} k^n \delta(t - nT) \tag{C.32}$$

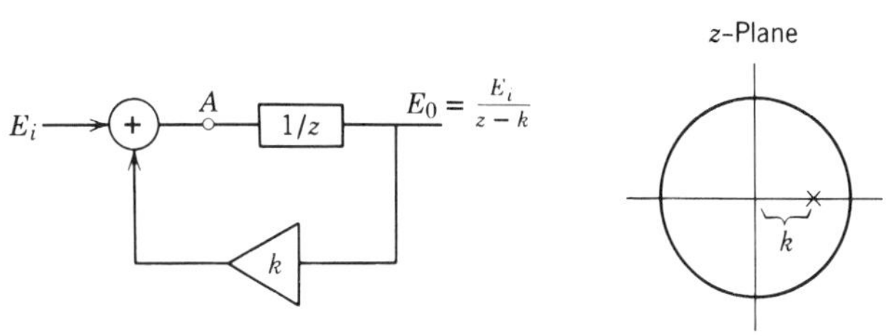

Figure C.9 Video sweep integrator.

C.2 BILINEAR TRANSFORMS

Conventional filter design techniques can be applied to sampled-data type filters through the use of bilinear transformations that map the pole-zero pattern of the conventional filter into the z-plane. The first step in this procedure is to locate the poles and zeros of a suitable continuous data filter in the p-plane (Laplace transformation plane), which provides a desired frequency characteristic. This filter serves as the basis for the design of the sampled-data filter. The poles and zeros are then mapped into the z-plane by using the appropriate transformation. The z-plane representation is a sampled-data filter having the desired characteristics.

Under the bilinear transformation, the mapping from the p-plane to the z-plane is one to one, whereby the number of poles and zeros in either plane remains the same. The entire imaginary axis of the p-plane becomes the unit circle of the z-plane. A scaling factor (Ω) is employed that allows the adjustment of the location of one significant point in the filter's transfer function (e.g., the corner frequency) as it is mapped into the z-plane. Although the frequency axis is distorted in mapping from the p-plane to the z-plane, these transformations preserve ripple characteristics such as maximally flat or equal-ripple amplitude responses. Thus, digital filters having responses such as Butterworth, Chebyschev, or elliptic filters may be synthesized by first designing a conventional version of the filter in the p-plane and then mapping the poles and zeros onto the z-plane. Figure C.10 depicts the mapping of a three-pole Butterworth low-pass filter into an MTI canceller design or video integrator design [6].

The first bilinear transformation useful to synthesize an MTI filter from a low-pass conventional filter design is given by

$$z = \frac{p + \Omega}{p - \Omega} \tag{C.33}$$

where z is the pole or zero location in the z-plane, p is the pole or zero location of the conventional filter in the p-plane, and Ω is a scale factor related to the filter's cut-off frequency. The inverse transformation from the z-plane into the p-plane is given by

$$p = \Omega \frac{z + 1}{z - 1} \tag{C.34}$$

which provides a mapping of poles and zeros in the p-plane into the z-plane. The scale factor is determined by adjusting the 3 dB frequency point in the p-plane ($+j$ on the imaginary axis) to map into the point where the angle $\phi_0 = \omega_{c_0}T$ intersects the unit circle in the z-plane (see Figure 6.31). The resulting scale factor is given by

$$\Omega = \tan(\phi_0/2) \tag{C.35}$$

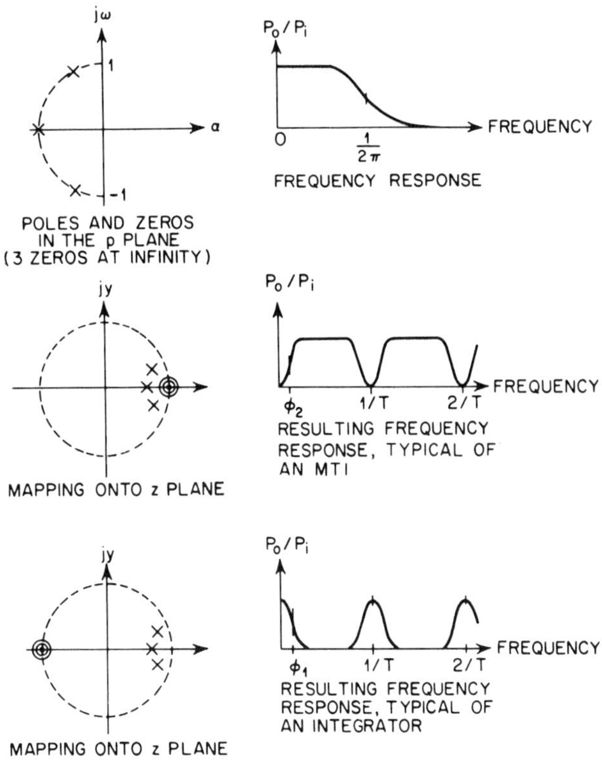

Figure C.10 Bilinear mapping of conventional filter into the z-plane.

where $\phi_0 = \omega_{c_0}T$, and ω_{c_0} is the desired 3 dB frequency of the filter. Note that this transformation is also useful for synthesizing a sampled-data integrator from a conventional high-pass filter design. This transformation forces the points $j\omega = \pm 1$ in the p-plane to map on the unit circle in the z-plane at an angle $\mp\ \phi_0$, which represents $f_{c_0} = \phi_0/2\pi T$.

The second bilinear transformation allows the synthesis of an MTI canceller from a conventional high-pass filter design or the synthesis of a video integrator from a conventional low-pass filter design. This transformation can be obtained from the first by noting that it involves a frequency translation from a low-pass to a high-pass filter in the conventional filter domain. A low-pass to high-pass filter design is accomplished through the transformation:

$$p^* = \frac{1}{p} \tag{C.36}$$

where p is the low-frequency poles or zeros and p^* is the transformed high-frequency poles or zero. The resulting second bilinear transformation is given by

$$z = \frac{\Omega^* + p^*}{\Omega^* - p^*} \tag{C.37}$$

$$p^* = \Omega^* \frac{z - 1}{z + 1} \tag{C.38}$$

where the scale factor is

$$\Omega^* = \cot\frac{\phi_1}{2} \tag{C.39}$$

This transformation forces the points $j\omega = \pm 1$ in the p-plane to map on the unit circle in the z-plane at an angle $\pm\phi_1$, which represents $f_{c_0} = \phi_1/2\pi T$ in the z-plane.

REFERENCES

1. J. Truxal, *Automatic Feedback Control System Synthesis,* McGraw-Hill, New York, 1955.
2. D. Linden and B. Steinberg, "Synthesis of Delay Line Networks," *IRE Trans.,* Vol. ANE-4, No. 1, March 1957.
3. H. Urkowitz, "Analysis and Synthesis of Delay Line Periodic Filters," *IRE Trans.,* Vol. CT-4, No. 2, June 1957.
4. R. Berkowitz, *Modern Radar,* John Wiley and Sons, New York, 1965.
5. W. White and A. Ruvin, "Recent Advances in the Synthesis of Comb Filters," *IRE Nat. Conv. Record,* 1957.
6. J. Shreve, "Digital Signal Processing," Chapter 35, in M.I. Skolnik, ed., *Radar Handbook,* 1st Ed., McGraw-Hill, New York, 1970.

Index

The Artech House Radar Library

David K. Barton, *Series Editor*

Active Radar Electronic Countermeasures by Edward J. Chrzanowski

Airborne Pulsed Doppler Radar by Guy V. Morris

AIRCOVER: Airborne Radar Vertical Coverage Calculation Software and User's Manual by William A. Skillman

Analog Automatic Control Loops in Radar and EW by Richard S. Hughes

Aspects of Modern Radar by Eli Brookner, *et al.*

Aspects of Radar Signal Processing by Bernard Lewis, Frank Kretschmer, and Wesley Shelton

Bistatic Radar by Nicholas J. Willis

Detectability of Spread-Spectrum Signals by Robin A. Dillard and George M. Dillard

Electronic Countermeasure System Design by Richard J. Wiegand

Electronic Homing Systems by M.V. Maksimov and G.I. Gorgonov

Electronic Intelligence: The Analysis of Radar Signals by Richard G. Wiley

Electronic Intelligence: The Interception of Radar Signals by Richard G. Wiley

EREPS: Engineer's Refractive Effects Prediction System Software and User's Manual, developed by NOSC

Handbook of Radar Measurement by David K. Barton and Harold R. Ward

High Resolution Radar by Donald R. Wehner

High Resolution Radar Cross-Section Imaging by Dean L. Mensa

Interference Suppression Techniques for Microwave Antennas and Transmitters by Ernest R. Freeman

Introduction to Electronic Warfare by D. Curtis Schleher

Introduction to Sensor Systems by S.A. Hovanessian

Logarithmic Amplification by Richard Smith Hughes

Modern Radar System Analysis by David K. Barton

Modern Radar System Analysis Software and User's Manual by David K. Barton and William F. Barton

Monopulse Principles and Techniques by Samuel M. Sherman

Monopulse Radar by A.I. Leonov and K.I. Fomichev

MTI and Pulsed Doppler Radar by D. Curtis Schleher

Multifunction Array Radar Design by Dale R. Billetter

Multisensor Data Fusion by Edward L. Waltz and James Llinas

Multiple-Target Tracking with Radar Applications by Samuel S. Blackman

Multitarget-Multisensor Tracking: Advanced Applications, Yaakov Bar-Shalom, ed.

Over-The-Horizon Radar by A.A. Kolosov, et al.

Principles and Applications of Millimeter-Wave Radar, Charles E. Brown and Nicholas C. Currie, eds.

Principles of Modern Radar Systems by Michel H. Carpentier

Pulse Train Analysis Using Personal Computers by Richard G. Wiley and Michael B. Szymanski

Radar and the Atmosphere by Alfred J. Bogush, Jr.

Radar Anti-Jamming Techniques by M.V. Maksimov, *et al.*

Radar Cross Section by Eugene F. Knott, *et al.*

Radar Detection by J.V. DiFranco and W.L. Rubin

Radar Evaluation Handbook by David K. Barton, *et al.*

Radar Evaluation Software by David K. Barton and William F. Barton

Radar Propagation at Low Altitudes by M.L. Meeks

Radar Range-Performance Analysis by Lamont V. Blake

Radar Reflectivity Measurement: Techniques and Applications, Nicholas C. Currie, ed.

Radar Reflectivity of Land and Sea by Maurice W. Long

Radar System Design and Analysis by S.A. Hovanessian

Radar Technology, Eli Brookner, ed.

Receiving Systems Design by Stephen J. Erst

Radar Vulnerability to Jamming by Robert N. Lothes, Michael B. Szymanski, and Richard G. Wiley

RGCALC: Radar Range Detection Software and User's Manual by John E. Fielding and Gary D. Reynolds

SACALC: Signal Analysis Software and User's Guide by William T. Hardy

Secondary Surveillance Radar by Michael C. Stevens

SIGCLUT: Surface and Volumetric Clutter-to-Noise, Jammer and Target Signal-to-Noise Radar Calculation Software and User's Manual by William A. Skillman

Signal Theory and Random Processes by Harry Urkowitz

Solid-State Radar Transmitters by Edward D. Ostroff, *et al.*

Space-Based Radar Handbook, Leopold J. Cantafio, ed.

Spaceborne Weather Radar by Robert M. Neneghini and Toshiaki Kozu

Statistical Theory of Extended Radar Targets by R.V. Ostrovityanov and F.A. Basalov

The Scattering of Electromagnetic Waves from Rough Surfaces by Peter Beckmann and Andre Spizzichino

VCCALC: Vertical Coverage Plotting Software and User's Manual by John E. Fielding and Gary D. Reynolds